Progress in Mathematics
Volume 233

Guy David

Singular Sets of Minimizers for the Mumford-Shah Functional

Birkhäuser Verlag
Basel · Boston · Berlin

Author:

Guy David
Mathématiques, Bâtiment 425
Université de Paris-Sud
91405 Orsay Cedex
France
guy.david@math.u-psud.fr

2000 Mathematics Subject Classification 49K99, 49Q20

A CIP catalogue record for this book is available from the Library of Congress,
Washington D.C., USA

Bibliographic information published by Die Deutsche Bibliothek
Die Deutsche Bibliothek lists this publication in the Deutsche Nationalbibliografie;
detailed bibliographic data is available in the Internet at <http://dnb.ddb.de>.

ISBN 3-7643-7182-X Birkhäuser Verlag, Basel – Boston – Berlin

© 2005 Birkhäuser Verlag, P.O. Box 133, CH-4010 Basel, Switzerland
Part of Springer Science+Business Media
Printed on acid-free paper produced of chlorine-free pulp. TCF ∞
Printed in Germany
ISBN-10: 3-7643-7182-X
ISBN-13: 978-3-7643-7182-1

9 8 7 6 5 4 3 2 1   www.birkhauser.ch

Ferran Sunyer i Balaguer (1912–1967) was a self-taught Catalan mathematician who, in spite of a serious physical disability, was very active in research in classical mathematical analysis, an area in which he acquired international recognition. His heirs created the Fundació Ferran Sunyer i Balaguer inside the Institut d'Estudis Catalans to honor the memory of Ferran Sunyer i Balaguer and to promote mathematical research.

Each year, the Fundació Ferran Sunyer i Balaguer and the Institut d'Estudis Catalans award an international research prize for a mathematical monograph of expository nature. The prize-winning monographs are published in this series. Details about the prize and the Fundació Ferran Sunyer i Balaguer can be found at

http://www.crm.es/FSBPrize/ffsb.htm

**This book has been awarded the
Ferran Sunyer i Balaguer 2004 prize.**

The members of the scientific commitee of the 2004 prize were:

Hyman Bass
    University of Michigan

Antonio Córdoba
    Universidad Autónoma de Madrid

Paul Malliavin
    Université de Paris VI

Joseph Oesterlé
    Université de Paris VI

Oriol Serra
    Universitat Politècnica de Catalunya, Barcelona

Ferran Sunyer i Balaguer Prize winners since 1996:

1996   V. Kumar Murty and M. Ram Murty
Non-vanishing of L-Functions and Applications, PM 157

1997   Albrecht Böttcher and Yuri I. Karlovich
Carleson Curves, Muckenhoupt Weights, and Toeplitz Operators, PM 154

1998   Juan J. Morales-Ruiz
Differential Galois Theory and Non-integrability of Hamiltonian Systems, PM 179

1999   Patrick Dehornoy
Braids and Self-Distributivity, PM 192

2000   Juan-Pablo Ortega and Tudor Ratiu
Hamiltonian Singular Reduction, PM 222

2001   Martin Golubitsky and Ian Stewart
The Symmetry Perspective, PM 200

2002   André Unterberger
Automorphic Pseudodifferential Analysis and Higher Level Weyl Calculi, PM 209

        Alexander Lubotzky and Dan Segal
Subgroup Growth, PM 212

2003   Fuensanta Andreu-Vaillo, Vincent Caselles and José M. Mazón
Parabolic Quasilinear Equations Minimizing Linear Growth Functionals, PM 223

2004   Guy David
Singular Sets of Minimizers for the Mumford-Shah Functional, PM 233

Contents

Contents

F Global Mumford-Shah Minimizers in the Plane

G Applications to Almost-Minimizers $(n = 2)$

H Quasi- and Almost-Minimizers in Higher Dimensions

I Boundary Regularity

Foreword

As this book is published, the study of the Mumford-Shah functional is at a curious stage. There was a quite rapid progress a few years ago, in particular with the work of A. Bonnet, and at the same time the most famous question in the subject, the Mumford-Shah conjecture, is still open. Recall that this conjecture says that in dimension 2, the singular set of a reduced minimizer of the functional is locally a C^1 curve, except at a finite number of points.

In this respect, it is probable that some new ideas are needed, but it seems equally likely that many of the recently developed tools will be useful. For instance, it now would seem quite awkward to try to prove the Mumford-Shah conjecture directly, instead of using blow-up limits and working on global minimizers in the plane.

The official goal of this book is to take the optimistic view that we first need to digest the previous progress, and then things will become easier. So we shall try to describe a lot of the available machinery, in the hope that a large part of it will be used, either for the Mumford-Shah problem, or in some other context.

From the author's point of view, the main reason why the Mumford-Shah functional is important is not really its connection with image segmentation, but the fact that it is a very good model for a whole class of problems with free boundaries and a length or surface term, and that there are very interesting mathematical questions connected with it. Hopefully the techniques discovered for the Mumford-Shah functional will be useful somewhere else too, just as some of the most significant improvements in Mumford-Shah Theory come from different areas.

It should probably be stressed at the outset that there is a life outside of the Mumford-Shah conjecture; there are many other interesting (and perhaps easier) questions related to the functional, in particular in dimension 3; we shall try to present a few in the last section of the book.

The project of this book started in a very usual way: the author taught a course in Orsay (fall, 1999), and then trapped himself into writing notes, that eventually reached a monstrous size. What hopefully remains from the initial project is the will to be as accessible and self-contained as possible, and not to treat every aspect of the subject. In particular, there is an obvious hole in our treatment: we shall almost never use or mention the bounded variation approach, even though this approach is very useful, in particular (but not only) with existence results. The author agrees that this attitude of avoiding BV looks a lot like ignoring progress, but he has a good excuse: the BV aspects of the theory have been treated very well in the book of Ambrosio, Fusco, and Pallara [AFP3], and there would be no point in doing the same thing badly here.

Part A will be a general presentation of the Mumford-Shah functional, where we shall discuss its origin in image segmentation, existence and nonuniqueness of minimizers, and give a quick presentation of the Mumford-Shah conjecture and some known results. We shall also give slightly general and complicated (but I claim natural) definitions of almost-minimizers and quasiminimizers. Incidentally, these are slightly different from definitions with the same names in other sources.

Part B reviews simple facts on the Sobolev spaces $W^{1,p}$ that will be needed in our proofs. These include Poincaré estimates, boundary values and traces on planes and spheres, and a corresponding welding lemma. We shall also discuss the existence of functions u that minimize energy integrals $\int_U |\nabla u|^2$ with given boundary values. Thus this part exists mostly for self-containedness.

Part C contains the first regularity results for minimizers and quasiminimizers in dimension 2, i.e., local Ahlfors-regularity, some useful Carleson measure estimates on $|\nabla u|^p$, the projection and concentration properties, and local uniform rectifiability. The proofs are still very close to the original ones.

Part D is a little more original (or at least was when the project started), but is very much inspired by the work of Bonnet. The main point is to use the concentration property of Dal Maso, Morel, and Solimini to prove that limits of minimizers or almost-minimizers are still minimizers or almost-minimizers (actually, with a small additional topological constraint that comes from the normalization of constants). Bonnet did this only in dimension 2, to study blow-up limits of Mumford-Shah minimizers, but his proof is not really hard to adapt. The results of Part D were also proved and published by Maddalena and Solimini [MaSo4] (independently (but before!), and with a different approach to the concentration property).

Part E contains the C^1-regularity almost-everywhere of almost-minimizers, in dimension 2 only. The proof is not really new, but essentially unreleased so far. As usual, the central point is a decay estimate for the normalized energy $\frac{1}{r} \int_{B(x,r) \setminus K} |\nabla u|^2$; in the argument presented here, this decay comes from a variant Bonnet's monotonicity argument. See Sections 47–48.

In Part F we try to give a lot of properties of global minimizers in the plane (these include the blow-up limits of standard Mumford-Shah minimizers). This part is already a slight mixture of recent results (not always given with full detailed proofs) and some mildly new results. The main tools seem to be the results of Section D (because taking limits often simplifies things), Bonnet's monotonicity argument and a variant (which in the best cases play the role of the standard monotonicity result for minimal surfaces), Léger's formula (63.3) (which allows us to compute u given K), and of course some work by hand.

In Part G we return briefly to almost-minimizers in a domain and use Part F to derive a few additional regularity results. We also check that the standard Mumford-Shah conjecture would follow from its counterpart for global minimizers in the plane.

We decided to wait until Part H for a quicker and less complete description of the situation in higher dimensions. For instance, the result of Ambrosio, Fusco, and Pallara on C^1 regularity almost-everywhere is discussed, but not entirely proved.

Part I contains a description of the regularity of K near the boundary, which we decided to keep separate to avoid confusion; however, most of our inside regularity results still hold at the boundary (if our domain Ω is C^1, say). In dimension 2, we even get a good description of K near $\partial\Omega$, as a finite union of C^1 curves that meet $\partial\Omega$ orthogonally.

We conclude with a small isolated section of questions.

This book is too long, and many arguments look alike (after all, we spend most of our time constructing new competitors). Even the author finds it hard to find a given lemma. To try to ameliorate this, some small arguments were repeated a few times (usually in a more and more elliptic way), and a reasonably large index is available. It is hard to recommend a completely linear reading, but probably the last parts are easier to read after the first ones, because similar arguments are done faster by the end.

To make the book look shorter locally, references like (7) will refer to Display (7) in the current section, and (3.7) will refer to (7) in Section 3. The number of the current section is visible in the running title on the top of each page.

Notation

We tried to make reasonable choices; here are just a few:

C is a positive, often large constant that may vary from line to line,

$B(x, r)$ is the open ball centered at x with radius r,

λB is the ball with same center as B, and radius λ times larger,

ω_n is the Lebesgue measure of the unit ball in $\mathbb{R}^n$,

$\widetilde{\omega}_{n-1}$ is the H^{n-1}-measure of the unit sphere in $\mathbb{R}^n$,

$\square$ signals the end of a proof,

$\subset\subset$ means relatively compact in.

See the index for a few other symbols, that come with a definition.

Acknowledgments

I would like to thank the many people who helped with the preparation of this
book, for instance by answering questions or suggesting improvements; here is
only a short list: Alano Ancona, Alexis Bonnet, Thierry De Pauw, David Jerison,
Jean-Christophe Léger, Noël Lohoué, Francesco Maddalena, Hervé Pajot, Séverine
Rigot, Anthony Siaudeau, Sergio Solimini. I also wish to thank Antoinette Bardot,
who kindly started to type the manuscript when I knew nothing about TEX (this
looks like a very remote time now!).

Many thanks are due to the Institut Universitaire de France (and at the same
time my colleagues in Orsay), which gave me lots of very useful extra time, and
to the Fundació Ferran Sunyer i Balaguer. The author is partially supported by
the HARP (Harmonic Analysis and Related Problems) European network.

A. Presentation of the Mumford-Shah Functional

In this first part we want to give a general presentation of the Mumford-Shah functional. We shall define the functional and rapidly discuss some basic issues like the existence of minimizers, the lack of uniqueness in general, and the fact that the functional becomes much easier to study when the singular set K is fixed. We shall also present the Mumford-Shah conjecture on the regularity of minimizers in dimension 2, and give a few hints on the contents of this book. The part will end with two sections of definitions of almost-minimizers and quasiminimizers, which we think are reasonably important.

1 The Mumford-Shah functional and image segmentation

In this section we want to describe the origin of the Mumford-Shah problem, in connection with the issue of image segmentation. Part of this description is fairly subjective, and this introduction may not reflect much more than the author's personal view on the subject.

Consider a simple domain $\Omega \subset \mathbb{R}^n$. For image segmentation, the most important case is when $n = 2$, and Ω may as well be a rectangle. Also let $g \in L^\infty(\Omega)$ be given. We think of g as a representation of an image; we shall take it real-valued to simplify the exposition, but vector values are possible (for color pictures, or textures, etc.) and would lead to a similar discussion.

The point of image segmentation is to replace g with a simpler function (or image) u which captures "the main features" of g. A very natural idea is to define a functional that measures how well these two contradictory constraints (simplicity and good approximation of g) are satisfied by candidates u, and then minimize the functional.

Of course there are many possible choices of functionals, but it seems that most of those which have been used in practice have the same sort of structure as the Mumford-Shah functional studied below. See [MoSo2] for a thorough description of this and related issues.

The Mumford-Shah functional was introduced, I think, in 1985 [MuSh1], but the main reference is [MuSh2]. The good approximation of g by u will be measured in the simplest way, i.e., by

$$A = \int_\Omega |u - g|^2. \tag{1}$$

This is a fairly reasonable choice, at least if we don't have any information a priori on which sort of image g we have. (We may return to this issue soon.) Of course minor modifications, such as using an L^p-norm instead of L^2, or integrating $|u - g|^2$ against some (slowly varying) weight are possible, and they would not really change the mathematics in this book. Note that we do not want to imply here that images are well described by L^2 functions. (L^∞ functions with bounded variation, for instance, would be better, because of the importance of edges.) We just say that for image segmentation we prefer to use a weak norm (like the L^2 norm) in the approximation term. If nothing else, it should make the process less dependent on noise.

Let us now say what we shall mean by a simple function u. We want to authorize u to have singularities (mainly, jumps) along a closed set K, but we want K to be as "simple" as possible. For the Mumford-Shah functional, simple will just mean short: we shall merely measure

$$L = H^{n-1}(K), \tag{2}$$

the Hausdorff measure of dimension $n - 1$ of K (if we work in $\Omega \subset \mathbb{R}^n$). See the next section for a definition; for the moment let us just say that $H^{n-1}(K)$ is the same as the surface measure of K (length when $n = 2$) when K is a reasonably smooth hypersurface (which we do *not* assume here).

The reader may be surprised that we confuse length with simplicity. The objection is perhaps a little less strong when everything is discretized, but also one of the good features of the Mumford-Shah functional is that, for minimizers, K will turn out to have some nontrivial amount of regularity, which cannot be predicted immediately from the formula but definitely makes its choice more reasonable.

Here also many other choices are possible. A minor variant could be to replace L with $\int_K a(x)dH^{n-1}(x)$ for some smooth, positive function a; we shall try to accommodate this variant with some of the definitions of almost-minimizers below. One could also replace H^{n-1} with some roughly equivalent measurement of surface measure, which would not be isotropic but where for instance horizontal and vertical directions would be privileged. This may look more like what happens when one discretizes; in this case we cannot expect K to be C^1 for minimizers, but some of the weaker regularity properties (like local Ahlfors-regularity and uniform rectifiability) will remain true. And this is one of the points of the definition of quasiminimizers in Section 7.

In some cases, one requires K to be more regular directly, by replacing L with the integral over K (against Hausdorff measure) of some function of the curvature of K. We shall not study these variants, but they make sense: after all, if you want K to be fairly smooth, you may as well require this up front instead of trying to get it too indirectly.

So we allow u to jump across the simple (or short) singular set K, but away from K we want u to be reasonably smooth. Here again we shall measure this in the simplest way: we shall require u to have one derivative in $L^2(\Omega \setminus K)$, and use

$$R = \int_{\Omega \setminus K} |\nabla u|^2 \tag{3}$$

as a measure of the smoothness of u.

Taking other powers than 2 does not seem to have obvious advantages here, and it would make some of the mathematics substantially more complicated. Altogether we arrive at the functional

$$J(u, K) = a \int_{\Omega} |u - g|^2 + bH^{n-1}(K) + c \int_{\Omega \setminus K} |\nabla u|^2, \tag{4}$$

where a, b, c are positive constants, and the competitors are pairs (u, K) for which

$$K \subset \Omega \text{ is closed in } \Omega, \tag{5}$$
$$H^{n-1}(K) < +\infty, \tag{6}$$

and

$$u \in W^{1,2}(\Omega \setminus K) \tag{7}$$

(which means that the derivative of u lies in L^2 on the domain $\Omega \setminus K$, see Definition 2.2). We could have required instead that $u \in C^1(\Omega \setminus K)$; it turns out that this makes no difference because when (u, K) minimizes J (or even when u minimizes $J(u, K)$ for a fixed K), u is automatically C^1 on $\Omega \setminus K$. (See Corollary 3.19.)

There are still many other variants of (4) that we did not mention. For instance we could replace $H^{n-1}(K)$ with the integral on K (against H^{n-1}) of some function of the jump of u across K (which can be defined correctly by changing contexts slightly). See [MoSo2] for details.

What are the chances that the Mumford-Shah functional will give good segmentations? This question is not precise enough, because we did not say what we mean by capturing the main features of an image g. We should definitely distinguish image segmentation from compression, where the goal would be to keep as much of g as possible within a given amount of storage room. Here we are happy to forget some information about g, make it simple, and hopefully still retain the information that we want to use for a later purpose. So everything seems to depend on what we want to do with our segmentation. We probably won't be able to use a Mumford-Shah segmentation directly to answer elaborate questions about our image, but it may help answering simple questions (like, where are the main parts of the picture) before we do a further analysis.

Using the functional (4) has the a priori advantage that we don't need to know anything about our function g to do it. On the other hand, precisely because of this, we cannot expect the functional to take any intelligent decision that would require some a priori knowledge about g. It seems that a current tendency in image segmentation is to try to give the computer as much a priori knowledge as possible, often on statistical bases, before we do the image processing. In spite of all this, the Mumford-Shah functional has been extensively used for image segmentation, at least as a reference method.

We should also mention that there are nontrivial difficulties with the actual implementation and discretization of the Mumford-Shah functional; it is often much easier to look for local minima (that is, with respect to small modifications like adding or subtracting line segments), or use approximations of the Mumford-Shah functional by other functionals with better convexity properties. This is an interesting issue, but will be ignored in this book.

The Mumford-Shah functional still has a few good features that we should mention here. As was said before, from looking at (4) and (2) we should only expect K to be reasonably short (when $n = 2$) when (u, K) minimizes J. But it turns out that K is also fairly regular. This was expected from the start by Mumford and Shah: they conjectured in [MuSh2] that modulo a set of H^1-measure zero which is irrelevant, K is a finite union of C^1-curves. See Section 6 for details. Even though we do not know this much yet, K has lots of similar, barely weaker properties; this is the main point of this book.

This is a good thing. First, the set K is an important piece of the image segmentation: we may want to use it as a rough, cartoon-like, description of the edges in our image g. Thus it is more pleasant to know that it looks like a fairly regular collection of curves (or boundaries), instead of some ugly Cantor set.

We also like to know that K is regular, because it is a good hint that the functional is not too sensitive to noise. Indeed we may think of noise as producing lots of little singularities in random places. In this respect, it is good to know that the functional decides that they are not sufficiently coherently organized, and does not pick them up. As was alluded to above, choosing a fairly rudimentary norm in the definition of our approximation term A probably helps getting minimal segmentations that are less sensitive to noise. Notice the difference with the stupider algorithm where we would try to find edges in a picture g simply by computing the gradient of g and keeping the places where it is largest.

Observations like these were probably quite comforting to the people that actually used the functional for image segmentation, and this may have been one of the main incentives for the theoretical study of the functional and its minimizers. We cannot use this excuse any more: the subtle difference between the Mumford-Shah conjecture (which we want to prove) and what we already know about minimizers will never make a big difference for actual image segmentation.

The author's personal point of view is that the theoretical aspects of the study of the functional are much more interesting now than its applications to image processing, even though a large part of the beauty of the problems comes from the geometry and the close relation with the initial image processing setting. For the rest of this book, we shall almost entirely forget about image segmentation and concentrate on the (theoretical) study of minimizers for J, and in particular on the regularity properties of the singular set K.

2 Definition of the functional

First we want to give a more detailed definition of the Mumford-Shah functional. Let a domain $\Omega \subset \mathbb{R}^n$ and a bounded function $g \in L^\infty(\Omega)$ be given. We first define the set of admissible pairs (or competitors) by

$$\mathcal{A} = \mathcal{A}(\Omega) = \{(u, K)\,;\; K \subset \Omega \text{ is relatively closed and } u \in W^{1,2}_{\text{loc}}(\Omega \backslash K)\}. \quad (1)$$

By relatively closed we mean that K is closed in Ω, or equivalently that K is the intersection of Ω with a closed subset of $\mathbb{R}^n$. The definition of $W^{1,2}_{\text{loc}}$ is as follows.

Definition 2. *Let $1 \leq p \leq +\infty$ and an open set $U \subset \mathbb{R}^n$ be given. We denote by $W^{1,p}_{\text{loc}}(U)$ the set of functions $f \in L^1_{\text{loc}}(U)$ whose partial derivatives (of order 1, and in the sense of distributions) lie in $L^p_{\text{loc}}(U)$. This means that there exist functions $f_1, \ldots, f_n \in L^p_{\text{loc}}(U)$ such that*

$$\int_U f(x) \frac{\partial \varphi}{\partial x_i}(x) dx = -\int_U f_i(x)\varphi(x) dx \quad (3)$$

for $1 \leq i \leq n$ and all functions $\varphi \in C_c^\infty(U)$. The Sobolev space $W^{1,p}(U)$ is just the set of functions $f \in W^{1,p}_{\text{loc}}$ for which $f_1, \ldots, f_n \in L^p(U)$. Note that we do not require that $f \in L^1(U)$.

Here we denoted by $L^p_{\text{loc}}(U)$ the set of (measurable) functions g on U such that $\int_K |g(x)|^p dx < +\infty$ for every compact set $K \subset U$, with the usual modification when $p = +\infty$. We also denoted by $C_c^\infty(U)$ the set of C^∞-functions with compact support in U (test-functions).

In our definition of $W^{1,p}_{\text{loc}}$ and $W^{1,p}$ we included explicitly the demand that $f \in L^1_{\text{loc}}(U)$. This implies that f defines a distribution on U, by

$$\langle f, \varphi \rangle = \int_U f(x)\varphi(x) dx \quad \text{for } \varphi \in C_c^\infty(U). \quad (4)$$

A possibly more natural definition would have only required that f be a distribution on U instead. We decided to avoid this, first to avoid talking too much about distributions, and also because the two definitions coincide and we would have had to check that distributions with derivatives in L^p_{loc} lie in L^1_{loc} (by Sobolev).

Of course the partial derivatives $\frac{\partial f}{\partial x_i}$ exist (as distributions) as soon as $f \in L^1_{\text{loc}}(U)$; they are defined by

$$\left\langle \frac{\partial f}{\partial x_i}, \varphi \right\rangle = -\int_U f(x) \frac{\partial f}{\partial x_i}(x) dx \quad \text{for } \varphi \in C_c^\infty(U). \quad (5)$$

Thus (3) says that $\frac{\partial f}{\partial x_i} = f_i$. We shall find it convenient to denote by ∇f the vector-valued function with coordinates $\frac{\partial f}{\partial x_i}, 1 \le i \le n$.

Next we define the Hausdorff measure.

Definition 6. *For $K \subset \mathbb{R}^n$ and $d > 0$, set*

$$H^d(K) = \sup_{\varepsilon > 0} H^d_\varepsilon(K), \tag{7}$$

where

$$H^d_\varepsilon(K) = c_d \inf \left\{ \sum_{i=1}^\infty (\text{diam } A_i)^d \right\}, \tag{8}$$

where the infimum is taken over all countable families $\{A_i\}_{i=1}^\infty$ of closed sets A_i such that

$$K \subset \bigcup_{i=1}^\infty A_i \quad \text{and } \text{diam } A_i \le \varepsilon \text{ for all } i. \tag{9}$$

Here the constant c_d is chosen so that H^d coincides with the Lebesgue measure on d-planes; its precise value won't really matter here. Note that $H^d_\varepsilon(K)$ is a nonincreasing function of ε; thus the sup in (7) is a limit. It is well known that (the restriction to Borel sets) of H^d is a Borel measure; in fact H^d is a metric outer measure. [This is not the case of H^d_ε when ε is fixed, because H^d_ε is not even additive on disjoint sets; this is why the limit in (7) is needed if we want a measure.] See [Mat2] for details.

Remark 10. When d is an integer and K is contained in a C^1-surface of dimension d, $H^d(K)$ coincides with its d-dimensional surface measure. Since we don't know in advance that the good competitors for the Mumford-Shah functional are smooth, we have to use something like the Hausdorff measure.

Remark 11. If $K \subset \mathbb{R}^n$ and $f : K \to \mathbb{R}^m$ is Lipschitz with constant $\le M$ (that is, if $|f(x) - f(y)| \le M|x - y|$ for $x, y \in K$), then

$$H^d(f(K)) \le M^d H^d(K). \tag{12}$$

This follows directly from the definitions: if $\{A_i\}$ covers K, $\{f(A_i)\}$ covers $f(K)$.

Notice also that if K is a Borel set and $H^d(K) < +\infty$, then $H^\lambda(K) = 0$ for all $\lambda > d$. See Exercise 20.

We are now ready to define the functional J: given $\Omega \subset \mathbb{R}^n$ and $g \in L^\infty(\Omega)$ as above, as well as constants $a, b, c > 0$, set

$$J(u, K) = a \int_{\Omega/K} |u - g|^2 + bH^{n-1}(K) + c \int_{\Omega/K} |\nabla u|^2 \tag{13}$$

for all pairs $(u, K) \in \mathcal{A}$.

This is slightly different from what we did in Section 1: here we decided to take a rather large class $\mathcal{A}$ of competitors, and now $J(u, K) = +\infty$ for some of them. Of course we are still mostly interested in the restriction of J to the smaller class of pairs $(u, K) \in \mathcal{A}$ such that $J(u, K) < +\infty$. And we shall always assume that this class is not empty. [This is why Ω is often assumed to be bounded in the literature, and also partly why it is customary to take $g \in L^\infty$.]

Remark 14. The precise form of the first term of the functional is not too important; we could also have taken an L^p-norm, $1 \leq p < +\infty$. Similarly, the assumption that $g \in L^\infty$ is not vital. We shall see in Section 7 definitions of almost-minimizer that accommodate these variations. Changing the power 2 in the last term of J is more dangerous; at least quite a few of the proofs below break down in this case.

Note that taking $c = +\infty$ in (13) corresponds to the variant of the Mumford-Shah functional where u is forced to be constant on each of the connected components of Ω/K. See [MoSo2] and its references for more information on this variant.

To end this section, let us see why we can always reduce to $a = b = c = 1$ in the definition (13) of J.

Remark 15 on homogeneity. The three terms of the functional are obviously invariant by joint translations of Ω, g, u, K, but they have different homogeneities with respect to dilations. We shall see later that this has very important consequences on the way J chooses its minimizers, but for the moment we just want to see how to use this fact to normalize out the constants in (13).

Let Ω and g be as before and let $\lambda > 0$ be given. Set $\widetilde{\Omega} = \lambda\Omega$, $\widetilde{g}(x) = g(\frac{x}{\lambda})$, and call $\widetilde{J}$ the analogue of J on the domain $\widetilde{\Omega}$, defined with the function $\widetilde{g}$ and the coefficients

$$\widetilde{a} = \lambda^{-n}a, \ \ \widetilde{b} = \lambda^{1-n}b, \ \text{ and } \ \widetilde{c} = \lambda^{2-n}c. \tag{16}$$

For $(u, K) \in \mathcal{A}$ we set $\widetilde{K} = \lambda K$ and $\widetilde{u}(x) = u(\frac{x}{\lambda})$. Then $(\widetilde{u}, \widetilde{K}) \in \widetilde{\mathcal{A}}$ (the analogue of $\mathcal{A}$ on $\widetilde{\Omega}$), and

$$\widetilde{J}(\widetilde{u}, \widetilde{K}) = \widetilde{a} \int_{\widetilde{\Omega}/\widetilde{K}} |\widetilde{u} - \widetilde{g}|^2 + \widetilde{b}H^{n-1}(\widetilde{K}) + \widetilde{c} \int_{\widetilde{\Omega}/\widetilde{K}} |\nabla\widetilde{u}|^2 \tag{17}$$

$$= \lambda^n \, \widetilde{a} \int_{\Omega/K} |u - g|^2 + \lambda^{n-1} \, \widetilde{b}H^{n-1}(K) + \lambda^{n-2} \, \widetilde{c} \int_{\Omega/K} |\nabla u|^2 = J(u, K).$$

Now keep $\widetilde{\Omega}$ and $\widetilde{K}$, but replace $\widetilde{g}$ with $g^* = \mu\widetilde{g}$ and $\widetilde{u}$ with $u^* = \mu\widetilde{u}$ for some $\mu > 0$. Then

$$J(u, K) = J^*(u^*, \widetilde{K}), \tag{18}$$

where J^* corresponds to $\widetilde{\Omega}$, g^*, and the constants

$$a^* = \mu^2\widetilde{a} = \mu^2\lambda^{-n}a, \ \ b^* = \widetilde{b} = \lambda^{1-n}b, \ \text{ and } \ c^* = \mu^2\widetilde{c} = \mu^2\lambda^{2-n}c. \tag{19}$$

We can choose the constants λ and μ so that $a^* = b^* = c^*$. Thus, modulo a composition with a dilation and a multiplication of all functions by a constant, we can always reduce the study of J to the case where $a = b = c = 1$. (The last condition comes for free, because multiplying J by a constant does not change things seriously.) Because of this, we shall take $a = b = c = 1$ for the rest of this book. Of course, we cannot do this and normalize $||g||_\infty$ at the same time, but this will not matter.

We shall see later another facet of the fact that the three terms of J have different homogeneities: the approximation term of the functional only gives very small contributions at small scales (compared with the other two). This is why it only makes marginal appearances when we prove regularity properties for minimizers. Note that this is not bad for image processing: we actually want g to play an important role, but only at large scales.

Exercise 20. Let K be a Borel set such that $H^d(K) < +\infty$, and let $\lambda > d$ be given. Show that $H^\lambda_\varepsilon(K) \leq C\varepsilon^{\lambda-d}H^d_\varepsilon(K)$, and then that $H^\lambda(K) = 0$.

3   Minimizing in u with K fixed

In this section we want to show that when the closed set $K \subset \Omega$ is given, there is a unique function $u \in W^{1,2}(\Omega \setminus K)$ for which $J(u, K)$ is minimal. That is, assuming that $J(v, K) < +\infty$ for some $v \in W^{1,2}(\Omega \setminus K)$. Thus, when we try to minimize the Mumford-Shah functional, the only problem will be the determination of the free boundary K.

Since K will be fixed, let us even remove it from our notation, as follows. Let the open set $V \subset \mathbb{R}^n$ be given (think about $\Omega \setminus K$), and let $g \in L^2_{loc}(V)$ be given also. Set

$$J(u) = \int_V |u - g|^2 + \int_V |\nabla u|^2 \tag{1}$$

for $u \in W^{1,2}(V)$. We shall assume that

$$\text{there exists } v \in W^{1,2}(V) \text{ such that } J(v) < +\infty, \tag{2}$$

since otherwise nothing interesting can be said.

Proposition 3.   *There is a unique function* $u_0 \in W^{1,2}(V)$ *such that*

$$J(u_0) = \inf \{J(u) \, ; u \in W^{1,2}(V)\}. \tag{4}$$

This is a simple convexity result. Recall that for $u, v \in L^2$,

$$\left\|\tfrac{u+v}{2}\right\|_2^2 + \left\|\tfrac{u-v}{2}\right\|_2^2 = \frac{1}{2}\|u\|_2^2 + \frac{1}{2}\|v\|_2^2, \tag{5}$$

and so

$$\left\|\tfrac{u+v}{2}\right\|_2^2 = \frac{1}{2}\|u\|_2^2 + \frac{1}{2}\|v\|_2^2 - \left\|\tfrac{u-v}{2}\right\|_2^2 . \tag{6}$$

Let $u, v \in W^{1,2}(V)$ be such that $J(u), J(v) < \infty$, and apply (6) to $u - g$ and $v - g$, and then to ∇u and ∇v; we get that

$$J\!\left(\frac{u+v}{2}\right) = \frac{1}{2}J(u) + \frac{1}{2}J(v) - \frac{1}{4}\int_V |u-v|^2 - \frac{1}{4}\int_V |\nabla(u-v)|^2 \tag{7}$$

$$= \frac{1}{2}J(u) + \frac{1}{2}J(v) - \frac{1}{4}\|u-v\|_H^2,$$

where we set

$$\|w\|_H^2 = \int_V \{|w|^2 + |\nabla w|^2\} \tag{8}$$

for $w \in L^2(W) \cap W^{1,2}(V)$. Set

$$m = \inf\{J(u)\,;\, u \in W^{1,2}(V)\} < +\infty \tag{9}$$

(by (2)). If we choose $u, v \in W^{1,2}(V)$ such that $J(u), J(v) \le m + \varepsilon$, then

$$m \le J\!\left(\frac{u+v}{2}\right) \le m + \varepsilon - \frac{1}{4}\|u-v\|_H^2 \tag{10}$$

by (9) and (7), whence

$$\|u-v\|_H^2 \le 4\varepsilon. \tag{11}$$

Let $\{u_j\}_{j\ge 1}$ be a minimizing sequence (i.e., $\lim_{j\to+\infty} J(u_j) = m$). We may assume that $J(u_j) < +\infty$ for all j; then $w_j = u_j - u_1$ lies in the Hilbert space $H = L^2(V) \cap W^{1,2}(V)$. Moreover (11) says that $\{w_j\}$ is a Cauchy sequence in H; let w denote its limit and set $u_0 = u_1 + w$. Then

$$J(u_0) = \|u_0 - g\|_2^2 + \|\nabla u_0\|_2^2 = \|(u_1 - g) + w\|_2^2 + \|\nabla u_1 + \nabla w\|_2^2 \tag{12}$$

$$= \lim_{j\to+\infty} \{\|(u_1 - g) + w_j\|_2^2 + \|\nabla u_1 + \nabla w_j\|_2^2\} = \lim_{j\to+\infty} J(u_j) = m,$$

by the continuity of L^2-norms. This proves the existence in Proposition 3; the uniqueness follows from (11) with $\varepsilon = 0$. $\qquad\square$

Remark 13. Proposition 3 and its proof also work for functionals

$$J(u) = \int_V |u - g|^p + \int_V |\nabla u|^q, \tag{14}$$

with $1 < p, q < +\infty$, $g \in L_{\mathrm{loc}}^p(V)$, and the analogue of (2). The point is that the parallelogram identity (5) can be replaced by the fact that the L^p-norms are uniformly convex for $1 < p < +\infty$.

Next we want to say a few words about the regularity of the minimizer for J (which we now call u). So let J be as in (1), and let u be such that

$$J(u) = \inf\{J(v)\,;\, v \in W^{1,2}(V)\} < +\infty. \tag{15}$$

Proposition 16. *If* (15) *holds, then* $\Delta u = u - g$ *(in the sense of distributions) on* V. *This means that*

$$\int_V u(x)\Delta\varphi(x) = \int_V [u(x) - g(x)]\varphi(x) \ \text{for} \ \varphi \in C_c^\infty(V). \tag{17}$$

To prove this let $\varphi \in C_c^\infty(V)$ be given, and set $A(\lambda) = J(u + \lambda\varphi)$ for $\lambda \in \mathbb{R}$. This is a quadratic function of λ, and its derivative at $\lambda = 0$ is

$$A'(0) = \frac{\partial}{\partial\lambda}\Big\{\langle u - g + \lambda\varphi, u - g + \lambda\varphi\rangle_2 + \langle\nabla u + \lambda\nabla\varphi, \nabla u + \lambda\nabla\varphi\rangle_2\Big\}(0)$$

$$= 2\langle u - g, \varphi\rangle_2 + 2\langle\nabla u, \nabla\varphi\rangle_2 = 2\int_V (u - g)\varphi + 2\int_V \nabla u \cdot \nabla\varphi \tag{18}$$

$$= 2\int_V (u - g)\varphi + 2\sum_{i=1}^n \int_V \frac{\partial u_i}{\partial x_i}\frac{\partial\varphi_i}{\partial x_i} = 2\int_V (u - g)\varphi - 2\int_V u\Delta\varphi,$$

where the last line comes from the definition of $\frac{\partial u_i}{\partial x_i}$ in (2.3) and uses the fact that φ has compact support in V. Since $A(\lambda)$ is minimal for $\lambda = 0$, $A'(0) = 0$ and (17) holds. The proposition follows. □

Corollary 19. *If* $g \in L^\infty(V)$ *and* u *satisfies* (15), *then* $u \in C_{\text{loc}}^1(V)$.

Here we used the strange notation $C_{\text{loc}}^1(V)$ to stress the fact that we do not imply any regularity up to the boundary. Corollary 19 is a consequence of Proposition 16 and simple elliptic regularity theory. Much more is true (especially if g is more regular), but we shall never need to know that.

Let us sketch a proof. We shall need to know that u is bounded and

$$||u||_\infty \leq ||g||_\infty. \tag{20}$$

To check this, let

$$u^*(x) = \text{Max}\big\{ -||g||_\infty, \text{Min}(||g||_\infty, u(x))\big\} \tag{21}$$

be the obvious truncation of u. We need to know that $u^* \in W^{1,2}(V)$ and $|\nabla u^*(x)| \leq |\nabla u(x)|$ almost everywhere. This follows from the following (classical) lemma, which will be proved later. (See Proposition 11.15.)

Lemma 22. *If* V *is a domain and* $u \in W^{1,2}(V)$, *then* $u^+ = \text{Max}(u, 0)$ *also lies in* $W^{1,2}(V)$, *and* $|\nabla u^+(x)| \leq |\nabla u(x)|$ *almost everywhere.*

Thus $\int_V |\nabla u^*|^2 \leq \int_V |\nabla u|^2$, and since $\int_V |u^* - g|^2 \leq \int_V |u - g|^2$ by (21), we get that $J(u^*) \leq J(u)$. Since u is the unique minimizer for J, $u^*(x) = u(x)$ almost everywhere, and (20) holds.

We may now return to the regularity of u in V. Let $x_0 \in V$ be given. Let φ be a bump function with compact support in V, and such that $\varphi(x) \equiv 1$ in a neighborhood of x_0. Set $b = (u - g)\varphi$ and let w be such that $\Delta w = b$. We shall produce such a w shortly.

Because of Proposition 16, $\Delta(u - w) = (u - g)(1 - \varphi) \equiv 0$ in a neighborhood of x_0. Thus $u - w$ is harmonic (and hence C^∞) near x_0, and it is enough to check that w is C^1 near x_0.

Now let us compute a reasonable choice of w. Let G denote the Green function (i.e., the fundamental solution of $\Delta G = \delta_0$). We know that

$$G(x) = \begin{cases} c \operatorname{Log} \dfrac{1}{|x|} & \text{if } n = 2, \\ c_n |x|^{-n+2} & \text{if } n \geq 3. \end{cases} \tag{23}$$

We can take $w = G * b$. Note that this is a nice continuous function because b is bounded with compact support, and $\Delta w = b$ by the theory of distributions. We also get that $\nabla w = \nabla G * b$. Here ∇G can be computed directly from (23). It is a vector-valued function which is smooth away from the origin and homogeneous of degree $-n + 1$. In particular, $\nabla G \in L^p_{\text{loc}}(\mathbb{R}^n)$ for $p < \frac{n}{n-1}$.

Since b is compactly supported, the values of $\nabla w = \nabla G * b$ near x_0 do not depend on the values of ∇G out of some big ball $B(0, R)$, and we can replace ∇G with $H = \mathbf{1}_{B(0,R)} \nabla G$. Then $H \in L^1(\mathbb{R}^n)$, and $H * b$ is bounded. It is even continuous, because $\|H(\cdot + t) - H(\cdot)\|_{L^1}$ tends to 0 when t tends to 0.

Altogether, ∇w is bounded and continuous near x_0, and Corollary 19 follows. $\qquad\square$

Of course our conclusion that $u \in C^1_{\text{loc}}(V)$ is not optimal. For instance, since $u - g$ is bounded, we get that $b = (u - g)\varphi \in L^p$ for all $p < +\infty$, hence $w = \Delta^{-1} b \in W^{2,p}$ for $p < +\infty$. The Sobolev injections then give a better regularity than C^1. Our assumption that g is bounded could be weakened too.

4 H^{n-1} is not semicontinuous but J has minimizers

In this section we just want to explain rapidly why the most logical method to produce minimizers for the Mumford-Shah functional fails, and how one can prove existence in spite of that. We shall return to this in Section 36, with some more details.

We have seen in the previous section that for each choice of a closed set $K \subset \Omega$ there is a unique $u \in W^{1,2}(\Omega \setminus K)$ for which $J(u, K)$ is minimal (that is, assuming that there is at least one $v \in W^{1,2}(\Omega \setminus K)$ for which $J(u, K) < +\infty$). Thus finding minimizers for J is only a problem about K.

Here is how we would like to proceed. First call

$$m = \inf \{ J(u, K) ; (u, K) \in \mathcal{A} \}, \tag{1}$$

where $\mathcal{A}$ is the set of admissible pairs (as in (2.1)). Of course we assume that $m < +\infty$ because otherwise the problem is not interesting. Let $\{(u_k, K_k)\}_{k \geq 0}$ be a minimizing sequence, i.e., a sequence of admissible pairs for which

$$\lim_{k \to +\infty} J(u_k, K_k) = m. \tag{2}$$

We wish to use this sequence to produce a minimizer, and so we want to extract a subsequence that converges. First, we can extract a subsequence (which we shall still denote the same way) such that $\{K_k\}$ converges to some relatively closed subset K of Ω. The convergence is for the Hausdorff metric on every compact subset of Ω. The existence of a subsequence that converges is a fairly straightforward exercise on completely bounded spaces and extractions of subsequences; we shall give more details in Section 34.

After this, we want to choose a new subsequence such that (after extraction) $\{u_k\}$ converges to some limit $u \in L^2(\Omega \setminus K) \cap W^{1,2}(\Omega \setminus K)$, and

$$\int_{\Omega \setminus K} \left[|u - g|^2 + |\nabla u|^2 \right] \leq \liminf_{k \to +\infty} \int_{\Omega \setminus K_k} \left[|u_k - g|^2 + |\nabla u_k|^2 \right]. \tag{3}$$

We shall see in Section 37 that this can be done rather automatically, just with the knowledge that for each closed ball $B \subset \Omega \setminus K$, there is a constant $C(B)$ such that $\int_B |\nabla u_k|^2 \leq C(B)$ for k large enough. Note that B does not meet K_k for k large, because it is compactly contained in $\Omega \setminus K$ and $\{K_k\}$ converges to K. The existence of the bound $C(B)$ on $\int_B |\nabla u_k|^2$ comes directly from (2).

Let us rapidly sketch a more brutal way to get a new subsequence for which $\{u_k\}$ converges and (3) holds. We may assume that for each k the function u_k was chosen to minimize $J(u_k, K_k)$ for the given K_k. Because of (3.20), the functions u_k are uniformly bounded (by $||g||_\infty$). By Corollary 3.19, u_k is also C^1 away from K_k. Assume for simplicity that Ω is bounded; then the proof of Corollary 3.19 also give bounds on $|\nabla u_k(x)|$ that depend only on $||g||_\infty$, Ω, and the distance from x to $K_k \cup \partial\Omega$. These estimates allow us to extract a subsequence so that, for each compact subset H of $\Omega \setminus K$, $\{u_k\}$ converges to u uniformly on H and $\{\nabla u_k\}$ converges to ∇u in $L^2(H)$. Then $u \in W^{1,2}(\Omega \setminus K)$ and (3) holds by Fatou's lemma.

So far, everything works out fine (except that we would have to check the details). To prove that (u, K) minimizes J, it would now be enough to verify that (possibly for a new subsequence)

$$H^{n-1}(K) \leq \liminf_{k \to +\infty} H^{n-1}(K_k). \tag{4}$$

Unfortunately, (4) is wrong for general sets: even for $n = 2$ it is very easy to find sequences $\{K_k\}$ that converge to limits K with much larger Hausdorff measure. For instance, choose for K_k a union of small intervals in $[0, 1]$. We can manage so that $H^1(K_k) = \alpha_k$ for any $\alpha_k < 1$ given in advance, and at the same time K_k is 2^{-k}-dense in $[0, 1]$. Then $\{K_k\}$ converges to $[0, 1]$ (for the Hausdorff metric), even though $\lim_{k\to+\infty} H^1(K_k) = \lim_{k\to\infty} \alpha_k$ could be any number smaller than 1.

The Hausdorff measure is not semicontinuous in the other direction either: $H^{n-1}(K)$ can be much smaller than $\liminf_{k\to\infty} H^{n-1}(K_k)$. Take $n = 2$ and for K_k the graph of the function $2^{-n} \sin 2^n x$ on $[0,1]$, for instance. This second bad behavior does not matter to us, though.

Let us return to our approach to the existence of minimizers. It would seem that it fails miserably, but the situation is not that desperate. We can still hope that we can find a minimizing sequence $\{(u_k, K_k)\}$ as before, but for which (4) holds.

This is actually possible. The first argument to this effect was given by Dal Maso, Morel, and Solimini [DMS1, 2], but only in the case when $n = 2$. This was later extended to dimensions $n \geq 3$ by Maddalena and Solimini [MaSo3]. The point is to show that we can choose sets K_k with a special regularity property, the "uniform concentration property", which itself yields (4). This is similar, but a little harder, to proving that for actual minimizers of the functional, the set K satisfies the same concentration property. We shall return to this approach later; see in particular Section 36. For the moment let us just observe that for this approach, proving regularity properties for minimizers (or approximate minimizers) can also be helpful when we try to get existence results.

To end this section, let us say a few words about the (by now) standard approach to existence. The idea is to transpose the problem to a slightly different setting, solve it there, and return to the initial formulation. In the new setting u is a function in SBV (special bounded variation) and $K = K_u$ denotes the singular set of u. SBV is a subset of BV, the class of functions whose derivative Du is a finite (vector-valued) measure. For general BV functions Du is composed of three parts: a part which is absolutely continuous with respect to the Lebesgue measure (the gradient part), a part which comes from jumps across a singular set K_u, and a third, more diffuse, singular measure. In dimension 1, a typical example of this third part is the derivative of the function called "devil's staircase", or Lebesgue's function, a derivative that is supported on a Cantor set. SBV functions are BV functions for which the third part of Du vanishes. See for instance [Gi] or [AFP3]. In this different setting it is easier to prove that the analogue of the Mumford-Shah functional has minimizers. This was proved by L. Ambrosio [Am]; one of the points of the whole approach is that BV is locally compact (basically, because we can always extract weakly convergent subsequences from sequences of measures with total mass $\leq C$). Then one still needs to show that minimizers for the sister functional on SBV yield minimizers for the Mumford-Shah functional. This was done in [DCL]. The whole argument extends to the variants

$$J(u, K) = \int_{\Omega\backslash K} |u - g|^p + \int_{\Omega\backslash K} |\nabla u|^q + H^{n-1}(K), \tag{5}$$

with $1 < p, q < +\infty$; the proof for the second part of the argument can be found in [CaLe2]. See also [AFP3].

Let us also note that the proof gives minimizers (u, K) for which K is automatically rectifiable, because it is a singular set coming from a SBV function.

The general method that we rapidly described here (that is, state the problem on a space like BV where there is compactness and then try to go back) is quite popular and fruitful in geometric measure theory and calculus of variations. For more details on the BV approach to the Mumford-Shah functional (and others), the best reference is almost surely the book [AFP3].

In the present text we shall try to take the opposite point of view and do almost everything without using BV. Because of this bias, we shall have to admit the existence of minimizers, all the way up to Section 36 where we shall at least present an alternative proof. In the meantime, we shall mostly concentrate on the regularity properties of K when (u, K) minimizes J.

5   Simple examples, no uniqueness, but few butterflies in Hong-Kong

In general minimizers for the Mumford-Shah functional are not unique, even if we take into account reasonable equivalence relations between admissible pairs (for instance, to allow us to add a set of H^{n-1}-measure 0 to K). This is not surprising; we shall see a few examples where the following situation occurs naturally. There are only two pairs that can minimize the functional J (typically, one with $K = \emptyset$ and another one with a substantial K). And if in addition some parameters are chosen right, these two competitors give the same result. Then J has exactly two minimizers, and we do not have uniqueness.

Of course, this lack of uniqueness also says that minimizers do not depend continuously on the data g. This does not necessarily mean that minimizers will always depend very wildly on g, though; see the comment about butterflies in Subsection **d**.

Here we shall try to give a few simple examples where there is no uniqueness, and where something can be actually computed. As it turns out, it is often the case that we can easily guess what the minimizers must be, but it is much more painful to prove our guesses. This is one of the very amusing (but sometimes a little irritating) properties of the Mumford-Shah functional.

There is a very nice way to deal with this problem and show that some presumed minimizers are actually minimizers. It is called (finding a) calibration, and when it works it gives slick and short proofs. We shall not use calibrations here, but the reader is welcome to consult [ABD1, 2].

a. Mumford-Shah functionals in dimension 1

Mumford-Shah minimizers are much easier to study on the line, and they already give a good idea of some of the features of higher dimensions. In this context Ω is an open interval (we shall always take connected domains Ω, because otherwise we could study the functional independently on each component). Recall that $\mathcal{A}$ is the set of pairs (u, K), where K is a closed set in Ω and u is defined on $\Omega \setminus K$, and its restriction to each of the open intervals I that compose $\Omega \setminus K$ has a derivative $u' \in L^2_{\text{loc}}(I)$. Here we mean this in the sense of distributions, but it is easy to check that it is the same as saying that on I, u is the indefinite integral of some function in L^2_{loc}, which turns out to be the distribution derivative u'. See Exercise 59.

It will not be much harder to work with the slightly more general functional

$$J(u, K) = \int_{\Omega \setminus K} |u(x) - g(x)|^2 a(x)\, dx + \int_{\Omega \setminus K} |u'(x)|^2 b(x) dx + \sum_{z \in K} c(z), \quad (1)$$

where g is a given function in $L^2_{\text{loc}}(\Omega)$, say, and a, b, c are three bounded positive functions defined on Ω. To simplify the discussion, let us also assume that c is bounded from below and that

$$m = \inf \{ J(u, K) \,;\, (u, K) \in \mathcal{A} \} \text{ is finite.} \quad (2)$$

Let us restrict our attention to pairs in $\mathcal{A}_1 = \{(u, K) \in \mathcal{A} \,;\, J(u, K) \leq m + 1\}$. Since c is bounded from below, K has a bounded number of elements. If we fix the set K, J is the constant $\sum_{z \in K} c(z)$, plus the sum of a finite number of pieces $J_{s,t}$ that come from the connected components (s, t) of $\Omega \setminus K$ and are defined by

$$J_{s,t}(u) = \int_s^t \left\{ |u(x) - g(x)|^2 a(x) + |u'(x)|^2 b(x) \right\} dx \quad (3)$$

for $s, t \in \overline{\Omega}$, $s \leq t$, and $u \in W^{1,2}_{\text{loc}}(s, t)$. Set

$$m_{s,t} = \inf \{ J_{s,t}(u) \,;\, u \in W^{1,2}_{\text{loc}}(s, t)\}. \quad (4)$$

Let us spend some time to check that

$$m_{s,t} \text{ is a continuous function of } s \text{ and } t \text{ in } \Omega. \quad (5)$$

Here we restrict our attention to $s, t \in \Omega$ because we do not seem to have much control on g and the competitors u near the endpoints. [We are trying not to add nondegeneracy assumptions on the functions a and b, and here we are paying for this.] However the proof below would also give the continuity at the endpoints of Ω if we assumed that $\int_\Omega g(x)^2 a(x) dx < +\infty$, for instance.

Let us fix a compact interval $\Omega_0 \subset \Omega$, and restrict to $s, t \in \Omega_0$. Also let $\varepsilon > 0$ be given. Since $g \in L^2(\Omega_0)$, we can find $\lambda > 0$ such that

$$\int_{\{x \in \Omega_0 ; |g(x)| \geq \lambda\}} g(x)^2 a(x) dx \leq \varepsilon. \quad (6)$$

Now let $s, t \in \Omega_0$ be given, with $s \leq t$. Let us even assume that $s < t$; the construction when $s = t$ is easier, and will be left to the reader. Let $u \in W^{1,2}_{\text{loc}}(s, t)$ be such that $J_{s,t}(u) \leq m_{s,t} + \varepsilon$. First we replace u with the truncation $u_\lambda = \text{Min}\,(-\lambda, \text{Max}\,(\lambda, u))$. It is easy to see that $u_\lambda \in W^{1,2}_{\text{loc}}(s, t)$ and $|u'_\lambda(x)| \leq |u'(x)|$ almost-everywhere (but if you find this hard, see Proposition 11.15). Hence

$$\int_s^t |u'_\lambda(x)|^2 b(x)dx \leq \int_s^t |u'(x)|^2 b(x)dx. \tag{7}$$

Besides,

$$\int_s^t |u_\lambda(x) - g(x)|^2 \, a(x)dx$$

$$\leq \int_{\{x \in (s,t); |g(x)| \leq \lambda\}} |u_\lambda(x) - g(x)|^2 a(x)dx + \int_{\{x \in \Omega_0; |g(x)| > \lambda\}} [|\lambda| + |g(x)|]^2 a(x)dx$$

$$\leq \int_s^t |u(x) - g(x)|^2 a(x)dx + 4 \int_{\{x \in \Omega_0; |g(x)| > \lambda\}} g(x)^2 a(x)dx$$

$$\leq \int_s^t |u(x) - g(x)|^2 a(x)dx + 4\varepsilon \tag{8}$$

because $|u_\lambda(x) - g(x)| \leq |u(x) - g(x)|$ when $|g(x)| \leq \lambda$ and by (6). Then $J_{s,t}(u_\lambda) \leq J_{s,t}(u) + 4\varepsilon \leq m_{s,t} + 5\varepsilon$.

Next we want to use u_λ to construct a good competitor for $J_{s',t'}$ when s' and t' are near s and t. Since we did not dare to require a lower bound for $b(x)$, we do not know for sure that $u_\lambda \in W^{1,2}(s, t)$ (even though $\int_s^t |u'_\lambda(x)|^2 b(x)dx < +\infty$), and so $u_\lambda(s)$ and $u_\lambda(t)$ may not be defined. So we choose s_1 and t_1 (very close to s and t, see later) such that $s < s_1 < t_1 < t$, and now u_λ is continuous on $[s_1, t_1]$ because $u_\lambda \in W^{1,2}_{\text{loc}}(s, t)$. In particular, $u_\lambda(s_1)$ and $u_\lambda(t_1)$ are defined. We define v on Ω_0 by $v(x) = u_\lambda(s_1)$ for $x \leq s_1$, $v(x) = u_\lambda(x)$ for $s_1 \leq x \leq t_1$, and $v(x) = u_\lambda(t_1)$ for $x \geq t_1$. It is easy to see that $v \in W^{1,2}_{\text{loc}}(\Omega_0)$. Moreover,

$$\int_{\Omega_0} |v'(x)|^2 b(x)dx = \int_{s_1}^{t_1} |u'_\lambda(x)|^2 b(x)dx \leq \int_s^t |u'(x)|^2 b(x)dx, \tag{9}$$

by (7). Now consider $s', t' \in \Omega_0$, with $s' \leq s \leq t \leq t'$, and set $\Delta = [s', t'] \setminus [s_1, t_1]$. Then

$$\int_{s'}^{t'} |v(x) - g(x)|^2 a(x)dx = \int_{s_1}^{t_1} |u_\lambda(x) - g(x)|^2 a(x)dx + \int_\Delta |v(x) - g(x)|^2 a(x)dx$$

$$\leq \int_s^t |u(x) - g(x)|^2 a(x)dx + 4\varepsilon + \int_\Delta [\lambda + |g(x)|]^2 a(x)dx \tag{10}$$

by (8). We already chose λ (depending on ε), and now we can use the integrability of $g(x)^2 a(x)$ on Ω_0 once again, to find $\delta > 0$ such that $\int_\Delta [\lambda + |g(x)|]^2 a(x)dx \leq \varepsilon$

for any set $\Delta \subset \Omega_0$ whose Lebesgue measure is less than 2δ. Thus, if we choose s_1 and t_1 above so close to s and t that $(s_1 - s) + (t - t_1) \leq \delta$, then (9) and (10) imply that

$$J_{s',t'}(v) \leq J_{s,t}(u) + 5\varepsilon \leq m_{s,t} + 6\varepsilon \qquad (11)$$

for all pairs $(s',t') \in \Omega_0^2$ such that $s' \leq s \leq t \leq t'$ and $(s - s') + (t' - t) \leq \delta$.

Note that λ, and then δ, depend only on ε and Ω_0. Since the function $m_{s,t}$ is nondecreasing in t and nonincreasing in s, its continuity on $\Omega_0 \times \Omega_0$ follows from (11), which proves (5).

A direct consequence of our definitions is that

$$m = \inf \left\{ \sum_{i=1}^{N-1} c(t_i) + \sum_{i=1}^{N} m_{t_{i-1},t_i} \right\}, \qquad (12)$$

where the infimum is taken over all finite sequences $\{t_i\}_{0 \leq i \leq N}$ such that $t_0 < t_1 < \cdots < t_N$ and t_0 and t_N are the endpoints of Ω. Moreover, we know that we can restrict to $N \leq C$, because c is bounded from below. If in addition c is lower semicontinuous (which means that $c(x) \leq \liminf_{k \to +\infty} c(x_k)$ whenever $\{x_k\}$ converges to x), then (5) allows us to find a finite sequence $\{t_i\}_{0 \leq i \leq N}$ such that

$$m = \sum_{i=1}^{N-1} c(t_i) + \sum_{i=1}^{N} m_{t_{i-1},t_i}.$$

Because of this, it is enough to study the individual functionals $J_{s,t}$, which is substantially easier than what happens in higher dimensions. In particular, the same convexity argument as in Section 3 gives the existence (and uniqueness) of minimizers for $J_{s,t}$ if $m_{s,t} < +\infty$ and under mild nondegeneracy conditions on the functions a and b (for instance, if they are locally bounded from below). The existence of minimizers for J follows.

Before we get to counterexamples, let us record that minimizers for $J_{s,t}$ satisfy a differential equation with Neumann conditions at the endpoints. To simplify the notation, let us only consider bounded intervals.

Lemma 13. *If b is of class C^1 and does not vanish on $[s,t]$, and if $u \in W_{\text{loc}}^{1,2}(s,t)$ is such that $J_{s,t}(u) = m_{s,t} < +\infty$, then u is of class C^1 on $[s,t]$, $u'' \in L^2((s,t))$,*

$$b(x)u''(x) + b'(x)u'(x) = a(x)(u(x) - g(x)) \quad \text{on } (s,t),$$
$$\text{and } u'(s) = u'(t) = 0. \qquad (14)$$

First observe that $u' \in L^2(s,t)$, because $J_{s,t}(u) < +\infty$ and b is bounded from below. Similarly, $a(u - g) \in L^2(s,t)$, because a is bounded and $\int_s^t a(u-g)^2 \leq J_{s,t}(u) < |\infty$. Let us proceed as in Proposition 3.16, pick any smooth function φ (not necessarily compactly supported in (s,t) yet), and compare u with its competitors $u + \lambda\varphi$, $\lambda \in \mathbb{R}$. We get the scalar product condition

$$\int_s^t \left[a(x)(u(x) - g(x))\varphi(x) + b(x)u'(x)\varphi'(x) \right] dx = 0, \qquad (15)$$

as in (3.18). When we take φ compactly supported in (s,t), we get that $(bu')' = a(u-g)$ in the sense of distributions. But this forces bu' to be the indefinite integral of $a(u-g)$ (which lies in $L^2(s,t)$, as we just noticed); see Exercise 59. In particular bu' is bounded, and hence also u', because b is bounded from below. Now $u' = b^{-1}(bu')$ and b^{-1} is C^1, so $u'' = b^{-1}(bu')' - b'b^{-2}bu' = b^{-1}a(u-g) - b'b^{-1}u'$, both in the sense of distributions and pointwise almost everywhere (see Exercise 59 again). The first part of (14) follows.

Next, u' is continuous on $[s,t]$, because it is the indefinite integral of $u'' \in L^2$. In particular, $u'(s)$ and $u'(t)$ are well defined. To prove that $u'(t) = 0$, we apply (15) with a smooth nondecreasing function φ such that $\varphi(x) = 0$ for $x \leq t'$ and $\varphi(x) = 1$ for $x \geq t$, where $t' < t$ is very close to t. The contribution $\int_s^t a(x)(u(x) - g(x))\varphi(x)dx$ to (15) tends to zero as t' tends to t (because φ is bounded and $a(u-g) \in L^2$), while $\int_s^t b(x)u'(x)\varphi'(x)dx$ tends to $b(t)u'(t)$. This proves that $b(t)u'(t) = 0$; hence $u'(t) = 0$ because $b(t) > 0$. Of course $u'(s) = 0$ for similar reasons, and Lemma 13 follows. $\square$

Example 16. Take $n = 1$, $\Omega = [-1,1]$, $a \equiv b \equiv c \equiv 1$, and $g_\lambda = \lambda\mathbf{1}_{[0,1]}$ for $\lambda \geq 0$. It is clear now that there are only two possible minimizers. The first one is given by $K = \{0\}$ and $u = g_\lambda$, and then $J(u,K) = 1$. The second one is for $K = \emptyset$ and $u = u_\lambda$, where u_λ is the only minimizer of $J_{-1,1}$ in (3). We could easily use Lemma 13 to compute u_λ, but the main point is that $u_\lambda = \lambda u_1$, so that $J(u_\lambda,\emptyset) = A\lambda^2$, where $A = \int[(u_1-g_1)^2 + (u_1')^2]$ is a fixed positive constant.

For $\lambda^2 < A^{-1}$, the only minimizer for J is $(u_\lambda,\emptyset)$, while for $\lambda^2 > A^{-1}$ the only minimizer is $(g_\lambda, \{0\})$. And for $\lambda^2 = A^{-1}$, we have exactly two minimizers. Thus uniqueness and smooth dependence on parameters fail in a very simple way. The principle behind all our other examples will be the same.

b. Simple examples in dimension 2

Here we return to the usual Mumford-Shah functional in (2.13), with $a = b = c = 1$. We start with the product of a line segment with Example 16.

Example 17. (A white square above a black square.) This is taken from [DaSe5] but the example could be much older. We take $\Omega = (0,1) \times (-1,1) \subset \mathbb{R}^2$ and $g = g_\lambda = \lambda\mathbf{1}_{\Omega^+}$, where $\Omega^+ = (0,1) \times (0,1)$ and λ is a positive parameter. We shall denote by $L = (0,1) \times \{0\}$ the singular set of g_λ.

There are two obvious candidates to minimize J. The first one is the pair (g_λ, L) itself. The second one is $(\tilde{u}_\lambda, \emptyset)$, where $\tilde{u}_\lambda(x,y) = u_\lambda(y)$ and u_λ is the C^1-function on $(-1,1)$ for which

$$I = \int_{(-1,1)} \left|u_\lambda(y) - \lambda\mathbf{1}_{(0,1)}(y)\right|^2 + |u_\lambda'(y)|^2 \tag{18}$$

is minimal. Thus u_λ is the same as in Example 16. Set

$$m = \text{Min}(J(g_\lambda, L), J(\tilde{u}_\lambda, \emptyset)); \tag{19}$$

we want to check that

$$J(u, K) \geq m \text{ for all } (u, K) \in \mathcal{A}. \tag{20}$$

So let $(u, K) \in \mathcal{A}$ be an admissible pair (see (2.1) for the definition). We may as well assume that u minimizes J with the given set K. This will not really change the proof; it will only make it a little more pleasant because we know that u is C^1 on $\Omega \setminus K$. Set

$$X = \Big\{ x \in (0, 1) \, ; \, K \cap [\{x\} \times (-1, 1)] = \emptyset \Big\}. \tag{21}$$

For $x \in X$, the function $y \to u(x, y)$ is C^1 on $(-1, 1)$; then

$$\int_{X \times (-1,1)} |u - g_\lambda|^2 + |\nabla u|^2 \tag{22}$$

$$\geq \int_X \Big\{ \int_{(-1,1)} |u(x, y) - \lambda \mathbf{1}_{(0,1)}(y)|^2 + \Big| \frac{\partial u}{\partial y}(x, y) \Big|^2 dy \Big\} dx \geq H^1(X) \, I,$$

where I is as in (18). On the other hand, $(0, 1) \setminus X = \pi(K)$, where π is the orthogonal projection on the first axis. Since π is 1-Lipschitz, Remark 2.11 says that $H^1(K) \geq H^1((0, 1) \setminus X) = 1 - H^1(X)$. We can combine this with (22) and get that

$$J(u, K) \geq \int_{X \times (-1,1)} \Big[|u - g_\lambda|^2 + |\nabla u|^2 \Big] + H^1(K) \geq H^1(X) \, I + [1 - H^1(X)]. \tag{23}$$

Thus $J(u, K)$ is a convex combination of I and 1. Since $I = J(\tilde{u}_\lambda, \emptyset)$ and $1 = J(g_\lambda, L)$, (20) follows from (23) and (19).

With just a little more work, we could prove that if $J(u, K) = m$, then $(u, K) = (\tilde{u}_\lambda, \emptyset)$ or $(u, K) = (g_\lambda, L)$, modulo adding a set of measure 0 to K.

Thus at least one of our two favorite candidates is a minimizer for J. Since I is obviously proportional to λ^2, there is a value of λ for which $I = 1$, and for this value J has two minimizers. Of course one could object that this example is very special, and that we are using the two vertical sides of $\partial\Omega$, so we shall give two other ones.

Example 24. Take $g = \lambda \mathbf{1}_{B(0,1)}$, in the domain $\Omega = \mathbb{R}^2$ or $\Omega = B(0, R) \subset \mathbb{R}^2$, where $\lambda > 0$ and $R > 1$ are given constants.

It is tempting here to use the symmetry of the problem to compute things almost explicitly. By an averaging argument similar to the proof of (20) above (but radial), we can show that for each $(u, K) \in \mathcal{A}$ there is a radial competitor (u^*, K^*) such that $J(u^*, K^*) \leq J(u, K)$ (see [DaSe5], page 164, for details). Here radial means that K^* is a union of circles centered at the origin, and $u^*(\rho e^{i\theta})$ is just a function of ρ. Then the choice can be reduced to two competitors, as follows.

Lemma 25. *If (u, K) is a radial minimizer for J, then $K = \partial B(0, 1)$ and $u = g$, or*

$$K = \emptyset \ \text{ and } \ u = u_0 \ \text{ is the function of } W^{1,2}(\Omega)$$

$$\text{that minimizes } \int_\Omega |u - g|^2 + |\nabla u|^2. \tag{26}$$

We shall prove this when $\Omega = B(0, R)$, but the argument for $\Omega = \mathbb{R}^2$ would be similar.

Some choices of K^* are clearly impossible. For instance, including in K^* a circle $\partial B(0, r)$ such that $r > 1$ does not make sense, because $\partial B(0, 1)$ is shorter and does at least as well in the approximation term. Also, it does not make sense to have two circles $\partial B(0, r)$ with $r \leq 1$ in K^*, since the largest one alone does equally well in the approximation term. Thus the competition reduces to the candidate in (26) and pairs (u_r, K_r), where $0 < r \leq 1$, $K_r = \partial B(0, r)$, and u_r is the radial function on $\Omega \setminus K_r$ which is equal to λ on $B(0, r)$ and minimizes

$$E_r = \int_{\Omega \setminus \overline{B}(0, r)} |u_r - g|^2 + |\nabla u_r|^2. \tag{27}$$

We want to show that $0 < r < 1$ does not lead to a minimizer, and the main step will be the following.

Sublemma 28. E_r *is a concave function of r.*

The functions u_r are not so explicit, so we shall use the differential equation that they satisfy to study their variations. Eventually we shall compute the derivative of E_r in terms of $u_r(r)$, and the sublemma will follow.

First write the radial function $u_r(x)$ as $\lambda v_r(|x|)$. Of course $v_r(x) = 1$ on $[0, r)$, but what we care about is its values on (r, R). We write the integrals in polar coordinates, and get that (the restriction to (r, R) of) v_r is the function in $W^{1,2}(r, R)$ which minimizes

$$\overline{E}_r = \int_r^R \Big[x \left[v_r(x) - \mathbf{1}_{[0,1]}(x) \right]^2 + x\, v_r'(x)^2 \Big]\, dx, \tag{29}$$

and then $E_r = 2\pi\lambda^2 \overline{E}_r$. We want to show that for r fixed,

$$v_r(x) \text{ is a nonincreasing function of } x, \tag{30}$$

and that it is a nonincreasing function of r as well. Figure 1 shows what v_r could look like for different values of r.

First apply Lemma 13, with $(s, t) = (r, R)$ and $a(x) = b(x) = x$. We get that

$$xv_r''(x) + v_r'(x) = x[v_r(x) - \mathbf{1}_{[0,1]}(x)] \ \text{ on } (r, R), \quad \text{and } v_r'(r) = v_r'(R) = 0. \tag{31}$$

Lemma 13 also tells us that v_r and v_r' are continuous on $[r, R]$, and the first part of (31) also says that xv_r' is the indefinite integral of $x[v_r - \mathbf{1}_{[0,1]}]$ (this is even

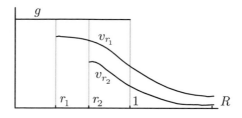

Figure 1

how we proved it in Lemma 13). Thus the variations of xv'_r will depend on the sign of $v_r - \mathbf{1}_{[0,1]}$.

Note that $0 \le v_r(x) \le 1$ on $[r, R]$, by uniqueness of u_r or v_r and because otherwise $\mathrm{Max}(0, \mathrm{Min}(v_r, 1))$ would give a smaller integral in (29). (See Proposition 3.3 and (3.20) for details.) We could also get this from (31).

Now (31) says that $xv'_r(x)$ decreases for $x \le r$, and increases for $x \ge r$. Since its boundary values are 0, we see that $v'_r(x) \le 0$ everywhere, and (30) follows.

Next let $0 < r_1 < r_2 \le r$ be given, and set $w = v_{r_1} - v_{r_2}$ on $[r_2, R]$. Then

$$(xw')'(x) = xw''(x) + w'(x) = xw(x) \text{ on } [r_2, R],$$
$$w'(r_2) \le 0, \text{ and } w'(R) = 0, \tag{32}$$

by (31) and the discussion above. Let us check that

$$w(x) \ge 0 \text{ and } w'(x) \le 0 \text{ on } [r_2, R]. \tag{33}$$

Let x_0 be the first point of $[r_2, R]$ where $w(x_0) \le 0$ or $w'(x_0) = 0$. This point exists, because (32) says that $w'(R) = 0$. Note that $w'(x_0) \le 0$ even if $w(x_0) \le 0$ is reached first; indeed either $x_0 = r_2$ and $w'(x_0) \le 0$ by (32), or else $w'(x_0) > 0$ would lead to a contradiction because it would imply that x_0 is not the first point where $w(x_0) \le 0$.

Let us check that $w(x_0) \ge 0$. Indeed, if $w(x_0) < 0$, (32) says that $xw'(x)$ is strictly decreasing for some time after x_0, which forces $w < 0$ and $w' < 0$ there (because $w'(x_0) \le 0$). Moreover, $xw'(x)$ stays decreasing as long as $w(x) < 0$, which means forever (because w cannot increase again before w' itself becomes nonnegative). This is impossible because $w'(R) = 0$. So $w(x_0) \ge 0$.

Note that the same argument also excludes the case when $w(x_0) = 0$ and $w'(x_0) < 0$. Since the case when $w(x_0) > 0$ and $w'(x_0) < 0$ is impossible by definition of x_0, we get that $w'(x_0) < 0$ is impossible, and so $w'(x_0) = 0$.

In the unlikely case when $w(x_0) = w'(x_0) = 0$, we get that $w(x) = 0$ on the whole $[r_1, R]$, because 0 is the only solution of $xw'' + w' = xw$ with these Cauchy data at x_0. Then (33) holds trivially.

Thus we may assume that $w(x_0) > 0$. If $x_0 < R$, (32) says that $xw'(x)$ is strictly increasing for some time after x_0, which forces the functions w and w' to be positive there. And then $xw'(x)$ stays increasing as long as this is the case, which therefore is forever. This is impossible, because $w'(R) = 0$. So $x_0 = R$, and (33) holds by definition of x_0.

Next we want to check that

$$w(r_2) \leq C|w'(r_2)|, \tag{34}$$

where we allow C to depend on R and a lower bound for r_2. Set $A = |w'(r_2)|$. Since $xw'(x)$ is nondecreasing (by (32) and (33)), $r_2 w'(r_2) \leq xw'(x) \leq Rw'(R) = 0$ on $[r_2, R]$. Thus $|w'(x)| \leq x^{-1} r_2 A \leq A$. Take $C \geq 2(R - r_2)$ (to be chosen soon), and suppose that (34) fails. Then $w(x) \geq w(r_2) - A|x - r_2| \geq CA - (R - r_2)A \geq CA/2$ for $x \in [r_2, R]$. Now $(xw')'(x) \geq Cr_2A/2$ by (32), so $0 = Rw'(R) \geq r_2 w'(r_2) + CAr_2(R - r_2)/2 = -r_2 A + CAr_2(R - r_2)/2$. This is clearly impossible if C is large enough; (34) follows.

We are now ready to compute the derivative of E_r at the point r_1. Take $r_2 > r_1$, very close to r_1, and let us first evaluate the part of $\overline{E}_{r_1} - \overline{E}_{r_2}$ that lives on $[r_2, R]$. Set

$$\Delta = \int_{r_2}^{R} \left\{ [v_{r_1}(x) - \mathbf{1}_{[0,1]}(x)]^2 + v'_{r_1}(x)^2 - [v_{r_2}(x) - \mathbf{1}_{[0,1]}(x)]^2 - v'_{r_2}(x)^2 \right\} x \, dx$$

$$= \int_{r_2}^{R} \left\{ [v_{r_1(x)} - \mathbf{1}_{[0,1]}(x)]^2 + v'_{r_1}(x)^2 \right\} x \, dx - \overline{E}_{r_2}. \tag{35}$$

[See (29) for the definition of $\overline{E}_r$.] Also set $f_i(x) = v_{r_i}(x) - \mathbf{1}_{[0,1]}(x)$ and $g_i(x) = v'_{r_i}(x)$ for $i = 1, 2$, so that

$$\Delta = \langle f_1, f_1 \rangle + \langle g_1, g_1 \rangle - \langle f_2, f_2 \rangle - \langle g_2, g_2 \rangle$$
$$= \langle f_1 - f_2, f_1 + f_2 \rangle + \langle g_1 - g_2, g_1 + g_2 \rangle, \tag{36}$$

where we set $\langle f, g \rangle = \int_{r_2}^{R} f(x) g(x) x \, dx$ to condense notation.

For each $\lambda \in \mathbb{R}$, the function $h_\lambda = v_{r_2} + \lambda(v_{r_1} - v_{r_2})$ is of class C^1 on $[r_2, R]$, so it is a competitor in the definition of v_{r_2} (i.e., the fact that it minimizes $\overline{E}_{r_2}$ in (29)). Because of this, the derivative at $\lambda = 0$ of $\int_{r_2}^{R} \{ [h_\lambda(x) - \mathbf{1}_{[0,1]}(x)]^2 + h'_\lambda(x)^2 \} x \, dx$ vanishes. And this yields $\langle f_1 - f_2, f_2 \rangle + \langle g_1 - g_2, g_2 \rangle = 0$. [The computations are the same as in Proposition 3.16 and Lemma 13 above.] Because of this, (36) becomes

$$\Delta = \langle f_1 - f_2, f_1 - f_2 \rangle + \langle g_1 - g_2, g_1 - g_2 \rangle. \tag{37}$$

Since $f_1 - f_2 = v_{r_1} - v_{r_2} = w$ and $g_1 - g_2 = v'_{r_1} - v'_{r_2} = w'$,

$$||f_1 - f_2||_\infty + ||g_1 - g_2||_\infty = ||w||_\infty + ||w'||_\infty \leq w(r_2) + r_2^{-1}||xw'||_\infty \tag{38}$$
$$= w(r_2) + |w'(r_2)| \leq C|w'(r_2)| = C|v'_{r_1}(r_2)|$$

by (33), because $xw'(x)$ is nondecreasing and $w'(R) = 0$ (by (32) and (33)), by (34), and because $v'_{r_2}(r_2) = 0$. Here again C may depend on R and a lower bound for r_1.

Now $v'_{r_1}(r_1) = 0$ (by (31)) and $(xv'_{r_1})'$ is bounded (again by (31) and because $|v_{r_1}| \leq 1$), so $|v'_{r_1}(r_2)| \leq C(r_2 - r_1)$. Altogether $|\Delta| \leq C(r_2 - r_1)^2$, by (37) and (38).

Recall from (29) and (35) that

$$\overline{E}_{r_1} - \overline{E}_{r_2} = \Delta + \int_{r_1}^{r_2} \left\{ [v_{r_1}(x) - \mathbf{1}_{[0,1]}(x)]^2 + v'_{r_1}(x)^2 \right\} x\, dx, \qquad (39)$$

and the integral is equivalent to $(r_2 - r_1)r_1[1 - v_{r_1}(r_1)]^2$ because $v'_{r_1}(r_1) = 0$ and $r_1 < 1$. This proves that $\overline{E}_r$ has a half-derivative on the right at the point r_1, which is equal to $-r_1[1 - v_{r_1}(r_1)]^2$.

The computation of the half-derivative on the left is similar. This time we compute it at r_2, to be able to keep the same notation. Observe that for $x \in (r_1, r_2)$, $|v'_{r_1}(x)| \leq |v'_{r_1}(r_1)| + C(r_2 - r_1) = C(r_2 - r_1)$, because $(xv'_{r_1})'$ is bounded. Similarly, $|v_{r_1}(x) - v_{r_2}(r_2)| \leq |v_{r_1}(x) - v_{r_1}(r_2)| + |v_{r_1}(r_2) - v_{r_2}(r_2)|$. The first term is at most $C(r_2 - r_1)$, because $|v'_{r_1}(\xi)| \leq C$ in (r_1, r_2) (recall that $v'_{r_1}(r_1) = 0$ and $(xv'_{r_1})'$ is bounded, see above (39)). The second term is $|w(r_2)| \leq C(r_2 - r_1)$, by (38) and the discussion that follows. Altogether $\overline{E}_r$ has a half-derivative on the left at r_2, equal to $-r_2[1 - v_{r_2}(r_2)]^2$. Thus we have a full derivative, and

$$E_r = 2\pi\lambda^2\overline{E}_r \text{ is differentiable on } (0,1], \text{ and } \frac{\partial E_r}{\partial r} = -2\pi\lambda^2 r[1 - v_r(r)]^2. \quad (40)$$

To complete the proof of Sublemma 28, note that $1 \geq v_{r_1}(r_1) \geq v_{r_1}(r_2) \geq v_{r_2}(r_2)$ for $0 < r_1 < r_2 \leq r$, by (30) and because $v_{r_1} \geq v_{r_2}$ (by (33)). Thus $[1 - v_r(r)]^2$ is nondecreasing, and the derivative of E_r is nonincreasing. Sublemma 28 follows. □

Recall from the few lines before Sublemma 28 that the only possible radial minimizers are the one given by (26) and the pairs (u_r, K_r) associated with $K_r = \partial B(0, r)$, $0 < r \leq 1$. Also, $J(u_r, K_r) = H^1(K_r) + E_r$ (by the definitions (2.13) and (27), and because $u_r = \lambda$ in $B(0, r)$). Thus $J(u_r, K_r)$ also is a concave function of r, and its minimum is reached for $r = 1$ or $r = 0$, which correspond to the two choices in Lemma 25. If the minimum is also reached at any other point, then $J(u_r, K_r)$ is constant, hence $E_r = J(u_r, K_r) - 2\pi r = 2\pi(1 - r)$ and its derivative is -2π. This is incompatible with (40), which says that the derivative vanishes at $r = 0$. This proves Lemma 25. □

The proof of Lemma 25 also gives the existence of minimizers for J. Indeed, for each pair $(u, K) \in \mathcal{A}$, there is a radial pair that does better, and even a radial pair with at most one circle. Thus it is enough to look for a minimizer in the class of radial pairs considered in Lemma 25. But Sublemma 28 allows us to reduce to two competitors, and the existence follows.

We now return to our story about uniqueness. We keep the same domain $\Omega = B(0, R)$ (possibly, with $R = +\infty$) and function $g = g_\lambda = \lambda \mathbf{1}_{B(0,1}$ as before, but now we worry about how things depend on λ. Call J_λ the corresponding functional.

There are two interesting competitors. The first one is $(\partial B(0,1), g_\lambda)$, and of course $J_\lambda(\partial B(0,1), g_\lambda) = 2\pi$. The second one is the pair $(\emptyset, u_0)$ described in (26). Note that $u_0 = u_{0,\lambda}$ depends on λ, but the dependence is very simple: $u_{0,\lambda} = \lambda u_{0,1}$ (by linearity and uniqueness). Thus $J_\lambda(\emptyset, u_{0,\lambda}) = \lambda^2 \alpha$, with $\alpha = J_1(\emptyset, u_{0,1})$.

For λ small, the pair $(\emptyset, u_{0,\lambda})$ is better than $(\partial B(0,1), g_\lambda)$, and Lemma 25 (together with a closer look at the reduction to radial competitors) says that $(\emptyset, u_{0,\lambda})$ is the only minimizer (modulo trivial equivalences, like adding a closed set of measure zero to K). For λ large, we get that $(\partial B(0,1), g_\lambda)$ is the only minimizer. And when $\alpha \lambda^2 = 2\pi$, we get two different minimizers.

Remark 41. Even though we used the invariance under rotations to reduce to the one-dimensional Mumford-Shah functional, the argument above is not very specific. We used the fact that we could reduce the competition to very few pairs (u, K) (first, a one-parameter family, and then only two) to make the argument simpler, but the most important features of our example were the following. When λ goes from 0 to large values, K starts from the empty set, and eventually has to be quite large. But also, there is no way that it will make this transition continuously. The point is that very small sets K are forbidden.

In Example 24, this is because little circles are not allowed. One can see this because (40) says that the derivative of E_r vanishes at the origin, so $E_0 - E_r$ is of the order of r^2 for r small, which is not enough to compensate the $2\pi r$ that you lose in the length term when you add $\partial B(0, r)$ to K.

In this example, we could also have proved the existence of a $\lambda_0 > 0$ for which there are two very different minimizers, without using Lemma 25. This would have forced us to check two additional things. First, there is at least one radial minimizer for each J_λ, but also, if $\{\lambda_k\}$ is a sequence that tends to some λ_0, and if (u_k, K_k) minimizes J_{λ_k} for each k, then there is a subsequence of $\{(u_k, K_k)\}$ that converges, and the limit is a minimizer for λ_0. We would apply this to the infimum λ_0 of values where there is a minimizer with $K \neq \emptyset$, to get a first minimizer with $K = \emptyset$, and a different one that comes from limits of nonempty (and therefore large) sets. In the case of radial minimizers with $K_k = \partial B(0, r_k)$, all this would be reasonably easy to do.

The point of Example 43 below is to show that this is a general phenomenon.

Example 42. We can take $\Omega = \mathbb{R}^2$ or $\Omega = B(0, R)$, and $g = g_r = \lambda \mathbf{1}_{B(0,r)}$, where this time λ is fixed and we let r vary. If λ is chosen large enough, then there is $r_0 \in (0, R)$ such that $K = \emptyset$ does not give a minimizer for $r = r_0$, and the same sort of argument as in Example 24 shows that there is a value of $r \leq r_0$ for which J has two different minimizers.

c. Description of a more general example

The point of the next example is to show that the lack of uniqueness of minimizers for J is a quite general phenomenon. It will also serve as an advertisement for other parts of this book, because we shall discover that complete justifications of simple little facts may require a nontrivial amount of machinery (some of which is included later). Examples in the same spirit were already given in [MoSo2], Theorem 15.49 on page 198.

Example 43. Let $\Omega \subset \mathbb{R}^n$ be a fixed domain; some constraints will arise later. Also let g_t, $t \in [0,1]$, be a one-parameter family of bounded functions on Ω, with $\sup_{t \in [0,1]} ||g_t||_\infty \leq C$. Denote by J_t the Mumford-Shah functional with initial image $g = g_t$ and constants $a = b = c = 1$ (see (2.13)). Thus

$$J_t(u,K) = \int_{\Omega/K} |u - g_t|^2 + H^{n-1}(K) + \int_{\Omega/K} |\nabla u|^2 \tag{44}$$

for $(u,K) \in \mathcal{A}$ (see (2.1)). We shall assume that $g_t \in L^2(\Omega)$ for each t (which does not follow from $g_t \in L^\infty$ if Ω is not bounded). Then

$$m(t) = \inf \{J_t(u,K) \,;\, (u,K) \in \mathcal{A}\} \leq ||g_t||^2_{L^2(\Omega)} < +\infty \tag{45}$$

because $m(t) \leq J(\emptyset, 0) = ||g_t||^2_2$. We also assume that

$$t \to g_t \text{ is a continuous mapping from } [0,1] \text{ to } L^2(\Omega). \tag{46}$$

We are interested in conditions that force jumps, or absence of uniqueness, for minimizers of J_t. Thus we also demand that

$$g_0 = 0 \quad \text{and} \quad \inf \{J_1(u, \emptyset) \,;\, u \in W^{1,2}_{\text{loc}}(\Omega)\} > m(1), \tag{47}$$

where the last condition forces minimizers for J_1 to have some nontrivial amount of K.

A typical example would be to take a nice set $A \subset \Omega$, with $|A| > 0$ and $H^{n-1}(\partial A) < 1$, and to take $g_t = \alpha t \mathbf{1}_A$. If the constant α is large enough, (47) holds because $m(1) \leq J_1(\partial A, g_1)$, while the infimum in (47) is of the order of α^2.

Before we get to minimizers, let us already observe that

$$m(t) \text{ is a continuous function of } t. \tag{48}$$

Indeed let s, $t \in [0,1]$, and $(u,K) \in \mathcal{A}$ be given. Then

$$J_t(u,K) - J_s(u,K) = \int_{\Omega \backslash K} [|u - g_t|^2 - |u - g_s|^2] = \int_{\Omega \backslash K} [g_s - g_t][2u - g_t - g_s]. \tag{49}$$

If $J_s(u,K) \leq m(s) + \varepsilon$, then $\int_{\Omega \backslash K} |u - g_s|^2 \leq m(s) + \varepsilon$,

$$||2u - g_t - g_s||_2 \leq 2||u - g_s||_2 + ||g_s - g_t||_2 \leq 2(m(s) + \varepsilon)^{1/2} + ||g_t - g_s||_2 , \tag{50}$$

and (49) says that

$$J_t(u, K) \leq J_s(u, K) + ||g_t - g_s||_2 \left[2(m(s) + \varepsilon)^{1/2} + ||g_t - g_s||_2 \right]. \qquad (51)$$

Since $m(t) \leq J_t(u, K)$ and we can do this with ε as small as we want, we get that

$$m(t) \leq m(s) + ||g_t - g_s||_2 \left[(2m(s))^{1/2} + ||g_t - g_s||_2 \right]. \qquad (52)$$

The symmetric estimate would be obtained the same way, and (48) follows.

The most obvious candidate (for minimizing J_t) is the pair $(v_t, \emptyset)$, where v_t denotes the function in $W^{1,2}(\Omega)$ for which

$$I(t) = \int_\Omega |v_t - g_t|^2 + |\nabla v_t|^2 \qquad (53)$$

is minimal. Note that v_t is unique (by Proposition 3.3), and $I(t)$ is also a continuous function of t, by the proof of (48). Our assumption (47) says that $m(1) < I(1)$, so $(v_t, \emptyset)$ does not minimize J_t for $t = 1$ (or even t close to 1, by continuity). The first step of our argument will be to check that we can find $t_0 > 0$ such that

$$(v_t, \emptyset) \text{ is the unique minimizer for } J_t \text{ when } 0 \leq t < t_0. \qquad (54)$$

Here uniqueness is to be understood modulo trivial equivalence: the other minimizers are $(\tilde{v}_t, K)$, where K is a closed subset of Ω such that $H^{n-1}(K) = 0$, and $\tilde{v}_t$ is the restriction of v_t to $\Omega \setminus K$. With the terminology of Section 8, $(v_t, \emptyset)$ is the only "reduced" minimizer.

Actually, the proof will be more important than (54) itself. The main ingredient will be the local Ahlfors-regularity of K when (u, K) is a reduced minimizer. Let us not give a precise definition here, but only say two things: a reduced minimizer is a minimizer (u, K) for which K is minimal with the given u; for each minimizer (u, K), there is an equivalent reduced minimizer (u', K') (i.e., for which $K' \subset K$ and u is the restriction of u' to $\Omega \setminus K$). See Section 8 for details.

What about local Ahlfors-regularity? We shall see in Section 18 (and discuss the proof later) that if (u, K) is a reduced minimizer of the Mumford-Shah functional (in any domain, and associated to any bounded function g), then

$$C_0^{-1} r^{n-1} \leq H^{n-1}(K \cap B(x, r)) \leq C_0 r^{n-1} \qquad (55)$$

for all $x \in K$ and $r \leq 1$ such that $B(x, r) \subset \Omega$, where C_0 is a constant that depends only on n and $||g||_\infty$. This is a theorem of Dal Maso, Morel, and Solimini [DMS] when $n = 2$, and Carriero and Leaci [CaLe2] for $n > 2$. Also see [So2] for a proof without SBV.

Let us return to our uniformly bounded functions g_t and the corresponding J_t. If $\Omega = \mathbb{R}^n$, we can immediately apply this to any point of K and $r = 1$. We get that

$$H^{n-1}(K) \geq C_0^{-1} \text{ when } (u, K) \text{ is a (reduced) minimizer for } J_t \text{ and } K \neq \emptyset. \qquad (56)$$

If $\Omega \neq \mathbb{R}^n$, we can still prove (56) under reasonable assumptions on Ω, but we need a boundary version of the local Ahlfors-regularity result to get (56). Indeed, we could imagine reduced minimizers (u, K) for J_t such that K is not empty, but stays very close to $\partial\Omega$, so that we cannot apply (55) with a large radius.

If Ω is bounded and smooth (C^2 would be more than enough), then we also have (55) for any $x \in K$ (even when $B(x, r)$ is not contained in Ω), and hence also (56). The best way to see this is probably to check that the proofs of [DMS2] and [CaLe2] go through. We shall give in Section 79 a rapid proof of (55) when Ω is a Lipschitz domain (with small constants in dimension $n > 2$).

Anyway, let us assume that we have (56). Then we know that if we ever switch from $(v_t, \emptyset)$ to some other minimizer, this will have to be in a discontinuous way: any K that shows up has to be fairly large. Here we expressed largeness in terms of Hausdorff measure, but (55) tells us that Hausdorff measure controls other quantities (like the diameter, or how many disjoint balls of a given size can meet K).

We are not finished yet; we need to know that

for each $t \in [0, 1]$, we can find a reduced minimizer (u_t, K_t) for J_t. (57)

This is the existence result that was rapidly discussed in Section 4. We can easily deduce (54) from this and (56). Indeed, $m(t) \leq I(t) < C_0^{-1}$ for t small, so (56) says that there is no reduced minimizer with $K \neq \emptyset$. Since $(v_t, \emptyset)$ is the only other possible minimizer, (54) holds.

Let us concentrate our attention on what happens near

$$T = \sup\{t \in [0, 1] \,;\, (v_t, \emptyset) \text{ is a minimizer for } J_t\}. \tag{58}$$

We know that $0 < T < 1$, by (54) and (47). For $t > T$, let (u_t, K_t) be a reduced minimizer for J_t (provided to us by (57)). By definition of T, K_t is not empty, so (56) says that $H^{n-1}(K) \geq C_0^{-1}$.

We want to produce a minimizer (u_T, K_T) for J_T, with $K_T \neq \emptyset$, and for this we try to follow the argument suggested in Section 4. That is, we first take a decreasing sequence $\{t_k\}$ that converges to T, and for which the sets K_{t_k} converge to some limit K_T. For the convergence of closed sets in Ω, we use the Hausdorff distance, see Section 34. Then we extract a subsequence for which the functions u_{t_k} converge to some limit u_T, uniformly on compact subsets of $\Omega \setminus K_T$. This is easy too. But then we need to know that (u_T, K_T) is a reduced minimizer for J_T, which is somewhat harder. This uses the continuity of $m(t)$ and Fatou's lemma (for the convergence of the integral terms), but more importantly the lower semicontinuity of the Hausdorff measure restricted to our sequence K_{t_k}. Which in turn uses the uniform concentration property of [DMS]. See Section 4 for a preview, Section 25 for the uniform concentration property in dimension 2, Section 35 for the relevant semicontinuity result, and Sections 36–38 for the limiting argument. Altogether, the desired result is a consequence of Exercise 38.102.

Once we get our minimizer (u_T, K_T), we are almost finished. We have to check that K_T is not empty. This is not excluded a priori, because we could imagine that the K_{t_k} go to the boundary $\partial\Omega$ and disappear at the limit, but this does not happen. One way to see this is to look more carefully at the proof of minimality for (u_T, K_T) and observe that we must have $H^{n-1}(K_T) = \liminf_{k\to+\infty} H^{n-1}(K_{t_k})$, because if there was a strict inequality, our proof would show that (u_{t_k}, K_{t_k}) was not a minimizer. [The point is that we use (u_T, K_T) to construct a competitor for (u_{t_k}, K_{t_k}).] This proof yields $H^{n-1}(K_T) \geq C_0^{-1}$ directly. Another way to proceed is to show that (since we assumed Ω to be bounded and regular) K_{t_k} cannot be contained in a very thin neighborhood of $\partial\Omega$. The idea is that if this was the case, we could compose u_{t_k} with a deformation of $\overline{\Omega}$ that sends this thin neighborhood to $\partial\Omega$. We could do this so that the integral terms of J_{t_k} would be increased only slightly, while we could get rid of K_{t_k} altogether. The contradiction would then come from (56). See Lemma 79.3 for a similar argument. So some part of K_{t_k} stays reasonably far from the boundary, and the limit K_T is not empty.

Once all of this is done, we get a reduced minimizer (u_T, K_T) for J_T such that $K \neq \emptyset$, and even $H^{n-1}(K_T) \geq C_0^{-1}$ (by (56)). The pair $(v_t, \emptyset)$ is also a minimizer for J_T, by definition of T and because $I(t)$ and $m(t)$ are both continuous. So we have at least two fairly different minimizers for J_T. Note also that at the point T, the set of (all the) minimizers for J_t also has a nontrivial jump.

It is probable that the proof described above is not optimal, but the author doubts that there is a simple proof of nonuniqueness in the general case of Example 43. The reader is welcome to try; it is a good exercise.

d. How many Hong-Kong butterflies?

The title refers to the nasty Hong-Kong butterflies that are known to cause storms in New York (or the too often forgotten good ones that prevent weather disasters). If we start from a function g for which there are two different minimizers and add to g a small perturbation (the butterfly), this can change the set of minimizers dramatically. But does this really happen often?

All the examples above are based on the same principle: when we minimize the Mumford-Shah functional associated to slowly varying functions g_t, the corresponding (set of) minimizer(s) may have jumps at some points. Here we singled out situations where we jump from the empty set to reasonably large sets, but one can easily imagine other types of jumps or lack of uniqueness.

For instance, one should be able to come up with situations where the lack of uniqueness comes from a rupture of symmetry. That is, Ω and g would have some symmetry, and there would be different minimizers for the functional that would not have the same symmetries as Ω and g, but would be obtained from each other by the missing symmetries. Figure 2 suggests an initial image and domain (a square paved with four black and white tiles) and two likely minimizers. If (u, K) is a minimizer and the square is large enough, K should not be empty;

the argument is the same as in the examples above. We also know that taking $u = g$ and K equal to two perpendicular line segments is not optimal, by the same argument as in the next section, near Figure 6.2. Then the two sets K in Figure 2.b and 2.c seem optimal, and in particular there is no special need for a piece of curve that would separate the two main parts of the big component of $\Omega \setminus K$, because we expect u and K to be symmetric with respect to the diagonals. Note that we are saying "likely minimizers" here, because it is not clear that we would be able to prove that the minimizers of J are as in the picture. Also, the rupture of symmetry here is with respect to a joint rotation of 90° and replacement of g with $1 - g$, for instance; one could look for examples with more obvious ruptures of symmetry.

Similarly, it would be amusing to have examples where different Mumford-Shah minimizers correspond to different logical interpretations of a picture.

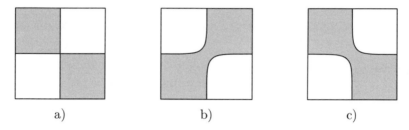

Figure 2. a) A function g b) A minimizer   c) Another minimizer

How often can we get two (or more) different minimizers? The author (at least) does not know, but would suspect that this is fairly rare. One possible statement is that, generically, there is a unique minimizer. ["Generically" seems natural, because it is perhaps artificial to put a measure on the set of functions g, and the notion of forcing uniqueness by adding a small perturbation makes sense.] Maybe this is not even so hard to prove.

Similarly, could it be that the Mumford-Shah functional never has two different minimizers that are very close to each other? Of course this would have to be made precise.

If all this was true, it would also mean that we don't need to be too concerned about our lack of uniqueness (and maybe even dependence on parameters); maybe they only correspond to rare situations of ambiguous images. Of course the lack of uniqueness is not in itself bad for image processing: we already know that for some images there are more than one equally good ways to interpret them. It is a sign that finding exact minimizers may be hard, though. But of course finding some sort of a local minimizer is often good enough in practice.

Exercise 59. Let I be an open integral, and f be a locally integrable function on I such that $f' \in L^1_{\text{loc}}(I)$ in the sense of distributions. This means that there is a locally integrable function f' on I such that $\int_I f(x)\varphi'(x)dx = -\int_I f'(x)\varphi(x)dx$ for all C^∞ functions φ with compact support in I. Let $z \in I$ be given.

1. For each $g \in L^1_{\text{loc}}(I)$, set $\widetilde{g}(x) = \int_z^x g(t)dt$ for $x \in I$. Show that $\widetilde{g}$ is locally integrable and that its derivative (in the sense of distributions) is g. [Hint: Fubini.]

2. Show that $f(x) = \int_z^x f'(t)dt + C$ on I.

3. Let h be another function such that $h' \in L^1_{\text{loc}}(I)$ in the sense of distributions. Show that $(fh)' \in L^1_{\text{loc}}(I)$, with $(fh)' = fh' + f'h$. [Hint: more Fubini.] The case when h is smooth is easier.

Exercise 60. Complete the verification of Example 42 (using Lemma 25).

Exercise 61. Try to follow the argument of Subsection **c** in the special case of Examples 24 and 42.

Exercise 62. Guess what could be a one-parameter family of functions g for which the Mumford-Shah functional has two minimizers. Here again, a verification may be very hard.

Exercise 63. Consider $g = \lambda \mathbf{1}_{B(0,1)}$ in the domain $\Omega = B(0, R)$, where λ is chosen so that $K = \emptyset$ and $K = \partial B(0,1)$ give two minimizers for J (as in Example 24). Try to guess what minimizers of J look like when we replace g with $g \pm \alpha \mathbf{1}_{B(z,\rho)}$, with α and ρ very small (and depending on the position of z in $B(0, R)$, and the sign). The author has a vague guess, but he does not claim he can justify it.

6   The Mumford-Shah conjecture and some known results

Let $\Omega \subset \mathbb{R}^n$ and $g \in L^\infty(\Omega)$ be as in the previous sections, and assume that

$$m = \inf \{J(u, K); (u, K) \in \mathcal{A}\} < +\infty, \tag{1}$$

where $\mathcal{A}$ (the set of admissible pairs) is as in (2.1) and J is the Mumford-Shah functional, as in (2.13) but with $a = b = c = 1$. (We have seen in Remark 2.15 how to reduce to this choice.)

We said in the previous sections that J has (not necessarily unique) minimizers. Let (u, K) be one of them (i.e., suppose that $J(u, K) = m$). Note that K determines u, and u is reasonably regular on $\Omega \setminus K$ (see Section 3), so we shall naturally concentrate on the regularity properties of K first. Once we get some amount of regularity for K, we may consider getting higher regularity for both K and u, but this will not be our main concern.

If we want to have clean statements on K, we shall need to restrict to *reduced* minimizers. Indeed, if Z is any closed subset of Ω such that $H^{n-1}(Z) = 0$, we can add Z to K, keep the same function u (on $\Omega \setminus (K \cup Z)$), and we get another minimizer for J. We want to avoid this because Z could be really ugly, and it is useless.

Definition 2. *A* <u>*reduced minimizer*</u> *for* J *is a pair* $(u, K) \in \mathcal{A}$ *such that* $J(u, K) = m$ *(with* m *as in (1)) and, whenever* $\widetilde{K} \subset K$ *is another relatively closed subset of* Ω *with* $\widetilde{K} \neq K$, u *does not have any extension in* $W^{1,2}_{\mathrm{loc}}(\Omega \setminus \widetilde{K})$.

See Section 8 for a slightly more general definition.

Note that since $H^{n-1}(K) < +\infty$, K has zero Lebesgue measure in $\mathbb{R}^n$ (see Exercise 2.20); thus any extension of u would have to coincide with u almost everywhere. So we can simplify our last statement and just say that if $\widetilde{K} \subset \Omega$ is (relatively) closed and strictly contained in K, then $u \notin W^{1,2}_{\mathrm{loc}}(\Omega \setminus \widetilde{K})$.

We shall see in Section 8 that for each minimizer (u, K) there is an equivalent reduced minimizer $(\widetilde{u}, \widetilde{K})$, i.e., one for which $\widetilde{K} \subset K$ and u is the restriction of $\widetilde{u}$. Thus we can concentrate on reduced minimizers.

The Mumford-Shah conjecture [MuSh2] is that, when n= 2, Ω is nice enough, and (u, K) is a reduced minimizer for J,

$$K \text{ is a finite union of } C^1 \text{ arcs} . \tag{3}$$

This is just a minimal way of asking for more. That is, once we know (3), we can prove more. First, it is not too hard to see that if (3) holds,

$$\begin{array}{c} \text{the arcs that compose } K \text{ can only meet at their} \\ \text{extremities, by sets of 3, and with angles of } 120°. \end{array} \tag{4}$$

The main point is the following. If some of the C^1 arcs that compose K meet at $x \in K$ in a different way, then for each small disk $B = B(x, \varepsilon)$ we can replace $K \cap B$ with a union P of line segments with the same endpoints on $K \cap \partial B$, so that $H^1(P) \leq H^1(K \cap B) - \alpha\varepsilon$, where $\alpha > 0$ is a small constant that depends on the angles (see the pictures below). Then we replace u in B with a function v with jumps on P (instead of $K \cap B$). That is, we consider the restriction of u to $\partial B \setminus K$, and extend it to $B \setminus P$. We get the desired contradiction with the minimality of (u, K) if we can find v so that $\int_{B \setminus P} |\nabla u|^2 + |u - g|^2 \leq \alpha\varepsilon/2$, so that we do not lose in the two other terms of the functional what we won in length. The verifications are not very difficult, but take some time. The most straightforward route seems to prove first that

$$\int_{B(x,t) \setminus K} |\nabla u|^2 \leq C\, t^{1+\tau} \tag{5}$$

for t small. The estimates on v are then fairly easy. For the decay estimate (5), the simplest is probably to prove it first for energy-minimizing functions on $\Omega \setminus K$, because they are easier to control. If we have $C^{1+\varepsilon}$-regularity for K near x, we can use the conformal invariance of energy integral, as in Section 17 and Remark 17.17. Otherwise, we can use Bonnet's decay estimate from Section 47. Once we have (5) for energy-minimizing functions, there is a small trick to get (5) itself. See Exercises 11 and 12 for a few more details, and Exercise 55.11 for a proof.

The property (4) is a little shocking when we think about image processing. First, it means that J does not like corners. To a segmentation with a corner (as in Figure 1.a), it will prefer smoothed out versions like the one in Figure 1.b. [The little faces show whether the functional likes the segmentation.] It is somewhat harder to show that J will also refuse segmentations like the one suggested in Figure 1.c, but this can be done. For this, it seems that the best is to use blow-up limits. We shall see later (Proposition 37.8) that we can find dilations of the pair (u, K) at x (see near (9) below), that converge to some limit (v, G). Then (v, G) is a "reduced global minimizer in the plane" (see Proposition 40.9). In the situations suggested in Figures 1.c and 2.d, G separates the plane in at least two connected components, and we shall see in Section 66 that this forces G to be a line or a propeller (three half-lines with the same origin, that make 120° angles).

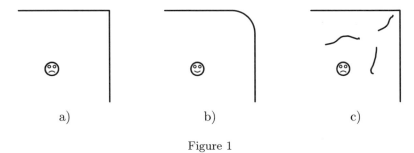

Figure 1

Similarly J does not like crossings. Figure 2.a may look reasonable, but Figures 2.b and 2.c will do better. In this case also we cannot trick J by adding lots of little connected components like the ones in Figure 2.d.

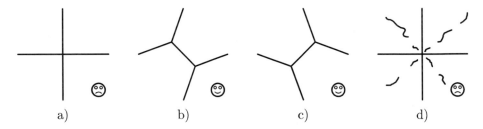

Figure 2

Finally, the fact that J wants to deform T-junctions like the one in Figure 3.a into symmetric junctions those like in Figure 3.b is not very satisfactory. Recall that T-junctions do carry some important pieces of information, in particular because we can often infer that the T-junction arises because some object is in front of another one and hides it partially, and that its boundary corresponds to the top branch of the T. Thus, in the case of Figure 3.a, we could infer that the arc of the circle is closer than the line segment; Figure 3.b is less clear. Perhaps this is not extremely bad news for image processing, though, because all these things may happen at very small scales.

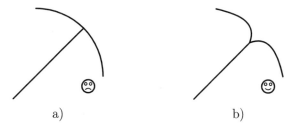

a)   b)

Figure 3

A second remark about (3) is that it does not give the optimal regularity. Actually, the choice of C^1 was made in [MuSh2] because this is the condition on K that we need to get more regularity easily. It turns out that if (3) holds,

$$\text{each of the } C^1 \text{ arcs that compose } K \text{ is at least } C^{1,1},$$
$$\text{except perhaps at its extremities.} \tag{6}$$

This was proved in [Bo]. By $C^{1,1}$ we mean that the tangent vector to the curve at the point x is a Lipschitz function of x; thus the curve is almost C^2. The fact that our curves are $C^{1,\alpha}$ (i.e., that the tangent vector is Hölder-continuous with exponent α) for $0 < \alpha < 1$ is substantially easier, and could almost be done by hand. On the other hand, the precise behavior of K near a tip (i.e., the free end of a C^1 curve) is not known yet. As far as the author can tell, it is not even known whether the curve is C^1 at the tip. See Section 69 for other details about tips.

When K is a nice curve, one can write down an equation that relates the curvature of K to the jump of $|\nabla u|^2$ across K and the function g. The optimal regularity that one can prove on K will then depend on the regularity of g. By contrast, it is amusing to note that knowing additional regularity properties of g never helped (so far) to prove reasonably low regularity properties of K (like those in (3) above, or the results mentioned below).

Whether (3) always holds is still a conjecture, but fortunately there are lots of partial results, some of them listed below. Most people think that the conjecture is true; the author is even ready to bet that in the description (3), there is a lower bound on the diameter of each C^1 curve (and hence an upper bound on the number of curves) that depends only on $||g||_\infty$ and the nice domain Ω.

There is no precise analogue of the Mumford-Shah conjecture in higher dimensions. Even when $n = 3$, we do not have a list of all the asymptotic behaviors of K and u near a point of Ω that we should expect. We shall return to this issue in Section 76, and try to give a reasonable partial conjecture in dimension 3.

Let us now turn to what we know. In this book we shall mostly concentrate on the dimension $n = 2$ and the behavior of K inside Ω, but surprisingly the behavior of K near $\partial\Omega$ is easier to deal with. It turns out that if when $n = 2$ and Ω is bounded and C^1, there is a neighborhood of $\partial\Omega$ where K is composed of a finite number of C^1 curves that meets $\partial\Omega$ perpendicularly. See [MaSo4] or Section 79.

A very useful property of K (when (u, K) is a reduced minimizer of J) is its local <u>Ahlfors-regularity</u>: there is a constant C, that depends only on $\|g\|_\infty$ and the dimension n, such that for $x \in K$ and $0 < r < 1$ such that $B(x, r) \subset \Omega$,

$$C^{-1} r^{n-1} \leq H^{n-1}(K \cap B(x, r)) \leq C r^{n-1}. \tag{7}$$

This was proved in [DMS] when $n = 2$, and in [CaLe2] when $n \geq 2$. We shall give two proofs of (7). The first one (in Sections 18–22) will be based on [DMS]; it works only when $n = 2$, but has the advantage of not using a compactness argument. The second one (in Section 72) looks more like the one from [CaLe1], except that it was translated back from the BV setting by Siaudeau [Si]; the presentation will be slightly more elliptic. See [So2] for a third one, based on the bisection property.

There are a few "intermediate" regularity properties of K when (u, K) is a reduced minimizer for J, like the property of projections of [DiKo] when $n = 2$ (see Section 24) and the uniform concentration property of [DMS] (see Section 25 when $n = 2$, and a few comments in Remark 25.5 and Remark 75.17 when $n > 2$) that we do not wish to explain here; let us just quote two main properties of K.

First, K is locally <u>uniformly rectifiable</u>. When $n = 2$, this was proved in [DaSe4]; it means that for every choice of $x \in K$ and $0 < r \leq 1$ such that $B(x, r) \subset \Omega$, there is an Ahlfors-regular curve Γ with constant $\leq C$ that contains $K \cap B(x, r)$. Moreover, C depends only on $\|g\|_\infty$. Let us not define Ahlfors-regular curves here, but refer to Definition 27.3 and say that they are connected sets that satisfy the analogue of (7) with $n = 2$. They have the same sort of regularity as graphs of Lipschitz functions, but form a strictly larger class mainly because they are allowed to cross themselves. When $n \geq 3$, local uniform rectifiability was proved in [DaSe6]; the definition is somewhat more complicated, but the main point is still that $K \cap B(x, r)$ is contained in a "hypersurface" with fairly decent parameterization. We shall only prove this result in detail when $n = 2$; see Sections 27–32. Then see Sections 73 and 74 for hints about higher dimensions.

We also know that for H^{n-1}-almost every $x \in K$ we can find $t > 0$ such that

$$K \cap B(x, t) = \Gamma \cap B(x, t) \text{ for some } C^1 \text{ (embedded) hypersurface } \Gamma. \tag{8}$$

More precisely, there is a constant $C > 0$ that depends only on $\|g\|_\infty$ and n, such that if $x_0 \in K$, $0 < r \leq 1$, and $B(x_0, r) \subset \Omega$, we can find $x \in K \cap B\left(x_0, \frac{r}{2}\right)$ and $t \in \left[\frac{r}{C}, \frac{r}{4}\right]$ such that $B(x, t)$ satisfies (8). This is more precise, because it implies that the singular subset of K composed of the points x for which (8) fails for all $t > 0$ is actually of Hausdorff dimension $< n - 1$. This was proved in [Da5] when $n = 2$, and in [AFP1] when $n \geq 2$, completed by [Ri1] for the more precise statement. Here again, we shall only prove it for $n = 2$; this will be the main point of Part E. See Section 75 for a few comments about the proof in higher dimensions, and [MaSo2] for a different approach to [Ri1].

At the present time it seems that the most promising approach to the conjecture is through blow-ups and monotonicity formulae. Take $n = 2$, and let $x_0 \in K$ be given. Also choose a sequence $\{t_n\}$ of positive numbers that tends to 0. Define pairs (u_n, K_n) by $K_n = t_n^{-1}(K - x_0)$ and

$$u_n(x) = t_n^{-1/2} u(x_0 + t_n x) \quad \text{for } x \in \Omega_n \setminus K_n, \tag{9}$$

where $\Omega_n = t_n^{-1}(\Omega - x_0)$. From each such sequence, it is possible to extract subsequences that converge to limits (u_∞, K_∞), where K_∞ is a closed set in $\mathbb{R}^2$ and $u_\infty \in W^{1,2}_{\text{loc}}(\mathbb{R}^2 \setminus K_\infty)$. (The notion of convergence is a little like in Section 4, and will be discussed in Section 36.)

A. Bonnet [Bo] proved that each such limit (u_∞, K_∞) is a "global minimizer in the plane". See the beginning of Section 54 for the definition and Section 38 for the proof. This is interesting because the term $\int |u - g|^2$ is no longer present in the definition of a global minimizer, and we can expect a much simpler description of those. Actually, there is even a short list of four types of global minimizers (where K is empty, or a line, or three half-lines meeting with $120°$ angles, or a half-line), and we can conjecture that all global minimizers are in the short list. We shall see in Section 71 that this new conjecture implies the old one (i.e., (3)). See Section 54 for additional details about all this.

Bonnet also proved that when (u_∞, K_∞) is a (reduced) global minimizer in the plane and K_∞ is connected, (u_∞, K_∞) is indeed of one of the four types in the short list. [See more details in Section 61.] This allowed him to prove that when (u, K) is a reduced minimizer for J and K_0 is an *isolated* connected component of K, then K_0 is a finite union of C^1 curves (as in (3)). Thus (3) holds when K has only finitely many connected components.

Some other cases of the new conjecture are known; see for instance Section 64 for the case when K_∞ is contained in a countable union of lines and Section 66 for the case when $\mathbb{R}^2 \setminus K_\infty$ is not connected or K_∞ is symmetric with respect to the origin. Also see Section 63 for a formula that allows us to compute u_∞ given K_∞, and Part G for local consequences.

Remark 10. The various regularity properties that we can prove for singular sets of minimizers of the Mumford-Shah functional J give nice justifications for the use of J for image segmentation, because they say that J will automatically select segmentations with nice-looking boundaries. The Ahlfors-regularity property (7), for instance, is a good indication of resistance to noise (because noise is not Ahlfors-regular); similarly, the reader will probably be happy to know that J does not select sets K like the Cantor set in Exercise 18.26.

Exercise 11. Suppose (u, K) is a Mumford-Shah minimizer, that $x \in K$, and that there is a small ball centered at x where K is the union of three C^1 curves that meet at x with angles other than $120°$. Also assume that we have (5). Let t be small.

1. Find $\rho \in (t/2, t)$ such that $\int_{\partial B(x,\rho) \backslash K} |\nabla u|^2 \leq Ct^\tau$. Set $B = B(x, \rho)$.
   Let $x_0 \in B(x, \rho/2)$ be such that the union P of the tree segments from x_0 to the points of $K \cap \partial B$ is substantially shorter than $K \cap B$. Call u_0 the restriction of u to $\partial B \backslash K$.

2. Find an extension v of u_0 to $\overline{B} \backslash P$ such that $\int_{B \backslash G} |\nabla v|^2 \leq Ct^{1+\tau}$. You may use the fact that u_0 is C^1 if you want; if in trouble see Lemma 22.16 or even Exercise 20.49.

3. Check that you can choose v so that $||v||_\infty \leq ||u||_\infty \leq ||g||_\infty$, so $\int_{B \backslash P} |u - g|^2 \leq Ct^2$, by (3.20). Find a contradiction.

Exercise 12. Suppose that (u, K) is a Mumford-Shah minimizer, that $x \in K$ is given, and that we can prove the following: for t small enough, we can find a function v_t that coincides with u on $\partial B(x,t) \backslash K$, lies in $W^{1,2}(B(x,t) \backslash K)$, and minimizes $\int_{B(x,t) \backslash K} |\nabla v_t|^2$ under this constraint, and in addition $\int_{B(x,at) \backslash K} |\nabla v_t|^2 \leq Ca^{1+\tau} \int_{B(x,t) \backslash K} |\nabla v_t|^2$ for $a < 1$.

1. Show that $\nabla(u - v_t)$ is orthogonal to ∇v_t in $L^2(B(x,t) \backslash K)$. [Hint: $v_t + \lambda(u - v_t)$ is a competitor in the definition of v_t.]

2. Check that (we can choose v_t so that) $||v_t||_\infty \leq ||g||_\infty$ and that

$$\int_{B(x,t) \backslash K} |\nabla u|^2 \leq \int_{B(x,t) \backslash K} |\nabla v_t|^2 + 4\pi t^2 ||g||_\infty^2.$$

3. Show that

$$\int_{B(x,at) \backslash K} |\nabla u|^2 \leq Ca^{1+\tau} \int_{B(x,t) \backslash K} |\nabla v_t|^2 + Ct^2 ||g||_\infty^2$$
$$\leq Ca^{1+\tau} \int_{B(x,t) \backslash K} |\nabla u|^2 + Ct^2 ||g||_\infty^2.$$

4. Fix a small a and deduce (5) (maybe with a different τ) from this. We shall see variants of this argument a few times.

7 Many definitions of almost- and quasiminimizers

Even if we only want to study minimizers of the Mumford-Shah functional, it is often convenient to forget about the function g and use the slightly more general notions of almost- and quasiminimizers defined below. It will also be good to know that our arguments work for analogues of the Mumford-Shah functional with slowly varying coefficients, or that we can treat Dirichlet problems (i.e., where the values of u on $\partial\Omega$, the boundary of our (given) open set in $\mathbb{R}^n$, are fixed). In this section we want to give many definitions of weaker notions of minimality, and prove the easy inclusions between the corresponding classes. The author is convinced that the right choice of definitions in this context may be important, and realizes that there is a compromise to be found between relatively simple and natural definitions, and definitions that are sufficiently general to be applied in many cases. Possibly the choice of definitions made in this book is a little too complicated, but hopefully the reader will find the notions of almost- and quasiminimizers below natural. They should give a good control of what happens to the singular set K, and perhaps a little less good description of problems related to u; but we do this on purpose because we think the dependence on u is often a simpler problem.

We shall distinguish between almost-minimizers, where the pair (u, K) looks more and more like a local minimizer of $H^1(K) + \int_{\Omega \setminus K} |\nabla u|^2$ when we look at smaller and smaller scales, and quasiminimizers that are more scale invariant. Unfortunately, these names can be found in the literature with different meanings and the reader should be a little careful about this. In particular, what we call almost-minimizers here is often called quasiminimizers, and the (more general) notion of quasiminimizers here does not seem to exist in other references.

Additional modifications to the basic definitions are needed, because we often want the notions to be local (i.e., we only want to compare our pair (u, K) with competitors that coincide with (u, K) on the boundary of the domain Ω, for instance). Also, there is a notion of topological almost- or quasiminimizer where we only allow competitors (v, G) of (u, K) for which G still has some of the separating properties as K near $\partial \Omega$, because this seems to be the right notion when we take limits. Finally, we would like to allow ourselves to replace the usual gradient by its images under linear transformations, and this will lead to linearly distorted versions of all our definitions.

a. Almost-minimizers

Many differences between our classes will come from conditions that we impose on competitors for a given pair $(u, K) \in \mathcal{A}$ before we allow the comparison. Recall from (2.1) that

$$\mathcal{A} = \mathcal{A}(\Omega) = \big\{ (u, K); K \subset \Omega \text{ is (relatively) closed in } \Omega \tag{1}$$
$$\text{and } u \in W^{1,2}_{\text{loc}}(\Omega \setminus K) \big\}.$$

Definition 2. *Let $(u, K) \in \mathcal{A}$ and a ball $B(x, r)$ be given. A <u>competitor for (u, K)</u> <u>in $B(x, r)$</u> is a pair $(\widetilde{u}, \widetilde{K}) \in \mathcal{A}$ such that*

$$\widetilde{K} \setminus \overline{B}(x, r) = K \setminus \overline{B}(x, r) \tag{3}$$

and

$$\widetilde{u}(z) = u(z) \text{ for almost-every } z \in \Omega \setminus (K \cup \overline{B}(x, r)). \tag{4}$$

Thus we just require $(\tilde{u}, \tilde{K})$ to coincide with (u, K) out of $\overline{B}(x,r)$. (Taking the closure will make the definition easier to use, because we often want to modify things on $\partial B(x,r)$.)

For our definition of almost-minimizers we need a nonnegative function $h : (0, +\infty) \to [0, +\infty]$ such that

$$h \text{ is nondecreasing and } \lim_{r \to 0^+} h(r) = 0. \tag{5}$$

The fact that h is required to be nondecreasing is just here for convenience. We decided to authorize $h(r) = +\infty$ for r large to stress the fact that our almost-minimality condition often does not give much information at large scales.

Definition 6. *The pair $(u, K) \in \mathcal{A}$ is an* <u>*almost-minimizer with gauge function h*</u> *if*

$$\int_{\Omega \cap B(x,r) \setminus K} |\nabla u|^2 + H^{n-1}(K \cap \overline{B}(x,r))$$
$$\leq \int_{\Omega \cap B(x,r) \setminus \tilde{K}} |\nabla \tilde{u}|^2 + H^{n-1}(\tilde{K} \cap \overline{B}(x,r)) + h(r) r^{n-1} \tag{7}$$

for all balls $B(x,r)$ and all choices of competitors $(\tilde{u}, \tilde{K})$ for (u, K) in $B(x,r)$.

Note that this is called quasi-minimizer in [AFP] or [MaSo], for instance. The main example is minimizers of the Mumford-Shah functional (or minor variants), as follows.

Proposition 8. *Let $\Omega \subset \mathbb{R}^n$ be open, and let $g \in L^\infty(\Omega)$ and $1 \leq p < +\infty$ be given. Set*

$$J(u, K) = H^{n-1}(K) + \int_{\Omega \setminus K} |u - g|^p + \int_{\Omega \setminus K} |\nabla u|^2 \tag{9}$$

for $(u, K) \in \mathcal{A}$. Let $(u, K) \in \mathcal{A}$ be such that

$$J(u, K) = \inf \{J(v, G)\,;\, (v, G) \in \mathcal{A}\} < +\infty. \tag{10}$$

Then (u, K) is an almost-minimizer, with the gauge function $h(r) = 2^p \omega_n ||g||_\infty^p\, r$. Here ω_n denotes the Lebesgue measure of the unit ball in $\mathbb{R}^n$.

To prove the proposition, let a minimizer (u, K) be given. First we want to check that

$$||u||_\infty \leq ||g||_\infty. \tag{11}$$

To see this, set $u^*(z) = \text{Max}\{-||g||_\infty, \text{Min}[||g||_\infty, u(z)]\}$. Then Lemma 3.22 (see Proposition 11.15 for its proof) says that $u^* \in W^{1,2}_{\text{loc}}(\Omega \setminus K)$ and $|\nabla u^*(z)| \leq |\nabla u(z)|$ almost-everywhere. Thus $(u^*, K) \in \mathcal{A}$ and $J(u^*, K) \leq J(u, K)$, with a strict inequality if u^* is not equal to u almost-everywhere (because $\int_{\Omega \setminus K} |u^* - g|^p$ would be strictly smaller). Since $J(u, K)$ is minimal, (11) follows.

Now let $(\widetilde{u}, \widetilde{K})$ be a competitor for (u, K) in some $B(x, r)$. We want to prove that (7) holds, with $h(r) = 2^p \omega_n ||g||_\infty^p r$. Set

$$v(z) = \text{Max}\{-||g||_\infty, \text{Min}(||g||_\infty, \widetilde{u}(z))\}$$

for $z \in \Omega \setminus \widetilde{K}$. By Lemma 3.22 again, $v \in W^{1,2}_{\text{loc}}(\Omega \setminus \widetilde{K})$ and so $(v, \widetilde{K}) \in \mathcal{A}$. Hence

$$J(u, K) \leq J(v, \widetilde{K}), \tag{12}$$

by (10). Now the pair $(\widetilde{u}, \widetilde{K})$ coincides with (u, K) out of $\overline{B}(x, r)$ (by Definition 2), and so $|\widetilde{u}(z)| \leq ||g||_\infty$ a.e. out of $\overline{B}(x, r)$ (by (11)). Hence $v(z) = \widetilde{u}(z) = u(z)$ a.e. out of $\overline{B}(x, r)$, and in fact the pairs $(v, \widetilde{K})$ and (u, K) coincide out of $\overline{B}(x, r)$. Thus everything in (12) happens in $\overline{B}(x, r)$, and

$$\int_{\Omega \cap B(x,r) \setminus K} |\nabla u|^2 + H^{n-1}(K \cap \overline{B}(x, r))$$

$$\leq \int_{\Omega \cap B(x,r) \setminus \widetilde{K}} |\nabla v|^2 + H^{n-1}(\widetilde{K} \cap \overline{B}(x, r))$$

$$+ \int_{\Omega \cap B(x,r) \setminus \widetilde{K}} |v - g|^p - \int_{\Omega \cap B(x,r) \setminus K} |u - g|^p$$

$$\leq \int_{\Omega \cap B(x,r) \setminus \widetilde{K}} |\nabla \widetilde{u}|^2 + H^{n-1}(\widetilde{K} \cap \overline{B}(x, r)) + \int_{\Omega \cap B(x,r) \setminus \widetilde{K}} |v - g|^p \tag{13}$$

because $|\nabla v(z)| \leq |\nabla \widetilde{u}(z)|$ almost-everywhere. Now

$$\int_{\Omega \cap B(x,r) \setminus \widetilde{K}} |v - g|^p \leq 2^p ||g||_\infty^p \omega_n r^n$$

because $|v - g| \leq 2||g||_\infty$ almost-everywhere, and (7) follows from (13). This proves Proposition 8. □

Note that our hypothesis that $g \in L^\infty$ was very convenient here. Probably something like Proposition 8 stays true with weaker hypotheses on g, but I did not check.

If we are only interested in the behavior of K far from $\partial\Omega$, the following weaker notion of almost-minimality will be enough; it has the advantage of being satisfied by pairs (u, K) that minimize the Mumford-Shah functional with prescribed boundary values on $\partial\Omega$ (by the proof of Proposition 8).

Definition 14. *The pair $(u, K) \in \mathcal{A}$ is a local almost-minimizer (with gauge function h) if (7) holds for all choices of balls $B(x, r)$ such that $\overline{B}(x, r) \subset \Omega$ and all competitors $(\widetilde{u}, \widetilde{K})$ for (u, K) in $B(x, r)$.*

Thus the definition is almost the same, but we only allow perturbations that live in compact balls in Ω. Here we could have been a little more generous and have allowed competitors $(\widetilde{u}, \widetilde{K})$ for (u, K) in $B(x, r)$ even when $\overline{B}(x, r)$ is not contained in Ω, but with the only restriction that $(\widetilde{u}, \widetilde{K})$ coincides with (u, K) out of a compact subset of Ω (that could depend on the competitor). It is not too clear which should be the best definition, but Definition 14 looks a little bit simpler to use.

We are not finished with almost-minimizers yet. As we shall see in Section 38, limits of almost-minimizers, and even limits of minimizers under blow ups, are not always almost-minimizers, but merely "topological almost-minimizers", where we keep a similar definition, but put an additional constraint on competitors.

Definition 15. *Let $(u, K) \in \mathcal{A}$ and a ball $B(x, r)$ be given. A <u>topological competitor</u> for (u, K) in $B(x, r)$ is a competitor for (u, K) in $B(x, r)$ (as in Definition 2) which also satisfies the following topological condition:*

$$\text{if } y, z \in \Omega \setminus [K \cup \overline{B}(x, r)] \text{ are separated by } K \text{ in } \Omega,$$
$$\text{then } \widetilde{K} \text{ also separates them in } \Omega. \tag{16}$$

By (16) we mean that if $y, z \in \Omega \setminus [K \cup \overline{B}(x, r)]$ do not lie in the same connected component of $\Omega \setminus K$, then they also lie in different connected components of $\Omega \setminus \widetilde{K}$. Thus some competitors will not be allowed, when they introduce new holes in K that would put in communication different components of $\Omega \setminus K$. For instance, if $\Omega = \mathbb{R}^2$ and K is a line, no proper subset of K will be allowed.

We shall most often use this definition when $\overline{B}(x, r) \subset \Omega$ (and (u, K) is a local topological almost-minimizer as in the next definition); the first exception to that rule will not come before Section 39, when we study limits of almost-minimizers near the boundary.

From time to time, we shall need to prove that a given competitor is a topological competitor. See Lemmas 20.10 and 45.8 for sufficient conditions.

Definition 17. *The pair $(u, K) \in \mathcal{A}$ is a <u>local topological almost-minimizer</u> with gauge function h if (7) holds whenever $\overline{B}(x, r) \subset \Omega$ and $(\widetilde{u}, \widetilde{K})$ is a topological competitor for (u, K) in $B(x, r)$. It is a <u>topological almost-minimizer</u> with gauge function h if (7) holds for all balls $B(x, r)$ and all topological competitors $(\widetilde{u}, \widetilde{K})$ for (u, K) in $B(x, r)$.*

The notions of local almost-minimizer and local topological almost-minimizer localize well: if (u, K) is a local topological almost-minimizer in a domain Ω, and Ω_1 is an open set contained in Ω, then (u, K) is a local topological almost-minimizer in Ω_1, with the same gauge function (and similarly for local almost-minimizers). See Exercise 58 (for the topological case).

Probably Definitions 15 and 17 are the right ones for the local almost-minimizers. For plain almost-minimizers, the situation is less clear. In fact, the author spent a long time hesitating between various definitions, and settled for the ones above mostly because they are the simplest.

Just as for local minimizers, a perhaps more natural notion would allow topological competitors in compact sets $T \subset \Omega$ (and not only balls), with a definition like Definition 15, and where we would replace r in (7) with $\operatorname{diam}(T)/2$. This would make our condition more nicely invariant under bilipschitz changes of domains, for instance. Hopefully, it would not make major differences in the various theorems below, but we did not check.

A defect of our definition will arise in the next section, when we wonder about the following question: for each local topological almost-minimizer, is there a topologically reduced local topological almost-minimizer that is equivalent to it? For this it would seem easier to use a slightly stronger notion of local topological almost-minimizer, associated to the following less restrictive definition of topological competitor in $B(x,r)$. We would allow all the competitors $(\widetilde{u}, \widetilde{K})$ in $B(x,r)$ for which there is a compact set $T \subset \Omega$ such that $\widetilde{K}$ separates y from z in Ω as soon as $y, z \in \Omega \setminus [\overline{B}(x,r) \cup T \cup K]$ are separated by K. [Recall that in (16) we required this for all $y, z \in \Omega \setminus [\overline{B}(x,r) \cup K]$.] See Definitions 8.19 and 8.21. But apparently this other definition would behave less nicely when we try to take limits. We shall be more explicit about this issue in the next section (see Remark 8.18 and Proposition 8.22).

For many results, the fact that our gauge function $h(r)$ tends to 0 when r tends to 0 is not really needed, and it is enough that $h(r)$ stays small enough for r small. We do not give a definition, because quasiminimizers will be even more general.

b. Quasiminimizers

Almost-minimizers are not yet general enough for us, so let us also define quasiminimizers. We prefer to use the term "quasi" instead of "almost", because constants do not tend to 0; however the reader should be warned again that our terminology is not standard. In fact, the next definition probably does not appear anywhere else in the literature. The main motivation for it is that it is sufficiently general to include Example 23 below, and yet it still gives good rectifiability properties for K. The proofs will often be the same.

The definition will use two constants $a \geq 0$ (small) and $M \geq 1$, as well as some additional notation. Let $(\widetilde{u}, \widetilde{K})$ be a competitor for (u, K) in some ball $B(x,r)$. Set

$$E = \int_{\Omega \cap B(x,r) \setminus K} |\nabla u|^2, \quad \widetilde{E} = \int_{\Omega \cap B(x,r) \setminus \widetilde{K}} |\nabla \widetilde{u}|^2, \tag{18}$$

and then

$$\delta E = \operatorname{Max} \{(\widetilde{E} - E), M(\widetilde{E} - E)\}. \tag{19}$$

This may look a little weird, but recall that $\widetilde{E} - E$ may be positive or negative, and we want different formulae for these cases. Our quasiminimality condition is that

$$H^{n-1}(K \setminus \widetilde{K}) \leq M H^{n-1}(\widetilde{K} \setminus K) + \delta E + a r^{n-1}. \tag{20}$$

The reader may find this definition slightly strange, in particular because the two terms of the functional are not treated the same way. Here we put the focus on the variations of K, which we consider more important, but another reason for this different treatment is that the energy term is a little less local. That is, it seems hard to come up with analogues of $\widetilde{K} \setminus K$ and $K \setminus \widetilde{K}$ for the energy term. At least note that when $M = 1$, (20) is the same thing as (7), with $h(r)$ replaced with a. Taking $M > 1$ will give more flexibility (for the verification that (u, K) is a quasiminimizer).

Definition 21. *Let $(u, K) \in \mathcal{A}$ and constants $r_0 > 0$, $M \geq 1$, $a \geq 0$ be given. We say that (u, K) is an (r_0, M, a)-quasiminimizer if (20) holds whenever $(\widetilde{u}, \widetilde{K})$ is a competitor for (u, K) in $B(x, r)$ and $r \leq r_0$. If we only require (20) when $\overline{B}(x, r) \subset \Omega$ and $r \leq r_0$, we say that (u, K) is a local (r_0, M, a)-quasiminimizer. If we only require (20) when $(\widetilde{u}, \widetilde{K})$ is a topological competitor for (u, K) in $B(x, r)$ for some ball $B(x, r)$ such that $\overline{B}(x, r) \subset \Omega$ and $r \leq r_0$, we say that (u, K) is a local topological (r_0, M, a)-quasiminimizer.*

As above for local almost-minimizers, the restriction to a smaller domain of a local quasiminimizer is a local quasiminimizer with the same constants, and the same thing holds with local topological almost-minimizers. See Exercise 58.

As for topological almost-minimizers, there are other possible definitions of topological quasiminimizers (based on variants in the definition of topological competitors), and it is not obvious that we chose the best one. We tried to pick the simplest (and most general) one. See Remark 8.18 and the definitions that follow it.

Remark 22. If (u, K) is an almost-minimizer, then for each $a > 0$ it is also an $(r_0, 1, a)$-quasiminimizer for r_0 small enough (so that $h(r_0) \leq a$). This follows from the remark just after (20). The same remark applies to local and topological almost-minimizers. Note also that our classes of quasiminimizers get larger when M and a get larger. In typical statements, M will be given, and we shall prove that good properties of quasiminimizers hold for a small enough.

The main justification for the definition above is the following type of examples.

Example 23. Let $\Omega \subset \mathbb{R}^n$ (open), $g \in L^\infty(\Omega)$, and $1 \leq p < +\infty$ be given. Also let $b_1, b_2, b_3 \in L^\infty(\Omega)$ be positive Borel functions, and assume that

$$1 \leq b_1(x) \leq M \text{ and } 1 \leq b_2(x) \leq M \text{ for } x \in \Omega, \tag{24}$$

and

$$b_2 \text{ is uniformly continuous on } \Omega. \tag{25}$$

Define a functional J on the usual set $\mathcal{A}$ of admissible pairs, by

$$J(u, K) = \int_K b_1(x) \, dH^{n-1}(x) + \int_{\Omega \setminus K} b_2(x) |\nabla u(x)|^2 \, dx$$

$$+ \int_{\Omega \setminus K} b_3(x) |u(x) - g(x)|^p \, dx. \tag{26}$$

See Remark 42 below for a slightly more general formula and comments on our assumptions. Without further information on all our data, we cannot be sure that J has minimizers, but let us not worry about that. The next proposition says that any minimizer for J is automatically a quasiminimizer.

Proposition 27. *Let b_1, b_2, b_3, and J be as above, and let $(u, K) \in \mathcal{A}$ be such that*

$$J(u, K) = \inf \{ J(v, G) \, ; \, (v, G) \in \mathcal{A} \} < +\infty. \tag{28}$$

Then for every $a > 0$ there is an $r_0 > 0$ such that (u, K) is an (r_0, M, a)-quasiminimizer.

Here M is as in (24), and r_0 depends only on M, $||g||_\infty$, $||b_3||_\infty$, a, n, p, and the modulus of continuity which is implicit in (25).

The proof will be similar to what we did for Proposition 8. We still have that

$$||u||_\infty \leq ||g||_\infty, \tag{29}$$

because otherwise we may truncate u and strictly decrease the value of $J(u, K)$. Next we want to check that for every ball $B(x, r)$,

$$H^{n-1}(K \cap \overline{B}(x, r)) + \int_{\Omega \cap B(x,r) \setminus K} |\nabla u|^2 \leq M \, \widetilde{\omega}_{n-1} \, r^{n-1} + ||b_3||_\infty \, \omega_n \, ||g||_\infty^p \, r^n, \tag{30}$$

where $\widetilde{\omega}_{n-1} = H^{n-1}(\partial B(0, 1))$ and ω_n is the Lebesgue measure of $B(0, 1)$.

To see this we define a new pair $(\widetilde{u}, \widetilde{K})$. We take $\widetilde{K} = [K \setminus \overline{B}(x, r)] \cup [\Omega \cap \partial B(x, r)]$, keep $\widetilde{u} = u$ out of $\overline{B}(x, r) \cup K$, and take $\widetilde{u} = 0$ on $\Omega \cap B(x, r)$. It is easy to see that $(\widetilde{u}, \widetilde{K}) \in \mathcal{A}$, because we only introduced new discontinuities of $\widetilde{u}$ on $\Omega \cap \partial B(x, r)$, which we added to $\widetilde{K}$. Then $J(u, K) \leq J(\widetilde{u}, \widetilde{K})$. Since the left-hand side of (30) is less than the contribution of $\overline{B}(x, r)$ to the first two terms of $J(u, K)$, and the right-hand side is an easy upper bound for the contribution of $\overline{B}(x, r)$ to $J(\widetilde{u}, \widetilde{K})$, (30) follows.

Now let $(\widetilde{u}, \widetilde{K})$ be a competitor for (u, K) in some ball $B(x, r)$. Set

$$v = \text{Max}\{-||g||_\infty, \text{Min}(||g||_\infty, \widetilde{u})\}; \tag{31}$$

then $v = \widetilde{u} = u$ out of $\overline{B}(x, r) \cup K$, by (29) and because $\widetilde{u} = u$ out of $\overline{B}(x, r) \cup K$. Also, $v \in W^{1,2}_{\text{loc}}(\Omega \setminus \widetilde{K})$ and

$$|\nabla v(y)| \leq |\nabla \widetilde{u}(y)| \tag{32}$$

almost everywhere, by Lemma 3.22. In particular, $(v, \widetilde{K}) \in \mathcal{A}$ and hence

$$J(v, \widetilde{K}) \geq J(u, K), \tag{33}$$

by (28). From (24) we deduce that

$$H^{n-1}(K \setminus \widetilde{K}) - M H^{n-1}(\widetilde{K} \setminus K) \leq \int_K b_1(y) \, dH^{n-1}(y) - \int_{\widetilde{K}} b_1(y) \, dH^{n-1}(y). \tag{34}$$

Next set $\tau(r) = \sup\{|b_2(y) - b_2(z)|; |y - z| \le r\}$; (25) says that $\tau(r)$ tends to 0 when r tends to 0. Also,

$$\left| b_2(x)E - \int_{\Omega \cap B(x,r) \setminus K} b_2(y)|\nabla u(y)|^2 \right| \le \tau(r)\, E \qquad (35)$$

(with E as in (18)), and similarly

$$\left| b_2(x)\widetilde{E} - \int_{\Omega \cap B(x,r) \setminus \widetilde{K}} b_2(y)|\nabla \widetilde{u}(y)|^2 \right| \le \tau(r)\, \widetilde{E}. \qquad (36)$$

First assume that $\widetilde{E} \ge 2\eta$, where η denotes the right-hand side of (30). Then $\widetilde{E} - E \ge 2\eta - E \ge \eta$ by (30), hence $\delta E \ge \eta$ (see (19)). Also, $H^{n-1}(K \setminus \widetilde{K}) \le H^{n-1}(K \cap \overline{B}(x,r)) \le \eta$ (by (30) again). Thus (20) holds trivially in this case.

So we can assume that $\widetilde{E} \le 2\eta$. Then

$$\int_{\Omega \setminus \widetilde{K}} b_2|\nabla v|^2 - \int_{\Omega \setminus K} b_2|\nabla u|^2 \le \int_{\Omega \setminus \widetilde{K}} b_2|\nabla \widetilde{u}|^2 - \int_{\Omega \setminus K} b_2|\nabla u|^2 \qquad (37)$$

$$\le b_2(x)(\widetilde{E} - E) + \tau(r)(\widetilde{E} + E) \le \delta E + \tau(r)(\widetilde{E} + E) \le \delta E + 3\eta\tau(r)$$

by (32), because $\widetilde{u} = u$ out of $\overline{B}(x,r)$, by (35) and (36), and then by (19), (24), (30), and our definition of η. Finally

$$\int_{\Omega \setminus \widetilde{K}} b_3|v - g|^p - \int_{\Omega \setminus K} b_3|u - g|^p \le \int_{\Omega \cap B(x,r) \setminus \widetilde{K}} b_3|v - g|^p$$
$$\le 2^p \, ||b_3||_\infty \, ||g||_\infty^p \, \omega_n \, r^n, \qquad (38)$$

because $v = u$ out of $\overline{B}(x,r)$ and $||v||_\infty \le ||g||_\infty$ (by (31)).

Let us now add up (34), (37), and (38). We get that

$$H^{n-1}(K \setminus \widetilde{K}) - M\, H^{n-1}(\widetilde{K} \setminus K) + J(v, \widetilde{K}) - J(u, K)$$
$$\le \delta E + 3\eta\tau(r) + 2^p||b_3||_\infty \, ||g||_\infty^p \, \omega_n \, r^n. \qquad (39)$$

Since $J(v, \widetilde{K}) - J(u, K) \ge 0$ by (33), (39) implies that

$$H^{n-1}(K \setminus \widetilde{K}) \le M\, H^{n-1}(\widetilde{K} \setminus K) + \delta E + \eta_1, \qquad (40)$$

with

$$\eta_1 = 3\eta\tau(r) + 2^p||b_3||_\infty||g||_\infty^p \, \omega_n \, r^n \qquad (41)$$
$$= 3M\widetilde{\omega}_{n-1}\tau(r)r^{n-1} + \{3\tau(r) + 2^p\}||b_3||_\infty||g||_\infty^p \, \omega_n \, r^n$$

because η is the right-hand side of (30). Now (20) follows from (40) as soon as $\eta_1 \le a r^{n-1}$. This is the case for r small enough, because $\tau(r)$ tends to 0. In other words, we can find $r_0 > 0$ such that (20) holds for $r < r_0$; this completes our proof of Proposition 27. $\qquad \square$

Remark 42. Proposition 27 applies with the same proof to functionals that are a little more general than the ones defined in (26). We can replace the term $\int_K b_1(x) dH^{n-1}(x)$ by less isotropic variants of the form

$$\int_K b_1(x, \tau(x)) \, dH^{n-1}(x) \tag{43}$$

where $\tau(x)$ denotes a unit normal vector to the approximate tangent plane to K at x, and $b_1(x, \tau)$ is an even function of the second variable (i.e., $b_1(x, -\tau) = b_1(x, \tau)$).

Here we do not wish to discuss too much the proper way to define the integral in (43). One way to do so would be to define J only for pairs $(u, K) \in \mathcal{A}$ such that K is rectifiable, so that we can talk about the approximate tangent plane to K at H^{n-1}-almost every point $x \in K$. This would possibly make it more unpleasant to prove existence of minimizers, and the reader may be shocked that we restrict to rectifiable sets a priori. Incidentally, see Exercise 24.35 for a failed attempt to reduce to rectifiable sets immediately.

One could also use a weak formulation where (u, K) is replaced by a function of "special bounded variation" (SBV), and (43) becomes an integral on the corresponding singular set. This is probably more pleasant if we want to prove existence results, and the rectifiability issue is nicely hidden in the fact that singular sets of SVB functions are automatically rectifiable.

But we could define directly variants of Hausdorff measures adapted to the function $b_1(x, \tau)$. That is, we would still use coverings as in Definition 2.6, but sum something else than a power of the diameters. See [Mat2] for definitions of generalized Hausdorff measures.

On the other hand, we should perhaps not spend too much time on finding the right definition of J when K is not rectifiable, since we shall see later that for reduced quasiminimizers, K is (more than) rectifiable.

Remark 44. In all the discussion above, we carefully avoided the issue of existence of minimizers for J, but merely talked about properties of minimizers. For existence, we would probably need to add convexity assumptions on $b_1(x, \cdot))$ and lower semicontinuity assumptions on everything. However Remark 42 and regularity results for quasiminimizers could help define a functional and find minimizers, by arguments similar to the existence method rapidly described at the end of Section 4. That is, we would use quasiminimality to get minimizing sequences with good regularity properties, and lower semicontinuity of (variants of) the Hausdorff measure would be easier to get.

c. Almost- and quasiminimizers with linear distortion

This subsection may easily be skipped on a first reading.

The nonisotropic variants of J discussed in Remark 42 are not as weird as it may seem. For instance, if we start from the usual Mumford-Shah functional and transport it by a smooth change of variables, we immediately get a functional like this. We only get an isotropic function b_1 if the change of variables is conformal; this may almost be a reasonable constraint in dimension 2, but certainly not in general (because there are so few conformal mappings). Also, one can argue that many natural problems are nonisotropic.

One may also need to consider nonisotropic variants of the energy term in the functional J, for the same good reasons. In this respect, the way we treated energy so far is probably too rigid. For instance, if we replace the energy term in (26) with a nonisotropic variant like $\int_{\Omega \setminus K} |A(x)\nabla u(x)|^2 dx$, where A is a smooth matrix-valued function, with $A(x)$ and $A(x)^{-1}$ uniformly bounded, we cannot say that minimizers for the corresponding functional are quasiminimizers.

Even with isotropic functions b_1 and b_2, we should expect more than the quasiminimality property of Proposition 27. We do not get almost-minimality because the relative weight of the energy term, compared to the Hausdorff measure term, varies from point to point. But if b_1 and b_2 are smooth, we should expect the same sort of regularity as we get for almost-minimizers. The issue may have some importance, because we shall get C^1 regularity for almost-minimizers, but not for quasiminimizers.

A good treatment of this issue would probably force us to replace energy integrals with expressions like $\int_{\Omega \setminus K} |A(x)\nabla u(x)|^2 dx$, and then use elliptic differential operators wherever we use the Laplacian, but the author refuses to do this. Instead we shall take the constant coefficients approach, which will allow us to use harmonic functions essentially as before. And since we already distort the energy integrals, we shall distort the Hausdorff measure as well.

Let us now say how we generalize our definitions of almost- and quasiminimizers. We start with plain almost-minimizers.

Definition 45. *The pair $(u, K) \in \mathcal{A}$ is an almost-minimizer with gauge function h and linear distortion $\leq L$ if for every ball $B(x, r)$ we can find invertible $n \times n$ matrices A_1 and A_2, with*

$$\text{Max}\left\{\|A_1\|, \|A_2\|, \|A_1^{-1}\|, \|A_2^{-1}\|\right\} \leq L, \tag{46}$$

such that

$$H^{n-1}(A_1(K \cap \overline{B}(x, r))) + \int_{\Omega \cap B(x,r) \setminus K} |A_2 \nabla u|^2 \tag{47}$$

$$\leq H^{n-1}(A_1(\widetilde{K} \cap \overline{B}(x, r))) + \int_{\Omega \cap B(x,r) \setminus \widetilde{K}} |A_2 \nabla \widetilde{u}|^2 + h(r) r^{n-1}$$

whenever $(\widetilde{u}, \widetilde{K})$ is a competitor for (u, K) in $B(x, r)$.

Here we identified A_1 with the associated linear mapping to save notation. There are probably lots of more general ways to do the distortion of Hausdorff measure, but let us not worry.

We define <u>local almost-minimizers with linear distortion $\leq L$</u> the same way, except that we only require (47) for balls $B(x,r)$ such that $\overline{B}(x,r) \subset \Omega$. For <u>topological almost-minimizers with linear distortion $\leq L$</u> and their local variant, we only require (47) for topological competitors in $B(x,r)$.

Example 48. Let $\Omega \subset \mathbb{R}^n$ be open, and consider the functional

$$J(u,K) = \int_K b_1(x)\,dH^{n-1}(x) + \int_{\Omega \setminus K} |B_2(x)\nabla u(x)|^2 dx$$

$$+ \int_{\Omega \setminus K} b_3(x)|u(x) - g(x)|^p dx, \qquad (49)$$

defined on the usual set $\mathcal{A}$ of admissible pairs, and where $1 \leq p < +\infty$, $g \in L^\infty$, b_1 and b_3 are positive bounded functions on Ω, and B_2 is a bounded, $n \times n$ matrix-valued function. We assume that

$$1 \leq b_1(x) \leq L, \ \|B_2(x)\| \leq L, \ \text{and} \ \|B_2(x)^{-1}\| \leq L \text{ everywhere}, \qquad (50)$$

and that

$$b_1 \text{ and } B_2 \text{ are uniformly continuous on } \Omega. \qquad (51)$$

Then every minimizer for J is an almost-minimizer with gauge function h and linear distortion $\leq L$, where h can easily be computed in terms of L, $\|b_3\|_\infty$, $\|g\|_\infty$, and the modulus of continuity of b_1 and B_2. [In particular, h tends to 0 at 0.]

The verification is the same as for Proposition 27; we take $A_1 = b_1(x)I$ and $A_2 = B_2(x)$ in Definition 45, and then use analogues of (35) and (36) (both for the energy and the Hausdorff measure terms) to reduce to constant coefficients. Note that we do not need the linear distortion of Hausdorff measure here; we could also take $A_1 = I$ and $A_2 = B_2(x)/b_1(x)$ in Definition 48 (and just get a worse constant L^2).

Let us also define quasiminimizers with linear distortion. We shall only need to distort the gradient, because our definition of quasiminimizers is already flexible enough with respect to the Hausdorff measure.

Definition 52. *The pair $(u,K) \in \mathcal{A}$ is called a (r_0, M, a)-quasiminimizer with linear distortion $\leq L$ if for every ball $B(x,r)$ such that $r \leq r_0$, we can find an invertible $n \times n$ matrix A, with*

$$\|A\| \leq L \quad and \quad \|A^{-1}\| \leq L, \qquad (53)$$

such that

$$H^{n-1}(K \setminus \widetilde{K}) \leq M H^{n-1}(\widetilde{K} \setminus K) + \delta E_A + a r^{n-1} \qquad (54)$$

whenever $(\widetilde{u}, \widetilde{K})$ is a competitor for (u,K) in $B(x,r)$, and where we set

$$E_A = \int_{\Omega \cap B(x,r) \setminus K} |A\nabla u|^2, \quad \widetilde{E}_A = \int_{\Omega \cap B(x,r) \setminus \widetilde{K}} |A\nabla \widetilde{u}|^2 \qquad (55)$$

and

$$\delta E_A = \text{Max}\{(\widetilde{E}_A - E_A), M(\widetilde{E}_A - E_A)\}. \tag{56}$$

Local, topological, and local topological (r_0, M, a)-quasiminimizers with linear distortion $\leq L$ are defined in a similar way, with the same restrictions as before on competitors.

The reason why we dare to introduce almost- and quasiminimizers with linear distortion is that most of our results for almost- and quasiminimizers will still hold with linear distortion. The proofs will need to be modified a little bit, but not too much, and most of the time the constants will just get multiplied by a power of L. The main point (for the distortion of energy, say) is that the functions that minimize $\int |A\nabla u|^2$ locally are just as good as harmonic (and can be obtained from harmonic functions by a linear change of variable, see Exercise 59).

Since we do not want the reader to be overwhelmed with notation, we shall give all our proofs in the standard case (without distortion), and add a few remarks here and there about linear distortion, typically when there is a difficulty with the proof, or the generalization of the statement is not obvious. The reader may use the entry "linear distortion" in the index to find such remarks.

Exercise 57. We want to check that if (u, K) is a local topological almost- or quasiminimizer in Ω, and we add a constant to u on one of the connected components of $\Omega \setminus K$, then we get another pair with the same topological almost- or quasiminimality properties. So let $(u, K) \in \mathcal{A}$ be given, let W_0 be one of the components of $\Omega \setminus K$, let $A \in \mathbb{R}$ be given, and define a function v on $\Omega \setminus K$ by $v(x) = u(x)$ for $x \in \Omega \setminus [K \cup W_0]$ and $v(x) = u(x) + A$ for $x \in W_0$.

  1. Show that $v \in W^{1,2}_{\text{loc}}(\Omega \setminus K)$.

Now let $B = B(x, r)$ be such that $\overline{B} \subset \Omega$, and let $(\widetilde{v}, \widetilde{K})$ be a topological competitor for (v, K) in B. We want to construct a competitor for (u, K).

  2. Call $\mathcal{B}$ the set of connected components of $\Omega \setminus \widetilde{K}$, and then set $\mathcal{B}_1 = \{V \in \mathcal{B}; V \subset B\}$, $\mathcal{B}_2 = \{V \in \mathcal{B}; V \text{ meets } \partial B\}$, and $\mathcal{B}_3 = \{V \in \mathcal{B}; V \subset \Omega \setminus B\}$. Check that $\mathcal{B} = \mathcal{B}_1 \cup \mathcal{B}_2 \cup \mathcal{B}_3$.

  3. Check that for each $V \in \mathcal{B}_2$ and $x \in V \cap \partial B$, there is a connected component $W(x)$ of $\Omega \setminus K$ that contains $B(x, \rho) \setminus \overline{B}$ for ρ small.

  4. Check that $W(x)$ depends only on V. Show that for $V \in \mathcal{B}$, $V \setminus \overline{B}$ meets at most one component of $\Omega \setminus K$.

Now we are ready to define $\widetilde{u}$. We do this component by component. If $V \in \mathcal{B}$ meets $W_0 \setminus \overline{B}$, we set $\widetilde{u}(x) = \widetilde{v}(x) - A$ for $x \in V$. Otherwise, we keep $\widetilde{u} = \widetilde{v}$ on V.

  5. Show that $\widetilde{u} \in W^{1,2}_{\text{loc}}(\Omega \setminus \widetilde{K})$ and that $\widetilde{u} = u$ on $\Omega \setminus [K \cup \overline{B}]$.

  6. Check that $(\widetilde{u}, \widetilde{K})$ is a topological competitor for (u, K) in B. Note that $\nabla \widetilde{u} = \nabla \widetilde{v}$ everywhere on $\Omega \setminus \widetilde{K}$.

7. Show that if (u, K) is a local topological almost-minimizer on Ω with some gauge function h, so is (v, K). And similarly for local topological (r_0, M, a)-quasiminimizers.

Exercise 58 (Localization of the topological condition). Let $\Omega_1 \subset \Omega$ be open sets, and let $(u, K) \in \mathcal{A}(\Omega)$ be an acceptable pair in Ω. Of course the restriction of (u, K) to Ω_1 lies in $\mathcal{A}(\Omega_1)$ too.

1. Let B be a ball such that $\overline{B} \subset \Omega_1$, and let (v, G) be a topological competitor for (u, K) in B, seen in the domain Ω_1. Extend (v, G) to Ω, by taking $(v, G) = (u, K)$ in $\Omega \setminus \Omega_1$. Check that $(v, G) \in \mathcal{A}(\Omega)$; thus it is a competitor for (u, K) in B (in Ω).

2. We want to see that (v, G) is also a topological competitor for (u, K) in B, in Ω. So we assume that we can find points y and z in $\Omega \setminus [\overline{B} \cup K]$ that lie in the same component of $\Omega \setminus G$, but in different components of $\Omega \setminus K$.

    a. Let B_1 be a ball such that $\overline{B} \subset B_1$, $\overline{B_1} \subset \Omega_1$, and y, z lie out of $\overline{B_1}$. Find a path γ from y to z that meets ∂B_1 only finitely many times.

    b. Find two consecutive points of $\gamma \cap \partial B_1$ that lie in different components of $\Omega \setminus K$.

    c. Show that the arc of γ between these points stays in B_1.

    d. Find a contradiction and conclude.

3. Suppose (u, K) is a local topological almost-minimizer in Ω. Show that it is a local topological almost-minimizer in Ω_1, with the same gauge function.

4. Suppose (u, K) is a local topological (r_0, M, a)-quasiminimizer in Ω. Show that it is a local topological (r_0, M, a)-quasiminimizer in Ω_1.

Exercise 59. Let $D \subset \mathbb{R}^n$ be open, and let A be an invertible $n \times n$ matrix. Call $\widetilde{A}$ the transposed matrix, set $V = \widetilde{A}^{-1}(D)$ and, for $u \in C^1(D)$, set $v = u \circ \widetilde{A}$ on V.

1. Check that $\nabla v(x) = A \nabla u(y)$, with $y = \widetilde{A} x$, and that

$$\int_D |A \nabla u|^2 = |\det(A)| \int_V |\nabla v|^2.$$

2. Suppose that $\int_D |A \nabla u|^2 \leq \int_D |A \nabla w|^2$ for every C^1 function w on D such that $w(y) = u(y)$ out of some compact subset of D. Show that v is harmonic on V. Hint: Lemma 15.17.

Exercise 60. Let φ be a C^1 diffeomorphism from Ω to Ω', and assume that the derivatives of φ and φ^{-1} are bounded and uniformly continuous. Let (u, K) be an almost-minimizer with linear distortion in Ω, and set $G = \varphi(K)$ and $v = u \circ \varphi^{-1}$ on $\Omega' \setminus G$. Show that (v, G) is an almost-minimizer with linear distortion in Ω'. Also prove a similar statement for quasiminimizers, local and topological almost-minimizers and quasiminimizers (all with linear distortion).

8 Reduced minimizers and pairs, coral pairs

a. Without topology

We have rapidly seen in Section 6 why it is interesting to replace minimizers of the Mumford-Shah functional, for instance, by reduced minimizers. Here we verify that for each minimizer (and in fact for each admissible pair) (u, K), we can find an equivalent pair $(\widetilde{u}, \widetilde{K})$ which is reduced (i.e., for which $\widetilde{K}$ is minimal with the given function). In later subsections, we shall try to do this in the context of topological competitors. We shall also introduce a slightly different notion, corality, which may turn out to be just as useful as reduction and a little more convenient in that context.

Recall from (2.1) that we denote by $\mathcal{A} = \{(u, K);\ K \subset \Omega$ is closed in Ω and $u \in W^{1,2}_{\text{loc}}(\Omega \setminus K)\}$ the set of admissible pairs associated to any given open set $\Omega \subset \mathbb{R}^n$.

Definition 1. *We say that the pair $(u, K) \in \mathcal{A}$ is reduced when there is no other pair $(\widetilde{u}, \widetilde{K}) \in \mathcal{A}$ such that $\widetilde{K} \subset K$, $\widetilde{K} \neq K$, and $\widetilde{u}$ is an extension of u.*

In general, we consider sets K for which $H^{n-1}(K)$ is locally finite, hence $H^n(K) = 0$ (see Exercise 2.20). Then u is defined almost-everywhere, and the definition just says that u does not lie in $W^{1,2}_{\text{loc}}(\Omega \setminus \widetilde{K})$ for any closed set $\widetilde{K}$ in Ω that is strictly contained in K. Actually, when we deal with topological almost (or quasi-) minimizers, we shall have to modify this definition. See Remark 6 below and the second and third parts of this section.

Proposition 2. *For each $(u, K) \in \mathcal{A}$ we can find a reduced pair $(\widetilde{u}, \widetilde{K}) \in \mathcal{A}$ such that $\widetilde{K} \subset K$ and $\widetilde{u}$ is an extension of u.*

Before we prove the proposition, let us say why it allows us to reduce our study of minimizers, almost-minimizers, and quasiminimizers to reduced pairs. For all these categories of objects, $H^{n-1}(K)$ is locally finite in Ω, hence K has zero Lebesgue measure and $(\widetilde{u}, \widetilde{K})$ is at least as good as (u, K) (the integrals on $\Omega \setminus K$ are the same for u and $\widetilde{u}$). Thus $(\widetilde{u}, \widetilde{K})$ is a minimizer (or almost-, or quasi-) just like (u, K). We even claim that

$$H^{n-1}(K \setminus \widetilde{K}) = 0, \tag{3}$$

so that $(\widetilde{u}, \widetilde{K})$ is really equivalent to (u, K). (In the case of quasiminimizers, we have to assume a to be small enough, though; this is natural even if $n = 2$, because for a large we could add one or two useless lines to K.) In the case of minimizers, (3) is clear because otherwise $J(\widetilde{u}, \widetilde{K}) < J(u, K)$. For almost- and quasiminimizers, we have to use the fact that for H^{n-1}-almost all $x \in K \setminus \widetilde{K}$,

$$\limsup_{r \to 0} \left\{ r^{1-n} H^{n-1}([K \setminus \widetilde{K}] \cap B(x, r)) \right\} \geq c_n \tag{4}$$

for some geometric constant $c_n > 0$ (see Lemma 19.1). When (4) holds for some x, the analogue of (7.7) with $\widetilde{K}' = [K \setminus B(x,r)] \cup [\widetilde{K} \cap B(x,r)]$ does not hold for every small r, and neither does (7.20) if $a < c_n$. This proves that (4) never holds, and (3) follows.

Let us now prove Proposition 2. Let $(u, K) \in \mathcal{A}$ be given. For $k \geq 0$, set

$$\mathcal{R}_k = \left\{ B(x,r); |x| \leq 2^k, 2^{-k-1} < r \leq 2^{-k} \text{ and } B(x,r) \subset \Omega \right\}. \tag{5}$$

Define a sequence of admissible pairs $(u_j, K_j) \in \mathcal{A}$, as follows. Start with $(u_0, K_0) = (u, K)$. If (u_j, K_j) has already been chosen, and is a reduced pair, just stop the construction. Otherwise, since (u_j, K_j) is not reduced we can find a point $x_j \in K_j$ and a small ball B_j centered at x_j, such that u_j has an extension $u_j^* \in W_{\text{loc}}^{1,2}((\Omega \setminus K_j) \cup B_j)$. We can always replace B_j with a smaller ball if needed, and so we can make sure that $B_j \in \mathcal{R}_k$ for some k. Let us choose x_j and B_j like this, and with k as small as possible. Then set $K_{j+1} = K_j \setminus B_j$ and $u_{j+1} = u_j^*$. Clearly $(u_{j+1}, K_{j+1}) \in \mathcal{A}$ if $(u_j, K_j) \in \mathcal{A}$.

Our construction gives pairs $(u_j, K_j) \in \mathcal{A}$ such that $K_j \subset K$ and u_j is an extension of u. If the construction stops, the last pair (u_j, K_j) is reduced and we are done. So let us assume that it does not stop. Take $\widetilde{K} = \bigcap_j K_j$ and let $\widetilde{u}$ be the function on $\Omega \setminus \widetilde{K}$ that coincides (almost-everywhere) with u_j on $\Omega \setminus K_j$ for each j. Clearly $\widetilde{K}$ is closed and contained in K, and $\widetilde{u}$ is an extension of u. Moreover $\widetilde{u} \in W_{\text{loc}}^{1,2}(\Omega \setminus \widetilde{K})$, because for each $x \in \Omega \setminus \widetilde{K}$ we can find $r > 0$ such that $\overline{B}(x,r) \subset \Omega \setminus \widetilde{K}$, then $\overline{B}(x,r)$ does not meet K_j for j large enough, and, for these j, u coincides with $u_j \in W_{\text{loc}}^{1,2}(B(x,r))$ in $B(x,r)$. Thus $(\widetilde{u}, \widetilde{K}) \in \mathcal{A}$.

We still need to check that $(\widetilde{u}, \widetilde{K})$ is reduced. If not, we can find $x \in \widetilde{K}$ and $r > 0$ such that $B(x,r)$ lies in some $\mathcal{R}_k$ and $\widetilde{u}$ has an extension u^* in $W^{1,2}((\Omega \setminus \widetilde{K}) \cup B(x,r))$. At each stage of the construction we could have selected $B_j = B(x,r)$ and we did not. This means that there was an equally good choice, i.e., that $B_j \in \mathcal{R}_{k'}$ for some $k' \leq k$. But this does not make sense: we cannot find infinitely many balls B_j centered in $\overline{B}(0, 2^k)$, with radii $\geq 2^{-k+1}$, and such that the center of B_j is not contained in any of the previous balls. This contradiction proves that $(\widetilde{u}, \widetilde{K})$ is reduced; Proposition 2 follows. $\square$

b. Topologically reduced pairs

Remark 6. When we deal with topological almost (or quasi-) minimizers, Definition 2 does not give the appropriate notion of reduction. For instance, Ω could be the plane, K a line, and u the constant function equal to 0 on $\Omega \setminus K$. Then u extends to the whole plane, but it is not fair to remove any piece of K, because this would give competitors of (u, K) that are not topological. We want to say that (u, K) is topologically reduced, even though it is not reduced as in Definition 2. Here again, it is not too clear what the best definition of reduction should be. We shall try to use the following one, which at least is reasonably simple.

Definition 7. *We say that the pair* $(u, K) \in \mathcal{A}$ *is* topologically reduced *when for all choices of balls* $B(x, r)$ *such that* $\overline{B}(x, r) \subset \Omega$ *and all topological competitors* $(\widetilde{u}, \widetilde{K})$ *for* (u, K) *in* $B(x, r)$ *such that* $\widetilde{K} \subset K$ *and* $\widetilde{u}$ *is an extension of* u, *we have that* $\widetilde{K} = K$.

See Definition 7.15 for topological competitors. Here also we can replace most pairs $(u, K) \in \mathcal{A}$ with topologically reduced pairs, but the statements will be more complicated. Also see Subsection c for a possible way to avoid the issue. In practice, most of the almost- and quasiminimizers will come as limits of reduced pairs, and will be topologically reduced because they are coral (see Remark 31). So the question of replacing pairs with topologically reduced pairs will not really arise, and the reader may find it more pleasant to go directly to the beginning of Subsection c.

Proposition 8. *Let* $(u, K) \in \mathcal{A}$ *be such that*

$$\text{if } (\widetilde{u}, \widetilde{K}) \text{ is a topological competitor for } (u, K) \text{ in some ball } B \subset\subset \Omega, \\ \widetilde{K} \subset K, \text{ and } \widetilde{u} \text{ is an extension of } u, \text{ then } H^{n-1}(K \setminus \widetilde{K}) = 0. \tag{9}$$

Then we can find a topologically reduced pair $(\widetilde{u}, \widetilde{K}) \in \mathcal{A}$ *such that* $\widetilde{K} \subset K$, $\widetilde{u}$ *is an extension of* u, *and* $H^{n-1}(K \setminus \widetilde{K}) = 0$.

See Remark 18 for an example that shows that something like (9) is needed if we want to replace local topological almost-minimizers with reduced ones.

Fortunately, (9) will automatically be satisfied when (u, K) is a (local topological) minimizer for a functional like the ones in (7.9) or (7.26), and for most quasiminimizers. See Remark 18 and Proposition 22. For other cases, let us just say that for most purposes, it will be enough to restrict to coral pairs (instead of topologically reduced ones). See Subsection **c**.

Some additional properties of the set $\widetilde{K}$ that we obtain in Proposition 8 will show up in the proof, but the author does not know which ones should be considered important. See Lemma 16, for instance.

It would have been nicer if we could say that $(\widetilde{u}, \widetilde{K})$ is a topological competitor for (u, K), but this is not the case because we may have to remove a sequence of small sets that accumulate on $\partial\Omega$. The best that we can say is that

$$\widetilde{K} = \bigcap_j K_j, \tag{10}$$

where $K_0 = K$ and

$$K_{j+1} = K_j \setminus B_j \tag{11}$$

for $j \geq 0$, where B_j is a ball centered on K_j such that $\overline{B}_j \subset \Omega$ and

$$(u_{j+1}, K_{j+1}) \text{ is a topological competitor} \\ \text{for } (u_j, K_j) \text{ in some ball } B'_j \supset B_j, \tag{12}$$

where we denote by u_j the restriction of $\widetilde{u}$ to $\Omega \setminus K_j$, and we also have that $\overline{B'_j} \subset \Omega$.

The construction will follow the same lines as for Proposition 2, except that we only consider balls B_j such that

u has an extension $u_j^* \in W_{\text{loc}}^{1,2}(B_j \cup (\Omega \setminus K))$, and $(u_j^*, K \setminus B_j)$ is a $\qquad$ (13)

topological competitor for (u, K) in some B_j' such that $B_j \subset B_j' \subset\subset \Omega$.

We start from the pair $(u_0, K_0) = (u, K)$, and construct $\{(u_j, K_j)\}_{j>0}$ by induction. Suppose we already defined (u_j, K_j), so that $K_j \subset K$ and u_j is an extension of u. If we cannot find any B_j centered on K_j such that (13) holds, we stop and take $(\widetilde{u}, \widetilde{K}) = (u_j, K_j)$. Otherwise, we choose B_j so that (13) holds and $B_j \in \mathcal{R}_k$ for the smallest possible value of k (see (5)). Then we take $K_{j+1} = K_j \setminus B_j$. We also need to define u_{j+1} on $\Omega \setminus K_{j+1}$. Note that

$$H^{n-1}(K \cap B_j) = 0, \qquad (14)$$

by (13) and (9). So u was already defined almost-everywhere on B_j, and we can keep $u_{j+1}(x) = u_j(x) = u(x)$ almost-everywhere on B_j. In addition, $u_{j+1} \in W_{\text{loc}}^{1,2}(\Omega \setminus K_{j+1})$ because (13) says that it lies in $W_{\text{loc}}^{1,2}(B_j)$, and the induction hypothesis says that $u_j \in W_{\text{loc}}^{1,2}(\Omega \setminus K_j)$. Thus $(u_{j+1}, K_{j+1}) \in \mathcal{A}$.

If the construction does not stop, take $\widetilde{K} = \bigcap_j K_j$. This is still a closed set in Ω, and $H^{n-1}(K \setminus \widetilde{K}) = 0$ because $H^{n-1}(K \cap B_j) = 0$ for each j. This allows us to keep $\widetilde{u} = u$ almost-everywhere on $\Omega \setminus \widetilde{K}$. Moreover, $\widetilde{u} \in W_{\text{loc}}^{1,2}(\Omega \setminus \widetilde{K})$, because this is a local property (see the argument in Proposition 2). Thus $(\widetilde{u}, \widetilde{K}) \in \mathcal{A}$.

Next we want to check that

if x, y lie in different connected components of $\Omega \setminus K_j$,

then they also lie in different components of $\Omega \setminus K_{j+1}$. $\qquad$ (15)

We start with the case when x and y lie in $B_j \setminus K_j$, and verify that x and y must lie in the same component of $\Omega \setminus K_j$. Let $\rho > 0$ be so small that $B(x, \rho) \cup B(y, \rho) \subset B_j \setminus K_j$, and consider all the line segments that are parallel to $[x, y]$ and go from $B(x, \rho)$ to $B(y, \rho)$. If all these segments met K, then the orthogonal projection of $K \cap B_j$ on any hyperplane perpendicular to $[x, y]$ would contain a ball of radius ρ. This is impossible because of (14) and Remark 2.11, so at least one segment does not meet K, and K does not separate x from y in Ω.

Next suppose that $x \in B_j$ and $y \notin B_j$, and that they lie in the same component of $\Omega \setminus K_{j+1}$. We want to show that $x \in V_y$, where V_y denotes the connected component of $\Omega \setminus K_j$ that contains y. Let γ be a path from x to y in $\Omega \setminus K_{j+1}$; also let z denote the first point of $\gamma \cap \overline{B}_j$ when we go from y to x along γ. Since K_j and K_{j+1} coincide out of B_j, the arc of γ between y and z does not meet K_j, and therefore $z \in V_y$. Since $z \notin K_j$, there is a small ball B centered at z that does not meet K_j, and then $B \subset V_y$. Since some points of B lie in $B_j \setminus K_j$, we get that $x \in V_y$, by our previous case.

We are left with the case when x and y both lie out of B_j. Suppose that they lie in different components of $\Omega \setminus K_j$, but in the same component of $\Omega \setminus K_{j+1}$; we want to find a contradiction. Let γ be a path from x to y in $\Omega \setminus K_{j+1}$. Note that γ meets K_j. Since $K_j = K_{j+1}$ out of B_j, γ meets B_j. Let z be the first point of $\gamma \cap \overline{B}_j$ when we go along γ from y to x. Then z lies in the same component of $\Omega \setminus K_j$ as y (because $K_j = K_{j+1}$ out of B_j). Similarly, if z' is the last point of $\gamma \cap \overline{B}_j$ when we go from y to x, then z' lies in the same component of $\Omega \setminus K_j$ as x. But z' and z also lie in the same component of $\Omega \setminus K_j$, by our first case (applied to points of $B_j \setminus K_j$ very close to z and z'). This proves the desired contradiction, and (15) follows.

Note that (12) follows at once from (15). We still need to check that $(\widetilde{u}, \widetilde{K})$ is topologically reduced, and this will complete our proof of Proposition 8 (and of all our statements so far). The following will be useful.

Lemma 16. *Suppose $\overline{B}(x,r) \subset \Omega$ and B is a ball such that $\overline{B} \subset B(x,r)$, and let (u^*, K^*) be a topological competitor for $(\widetilde{u}, \widetilde{K})$ in B. Set $\widehat{K} = [K^* \cap B(x,r)] \cup [K \setminus B(x,r)]$; observe that $K^* \subset \widehat{K}$, because $K^* \setminus B(x,r) = \widetilde{K} \setminus B(x,r) \subset K \setminus B(x,r)$. Still call u^* the restriction of u^* to $\Omega \setminus \widehat{K}$. Then $(u^*, \widehat{K})$ is a topological competitor for (u, K) in $B(x,r)$.*

First, $\widehat{K}$ is closed in Ω: since $K \setminus B(x,r)$ is obviously closed, the only problem could come from a limit of points of $K^* \cap B(x,r)$. Such a limit either lies in $K^* \cap B(x,r)$, or in $K^* \cap \partial B(x,r) = \widetilde{K} \cap \partial B(x,r) \subset K \setminus B(x,r)$ (because $K^* = \widetilde{K}$ out of $\overline{B} \subset B(x,r)$, and $\widetilde{K} \subset K$). Since $K^* \subset \widehat{K}$, $u^* \in W_{\text{loc}}^{12}(\Omega \setminus \widehat{K})$, and hence $(u^*, \widehat{K}) \in \mathcal{A}$. Notice that $u^* = \widetilde{u}$ out of $\overline{B}$ and $\widetilde{u} = u$ on $\Omega \setminus K$, so $u^* = u$ on $\Omega \setminus (K \cup B(x,r))$.

We still need to check that $(u^*, \widehat{K})$ is a *topological* competitor for (u, K) in $B(x,r)$. Let y, $z \in \Omega \setminus [\overline{B}(x,r) \cup K]$ be given, and suppose that they lie in different components of $\Omega \setminus K$. We want to show that $\widehat{K}$ also separates y from z in Ω. Multiple applications of (15) show that for each j, y and z lie in different components of $\Omega \setminus K_j$. Then $\widehat{K}$ also separates them (otherwise, there is a path $\gamma \subset \Omega \setminus \widehat{K}$ that goes from y to z, this path meets every K_j, i.e., each compact set $\gamma \cap K_j$ is nonempty, hence also the intersection $\gamma \cap \widetilde{K}$, a contradiction). Next, K^* separates y from z, by (7.16) and because (u^*, K^*) is a topological competitor for $(\widetilde{u}, \widetilde{K})$ in $B \subset\subset B(x,r)$. Finally, $\widehat{K}$ also separates y from z, because $K^* \subset \widehat{K}$. Lemma 16 follows. $\qquad\square$

It is now easy to check that $(\widetilde{u}, \widetilde{K})$ is topologically reduced. Suppose $\overline{B} \subset \Omega$, let (u^*, K^*) be a topological competitor for $(\widetilde{u}, \widetilde{K})$ in B, and suppose that K^* is strictly contained in $\widetilde{K}$ and u^* is an extension of $\widetilde{u}$. We want to get a contradiction. Choose $B(x,r) \subset\subset \Omega$ so that $\overline{B} \subset B(x,r)$. Lemma 16 says that

$$(u^*, \widehat{K}) \text{ is a topological competitor for } (u, K) \text{ in } B(x,r), \qquad (17)$$

with $\widehat{K} = [K^* \cap B(x,r)] \cup [K \setminus B(x,r)]$.

Pick a point $z \in \widetilde{K} \setminus K^*$. Note that $z \in \overline{B} \subset B(x, r)$, because (u^*, K^*) is a topological competitor for $(\widetilde{u}, \widetilde{K})$ in B. Let B' be a small ball centered at z, contained in $B(x, r)$, and that does not meet K^*. Then $\widehat{K} \subset K \setminus B'$ (because $K^* \subset \widetilde{K} \subset K$), so $(u^*, K \setminus B')$ is a topological competitor for (u, K) in $B(x, r)$, by (17). [Note that (7.16) is stronger for $\widehat{K}$ than for the larger $K \setminus B'$.] And the restriction to $\Omega \setminus (K \setminus B')$ of u^* is an extension of u (recall that u^* extends $\widetilde{u}$ which extends u). So B' satisfies (13), and it would surely have been a better choice than some of the late B_j, because they cannot all lie in the same $\mathcal{R}_k$ as B', or an earlier one. [See the same argument in the proof of Proposition 2.] This contradiction proves that $(\widetilde{u}, \widetilde{K})$ is topologically reduced, and Proposition 8 follows. $\qquad\square$

Remark 18. The main point of Proposition 8 is to convince the reader that we may restrict our attention to reduced topological almost-minimizers and quasiminimizers. However, our additional condition (9) is not automatically satisfied, even for local topological almost-minimizers.

Let us try to explain what may go wrong. Take $\Omega = \mathbb{R}^2$, K equal to a circle of (large) radius R that goes through the origin, and $u = 0$ everywhere. Obviously, $(u, K) \in \mathcal{A}$.

Now let $r > 0$ be small, and consider the pair $(\widetilde{u}, \widetilde{K})$, where $\widetilde{K} = K \setminus B(0, r)$ and $\widetilde{u} = 0$. This pair is a competitor for (u, K) in $B(0, r)$, and it is a topological competitor for (u, K) in $B(0, 2R)$ (because $\Omega \setminus [K \cup \overline{B}(0, 2R)] = \mathbb{R}^2 \setminus \overline{B}(0, 2R)$ is connected). In particular, the pair (u, K) is not topologically reduced, and also (9) does not hold. But this concerns the situation at a large scale, and things are different at scales smaller than R.

If B is a ball such that $(\widetilde{u}, \widetilde{K})$ is a topological competitor for (u, K) in B, then B must contain K. Indeed, otherwise we can find two points y, z that lie out of $B \cup K$, and on different sides of K. These two points are separated by K in Ω, but not by $\widetilde{K}$, hence $(\widetilde{u}, \widetilde{K})$ is not a topological competitor for (u, K) in B.

Also, (u, K) is a local topological almost-minimizer (and quasiminimizer too). More precisely, let h be any gauge function such that $\lim_{R \to +\infty} h(R) > 2\pi$ and for which there is a positive constant b such that $h(t) \geq bt$ for $t \leq 1$. Then (u, K) is a local topological almost-minimizer in $\mathbb{R}^2$ with this gauge function as soon as R is large enough. To check this, we consider a topological competitor $(\widetilde{u}, \widetilde{K})$ for (u, K) in some ball B. Call $r(B)$ the radius of B. If $r(B) \geq R/10$, and we chose R so large that $h(R/2\pi) \geq 2\pi$, the almost-minimizing condition (7.7) holds simply because the left-hand side is $H^1(K \cap \overline{B}) \leq 2\pi r(B) \leq h(r(B)) r(B)$. So we are left with the case when $r(B) < R/10$. The same sort of argument as we did above for $K \setminus B(0, r)$ says that $\widetilde{K}$ must keep the same separation properties as K (that is, we cannot open holes in K). Then the best way we can pick $\widetilde{K}$ is to replace the arc of circle $K \cap B$ with the line segment with the same endpoints, and a small computation shows that (7.7) holds if R is large enough, depending on the constant b above.

Thus (u, K) is a local topological almost-minimizer, with h essentially as small as we want. By Remark 7.22, it is also a topological almost quasiminimizer, with constants $M = 1$, a as small as we want, and r_0 as large as we want (but of course we have to take R large enough, depending on these data).

Altogether, we have examples of local topological almost- and quasiminimizer (u, K) that are not topologically reduced, do not satisfy (9), and for which there is no really equivalent reduced topological almost- or quasiminimizer.

In the example above, we can find a reduced almost- or quasiminimizer $(\widetilde{u}, \widetilde{K})$ for which $\widetilde{K} \subset K$ and $\widetilde{u}$ is an extension of u (take $\widetilde{K} = \emptyset$), but it is easy to modify u so that even this becomes wrong. Instead of taking $u = 0$, we choose u smooth, still constant on $B(0, r)$, but with a small positive jump across $K_1 = K \setminus B(0, r)$. If u is small and smooth enough, the almost-minimality properties of (u, K) are preserved. But now u only extends to $\mathbb{R}^2 \setminus K_1$, and the pair (u, K_1) is no longer an almost-minimizer, because it does not compare favorably with a pair with an even slightly shorter $\widetilde{K}$ and a slightly modified $\widetilde{u}$ near the two points of $K \cap \partial B(0, r)$.

One can try to blame this failure of (9) and the existence of equivalent topologically reduced almost- and quasiminimizers on our definition of topological competitors, and the definitions of local topological almost- and quasiminimizers that ensue. We give below a slightly different set of definitions for which this problem does not arise. We mention it mostly for completeness, but warn the reader that the notions of almost- and quasiminimizers that follow are a little more restrictive, and have the following two defects. First, limits of strong local topological almost-minimizers are local topological almost-minimizers (see Theorem 38.3), but are not always strong local topological almost-minimizers. See Remark 38.63. Also, it is no longer true that when you start from a strong local topological almost-minimizers and add a constant to u in one of the components of $\Omega \setminus K$, you get another strong local topological almost-minimizer. See Exercise 32 and compare with the positive result in Exercise 7.57.

Probably the best way to deal with examples like the big circle above is to replace (u, K) by an equivalent coral pair (see Subsection c below), and not worry too much about reduction.

Definition 19. *Let $(u, K) \in \mathcal{A}$ and a ball $B(x, r)$ be given, with $\overline{B}(x, r) \subset \Omega$. A weak topological competitor for (u, K) in $B(x, r)$ is a competitor for (u, K) in $B(x, r)$ (as in Definition 7.2) which also satisfies the following topological condition:*

$$\text{there is a compact set } T \subset \Omega \text{ such that if } y, z \in \Omega \setminus [K \cup T \cup \overline{B}(x,r)] \qquad (20)$$
$$\text{are separated by } K \text{ in } \Omega, \text{ then } \widetilde{K} \text{ also separates them in } \Omega.$$

Note that (20) is weaker than (7.16). When we replace topological competitors with weak topological competitors in Definitions 7.17 and 7.21, we get the following.

Definition 21. *Let the pair $(u, K) \in \mathcal{A}$ be given. We say that (u, K) is a strong local topological almost-minimizer in Ω, with gauge function h, if (7.7) holds whenever $\overline{B}(x, r) \subset \Omega$ and $(\widetilde{u}, \widetilde{K})$ is a weak topological competitor for (u, K) in $B(x, r)$. We say that (u, K) is a strong local topological (r_0, M, a)-quasiminimizer in Ω when (7.20) holds whenever $\overline{B}(x, r) \subset \Omega$, $r \leq r_0$, and $(\widetilde{u}, \widetilde{K})$ is a weak topological competitor for (u, K) in $B(x, r)$.*

Notice that in the example above where K is a big circle, (u, K) is not a strong local topological almost- or quasiminimizer any more, because this time $(\widetilde{u}, \widetilde{K})$ is a weak topological competitor for (u, K) already in the ball $B(0, r)$. Of course this requires $h(r)$

or a to be small enough, but this is all right. Note also that the difference between the two definitions is not so much in the total list of topological competitors; that list would even be exactly the same if we only considered compact sets T in (20) that are balls. The difference comes from the size of the ball where we are allowed to do the comparison, which determines the value of $h(r)r^{n-1}$ in (7.7) or ar^{n-1} in (7.20). In particular, if we were working with the gauge function $h(r) = 0$ or the constant $a = 0$, we would see essentially no difference between the two definitions.

Maybe we should say again that all these subtle differences do not concern minimizers of functionals like the ones in (7.9) or (7.26), or even topological minimizers of such functionals (with appropriate definitions), because for them we already have (9) (and the two definitions are essentially the same anyway).

Proposition 22. *Let (u, K) be a strong local topological almost-minimizer in Ω with gauge function h (respectively, a strong local topological (r_0, M, a)-quasiminimizer in Ω, with a small enough (depending on n)). Then (u, K) satisfies (9), and the pair $(\widetilde{u}, \widetilde{K})$ given by Proposition 8 is a (topologically reduced) strong local topological almost minimizer in Ω with the gauge function h^{+} defined by $h^{+}(r) = \lim_{\rho \to r+} h(\rho)$ (respectively, a strong local topological (r_1, M, a)-quasiminimizer in Ω for any $r_1 < r_0$).*

When we say topologically reduced local topological almost- or quasiminimizer, we just mean an almost- or quasiminimizer that is topologically reduced as a pair.

Let us first check (9). Let $(\widetilde{u}, \widetilde{K})$ be a topological competitor for (u, K) in some ball $B \subset\subset \Omega$. It is even enough to assume that $(\widetilde{u}, \widetilde{K})$ is a weak topological competitor in B. Also assume that $\widetilde{K} \subset K$ and $\widetilde{u}$ is an extension of u, and that $H^{n-1}(K \setminus \widetilde{K}) > 0$. We want to get a contradiction. We shall do this when (u, K) is a quasiminimizer; the case of an almost-minimizer is even simpler.

By Lemma 19.1, we can find $x \in B$ such that (4) holds. In particular, we can find $r > 0$ so small that $r < r_0$ (with r_0 as in the definition of quasiminimizers), $B(x, r) \subset B$, and

$$H^{n-1}(B(x, r) \cap [K \setminus \widetilde{K}]) \geq c_n\, r^{n-1}/2. \tag{23}$$

Set $K^* = [\widetilde{K} \cap B(x, r)] \cup [K \setminus B(x, r)]$. Note that K^* is closed in Ω and $\widetilde{K} \subset K^* \subset K$. Then $(\widetilde{u}, K^*)$ is a weak topological competitor for (u, K) in $B(x, r)$ (for (20), take the compact set T given by the fact that $(\widetilde{u}, \widetilde{K})$ is a weak topological competitor for (u, K) in B, and note that K^* separates y from z as soon as $\widetilde{K}$ does). In this case, (7.20) reduces to $H^{n-1}(K \setminus K^*) \leq ar^{n-1}$. Since $K \setminus K^* = [K \cap B(x, r)] \setminus [\widetilde{K} \cap B(x, r)]$, we get the desired contradiction with (23) as soon as $a < c_n/2$. Thus (9) holds.

Now Proposition 8 applies, and gives a topologically reduced pair $(\widetilde{u}, \widetilde{K})$, which we may consider equivalent to (u, K). Note in particular that $H^{n-1}(K \setminus \widetilde{K}) = 0$. We still need to check that $(\widetilde{u}, \widetilde{K})$ is a strong local topological almost- or quasiminimizer.

If we just want to show that it is a local topological almost- or quasiminimizer, we can use Lemma 16 as it is: if (u^*, K^*) is a topological competitor for $(\widetilde{u}, \widetilde{K})$ in some ball B, and if $B(x, r) \subset\subset \Omega$ is a slightly larger ball, then $(u^*, \widehat{K})$ is a topological competitor for (u, K) in $B(x, r)$, where we set $\widehat{K} = [K^* \cap B(x, r)] \cup [K \setminus B(x, r)]$. When we apply the definition of almost- or quasiminimality to $(u^*, \widehat{K})$ and (u, K) in $B(x, r)$, we also get (7.7) or (7.20) for (u^*, K^*) and $(\widetilde{u}, \widetilde{K})$ in $B(x, r)$, because $H^{n-1}(K \setminus \widehat{K}) = H^{n-1}(K^* \setminus \widehat{K}) = 0$.

The desired result, i.e., (7.7) or (7.20) for (u^*, K^*) and $(\tilde{u}, \tilde{K})$ in B, follows by letting $B(x, r)$ tend to B.

Since we required strong almost- or quasiminimality, we have to check that Lemma 16 also holds with topological competitors replaced with weak topological competitors. This is easy; we leave the details. This completes our proof of Proposition 22. $\square$

One could object that if we consider strong topological almost- or quasiminimizers, one should use weak competitors in the definition of reduced pairs. It is fairly easy to verify that Proposition 8 stays valid, with only minor modifications in the proof, but let us not bother the reader with this.

In the future we shall feel free to restrict our study of local topological almost- or quasiminimizers to topologically reduced objects. With the definitions of Section 7, this is not entirely justified, because of the example in Remark 18, but the author feels that this is not a sufficient reason to introduce more complicated definitions. If we ever need to study nonreduced local topological almost- or quasiminimizers that do not satisfy the stronger property of Definition 21, we shall simply have to do this by hand, or use the notion of coral pairs below.

c. Coral pairs

It is probable that in the context of topological almost- or quasiminimizers, our attempt to reduce to topologically reduced pairs was uselessly ambitious. Here we describe a first cleaning of pairs in $\mathcal{A}$ that is a little less natural, but is probably just as useful.

Definition 24. *Let K be a (relatively) closed subset of Ω. We shall call core of K the closed support of the restriction of H^{n-1} to K, i.e., the set*

$$\operatorname{core}(K) = \left\{ x \in K \,;\, H^{n-1}(K \cap B(x, r)) > 0 \text{ for all } r > 0 \right\}. \tag{25}$$

If $\operatorname{core}(K) = K$, we shall say that K is coral. If $(u, K) \in \mathcal{A}$ and K is coral, we shall say that (u, K) is coral.

The advantage of the notion is that it is very easy to reduce to coral pairs, and most of our regularity results on K when (u, K) is reduced will easily extend to coral pairs.

Let $(u, K) \in \mathcal{A}$ be given. Let us check that $\operatorname{core}(K)$ is closed in Ω and

$$H^{n-1}(K \setminus \operatorname{core}(K)) = 0. \tag{26}$$

Indeed set $V = \Omega \setminus \operatorname{core}(K)$. Then

$$V = \left\{ x \in \Omega \,;\, H^{n-1}(K \cap B(x, r)) = 0 \text{ for some } r > 0 \right\},$$

V is obviously open, and it is also very easy to cover it with countably many balls B_j such that $H^{n-1}(K \cap B_j) = 0$.

Since $H^{n-1}(K \setminus \mathrm{core}\,(K)) = 0$, we can still see u as defined (almost-every-where) on $\Omega \setminus \mathrm{core}\,(K)$. Let us assume that

$$u \in W^{1,2}(B \setminus K) \text{ for every ball } B \text{ such that } \overline{B} \subset \Omega. \qquad (27)$$

This is a very reasonable assumption, even though we only require that $u \in W^{1,2}_{\mathrm{loc}}(B \setminus K)$ in our definition of $\mathcal{A}$. In particular, every local almost- or quasimin-imizer satisfies (27), by a very simple truncation argument: add ∂B to K, remove $K \cap B$, and replace u with a constant inside B. See Lemma 18.19 for a proof (and Remark 7.22 if you want to get the case of almost-minimizers without even checking that the argument goes through). Also note that it is enough to prove (27) for balls of radius less than r_0.

We claim that if (27) holds, then $(u, \mathrm{core}\,(K)) \in \mathcal{A}$. Since $\mathrm{core}\,(K)$ is closed in Ω, this just means that $u \in W^{1,2}_{\mathrm{loc}}(\Omega \setminus \mathrm{core}\,(K))$. In fact, we even have that

$$u \in W^{1,2}(B \setminus \mathrm{core}\,(K)) \text{ for every ball } B \text{ such that } \overline{B} \subset \Omega. \qquad (28)$$

To see this, we need Proposition 10.1 below on the removability of sets of vanishing H^{n-1}-measure for functions in $W^{1,p}_{\mathrm{loc}}$. Let B be as in (28), and apply Proposition 10.1 to u on the domain $\Omega' = B \setminus \mathrm{core}\,(K)$. Note that $K' = K \cap \Omega'$ is closed in Ω', has vanishing H^{n-1}-measure by (26), and that $u \in W^{1,2}(\Omega' \setminus K')$ by (27). Then Proposition 10.1 says that $u \in W^{1,2}(\Omega')$, which proves (28) and our claim.

As before for reduction, we cannot hope that $(u, \mathrm{core}\,(K))$ is a competitor for (u, K), because $K \setminus \mathrm{core}\,(K)$ may accumulate near $\partial\Omega$. So we cannot really say that it is a topological competitor either. However, it is topologically equivalent to (u, K) in the sense that

$$\begin{aligned} &\text{if } y, z \in \Omega \setminus K \text{ lie in different connected components} \\ &\quad \text{of } \Omega \setminus K, \text{ then } \mathrm{core}\,(K) \text{ also separates them in } \Omega. \end{aligned} \qquad (29)$$

The proof goes roughly as for (15). Assume that $\mathrm{core}\,(K)$ does not separate y from z in Ω, and let $\gamma \subset \Omega \setminus \mathrm{core}\,(K)$ be a path from y to z. Since $\mathrm{core}\,(K)$ is closed, we can move γ a tiny bit, and replace it with a piecewise affine path, i.e., the juxtaposition of consecutive line segments $L_1, \ldots, L_m$. Call a_j the initial endpoint of L_j and b_j its final endpoint; thus $a_1 = y$, $a_j = b_{j-1}$ for $j \geq 1$, and $b_m = z$. We may assume that $m \geq 1$ (otherwise, add a vertex in the middle of $[x, y]$). We can replace $b_1 = a_2$ with any point in a small ball, without altering the fact that $\gamma \subset \Omega \setminus \mathrm{core}\,(K)$. Because of (26) and by an easy general position argument, we can pick our new choice of b_1 so that $[a_1, b_1]$ does not meet K any more. Then we replace b_2 with a new point, so close to the previous one that $[a_2, b_2] \cup [b_2, b_3] \subset \Omega \setminus \mathrm{core}\,(K)$, and such that $[a_2, b_2]$ no longer meets K. After a finite number of modifications, we get a path in $\Omega \setminus K$ that goes from y to z. Notice that for the last modification, when we replace a_m, we need to make sure at the same time that $[a_{m-1}, a_m]$ and $[a_m, z]$ do not meet K, but the same general position argument works. This proves (29).

Because of (29), $\widetilde{K} = \mathrm{core}\,(K)$ satisfies all the requirements of Proposition 16, except for the fact that $(u, \widetilde{K})$ is not reduced, but merely coral. Also, Lemma 16 holds with this choice of $\widetilde{K}$. If in addition (u, K) is a local topological almost-minimizer with gauge function h, or a local topological (r_0, M, a)-quasiminimizer with a small enough, then the same thing holds for $(u, \mathrm{core}\,(K))$, except that we have to replace $h(r)$ with $h^+(r) = \lim_{\rho \to r^+} h(\rho)$ in the first case, and replace r_0 with any smaller r_1 in the second case. The proof is the same as in the last lines of Proposition 22: we apply Lemma 16 and find that the almost- or quasiminimality condition for $(u, \mathrm{core}\,(K))$ is the same as for (u, K), because of (26).

Let us try to summarize the situation. We would have liked to replace all local topological almost- or quasiminimizers with equivalent topologically reduced pairs, but this is only possible most of the time. For the other cases, we can still replace K with its core, and get a coral almost- or quasiminimizer. Although this pair is not a topological competitor for the initial pair, it can be considered equivalent because of (26), (29) and the fact that it satisfies the conclusion of Lemma 16. And we shall see that most of the properties of topologically reduced local topological almost- or quasiminimizers are still satisfied by coral ones.

Remark 30. Reduced local almost- or quasiminimizers, and topologically reduced local topological almost- or quasiminimizers are coral. Indeed, suppose for a minute that this is not the case for (u, K). We can find a small ball $B \subset\subset \Omega$ centered on K and such that $H^{n-1}(K \cap B) = 0$. Then $u \in W^{1,2}(B)$ because (27) holds for local topological almost- and quasiminimizers, and by Proposition 10.1. Hence $(u, K \setminus B)$ is a competitor for (u, K) in B. It is a topological competitor, by (29). And this contradicts Definition 7 or Definition 1.

Remark 31. When (u, K) is a local topological minimizer of a functional like the ones in (7.9) or (7.26), corality is clearly a good enough notion, because it implies topological reduction.

Indeed suppose that (u, K) is coral, and let $(\widetilde{u}, \widetilde{K})$ be a topological competitor for (u, K) such that $\widetilde{K} \subset K$ and $\widetilde{u} = u$ on $\Omega \setminus K$ (as in Definition 7). Then $H^{n-1}(K \setminus \widetilde{K}) = 0$, because (u, K) is a local topological minimizer. If x is a point of $K \setminus \widetilde{K}$, then $\mathrm{dist}(x, K) > 0$ because K is closed, hence $H^{n-1}(K \cap B(x, t)) = H^{n-1}([K \setminus \widetilde{K}] \cap B(x, t)) = 0$ for t small, and $x \notin \mathrm{core}\,(K)$. Such a point does not exist if (u, K) is coral. This proves that $\widetilde{K} = K$, hence (u, K) is topologically reduced.

Similarly, if (u, K) is a coral local minimizer of one of the functionals in (7.9) or (7.26), then it is reduced. The proof is the same.

Exercise 32. Let K be a big circle as in Remark 18, and u be constant on each component of $\mathbb{R}^2 \setminus K$. Check that the fact that (u, K) is a strong local topological almost-minimizer with a given gauge function, for instance, depends on the jump of u on K. Thus the notion is not stable under the operation of adding a constant to u in a component of $\Omega \setminus K$. Compare with the situation for local topological almost-minimizers and Exercise 7.57.

Exercise 33. Suppose (u, K) is a reduced (respectively topologically reduced, respectively coral) pair in Ω, and $\Omega_1 \subset \Omega$ is a smaller domain. Show that (u, K) is reduced (respectively topologically reduced, respectively coral) in Ω_1. Hint: Exercise 7.58.

... ...
... ...

B. Functions in the Sobolev Spaces $W^{1,p}$

In the next six sections we want to record elementary properties of functions in $W_{\text{loc}}^{1,p}$ that will be used later. These include absolute continuity on almost every line, traces and limits on hyperplanes, Poincaré estimates, and a few others. These sections are present for self-containedness, but most readers should be able to go through them very rapidly, or skip them altogether. One may also consult [Zi] for a much more thorough treatment of these topics.

We shall also include in Sections 15–17 a few standard facts on functions f that (locally) minimize the energy $\int |\nabla f|^2$ with given boundary data. Things like harmonicity, the Neumann condition at the boundary, or conformal invariance in dimension 2. These sections also should be skippable to a large extent.

9 Absolute continuity on lines

We start our study of the Sobolev spaces $W^{1,p}$ with the absolute continuity of Sobolev functions on almost-every line.

Let $\Omega \subset \mathbb{R}^n$ be open. Recall from Definition 2.2 that $W^{1,p}(\Omega)$ is the set of functions $f \in L_{\text{loc}}^1(\Omega)$ whose partial derivatives $\frac{\partial f}{\partial x_i}$ (in the sense of distributions) lie in $L^p(\Omega)$. [Note that we do not require that $f \in L^p(\Omega)$, as many authors do.] Similarly, $W_{\text{loc}}^{1,p}(\Omega)$ is the set of functions $f \in L_{\text{loc}}^1(\Omega)$ such that $\frac{\partial f}{\partial x_i} \in L_{\text{loc}}^p(\Omega)$ for $1 \leq i \leq n$.

We start with the simple case when $n = 1$.

Proposition 1. *Let $I \subset \mathbb{R}$ be an open interval and $f \in W_{\text{loc}}^{1,1}(I)$. Denote by $f' \in L_{\text{loc}}^1(I)$ the derivative of f (in the sense of distributions). Then*

$$f(y) - f(x) = \int_x^y f'(t)\,dt \quad \text{for } x, y \in I. \tag{2}$$

Moreover, f is differentiable almost-everywhere on I, and its differential is equal to f' almost-everywhere.

This is what we shall mean by absolute continuity on I. To prove the proposition, fix $x \in I$ and set $F(y) = \int_x^y f'(t)\,dt$ for $y \in I$. Since F is continuous, it defines a distribution on I and we can talk about its derivative F'. We want to show that $F' = f'$, so let $\varphi \in C_c^\infty(I)$ be given, and let $[x_1, x_2]$ denote a compact interval in I that contains the support of φ. Then

$$\langle F', \varphi \rangle = -\int_I F(y)\,\varphi'(y)\,dy = -\int_I [F(y) - F(x_1)]\,\varphi'(y)\,dy \tag{3}$$

$$= -\int_{x_1}^{x_2} \left\{ \int_{x_1}^{y} f'(t)\,dt \right\} \varphi'(y)\,dy = -\int_{x_1}^{x_2} \int_{x_1}^{x_2} f'(t)\,\varphi'(y)\,\mathbf{1}_{\{t \le y\}}(t, y)\,dy\,dt$$

$$= \int_{x_1}^{x_2} f'(t)\,\varphi(t)\,dt = \langle f', \varphi \rangle,$$

by definition of F', because $\int \varphi'(y)\,dy = 0$, and by Fubini. Thus $(F - f)' = 0$ distributionwise, and it is easy to see that it is constant (exercise!). This proves (2).

For the remaining part of Proposition 1, we apply the Lebesgue differentiation theorem (see for instance [Ru] or [Mat2]) to the function $f' \in L_{\text{loc}}^1(I)$, to get that

$$\lim_{\varepsilon \to 0^+} \frac{1}{2\varepsilon} \int_{y-\varepsilon}^{y+\varepsilon} |f'(t) - f'(y)|\,dt = 0 \tag{4}$$

for almost-every $y \in I$. Proposition 1 follows, because (4) implies that f is differentiable at y, with derivative $f'(y)$. $\qquad\qquad\qquad\qquad\qquad\qquad\qquad\qquad\Box$

We may now consider larger dimensions.

Proposition 5. *Let* $\Omega = \Omega_1 \times \Omega_2 \subset \mathbb{R}^m \times \mathbb{R}^n$ *be a product of open sets, and let* $f \in W^{1,1}(\Omega)$. *Denote by* $(x, y) \in \Omega_1 \times \Omega_2$ *the points of* Ω, *and call* $f_j = \frac{\partial f}{\partial y_j} \in L^1(\Omega)$, $1 \le j \le n$, *the derivatives of* f *with respect to the last variables. For almost-every* $x \in \Omega_1$, *the function* F_x *defined by* $F_x(y) = f(x, y)$ *lies in* $W^{1,1}(\Omega_2)$, *and* $\frac{\partial F_x}{\partial y_j} = f_j(x, \cdot)$ *(distributionwise) for* $j = 1, \dots, n$.

Comments. We shall mostly be interested in the case when $n = 1$ (and y is any of the coordinate functions). Our statement concerns functions in $W^{1,1}$, but it is easy to localize (to treat functions in $W_{\text{loc}}^{1,1}$). Also, we get a similar statement for functions in $W_{\text{loc}}^{1,p}$, $p > 1$, because these functions lie in $W_{\text{loc}}^{1,1}$ and our statement says how to compute $\frac{\partial F_x}{\partial y_j}$.

In the statement of Proposition 5 (and similar ones later) we assume that a representative in the class of $f \in L_{\text{loc}}^1$ has been chosen, and similarly for the functions f_j, and our statements concern these representatives. Thus the bad sets where $F_x \notin W^{1,1}(\Omega_2)$ or $\frac{\partial F_x}{\partial y_j} \ne f_j(x, \cdot)$ may depend on our choice of representatives, but this is all right.

The proof of Proposition 5 will be an exercise on distributions and Fubini. By definition,

$$\int_\Omega f(x, y)\,\frac{\partial \Phi}{\partial y_j}(x, y)\,dx\,dy = \left\langle f, \frac{\partial \Phi}{\partial y_j} \right\rangle = -\langle f_j, \Phi \rangle = -\int_\Omega f_j(x, y)\,\Phi(x, y)\,dx\,dy$$

$$\tag{6}$$

for all test functions $\Phi \in C_c^\infty(\Omega)$. We only want to apply this to decomposed functions of the form $\varphi(x)\psi(y)$, with $\varphi \in C_c^\infty(\Omega_1)$ and $\psi \in C_c^\infty(\Omega_2)$; then (6) becomes

$$\int_\Omega f(x,y)\,\varphi(x)\,\frac{\partial\psi}{\partial y_j}(y)\,dx\,dy = -\int_\Omega f_j(x,y)\,\varphi(x)\,\psi(y)\,dx\,dy. \tag{7}$$

Since f and f_j are locally integrable on Ω, Fubini says that for each compact set $K \subset \Omega_2$,

$$\int_K \{|f(x,y)| + |f_j(x,y)|\}\,dy < +\infty \tag{8}$$

for almost every $x \in \Omega_1$. In fact, since we can cover Ω_2 by countably many compact subsets, we can find a set $Z \subset \Omega_1$, of measure zero, such that for $x \in \Omega_1 \setminus Z$, (8) holds for all compact subsets $K \subset \Omega_2$.

Fix any $\psi \in C_c^\infty(\Omega_2)$. For each $x \in \Omega_1 \setminus Z$ we can define

$$A(x) = \int_{\Omega_2} f(x,y)\,\frac{\partial\psi}{\partial y_j}(y)\,dy \tag{9}$$

and

$$B(x) = -\int_{\Omega_2} f_j(x,y)\,\psi(y)\,dy\,. \tag{10}$$

By Fubini, we even know that A and B are locally integrable on Ω_1. Moreover, (7) says that

$$\int_{\Omega_1} A(x)\,\varphi(x)\,dx = \int_{\Omega_1} B(x)\,\varphi(x)\,dx \tag{11}$$

for all $\varphi \in C_c^\infty(\Omega_1)$. Hence $A(x) = B(x)$ almost-everywhere.

Let us apply this argument for $\psi \in \mathcal{D}$, where $\mathcal{D} \subset C_c^\infty(\Omega_2)$ is a countable set of test functions. For each $\psi \in \mathcal{D}$ we get functions A and B and a set $Z(\psi)$ of measure zero such that $A(x) = B(x)$ on $\Omega_1 \setminus Z(\psi)$. Set $Z_1 = Z \cup \left(\bigcup_{\psi \in \mathcal{D}} Z(\psi) \right)$. Thus Z_1 is still negligible.

Let $x \in \Omega_1 \setminus Z_1$ be given. We know that

$$\int_{\Omega_2} f(x,y)\,\frac{\partial\psi}{\partial y_j}(y)\,dy = -\int_{\Omega_2} f_j(x,y)\,\psi(y)\,dy \tag{12}$$

for all $\psi \in \mathcal{D}$, because $A(x) = B(x)$ with the notation above. Let us assume that we chose $\mathcal{D}$ so large that, for each $\psi \in C_c^1(\Omega_2)$, we can find a sequence $\{\psi_j\}$ in $\mathcal{D}$ such that the ψ_j have a common compact support in Ω_2 and $\{\psi_j\}$ converges to ψ for the norm $\|\psi\|_\infty + \|\nabla\psi\|_\infty$. (This is easy to arrange.) Then (12) extends to all $\psi \in C_c^1(\Omega_2)$, because since $x \notin Z$, (8) holds for all compact sets $K \subset \Omega_2$.

Continue with $x \in \Omega_1 \setminus Z_1$ fixed. Set $F_x(y) = f(x,y)$, as in the statement of the proposition. Note that $f_j(x,\cdot) \in L_{\mathrm{loc}}^1(\Omega_2)$ because (8) holds for all compact sets $K \subset \Omega_2$. Since (12) holds for all $\psi \in C_c^\infty(\Omega_2)$ we get that $\frac{\partial F_x}{\partial y_j} = f_j(x,\cdot)$. To complete the proof of Proposition 5 we just need to observe that $f_j(x,\cdot) \in L^1(\Omega_2)$ (and not merely $L_{\mathrm{loc}}^1(\Omega_2)$). This is the case, by Fubini. $\qquad\square$

Corollary 13. *If $\Omega = \Omega_1 \times I$ for some open interval I, and $f \in W^{1,p}_{\mathrm{loc}}(\Omega)$ for some $p \in [1, +\infty)$, then for almost every $x \in \Omega_1$, $f(x, \cdot)$ is absolutely continuous on I (i.e., satisfies the hypotheses and conclusions of Proposition 1). Furthermore, its derivative equals $\frac{\partial f}{\partial y}(x, y)$ almost-everywhere.*

This is easy to prove. First note that we can restrict to $p = 1$, because $W^{1,p}_{\mathrm{loc}} \subset W^{1,1}_{\mathrm{loc}}$ anyway. Since we only assumed f to be locally in $W^{1,p}$, we need a small localization argument. Cover Ω_1 by a countable collection of relatively compact open subsets $\Omega_{1,k} \subset\subset \Omega_1$, and similarly cover I by an increasing sequence of open intervals $I_k \subset\subset I$. Then $f \in W^{1,1}(\Omega_{1,k} \times I_k)$ for each k, and Proposition 5 says that for almost every $x \in \Omega_{1,k}$, $f(x, .)$ is absolutely continuous on I_k, with a derivative that coincides with $\frac{\partial f}{\partial y}(x, .)$ almost-everywhere. The conclusion follows: when $x \in \Omega_1$ does not lie in any of the exceptional sets in the $\Omega_{1,k}$, $f(x, .)$ is absolutely continuous on I, still with the same derivative.   □

Corollary 14. *Suppose $f \in W^{1,p}_{\mathrm{loc}}(\Omega)$ for some $p \geq 1$ and some open set $\Omega \subset \mathbb{R}^n$. Let $v = (v_i) \in \mathbb{R}^n$ be given. For almost every $x \in \Omega$, f is differentiable at x in the direction v, and the differential is $\sum_{i=1}^n v_i \frac{\partial f}{\partial x_i}(x)$.*

Here $\frac{\partial f}{\partial x_i}$ denotes the distributional derivative of f. Note that the distribution $\frac{\partial f}{\partial v}$ makes sense, and is equal to $\sum_i v_i \frac{\partial f}{\partial x_i}$. Corollary 14 is now a direct consequence of Corollary 13 after a linear change of coordinates.   □

We can apply Corollary 14 to all the vectors v with rational coordinates at the same time, and get that for almost all $x \in \Omega$, f is differentiable at x in all the directions v with rational coordinates.

When f is Lipschitz, one can prove more than this: f is (truly) differentiable at x for almost all $x \in \Omega$. This is Rademacher's theorem; see for instance [Mat2].

Exercise 15. Let $f \in W^{1,p}_{\mathrm{loc}}(\Omega)$, $1 \leq p < +\infty$, and $g \in C^1_{\mathrm{loc}}(\Omega)$ be given. Show that $fg \in W^{1,p}_{\mathrm{loc}}(\Omega)$, and that $\nabla(fg) = f\nabla g + g\nabla f$ almost-everywhere. Hint: start with $g \in C^\infty_c$ and a single $\frac{\partial f}{\partial x_i}$.

Exercise 16. Let $f \in W^{1,p}_{\mathrm{loc}}(\Omega_1)$ and $g \in W^{1,p}_{\mathrm{loc}}(\Omega_2)$ be given. Show that the function F defined on $\Omega_1 \times \Omega_2$ by $F(x, y) = f(x)g(y)$ lies in $W^{1,p}_{\mathrm{loc}}(\Omega_1 \times \Omega_2)$, and compute its partial derivatives.

Exercise 17. Let $f \in L^1_{\mathrm{loc}}(\mathbb{R}^n)$ be given.

1. Show that if $\nabla f \in L^\infty(\mathbb{R}^n)$ (distributionwise), then f is Lipschitz and

$$|f(x) - f(y)| \leq ||\nabla f||_\infty |x - y| \text{ for } x, y \in \mathbb{R}^n. \qquad (16)$$

Hint: reduce to one dimension and try test functions that approximate $\mathbf{1}_{[x,y]}$.

2. Show the converse: if f is Lipschitz with constant $\leq C$, then $\nabla f \in L^\infty(\mathbb{R}^n)$ (distributionwise) and $||\nabla f||_\infty \leq C$. Hint: the problem is local, and we can get that $\nabla f \in L^2_{\mathrm{loc}}$ using the Riesz representation theorem. Then $\nabla f \in L^\infty$ because of Corollary 14.

10 Some removable sets for $W^{1,p}$

Proposition 1. *Let $\Omega \subset \mathbb{R}^n$ be open, and let $K \subset \Omega$ be a relatively closed subset such that*

$$H^{n-1}(K) = 0. \tag{2}$$

If $1 \leq p \leq +\infty$ and $f \in W^{1,p}(\Omega \setminus K)$, then $f \in W^{1,p}(\Omega)$, with the same derivative.

Note that we do not need to extend f to Ω, because K has vanishing Lebesgue measure by (2) (see Exercise 2.20). For the same reason, the derivatives $\frac{\partial f}{\partial x_i}$ are initially defined almost-everywhere on $\Omega \setminus K$, but we can easily (and uniquely) extend them to Ω.

Proposition 1 says that closed sets of H^{n-1}-measure zero are removable for $W^{1,p}$ functions. There are lots of other removable sets, but Proposition 1 will be enough here.

It is enough to prove the proposition with $p = 1$, because the derivative of f on Ω is the same as on $\Omega \setminus K$. The proof will be somewhat easier in the special case when we know that

$$f \in L^1_{\mathrm{loc}}(\Omega), \tag{3}$$

which we shall treat first. (Note that (3) is not automatic: we only know that f lies in $L^1_{\mathrm{loc}}(\Omega \setminus K)$, and a priori it may have monstrous singularities near K.) In our applications to Mumford-Shah minimizers, this special case would often be enough because f is bounded.

So let us assume that $f \in W^{1,1}(\Omega \setminus K)$ and that (3) holds. All we have to do is show that the distributional derivatives $\frac{\partial f}{\partial x_j}$ on $\Omega \setminus K$ also work on Ω. Let us only do this for $j = 1$. We want to show that for all $\varphi \in C_c^\infty(\Omega)$,

$$\int_\Omega f(x) \frac{\partial \varphi}{\partial x_1} \, dx = - \int_\Omega \frac{\partial f}{\partial x_1}(x)\, \varphi(x)\, dx. \tag{4}$$

Clearly the problem is local: it is enough to check (4) when φ is supported in a small rectangle $R \subset\subset \Omega$ (because we can cut the original φ into finitely many small pieces).

So let us assume that φ is supported in $R = I \times J$, where I is a (bounded) open interval and J is a (bounded) product of open intervals. Denote by $(x, y) \in I \times J$ the points of R, and call $\pi : (x, y) \to y$ the projection on the hyperplane that contains J. Set

$$Z = \pi(K \cap \overline{R}). \tag{5}$$

Then $H^{n-1}(Z) = 0$, by (2) and because π is Lipschitz (see Remark 2.11). Also, Z is closed if we took R relatively compact in Ω. Set $F_y(x) = f(x, y)$ for $y \in J \setminus Z$ and $x \in I$. Since $f \in W^{1,1}(I \times (J \setminus Z))$, Corollary 9.13 says that for almost-every $y \in J \setminus Z$, F_y is absolutely continuous on I, with derivative

$$F_y'(x) = \frac{\partial f}{\partial x}(x, y) \text{ for almost every } x \in I. \tag{6}$$

Call $J' \subset J \setminus Z$ the set of full measure for which this holds. Then

$$\int_R f(x,y) \frac{\partial \varphi}{\partial x}(x,y)\, dx\, dy = \int_{J'} \left\{ \int_I f(x,y) \frac{\partial \varphi}{\partial x}(x,y)\, dx \right\} dy$$

$$= \int_{J'} \left\{ -\int_I F_y'(x) \varphi(x,y)\, dx \right\} dy$$

$$= -\int_{J'} \int_I \frac{\partial f}{\partial x}(x,y) \varphi(x,y)\, dx\, dy = -\int_\Omega \frac{\partial f}{\partial x} \varphi, \qquad (7)$$

by Proposition 9.1 and (6). This is just (4) with different notations; our verification is now complete in the special case when (3) holds.

The remaining case will be treated by induction on the dimension n. We want to show that when R is a product of bounded open intervals, $K \subset \overline{R}$ is a compact set that satisfies (2), and $f \in W^{1,1}(R \setminus K)$, we can find $z_0 \in R \setminus K$ such that

$$\int_{R \setminus K} |f(z) - f(z_0)| \le C_0 \int_{R \setminus K} |\nabla f|, \qquad (8)$$

with a constant C_0 that depends only on n and the various sidelengths of R. Here ∇f denotes the vector-valued L^1 function whose coordinates are the (distributional) derivatives $\frac{\partial f}{\partial z_i}(z)$.

Notice that our statement makes sense and is trivial when $n = 1$. Indeed $H^0(K) = 0$ means that K is empty, and then (8) follows from (9.2). Let us assume now that $n \ge 2$, that our statement holds for $n-1$, and let R, K, f be as in the statement. Write $R = I \times J$, where I is an interval, and call $(x,y) \in I \times J$ the generic point of R.

Set $Z = \pi(K) \subset \overline{J}$, as in (5); we still have that Z is closed and $H^{n-1}(Z) = 0$. Also set $F_y(x) = f(x,y)$ for $x \in I$ and $y \in J \setminus Z$; as before, there is a set $J' \subset J \setminus Z$ of full measure such that for $y \in J'$, F_y is absolutely continuous on I and (6) holds. Set

$$J_0 = \left\{ y \in J'; \int_I \left| \frac{\partial f}{\partial x}(x,y) \right| dx \le a \right\}, \qquad (9)$$

with $a = \dfrac{2}{|J|} \displaystyle\int_{R \setminus K} |\nabla f|$. Then

$$a |J' \setminus J_0| \le \int_{J' \setminus J_0} \left\{ \int_I \left| \frac{\partial f}{\partial x}(x,y) \right| dx \right\} dy \le \int_{R \setminus K} |\nabla f| = \frac{a|J|}{2}$$

by (9), Fubini, and Chebyshev, so $|J' \setminus J_0| \le \frac{1}{2}|J|$ and

$$|J_0| \ge \frac{1}{2}|J|. \qquad (10)$$

Observe that for $y \in J_0$, Proposition 9.1 says that F_y is absolutely continuous and has total variation $\le a$. Hence

$$|f(x,y) - f(x',y)| = |F_y(x) - F_y(x')| \le a \quad \text{for } x, x' \in I \text{ and } y \in J_0. \qquad (11)$$

Now we want to do slices in the other direction. For each $x \in I$, set $K^x = \{y \in \bar{J} ; (x,y) \in K\}$ and $f^x(y) = f(x,y)$ for $y \in J \setminus K^x$. Proposition 9.5 says that there is a set $I' \subset I$ of full measure such that, for $x \in I'$, $f^x \in W^{1,1}(J \setminus K^x)$, and its derivatives in the y-directions coincide with the restrictions $\frac{\partial f}{\partial y_j}(x, \cdot)$. To be fair, we cannot apply Proposition 9.5 directly because $R \setminus K$ is not a product of domains, but we can cover it by countably many small cubes and apply the proposition on each one separately. We get that $f^x \in W^{1,1}_{\text{loc}}(J \setminus K^x)$ for almost all x, and the fact that the derivatives lie in L^1 for almost every x follows easily from Fubini. Next we want to apply our induction hypothesis, and so we want to check that

$$H^{n-2}(K^x) = 0 \text{ for almost every } x \in I. \tag{12}$$

This looks a little like Fubini, but we shall just use definitions with coverings. For each $\varepsilon > 0$ we can cover K by sets A_j, $j \geq 0$, with

$$\sum_{j=0}^{\infty} (\text{diam } A_j)^{n-1} \leq \varepsilon. \tag{13}$$

(See Definition 2.6 for the Hausdorff measure.) We can even assume that each A_j is a cube with sides parallel to the axes, because we could replace A_j with a cube that contains it and has a diameter at most $\sqrt{n}$ times larger.

Call π the orthogonal projection on the line that contains I and π^* the orthogonal projection on the hyperplane that contains J. For each $x \in I$, K^x is covered by the projections $\pi^*(A_j)$, $j \in \mathcal{D}(x)$, where $\mathcal{D}(x)$ denotes the set of indices $j \geq 0$ such that $\pi(A_j)$ contains x. Notice that all the sets $\pi^*(A_j)$ have diameters $\leq \varepsilon^{1/n-1}$, by (13). Hence

$$H^{n-2}_\delta(K^x) \leq c_{n-2} \sum_{j \in \mathcal{D}(x)} (\text{diam } A_j)^{n-2} \tag{14}$$

for $x \in I$, as soon as $\delta \geq \varepsilon^{1/n-1}$. (See the definition (2.8).) Now

$$\int_I H^{n-2}_\delta(K^x)\, dx \leq c_{n-2} \int_I \sum_{j \in \mathcal{D}(x)} (\text{diam } A_j)^{n-2}\, dx$$

$$\leq c_{n-2} \sum_{j=0}^{\infty} (\text{diam } A_j)^{n-2} \int_{\pi(A_j)} dx \leq c'_{n-2} \sum_j (\text{diam } A_j)^{n-1} \leq C\varepsilon. \tag{15}$$

Since this holds for all $\varepsilon > 0$, we get that $H^{n-2}_\delta(K^x) = 0$ for almost-every $x \in I$ and each $\delta > 0$; (12) follows by taking a countable intersection of bad sets.

Because of (12), we can apply our induction hypothesis to $f^x \in W^{1,1}(J \setminus K^x)$ for almost every $x \in I'$. We get a set $I'' \subset I'$ of full measure such that, for $x \in I''$, we can find $y_x \in J \setminus K^x$ such that

$$\int_{J \setminus K^x} |f^x(y) - f^x(y_x)| \leq C_0 \int_{J \setminus K^x} |\nabla f^x|. \tag{16}$$

From (16) and the triangle inequality we deduce that

$$\int_{J\setminus K^x}\int_{J\setminus K^x}|f^x(y)-f^x(y')|\leq 2C_0|J|\int_{J\setminus K^x}|\nabla f^x|. \qquad (17)$$

Then we integrate with respect to $x\in I''$ and get that

$$\int_I\int_J\int_J|f(x,y)-f(x,y')|\,dx\,dy\,dy'\leq 2C_0|J|\int_{R\setminus K}|\nabla f|, \qquad (18)$$

where we added a set of measure zero to the domains of integration to simplify the notation. By (18), Chebyshev (applied to the variable y'), and (10) we can find $y_0\in J_0$ such that

$$\int_I\int_J|f(x,y)-f(x,y_0)|\,dx\,dy\leq 4C_0\int_{R\setminus K}|\nabla f|. \qquad (19)$$

Pick any $x_0\in I$, apply (11) with $x'=x_0$ and $y=y_0$, and integrate with respect to x; this yields

$$\int_I|f(x,y_0)-f(x_0,y_0)|\,dx\leq a|I|=\frac{2|I|}{|J|}\int_{R\setminus K}|\nabla f|. \qquad (20)$$

Then integrate this with respect to y and add to (19); this gives

$$\iint_R|f(x,y)-f(x_0,y_0)|\,dx\,dy\leq\{4C_0+2|I|\}\int_{R\setminus K}|\nabla f|, \qquad (21)$$

where the apparently strange homogeneity comes from the fact that C_0 scales like a length. This completes the proof by induction of our claim (8).

    We may now finish the proof of Proposition 1: if $f\in W^{1,1}(\Omega\setminus K)$ is as in the statement of the proposition, our claim (8) says that $f\in L^1(R\setminus K)$ for every product of intervals $R\subset\subset\Omega$, and so f satisfies (3) and we can conclude as before. $\qquad\square$

Exercise 22. Suppose K is closed in the open set $\Omega\subset\mathbb{R}^n$, and there are n independent vectors v_i such that $H^{n-1}(\pi_i(K))=0$, where π_i denotes the orthogonal projection onto the vector hyperplane perpendicular to v_i. Let $p\in[1,+\infty)$ be given.

   1. Show that if $f\in W^{1,p}(\Omega\setminus K)\cap L^1_{\mathrm{loc}}(\Omega)$, then $f\in W^{1,p}(\Omega)$, with the same derivative.

   2. Suppose in addition that $n=2$. Show that if $f\in W^{1,p}(\Omega\setminus K)$, then $f\in W^{1,p}(\Omega)$, with the same derivative.

11   Composition with other functions

In this section we just want to check the stability of $W^{1,p}$ under composition with smooth enough functions. We start with two easy approximation lemmas.

Lemma 1. *Let $\Omega \subset \mathbb{R}^n$ be open, $1 \le p < +\infty$, and $f \in W^{1,p}_{\text{loc}}(\Omega)$. For each compact set $K \subset \Omega$ we can find a sequence of smooth functions $f_k \in C^\infty_c(\Omega)$ such that $\lim_{k\to\infty} f_k = f$ in $L^1(K)$ and $\lim_{k\to\infty} \nabla f_k = \nabla f$ in $L^p(K)$.*

This is standard. We can take $f_k = f * \varphi_k$, where $\{\varphi_k\}$ is any reasonable approximation of the Dirac mass at 0. For k large enough, f_k is well defined in a neighborhood of K; $\{f_k\}$ converges to f in $L^1(K)$ because $f \in L^1(K')$ for some slightly larger compact neighborhood $K' \subset \Omega$ (recall that $f \in L^1_{\text{loc}}(\Omega)$); $\{\nabla f_k\}$ converges to ∇f in $L^p(K)$ because $\nabla f_k = \nabla f * \varphi_k$ (exercise!) and $\nabla f \in L^p(K')$; finally we can multiply f_k by a smooth cut-off function to make it compactly supported in Ω.   □

Lemma 2. *Let $V \subset \mathbb{R}^n$ be open, $f \in L^1(V)$, and let $\{f_k\}$ be a sequence in $W^{1,p}(V)$ for some $p \ge 1$. Suppose that $f = \lim_{k\to\infty} f_k$ in $L^1(V)$ and $\{\nabla f_k\}$ converges in $L^p(V)$ to some limit g. Then $f \in W^{1,p}(V)$ and $\nabla f = g$.*

We gave V a different name because in practice it will often be a small, relatively compact in Ω, neighborhood of some point of Ω. The verification is immediate: if $\varphi \subset C^\infty_c(V)$,

$$\left\langle \frac{\partial f}{\partial x_i}, \varphi \right\rangle = -\left\langle f, \frac{\partial \varphi}{\partial x_i} \right\rangle = -\lim_{k\to+\infty} \int_V f_k(x) \frac{\partial \varphi}{\partial x_i}(x)dx \tag{3}$$

$$= \lim_{k\to+\infty} \int_V \frac{\partial f_k}{\partial x_i}(x)\,\varphi(x)\,dx = \int_V g_i(x)\,\varphi(x)\,dx = \langle g_i, \varphi \rangle,$$

where we called g_i the ith coordinate of g. There is no difficulty with the limits because $\frac{\partial \varphi}{\partial x_i}$ and φ are both bounded and compactly supported. Actually, the weak convergence of $\{f_k\}$ to $f \in L^1_{\text{loc}}$ and of $\{\nabla f_k\}$ to $g \in L^p$ would have been enough. Lemma 2 follows.   □

Next we consider smooth changes of variables.

Proposition 4. *Let $\Phi : \Omega_1 \to \Omega_2$ be a C^1-diffeomorphism and $f : \Omega_2 \to \mathbb{R}$ lie in $W^{1,p}_{\text{loc}}(\Omega_2)$ for some $p \ge 1$. Then $f \circ \Phi \in W^{1,p}_{\text{loc}}(\Omega_1)$ and*

$$\frac{\partial}{\partial x_i}(f \circ \Phi) = \sum_j \frac{\partial \Phi_j}{\partial x_i} \cdot \frac{\partial f}{\partial y_j} \circ \Phi. \tag{5}$$

To prove the proposition we may restrict to $p = 1$, because the right-hand side of (5) clearly lies in $L^p_{\text{loc}}(\Omega_2)$ when $\nabla f \in L^p_{\text{loc}}(\Omega_1)$ and Φ is C^1. Also note that the result is trivial when f is smooth. Now let $f \in W^{1,1}_{\text{loc}}(\Omega_2)$ be given, and let $V \subset\subset \Omega_2$ be any relatively compact open subset of Ω_2. Let $\{f_k\}$ be a sequence as in Lemma 1, with $K = \overline{V}$. Set $V_1 = \Phi^{-1}(V)$ and $h_k = f_k \circ \Phi$ on V_1. Then $\{h_k\}$

converges to $f \circ \Phi$ in $L^1(V_1)$. Also, $\frac{\partial h_k}{\partial x_i} = \sum_j \frac{\partial \Phi_j}{\partial x_i} \cdot \frac{\partial f_k}{\partial y_j} \circ \Phi$ (by (5) in the smooth case), and so $\left\{ \frac{\partial h_k}{\partial x_i} \right\}$ converges to the right-hand side of (5) in $L^1(V)$. Now (5) (in the set V) and the proposition follow from Lemma 2.   □

Proposition 4 allows us to distort the results of Section 9 by changes of variables. For instance, we get that functions in $W^{1,1}_{loc}(\mathbb{R}^2)$ are absolutely continuous on almost every circle centered at the origin. We may need later the following slight improvement of this.

Corollary 6. *Suppose K is a closed subset of $B(0, R)$ in $\mathbb{R}^2$, and $f \in W^{1,1}(B(0, R) \setminus K)$. Then $f \in W^{1,1}(\partial B(0, r) \setminus K)$ for almost every $r \in (0, R)$, and its (tangential) derivative can be computed from the restriction of ∇f to $\partial B(0, r) \setminus K$ in the obvious way.*

Thus, for almost every $r < R$, f is absolutely continuous on each of the open arcs that compose $\partial B(0, r) \setminus K$. The analogue of Corollary 6 in dimensions $n > 2$ is also true, with almost the same proof. The proof is easy. First, it is enough to check that $f \in W^{1,1}_{loc}(\partial B(0, r) \setminus K)$ for almost every $r < R$, with a derivative given by the restriction of ∇f, because the average L^1-norm of the derivative will be controlled by Fubini. Thus we can localize and reduce to a region like $S = \{ re^{i\theta} ; r_0 < r < R \text{ and } \theta_0 < \theta < \theta_0 + 2\pi \}$. Then we can use polar coordinates and reduce to a rectangle $R = \phi(S)$, with the closed set $K' = \phi(K \cap S)$. Once again, we localize: we cover $R \setminus K'$ with countably many open rectangles T_j, and apply Proposition 9.5 to get that the restriction of $f \circ \phi^{-1}$ to almost every vertical line in a T_j is absolutely continuous, with a derivative that can be computed from ∇f. Then we return to circles via ϕ^{-1}, and notice that for almost every $r < R$, $\partial B(0, r) \setminus K$ is covered by open arcs of circles where f is absolutely continuous. The conclusion follows, because any compact interval in a component of $\partial B(0, r) \setminus K$ is then covered by finitely many such intervals.   □

Now we want to study $g \circ f$ when $f \in W^{1,p}_{loc}(\Omega)$ and g is Lipschitz. We start with the case when g is smooth.

Lemma 7. *Suppose $f \in W^{1,p}_{loc}(\Omega)$ for some open set $\Omega \subset \mathbb{R}^n$ and $1 \leq p \leq +\infty$. Let $g : \mathbb{R} \to \mathbb{R}$ be a C^1 function, and assume that g' is bounded. Then $g \circ f \in W^{1,p}_{loc}(\Omega)$ and*

$$\frac{\partial}{\partial x_i}(g \circ f) = g' \circ f \cdot \frac{\partial f}{\partial x_i} \quad \text{for } 1 \leq i \leq n. \tag{8}$$

Here again the result is trivial when f is smooth. Otherwise, let $V \subset\subset \Omega$ be any relatively compact open set, and let $\{f_k\}$ be a sequence of smooth functions, with the properties of Lemma 1 (for $K = \overline{V}$). Set $h_k = g \circ f_k$. Then

$$
\begin{aligned}
\|h_k - g \circ f\|_{L^1(V)} &= \int_V |g(f_k(x)) - g(f(x))| \, dx \\
&\leq \|g'\|_\infty \int_V |f_k(x) - f(x)| \, dx \leq \|g'\|_\infty \|f_k - f\|_{L^1(V)},
\end{aligned}
\tag{9}
$$

which tends to 0. Also, we can extract a subsequence from $\{f_k\}$, which we shall still denote by $\{f_k\}$ for convenience, such that $\{f_k\}$ and $\{\nabla f_k\}$ converge to f and ∇f almost-everywhere on V, and even

$$\sum_k ||\nabla f_{k+1} - \nabla f_k||_{L^1(V)} < +\infty. \tag{10}$$

We want to apply the dominated convergence theorem, so we set

$$F(x) = \sum_k |\nabla f_{k+1}(x) - \nabla f_k(x)|.$$

Then $F \in L^1(V)$ by (10), and also

$$|\nabla f_k| \leq |\nabla f| + F \quad \text{almost-everywhere.} \tag{11}$$

Set $G = (g' \circ f)\nabla f$; thus G is the vector-valued function whose coordinates are the right-hand sides of (8). Also,

$$G(x) = \lim_{k \to \infty} \left[(g' \circ f_k)\nabla f_k \right](x) = \lim_{k \to \infty} \nabla h_k(x) \tag{12}$$

almost-everywhere, because g' is continuous. Since for almost every $x \in V$

$$|G(x) - \nabla h_k(x)| \leq |G(x)| + |\nabla h_k(x)| \tag{13}$$
$$\leq ||g'||_\infty \{ |\nabla f| + |\nabla f_k| \} \leq ||g'||_\infty \{ 2|\nabla f| + F \}$$

by (11), we can apply the dominated convergence theorem and get that

$$G = \lim_{k \to \infty} \nabla h_k \quad \text{in } L^1(V). \tag{14}$$

Now we can apply Lemma 2, and get that $g \circ f \in W^{1,1}(V)$ and (8) holds in V. This proves (8) in Ω as well, and now $g \circ f \in W^{1,p}_{\text{loc}}(\Omega)$ because the right-hand side of (8) lies in $L^p_{\text{loc}}(\Omega)$. Lemma 7 follows. $\qquad \square$

Proposition 15. *Suppose that $f \in W^{1,p}_{\text{loc}}(\Omega)$ for some open set $\Omega \subset \mathbb{R}^n$ and some $p > 1$, and let $g : \mathbb{R} \to \mathbb{R}$ be such that*

$$|g(x) - g(y)| \leq |x - y| \quad \text{for } x, y \in \mathbb{R}. \tag{16}$$

Then $g \circ f \in W^{1,p}_{\text{loc}}(\Omega)$ and

$$|\nabla(g \circ f)(x)| \leq |\nabla f(x)| \quad \text{almost-everywhere on } \Omega. \tag{17}$$

Here we took $p > 1$ just to have a simpler proof. See [Zi] for the case when $p = 1$ and more details. Set $h = g \circ f$. Note that $h \in L^1_{\text{loc}}(\Omega)$ because $|h(x)| \leq |g(0)| + |f(x)|$ (by (16)). We want to check first that $\nabla h \in L^p_{\text{loc}}(\Omega)$. For this it is enough to show that for every compact subset K of Ω, there is a constant C_K such that

$$\left| \left\langle \frac{\partial h}{\partial x_i}, \varphi \right\rangle \right| =: \left| \int_\Omega h \frac{\partial \varphi}{\partial x_i} \right| \leq C_K \|\varphi\|_q \tag{18}$$

for all $\varphi \in C^\infty_c(\Omega)$ supported in K and $1 \leq i \leq n$, where q is the conjugate exponent to p. Indeed if we can prove (18), the Riesz representation theorem says that for each K the action of $\frac{\partial h}{\partial x_i}$ on $C^\infty_c(K)$ is given by an L^p-function. Then $\frac{\partial h}{\partial x_i} \in L^p_{\text{loc}}(\Omega)$.

In the special case when g is of class C^1, Lemma 7 applies and $\left| \frac{\partial h}{\partial x_i} \right| = |g' \circ f| \left| \frac{\partial f}{\partial x_i} \right| \leq \left| \frac{\partial f}{\partial x_i} \right|$ almost-everywhere, by (8) and (16). Then (18) holds, with $C_K = \left\{ \int_K |\nabla f|^p \right\}^{1/p}$. In the general case we can easily find a sequence $\{g_k\}$ of C^1 functions that converges pointwise to g and such that $\|g'_k\|_\infty \leq M$ for some M (and all k). Then

$$\left| \int_\Omega (g_k \circ f) \frac{\partial \varphi}{\partial x_i} \right| \leq M \left\{ \int_K |\nabla f|^p \right\}^{1/p} \|\varphi\|_q \tag{19}$$

for all $\varphi \in C^\infty_c(\Omega)$ with support in K, because of the C^1 case that we already know about. Also,

$$\int_\Omega h \frac{\partial \varphi}{\partial x_i} = \int_\Omega (g \circ f) \frac{\partial \varphi}{\partial x_i} = \lim_{k \to +\infty} \int (g_k \circ f) \frac{\partial \varphi}{\partial x_i} \tag{20}$$

for $\varphi \in C^\infty_c(\Omega)$; for the convergence, notice that $|g_k \circ f(x)| \leq |g_k(0)| + M |f(x)|$ and $\frac{\partial \varphi}{\partial x_i}$ is bounded with compact support, so we can apply the dominated convergence theorem. Now (18) follows from (19) and (20), and $h \in W^{1,p}_{\text{loc}}(\Omega)$ in all cases.

We still need to check (17). Since f and $h = g \circ f$ both lie in $W^{1,p}_{\text{loc}}(\Omega)$, we can apply Corollary 9.14 to them. For almost-every $x \in \Omega$, they have differentials in every direction v with rational coordinates, which we shall denote by $\frac{\partial f}{\partial v}(x)$ and $\frac{\partial h}{\partial v}(x)$. Furthermore,

$$\frac{\partial f}{\partial v}(x) = \nabla f(x) \cdot v \quad \text{and} \quad \frac{\partial h}{\partial v}(x) = \nabla h(x) \cdot v \tag{21}$$

almost-everywhere, where ∇f and ∇h denote the distributional gradients.

When $\frac{\partial f}{\partial v}(x)$ and $\frac{\partial h}{\partial v}(x)$ both exist, we have that $\left| \frac{\partial h}{\partial v}(x) \right| \leq \left| \frac{\partial f}{\partial v}(x) \right|$ because $h = g \circ f$ and g is 1-Lipschitz. Then (21) says that for almost all $x \in \Omega$, $|\nabla h(x) \cdot v| \leq |\nabla f \cdot v|$ for every vector v with rational coordinates; (17) follows at once. This completes our proof of Proposition 15. $\square$

Exercise 22. Extend the result of Exercise 9.15 (on the fact that the product fg lies in $W^{1,p}_{\text{loc}}(\Omega)$, with the natural formula for the gradient) to the case when f and g lie in $W^{1,p}_{\text{loc}}(\Omega) \cap L^\infty_{\text{loc}}(\Omega)$.

12   Poincaré estimates for $f \in W^{1,p}(\Omega)$

In this section we try to describe how to prove Sobolev and Poincaré estimates for functions $f \in W^{1,p}(\Omega)$, $p \geq 1$. We want to do this because the proofs will be used in the next sections, in particular to define the restrictions or boundary values of f on (reasonably smooth) hypersurfaces. Our plan is to derive the various estimates when f is of class C^1 first, and then use limiting arguments.

    Let $\Omega \subset \mathbb{R}^n$ be open, and let f be defined and C^1 on Ω. We shall first estimate $|f(x) - f(y)|$ in terms of integrals of $|\nabla f|$. Some notation will be useful. Fix a constant $\alpha \in (0,1]$. For most of this section, $\alpha = 1$ will be a good choice, but we want to keep the option of taking α small later. For $x, y \in \mathbb{R}^n$, define a hyperdisk $D(x,y)$ by

$$D(x,y) = D_\alpha(x,y) = \left\{ z \in \mathbb{R}^n : |z-x| = |z-y| \text{ and } \left| z - \frac{x+y}{2} \right| \leq \alpha |x-y| \right\}, \quad (1)$$

and then set

$$C(x,y) = C_\alpha(x,y) = \text{conv}\left(\{x\} \cup \{y\} \cup D(x,y) \right) \quad (2)$$

(the intersection of two convex cones; see Figure 1).

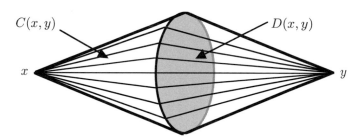

Figure 1

    For each $\ell \in D(x,y)$ denote by Γ_ℓ the path from x to y whose support is $[x, \ell] \cup [\ell, y]$. Also call $\gamma_\ell : [x,y] \to C(x,y)$ the parameterization of Γ_ℓ by its projection on $[x,y]$. Let us assume that

$$C(x,y) \subset \Omega. \quad (3)$$

The fundamental theorem of calculus says that

$$|f(x) - f(y)| \leq \int_{\Gamma_\ell} |\nabla f| dH^1 \quad (4)$$

for each $\ell \in D(x,y)$, and we can average over $\ell \in D(x,y)$ to get that

$$|f(x) - f(y)| \leq H^{n-1}(D(x,y))^{-1} \int_{D(x,y)} \int_{[x,y]} |\nabla f(\gamma_\ell(u))| |\gamma_\ell'(u)| \, du \, d\ell, \quad (5)$$

where $d\ell$ denotes the restriction of H^{n-1} to $D(x,y)$. The change of variable $(\ell, u) \to \gamma_\ell(u)$ yields

$$|f(x) - f(y)| \leq \int_{C(x,y)} |\nabla f(z)|\theta(z)\,dz, \tag{6}$$

where dz denotes the Lebesgue measure and $\theta(z)$ comes from the Jacobian determinant. A simple computation shows that

$$\theta(z) \leq C(n,\alpha)\, w_{x,y}(z), \tag{7}$$

with

$$w_{x,y}(z) = \mathbf{1}_{C(x,y)}(z)\{|z - x|^{1-n} + |z - y|^{1-n}\}. \tag{8}$$

The Sobolev estimates only hold when $p > n$, which is slightly unfortunate for us here because we shall mostly operate with $p = 2$ and $n \geq 2$. When $p > n$ we can use Hölder's inequality to estimate the integral in (6), because if $q = \frac{p}{p-1}$ denotes the conjugate exponent, then $q < \frac{n}{n-1}$, $|z|^{q(1-n)}$ is locally integrable, and so

$$\|w_{x,y}\|_q \leq 2\left\{\int_{|z-x|\leq 2|x-y|} |z - x|^{q(1-n)}\right\}^{1/q} \leq C|x - y|^{(1-n)+\frac{n}{q}} = C|x - y|^{\frac{p-n}{p}}. \tag{9}$$

Thus (6) implies that

$$|f(x) - f(y)| \leq C\|\nabla f\|_{L^p(C(x,y))}\|w_{x,y}\|_q \leq C|x-y|^{\frac{p-n}{p}}\left\{\int_{C(x,y)} |\nabla f|^p\right\}^{1/p}. \tag{10}$$

Thus for $p > n$ functions in $W^{1,p}_{\text{loc}}$ are (locally) Hölder-continuous. This fails for $p \leq n$, and for instance functions in $W^{1,2}_{\text{loc}}(\mathbb{R}^2)$ are not necessarily continuous. Take for example $f(x) = \text{Log Log}\frac{1}{|x|}$ near 0 in $\mathbb{R}^2$.

When $p \leq n$ we can still deduce from (6) that functions of $W^{1,p}_{\text{loc}}(\Omega)$ lie in $L^r_{\text{loc}}(\Omega)$ for $r < \frac{np}{n-p}$. This uses the fact that $w \in L^q$ for $q < \frac{n}{n-1}$, that convolution maps $L^p \times L^q$ into L^r when $\frac{1}{p} + \frac{1}{q} = \frac{1}{r} + 1$, and also a small limiting argument to get rid of our assumption that f is C^1. For the purposes of most of this book, we shall be happy to settle for $r = 1$ and Poincaré estimates, as follows.

Proposition 11. *If $\Omega \subset \mathbb{R}^n$ is open, $f \in W^{1,1}_{\text{loc}}(\Omega)$, and $B \subset \Omega$ is a ball, then*

$$\int_B |f(x) - m_B f|\,dx \leq C_n |B|^{1/n} \int_B |\nabla f|, \tag{12}$$

where $m_B f = \frac{1}{|B|}\int_B f(x)\,dx$ denotes the mean value of f on B.

Our statement does not exclude the case when (B is not relatively compact in Ω and) the right-hand side of (12) is infinite, but of course this case is trivial.

Let $B_1 = \frac{1}{10}B$ denote the ball with the same center and a radius 10 times smaller. We want to prove that

$$\int_B |f(x) - m_{B_1}f|\,dx \le C_n |B|^{1/n} \int_B |\nabla f|. \tag{13}$$

This will imply (12) at once, because $|m_B f - m_{B_1}f| \le \frac{1}{|B|}\int_B |f(x) - m_{B_1}f|$ by the triangle inequality, and it will be slightly more convenient to prove.

We start with the easier case when f is C^1. Notice that when $x \in B$ and $y \in B_1$ the set $C(x,y)$ is contained in B and (6) holds. Thus

$$\int_B |f(x) - m_{B_1}f|\,dx \le \frac{1}{|B_1|}\int_B \int_{B_1} |f(x) - f(y)|\,dy\,dx \tag{14}$$

$$\le \frac{C}{|B_1|}\int_B \int_{B_1} \int_{C(x,y)} \{|z-x|^{1-n} + |z-y|^{1-n}\}|\nabla f(z)|\,dz\,dy\,dx$$

$$\le \frac{C}{|B_1|}\int_B |\nabla f(z)|\left[\int_B \int_{B_1} \{|z-x|^{1-n} + |z-y|^{1-n}\}\,dy\,dx\right]dz$$

$$\le C|B|^{1/n}\int_B |\nabla f(z)|\,dz,$$

as needed. In the general case when $f \in W^{1,1}_{\text{loc}}(\Omega)$ only, note that it is enough to prove that

$$\int_{B'} |f(x) - m_{B_1}f|\,dx \le C_n |B'|^{1/n}\int_B |\nabla f| \tag{15}$$

for every ball $B' \subset\subset B$. The point of using this variant is that, since B' is relatively compact in Ω, Lemma 11.1 says that we can find a sequence of smooth functions $f_k \in C^1_c(\Omega)$ such that $\lim_{k \to +\infty} f_k = f$ and $\lim_{k \to +\infty} \nabla f_k = \nabla f$ in $L^1(B')$. Since f_k is smooth,

$$\int_{B'} |f_k(x) - m_{B_1'}f_k|\,dx \le C_n |B'|^{1/n}\int_{B'} |\nabla f_k|, \tag{16}$$

where we set $B_1' = \frac{1}{10}B'$. We can even replace $m_{B_1'}f_k$ with $m_{B_1}f_k$ in (16), either by modifying slightly the proof of (14) above, or by the same argument that allowed us to deduce (12) from (13). Now (15) follows from (16) by taking limits in k (with B' fixed), and Proposition 11 follows from (15). $\qquad\square$

Corollary 17. *Let $\Omega \subset \mathbb{R}^n$ be a (connected) domain, and let $\{f_k\}_{k \ge 0}$ be a sequence in $W^{1,p}(\Omega)$ for some $p \ge 1$. Assume that*

$$\{\nabla f_k\} \text{ converges in } L^p(\Omega) \text{ to some limit } h \in L^p(\Omega), \tag{18}$$

and that we can find a ball $B \subset \Omega$ such that

$$\lim_{k \to +\infty} m_B f_k \text{ exists.} \tag{19}$$

Then $\{f_k\}$ converges in $L^1_{loc}(\Omega)$ to some limit f, $f \in W^{1,p}(\Omega)$, and $\nabla f = h$.

First observe that if $f \in W^{1,1}_{loc}(\Omega)$ and B_1, B are balls such that $B_1 \subset B \subset \Omega$, then

$$|m_{B_1} f - m_B f| \leq \frac{1}{|B_1|} \int_{B_1} |f(x) - m_B f| \, dx \leq C_n |B_1|^{-1} |B|^{1/n} \int_B |\nabla f|, \tag{20}$$

by (12). Since

$$\lim_{k, \ell \to +\infty} \int_B |\nabla f_k - \nabla f_\ell| = 0 \tag{21}$$

by (18) and Hölder, (20) implies that

$$\lim_{k, \ell \to +\infty} |m_{B_1}(f_k - f_\ell) - m_B(f_k - f_\ell)| = 0 \tag{22}$$

when $B_1 \subset B \subset \Omega$. In particular, (19) holds for B if and only if it holds for B_1 (because this happens if and only if $m_B(f_k - f_\ell)$ tends to 0).

Set $V = \{x \in \Omega; (19) \text{ holds for some open ball } B \subset \Omega \text{ that contains } x\}$. Obviously V is open. Also, if $x \in \Omega \setminus V$ and $B \subset \Omega$ is any open ball that contains x, then (19) fails for B, hence also for all the open balls $B' \subset \Omega$ that meet B because we can apply the observation above to a ball $B_1 \subset B \cap B'$ to find that (19) fails for B_1, then for B'. This proves that B does not meet V; hence V is also closed in Ω. Since Ω is connected and V is not empty by assumption, $V = \Omega$. Moreover, for every ball $B \subset \Omega$, the center of B lies in V, so we can find a ball B' that contains it and satisfies (19), hence (19) also holds for some ball $B_1 \subset B \cap B'$, and finally B satisfies (19).

From (12) and (21) we deduce that the sequence $\{f_k(\cdot) - m_B f_k\}$ converges in $L^1(B)$ for every open ball $B \subset \Omega$. Since B satisfies (19), $\{f_k\}$ also converges in $L^1(B)$. By a trivial covering argument, $\{f_k\}$ converges in $L^1(K)$ for every compact set $K \subset \Omega$, i.e., $\{f_k\}$ converges in $L^1_{loc}(\Omega)$.

Call f the limit of $\{f_k\}$. Lemma 11.2 says that $f \in W^{1,p}(B)$ for every open ball $B \subset \Omega$, and that $\nabla f = h$ on B. Corollary 17 follows. $\quad\square$

The following slight improvement of Proposition 11 will be used in Section 38.

Proposition 23. For each $r \in [2, \frac{2n}{n-2})$, we can find a constant $C(n, r)$ such that if $f \in W^{1,2}(B)$ for some ball $B \subset \mathbb{R}^n$, then $f \in L^r(B)$ and

$$\left\{ \frac{1}{|B|} \int_B |f(x) - m_B f|^r \right\}^{1/r} \leq C(n, r) |B|^{1/n} \left\{ \frac{1}{|B|} \int_B |\nabla f(x)|^2 \right\}^{1/2}. \tag{24}$$

By homogeneity, it is enough to prove this when $B = B(0,1)$. Let us proceed as in Proposition 11 and start with the case when f is C^1. Set $B_1 = \frac{1}{10}B$, fix $x \in B$, and integrate (6) with respect to $y \in B_1$. We get that

$$|f(x) - m_{B_1}f| \le C \int_{y \in B_1} \int_{z \in C(x,y)} |\nabla f(z)|\, \omega_{x,y}(z)\, dy dz, \qquad (25)$$

by (7). Note that $C(x,y) \subset B$ when $x \in B$ and $y \in B_1$. When we integrate $\omega_{x,y}(z)$ with respect to y, we get less than $C(1 + |z - x|^{1-n})$. Thus

$$|f(x) - m_{B_1}f| \le C\left[(\mathbf{1}_B |\nabla f|) * h\right](x), \qquad (26)$$

where $h(u) = (1 + |u|^{1-n})\mathbf{1}_{2B}(u)$. Set $q = \frac{2r}{2+r}$. Note that q is an increasing function of r, so $q < 2\frac{2n}{n-2}(2 + \frac{2n}{n-2})^{-1} = \frac{4n}{n-2}\frac{n-2}{4n-4} = \frac{n}{n-1}$. Hence $h \in L^q(\mathbb{R}^n)$. Also, $\frac{1}{2} + \frac{1}{q} = \frac{1}{r} + 1$, so the convolution maps $L^2 \times L^q$ to L^r, and

$$\left\{\int_B |f(x) - m_{B_1}f|^r\right\}^{1/r} \le C\left\{\int_B |\nabla f|^2\right\}^{1/2}. \qquad (27)$$

So far we assumed that f is C^1. Suppose now that this is not the case, and let $B' = B(0, r')$ be given, with $r' \in (0,1)$ close to 1. We can use Lemma 11.1 to find a sequence $\{f_k\}$ in $C_c^1(B)$ that tends to f in $L^1(B')$ and for which $\{\nabla f_k\}$ converges to ∇f in $L^2(B')$. Notice that

$$\left\{\int_{B'} |f_k(x) - m_{B_1'}f_k|^r\right\}^{1/r} \le C\left\{\int_{B'} |\nabla f_k|^2\right\}^{1/2}, \qquad (28)$$

with $B_1' = \frac{1}{10}B'$, by our proof of (27). When we let k tend to $+\infty$, we get the analogue of (27) for B', by Fatou. [Recall that we may always extract a subsequence of $\{f_k\}$ that converges almost-everywhere.] Now (27) for B itself follows by letting r' tend to 1.

So (27) holds in all cases; (24) follows, because we can easily control $|m_B f - m_{B_1}f|$ in terms of $\{\int_B |\nabla f_k|^2\}^{1/2}$, by (27). This proves Proposition 23. $\square$

Exercise 29. Show that if $f \in W^{1,p}(B)$ and $r < \frac{np}{n-p}$, then

$$\left\{\frac{1}{|B|}\int_B |f(x) - m_B f|^r\right\}^{1/r} \le C(n,p,r)\,|B|^{1/n}\left\{\frac{1}{|B|}\int_B |\nabla f(x)|^p\right\}^{1/p}. \qquad (30)$$

Also check the equivalent estimate

$$\left\{\frac{1}{|B|^2}\int_B \int_B |f(x) - f(y)|^r\right\}^{1/r} \le C(n,p,r)\,|B|^{1/n}\left\{\frac{1}{|B|}\int_B |\nabla f(x)|^p\right\}^{1/p}. \qquad (31)$$

13   Boundary values and restrictions to hypersurfaces

Let $f \in W^{1,p}(\Omega)$ for some $p \geq 1$. We want to study the existence of boundary values of f on a reasonably smooth piece of $\partial\Omega$ (say, Lipschitz or C^1). In the following, our attitude will be to take a naive approach and get some reasonable information. See [Zi] for a more general and systematic point of view.

When $p > n$, it is possible to use the proof of the Sobolev inequalities in Section 12 to see that f is continuous up to the boundary. In general, though, we cannot expect as much, because f may have (mild) singularities everywhere. See Exercise 36. So we shall have to settle for limits almost-everywhere in transverse directions. The same thing happens when we want to define the restriction of f to a hypersurface inside Ω. Note that in both cases the problem is local; thus we can easily reduce to the following setting.

Let $D \subset \mathbb{R}^{n-1}$ be an open hyperdisk, and let $A : D \to \mathbb{R}$ be a Lipschitz function. We assume that

$$|A(x) - A(y)| \leq M |x - y| \text{ for } x, y \in D \tag{1}$$

and

$$0 < A(x) < T_0 \text{ for } x \in D, \tag{2}$$

where M and T_0 are positive constants. (This is mostly for definiteness.) We consider the cylindrical domain

$$V = \big\{(x,t) \in D \times (0,T)\,; A(x) < t < T\big\} \subset \mathbb{R}^n, \tag{3}$$

where $T > T_0$ is another constant (see Figure 1).

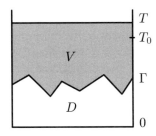
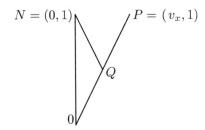

Figure 1.   Figure 2. The set Z

For each function $f \in W^{1,1}(V)$ we can define a function f^* on the graph

$$\Gamma = \{(x, A(x)); x \in D\} \tag{4}$$

in the following simple way. Set $L(x) = \{(x,t); A(x) < t < T\}$ for $x \in D$. By Proposition 9.5, the restriction of f to $L(x)$ is (locally) absolutely continuous for almost-every $x \in D$, and its derivative is the restriction of the (distributional) derivative $\frac{\partial f}{\partial t}$ to $L(x)$. By Fubini, $\int_{L(x)} \left| \frac{\partial f}{\partial t} \right| dH^1 < +\infty$ for almost-every x, which means that the restriction of f to $L(x)$ has finite total variation. In particular we can define

$$f^*(x, A(x)) = \lim_{t \to A(x)^+} f(x, t) \tag{5}$$

for almost-every $x \in D$. Notice that for almost-every $x \in D$ and all $t \in (A(x), T)$,

$$f(x, t) - f^*(x, A(x)) = \int_{A(x)}^{t} \frac{\partial f}{\partial t} (x, u) \, du, \tag{6}$$

by absolute continuity. (See Proposition 9.1.)

At this point we have a definition of the boundary value f^* of f on Γ, but it is clearly not satisfactory without further checking: we do not know yet that our definition is reasonably invariant under smooth changes of variables and, even worse, that f^* would not be completely different if we had chosen slightly different sets of coordinates in $\mathbb{R}^n$ (i.e., if we had taken limits in (5) in slightly different directions). Let us at least address this last issue carefully; we shall say a few words about smooth changes of variables later.

Let V and $f \in W^{1,1}(V)$ be as before, and let $v = (v_x, 1)$ be a vector in $\mathbb{R}^n$, thus pointing upward, and such that $|v_x| < \frac{1}{M}$. By the same argument as above,

$$f^v(z) = \lim_{\varepsilon \to 0^+} f(z + \varepsilon v) \tag{7}$$

exists for H^{n-1}-almost all $z \in \Gamma$, and

$$f(z + \varepsilon v) - f^v(z) = \int_0^\varepsilon \frac{\partial f}{\partial v} (z + sv) \, ds \tag{8}$$

for H^{n-1}-almost all $z \in \Gamma$ and all $\varepsilon > 0$ such that $[z, z + \varepsilon v] \subset V$, and where we set $\frac{\partial f}{\partial v} = \nabla f \cdot v$.

Lemma 9. *For all choices of V, f, v as above, $f^v(z) = f^*(z)$ for H^{n-1}-almost every $z \in \Gamma$.*

First let Z be as in Figure 2; that is, call 0 the origin, set $N = (0, 1)$, $P = (v_x, 1) = v$, $Q = \left(\frac{v_x}{2}, \frac{1}{2} \right)$, and then $Z = (0, N] \cup (0, P] \cup [N, Q] \subset \mathbb{R}^n$. Observe that the three segments that compose Z have slopes larger than M, because $|v_x| < \frac{1}{M}$.

Let $\varepsilon > 0$ be given (small). Set

$$\Gamma_\varepsilon = \{(x, A(x)) \in \Gamma; x + \varepsilon v_x \in D\} \tag{10}$$

and $Z_\varepsilon(z) = z + \varepsilon Z$ for $z \in \Gamma_\varepsilon$. Because of (1) in particular, $Z_\varepsilon(z) \subset V$ for all $z \in \Gamma_\varepsilon$, at least if we take $\varepsilon < T - T_0$. We claim that

$$\int_{\Gamma_\varepsilon} \left\{ \int_{Z_\varepsilon(z)} |\nabla f| dH^1 \right\} dH^{n-1} \le C \int_{V_\varepsilon} |\nabla f|, \tag{11}$$

where $V_\varepsilon = \bigcup_{z \in \Gamma_\varepsilon} Z_\varepsilon(z) \subset V$ looks like a thin stripe of V above Γ.

To prove this, we can cut the integral on the left-hand side of (11) into three parts, each one coming from one of the three segments that compose Z. Let L be one of these segments. The application from $\Gamma_\varepsilon \times (\varepsilon L)$ into V_ε which maps (z, ℓ) to $z + \ell$ is bilipschitz (onto its image), because the slope of L is larger than M (and by (1)). Our claim (11) follows from this.

Let L be, as before, one of the three line segments that compose Z. Proposition 9.1 says that f is absolutely continuous on $z + \varepsilon L$ for almost-every $z \in \Gamma_\varepsilon$. (This uses the fact that the slope of L is larger than M, as before.) Thus f is absolutely continuous on $Z_\varepsilon(z)$ for almost-every $z \in \Gamma_\varepsilon$; as usual Proposition 9.1 also says that the derivative of f along $Z_\varepsilon(z)$ can be computed from the restriction of ∇f to $Z_\varepsilon(z)$. Hence, if

$$\Delta_\varepsilon(z) = \sup\left\{ f(w); w \in Z_\varepsilon(z) \right\} - \inf\left\{ f(w); w \in Z_\varepsilon(z) \right\} \qquad (12)$$

denotes the oscillation of f on $Z_\varepsilon(z)$, then

$$\Delta_\varepsilon(z) \leq \int_{Z_\varepsilon(z)} |\nabla f| \, dH^1 \qquad (13)$$

for almost all $z \in \Gamma_\varepsilon$. (Note that $Z_\varepsilon(z)$ is connected, because of the third segment $[NQ]$ in Z.) Now

$$\int_{\Gamma_\varepsilon} \Delta_\varepsilon(z) \, dH^{n-1}(z) \leq C \int_{V_\varepsilon} |\nabla f|, \qquad (14)$$

by (11). Since $|f^v(z) - f^*(z)| \leq \Delta_\varepsilon(z)$ almost-everywhere on Γ_ε (by (5), (7), and our definition of Z), we get that

$$\int_{\Gamma_\varepsilon} |f^v(z) - f^*(z)| \, dH^{n-1}(z) \leq C \int_{V_\varepsilon} |\nabla f|. \qquad (15)$$

The right-hand side of (15) tends to 0 with ε, because V_ε tends to the empty set and ∇f is integrable on V, and the left-hand side tends to $\int_\Gamma |f^v(z) - f^*(z)|$. Lemma 9 follows. $\qquad \square$

Remark 16. Our definition of f^* using vertical rays (or rays with enough slope, as authorized by Lemma 9) is far from perfect. One obvious defect it has is that it is not obviously invariant under smooth changes of variables. Indeed, images of parallel lines under such a smooth mapping Φ are no longer parallel lines, but they merely define a smooth vector field on a neighborhood of $\Phi(\Gamma)$. We can still use such curves (or vector fields) to define an analogue of $(f \circ \Phi^{-1})^*$ on $\Phi(\Gamma)$, and this gives a trace on $\Phi(\Gamma)$ that coincides almost-everywhere with the function that we would get if we used transverse lines (as in (5) or (7)) instead. The proof is rather easy; it is a minor variation of the argument for Lemma 9, so we shall kindly leave it to the reader.

We shall implicitly use a variant of this. We shall often work in the plane, with functions $f \in W^{1,2}(B(0, R) \setminus K)$ (where K is a closed set), and we shall be interested in the trace of f on $\partial B(0, r) \setminus K$ (for any given $r < R$). In such cases, we shall implicitly use polar coordinates, and define the traces as the radial limits of f along almost every ray (instead of using small domains where $\partial B(0, r)$ is Lipschitz, and taking limits along parallel lines).

Another reasonable definition of boundary value would be to take

$$f^{\sharp}(z) = \lim_{\varepsilon \to 0^+} \frac{1}{|B_\varepsilon(z)|} \int_{B_\varepsilon(z)} f, \tag{17}$$

where for instance $B_\varepsilon(z)$ denotes the ball centered at $z + (0, \varepsilon)$ and with radius $(10M + 10)^{-1}\varepsilon$, say. Because of Proposition 12.11, it is fairly easy to check that the limit in (17) exists for almost-every $z \in \Gamma$, and that $f^{\sharp}(z) = f^*(z)$ almost-everywhere. The precise choice of $B_\varepsilon(z)$ does not matter much either.

There is a more satisfactory approach to our problem of boundary values of functions in $W^{1,p}(\Omega)$, involving polar sets, capacities, etc. We refer the interested reader to [Zi].

Remark 18. Functions in $W^{1,p}(\Omega)$ (with $p \le n$) do not always have nontangential limits almost-everywhere, even if Ω is very regular. In fact, such functions do not even need to be bounded in any nontangential access region, because they may have little singularities all over the place. See Exercise 36.

However, this problem may disappear if we know more about f. For instance, if f is harmonic inside Ω, all the arguments above become substantially easier, and f has nontangential limits almost-everywhere on Γ (when Γ is a Lipschitz graph, as above).

Remark 19. Let $\Omega \subset \mathbb{R}^n$ be open, and let $\Gamma \subset \Omega$ be a piece of a Lipschitz graph. If $f \in W^{1,1}(\Omega \setminus \Gamma)$, then by the construction above f has "transverse" limits almost-everywhere on Γ, from both sides of Γ. If we know in addition that $f \in W^{1,1}(\Omega)$, then these transverse limits coincide almost-everywhere on Γ, because f is absolutely continuous on (almost-all) lines. Thus, for every reasonable set Γ of codimension 1, we are able to define f almost-everywhere on Γ. This is of course somewhat better than what we could do with any $f \in L^p$. Note that it is important in this discussion that Γ is *any* reasonable set of codimension 1; if we were willing to settle for almost-any such set, then we could use Proposition 9.5 and even get that the restriction of f to Γ lies in $W^{1,1}$.

The following lemma will be needed later.

Lemma 20. *Let the domain $V \subset \mathbb{R}^n$ be as in the beginning of this section, and let $B \subset V$ be a ball. Then there is a constant $C(V, B)$ such that if $f \in W^{1,1}(V)$ and f^* denotes the boundary extension of f on Γ given by (5),*

$$\int_{u \in B} \int_{z \in \Gamma} |f(u) - f^*(z)| \, dH^{n-1}(z) \, du \le C(V, B) \int_V |\nabla f|. \tag{21}$$

Note that from (21) we can easily deduce that

$$\frac{1}{H^{n-1}(\Gamma)}\int_{\Gamma}|f^*(z) - m_B f|dH^{n-1}(z) \leq C'(V,B)\int_V |\nabla f|. \qquad (22)$$

To prove the lemma, consider the rectangular domain $R = D \times (T_0, T) \subset V$ on the top of V, where T_0 is as in (2) an upper bound for the values of A. We claim that

$$\int_R |f(u) - m_R f|\, du \leq C(R)\int_R |\nabla f|. \qquad (23)$$

This can be proved like Proposition 12.11, or deduced from it. The constant $C(R)$ may be much larger than $\mathrm{diam}(R)$ if $T - T_0 << \mathrm{diam}\, D$, but this will not bother us. (In the proof, this shows up because we may have to take the parameter α in the definition (12.2) of $C(x,y)$ very small if we want to extend the proof without modifications, or because we may have to compose with an affine mapping $(x,t) \rightarrow (x, at)$ with a large to give R a normalized shape.)

Call π the orthogonal projection onto $\mathbb{R}^{n-1}$ (the hyperplane that contains D), and set $R_1 = R \cap \pi^{-1}(\pi(B))$. Then

$$|m_{R_1} f - m_R f| \leq C \int_R |\nabla f|, \qquad (24)$$

with a constant C that depends on B and R (but we shall no longer bother to say that), by (23). Also

$$|m_{R_1} f - m_B f| \leq C \int_V |\nabla f|. \qquad (25)$$

This can be proved like (23), but let us skip the details. Since Proposition 12.11 says that

$$\int_B |f - m_B f| \leq C \int_B |\nabla f|, \qquad (26)$$

we get that

$$\int_B |f(u) - m_R f|\, du \leq C \int_V |\nabla f|. \qquad (27)$$

Now we want to control $f^*(z)$. For $z \in \Gamma$, set $\ell(z) = V \cap \pi^{-1}(z)$. We know from (6) that for H^{n-1}-almost every $z \in \Gamma$,

$$|f^*(z) - f(u)| \leq \int_{\ell(z)} |\nabla f|dH^1 \quad \text{for } u \in \ell(z). \qquad (28)$$

When we average this over $\ell_1(z) = \ell(z) \cap R$ we get that

$$|f^*(z) - m_R f| \leq \frac{1}{H^1(\ell_1(z))}\int_{\ell_1(z)}|f(u) - m_R f|dH^1 + \int_{\ell(z)}|\nabla f|dH^1 \qquad (29)$$

almost-everywhere on Γ. Then we integrate (29) on Γ and get that

$$\int_{\Gamma} |f^*(z) - m_R f| \, dH^{n-1} \leq \frac{C}{T - T_0} \int_R |f(u) - m_R f| + C \int_V |\nabla f|, \qquad (30)$$

where the constant C comes from the fact that $dH^{n-1}(z)$ on Γ may be a little larger than $d\pi(z)$ on D.

Now (21), and hence Lemma 20, follow from (27), (30), and (23).   □

Corollary 31. *Let V be as in the beginning of this section, and let g be a measurable function on Γ. Set*

$$P_g = \left\{ f \in W^{1,1}(V) \, ; \, f^*(z) = g(z) \quad \text{for } H^{n-1}\text{-almost every } z \in \Gamma \right\}, \qquad (32)$$

where f^* still denotes the limit in (5). Then P_g is a (possibly empty) closed affine subspace of $W^{1,1}(V)$.

All we have to do is prove that P_g is closed. Let $\{f_n\}$ be a sequence in P_g, and assume that $\{\nabla f_n\}$ converges to some limit h in $L^1(V)$. Choose any ball $B \subset\subset V$. Since $f_m^* = f_n^*$, (22) says that $m_B(f_m - f_n)$ tends to 0 when m and n tend to $+\infty$. By Corollary 12.17, $\{f_n\}$ converges to some limit f in $L^1_{\text{loc}}(V)$, and $\nabla f = h$. Also,

$$\int_{\Gamma} |f^* - g| \, dH^{n-1} = \int_{\Gamma} |f^* - f_n^*| \, dH^{n-1}$$
$$\leq H^{n-1}(\Gamma) |m_B(f - f_n)| + C \int_V |\nabla(f - f_n)| \qquad (33)$$

by (22). The right-hand side tends to 0 by assumption; hence $f^*(z) = g(z)$ almost-everywhere on Γ, and $f \in P_g$. This proves the corollary.   □

Exercise 34. Let $f \in W^{1,1}(B(0,1))$ be given. Check that for almost-every $\xi \in \partial B(0,1)$, the restriction of f to the ray $\{r\xi \, ; \, 0 < r < 1\}$ is absolutely continuous, and $\lim_{r \to 1^-} f(r\xi) = f^*(\xi)$, where f^* is the trace on $\partial B(0,1)$ defined by (5) (in suitable local coordinates, with compatibility coming from Lemma 9).

Exercise 35. Prove that the limit in (17) exists and is equal to $f^*(z)$ for almost-every $z \in \Gamma$. You may also check that a different (but reasonable) choice of balls $B_\varepsilon(z)$ gives a function $\widetilde{f}^\sharp(z)$ that coincides with $f^\sharp$ almost-everywhere on Γ.

Exercise 36. [With thanks to A. Ancona for pointing out how easy it is.]

1. Check that $\text{Log}(\text{Log}(|x|^{-1})) \in W^{1,p}(B(0,1/2))$ in $\mathbb{R}^n$ for $p \leq n$.

2. Fix $p \leq n$. Show that for each ball B in $\mathbb{R}^n$ and each $\varepsilon > 0$ we can find a smooth function φ with compact support in B such that $\|\nabla \varphi\|_p \leq \varepsilon$ and $\|f\|_\infty \geq \varepsilon^{-1}$.

3. Find a (smooth) function f on $(0,1)^n$ such that $f \in W^{1,p}((0,1)^n)$ but that does not have nontangential limits at any point of $(0,1)^{n-1}$. The easiest is to make sure that f is not even bounded on any nontangential access sector $\{z \in (0,1)^n \, ; \, |z - x| \leq M|\pi(z) - \pi(x)|\}$, with $x \in (0,1)^{n-1}$ and $M > 0$, and where π denotes the orthogonal projection onto $\mathbb{R}^{n-1}$.

14   A simple welding lemma

In our constructions of competitors for the Mumford-Shah functional, we shall often define a function u differently on two pieces of a domain $\Omega \setminus K$, and we shall need to know that $u \in W^{1,2}_{\mathrm{loc}}(\Omega \setminus K)$. We give in this section a welding lemma that will be used for this purpose. Since our problem is local, we take the same sort of setting as in the last section.

Let $D \subset \mathbb{R}^{n-1}$ be a hyperdisk and $A : D \to \mathbb{R}$ a Lipschitz function. Set $\Gamma = \{(x, A(x)) \,;\, x \in D\} \subset \mathbb{R}^n$, and choose $T_1, T_2 \in \mathbb{R}$ such that

$$T_1 < \inf \{A(x) \,;\, x \in D\} \leq \sup\{A(x) \,;\, x \in D\} < T_2. \tag{1}$$

Consider the domains $V = D \times (T_1, T_2)$,

$$V_1 = \{(x,t) \in V \,;\, t < A(x)\}, \tag{2}$$

and

$$V_2 = \{(x,t) \in V \,;\, t > A(x)\}. \tag{3}$$

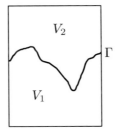

Figure 1

Lemma 4.   *Let V, V_1, V_2 be as above, and let $f \in L^1_{\mathrm{loc}}(V_1 \cup V_2)$ be given. Suppose that*

$$f_{|V_i} \in W^{1,1}(V_i) \ \text{ for } i = 1, 2, \tag{5}$$

and that

$$f^*(z) = \lim_{\varepsilon \to 0} f(z + \varepsilon e) \ \text{ exists for } H^{n-1}\text{-almost every } z \in \Gamma, \tag{6}$$

where $e = (0, \dots, 0, 1)$ denotes the vertical unit vector. Then $f \in W^{1,1}(V)$ and $\nabla f(x) = \nabla f_{|V_i}(x)$ almost-everywhere on each V_i.

Remark 7. We know from the previous section that the limits

$$f^*_{\pm}(z) = \lim_{\varepsilon \to 0^{\pm}} f(z + \varepsilon e)$$

exist almost-everywhere on Γ. Thus our new assumption in (6) is only that $f_+^*(z) = f_-^*(z)$ almost-everywhere. Of course in many applications f will just be continuous across Γ and (6) will be trivial.

Remark 8. If we also assume that $f_{|V_i} \in W^{1,p}(V_i)$ for some $p > 1$, then we get that $f \in W^{1,p}(V)$. This is obvious, since the lemma tells us how to compute $\nabla f(z)$.

To prove the lemma we want to check first that $f \in L^1(V)$. This will be useful, because so far we only know that $f \in L^1_{\mathrm{loc}}(V \setminus \Gamma)$, and this is not enough to say that f defines a distribution on V. To simplify some notations, define f also on Γ, by taking $f(z) = f^*(z)$ when $z \in \Gamma$ and the limit in (6) exists, and $f(z) = 0$ otherwise. Also define h on V by taking $h = \frac{\partial}{\partial t}(f_{|V_i})$ (the vertical derivative) on V_i, and $h = 0$ on Γ (say). Thus $h \in L^1(V)$ and

$$\int_{V_i} f \frac{\partial \varphi}{\partial t} = -\int_{V_i} h\varphi \text{ for } \varphi \in C_c^\infty(V_i). \tag{9}$$

From (13.6) and its analogue for V_2 we get that for almost-all $x \in D$,

$$f(x,t) = f^*(x, A(x)) + \int_{A(x)}^t h(x,u)\,du \text{ for } T_1 < t < T_2. \tag{10}$$

In particular, the restriction of f to almost-every vertical line is absolutely continuous. If we integrate (10) with respect to $x \in D$ and $t \in (T_1, T_2)$, we get that

$$\int_V |f(x,t)| \leq C \int_D |f^*(x, A(x))| + C \int_{V_1 \cup V_2} |\nabla f|, \tag{11}$$

where C depends on V, but we don't care. The right-hand side of (11) is finite, for instance by Lemma 13.20 or (13.22) (with any ball $B \subset\subset V_1$, say). Hence $f \in L^1(V)$, and f defines a distribution on V. All we have to do now is check that the (distributional) partial derivatives of f are given by the gradient of the restrictions $f_{|V_i}$, that is, that there is no weird part that shows up on Γ.

We start with the last derivative $\frac{\partial f}{\partial t}$. We want to show that $\frac{\partial f}{\partial t} = h$, where h was introduced a little before (9). This means that

$$\int_V f \frac{\partial \varphi}{\partial t} + \int_V h\varphi = 0 \text{ for all } \varphi \in C_c^\infty(V) \tag{12}$$

(and not only $\varphi \in C_c^\infty(V_1 \cup V_2)$). So let $\varphi \in C_c^\infty(V)$ be given. For each small $\varepsilon > 0$, choose a cut-off function $\chi_0 \in C^\infty(V)$ such that $0 \leq \chi_0 \leq 1$ on V,

$$\chi_0(z) = 1 \text{ on } \Gamma_\varepsilon = \{z \in V \,;\, \mathrm{dist}(z, \Gamma) \leq \varepsilon\}, \tag{13}$$

$$\chi_0(z) = 0 \text{ on } V \setminus \Gamma_{2\varepsilon}, \tag{14}$$

and

$$|\nabla \chi_0(z)| \leq C\varepsilon^{-1} \text{ everywhere on } V. \tag{15}$$

Such a function is easy to construct, for instance with partitions of identity and Whitney cubes. Also set $\chi_i = (1 - \chi_0)\mathbf{1}_{V_i}$ on V for $i = 1, 2$; note that χ_i is still C^∞ on V because of (13). Clearly $\chi_0 + \chi_1 + \chi_2 = 1$ on V.

Write our test-function φ as $\varphi_0 + \varphi_1 + \varphi_2$, with $\varphi_i = \chi_i \varphi$. Call A the left-hand side of (12); then $A = A_0 + A_1 + A_2$, with

$$A_i = \int_V f \frac{\partial \varphi_i}{\partial t} + \int_V h\varphi_i. \tag{16}$$

Note that for $i = 1, 2$, $\varphi_i = \chi_i \varphi \in C_c^\infty(V_i)$, because of (13). Hence $A_1 = A_2 = 0$, by (9). We still need to show that $A_0 = 0$. Write

$$A_0 = \int_V f\varphi \frac{\partial \chi_0}{\partial t} + \int_V f\chi_0 \frac{\partial \varphi}{\partial t} + \int_V h\chi_0 \varphi =: B_1(\varepsilon) + B_2(\varepsilon) + B_3(\varepsilon). \tag{17}$$

Notice that $\left| f\frac{\partial \varphi}{\partial t}\right| + |h\varphi| \in L^1(V)$, $|\chi_0| \le 1$ everywhere, and $\chi_0 \left(f\frac{\partial \varphi}{\partial t} + h\varphi\right)$ tends to 0 everywhere on $V \setminus \Gamma$ when ε tends to 0 (by (14)). Hence

$$\lim_{\varepsilon \to 0} \{B_2(\varepsilon) + B_3(\varepsilon)\} = 0, \tag{18}$$

for instance by the Lebesgue dominated convergence theorem. We are left with $B_1(\varepsilon)$. Set $a = f\varphi \frac{\partial \chi_0}{\partial t}$. By Fubini,

$$B_1(\varepsilon) = \int_{x \in D} \int_{T_1}^{T_2} a(x, t)\, dt\, dx. \tag{19}$$

For each $x \in D$, $\frac{\partial \chi_0}{\partial t}(x, t)$ vanishes outside an interval $I(x)$ with length $\le C\varepsilon$ that contains $A(x)$. This comes from (14) and the fact that A is Lipschitz. For almost-every $x \in D$, we can apply (10) and get that

$$|f(x, t) - f^*(x, A(x))| \le \int_{I(x)} |h(x, u)|\, du \tag{20}$$

for $t \in I(x)$. Then, if we set

$$\delta(x, t) = f(x, t)\varphi(x, t) - f^*(x, A(x))\,\varphi(x, A(x)), \tag{21}$$

$$|\delta(x, t)| \le |f(x, t)|\,|\varphi(x, t) - \varphi(x, A(x))| + |f(x, t) - f^*(x, A(x))|\,|\varphi(x, A(x))|$$

$$\le C\varepsilon|f(x, t)| + C\int_{I(x)} |h(x, u)|\, du \tag{22}$$

for $t \in I(x)$, because $\frac{\partial \varphi}{\partial t}$ is bounded and by (20). Let $t_0 \in (T_1, T_2)$ be given (to be chosen later). Then

$$|f(x, t)| \le |f(x, t_0)| + \int_{T_1}^{T_2} |h(x, u)|\, du, \tag{23}$$

by (10) again. Thus

$$|\delta(x,t)| \leq C\varepsilon|f(x,t_0)| + C\varepsilon \int_{T_1}^{T_2} |h(x,u)|\,du + C\int_{I(x)} |h(x,u)|\,du. \qquad (24)$$

Let us return to the inside integral in (19). Observe that if ε is small enough, $I(x) \subset\subset (T_1,T_2)$ and $\int_{I(x)} \frac{\partial\chi_0}{\partial t}(x,u)\,du = 0$. Thus we can remove a constant from the function $(f\varphi)(x,t)$ without changing the integral, and

$$\int_{T_1}^{T_2} a(x,t)\,dt = \int_{I(x)} \left(f\varphi\frac{\partial\chi_0}{\partial t}\right)(x,t)\,dt = \int_{I(x)} \delta(x,t)\frac{\partial\chi_0}{\partial t}(x,t)\,dt, \qquad (25)$$

by (21). Since $\left|\frac{\partial\chi_0}{\partial t}(x,t)\right| \leq C\varepsilon^{-1}$ (by (15)) and $H^1(I(x)) \leq C\varepsilon$, (24) yields

$$\left|\int_{T_1}^{T_2} a(x,t)\,dt\right| \leq C\varepsilon|f(x,t_0)| + C\varepsilon \int_{T_1}^{T_2} |h(x,u)|\,du + C\int_{I(x)} |h(x,u)|\,du, \qquad (26)$$

and then, by (19),

$$|B_1(\varepsilon)| \leq C\varepsilon \int_D |f(x,t_0)|\,dx + C\varepsilon \int_V |h| + C\int_{V_\varepsilon} |h|, \qquad (27)$$

where $V_\varepsilon = \bigcup_{x\in D} (\{x\} \times I(x))$ is contained in a thin stripe around Γ. Since $f \in L^1(V)$, we can choose t_0 so that $\int_D |f(x,t_0)|\,dx \leq C\int_V |f| < +\infty$. Thus the right-hand side of (27) tends to 0 when ε tends to 0, and so does $B_1(\varepsilon)$. Hence $A_0 = 0$, because of (17) and (18). This completes our proof of (12). In other words, $\frac{\partial f}{\partial t} = h$.

We still need to prove similar results for the other derivatives $\frac{\partial f}{\partial x_i}$, $1 \leq i \leq n-1$. Fortunately, we can reduce to the case of $\frac{\partial f}{\partial t}$ with a trick. Indeed, if $v = (v_i)$ is a unit vector, with $|v_i|$ sufficiently small for $1 \leq i \leq n-1$, we can apply the same argument as above to prove that the distribution $\frac{\partial f}{\partial v}$ (which exists because $f \in L^1(V)$) is given by integration against the L^1-function $\nabla f \cdot v$ on $V_1 \cup V_2$ which we get from our assumption (5). There is a small difference, because the vertical boundaries of V are no longer parallel to v, but this does not prevent us from doing the proof: since our problem is local, we can always restrict to smaller tubular domains (like V) contained in V, but with sides parallel to each other and to v. As for our hypothesis (6), we have to replace limits $f^*(z)$ on vertical rays with similar limits $f^v(z)$ on rays parallel to v. The existence of limits for almost-every $z \in \Gamma$ comes from Lemma 13.9 (which also says that $f^*(z) = f^v(z)$ almost-everywhere).

Thus we get that $\frac{\partial f}{\partial v} \in L^1(V)$, plus a description of $\frac{\partial f}{\partial v}$ in terms of the derivatives from (5), for every unit vector v that is close enough to e. The other derivatives, like the $\frac{\partial f}{\partial x_i}$, $1 \leq i \leq n-1$, are now obtained by linear combinations. This completes our proof of Lemma 4. $\qquad\square$

We shall often use the following combination of Lemma 4 and a spherical change of coordinates.

Corollary 28. *Let $K \subset \mathbb{R}^n$ be closed, and let $0 < r < R$ be given. Then for every $f \in W^{1,1}(B(0,R) \setminus [K \cup \partial B(0,r)])$, the radial limits $f^-(\xi) = \lim_{t \to 1-} (t\xi)$ and $f^+(\xi) = \lim_{t \to 1+} (t\xi)$ exist for H^{n-1}-almost every $\xi \in \partial B(0,r) \setminus K$. If in addition $f^+(\xi) = f^-(\xi)$ for H^{n-1}-almost every $\xi \in \partial B(0,r) \setminus K$, then $f \in W^{1,1}(B(0,R) \setminus K)$, with the same derivative.*

Let us start with a more local result. For each $\xi_0 \in \partial B(0,r) \setminus K$, choose a neighborhood $V = V(\xi_0)$ of ξ_0, so small that V does not meet K. We can also choose V so that, after a spherical change of coordinates, V becomes a rectangular domain $D \times (T_1, T_2)$, as in the beginning of this section. Then we have radial limits $f^-(\xi)$ and $f^+(\xi)$ for almost-every $\xi \in V \cap \partial B(0,r)$, by Proposition 11.4 (for the change of variable) and Proposition 9.5 (for the absolute continuity on almost all lines); the point is of course the same as for the definition (13.5) of vertical limits. Moreover, if $f^-(\xi) = f^+(\xi)$ almost-everywhere on $\xi \in V \cap \partial B(0,r)$, then $f \in W^{1,1}(V)$, this time by Proposition 11.4 and Lemma 4.

To complete the proof of the corollary, we just need to localize. Note that we can cover $\partial B(0,r) \setminus K$ by countably many sets $V(\xi_0)$, which gives the existence of radial limits almost everywhere on $\partial B(0,r) \setminus K$. Moreover, if $f^+(\xi) = f^-(\xi)$ almost everywhere on $\partial B(0,r) \setminus K$, then we get that $f \in W^{1,1}(V)$ for all V as above. This shows that $f \in W^{1,1}_{\text{loc}}(B(0,R) \setminus K)$, but since we know that the gradient of f on $B(0,R) \setminus K$ coincides with the previously known gradient on $B(0,R) \setminus [K \cup \partial B(0,r)]$, we immediately get that $f \in W^{1,1}(B(0,R) \setminus K)$. Corollary 28 follows. $\square$

15 Energy-minimizing functions with prescribed boundary values

Let $V \subset \mathbb{R}^n$ be an open set, $I \subset \partial V$ a piece of its boundary, and u a function on I. We shall often need to use the functions $f \in W^{1,2}(V)$ that minimize the energy $E = \int_V |\nabla f|^2$ under the constraint that the boundary values of f on I equal u. In this section we show that there is an essentially unique solution to this problem (unless no function $f \in W^{1,2}(V)$ has the given boundary values u), and that it is harmonic in V. More information on energy-minimizing functions will be given in the next two sections.

Let us be more specific about our Dirichlet problem. Let $V \subset \mathbb{R}^n$ be open, and let $I \subset \partial V$ be a (relatively) open subset of its boundary. We assume that for each $x_0 \in I$ there is an open neighborhood R of x_0 with the following properties. There is an isometry σ of $\mathbb{R}^n$ such that $\sigma(x_0) = 0$, $\sigma(R) = D \times (-T, T)$ is the product of a disk $D \subset \mathbb{R}^{n-1}$ centered at the origin and an interval, $\sigma(I \cap R) = \sigma(\partial V \cap R) = \Gamma$, where Γ is the graph of some Lipschitz function $A : D \to (-T, T)$, and $\sigma(V \cap R)$ is the part of $\sigma(R)$ above the graph. In other words, I is open in ∂V and locally I is a Lipschitz graph, with V on one side of I only. In most of our applications the dimension will be $n = 2$, V will be contained in a disk B, and I will be a finite union of arcs of ∂B.

The point of this hypothesis on I is that all functions $f \in W^{1,2}(V)$ have "transverse limits" almost-everywhere on I, by Section 13. More precisely, we can cover I by little boxes R where I is a Lipschitz graph and use formulae like (13.5) to define boundary values $f^*(z)$ almost-everywhere on $I \cap R$ for each R. This allows us to define f^* almost-everywhere on I, because Lemma 13.9 takes care of coherence issues. (That is, if R and R' meet, the two definitions of f^* on $I \cap R \cap R'$ coincide almost-everywhere.)

To each $f \in W^{1,2}(V)$ we shall systematically associate a choice of boundary values $f^*(z)$ on I, as above.

Next let u be a (measurable) function on I. To simplify things, we shall assume in later sections that

$$u \text{ is bounded}, \tag{1}$$

but this is not needed for the moment. Set

$$\mathcal{F} = \mathcal{F}(V, I, u) = \{f \in W^{1,2}(V)\,;\, f^*(x) = u(x) \ \ H^{n-1}\text{-almost-everywhere on } I\}. \tag{2}$$

Of course $\mathcal{F}$ can be empty if u is too wild; we shall assume that this is not the case, i.e., that

$$\mathcal{F} \neq \emptyset. \tag{3}$$

Next set

$$E(f) = \int_V |\nabla f|^2 \text{ for } f \in W^{1,2}(V), \tag{4}$$

and then

$$E = E(V, I, u) = \inf\{E(f)\,;\, f \in \mathcal{F}\}. \tag{5}$$

Proposition 6. *Let V, I, u be as above; in particular, $\mathcal{F} \neq \emptyset$. Then there is a function $f \in \mathcal{F}$ such that*

$$E(f) = E. \tag{7}$$

Also, f is almost unique: if $g \in \mathcal{F}$ is another function such that $E(g) = E$ and $\mathcal{O}$ is a connected component of V, then either

$$\mathcal{O} \text{ touches } I, \text{ and then } f = g \text{ on } \mathcal{O}, \tag{8}$$

or else

$$\mathcal{O} \text{ does not touch } I, \text{ and then } f \text{ and } g \text{ are constant on } \mathcal{O}. \tag{9}$$

Here we say that $\mathcal{O}$ touches I if there is a sequence $\{x_n\}$ in $\mathcal{O}$ that converges to some point of I. This is the same as saying that we can find $x_0 \in I$, and a rectangular domain R centered on x_0 as above (in the description of I) such that R meets $\mathcal{O}$ (or equivalently $R \cap V \subset \mathcal{O}$).

Before we prove the proposition, a few comments are in order.

Remark 10. We insisted on having V on only one side of I (locally), but this was mostly to make definitions simpler. If there is a portion J of I where V lies on both sides of J, we can give ourselves two functions u_1 and u_2 on J, and demand in the definition of $\mathcal{F}$ that the boundary values of f from each side of V coincide almost-everywhere on J with the corresponding u_j. We can also add to I some pieces of Lipschitz graphs Γ_j inside V, and demand that elements of $\mathcal{F}$ have "transverse" limits on these pieces equal almost-everywhere to a given function u. Because of the welding lemma in Section 14, this is the same thing as removing the Lipschitz graphs from V, and solving the previous problem on $V \setminus \bigcup_j \Gamma_j$ with $u_1 = u_2 = u$ on the Γ_j.

We shall prove the proposition component by component. This is possible because $f \in W^{1,2}_{\text{loc}}(V)$ if and only if $f_{|\mathcal{O}} \in W^{1,2}_{\text{loc}}(\mathcal{O})$ for every component $\mathcal{O}$ of V, and $E(f)$ is the sum of the energies $\int_{\mathcal{O}} |\nabla f|^2$. Also, $f^* = u$ almost-everywhere on I if and only if, for every component $\mathcal{O}$, $f^*_{|\mathcal{O}} = u$ almost-everywhere on the parts of I that are touched by $\mathcal{O}$.

This allows us to reduce to the case when V is connected. If $I = \emptyset$ (which corresponds to the case when $\mathcal{O}$ does not touch I), then $E = 0$ and the only solutions of $E(f) = E$ are constants. So we can assume that I is not empty. We want to use the convexity of $E(f)$. If $f, g \in W^{1,2}(V)$, the parallelogram identity says that

$$E\left(\frac{f+g}{2}\right) = \frac{1}{2}E(f) + \frac{1}{2}E(g) - \frac{1}{4}E(f-g). \tag{11}$$

Let us first deduce the uniqueness result in (8) from this. If $f, g \in \mathcal{F}$ are such that $E(f) = E(g) = E$, then $E(\frac{f+g}{2}) \geq E$ because $\frac{f+g}{2} \in \mathcal{F}$, and (11) says that $E(f-g) = 0$. Since V is now connected, $f - g$ is constant and then $f = g$ because $(f-g)^* = 0$ almost-everywhere on I. Here we use the fact that V touches I, to say that nonzero constant functions on V do not have limits on I that vanish almost-everywhere. This proves the uniqueness.

For the existence also we want to use (11). Let $\{f_k\}$ be a sequence in $\mathcal{F}$, with $\lim_{k \to +\infty} E(f_k) = E$. Let $\varepsilon > 0$ be given, and let k, ℓ be such that $E(f_k) \leq E + \varepsilon$ and $E(f_\ell) \leq E + \varepsilon$. Since $g = \frac{1}{2}(f_k + f_\ell) \in \mathcal{F}$,

$$E \leq E(g) = \frac{1}{2}E(f_k) + \frac{1}{2}E(f_\ell) - \frac{1}{4}E(f_k - f_\ell) \leq E + \varepsilon - \frac{1}{4}E(f_k - f_\ell), \tag{12}$$

and so $E(f_k - f_\ell) \leq 4\varepsilon$. Thus $\{\nabla f_k\}$ is a Cauchy sequence in $L^2(V)$, and

$$\{\nabla f_k\} \text{ converges in } L^2(V) \text{ to some limit } h. \tag{13}$$

Choose a point $x_0 \in I$, and let R be a rectangular neighborhood of x_0 with the properties listed at the beginning of this section (in the description of I). Also choose a ball $B \subset R \cap V$. We can apply Lemma 13.20 and its immediate consequence (13.22) to the domain $R \cap V$ and the ball B. We take $f = f_k - f_\ell$, so that $f^*(z) = 0$ almost-everywhere, and get that

$$|m_B(f_k - f_\ell)| \leq C \int_{R \cap V} |\nabla(f_k - f_\ell)|, \tag{14}$$

where C may depend on the geometry of $R \cap V$ and B, but this does not matter. By (13) and Cauchy–Schwarz, the right-hand side of (14) tends to 0, and so

$$\lim_{k \to +\infty} m_B f_k \text{ exists .} \tag{15}$$

Now we can apply Corollary 12.17, which tells us that $\{f_k\}$ converges in $L^1_{\text{loc}}(V)$ to some limit $f \in W^{1,2}(V)$, and that $\nabla f = h$. From (13) we deduce that

$$E(f) = \int_V |h|^2 = \lim_{k \to +\infty} E(f_k) = E;$$

thus Proposition 6 will follow as soon as we prove that $f \in \mathcal{F}$. In other words, we still need to check that $\mathcal{F}$ is closed in $W^{1,2}(V)$.

Pick any point $x_0 \in I$, and let R be an open rectangular neighborhood of x_0, as in the description of I (R may be different from the one we chose to show that $\{f_k\}$ converges). Set

$$\mathcal{F}_R = \{g \in W^{1,1}(R \cap V) \,;\, g^*(z) = u(z) \text{ almost-everywhere on } I \cap R\}. \tag{16}$$

We know from Corollary 13.31 that $\mathcal{F}_R$ is closed in $W^{1,1}(R \cap V)$. Since $\{f_k\}$ converges to f in $W^{1,2}(V)$, the convergence also holds in $W^{1,1}(R \cap V)$, and so (the restriction of) f lies in $\mathcal{F}_R$. Since we can do this for all x_0 and R, we get that $f^*(z) = u(z)$ almost-everywhere on I, and $f \in \mathcal{F}$. This completes our proof of Proposition 6. $\qquad \square$

Let us already state here the main property of our energy-minimizing functions: they are harmonic.

Lemma 17. *Let $V \subset \mathbb{R}^n$ be open and $f \in W^{1,2}_{\text{loc}}(V)$ satisfy the following property: for every ball $B \subset\subset V$ and every function $g \in W^{1,2}_{\text{loc}}(V)$ such that $f(x) = g(x)$ almost-everywhere on $V \setminus B$,*

$$\int_B |\nabla f|^2 \leq \int_B |\nabla g|^2. \tag{18}$$

Then f is harmonic in V.

Note that if V, I, u, $\mathcal{F}$ are as in Proposition 6 and $f \in \mathcal{F}$ is such that $E(f) = E$, then f satisfies the condition in the lemma (and hence is harmonic in V). This is an easy consequence of the definitions; indeed if $g(x) = f(x)$ almost-everywhere on $V \setminus B$ for some $B \subset\subset V$, then $g^* = f^*$ almost-everywhere on I.

The proof of Lemma 17 is even more standard than the rest of this section. Let $B \subset\subset V$ be given, and let $\varphi \in C_c^\infty(B)$. Try $g_\lambda = f + \lambda\varphi \in W_{\mathrm{loc}}^{1,2}(V)$. Since

$$\int_B |\nabla g_\lambda|^2 = \int_B |\nabla f|^2 + 2\lambda \int_B \nabla f \cdot \nabla \varphi + \lambda^2 \int_B |\nabla \varphi|^2 \tag{19}$$

is a quadratic function of λ, and (18) says that it is minimal for $\lambda = 0$, we get that

$$\int_B \nabla f \cdot \nabla \varphi = 0. \tag{20}$$

Since $\nabla \varphi$ is a (vector-valued) test function and ∇f is also the distributional gradient of $f \in W^{1,2}(B)$, $\int_B \nabla f . \nabla \varphi = - \int_B f \Delta \varphi$. Thus (20) says that $\int_B f \Delta \varphi = 0$ for all $\varphi \in C_c^\infty(B)$, another way of saying that $\Delta f = 0$ in B. We proved that f is harmonic in every ball $B \subset\subset V$; the lemma follows. $\qquad\square$

Remark 21. [Neumann condition at the boundary, first attempt.] Let V, I, u, $\mathcal{F}$ be as in Proposition 6, and let $f \in \mathcal{F}$ be such that $E(f) = E$. Then for every ball B such that $B \cap I = \emptyset$, and every $\varphi \in C_c^\infty(B)$,

$$\int_{B \cap V} \nabla f \cdot \nabla \varphi = 0. \tag{22}$$

The proof is the same as for (20); we do not need to know that $B \subset V$, just that $g_\lambda = g + \lambda\varphi \in \mathcal{F}$ for all λ.

If in addition $V \cap B$ is a nice piecewise C^1 open set and $f \in C^1(\overline{V \cap B})$, we can apply Green's theorem to the vector-valued function $\varphi \nabla f$ on $V \cap B$ and get that

$$\int_{\partial(V \cap B)} \varphi \frac{\partial f}{\partial n} = \int_{\partial(V \cap B)} \varphi \nabla f \cdot n = \int_{V \cap B} \mathrm{div}(\varphi \nabla f)$$
$$= \int_{V \cap B} \nabla \varphi \cdot \nabla f + \int_{V \cap B} \varphi \Delta f = 0, \tag{23}$$

by (22) and because $\Delta f = 0$. Here $\frac{\partial f}{\partial n}$ denotes the normal derivative of f (with unit normal pointing outwards). Note that $\varphi = 0$ near ∂B, so (23) is the same as

$$\int_{\partial V \cap B} \varphi \frac{\partial f}{\partial n} = 0 \tag{24}$$

and, since (24) holds for all $\varphi \in C_c^\infty(B)$, we get the Neumann condition

$$\frac{\partial f}{\partial n} = 0 \tag{25}$$

on $\partial V \cap B$.

Thus the Neumann condition (25) holds near every point of the boundary ∂V which has a neighborhood where ∂V is C^1, does not meet I, and where f is C^1 on $\overline{V}$. When this is not the case, we cannot use (25) directly, but we can still try to use (22) (which is a weak form of (25)), or use limiting arguments with smoother domains.

16   Energy minimizers for an increasing union of open sets

Let V, I, u, $\mathcal{F}$ be as in the beginning of the last section. Thus V is an open set in $\mathbb{R}^n$, I is a reasonably smooth open piece of ∂V, and u is a function on I such that $\mathcal{F} = \mathcal{F}(V, I, u)$ is not empty. In this section, we shall find it more convenient to assume that u is bounded, as in (15.1).

Let us assume that there is a nondecreasing sequence of open sets $V_k \subset V$ and a nondecreasing sequence of sets $I_k \subset I$ such that the pairs (V_k, I_k) also satisfy the hypotheses of Section 15,

$$V = \bigcup_k V_k, \text{ and } I = \bigcup_k I_k. \tag{1}$$

Set $\mathcal{F}_k = \mathcal{F}(V_k, I_k, u)$. Note that $\mathcal{F} \subset \mathcal{F}_k$, because $W^{1,2}(V) \subset W^{1,2}(V_k)$ and we add more constraints on the boundary values. Also set

$$E_k = \inf \left\{ \int_{V_k} |\nabla f|^2; f \in \mathcal{F}_k \right\}; \tag{2}$$

then $E_k \leq E = \inf\{E(f); f \in \mathcal{F}\}$. For similar reasons, $\{\mathcal{F}_k\}$ is a nonincreasing sequence and $\{E_k\}$ is nondecreasing.

Denote by f_k the function of $\mathcal{F}_k$ such that $\int_{V_k} |\nabla f_k|^2 = E_k$ and $f_k = 0$ on the connected components of V_k that do not touch I_k. (See Proposition 15.6.) Similarly let f denote the function in $\mathcal{F}$ such that $E(f) = E$ and $f = 0$ on the components of V that do not touch I.

Proposition 3. *With the notations and assumptions above,*

$$E = \lim_{k \to +\infty} E_k, \tag{4}$$

$$\{f_k\} \text{ converges to } f \text{ uniformly on every compact subset of } V, \tag{5}$$

and

$$\{\mathbf{1}_{V_k} \nabla f_k\} \text{ converges to } \nabla f \text{ in } L^2(V). \tag{6}$$

In the statement we added 1_{V_k} in (6) just because ∇f_k lives on the smaller V_k; we did not take such caution in (5) because every compact set $K \subset V$ is eventually contained in V_k.

To prove the proposition, first observe that

$$||f_k||_\infty \le ||u||_\infty < +\infty \tag{7}$$

for all k. This uses our assumption (15.1), the uniqueness of f_k, and the fact that if (7) did not hold we would be able to truncate f_k and get an equally good (or better) competitor. See the proof of (3.20) with Lemma 3.22, or directly Proposition 11.15.

Let $B \subset\subset V$ be a (relatively compact) open ball in V. By compactness, $\overline{B} \subset V_k$ for k large enough. Then Lemma 15.17 says that f_k is harmonic in B, and so

$$||\nabla f_k||_{L^\infty(\frac{1}{2}B)} \le C \tag{8}$$

for k large, with C independent of k (by (7)). Thus we can extract a subsequence of $\{f_k\}$ that converges uniformly on $\frac{1}{2}B$ to some limit g. We can even do this repeatedly with a sequence $\{B_j\}$ of balls such that every compact subset of V is contained in $\bigcup_j \left(\frac{1}{2}B_j\right)$. The diagonal process gives a subsequence, which we shall still denote by $\{f_k\}$, that converges uniformly on every compact subset of V to some limit g.

Note that g is harmonic on V (because for each ball $B \subset\subset V$, f_k is harmonic on B for k large, and $\{f_k\}$ converges uniformly on B). We also know that for each compact set $K \subset V$, $\{\nabla f_k\}$ converges to ∇g uniformly on K. Indeed K can be covered by finitely many balls B such that $3B \subset V$, and bounds on $||f_k - g||_{L^\infty(2B)}$ give bounds on $||\nabla f_k - \nabla g||_{L^\infty(B)}$ (by Harnack, as for (8)). Hence for K compact in V,

$$\int_K |\nabla g|^2 = \lim_{k \to +\infty} \int_K |\nabla f_k|^2 \le \lim_{k \to +\infty} E_k, \tag{9}$$

where the last limit exists because $\{E_k\}$ is nondecreasing. Since this holds for every K, we get that

$$E(g) =: \int_V |\nabla g|^2 \le \lim_{k \to +\infty} E_k \le E, \tag{10}$$

where the last inequality was already observed just after (2). Of course $g \in W^{1,2}(V)$, because it is harmonic in V and by (10). We also need to know that $g \in \mathcal{F}$, i.e., that

$$g^*(u) = u(x) \text{ almost-everywhere on } I. \tag{11}$$

Let $x_0 \in I$ be given. Then $x_0 \in I_k$ for k large enough. Fix such a k, and let R be a rectangular neighborhood of x_0, with the properties described at the beginning of Section 15 (with respect to I_k and V_k). Then $R \cap V_\ell = R \cap V = R \cap V_k$ and $R \cap I_\ell = R \cap I = R \cap I_k$ for $\ell > k$, because we cannot add anything to $R \cap V_k$ or $R \cap I_k$ and maintain the properties of the beginning of Section 15.

We know that $\{\nabla f_k\}$ converges to ∇g, uniformly on compact subsets of $R \cap V$. Also, $\{\nabla f_k\}$ is bounded in $L^2(R \cap V)$, because $\int_{R \cap V} |\nabla f_k|^2 = \int_{R \cap V_k} |\nabla f_k|^2 \leq E_k \leq E$ for k large. Hence

$$\lim_{k \to +\infty} \nabla f_k = \nabla g \ \text{ in } \ L^1(R \cap V) \tag{12}$$

(see Exercise 20 below). Since $\{f_k\}$ converges to g in $L^1_{\mathrm{loc}}(R \cap V)$ (and even uniformly on compact subsets), we have that

$$\lim_{k \to \infty} f_k = g \ \text{ in } \ W^{1,1}(R \cap V). \tag{13}$$

Now we can use Corollary 13.31: since

$$\mathcal{F}_R = \{f \in W^{1,1}(R \cap V) \, ; \, f^*(x) = u(x) \text{ almost-everywhere on } R \cap I\} \tag{14}$$

is closed and $f_k \in \mathcal{F}_R$ for k large, we get that $g \in \mathcal{F}_R$ too.

We proved that for all $x_0 \in I$, there is a neighborhood R of x_0 such that $g^*(x) = u(x)$ almost-everywhere on $R \cap I$. This proves (11).

Now we know that $g \in \mathcal{F}$. Then $E(g) \geq E$, by definition of E, and hence $E(g) = \lim_{k \to +\infty} E_k = E$, by (10). In particular, (4) holds.

From the uniqueness result in Proposition 15.6, we deduce that $g = f$ on all the components of V that touch I. If $\mathcal{O}$ is a component of V that does not touch I and $x \in \mathcal{O}$, then for all k large enough for V_k to contain x, the component of V_k that contains x is contained in $\mathcal{O}$, hence does not touch $I_k \subset I$, and so $f_k(x) = 0$. Thus $g = f$.

Let us now check (5). So far we only proved that we could extract a subsequence from $\{f_k\}$, which we still denoted by $\{f_k\}$, and which converges to $g = f$ uniformly on compact subsets of V. [See the few lines after (8).] But the same thing would hold if we started from any subsequence of $\{f_k\}$, so every subsequence contains a further subsequence that converges to f, uniformly on compact subsets of V. This forces $\{f_k\}$ itself to converge to f, as in (5).

We still need to check (6). Let $\varepsilon > 0$ be given. We can find a compact set $K \subset V$ such that

$$\int_K |\nabla f|^2 \geq E(f) - \varepsilon = E - \varepsilon. \tag{15}$$

For k large enough, $K \subset V_k$ and

$$\int_K |\nabla f - \nabla f_k|^2 \leq \varepsilon^2, \tag{16}$$

because $\{\nabla f_k\}$ converges to $\nabla g = \nabla f$ uniformly on K (by (5) and harmonicity). Then

$$\|\nabla f_k\|_{L^2(K)} \geq \|\nabla f\|_{L^2(K)} - \|\nabla(f_k - f)\|_{L^2(K)} \geq (E - \varepsilon)^{1/2} - \varepsilon, \tag{17}$$

by (15) and (16),

$$\int_{V\setminus K} \mathbf{1}_{V_k} |\nabla f_k|^2 = E_k - \|\nabla f_k\|^2_{L^2(K)} \le E - \left\{(E-\varepsilon)^{1/2} - \varepsilon\right\}^2 \le C\varepsilon, \qquad (18)$$

and finally

$$\int_V |\nabla f - \mathbf{1}_{V_k} \nabla f_k|^2 \le \int_K |\nabla f - \nabla f_k|^2 + \int_{V\setminus K} |\nabla f - \mathbf{1}_{V_k} \nabla f_k|^2$$
$$\le \varepsilon^2 + 2\int_{V\setminus K} |\nabla f|^2 + 2\int_{V\setminus K} |\mathbf{1}_{V_k} \nabla f_k|^2 \le C\varepsilon \qquad (19)$$

by (16), (15), and (18).

Thus (6) holds also, and our proof of Proposition 3 is complete. □

Exercise 20. Check our claim (12) above. Hint: for every $\varepsilon > 0$ we can find a compact set $K \subset R \cap V$ which is so large that $\int_{(R\cap V)\setminus K} |\nabla f_k| \le \varepsilon$ for k large (by Cauchy–Schwarz), and (hence) $\int_{(R\cap V)\setminus K} |\nabla g| \le \varepsilon$.

17   Conformal invariance and the Neumann condition

In this section we consider the same energy minimization problem as in Sections 15–16, but we restrict to dimension $n = 2$. We want to see how to use the invariance under conformal mappings of energy integrals like $E(f)$, to get information on the functions f that minimize $E(f)$ with given Dirichlet conditions. We start with two simple cases where we can compute everything.

Example 1. Take $V = B(0,1) \subset \mathbb{R}^2$ and $I = \partial B(0,1)$. Let $u \in L^\infty(I)$ be given. To simplify the discussion, we shall assume that $\mathcal{F} \ne \emptyset$, as in (15.3). [This is not automatic for $u \in L^\infty(I)$.] Denote by f the function in $\mathcal{F}$ such that

$$E(f) =: \int_V |\nabla f|^2 = \inf \left\{E(g)\,;\, g \in \mathcal{F}\right\}, \qquad (2)$$

as in Proposition 15.6. We know from Lemma 15.17 that f is harmonic in V. Also, f is bounded by $\|u\|_\infty$, by the uniqueness part of Proposition 15.6 and the usual truncation argument (see Lemma 3.22 and the argument just before). It is well known that bounded harmonic functions in the unit disk can be written as Poisson integrals of bounded functions. See for instance [Du], [Ga] page 18 (but on the upper half-plane), or [Ru], Theorem 17.10. This means that

$$f(r\cos\theta, r\sin\theta) = \int_0^{2\pi} \frac{1-r^2}{1 - 2r\cos(t-\theta) + r^2}\, g(t)\, dt \qquad (3)$$

for $r < 1$ and $\theta \in \mathbb{R}$, and some bounded function g. Moreover, for almost-every $\xi = (\cos\theta, \sin\theta) \in \partial B(0,1)$, f has a nontangential limit at ξ equal to $g(\theta)$. This means that

$$\lim_{\substack{z \to \xi \\ z \in \Gamma(\xi)}} f(z) = g(\theta), \tag{4}$$

where $\Gamma(\xi) = \{z \in B(0,1) \,;\, |z - \xi| < C(1 - |z|)\}$ is a conical region of approach to ξ, and we can choose the constant $C > 1$ as we like.

Since $f^*(\xi) = u(\xi)$ almost-everywhere on $\partial B(0,1)$, by definition of $\mathcal{F}$, we get that $g(\theta) = u(\cos\theta, \sin\theta)$ almost-everywhere. Thus f is just the harmonic extension of u, defined on $B(0,1)$ by convolution with the Poisson kernel.

When u is continuous, f has a continuous extension to $\overline{B}(0,1)$, which coincides with u on $\partial B(0,1)$. Then the way in which u is the boundary value of f is especially simple. When u is merely bounded, we still know that u is the nontangential limit of f almost-everywhere on $\partial B(0,1)$ (as in (4)), which is somewhat better than the sort of transversal limits we had in Section 13.

Note that we do not say here that $\mathcal{F} \neq \emptyset$ for every $u \in L^\infty(I)$, but only that if this is the case, then the minimizer f is the Poisson integral of u. In general, the Poisson integral of u is a harmonic function on $B(0,1)$ that tends to $u(x)$ nontangentially almost-everywhere on $\partial B(0,1)$, but its gradient may not lie in $L^2(B(0,1))$, and then $\mathcal{F} = \emptyset$. See Exercise 19.

Example 5. Take $V = \{(x,y) \in B(0,1) \,;\, y > 0\} \subset \mathbb{R}^2$, $I = \{(x,y) \in \partial B(0,1) \,;\, y > 0\}$, and let $u \in L^\infty(I)$ be given. Assume that $\mathcal{F} \neq \emptyset$ and let f denote the function in $\mathcal{F}$ that minimizes $E(f)$, as in (2) or Proposition 15.6. We know that f is unique, because V is connected and touches I.

Lemma 6. *The minimizer f is the restriction to V of the harmonic extension to $B(0,1)$ of the function $\widehat{u} \in L^\infty(\partial B(0,1))$ obtained from u by symmetry.*

More explicitly, we set $\widehat{u}(x,y) = u(x,y)$ when $(x,y) \in I$, $\widehat{u}(x,y) = u(x,-y)$ when $(x,-y) \in I$, and for instance $\widehat{u}(\pm 1, 0) = 0$. Then we call $\widehat{f}$ the harmonic extension of $\widehat{u}$, obtained by convolution of $\widehat{u}$ with the Poisson kernel (as in (3)). Our claim is that $f(x,y) = \widehat{f}(x,y)$ when $(x,y) \in V$.

To prove this, set $\widehat{V} = B(0,1)$, $\widehat{I} = \partial B(0,1)$, and define $\widehat{\mathcal{F}} = \mathcal{F}(\widehat{V}, \widehat{I}, \widehat{u})$ as in (15.2). We shall see soon (and without vicious circle) that $\widehat{\mathcal{F}} \neq \emptyset$. Also set $\widehat{E} = \inf\{\int_{B(0,1)} |\nabla g|^2 \,;\, g \in \widehat{\mathcal{F}}\}$. We know that we can find $\widehat{f} \in \widehat{\mathcal{F}}$ such that $\int_{B(0,1)} |\nabla \widehat{f}|^2 = \widehat{E}$, as in Example 1 above. Also, $\widehat{f}$ is symmetric because $\widehat{u}$ is, and by the uniqueness of $\widehat{f}$ (or because the Poisson kernel is symmetric). Therefore

$$E \leq \int_V |\nabla \widehat{f}|^2 = \frac{1}{2} \int_{B(0,1)} |\nabla \widehat{f}|^2 = \frac{1}{2} \widehat{E}. \tag{7}$$

On the other hand, let $g \in \mathcal{F}$ be given. Denote by $\widehat{g}$ the symmetric extension of g to $\widehat{V}$ (that is, keep $\widehat{g} = g$ on V and set $\widehat{g}(x,y) = g(x,-y)$ when $(x,-y) \in V$;

the values of $\widehat{g}$ on $(-1,1) \times \{0\}$ do not matter). Then $\widehat{g} \in W^{1,2}(\widehat{V})$, by the welding Lemma 14.4, and the fact that transversal limits g^* of g exist almost-everywhere on $(-1,1) \times \{0\}$ (by Section 13) and coincide with the transversal limits of $\widehat{g}$ from the other side (by symmetry). Also, $\widehat{g}^* = \widehat{u}$ almost-everywhere on $\widehat{I} = \partial B(0,1)$, because $g \in \mathcal{F}$ and by symmetry. Altogether, $\widehat{g} \in \widehat{\mathcal{F}}$. In particular, $\widehat{\mathcal{F}} \neq \emptyset$, as promised. Moreover $\widehat{E} \leq \int_{B(0,1)} |\nabla \widehat{g}|^2 = 2E(g)$ (by symmetry). If we take $g = f$, we get that $\widehat{E} \leq 2E$. Then (7) says that $\widehat{E} = 2E$, and we also get that $f = \widehat{f}_{|V}$, by (7) and the uniqueness of f. This proves the lemma. $\qquad\square$

Notice that $\widehat{f}$ is C^∞ on the unit ball because it is harmonic, and that $\frac{\partial \widehat{f}}{\partial y} = 0$ on $(-1,1) \times \{0\}$, by symmetry. Thus our minimizing function f has a C^∞ extension up to the boundary piece $(-1,1) \times \{0\}$, and satisfies the Neumann condition $\frac{\partial f}{\partial n} = 0$ there.

Next we want to see how we can use the conformal invariance of energy integrals to derive similar properties in different situations. This is the place where we shall use our assumption that $n = 2$. For the rest of this section, we shall identify $\mathbb{R}^2$ with the complex plane $\mathbb{C}$.

Lemma 8. *Let $\psi : V_1 \to V_2$ be a (bijective) conformal mapping between open sets, and let $f : V_2 \to \mathbb{R}$ be measurable. Then $f \in W^{1,2}(V_2)$ if and only if $f \circ \psi \in W^{1,2}(V_1)$, and in that case*

$$\int_{V_1} |\nabla(f \circ \psi)|^2 = \int_{V_2} |\nabla f|^2. \tag{9}$$

Since ψ and ψ^{-1} are both C^∞, the invariance of $W^{1,2}_{\mathrm{loc}}$ is a direct consequence of Proposition 11.4 on compositions. As for (9), note that

$$|\nabla(f \circ \psi)(z)|^2 = |\nabla f(\psi(z))|^2 \, |\psi'(z)|^2 \tag{10}$$

on V_1, because the differential of ψ acts as multiplication by $\psi'(z)$. Now $|\psi'(z)|^2$ is also the size of the Jacobian determinant, the two things cancel in the change of variables, and (9) follows. Of course all this is very special of conformal mappings in dimension 2. $\qquad\square$

Lemma 11. *Let V, I, and u be as in the beginning of Section 15, and let $f \in \mathcal{F}(V, I, u)$ minimize $E(f)$ (as in Proposition 15.6). Let $\psi : V \to \psi(V)$ be a conformal mapping. Assume that ψ has an extension to $V \cup I$ which is a C^1-diffeomorphism onto its image, and that $\psi(V)$, $\psi(I)$ satisfy the hypotheses at the beginning of Section 15. Then*

$$\mathcal{F}\big(\psi(V), \psi(I), u \circ \psi^{-1}\big) = \{g \circ \psi^{-1} \, ; \, g \in \mathcal{F}(V, I, u)\}, \tag{12}$$

and

$$\int_{\psi(V)} |\nabla(f \circ \psi^{-1})|^2 = \inf \left\{ \int_{\psi(V)} |\nabla h|^2 \, ; \, h \in \mathcal{F}\big(\psi(V), \psi(I), u \circ \psi^{-1}\big) \right\}. \tag{13}$$

This is fairly easy. We already know from Lemma 8 that $W^{1,2}(\psi(V)) = \{g \circ \psi^{-1}\,;\, g \in W^{1,2}(V)\}$. Thus (12) will follow as soon as we check that

$$(g \circ \psi^{-1})^*(z) = g^* \circ \psi^{-1}(z) \text{ almost-everywhere on } \psi(I) \tag{14}$$

for all $g \in W^{1,2}(V)$. Here the superscripts $*$ correspond to taking boundary values along transverse rays on our two domains V and $\psi(V)$, as in (13.5). What we need here is the invariance of our boundary values construction of Section 13 under C^1 changes of variables (up to the boundary). See Remark 13.16. Actually, if we just want to prove (13), we do not even need to do this, because we only need to check (14) when g minimizes $E(g)$ or $g \circ \psi^{-1}$ minimizes $\int_{\psi(V)} |\nabla h|^2$. In both cases g is harmonic in V and $g \circ \psi^{-1}$ is harmonic on $\psi(V)$ (the two things are equivalent, since ψ is conformal). Then $g \circ \psi^{-1}$, for instance, has nontangential limits almost-everywhere on $\psi(I)$, and (14) is true because the images by ψ of the access rays we used to define g^* are contained in the nontangential access regions to $\psi(I)$ in V.

Once we have (12), (13) is a direct consequence of (9). Lemma 11 follows. $\square$

For the next proposition we consider $V \subset \mathbb{R}^2$, $I \subset \partial V$, and $u \in L^\infty(I)$, with the same hypotheses as in the beginning of Section 15. Also let $f \in \mathcal{F}$ minimize $E(f)$ (i.e., be such that $E(f) = E$, as in Proposition 15.6).

Proposition 15.   *Suppose $x_0 \in \partial V \setminus I$ and $B = B(x_0, r)$ are such that $B \cap I = \emptyset$ and $\partial V \cap B$ is a $C^{1+\alpha}$-curve Γ that crosses B (for some $\alpha > 0$). If $V \cap B$ lies only on one side of Γ, f has a C^1 extension to $\overline{V} \cap B$ that satisfies the Neumann condition $\frac{\partial f}{\partial n} = 0$ on Γ. If $V \cap B$ lies on both sides of Γ, the restriction of f to each component $\mathcal{O}$ of $V \cap B = B \setminus \Gamma$ has a C^1 extension to $\mathcal{O} \cup \Gamma$, which satisfies the Neumann condition $\frac{\partial f}{\partial n} = 0$ on Γ.*

Here the value of $\alpha > 0$ does not matter. When we say that Γ crosses B we mean that it has two extremities in ∂B; it is not allowed to go out of B and come back again (Γ is connected).

Proposition 15 is a local result. That is, we only need to pick a component $\mathcal{O}$ of $V \cap B$ and a point $x \in \Gamma$, and check that $f_{|\mathcal{O}}$ has a C^1 extension to $\mathcal{O} \cup \Gamma$ in a small neighborhood of x. Once we have the C^1 extension, We could always derive the Neumann condition from Remark 15.21 (see in particular (15.25)), but our proof will give it anyway.

So let $\mathcal{O}$ and $x \in \Gamma$ be given. Choose a small ball B_1 centered at x and a simply connected domain $\mathcal{O}_1$ with $C^{1+\alpha}$ boundary such that

$$\mathcal{O} \cap B_1 \subset \mathcal{O}_1 \subset \mathcal{O} \cap B. \tag{16}$$

(See Figure 1.)

Let $\psi_1 : B(0,1) \to \mathcal{O}_1$ be a conformal mapping. Then ψ_1 has an extension to $\overline{B}(0,1)$ which is a C^1-diffeomorphism onto $\overline{\mathcal{O}}_1$. See for instance [Po], page 48. We can assume that $\psi_1(1) = x$, because otherwise we could compose ψ_1 with a rotation of $B(0,1)$.

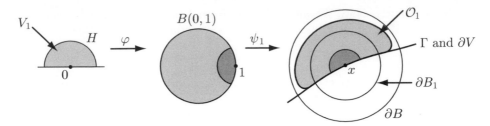

Figure 1

Next we want to replace $B(0,1)$ with the upper half-plane $\mathbb{H} = \{z \in \mathbb{C};\ \mathcal{I}m(z) > 0\}$. Define a conformal mapping $\varphi : \mathbb{H} \to B(0,1)$ by $\varphi(z) = \frac{i-z}{i+z}$, and set $\psi = \psi_1 \circ \varphi$. Now ψ is a conformal mapping from $\mathbb{H}$ to $\mathcal{O}_1$; it still has a C^1 extension to $\overline{\mathbb{H}}$, and $\psi(0) = x$.

Consider $V_1 = \mathbb{H} \cap B(0,\rho)$, where we choose ρ so small that $\psi(\overline{V}_1) \subset B_1$. We know that $\psi : \overline{V}_1 \to \psi(\overline{V}_1)$ is a C^1-diffeomorphism, and by (16) there is a neighborhood of x in which $\mathcal{O}$ and $\psi(V_1)$ coincide (and similarly for $\overline{\mathcal{O}}$ and $\psi(\overline{V}_1)$). Set $V_2 = \psi(V_1)$.

Cut ∂V_1 into $I_1 = \partial V_1 \cap \mathbb{H}$ and $L_1 = [-\rho, \rho] = \partial V_1 \setminus I_1$. We get a corresponding decomposition of $\partial V_2 = \partial(\psi(V_1)) = \psi(\partial V_1)$ into $I_2 = \psi(I_1)$ and $L_2 = \psi(L_1)$. Note that L_2 is an arc of $\Gamma \cap B_1$, while I_2 is an (analytic) arc in $\mathcal{O} \cap B_1$.

Set $u_2 = f$ on I_2, and $u_1 = f \circ \psi$ on I_1. Obviously the triple (V_1, I_1, u_1) satisfies the assumptions at the beginning of Section 15, and so does the triple (V_2, I_2, u_2). The restriction of f to V_2 is the function of $\mathcal{F}(V_2, I_2, u_2)$ that minimizes $\int_{V_2} |\nabla f|^2$: it has the right boundary values on I_2 by definition of u_2, and we cannot find a better competitor, because such a competitor would also give a better competitor than f in $\mathcal{F}(V, I, u)$. (This uses the welding Corollary 14.28.)

By Lemma 11, $h = f \circ \psi$ is the function of $\mathcal{F}(V_1, I_1, u_1)$ that minimizes $\int_{V_1} |\nabla h|^2$. But we know from Example 5 what it looks like: h is obtained from u_1 by symmetrization, Poisson extension to $B(0,\rho)$, and then restriction to V_1. In particular, $f \circ \psi$ has a C^1 extension to $D \cup (-\rho, \rho)$, and the normal derivative $\frac{\partial}{\partial n}(f \circ \psi)$ vanishes on $(-\rho, \rho)$. Hence f has a C^1 extension to $\psi(D \cup (-\rho, \rho))$, which we know contains a neighborhood of x in $\overline{\mathcal{O}}$. The Neumann condition also is preserved by composition, because ψ_2 preserves the normal direction (it is conformal in V_1 and C^1 up to the boundary).

This completes our proof of Proposition 15. Note that similar results hold in higher dimensions as well, with more standard proofs using P.D.E.'s. (But I did not check the amount of regularity for ∂V required for these arguments.) $\square$

Remark 17. We may need to know later what happens near points of $\partial V \setminus I$ where ∂V is only piecewise $C^{1+\alpha}$ for some $\alpha > 0$.

Let V, I, $\mathcal{F}$ be as in Section 15, and let $x_0 \in \partial V \setminus I$ and $r > 0$ be such that $B = B(x_0, r)$ does not meet I and $\partial V \cap B = \Gamma_1 \cup \Gamma_2$, where Γ_1 and Γ_2 are two $C^{1+\alpha}$ curves that go from x_0 to ∂B and only meet at x_0. When we say $C^{1+\alpha}$ curves, we mean all the way up to x_0, that is, that there is a choice of unit tangent $v_i(x)$ to Γ_i at x such that $|v_i(x) - v_i(y)| \leq C|x - y|^\alpha$.

Now let $\mathcal{O}$ be a component of $V \cap B$ (the only one if V is only on one side of $\Gamma_1 \cup \Gamma_2$), and call $\theta \in [0, 2\pi]$ the angle of Γ_1 and Γ_2 at x_0 inside $\mathcal{O}$. Thus we take $\theta < \pi$ if $\mathcal{O}$ is approximatively convex near x_0, and $\theta > \pi$ if $\mathcal{O}$ contains a big angular sector.

If $\theta = \pi$, we are in the situation of Proposition 15 (because of our definition of $C^{1+\alpha}$ curves). If $0 < \theta < \pi$, we claim that f has a C^1 extension to $\overline{\mathcal{O}} \cap B$, with $\frac{\partial f}{\partial n} = 0$ on $\Gamma_1 \cup \Gamma_2$, and even $Df(x_0) = 0$. If $\theta > \pi$, we claim that f has a C^1 extension to $[\overline{\mathcal{O}} \cap B] \setminus \{x_0\}$, with $\frac{\partial f}{\partial n} = 0$ on $\Gamma_1 \cup \Gamma_2 \setminus \{x_0\}$, but also that this extension is Hölder-continuous near x_0 and that $|Df(x)| \leq C|x - x_0|^{\frac{\pi}{\theta} - 1}$ near x_0.

This is not hard to check. Of course, the only difference with Proposition 15 is about what happens near x_0. Define $h : \mathcal{O} \to \mathbb{C}$ by $h(x) = (x - x_0)^{\frac{\pi}{\theta}}$, where we use the simple connectedness of $\mathcal{O}$ to define h, and any continuous determination of the power $\frac{\pi}{\theta}$ will do. Since Γ_1 and Γ_2 are $C^{1+\alpha}$, $\widetilde{V} = h(\mathcal{O})$ is piecewise $C^{1+\alpha}$ far from 0, and it is also $C^{1+\beta}$ near x_0 (with $\beta = \text{Min}(\alpha, \frac{\theta}{\pi}\alpha)$), because the two half-tangents of $h(\Gamma_1)$ and $h(\Gamma_2)$ at 0 go in opposite directions. Now set $\widetilde{I} = h(\partial \mathcal{O} \setminus (\overline{\Gamma}_1 \cup \overline{\Gamma}_2))$ and define $\widetilde{f} = f \circ h^{-1}$ on $\widetilde{V}$. It is easy to see that $\widetilde{V}$, $\widetilde{I}$, and $\widetilde{f}$ satisfy the conditions of Proposition 15. In particular, $\widetilde{f}$ has a C^1 extension to a neighborhood of 0 in $\overline{\mathcal{O}}$, and $D\widetilde{f}(x)$ is bounded there. If $\theta < \pi$, our claim follows because $f = \widetilde{f} \circ h$ and h has a vanishing derivative at x_0. If $\theta > \pi$, we get that $Df(x) \leq CDh(x) \leq C|x - x_0|^{\frac{\pi}{\theta} - 1}$, as promised.

Remark 18. We could have tried to get the minimizing function f in Proposition 15.6 by solving $\Delta f = 0$ on V with Dirichlet conditions $f = u$ on I and Neumann conditions $\frac{\partial f}{\partial n} = 0$ on $\partial V \setminus I$. When V is smooth enough, we know a posteriori that this is equivalent, because the minimizing function exists and satisfies the Dirichlet–Neumann conditions above, and in addition the solution to the Dirichlet–Neumann problem is unique. This last can be shown by computing $E(f) - E(g)$ when f is our minimizer and g is a solution of the Dirichlet–Neumann problem: if everything is regular enough, an integration by part shows that $E(f) = E(g)$, hence $g = f$ on components of V that touch I (see Exercise 20). A more logical way to get uniqueness for Dirichlet–Neumann solutions would be to prove the maximum principle in this context.

Although it is good to keep in mind that the two approaches to Proposition 15.6 exist, it seems to the author that the convexity approach taken in Section 15 is substantially simpler.

Exercise 19. Let J be any nontrivial arc of the unit circle, and set $u = \mathbf{1}_J$ on $\partial B(0, 1)$.

1. Call f the Poisson integral of u. Show that $\int_{B(0,1)} |\nabla f|^2 = +\infty$.

2. Deduce from this that the corresponding set $\mathcal{F}$ is empty, i.e., that there is no function $g \in W^{1,2}(B(0,1))$ whose radial limits are equal to u almost-everywhere on $\partial B(0,1)$.

3. Prove this last fact directly. Hint: let z denote an extremity of J, suppose g works, and show that g is absolutely continuous on almost every arc $\partial = \partial B(z, r) \cap B(0,1)$, with r small, and that $\int_{\partial} |\nabla g| \geq 1$.

Exercise 20. Let V, I, and u be as in Section 15, and assume that V, I, and u are smooth. Let f, g be C^1 on $\overline{V}$, harmonic on V, and satisfy the Dirichlet condition $f = g = u$ on I and the Neumann condition $\frac{\partial f}{\partial n} = \frac{\partial g}{\partial n} = 0$ on $\partial V \setminus \overline{I}$. Show that $\int_V |\nabla f|^2 = \int_V |\nabla g|^2$. Hint: write $|\nabla f|^2 - |\nabla g|^2 = \nabla(f - g) \cdot \nabla(f + g)$, and apply Green to $(f - g)\nabla(f + g)$. Also see similar integrations by parts in Section 21, and in particular (21.4). If we already know that the minimizer from Proposition 15.6 is smooth (for instance, by Proposition 15), we can apply this to show that any other solution to the Dirichlet–Neumann problem ($g = u$ on I and $\frac{\partial g}{\partial n} = 0$ on $\partial V \setminus \overline{I}$) is a minimizer too; this gives the essential uniqueness for the Dirichlet–Neumann problem.

C. Regularity Properties for Quasiminimizers

After our long series of generalities on Sobolev spaces and functions that minimize energy, we are ready to prove regularity results for K when (u, K) is a reduced minimizer of the Mumford-Shah functional, or an almost-minimizer, or even a quasiminimizer (with small enough constant a) locally in a domain Ω.

Recall that the regularity of u is not a serious issue in general, at least if we mean relatively low regularity away from K and $\partial \Omega$. For minimizers of the Mumford-Shah functional, u satisfies the elliptic equation $\Delta u = u - g$, and it is locally of class C^1 (see Proposition 3.16 and Corollary 3.19). Even if we only know that (u, K) is a (reduced) quasiminimizer, we can still show that u is locally Hölder-continuous on $\Omega \setminus K$, with exponent $1/2$. See Lemma 18.22 below.

In this part we shall concentrate on the dimension $n = 2$, even though most results generalize to higher dimensions. We shall try to give a rapid description of analogous results in higher dimensions later, in Sections 72–74. Also, we want to restrict to local properties (i.e., inside the domain) for the moment; boundary regularity will be treated in Part I.

The most important results of this part are the local Ahlfors-regularity of K (Theorem 18.16), the concentration property (Theorem 25.1, very important for the semicontinuity of the Hausdorff measure, see Sections 35–38), and the uniform rectifiability (Theorem 27.1, which says that K is contained in an Ahlfors-regular curve). This is essentially the best that we can prove for general quasiminimizers; there will be more regularity results in Part E, where we shall find big C^1 curves in K, but they will only apply to almost-minimizers with reasonably small gauge functions.

18   Local Ahlfors-regularity: the statements

Most of the statements in this section and the next ones hold in all dimensions n, but for the moment we shall only prove them in dimension 2, because the proofs are easier. Our assumptions will often be the same as in this section, i.e., that (u, K) is a reduced (or coral) local quasiminimizer, or a topologically reduced (or coral) topological local quasiminimizer in some domain $\Omega \subset \mathbb{R}^2$, with small enough constant a. See Standard Assumption 14 and Theorem 16 below.

Let us first give a simple statement with local almost-minimizers. Then we shall comment on the result and the notion of Ahlfors-regularity. The main statement will come later; see Theorem 16.

Theorem 1 [DMS], [CaLe2]. *There are constants $C = C(n)$ and $\varepsilon = \varepsilon(n) > 0$ such that if $\Omega \subset \mathbb{R}^n$ is open and (u, K) is a reduced local almost-minimizer in Ω, with gauge function h, then*

$$C^{-1} r^{n-1} \leq H^{n-1}(K \cap B(x, r)) \leq C r^{n-1} \tag{2}$$

for all $x \in K$ and $r > 0$ such that

$$B(x, r) \subset \Omega \ \text{ and } \ h(r) \leq \varepsilon. \tag{3}$$

Lots of comments are in order.

See Definitions 7.14 and 8.1 for local almost-minimizers (in Ω) and reduced pairs.

We do not need to ask explicitly that $h(\rho) \leq \varepsilon$ for $\rho \leq r$, because we required h to be nondecreasing (in (7.5)). Incidentally, we do not need our assumption that $\lim_{r \to 0} h(r) = 0$; see Theorem 16 below.

We could replace our hypothesis that (u, K) is reduced by the similar one that K be coral (see Definition 8.24), but at least something like this is needed. Otherwise we could add to K a thin set of H^{n-1}-measure zero, and the first part of (2) would become false for points x in the thin set.

Theorem 1 was proved by Dal Maso, Morel and Solimini [DMS] when $n = 2$, and by Carriero and Leaci [CaLe2] in higher dimensions (with a quite different proof, with a compactness argument). To be fair, the original statements are not exactly as in Theorem 1, but the proofs apply (even for $n > 2$). We shall only prove Theorem 1 in detail when $n = 2$, using a minor variant of the argument in [DMS]. For $n > 2$, we shall present the great lines of the proof of [CaLe1], as modified by A. Siaudeau [Si] to work with topological quasiminimizers. See Section 72. Also see [So2] for a different approach that works in all dimensions.

Remark 4. Theorem 1 applies to (reduced) minimizers of the Mumford-Shah functional (because they are almost-minimizers, by Proposition 7.8).

Remark 5. Theorem 1 also holds for topologically reduced (or coral) topological local almost-minimizers in Ω. (See Definitions 7.17, 8.7, and 8.24.) This seems a little harder to prove a priori, because we have to check the topological condition (7.16) each time we want to use a competitor for (u, K). However, the proof is not very different. Also, Theorem 1 and its topological variant extend to reduced (or coral) local quasiminimizers (with small enough constant a); see Theorem 16 below. Finally, Theorem 1 accepts linear distortion, as in Section 7.c; see Standard Assumption 14 and Remark 20.48.

Let us now discuss the condition (2). The upper bound in (2) is quite natural, and very easy to prove. First assume that $\overline{B}(x,r) \subset \Omega$, and consider the following pair $(\widetilde{u}, \widetilde{K})$. Set

$$\widetilde{K} = [K \cup \partial B(x,r)] \setminus B(x,r) \tag{6}$$

and

$$\begin{cases} \widetilde{u}(y) = u(y) \text{ for } y \in \Omega \setminus [K \cup \overline{B}(x,r)] \\ \widetilde{u}(y) = 0 \quad \text{ on } B(x,r). \end{cases} \tag{7}$$

Obviously $\widetilde{u} \in W^{1,2}_{loc}(\Omega \setminus \widetilde{K})$, because we can check this on each connected component of $\Omega \setminus \widetilde{K}$ independently, and $B(x,r)$ is not in the same component as the rest of $\Omega \setminus \widetilde{K}$. Thus $(\widetilde{u}, \widetilde{K}) \in \mathcal{A}$ (see (7.1)), and it is even a topological competitor for (u,K) in $B(x,r)$ (see Definitions 7.2 and 7.15, and Lemma 20.10 below if you have doubts about topology). Now (7.7) applies, and

$$\int_{B(x,r)\setminus K} |\nabla u|^2 + H^{n-1}(K \cap \overline{B}(x,r)) \leq w_{n-1} r^{n-1} + h(r) r^{n-1}, \tag{8}$$

where we set $w_{n-1} = H^{n-1}(\partial B(0,1))$, the surface measure of the unit sphere.

This proves the second inequality in (2), at least when $\overline{B}(x,r) \subset \Omega$. If we only have that $B(x,r) \subset \Omega$, we can apply (8) to all radii $r' < r$, and the second half of (2) follows in this case also. Notice that (8) only requires that $\overline{B}(x,r) \subset \Omega$, not that $x \in K$. It also works for topological almost-minimizers, as we have seen in the proof. Finally, if (u,K) is a (not just local) almost-minimizer, (8) still holds for all balls $B(x,r)$, not necessarily contained in Ω, with essentially the same proof.

The interesting part in Theorem 1 is therefore the lower bound. It says that K never gets too thin near a point $x \in K$, or else we may as well remove a piece of K near x and get a better competitor. In terms of applications of the Mumford-Shah functional to image segmentation, this is rather good news, because it means that the functional will not choose to represent little specs that are likely to come from noise, for instance.

Let us try to say rapidly why K is never too thin near a point x. We can compare (u,K) with a competitor $(\widetilde{u}, \widetilde{K})$ in some small ball $B(x,r)$, where $\widetilde{K} = K \setminus B(x,r)$ and $\widetilde{u}$ is the harmonic function on $B(x,r)$ that has the same boundary values on $\partial B(x,r)$ as u. Now $\int_{B(x,r)\setminus K} |\nabla u|^2$ may be smaller than $\int_{B(x,r)} |\nabla \widetilde{u}|^2$, because the boundary $K \cap B(x,r)$ allows u to have jumps that release some tension. The question is whether the gain in the energy term $\Delta E = \int_{B(x,r)} |\nabla \widetilde{u}|^2 - \int_{B(x,r)\setminus K} |\nabla u|^2$ that you get by keeping $K \cap B(x,r)$ is sufficient to compensate the additional term $H^{n-1}(K \cap B(x,r))$ in the surface term of the functional. It turns out that the answer is no, and even that ΔE is negligible compared to $H^{n-1}(K \cap B(x,r))$ when the latter is very small. A reasonable interpretation for this is that when $K \cap B(x,r)$ is very small, it is not sufficient to allow large discontinuities of u that would release enough tension. Our proof of

Theorem 1 in dimension 2 will make this manifest. As we shall see, there is an integration by parts that allows us to estimate ΔE in terms of some integral of the jump of u across K, and one of the key points will be that the jump is less than $CH^1(K \cap B(x,r))^{1/2}$.

It is difficult to resist the temptation of saying a few words about Ahlfors-regularity. In [DMS], (2) is called the elimination property, and in [So2] the uniform density property, but we shall refer to it as local Ahlfors-regularity. The name comes from [Ah], where L. Ahlfors studied curves with the property (2) (in a very different context of covering surfaces) and called them regular. Here is the current definition.

Definition 9. *Let $E \subset \mathbb{R}^n$ be a closed set, not reduced to a single point, and let $0 \leq d \leq n$ be given. We say that E is Ahlfors-regular of dimension d (or, in short, regular) if there is a constant $C \geq 0$ such that*

$$C^{-1}r^d \leq H^d(E \cap B(x,r)) \leq Cr^d \tag{10}$$

for all $x \in E$ and $0 < r < \text{diameter}(E)$.

Note that our definition allows both for bounded and unbounded regular sets. Here we restricted to sets that are contained in $\mathbb{R}^n$ for obvious reasons, but E may even be an (abstract) metric space. (In that case, though, it is a little more prudent to require (10) for $x \in E$ and $0 < r < 2$ diameter(E), to avoid the case of a set of diameter 1 composed of infinitely many copies of the same object, at mutual distances exactly equal to 1.)

In spite of what the name suggests, Ahlfors-regularity is not a notion of smoothness, even when d is an integer. For instance there are Cantor sets in the plane that are regular of dimension 1 (see Exercise 26). Instead we should view (10) as a strongly uniform way of demanding that E be d-dimensional. As we shall see in later sections, it is a very useful property to have, because it allows us to use lots of the usual techniques of analysis in $\mathbb{R}^d$. See for example Section 23 on Carleson measures.

It is not too surprising that with its simple definition Ahlfors-regularity showed up in a few different contexts before [DMS]. About fifty years after [Ah], it started to be used more and more in connection with problems of L^2-boundedness of some singular integral operators, like the Cauchy integral operator on curves and subsets of the complex plane. See for instance [Da4] for a description of such problems. The connection with geometric measure theory (and in particular rectifiability) also became stronger, and regularity can be seen as a uniform version of a standard hypothesis in this field, that the upper and lower densities at points of E be finite and positive. See for instance [DaSe3] concerning uniform rectifiability, and [DaSe7] for other connected topics. We shall actually meet uniform rectifiability later in this book.

Even in connection with minimizers (or at least, with minimal surfaces), the notion already appears in [Al]. See also [DaSe8,9] concerning the same "restricted sets" of F. Almgren, where the reference to Ahlfors-regularity may be more explicit. Of course there are certainly lots of other interesting occurrences of the notion.

We don't always need to estimate Hausdorff measures directly to show that a given set is Ahlfors-regular.

Lemma 11. *Let $E \subset \mathbb{R}^n$ be a closed set, not reduced to a point, and let μ be a (positive Borel) measure supported on E. If there is a constant $C \geq 1$ such that*

$$C^{-1} r^d \leq \mu(E \cap B(x,r)) \leq C r^d \text{ for } x \in E \text{ and } 0 < r < \text{diameter}(E), \quad (12)$$

then E is Ahlfors-regular of dimension d, and there is a constant C' such that

$$\frac{1}{C'} \mu(A) \leq H^d(E \cap A) \leq C' \mu(A) \text{ for all Borel sets } A \subset \mathbb{R}^n. \quad (13)$$

In other words, μ is equivalent to the restriction of H^d to E. Notice that there is no reasonable ambiguity about the notion of support, because E is closed and (12) holds. We wrote $\mu(E \cap B(x,r))$ in (12) instead of $\mu(B(x,r))$ just for additional stress, but the two are equal.

We shall not need this lemma, so let us leave its proof to the reader (see Exercise 25 below) and return to quasiminimizers.

Standard Assumption 14. In the next result, as well as in the next few sections, an open set $\Omega \subset \mathbb{R}$ will be given, often with $n = 2$, and we shall study a pair $(u, K) \in \mathcal{A}$ (see (2.1) or (7.1)). We shall assume that (u, K) is a reduced (or coral) local (r_0, M, a)-quasiminimizer in Ω, as in Definitions 7.21, 8.1, and 8.24, or that (u, K) is a topologically reduced (or coral) topological local (r_0, M, a)-quasiminimizer in Ω, as in Definitions 7.21 and 8.7. To save some space we shall just write $(u, K) \in RLQ(\Omega)$ in the first case, $(u, K) \in TTRLQ(\Omega)$ in the second case, and $(u, K) \in TRLQ(\Omega)$ when we allow both possibilities (i.e., most of the time). Most statements will require the constant a to be small enough, depending on n and possibly on M as well.

Recall from Remark 7.22 that if (u, K) is an almost-minimizer with gauge function h, then for every $a > 0$ it is also a $(r_0, 1, a)$-quasiminimizer, as long as we take r_0 so small that

$$h(r_0) \leq a. \quad (15)$$

The same thing is true with reduced, coral, local, topological almost-minimizers (or any combination). Thus our statements will also apply to almost-minimizers, and hence minimizers of the Mumford-Shah functional. They will also apply to minimizers of slightly more general functionals, as in Proposition 7.27 and Remark 7.42.

Also see Remarks 20.48, 25.28, and 27.6 for the extension of the results of this part to quasiminimizers with linear distortion.

Here is the quasiminimizer version of Theorem 1. Note that it contains Theorem 1.

Theorem 16. *For each choice of $n \geq 2$ and $M \geq 1$ we can find constants $a > 0$ and $C \geq 1$ such that if $\Omega \subset \mathbb{R}^n$ is open and $(u, K) \in TRLQ(\Omega)$ (as in Standard Assumption 14), with constants (r_0, M, a), then*

$$C^{-1} r^{n-1} \leq H^{n-1}(K \cap B(x, r)) \leq C r^{n-1} \qquad (17)$$

for all $x \in K$ and $r > 0$ such that

$$B(x, r) \subset \Omega \quad and \quad r \leq r_0. \qquad (18)$$

Correct attribution for Theorem 16 is a little complicated. If we are only concerned about local almost-minimizers, Theorem 16 reduces to Theorem 1, which was proved by Dal Maso, Morel and Solimini [DMS] when $n = 2$, and by Carriero and Leaci [CaLe2] in higher dimensions. Here we shall only prove Theorem 16 in full details in dimension 2. The case of higher dimensions will be treated briefly in Section 72, but for the full proof we refer the reader to [Si]. As far as the author knows, there is no prior appearance of this result (with our definition of quasiminimizers, and the topological condition). On the other hand, the proof presented below (of Theorem 16 in dimension 2) looks a lot like the original proof of [DMS], and similarly the proof of [Si] is strongly inspired by [CaLe1] (with a few translations to avoid talking about SBV, and a small additional argument to take care of the topological condition).

There is another proof of Theorem 1 in higher dimensions, by S. Solimini [So2]; it also uses direct comparisons with excised competitors, like the one in [DMS] (which we present here). Probably the proof would also give Theorem 16 with minor modifications, but I did not check.

The proof of Theorem 16 when $n = 2$ will keep us busy for the next four sections, but let us already verify the easy inequality (in slightly more generality, with balls that are not necessarily centered on K).

Lemma 19. *Let $(u, K) \in TRLQ(\Omega)$ (as in Standard Assumption 14) for some open set $\Omega \subset \mathbb{R}^n$. Let $B(x, r)$ be any open ball contained in Ω, and such that $r \leq r_0$. Then*

$$H^{n-1}(K \cap B(x, r)) + \int_{B(x,r) \setminus K} |\nabla u|^2 \leq (M w_{n-1} + a) r^{n-1}. \qquad (20)$$

Here again $w_{n-1} =: H^{n-1}(\partial B(0, 1))$. We shall often refer to (20) as the trivial estimate (we shall see why soon). The assumption that $B(x, r) \subset \Omega$ would not be needed if (u, K) was a quasiminimizer (topological or not, but not just local). Note also that we dropped our assumption that $x \in K$ (and we could have dropped the assumption that (u, K) is reduced or coral), which is only needed for the first half of (17).

It is enough to prove (20) when $\overline{B}(x,r) \subset \Omega$, because otherwise we could always prove (20) for $r' < r$ and go to the limit. Define a pair $(\widetilde{u}, \widetilde{K})$ as in (6) and (7). That is, take $\widetilde{K} = [K \cup \partial B(x,r)] \setminus B(x,r)$, set $\widetilde{u} = 0$ in $B(x,r)$, but do not change it outside of $\overline{B}(x,r)$. Then $(\widetilde{u}, \widetilde{K})$ is a topological competitor for (u,K) in $B(x,r)$, as before. Therefore we can apply the quasiminimality condition (7.20).

In the present situation (7.18) and (7.19) yield $E = \int_{B(x,r)\setminus K} |\nabla u|^2$, $\widetilde{E} = 0$, $\delta E = \widetilde{E} - E = -E$, and (7.20) says that

$$H^{n-1}(K \cap B(x,r)) + \int_{B(x,r)\setminus K} |\nabla u|^2 \leq M\, H^{n-1}(\widetilde{K} \setminus K) + a\,r^{n-1}$$
$$\leq M\, w_{n-1}\, r^{n-1} + a\, r^{n-1}, \tag{21}$$

as needed for (20). This proves the lemma, and of course the upper estimate in Theorem 16 follows. $\square$

An easy consequence of Lemma 19 is the local Hölder regularity of u on $\Omega \setminus K$ when $(u, K) \in TRLQ(\Omega)$, as follows.

Lemma 22. *Let $(u, K) \in TRLQ(\Omega)$ (as in Standard Assumption 14) for some open set $\Omega \subset \mathbb{R}^n$, with constants (r_0, M, a). Let B be any open ball contained in $\Omega \setminus K$. Then u is continuous on B, and even*

$$|u(x) - u(y)| \leq C|x - y|^{1/2} \quad \text{whenever } x, y \in B \text{ are such that } |x - y| \leq r_0, \tag{23}$$

for some constant C that depends only on n, M, and a.

Fix $(u, K) \in TRLQ(\Omega)$, and let B_1, B_2 be two balls such that $B_1 \subset B_2 \subset \Omega \setminus K$. Call r_i the radius of B_i, and first suppose that $r_2 \leq 10 r_1$ and $r_2 \leq r_0$. Also call m_i the mean value of u on B_i. Then

$$|m_1 - m_2| = \left| \frac{1}{|B_1|} \int_{B_1} [u(x) - m_2] dx \right| \leq 10^n \frac{1}{|B_2|} \int_{B_2} |u(x) - m_2| dx$$
$$\leq C r_2 \frac{1}{|B_2|} \int_{B_2} |\nabla u| \leq C r_2 \left\{ \frac{1}{|B_2|} \int_{B_2} |\nabla u|^2 \right\}^{1/2} \leq C r_2^{1/2} \tag{24}$$

by Poincaré (Proposition 12.11), Hölder, and (20).

If B_1, B_2 are still such that $B_1 \subset B_2 \subset \Omega \setminus K$ and $r_2 \leq r_0$ (but not necessarily $r_2 \leq 10 r_1$), we still have that $|m_1 - m_2| \leq C r_2^{1/2}$, because we can apply (24) to a series of intermediate balls with geometrically decreasing radii.

Because of this, if x is any point of B, the mean value of u on $B(x,r)$ has a limit $l(x)$ when r tends to 0. This limit coincides with $u(x)$ at every Lebesgue point of u (hence, almost-everywhere), and modulo redefining u on a set of measure zero (which is implicitly allowed by definitions), we can assume that $u(x) = l(x)$ on B.

To prove (23), consider x, $y \in B$, with $|x - y| \leq r_0$. First assume that we can find a ball $B_2 \subset B$ that contains x and y and whose radius is less than $|x - y|$, and take very small balls B_1 centered on x or y. We still have that $|m_1 - m_2| \leq Cr_2^{1/2} \leq C|x-y|^{1/2}$ for each choice of B_1, and (23) follows by taking limits. Even if we cannot find B_2 as above (because x and y are too close to ∂B), we can find additional points $x_1, y_1 \in B$, not to far, such that the argument above applies to the pairs (x, x_1), (x_1, y_1), and (y_1, y). This proves (23) in the general case. $\quad\square$

Exercise 25. (A proof of Lemma 11.) Let E and μ be as in the lemma.

1. Prove that $\mu(A) \leq C' H^d(E \cap A)$ for all Borel sets $A \subset \mathbb{R}^n$, just by definitions.
2. Prove that $H^d(E \cap B(x,r)) \leq C'' r^d$ for $x \in \mathbb{R}^n$ and $r > 0$. (Cover $E \cap B(x,r)$ by balls of radius ε for each small $\varepsilon > 0$. You will need less than $C\varepsilon^{-d} r^d$ of them, if you take their centers at mutual distances $\geq \frac{\varepsilon}{2}$.)
3. Let K be a compact subset of E. Check that for each $\delta > 0$ we can find $\varepsilon > 0$ such that $\mu(K^\varepsilon) \leq \mu(K) + \delta$, where K^ε is an ε-neighborhood of K.
4. Show that $H^d(K) \leq C''' \mu(K)$ for all compact subsets of E. (Cover K by balls of radius ε with centers on K and at mutual distances $\geq \frac{\varepsilon}{2}$.)
5. Complete the proof of Lemma 11. (Use the regularity of $H^d_{|E}$.)

Exercise 26. (A Cantor set of dimension 1.) Let $E_0 = [0,1]^2$ denote the unit square in $\mathbb{R}^2$. We want to construct a Cantor set $E \subset E_0$. We shall take $E = \bigcap_m E_m$, where $\{E_m\}$ is a decreasing sequence of (compact) sets. Each E_m is composed of 4^m squares of sidelength 4^{-m} (with sides parallel to the axes), and is obtained from E_{m-1} (when $m \geq 1$) by replacing each of the 4^{m-1} squares that compose it with 4 squares of size 4^{-m} contained in it and situated at its corners. This gives the set E suggested in Figure 1.

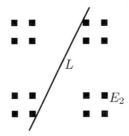

Figure 1

1. Show that E is Ahlfors-regular with dimension 1. It may be useful to construct a measure μ on E (either as weak limit of uniform measures on the sets E_n, or directly as the image of a measure on the set $\{0,1,2,3\}^{\mathbb{N}}$) and use Lemma 11. You may also try to estimate Hausdorff measures directly, but you may end up redoing parts of Exercise 25.
2. Try to show that there is no rectifiable curve Γ such that $E \subset \Gamma$. (A rectifiable curve is the support of a curve with finite length or, if you prefer, the image of a finite interval by a Lipschitz mapping.) Much more is true: $H^1(E \cap \Gamma) = 0$

for every rectifiable curve Γ, but this is hard to prove without knowledge of densities and the existence of tangents almost-everywhere on Γ.

3. Let L be a line of slope 2, and denote by π the orthogonal projection onto L. Show that there is a countable set of corners $D \subset E$ such that the restriction of π to $E \setminus D$ is a bijection onto an interval of length $\frac{\sqrt{5}}{2}$. This gives a complicated way to construct the measure μ in Question 1. Directions like L are exceptional: one can show that for almost-every $\theta \in [0, \pi]$, the measure $H^1(\pi_\theta(E))$ of the orthogonal projection of E onto a line of direction θ is equal to 0. (Do not try too hard to prove this directly.)

4. Show that E is removable for functions in $W^{1,p}$, i.e., that if $1 \leq p < +\infty$, Ω is an open set that contains E, and $f \in W^{1,p}(\Omega \setminus E)$, then $f \in W^{1,p}(\Omega)$. Use Exercise 10.22.

The Cantor set E is quite nice in many respects. It is probably the most standard example of a totally unrectifiable Ahlfors-regular set of dimension 1 in the plane. It was also the first example of an Ahlfors-regular set of dimension 1 with vanishing analytic capacity (Garnett and Ivanov), or where integration against the Cauchy kernel does not define a bounded operator on $L^2(E, dH^1)$.

19 A small density result

The following rudimentary variant of the Lebesgue differentiation theorem will be useful in the proof of Theorem 18.16. It is of course very standard.

Lemma 1. *Let $0 < d < n$ and a Borel set $K \subset \mathbb{R}^n$ be given. Suppose that $H^d(K) < +\infty$. Then*

$$\limsup_{r \to 0^+} \left\{ r^{-d} H^d(K \cap B(x, r)) \right\} \geq \frac{c_d}{2} \qquad (2)$$

for H^d-almost every $x \in K$.

Here c_d is the same positive constant as in (2.8), in the definition of the Hausdorff measure. Lemma 1 is just one among many density results that one can prove; see for instance [Mat2]. A somewhat singular feature of this one is that we can prove it without a covering lemma.

Let K be as in the lemma, and set

$$R_m = \left\{ x \in K \,;\, H^d(K \cap B(x, r)) \leq \frac{2c_d r^d}{3} \text{ for } 0 < r \leq 2^{-m} \right\} \qquad (3)$$

for $m \geq 1$. The exceptional set where (2) fails is contained in the union of the sets R_m, $m \geq 1$, and so it is enough to show that $H^d(R_m) = 0$ for all m.

So let $m \geq 1$ be given, and let $\eta > 0$ be small. By definition of $H^d(R_m)$, we can find a countable collection of sets E_i such that

$$R_m \subset \bigcup_i E_i, \tag{4}$$

$$\text{diam } E_i < 2^{-m} \text{ for all } i, \tag{5}$$

and

$$c_d \sum_i (\text{diam } E_i)^d \leq H^d(R_m) + \eta \tag{6}$$

(see Definition 2.6). Let us throw away the sets E_i that do not meet R_m. For the remaining ones, set $r_i = \text{diam } E_i$, and replace E_i with $B_i = \overline{B}(x_i, r_i)$, where we choose x_i in $E_i \cap R_m$. Note that $E_i \subset B_i$ by definition, and so

$$H^d(R_m) \leq \sum_i H^d(K \cap B_i) \leq \frac{2c_d}{3} \sum_i r_i^d \leq \frac{2}{3} H^d(R_m) + \frac{2\eta}{3}, \tag{7}$$

by (4), because $x_i \in R_m$ and $r_i < 2^{-m}$ (see (3) and (5)), and by (6).

Since (7) holds for every $\eta > 0$ and $H^d(R_m) \leq H^d(K) < +\infty$, we get that $H^d(R_m) = 0$. This proves the lemma. $\qquad \square$

20   Proof of local Ahlfors-regularity in dimension 2: the scheme

We are now ready to start the proof of the lower estimate in Theorem 18.16 when $n = 2$. Recall that the upper estimate was already proved in Lemma 18.19, and also that Theorem 18.1 is weaker than Theorem 18.16.

To prove the first inequality in (18.17), we shall proceed by contradiction. Let $\Omega \subset \mathbb{R}^2$ and $(u, K) \in TRLQ(\Omega)$ be given, as in Standard Assumption 18.14, and assume that we can find $y \in K$ and $t \leq r_0$ such that $B(y, t) \subset \Omega$ and

$$H^1(K \cap B(y, t)) \leq \alpha t, \tag{1}$$

where r_0 is as in the definition of quasiminimizers and $\alpha > 0$ is a very small constant that will be chosen near the end of the argument, depending on the quasiminimality constant M only. Our goal is to reach a contradiction.

We want to replace our pair (y, t) with a slightly better one.

Lemma 2.   *We can find* $x \in B\left(y, \frac{t}{2}\right)$ *and* $r \in \left(0, \frac{t}{2}\right)$ *such that*

$$H^1(K \cap B(x, r)) \leq 2\alpha r \tag{3}$$

and

$$H^1\left(K \cap B\left(x, \frac{r}{3}\right)\right) > \frac{2\alpha r}{3}. \tag{4}$$

The extra condition (4) will be useful in the argument, because we want to remove $K \cap B\left(x, \frac{r}{2}\right)$ from K, and we want to know that this reduces $H^1(K)$ substantially. Note that we still have that

$$0 < r < r_0 \text{ and } \overline{B}(x,r) \subset \Omega, \tag{5}$$

so replacing (y, t) with (x, r) will not cost us much.

To prove the lemma let us first notice that

$$u \in W^{1,2}(B(y,t) \setminus K). \tag{6}$$

Indeed $u \in W^{1,2}_{\text{loc}}(B(y,t) \setminus K)$ because $(u, K) \in \mathcal{A}$ (see (7.1)), and the trivial estimate (18.20) says that $\nabla u \in L^2(B(y,t) \setminus K)$. Next we want to check that

$$H^1\left(K \cap B\left(y, \frac{t}{2}\right)\right) > 0. \tag{7}$$

If (u, K) is coral, this is true simply because $y \in \text{core}\,(K)$, see (8.25). In the case of reduced or topologically reduced quasiminimizers, we could use Remark 8.30 that says that (u, K) is coral and conclude as before, but let us give a direct argument for the reader's convenience.

Set $B = B\left(y, \frac{t}{2}\right)$ for the next few lines. We know that $u \in W^{1,2}(B \setminus K)$, by (6), and if (7) fails Proposition 10.1 tells us that $u \in W^{1,2}(B)$. (Normally, we should say that u has an extension in $W^{1,2}(B)$, but here u is already defined almost-everywhere on B, so there is no real need to extend.) Set $\widetilde{K} = K \setminus B$. Then $\widetilde{K}$ is closed in Ω, and $u \in W^{1,2}_{\text{loc}}(\Omega \setminus \widetilde{K})$. In other words, $(u, \widetilde{K}) \in \mathcal{A}$. Also $(u, \widetilde{K})$ is a competitor for (u, K) in B, since we did not modify anything outside of B. Hence (u, K) is not a reduced pair (as in Definition 8.1), since $\widetilde{K}$ is strictly contained in K. (Recall that $y \in K \setminus \widetilde{K}$.) To see that (u, K) is not topologically reduced (as in Definition 8.7), and hence get the desired contradiction even in the case of topological local quasiminimizers, we still need to check that

$$(u, \widetilde{K}) \text{ is a topological competitor for } (u, K) \text{ in } B' \tag{8}$$

for some ball B' such that $\overline{B'} \subset \Omega$.

Set $\pi(x) = |x - y|$ for $x \in \mathbb{R}^2$; then π is 1-Lipschitz and hence

$$H^1\big(\pi(K \cap B(y,t))\big) \leq H^1(K \cap B(y,t)) \leq \alpha t, \tag{9}$$

by Remark 2.11 and (1). If $\alpha < \frac{1}{2}$ (which we can obviously assume), we can find $\rho \in \left(\frac{t}{2}, t\right)$ that does not lie in $\pi(K \cap B(y,t))$. This means that $\partial B(y, \rho)$ does not meet K. Now (8) will follow from the following general fact.

Lemma 10. *If $(u, K) \in \mathcal{A}$, B is a ball such that $\overline{B} \subset \Omega$, $(\widetilde{u}, \widetilde{K})$ is a competitor for (u, K) in B, and if in addition*

$$\partial B \setminus \widetilde{K} \text{ is contained in a single connected component of } \Omega \setminus K, \tag{11}$$

then $(\widetilde{u}, \widetilde{K})$ is a topological competitor for (u, K) in B.

To derive (8) from the lemma, just take $B = B(y, \rho)$ and note that (11) holds because $\partial B(y, \rho)$ is connected and contained in $\Omega \setminus K$.

To prove the lemma, let $y, z \in \Omega \setminus (K \cup \overline{B})$ be given, and assume that $\widetilde{K}$ does not separate y from z in Ω. We want to prove that K does not separate them in Ω either.

Let γ denote a path from y to z in $\Omega \setminus \widetilde{K}$. If γ does not meet $\overline{B}$, then $\gamma \subset \Omega \setminus K$ because K coincides with $\widetilde{K}$ out of $\overline{B}$, and we are happy. So let us assume that γ meets $\overline{B}$. Call y^* the first point of $\overline{B}$ that we meet when we go from y to z along γ. Note that γ does not meet K between y and y^* (excluded), because $K = \widetilde{K}$ out of $\overline{B}$. Also, $y^* \in \partial B \setminus \widetilde{K} \subset \Omega \setminus K$ (by (11)). Thus y^* lies in the same connected component of $\Omega \setminus K$ as y.

Similarly, call z^* the first point of $\overline{B}$ that we meet when we go from z to y along γ; then z^* lies in the same component of $\Omega \setminus K$ as z. Since $y^*, z^* \in \partial B \setminus \widetilde{K}$, (11) says that they also lie in the same component of $\Omega \setminus K$ as each other. Altogether, y and z lie in the same component of $\Omega \setminus K$, hence K does not separate them in Ω. Lemma 10 follows. $\qquad \square$

As we said earlier, (8) follows from Lemma 10. Thus (u, K) was not topologically reduced, and this contradiction completes our proof of (7). Note that this is the only place in the argument where we use our assumption that (u, K) is coral, reduced, or topologically reduced.

We may now return to Lemma 2 and our quest for (x, r). Lemma 19.1 (on upper densities) tells us that for H^1-almost every $x \in K \cap B\left(y, \frac{t}{2}\right)$,

$$\limsup_{\rho \to 0^+} \left\{ \rho^{-1} H^1(K \cap B(x, \rho)) \right\} \geq \frac{1}{2}. \tag{12}$$

Here we used the fact that when $d = 1$, the constant c_d that shows up in (19.2) (and the definition of the Hausdorff measure in (2.8)) is equal to 1. But the precise value does not matter much anyway.

We choose $x \in K \cap B\left(y, \frac{t}{2}\right)$ with the property (12). This is possible, by (7). Note that

$$\frac{2}{t} H^1\left(K \cap B\left(x, \frac{t}{2}\right)\right) \leq \frac{2}{t} H^1(K \cap B(y, t)) \leq 2\alpha, \tag{13}$$

because $x \in B\left(y, \frac{t}{2}\right)$ and by (1). Thus $r = \frac{t}{2}$ satisfies our first requirement (3), but of course it may not satisfy (4).

Set $\rho_n = 3^{-n} \frac{t}{2}$ for $n \geq 0$. We want to take $r = \rho_n$ for some n. Let us assume that this is not possible, and derive a contradiction with (12). Note that $r = \rho_0$ satisfies (3). Therefore it does not satisfy (4), and this means exactly that ρ_1 satisfies (3). But then ρ_1 does not satisfy (4), and ρ_2 satisfies (3). By induction, (3) holds for $r = \rho_n$ for every $n \geq 0$.

Now let $\rho > 0$ be small. Choose $n \geq 0$ so that $\rho_{n+1} < \rho \leq \rho_n$. Then

$$\rho^{-1} H^1(K \cap B(x, \rho)) \leq \rho^{-1} H^1(K \cap B(x, \rho_n))$$
$$\leq 2\alpha \rho^{-1} \rho_n \leq 2\alpha \rho_{n+1}^{-1} \rho_n = 6\alpha, \tag{14}$$

by (3) for $r = \rho_n$. Since this holds for all small $\rho > 0$, we get the desired contradiction with (12) (if $\alpha < \frac{1}{12}$). Thus we can choose $r = \rho_n$ for some n, and Lemma 2 follows. □

Next we want to choose a slightly better radius r_1, with the following properties. First,

$$\frac{r}{2} < r_1 < r \tag{15}$$

and

$$\partial B_1 \cap K = \emptyset, \tag{16}$$

where we set $B_1 = B(x, r_1)$. Also, we want to impose that

$$\begin{aligned} &u_{|\partial B_1} \in W^{1,2}(\partial B_1), \text{ its derivative on } \partial B_1 \\ &\text{is given by the restriction of } \nabla u \text{ to } \partial B_1, \end{aligned} \tag{17}$$

and

$$\int_{\partial B_1} |\nabla u|^2 \leq C_1. \tag{18}$$

The constant C_1 will be chosen soon. It will be allowed to depend on M, but not on the other constants (like r_0, a, and α). We shall try to keep this convention about constants C. Also, in future occurrences, we may just state (18) and consider (17) as implicit.

There is a last condition that we want to impose; it will be used later, to estimate a difference of energies. Set

$$J(r_1) = \int_{K \cap B_1} \text{dist}(z, \partial B_1)^{-1/2} \, dH^1(z). \tag{19}$$

We also demand that

$$J(r_1) \leq 20\alpha r^{1/2}. \tag{20}$$

Lemma 21. *We can choose $r_1 \in \left(\frac{r}{2}, r\right)$ with all the properties (16)–(20).*

To prove the lemma, set $I = \left(\frac{r}{2}, r\right)$, and first consider $I_1 = \{r_1 \in I; (16) \text{ fails}\}$. Then I_1 is contained in the image of $K \cap B(x, r)$ under the 1-Lipschitz mapping $\pi : z \to |x - z|$, and so

$$|I_1| = H^1(I_1) \leq H^1(K \cap B(x, r)) \leq 2\alpha r < \frac{r}{100} \tag{22}$$

if α is small enough, by Remark 2.11 and (3).

Next set $I_2 = \{r_1 \in I \setminus I_1; (18) \text{ fails}\}$ (here, we mean (18) alone). Then

$$|I_2| \leq C_1^{-1} \int_{I_2} \int_{\partial B(x, r_1)} |\nabla u|^2 \, dH^1 \, dr_1 \leq C_1^{-1} \int_{B(x, r) \setminus K} |\nabla u|^2 \leq C C_1^{-1} r, \tag{23}$$

by the trivial estimate (18.20). Here the constant C comes from (18.20); we can take $C = 2\pi M + a \le 2\pi M + 1$, for instance. Now choose $C_1 = 100C$; then $|I_2| \le \frac{r}{100}$.

Now let us check that (17) holds for almost-every $r_1 \in I \setminus (I_1 \cup I_2)$. First note that $I \setminus I_1$ is an open set and that for $r_1 \in I \setminus I_1$, K does not meet $\partial B(x, r_1)$. A polar change of coordinates (authorized by Proposition 11.4) tells us that the function v defined on $(I \setminus I_1) \times \mathbb{R}$ by $v(r_1, \theta) = u(r_1 e^{i\theta})$ lies in $W^{1,2}_{\text{loc}}((I \setminus I_1) \times \mathbb{R})$. Then Corollary 9.13 says that for almost every choice of $r_1 \in I \setminus I_1$, $v(r_1, \cdot)$ is absolutely continuous on $\mathbb{R}$, with a derivative that can be computed from the restriction of ∇u to $\partial B(x, r_1)$. If in addition $r_1 \notin I_2$, i.e., if (18) holds, then $u_{|\partial B(x,r_1)} \in W^{1,2}(\partial B(x, r_1))$ and (17) holds, as needed.

Finally set $I_3 = \{r_1 \in I; (20) \text{ fails}\}$. Then

$$|I_3| \le \{20\alpha r^{1/2}\}^{-1} \int_I J(t)\, dt \qquad (24)$$

by Chebyshev, and

$$
\begin{aligned}
\int_I J(t)\, dt &\le \int_I \int_{K \cap B(x,r)} \operatorname{dist}(z, \partial B(x,t))^{-1/2}\, dH^1(z)\, dt \\
&= \int_{K \cap B(x,r)} \left\{ \int_{\frac{r}{2}}^r |t - |z - x||^{-1/2}\, dt \right\} dH^1(z) \\
&\le \int_{K \cap B(x,r)} \left\{ 2 \int_0^r v^{-1/2}\, dv \right\} dH^1(z) \\
&\le 4 r^{1/2}\, H^1(K \cap B(x,r)) \le 8\alpha r^{3/2}
\end{aligned}
\qquad (25)
$$

by (19), Fubini, and (3). Thus $|I_3| \le \frac{8r}{20} = \frac{8|I|}{10}$. Lemma 21 follows from this and our previous estimates on the size of I_1 and I_2. $\qquad\square$

Choose r_1 and $B_1 = B(x, r_1)$ as in Lemma 21. Our next goal is to define a topological competitor (u_1, K_1) for (u, K) in B_1. Set

$$K_1 = K \setminus B_1, \qquad (26)$$

and denote by

$$v = u_{|\partial B_1} \qquad (27)$$

the restriction of u to ∂B_1. By (17) and Proposition 9.1, v is a continuous function on ∂B_1. Denote by u_1 the harmonic extension of v to $\overline{B}_1$, obtained by convolution with the Poisson kernel (as in Example 17.1). Thus u_1 is continuous on $\overline{B}_1$, harmonic in B_1, and equal to u and v on ∂B_1. We complete our definition of u_1 by setting

$$u_1(z) = u(z) \text{ for } z \in \Omega \setminus [K \cup B_1]. \qquad (28)$$

(Note that our two definitions of u_1 on ∂B_1 coincide.)

Lemma 29. *The pair (u_1, K_1) is a topological competitor for (u, K) in B_1.*

First we need to check that $(u_1, K_1) \in \mathcal{A}$ (as in (7.1)). Since K_1 is obviously closed, this reduces to verifying that

$$u_1 \in W^{1,2}_{\text{loc}}(\Omega \setminus K_1). \tag{30}$$

We know that $u \in W^{1,2}_{\text{loc}}(B_1)$ because it is harmonic. In fact, we even have that

$$u_1 \in W^{1,2}(B_1), \tag{31}$$

i.e., that $\int_{B_1} |\nabla u_1|^2 < +\infty$, because $v \in W^{1,2}(\partial B_1)$ (by (17), (18) and (27)). Let us not check this now, and refer to Lemma 22.16 below. (Also see Exercise 49 for a simpler proof.)

Next let $r_2 > r_1$ be such that $B_2 = B(x, r_2) \subset \Omega$, and K does not meet $B_2 \setminus B_1$. Such a radius exists, by (5), (15), (16) and because K is closed. Then

$$u_1 \in W^{1,2}(B_2 \setminus B_1) \tag{32}$$

by (28), because $u \in W^{1,2}_{\text{loc}}(B_2 \setminus K)$ (since $(u, K) \in \mathcal{A}$), and because $\int_{B_2 \setminus K} |\nabla u|^2 < +\infty$ by the trivial estimate (18.20).

Let us apply Corollary 14.28. By (31) and (32), $u_1 \in W^{1,2}(B_2 \setminus \partial B(x, r_1))$, so it has radial limits $u_1^+(\xi)$ (from outside of B_1) and $u_1^-(\xi)$ (from inside) at almost every point $\xi \in \partial B_1$. Also, u_1^- coincides with v on ∂B_1, because u_1 is the continuous harmonic extension of v to $\overline{B}_1$. Observe that

$$u_1^+(\xi) = v(\xi) \quad \text{for every } \xi \in \partial B(x, r_1), \tag{33}$$

because u is contibuous on $\Omega \setminus K$. More precisely, Lemma 18.22 says that we can replace u with a continuous function, which is equal to u almost-everywhere. We shall always assume that this was done. Now Corollary 14.28 says that $u_1 \in W^{1,2}(B_2)$, and (30) follows because we already know that $u_1 \in W^{1,2}_{\text{loc}}(\Omega \setminus [K \cup \overline{B}_1])$ (by (28) and because $(u, K) \in \mathcal{A}$).

Altogether $(u_1, K_1) \in \mathcal{A}$. Note that

$$\overline{B}_1 \subset \Omega, \tag{34}$$

by (5) and (15), and that (u_1, K_1) coincides with (u, K) out of B_1. So (u_1, K_1) is a competitor for (u, K) in B_1 (see Definition 7.2). To get the topological part, we can apply Lemma 10: (11) holds because $\partial B_1 \setminus K_1 = \partial B_1$ is connected and contained in $\Omega \setminus K$ (by (16)). This completes our proof of Lemma 29. $\square$

Recall from the statement of Theorem 18.16 that (u, K) is a topological local quasiminimizer on Ω. [Note that plain local quasiminimizers are automatically topological local quasiminimizers as well.] By Lemma 29, (34), and Definition 7.21, (7.20) holds for the pair (u_1, K_1) and the ball B_1. Set

$$E = \int_{B_1 \setminus K} |\nabla u|^2 \tag{35}$$

and

$$E_1 = \int_{B_1} |\nabla u_1|^2, \tag{36}$$

as in (7.18), and then

$$\delta E = \text{Max} \left\{ E_1 - E, M(E_1 - E) \right\} \tag{37}$$

as in (7.19); (7.20) says that

$$H^1(K \setminus K_1) < \delta E + a r \tag{38}$$

(because $K_1 \setminus K$ is empty, by (26)). Now

$$H^1(K \setminus K_1) = H^1(K \cap B_1) \geq H^1\left(K \cap B\left(x, \frac{r}{2}\right)\right) \geq \frac{2\alpha r}{3}, \tag{39}$$

by (26), (15), and (4). Thus if we choose $a \leq \frac{\alpha}{3}$, (38) says that $\delta E \geq \frac{\alpha r}{3}$, and hence

$$E_1 - E \geq \frac{\alpha r}{3M}. \tag{40}$$

Our plan is to prove an estimate on $E_1 - E$ which contradicts (40) (if α is small enough). Since E is hard to evaluate, we introduce a new function u_0 that will be easier to deal with. Call $\mathcal{F}$ the set of functions $f \in W^{1,2}(B_1 \setminus K)$ whose radial boundary values on ∂B_1 coincide with v (or u, by definition (27) of v) almost-everywhere. Note that $u \in \mathcal{F}$, so $\mathcal{F}$ is not empty. Thus the hypotheses of the beginning of Section 15 are satisfied and, by Proposition 15.6, there is a function $u_0 \in \mathcal{F}$ such that if we set

$$E_0 = \int_{B_1 \setminus K} |\nabla u_0|^2, \tag{41}$$

then

$$E_0 = \inf \left\{ \int_{B_1 \setminus K} |\nabla f|^2 ; f \in \mathcal{F} \right\}. \tag{42}$$

Proposition 15.6 also tells us that we can make u_0 unique, by deciding that

$$u_0 = 0 \text{ on the connected components of } B_1 \setminus K \text{ that do not touch } \partial B_1. \tag{43}$$

Set

$$\Delta E = E_1 - E_0 = \int_{B_1} |\nabla u_1|^2 - \int_{B_1 \setminus K} |\nabla u_0|^2. \tag{44}$$

Then $\Delta E \geq E_1 - E$, or equivalently $E \geq E_0$, because $u \in \mathcal{F}$. Hence

$$\Delta E \geq \frac{\alpha r}{3M}, \tag{45}$$

by (40). Our goal is to prove the following estimate, which obviously contradicts (45) if α is small enough.

Energy Lemma 46. *If a is small enough (depending on α and M), then*

$$\Delta E \leq C_2 \alpha^{3/2} r. \tag{47}$$

As usual, C_2 will be allowed to depend on M, but not on α or a.

Energy Lemma 46 will be proved in the next section; our proof of Theorem 18.16 in dimension 2 will be complete once we do that.

Remark 48. Our proof of Theorem 18.16 in dimension 2 extends to quasiminimizers with linear distortion (see Definition 7.52). The argument is the same up until the choice of B_1, but instead of choosing u_1 harmonic, we take the matrix $A = A(x, r_1)$ provided by Definition 7.52, call $\widetilde{A}$ the transpose of A, and choose u_1 so that $u_1(\widetilde{A}(x))$ is harmonic on $\widetilde{A}^{-1}(B_1)$. Thus u_1 minimizes $\int_{B_1} |A\nabla u_1|^2$ under the constraint that $u_1 = u$ on ∂B_1 (see Exercise 7.59). Similarly, u_0 is chosen so that $\int_{B_1 \setminus K} |A\nabla u_0|^2$ with the same boundary values, and $u_0 \circ \widetilde{A}$ is harmonic on $\widetilde{A}^{-1}(B_1 \setminus K)$.

The rest of the argument, including the proof of Energy Lemma 46 in the next section, goes as in the case without distortion. The point is that we can do the linear change of variable above, and integrate by parts and estimate our jumps on $\widetilde{A}^{-1}(B_1)$. Since A is bilipschitz, $H^1(\widetilde{A}^{-1}(K \cap B_1))$ is still as small as we want compared to r, which is all we really need to conclude. The fact that $\widetilde{A}^{-1}(B_1)$ is not exactly round does not matter (for (21.9) and similar estimates).

Exercise 49. Let B denote the unit ball in $\mathbb{R}^n$, and let $v \in W^{1,p}(\partial B)$ be given.

1. Find $u \in W^{1,p}(B)$ such that the radial limits of u equal v almost-everywhere on ∂B. Hint: take a smooth function φ on $[0, 1]$ which is 0 near 0 and 1 near 1, and try

$$u(r\theta) = \varphi(r) v(\theta) + (1 - \varphi(r)) m_{\partial B} v \tag{49}$$

for $r \leq 1$ and $\theta \in \partial B$, where $m_{\partial B} v$ is the mean value of v on ∂B. Also check that $\|\nabla u\|_{L^p(B)} \leq C \|\nabla v\|_{L^p(\partial B)}$.

2. Take $p = 2$ and $n = 2$ (to be allowed to apply the results of all the previous sections, but this is not really needed). Use the first question to show that the harmonic extension of v to B lies in $W^{1,2}(B)$, with energy less than $C \|\nabla v\|_{L^2(\partial B)}^2$. Compare with (31).

21   Energy estimates for the Ahlfors-regularity result $(n = 2)$

The goal of this section is to prove Energy Lemma 20.46. We want to start with a rapid description of how we would do it if we knew that

$$K \cap B_1 \text{ is a finite union of } C^2 \text{ curves that meet only transversally} \qquad (1)$$

(that is, with angles $\neq 0$). Later in the section we shall do a more rigorous proof, but the ideas will be the same.

The point of (1) is that we can integrate by parts on the domain $B_1 \setminus K$, but also that u_0 has C^1 extensions up to the C^2-arcs that compose $K \cap B_1$, and these extensions satisfy the Neumann condition $\frac{\partial u_0}{\partial n} = 0$. This description comes from Proposition 17.15 and Remark 17.17.

To compute ΔE (from (20.44)) we want to apply the Green formula to the vector-valued function

$$h = (u_1 - u_0) \nabla (u_1 + u_0) \qquad (2)$$

in the domain $B_1 \setminus K$. Note that

$$\mathrm{div}(h) = \nabla(u_1 - u_0) \cdot \nabla(u_1 + u_0) + (u_1 - u_0)\Delta(u_1 + u_0) = |\nabla u_1|^2 - |\nabla u_0|^2 \qquad (3)$$

because u_1 and u_0 are both harmonic in $B_1 \setminus K$. (See Lemma 15.17 concerning u_0.) Thus

$$\Delta E = \int_{B_1 \setminus K} |\nabla u_1|^2 - |\nabla u_0|^2 = \int_{\partial} (u_1 - u_0) \frac{\partial(u_1 + u_0)}{\partial n}, \qquad (4)$$

where ∂ denotes the boundary of $B_1 \setminus K$. The reader may be worried about the potential singularities of u_0 at the point of $\partial \cap B_1$ where the curves that compose $K \cap B_1$ meet. This is not a serious problem. These singularities are fairly weak (locally, $|\nabla u_0(z)| \leq C|z - z_0|^\beta$ for some $\beta > -\frac{1}{2}$) and one could also make them disappear locally by conformal changes of variables. See Remark 17.17. There is also a problem with the definition of $(u_1 - u_0)\frac{\partial(u_1+u_0)}{\partial n}$ on ∂B_1, which will be addressed later. For the moment, let us assume that u_1 and u_0 are sufficiently smooth near ∂B_1. Then the contribution of ∂B_1 to the integral in (4) is zero, because $u_1 = u_0 = v$ on ∂B_1. Thus

$$\Delta E = \int_{\partial \cap B_1} (u_1 - u_0) \frac{\partial(u_1 + u_0)}{\partial n} = \int_{\partial \cap B_1} (u_1 - u_0) \frac{\partial u_1}{\partial n}, \qquad (5)$$

because $\frac{\partial u_0}{\partial n} = 0$ on $K \cap B_1$. We can even simplify further. For each arc of curve γ that composes $K \cap B_1$, γ is counted twice in the boundary ∂, once for each access to γ from $B_1 \setminus K$. The values of the outward unit normal at each point of γ are opposite, and so $\frac{\partial u_1}{\partial n}$ also takes two opposite values at each point of γ. (Recall that

∇u_1 is well defined and continuous near $K \cap B_1$, because u_1 is harmonic on B_1.) Because of this, $\int_{\partial B_1} u_1 \frac{\partial u_1}{\partial n} = 0$, and

$$\Delta E = -\int_{\partial B_1} u_0 \frac{\partial u_1}{\partial n} = \int_{K \cap B_1} \text{Jump}(u_0) \frac{\partial u_1}{\partial n}. \tag{6}$$

To get the last integral we regrouped for almost-every point $z \in K \cap B_1$ the two occurrences of z in ∂. The jump $\text{Jump}(u_0)$ is the difference between the two boundary values of u_0 at z (computed with the sign that makes (6) true); it depends on a local orientation that we could put on K, but fortunately the product $\text{Jump}(u_0) \frac{\partial u_1}{\partial n}$ is well defined.

The formula (6) is a little less general than (5), because it depends on the local structure of K (the fact that $B_1 \setminus K$ lies on both sides of the arcs that compose its boundary), but it shows more clearly why we should not expect ΔE to be very large when $K \cap B_1$ is too small.

The homogeneity of our problem suggests that $\frac{\partial u_1}{\partial n} \leq C r^{-1/2}$ on average (see (9) for a precise estimate), and $\text{Jump}(u_0) \leq C r^{1/2}$. If things are merely like that, we can still get that ΔE is comparable to $H^1(K \cap B_1)$ (i.e., to αr), as in (20.45). But if $K \cap B_1$ is very small (like here), K does not separate B_1 well enough, and $\text{Jump}(u_0)$ cannot be more than $C(\alpha r)^{1/2}$, which leads to estimates like (20.47).

Our plan for the rest of this section is to prove estimates on $|\nabla u_1|$ and on the local oscillations of u_0 that fit with the comments we just made (and in particular would imply (20.47) if we had (6)), and then return to the estimates leading to (6) in a more rigorous way. We start with an estimate of $|\nabla u_1|$, which we state in a more general (but essentially equivalent) form, because we want to use it later.

Lemma 7. *Let $B = B(x, R)$ be a disk in the plane and $v \in W^{1,p}(\partial B)$ be given, with $1 \leq p < +\infty$. Also denote by v the harmonic extension of v to B, obtained by convolution with the Poisson kernel (as in Example 17.1, but with a different scaling). Then*

$$|\nabla v(z)| \leq C_p \, \text{dist}(z, \partial B)^{-1/p} \left\| \frac{\partial v}{\partial \tau} \right\|_p \quad \text{for } z \in B, \tag{8}$$

where $\left\| \frac{\partial v}{\partial \tau} \right\|_p$ denotes the L^p-norm on ∂B of the tangential derivative of v on ∂B.

We postpone the proof of Lemma 7 until Lemma 22.3 to avoid disrupting the flow of this section.

When we apply Lemma 7 to the function u_1, with $p = 2$ and $B = B_1$, we get that

$$|\nabla u_1(z)| \leq C \, \text{dist}(z, \partial B_1)^{-1/2} \quad \text{on } B_1, \tag{9}$$

because $\left\| \frac{\partial v}{\partial \tau} \right\|_2 \leq C_1^{1/2}$ by (20.27), (20.17) and (20.18).

Next we want to estimate the oscillation of u_0 near a point of $K \cap B_1$. For each $z \in K \cap B_1$, set

$$D^* = B(z, 20\alpha r) \tag{10}$$

and choose a radius $\rho = \rho(z) \in (10\alpha r, 20\alpha r)$ such that

$$\partial B(z, \rho) \cap B_1 \cap K = \emptyset \tag{11}$$

and

$$\int_{\partial B(z,\rho) \cap B_1} |\nabla u_0|^2 \leq I(z), \tag{12}$$

where we set

$$I(z) = (\alpha r)^{-1} \int_{D^* \cap B_1 \setminus K} |\nabla u_0|^2. \tag{13}$$

Such a radius exists, because $H^1(B_1 \cap K) \leq H^1(B(x,r) \cap K) \leq 2\alpha r$ (by (20.15) and (20.3)) and by the same argument as for Lemma 20.21. [That is, estimate the radial projection of $B_1 \cap K$ to get (11) and use Chebyshev to get (12).]

Lemma 14. $I(z) \leq C_3$, where as usual C_3 depends on M, but not on α or a.

To prove this we introduce a new competitor (u_2, K_2). Take

$$K_2 = K \cup \partial(B_1 \cap D^*), \tag{15}$$

and set

$$\begin{cases} u_2 = u & \text{on } \Omega \setminus (B_1 \cup K_2), \\ u_2 = u_0 & \text{on } B_1 \setminus (D^* \cup K), \text{ and} \\ u_2 = 0 & \text{on } B_1 \cap D^*. \end{cases} \tag{16}$$

The verification that $u_2 \in W^{1,2}_{\text{loc}}(\Omega \setminus K_2)$ can be done as for u_1 (in Lemma 20.29). Actually, the proof of (20.30) shows that $\tilde{u}_0 \in W^{1,2}_{\text{loc}}(\Omega \setminus K)$, where $\tilde{u}_0$ denotes the function on $\Omega \setminus K$ that is equal to u_0 on $B_1 \setminus K$ and to u on $\Omega \setminus (B_1 \cup K)$. The result for u_2 follows at once, because K_2 contains $\partial(B_1 \cap D^*)$ and $u_2 \in W^{1,2}_{\text{loc}}$ on both sides of this curve. Thus $(u_2, K_2) \in \mathcal{A}$.

Since we kept $(u_2, K_2) = (u, K)$ out of $\overline{B}_1$, and $\overline{B}_1 \subset \Omega$ by (20.34), (u_2, K_2) is a competitor for (u, K) in B_1. It is even a topological competitor, by Lemma 20.10, because $\partial B_1 \setminus K_2$ is contained in a single component of $\Omega \setminus K$ (namely, the component that contains ∂B_1), by (20.16). So we can use our hypothesis that (u, K) is a topological local quasiminimizer in Ω and apply (7.20). In the present situation (and with the notation of (7.18)),

$$E = \int_{B_1 \setminus K} |\nabla u|^2 \geq E_0 \tag{17}$$

by the comment just before (20.45) (and (20.35) for the coherence of notation), and

$$\widetilde{E} = \int_{B_1 \setminus K_2} |\nabla u_2|^2 = \int_{B_1 \setminus (K \cup D^*)} |\nabla u_0|^2$$

$$= E_0 - \int_{B_1 \cap D^* \setminus K} |\nabla u_0|^2 = E_0 - \alpha r I(z) \tag{18}$$

by (16), (20.41), and (13). Then

$$\delta E =: \operatorname{Max}\{(\widetilde{E} - E), M(\widetilde{E} - E)\} \leq -\alpha r I(z) \tag{19}$$

(compare with (7.19)) and (7.20) says that

$$\begin{aligned} 0 = H^1(K \setminus K_2) &\leq M H^1(K_2 \setminus K) + \delta E + ar_1 \\ &\leq M H^1(\partial(B_1 \cap D^*)) - \alpha r I(z) + ar \\ &\leq 40\pi M \alpha r - \alpha r I(z) + ar\,, \end{aligned} \tag{20}$$

by (15), (19) and (10). We already decided earlier to take $a \leq \frac{\alpha}{3}$, so we get that $I(z) \leq 40\pi M + \frac{1}{3}$, and Lemma 14 follows.   □

This will be the last time that we use quasiminimality in this proof. So our only constraint on a will be that $a \leq \frac{\alpha}{3}$, as required a little bit after (20.39).

We may now return to our point $z \in K \cap B_1$ and the radius $\rho = \rho(z)$. Set

$$D = B_1 \cap B(z, \rho) \subset D^*. \tag{21}$$

Note that

$$\partial D \cap K = \emptyset, \tag{22}$$

by (11) and (20.16) (which is needed in the rare case when $B(z, \rho)$ meets ∂B_1). Also, we claim that

$$u_0 \text{ is absolutely continuous on } \partial D, \tag{23}$$

and its derivative $\frac{\partial u_0}{\partial \tau}$ on ∂D satisfies

$$\int_{\partial D} \left| \frac{\partial u_0}{\partial \tau} \right| \leq H^1(\partial D)^{1/2} \left\{ \int_{\partial D} \left(\frac{\partial u_0}{\partial \tau} \right)^2 \right\}^{1/2} \leq C(\alpha r)^{1/2}. \tag{24}$$

If $\overline{B}(z, \rho) \subset B_1$, (23) is obvious because $\partial D = \partial B(z, \rho) \subset B_1 \setminus K$ (by (22)), and u_0 is harmonic there. Then (24) follows from (12), Lemma 14, and Cauchy–Schwarz. Let us decide to avoid the case when $\partial B(z, \rho)$ is tangent to ∂B_1 from inside, by choosing a different ρ if this happens.

Thus we are left with the case when $\partial B(z, \rho)$ crosses ∂B_1 transversally. In this case $\partial D = \partial_1 \cup \partial_2$, where $\partial_1 = \partial B_1 \cap \overline{B}(z, \rho)$ and $\partial_2 = \partial B(z, \rho) \cap B_1$. In (23) and (24) we implicitly extended u_0 to ∂B_1, by taking $u_0 = v = u$ there (see (20.27) and the definition of u_0 near (20.41)). We know that u_0 is absolutely continuous on ∂_1, with an estimate similar to (24), by (20.17) and (20.18). The same thing is true with ∂_2 (by the same argument as in our previous case). In particular, (24) will follow as soon as we prove (23). We already know (from the absolute continuity on ∂_2) that u_0 has limits at the two extremities of ∂_2. To prove (23), we only need to check that these limits are equal to the values of v at these points. There are a few ways to do this, each requiring a small amount of dirty work that we don't want to do in detail.

The first one is to observe that the property we want holds for almost-every choice of ρ (so that, assuming we chose ρ more carefully, we can take it for granted). This uses the fact that for almost all $\rho \in (10\alpha r, 20\alpha r)$ such that $B(z, \rho)$ meets ∂B_1, the restriction of u_0 to the arc ∂_2 has limits at both extremities (because $\int_{\partial_2} |\nabla u_0| dH^1 < +\infty$), and these limits coincide with the values of v at these points. This uses a variant of Lemma 13.9 that we did not actually prove, or Remark 13.16. The point is that the limits along the arcs of circles centered at z (and that meet ∂B_1) coincide almost-everywhere on ∂B_1 with the radial limits of u_0. Which are equal to v by definition of u_0.

Another option to show that, for almost-every choice of ρ, the limits of u_0 along circles $\partial B(x, \rho)$ coincide with the radial limits of u_0 at the corresponding points of ∂B_1, would be to use the fact that u_0 is harmonic, and so u_0 converges nontangentially almost everywhere (i.e., in little cones as in (17.4)).

We could also verify that u_0 is actually continuous on $\overline{B}_1$ in a neighborhood of ∂B_1. Let us give a rapid sketch of this approach. We first choose a radius $r_2 < r_1$, close to r_1, so that $B_1 \setminus B(x, r_2)$ does not meet K. We can also choose two radii L_1 and L_2 such that u_0 has limits (on ∂B_1) along L_1 and L_2 that coincide with v. [This is the case for almost all choices of radii, by definition of u_0.] Let R denote one of the two regions of $B_1 \setminus B(x, r_2)$ bounded by the radii L_i. Then u_0 is continuous on ∂R (by (20.16) and (20.17) in particular), and it is the harmonic extension of its values on ∂R inside R. If you already know how to take harmonic extensions of continuous functions on ∂R, you may proceed as in Example 17.1. Otherwise, you may compose with a conformal mapping ψ from the unit disk to R, to reduce to the case of the unit disk; this is possible by the proof of Lemma 17.11, and because ψ and ψ^{-1} extend continuously to the boundaries. Thus u_0 is continuous on $\overline{R}$. This gives the continuity of u_0 near ∂B_1, at least away from the L_i. But we could do the argument again with other choices of radii L_i, and get the continuity near any point of ∂B_1.

Let us not do more carefully any of the arguments suggested above, and consider that we have a sufficient proof of (23) and (24).

From (23), (24) and the connectedness of ∂D we deduce that there is a number $b = b(z)$ such that

$$|u_0(y) - b| \le C_4 (\alpha r)^{1/2} \text{ for } y \in \partial D. \tag{25}$$

(Take $b = u_0(y_0)$ for any $y_0 \in \partial D$.)

Denote by U_0 the connected component of $B_1 \setminus K$ that touches ∂B_1. We claim that

$$\partial D \subset U_0 \cup \partial B_1. \tag{26}$$

Indeed $\partial D \cap B_1$ is connected (by (21)), and does not meet K (by (22)), so it is contained in some component of $B_1 \setminus K$. Observe that diameter$(\partial D \cap B_1) \geq 10\alpha r$, by (21) and because $\rho \geq 10\alpha r$, so (26) will follow at once from the next lemma.

Lemma 27. *If U is a connected component of $B_1 \setminus K$ that does not touch ∂B_1, then* diameter$(U) \leq 3\alpha r$.

To see this, choose points $y_1, y_2 \in U$, with $|y_1 - y_2| > \frac{2}{3}$ diameter(U), and let $\gamma \subset U$ be a path from y_1 to y_2. Also call π the orthogonal projection onto the line L through y_1 and y_2. (See Figure 1.) Let us check that

$$\pi(K \cap B_1) \text{ contains the segment } (y_1, y_2). \tag{28}$$

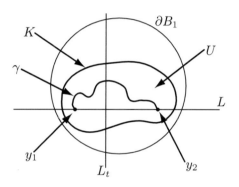

Figure 1

For $t \in (y_1, y_2)$, call L_t the line through t which is perpendicular to L, and set $L'_t = L_t \cap B_1$. Then γ meets L'_t, because y_1 and y_2 lie on opposite sides of L_t and $\gamma \subset B_1$. Thus L'_t meets U. Since both endpoints of L'_t lie in ∂B_1 (and hence away from U), L'_t must meet $\partial U \subset K$, which means that $t \in \pi(K \cap B_1)$. This proves (28). Then

$$\text{diameter}(U) < \frac{3}{2}|y_1 - y_2| \leq \frac{3}{2}H^1(\pi(K \cap B_1)) \tag{29}$$

$$\leq \frac{3}{2}H^1(K \cap B_1) \leq \frac{3}{2}H^1(K \cap B(x,r)) \leq 3\alpha r$$

by definition of y_1 and y_2, (28), Remark 2.11, (20.15), and (20.3). Lemma 27 and (26) follow. $\qquad\square$

Lemma 30. *Call $D' = (D \cap U_0) \cup \partial D$. Then*

$$|u_0(y) - b(z)| \leq C_4(\alpha r)^{1/2} \text{ for } y \in D', \tag{31}$$

where $b = b(z)$ and C_4 are still as in (25).

We already know the desired inequality for $y \in \partial D$, and we want to find out about $D \cap U_0$. Let us try a truncation argument. Define a new function $\tilde{u}_0$ on $B_1 \setminus K$, as follows. Keep $\tilde{u}_0 = u_0$ out of D, and replace u_0 on $D \setminus K$ with

$$\tilde{u}_0(y) = \mathrm{Min}\,\{b + C_4(\alpha r)^{1/2},\; \mathrm{Max}\,[b - C_4(\alpha r)^{1/2}, u_0(y)]\}. \tag{32}$$

By Lemma 3.22 (or Proposition 11.15), $\tilde{u}_0 \in W^{1,2}(D \setminus K)$, with smaller or equal energy. Because of (25), $\tilde{u}_0$ is continuous across $\partial D \cap B_1$ (recall that u_0 is harmonic near $\partial D \cap B_1$). Welding Lemma 14.4 applies (as in the proof of Corollary 14.28) and says that $\tilde{u}_0 \in W^{1,2}(B_1 \setminus K)$, again with smaller or equal energy. Note that $\tilde{u}_0$ has the same boundary values on ∂B_1 as u_0. (If ∂D meets ∂B_1, this uses (25) again.) In other words, $\tilde{u}_0 \in \mathcal{F}$, where $\mathcal{F}$ is the set of competitors that was used to define u_0. (See near (20.41).) By the uniqueness result in Proposition 15.6, $\tilde{u}_0 = u_0$ on U_0. Lemma 30 follows (compare with (32)). $\square$

We are almost ready to use the formula (6) (assuming that it holds) to get a good estimate on ΔE, but we want to have a better control on u_0 in the connected components of $B_1 \setminus K$ that do not touch ∂B_1. The easiest is to replace u_0 with an essentially equivalent function u_0^*. We keep $u_0^*(y) = u_0(y)$ on U_0, but modify the constant values of u_0 on the components of $B_1 \setminus K$ that do not touch ∂B_1. For each such component U, we pick a point $z_U \in \partial U \subset K \cap B_1$ and set

$$u_0^*(y) = b(z_U) \text{ for } y \in U, \tag{33}$$

where the constant $b(z_U)$ is the same as in (25) and (31), applied with the set $D = B_1 \cap B(z, \rho(z))$ that was constructed with $z = z_U$. Then our new function u_0^* has the same minimizing properties as u_0. In particular,

$$\int_{B_1 \setminus K} |\nabla u_0^*|^2 = \int_{B_1 \setminus K} |\nabla u_0|^2 = E_0, \tag{34}$$

and the same computations that yield (6), if they ever work, also give that

$$\Delta E = \int_{K \cap B_1} \mathrm{Jump}(u_0^*) \frac{\partial u_1}{\partial n}. \tag{35}$$

Now let $z \in K \cap B_1$ be given, and let us check that

$$|u_0^*(y) - b(z)| \leq 2C_4(\alpha r)^{1/2} \text{ for } y \in D \setminus K, \tag{36}$$

where again D is associated to z as above. When $y \in D \cap U_0$, this comes directly from Lemma 30, and we don't even need the constant 2. Otherwise, $y \in U$ for some other component of $B_1 \setminus K$, and $u_0^*(y) = b(z_U)$ for some $z_U \in \partial U$. Lemma 27 says that diameter$(U) \leq 3\alpha r$, so $|z_U - y| \leq 3\alpha r$. Note that y lies in $D = B(x, \rho) \cap B_1$ by assumption, and also in the set $D_U = B(z_U, \rho(z_U)) \cap B_1$ associated to z_U, because $|z_U - y| \leq 3\alpha < \rho(z_U)$. In particular, D_U meets D.

First suppose that ∂D_U meets $\overline{D}$, and let us check that it meets $D' = (D \cap U_0) \cup \partial D$. If it meets ∂D, we are happy. Otherwise, it meets D. Note that $\partial D_U \subset U_0 \cup \partial B_1$, by our proof of (26), and so ∂D_U meets $D \cap (U_0 \cup \partial B_1) = D \cap U_0$, as needed. So we can pick a point $y' \in \partial D_U \cap D'$. Then $|u_0(y') - b(z)| \leq C_4(\alpha r)^{1/2}$ by Lemma 27, and $|u_0(y') - b(z_U)| \leq C_4(\alpha r)^{1/2}$ by the analogue of (25) for ∂D_U. Thus $|u_0^*(y) - b(z)| = |b(z_U) - b(z)| \leq 2C_4(\alpha r)^{1/2}$, and (36) holds in this case.

If ∂D_U does not meet $\overline{D}$, then $\overline{D}$ is contained in D_U (because it is connected and meets D_U). In this case $\partial D \subset D_U \cap U_0$ (by (26) and because $D_U \subset B_1$), and $|b(z_U) - u_0(w)| \leq C_4(\alpha r)^{1/2}$ for $w \in \partial D$, by the analogue of Lemma 30 for D_U. Since $|b(z) - u_0(w)| \leq C_4(\alpha r)^{1/2}$ for all such w, by (25), we get (36) in this last case too. Thus (36) holds in all cases.

From (36) we deduce that

$$|\operatorname{Jump}(u_0^*)| \leq 4C_4(\alpha r)^{1/2} \text{ almost-everywhere on } K \cap B_1, \tag{37}$$

and then (if you trust the integrations by parts that lead to (6))

$$\begin{aligned}
\Delta E &\leq 4C_4(\alpha r)^{1/2} \int_{K \cap B_1} |\nabla u_1| dH^1 \\
&\leq C(\alpha r)^{1/2} \int_{K \cap B_1} \operatorname{dist}(z, \partial B_1)^{-1/2} dH^1(z) \\
&= C(\alpha r)^{1/2} J(r_1) \leq 20 C \alpha^{3/2} r
\end{aligned} \tag{38}$$

by (35), (9), (20.19) and (20.20). This is the estimate we want to prove for Energy Lemma 20.46, but our proof is not over yet because we did not check the integrations by parts that lead to (6) and (32).

Before we start with a rather long (and surpriseless) approximation argument, let us mention that there are actual ways to do the integrations by parts. And in a sense, this is what we shall check. We want to replace $B_1 \setminus K$ with smoother domains, integrate by parts there, and go to the limit. Let us first cover $K \cap B_1$ with small disks A_ℓ. Since we shall need to be a little more careful near ∂B_1, we decompose $K \cap B_1$ into the sets

$$K^m = \left\{ z \in K \cap B_1 \, ; \, 2^{-m} \leq \operatorname{dist}(z, \partial B_1) \leq 2^{-m+1} \right\}, \tag{39}$$

$m \in \mathbb{Z}$. Since $K \cap B_1$ lies at (strictly) positive distance from ∂B_1 (by (20.16)), there are only finitely many m for which K^m is not empty, and we shall only consider these. Let $k \geq 0$ be given. For each m as above, cover K^m by little disks A_ℓ, $\ell \in L_m$, so that

$$A_\ell = B(x_\ell, r_\ell), \text{ with } x_\ell \in K^m \text{ and } 0 < r_\ell \leq \operatorname{Min}(2^{-m-1}, 2^{-k}), \tag{40}$$

and

$$2 \sum_{\ell \in L_m} r_\ell < H^1(K^m) + 2^{-k-m}. \tag{41}$$

This can be done, by definition of the Hausdorff measure. (See Definition 2.6.) Also, we can take L_m finite, because K^m is compact. Set $L(k) = \bigcup_m L_m$. Then $L(k)$ also is finite, because we work with a finite set of integers m. Let us also modify some of our disks A_ℓ, if needed, so that

$$\text{the circles } \partial A_\ell \text{ always meet transversally (when they meet).} \qquad (42)$$

This is easily done, without ruining the other properties. Set

$$H_k = \bigcup_{\ell \in L(k)} \overline{A}_\ell \qquad (43)$$

and

$$V_k = B_1 \setminus H_k. \qquad (44)$$

Then

$$K \cap B_1 \subset \bigcup_{\ell \in L(k)} A_\ell \subset H_k \subset B_1 \qquad (45)$$

by construction, and so

$$\delta_k = \text{dist}(K \cap B_1, V_k) > 0 \qquad (46)$$

by compactness of $K \cap B_1$ and because each A_ℓ is open. Since we can construct our sets V_k by induction, we may also impose that $r_\ell \leq \frac{1}{2} \delta_k$ for all $\ell \in L(k+1)$, in addition to the constraint (40) for $k + 1$. (This amounts to replacing 2^{-k-1} with something smaller in (40).) Since each $\overline{A}_\ell$ is contained in a $2r_\ell$-neighborhood of $x_\ell \in K \cap B_1$, this is a simple way to make sure that

$$V_{k+1} \supset V_k \text{ for } k \geq 0, \qquad (47)$$

without taking subsequences. We already know from (45) that $H_k \subset B_1$; then (40) and the same argument as for (47) yield that

$$H_k \subset \left\{ y \in B_1; \text{dist}(y, K \cap B_1) \leq 2^{-k} \right\}. \qquad (48)$$

Hence $K \cap B_1 = \bigcap_k H_k$ and

$$B_1 \setminus K = \bigcup_k V_k \text{ (a nondecreasing union).} \qquad (49)$$

Set

$$\mathcal{F}_k = \Big\{ f \in W^{1,2}(V_k); \text{ the radial limits of } f \text{ on } \partial B_1$$
$$\text{are equal to } v \text{ almost-everywhere} \Big\}; \qquad (50)$$

notice that $\mathcal{F}_k \neq \emptyset$, for instance because $u_0 \in \mathcal{F}_k$. Then let $f_k \in \mathcal{F}_k$ be such that

$$\int_{V_k} |\nabla f_k|^2 = E(k) := \inf \left\{ \int_{V_k} |\nabla f|^2 ; f \in \mathcal{F}_k \right\}. \tag{51}$$

Thus f_k is the analogue of u_0 on V_k; see near (20.41). As before, f_k exists by Proposition 15.6, and we can make it unique by requiring that

$$f_k = 0 \text{ on the components of } V_k \text{ that do not touch } \partial B_1. \tag{52}$$

Let us now apply Proposition 16.3 to our increasing sequence of domains V_k and their limit $B_1 \setminus K$, with selected boundaries $I = I_k = \partial B_1$. The hypotheses are satisfied because ∂B_1 is smooth and lies at positive distance from K and the rest of the ∂V_k, v is bounded by (20.17) and (20.27), and $u_0 \in W^{1,2}(B_1 \setminus K)$ (so that $\mathcal{F}$ is not empty). We get that

$$E_0 = \lim_{k \to +\infty} E(k) \tag{53}$$

(with E_0 as in (20.41) and (20.42)), and

$$\{f_k\} \text{ converges to } u_0 \text{ uniformly on every compact subset of } B_1 \setminus K. \tag{54}$$

(See (16.5), and recall our convention (20.43) for the components of $B_1 \setminus K$ that do not touch ∂B_1.) We want to estimate the differences

$$\Delta_k = E_1 - E(k) = \int_{B_1} |\nabla u_1|^2 - E(k), \tag{55}$$

by the methods suggested above. Note that

$$\partial V_k = \partial B_1 \cup \partial H_k \tag{56}$$

by (44), and because H_k is compact and contained in B_1 (by (45)). From (43) it is clear that

$$\partial H_k \text{ is a finite union of (closed) arcs of circles,} \tag{57}$$

and these arcs always meet transversally (when they meet), by (42). It will be convenient to set

$$W_k = B_1 \setminus \partial H_k, \tag{58}$$

and then extend f_k to W_k by taking

$$f_k = 0 \text{ on the interior of } H_k. \tag{59}$$

Of course this does not increase the energy of f_k, and f_k is still the function of $W^{1,2}(W_k)$ whose boundary values on ∂B_1 coincide with v almost-everywhere and which minimizes the energy (with the normalization (52) and (59)). In particular, we can apply Proposition 17.15 to f_k and get that on each of the (open) arcs γ

that compose ∂H_k (minus a finite number of vertices), f_k has C^1 extensions up to γ from both sides, and these extensions satisfy the Neumann condition $\frac{\partial f_k}{\partial n} = 0$ on γ. As was discussed at the beginning of this section and in Remark 17.17, these extensions may have weak singularities at the vertices, but not to the point of disrupting our estimates. We could also smooth out ∂H_k without affecting the rest of the proof.

Also, we can integrate by parts on the piecewise smooth open set W_k. Let us not be that bold yet, and do it on the slightly smaller set $W_k^t = W_k \cap B(x,t)$, with $t < r_1$ very close to r_1. Consider the vector-valued function

$$h = (u_1 - f_k)\nabla(u_1 + f_k), \tag{60}$$

which is now sufficiently smooth near $\partial B(x,t)$. Since

$$\text{div}(h) = \nabla(u_1 - f_k) \cdot \nabla(u_1 + f_k) + (u_1 - f_k)\Delta(u_1 + f_k) = |\nabla u_1|^2 - |\nabla f_k|^2 \tag{61}$$

on W_k^t, Green says that

$$\int_{B(x,t)} |\nabla u_1|^2 - \int_{W_k^t} |\nabla f_k|^2 = \int_{W_k^t} \text{div}(h) = \int_{\partial W_k^t} (u_1 - f_k)\frac{\partial(u_1 + f_k)}{\partial n}. \tag{62}$$

When t tends to r_1^-, the left-hand side of (62) tends to Δ_k, by (51), (55), and because $\nabla f_k = 0$ on $W_k \setminus V_k$ (by (59)). Thus

$$\Delta_k = \int_{\partial_k} (u_1 - f_k)\frac{\partial(u_1 + f_k)}{\partial n} + \lim_{t \to r_1^-} J_k(t), \tag{63}$$

where $\partial_k = \partial W_k^t \setminus \partial B(x,t) = \partial W_k \cap B_1$ denotes the inside boundary of W_k^t (and W_k), and

$$J_k(t) = \int_{\partial B(x,t)} (u_1 - f_k)\frac{\partial(u_1 + f_k)}{\partial n}. \tag{64}$$

We want to check that

$$\lim_{t \to r_1^-} J_k(t) = 0, \tag{65}$$

and since we know from (63) that the limit exists, it will be enough to test this on any sequence $\{t_j\}$ that tends to r_1^-.

Maybe we should note that (65) is not completely clear because we do not know that u_1 and f_k are C^1 near ∂B_1. The situation would be somewhat better for minimizers of the Mumford-Shah functional, because in this case v is slightly better than C^1 on ∂B_1. Here we shall prove (65) in a fairly brutal way, using absolute continuity on lines and the existence of radial limits almost-everywhere.

From the definition of f_k as an element of $\mathcal{F}_k$ (see (50)) and the fact that u_1 is continuous on $\overline{B}_1$ and coincides with v on ∂B_1, we get that for almost every direction $\theta \in \partial B(0,1)$,

$$\lim_{t \to r_1^-} u_1(x + t\theta) = \lim_{t \to r_1^-} f_k(x + t\theta) = v(x + r_1^-\theta). \tag{66}$$

Set $\varphi = |\nabla u_1| + |\nabla f_k|$ on W_k, and let $t_0 < r_1$ be such that $B_1 \setminus B(x, t_0) \subset W_k$. Then

$$\int_{t_0}^{r_1} \varphi(x + t\theta) \, dt < +\infty \tag{67}$$

for almost-every $\theta \in \partial B(0, 1)$, because $\varphi \in L^2(W_k)$ and by Fubini. When (66) and (67) hold for some θ (and if in addition f_k is absolutely continuous along the corresponding ray), we can integrate derivatives along the ray $\{x + s\theta \,;\, t_0 < s < r_1\}$ all the way to ∂B_1, and get that

$$|u_1(x + t\theta) - f_k(x + t\theta)| \leq \int_t^{r_1} \varphi(x + s\theta) \, ds \tag{68}$$

for $t_0 < t < r_1$. Then

$$J_k(t) \leq r_1 \int_{\partial B(0,1)} \varphi(x + t\theta) \, |u_1(x + t\theta) - f_k(x + t\theta)| d\theta \tag{69}$$

$$\leq r_1 \int_{\partial B(0,1)} \varphi(x + t\theta) \int_t^{r_1} \varphi(x + s\theta) \, ds \, d\theta$$

by (64) and (68), and hence

$$\int_{t_0}^{r_1} J_k(t) \, dt \leq r_1 \int_{\partial B(0,1)} \int_{t_0}^{r_1} \int_t^{r_1} \varphi(x + t\theta) \varphi(x + s\theta) \, ds \, dt \, d\theta \tag{70}$$

$$\leq r_1 \int_{\partial B(0,1)} \left\{ \int_{t_0}^{r_1} \varphi(x + t\theta) \, dt \right\}^2 d\theta$$

$$\leq r_1(r_1 - t_0) \int_{\partial B(0,1)} \int_{t_0}^{r_1} \varphi(x + t\theta)^2 \, dt \, d\theta$$

$$\leq 2(r_1 - t_0) \int_{B_1 \setminus B(x, t_0)} \varphi^2$$

by Cauchy–Schwarz, a return from polar coordinates, and if $t_0 \geq \frac{r_1}{2}$. Thus

$$\lim_{t_0 \to r_1^-} (r_1 - t_0)^{-1} \int_{t_0}^{r_1} J_k(t) \, dt = 0, \tag{71}$$

because $\int_{B_1 \setminus B(x, t_0)} \varphi^2$ tends to 0 (since $\varphi \in L^2$).

It is now easy to use (71) and Chebyshev to construct sequences $\{t_j\}$ that tend to r_1^- and for which $J_k(t_j)$ tends to 0. This proves (65).

We can simplify (63) further, because $\frac{\partial f_k}{\partial n} = 0$ on ∂_k. Thus

$$\Delta_k = \int_{\partial_k} (u_1 - f_k) \frac{\partial u_1}{\partial n}. \tag{72}$$

Note that $\partial W_k \cap B_1 = \partial H_k$ (by (58) and the description of ∂H_k in (57)). Each point of ∂H_k, except perhaps for a finite number of vertices, is counted twice in ∂_k, with opposite values of the unit normal. The two contributions of $u_1 \frac{\partial u_1}{\partial n}$ cancel (because $u_1 \nabla u_1$ is defined and continuous in B_1), and

$$\Delta_k = -\int_{\partial_k} f_k \frac{\partial u_1}{\partial n} = \int_{\partial H_k} \text{Jump}(f_k) \frac{\partial u_1}{\partial n}, \qquad (73)$$

where in the last integral we grouped the two contributions of almost-every point of ∂H_k. As before, the signs of $\text{Jump}(f_k)$ and $\frac{\partial u_1}{\partial n}$ are not defined separately, but the product makes sense. Also, in the computations that lead to (73), we never used our conventions (52) and (59) that $f_k = 0$ on the connected components of W_k that do not touch ∂B_1. Thus, if f_k^* is another function on W_k such that

$$f_k^* = f_k \text{ on the component } W_{k,0} \text{ of } W_k \text{ that touches } \partial B_1 \qquad (74)$$

and

$$f_k^* \text{ is constant on each of the components of } W_k \text{ that do not touch } \partial B_1, \quad (75)$$

then $\Delta_k = E_1 - \int_{W_k} |\nabla f_k^*|^2$ (by (55) and (51)), and the computation above yields

$$\Delta_k = \int_{\partial H_k} \text{Jump}(f_k^*) \frac{\partial u_1}{\partial n}. \qquad (76)$$

To define an appropriate f_k^* and estimate its jump at the same time, cover $K \cap B_1$ by a finite number of sets $D_i = B_1 \cap B(z_i, \rho(z_i))$ like D in (21). Recall from (26) that

$$\partial D_i \subset U_0 \cup \partial B_1. \qquad (77)$$

Let us check that also

$$\partial D_i \subset W_{k,0} \cup \partial B_1 \text{ for } k \text{ large enough}, \qquad (78)$$

where $W_{k,0}$ still denotes the component of W_k that touches ∂B_1. Let $r_2 < r_1$ be such that

$$K \cap B_1 \subset B(x, r_2); \qquad (79)$$

then

$$H_k \subset B(x, r_2) \text{ for } k \text{ large}, \qquad (80)$$

by (48) and because $K \cap B_1$ is compact.

By (79), $\partial B(x, r_2) \subset U_0$. By (79), there is a path γ in $U_0 \cup \partial B_1$ that connects ∂D_i to $\partial B(x, r_2)$. Note that $(\partial D_i \cup \gamma) \cap \overline{B}(x, r_2)$ is a compact subset of U_0; hence (49) says that it is contained in V_k for k large enough. Then $\partial D_i \cup \gamma \subset V_k \cup \partial B_1$ also, because the remaining part $(\partial D_i \cup \gamma) \setminus \overline{B}(x, r_2)$ is controlled by (80). Our claim (78) follows, because $\partial D_i \cup \gamma$ is connected and meets $\partial B(x, r_2) \subset W_{k,0}$ (by (80)).

Set $D_i' = D_i \cap B(x, r_2)$. Then the D_i' still cover $K \cap B_1$, and hence

$$H_k \subset \bigcup_i D_i' \text{ for } k \text{ large,} \tag{81}$$

again by (48) and because $K \cap B_1$ is compact. Also, $\partial D_i' \subset W_{k,0}$ for k large, by (78) and because (80) forces $\partial B(x, r_2) \subset W_{k,0}$. Since $\partial D_i' \subset \partial D_i \cup [D_i \cap \partial B(x, r_2)] \subset \partial D_i \cup [D_i \cap U_0]$ (by (79)), Lemma 30 says that we can find a constant b_i such that

$$|u_0(y) - b_i| \leq C_4(\alpha r)^{1/2} \text{ for } y \in \partial D_i'. \tag{82}$$

Also, $\partial D_i' \subset [\partial D_i \cap \overline{B}(x, r_2)] \cup [D_i \cap \partial B(x, r_2)]$ is contained in U_0, by (77) and (79). Thus $\partial D_i'$ is a compact subset of $B_1 \setminus K$, and (54) says that $\{f_k\}$ converges to u_0 uniformly on $\partial D_i'$. In particular,

$$|f_k(y) - b_i| \leq 2C_4(\alpha r)^{1/2} \text{ on } \partial D_i' \tag{83}$$

for k large enough.

Let us now define f_k^* on the connected components of W_k that do not touch ∂B_1. Let U be such a component. Then $\partial U \subset \partial H_k$, and (81) says that for k large it is contained in the union of the D_i'. Pick a point $z_0 \in \partial U$, and then i such that $z_0 \in D_i'$. Then $U \cap D_i' \neq \emptyset$ (because U has points very close to z_0 and D_i' is open). Next $\partial D_i' \subset [\partial D_i \cap \overline{B}(x, r_2)] \cup [D_i \cap \partial B(x, r_2)] \subset W_{k,0}$, by (78) and (80). Since $W_{k,0}$ does not meet U, neither does $\partial D_i'$. But U is connected and meets $\partial D_i'$, so $U \subset D_i'$. Set $f_k^* = b_i$ on U. We can do this (independently) for each component U of W_k other than $W_{k,0}$. On $W_{k,0}$ we keep $f_k^* = f_k$.

We are ready to estimate the jump of f_k^*. Let us already check that for k large enough,

$$|f_k^*(y) - b_i| = |f_k(y) - b_i| \leq 2C_4(\alpha r)^{1/2} \text{ on } \overline{D}_i' \cap W_{k,0}. \tag{84}$$

We already know this for $y \in \partial D_i'$, by (83) and because $\partial D_i' \subset W_{k,0}$. Then we can apply the proof of Lemma 30 and get (84). The point is that if we truncate f_k on D_i' by a formula like (32), we get another competitor for f_k which is at least as good as f_k; this competitor must be equal to f_k on $W_{k,0}$ (by Proposition 15.6), and (84) follows.

Now we can prove that for k large,

$$|f_k^*(y) - b_i| \leq 4C_4(\alpha r)^{1/2} \text{ on } D_i' \setminus \partial H_k. \tag{85}$$

We already know this when $y \in W_{k,0}$, so let us assume that $y \in U$ for some other component U. Then $U \subset D_i'$, because U meets D_i' but not $\partial D_i' \subset W_{k,0}$. Also, $f_k^*(y) = b_j$ for some j that may be different from i, but for which D_j' contains U as well.

If $\partial D_j'$ meets $\overline{D}_i'$, then $|b_j - b_i| \leq 4C_4(\alpha r)^{1/2}$ by (84), the fact that $\partial D_j' \subset W_{k,0}$, and the analogue of (83) for j. Otherwise $\overline{D}_i' \subset D_j'$ (because D_i' meets

D'_j: their intersection contains U), and in particular $\partial D'_i \subset D'_j \cap W_{k,0}$. Then $|b_j - b_i| \leq 4C_4(\alpha r)^{1/2}$ by (83) and the analogue of (84) for j. In all cases we get the inequality in (85), because $f^*_k(y) = b_j$. This proves (85).

Since the D'_i cover $\partial H_k \subset H_k$ for k large (by (81)), we get that $\mathrm{Jump}(f^*_k) \leq 8C_4(\alpha r)^{1/2}$ almost-everywhere on ∂H_k, and (76) yields

$$\Delta_k \leq 8C_4(\alpha r)^{1/2} \int_{\partial H_k} |\nabla u_1| \leq C(\alpha r)^{1/2} \int_{\partial H_k} \mathrm{dist}(z, \partial B_1)^{-1/2} dH^1(z), \quad (86)$$

by (9). Now H_k is a union of closed disks $\overline{A}_\ell$, $\ell \in L(k)$, as in (43). Then ∂H_k is contained in the corresponding union of circles $\partial A_\ell = \partial B(x_\ell, r_\ell)$ (see (40)). The set $L(k)$ of indices ℓ is the union of sets L_m, and for these $x_\ell \in K^m$ lies at distance $\geq 2^{-m}$ from ∂B_1 and $r_\ell \leq 2^{-m-1}$. (See (39) and (40).) Hence $\mathrm{dist}(z, \partial B_1) \geq 2^{-m-1}$ on ∂A_ℓ when $\ell \in L_m$, and

$$\Delta_k \leq C(\alpha r)^{1/2} \sum_m \sum_{\ell \in L_m} 2^{m/2} H^1(\partial A_\ell) \leq C(\alpha r)^{1/2} \sum_m 2^{m/2} \sum_{\ell \in L_m} r_\ell$$
$$\leq C(\alpha r)^{1/2} \sum_m \left\{ 2^{m/2} H^1(K^m) + 2^{-k-\frac{m}{2}} \right\}, \quad (87)$$

by (41). We only sum on indices m for which K^m is not empty, which forces $2^{-m} \leq r_1$ (by (39)). Thus

$$\sum_m 2^{-k-\frac{m}{2}} \leq 2^{-k+2} r_1^{1/2} \leq 2^{-k+2} r^{1/2}. \quad (88)$$

Also,

$$\sum_m 2^{m/2} H^1(K^m) \leq 2 \sum_m \int_{K^m} \mathrm{dist}(z, \partial B_1)^{-1/2} dH^1(z)$$
$$\leq 4 \int_{K \cap B_1} \mathrm{dist}(z, \partial B_1)^{-1/2} dH^1(z) = 4J(r_1) \leq 80 \alpha r^{1/2} \quad (89)$$

by (39), (20.19) and (20.20). Altogether,

$$\Delta_k \leq C\alpha^{3/2} r + C 2^{-k+2} \alpha^{1/2} r \leq 2C\alpha^{3/2} r, \quad (90)$$

at least for k large.

Recall from (55) and (53) that $\Delta_k = E_1 - E(k)$ and $\lim_{k \to +\infty} E(k) = E_0$. Then (90) implies that $E_1 - E_0 \leq 2C\alpha^{3/2} r$, which is the same as (20.47). (See the definition (20.44).)

This completes our proof of Energy Lemma 20.46, Theorem 18.16, and Theorem 18.1 in the special case of dimension $n = 2$. $\qquad\square$

22   Simple estimates on the Poisson kernel

In this section we want to prove the estimates on Poisson extensions that we already used in Sections 20 and 21 (see (20.31) and Lemma 21.7) and intend to use later as well.

Let $B = B(x, R) \subset \mathbb{R}^2$, $1 \leq p < +\infty$, and $v \in W^{1,p}(\partial B)$ be given. Denote by u the harmonic extension of v to B, which is obtained by convolution of v with the Poisson kernel. More precisely, set

$$P_r(\theta) = \frac{1 - r^2}{1 - 2r \cos \theta + r^2} \quad \text{for } 0 \leq r < 1 \text{ and } \theta \in \mathbb{R}, \tag{1}$$

and then

$$u(x + r Re^{i\theta}) = \frac{1}{2\pi} \int_0^{2\pi} P_r(t - \theta) v(x + Re^{it}) \, dt \tag{2}$$

for $0 \leq r < 1$ and $\theta \in \mathbb{R}$ (and with the usual identification of $\mathbb{R}^2$ with $\mathbb{C}$).

Lemma 3. *Let B, v and u be as above; in particular $p \in [1, +\infty)$ and $v \in W^{1,p}(\partial B)$. Then*

$$|\nabla u(z)| \leq C_p \, \mathrm{dist}(z, \partial B)^{-1/p} \left\| \frac{\partial v}{\partial \tau} \right\|_p \quad \text{for } z \in B. \tag{4}$$

Here $\frac{\partial v}{\partial \tau}$ denotes the (tangential) derivative of v on ∂B, and $\left\| \frac{\partial v}{\partial \tau} \right\|_p$ its L^p-norm. Lemma 3 is the same as Lemma 21.7.

First note that it is enough to prove the lemma when $x = 0$ and $R = 1$. Indeed we can define $\tilde{v}$ on $\partial B(0, 1)$ by $\tilde{v}(y) = v(x + Ry)$, and then the harmonic extension of $\tilde{v}$ on $B(0, 1)$ is given by $\tilde{u}(y) = u(x + Ry)$ for $y \in B(0, 1)$. If we can prove the lemma for $\tilde{v}$ and $\tilde{u}$, then

$$|\nabla u(z)| = R^{-1} \left| \nabla \tilde{u} \left(\frac{z - x}{R} \right) \right| \leq C_p R^{-1} \, \mathrm{dist} \left(\frac{z - x}{R}, \partial B(0, 1) \right)^{-1/p} \left\| \frac{\partial \tilde{v}}{\partial \tau} \right\|_p$$
$$= C_p R^{-1} R^{1/p} \, \mathrm{dist}(z, \partial B)^{-1/p} \left\| \frac{\partial v}{\partial \tau} \right\|_p = C_p \, \mathrm{dist}(z, \partial B)^{-1/p} \left\| \frac{\partial v}{\partial \tau} \right\|_p \tag{5}$$

for $z \in B$, as needed.

So let us assume that $x = 0$ and $R = 1$, and let $z \in B$ be given. Call z^* the point of ∂B which is closest to z (for $z = 0$, take any point $z^* \in \partial B$). Also set

$$\delta = \mathrm{dist}(z, \partial B) \quad \text{and} \quad B(z) = B \left(z, \frac{\delta}{2} \right). \tag{6}$$

Observe that by Proposition 9.1, v is (equal almost-everywhere to) a continuous function on ∂B. We claim that

$$|u(y) - v(z^*)| \leq C_p \delta^{1 - \frac{1}{p}} \left\| \frac{\partial v}{\partial \tau} \right\|_p \quad \text{for } y \in B(z). \tag{7}$$

Before we prove the claim, we should note that it implies the lemma. Indeed,

$$|\nabla v(z)| \leq C \delta^{-1} \|u(\cdot) - v(z^*)\|_{L^\infty(B(z))}; \tag{8}$$

this can be seen for instance by writing $u(\cdot) - v(z^*)$ near z as the Poisson integral of its values on $\partial B\left(z, \frac{\delta}{3}\right)$, and then differentiating under the integral sign. Then (4) follows from (7) and (8).

To prove the claim we can assume (without loss of generality) that $z \geq 0$ and $z^* = 1$. Note that

$$|v(e^{it}) - v(z^*)| = \left| \int_0^t \frac{\partial v}{\partial \tau}(e^{is})\,ds \right| \leq |t|^{1 - \frac{1}{p}} \left\| \frac{\partial v}{\partial \tau} \right\|_p \tag{9}$$

by Proposition 9.1 and Hölder. Also note that

$$1 - 2r \cos\theta + r^2 = (1 - r\cos\theta)^2 + r^2 \sin^2\theta = \operatorname{dist}(re^{i\theta}, 1)^2, \tag{10}$$

so that

$$P_r(t - \theta) = (1 - r^2) \operatorname{dist}\left(r e^{i(t-\theta)}, 1\right)^{-2} = (1 - r^2) \operatorname{dist}\left(r e^{i\theta}, e^{it}\right)^{-2} \tag{11}$$

for $r < 1$ and $\theta, t \in \mathbb{R}$.

Let $y \in B(z)$ be given. Write $y = r e^{i\theta}$, and apply (2). Then

$$u(y) - v(z^*) = \frac{1}{2\pi} \int_{-\pi}^{\pi} P_r(t - \theta)[v(e^{it}) - v(z^*)]\,dt \tag{12}$$

because $\frac{1}{2\pi} P_r(\cdot)$ is 2π-periodic and its integral is 1. Note that

$$\begin{aligned} P_r(t - \theta) &= (1 - r^2) \operatorname{dist}\left(y, e^{it}\right)^{-2} \\ &\leq 4(1 - r^2) \operatorname{dist}\left(z, e^{it}\right)^{-2} \leq 12\delta \operatorname{dist}\left(z, e^{it}\right)^{-2} \end{aligned} \tag{13}$$

by (11), because $y \in B(z)$, by (6), and because $1 - r^2 = (1+r)(1-r) \leq 2(1-r) \leq 3\delta$. Now

$$\begin{aligned} |u(y) - v(z^*)| &\leq \frac{6\delta}{\pi} \int_{-\pi}^{\pi} \operatorname{dist}\left(z, e^{it}\right)^{-2} |v(e^{it}) - v(z^*)|\,dt \\ &\leq \frac{6\delta}{\pi} \left\| \frac{\partial v}{\partial \tau} \right\|_p \int_{-\pi}^{\pi} \operatorname{dist}(z, e^{it})^{-2} |t|^{1 - \frac{1}{p}}\,dt, \end{aligned} \tag{14}$$

by (12), (13) and (9). But

$$\int_{-\pi}^{\pi} \operatorname{dist}(z, e^{it})^{-2} |t|^{1 - \frac{1}{p}}\,dt \leq C \int_0^{\pi} \frac{t^{1 - \frac{1}{p}}}{\delta^2 + t^2}\,dt \leq C\delta^{-1/p} \int_0^{\infty} \frac{s^{1 - \frac{1}{p}}}{1 + s^2}\,ds, \tag{15}$$

where the last inequality comes from the change of variables $t = \delta s$. The last integral converges because $p < +\infty$, our claim (7) follows from (14) and (15), and this completes the proof of Lemma 3. $\qquad\square$

Lemma 16. *Let $B = B(x, R) \subset \mathbb{R}^2$, $1 < p \leq 2$, and $v \in W^{1,p}(\partial B)$ be given. Denote by u the harmonic extension of v to B, as in the beginning of this section. Then $u \in W^{1,2}(B)$, and*

$$\int_B |\nabla u|^2 \leq C_p R^{2 - \frac{2}{p}} \Big\{ \int_{\partial B} \Big| \frac{\partial v}{\partial \tau} \Big|^p \Big\}^{2/p}. \tag{17}$$

We already used Lemma 16 with $p = 2$ for (20.31), but it will be more useful later in its less trivial version where $p < 2$.

As before, it is enough to prove Lemma 16 when $x = 0$ and $R = 1$. Indeed, set $\widetilde{v}(y) = v(x + Ry)$ on $\partial B(0, 1)$ and call $\widetilde{u}(y) = u(x + Ry)$ its harmonic extension. If (17) holds for $\widetilde{v}$ and $\widetilde{u}$, then

$$\int_B |\nabla u|^2 = \int_{B(0,1)} |\nabla \widetilde{u}|^2 \leq C_p \Big\{ \int_{\partial B(0,1)} \Big| \frac{\partial \widetilde{v}}{\partial \tau} \Big|^p \Big\}^{2/p} = C_p R^{2 - \frac{2}{p}} \Big\{ \int_{\partial B} \Big| \frac{\partial v}{\partial \tau} \Big|^p \Big\}^{2/p} \tag{18}$$

by linear changes of variables, as needed.

To prove (17) when $x = 0$ and $R = 1$, note that if $y = r e^{i\theta}$,

$$u(y) = \frac{1}{2\pi} \int_0^{2\pi} P_r(t - \theta) v(e^{it}) \, dt = \frac{1}{2\pi} \int_0^{2\pi} (1 - r^2) \operatorname{dist}(y, e^{it})^{-2} v(e^{it}) \, dt, \tag{19}$$

by (2) and (11). Let us differentiate (with respect to y) under the integral sign. We get that

$$\nabla u(y) = \int_0^{2\pi} Q_y(t) \, v(e^{it}) \, dt = \int_{\theta - \pi}^{\theta + \pi} Q_y(t) \, v(e^{it}) \, dt \tag{20}$$

for some vector-valued, 2π-periodic function Q_y such that

$$|Q_y(t)| \leq C(1 - r) \operatorname{dist}(y, e^{it})^{-3}. \tag{21}$$

Also,

$$\int_0^{2\pi} Q_y(t) \, dt = 0, \tag{22}$$

since this is the gradient at y of the harmonic extension of $v \equiv 1$. Next we want to integrate by parts. Set

$$R_y(t) = \int_t^{\theta + \pi} Q_y(s) \, ds = - \int_{\theta - \pi}^t Q_y(s) \, ds \tag{23}$$

for $\theta - \pi \leq t \leq \theta + \pi$. Then (20) yields

$$\nabla u(y) = \int_{\theta - \pi}^{\theta + \pi} Q_y(t) v(e^{it}) \, dt = \int_{\theta - \pi}^{\theta + \pi} R_y(t) \frac{\partial v}{\partial \tau}(e^{it}) \, dt, \tag{24}$$

where the integrated terms vanish because $R_y(\theta \pm \pi) = 0$. Let us first estimate $R_y(\theta + t)$ for $0 \leq t \leq \pi$:

$$|R_y(\theta + t)| = \left| \int_{\theta+t}^{\theta+\pi} Q_y(s) \, ds \right| = \left| \int_t^\pi Q_y(\theta + s) \, ds \right| \tag{25}$$

$$\leq C(1-r) \int_t^\pi \operatorname{dist}\left(y, e^{i(\theta+s)}\right)^{-3} ds = C(1-r) \int_t^\pi \operatorname{dist}\left(r, e^{is}\right)^{-3} ds$$

$$\leq C(1-r) \int_t^{+\infty} ((1-r) + s)^{-3} \, ds \leq C(1-r)((1-r) + t)^{-2}.$$

We have a similar estimate for $-\pi \leq t \leq 0$:

$$|R_y(\theta + t)| = \left| -\int_{\theta-\pi}^{\theta+t} Q_y(s) \, ds \right| = \left| \int_{-\pi}^t Q_y(\theta + s) \, ds \right| \tag{26}$$

$$\leq C(1-r) \int_{-\pi}^t \operatorname{dist}(y, e^{i(\theta+s)})^{-3} \, ds \leq C(1-r)((1-r) + |t|)^{-2},$$

where we can almost copy down the two last lines of (25).

Set $h_r(t) = (1-r)[(1-r) + |t|]^{-2}$ for $t \in [-\pi, \pi]$. Then

$$|\nabla u(y)| = \left| \int_{\theta-\pi}^{\theta+\pi} R_y(t) \frac{\partial v}{\partial \tau}(e^{it}) \, dt \right| \tag{27}$$

$$= \left| \int_{-\pi}^\pi R_y(\theta + t) \frac{\partial v}{\partial \tau}(e^{i(\theta+t)}) \, dt \right| \leq C \int_{-\pi}^\pi h_r(t) \left| \frac{\partial v}{\partial \tau}(e^{i(\theta+t)}) \right| dt,$$

by (24) and (26). Recall that $y = re^{i\theta}$. When we fix r and let θ vary, (27) says that $|\nabla u(re^{i\theta})|$ is dominated by the convolution of $\left|\frac{\partial v}{\partial \tau}\right|$ with h_r. Now recall that convolution maps $L^p \times L^q$ into L^λ when $\frac{1}{p} + \frac{1}{q} = \frac{1}{\lambda} + 1$. Here we are interested in $\lambda = 2$ and hence $\frac{1}{q} = \frac{3}{2} - \frac{1}{p} \in (\frac{1}{2}, 1]$. Since $\|h_r\|_q^q = (1-r)^q \int_{-\pi}^\pi [(1-r) + |t|]^{-2q} dt \leq C(1-r)^{1-q}$, (27) leads to

$$\left\{ \int_\theta |\nabla u(re^{i\theta})|^2 \, d\theta \right\}^{1/2} \leq C \|h_r\|_q \left\| \frac{\partial v}{\partial \tau} \right\|_p \tag{28}$$

$$\leq C(1-r)^{\frac{1}{q}-1} \left\| \frac{\partial v}{\partial \tau} \right\|_p = C(1-r)^{\frac{1}{2}-\frac{1}{p}} \left\| \frac{\partial v}{\partial \tau} \right\|_p.$$

Finally,

$$\int_B |\nabla u|^2 \leq C \int_0^1 r (1-r)^{1-\frac{2}{p}} \left\| \frac{\partial v}{\partial \tau} \right\|_p^2 dr \leq C_p \left\| \frac{\partial v}{\partial \tau} \right\|_p^2, \tag{29}$$

because $1 - \frac{2}{p} > -1$. This completes our proof of Lemma 16. $\square$

Remark 30. We could also recover Lemma 3 from (24) and (26). This time choose p such that $\frac{1}{p} + \frac{1}{q} = 1$; Hölder gives that

$$
|\nabla u(re^{i\theta})| \leq \left\| \frac{\partial v}{\partial \tau} \right\|_p \|R_y(\cdot)\|_q \leq C \left\| \frac{\partial v}{\partial \tau} \right\|_p \|h_r\|_q
$$
$$
\leq C_p(1-r)^{\frac{1}{q}-1} \left\| \frac{\partial v}{\partial \tau} \right\|_p = C_p(1-r)^{-\frac{1}{p}} \left\| \frac{\partial v}{\partial \tau} \right\|_p,
\tag{31}
$$

which is the same thing as (4) with $x = 0$, $R = 1$ and $z = y = re^{i\theta}$.

The following generalization of Lemma 16 to higher dimensions will be used in Part E.

Lemma 32. *For each* $p \in (\frac{2(n-1)}{n}, 2]$, *there is a constant* $C(n,p)$ *such that if* $B = B(x,r)$ *is a ball in* $\mathbb{R}^n$ *and* $f \in W^{1,p}(\partial B)$, *then there is a function* $v \in W^{1,2}(B)$ *such that*

$$
\int_B |\nabla v|^2 \leq C(n,p)\, r^{n-\frac{2n}{p}+\frac{2}{p}} \left\{ \int_{\partial B} |\nabla f|^p \right\}^{2/p}
\tag{33}
$$

and the radial limits of v *on* ∂B *coincide with* f *almost-everywhere.*

By homogeneity, it is enough to prove the lemma for $x = 0$ and $r = 1$. Indeed, if we know (33) on the unit ball and $\widetilde{v}$ and $\widetilde{f}$ are the analogues of v and f on the unit ball (as for Lemma 16 and (18)),

$$
\int_B |\nabla v|^2 = r^{n-2} \int_{B(0,1)} |\nabla \widetilde{v}|^2 \leq C_p\, r^{n-2} \left\{ \int_{\partial B(0,1)} |\nabla \widetilde{f}|^p \right\}^{2/p}
$$
$$
= C_p\, r^{n-2}\, r^{2-2\frac{n-1}{p}} \left\{ \int_{\partial B} |\nabla f|^p \right\}^{2/p},
\tag{34}
$$

as needed. So let us assume that $B = B(0,1)$. The easiest will be to use a fake Poisson kernel. Set $\delta(z) = \mathrm{dist}(z, \partial B)$ for $z \in B$ and let $\varphi : B \times \partial B \to B$ be a smooth function such that

$$
\int_{\partial B} \varphi(z,y)\, dH^{n-1}(y) = 1 \text{ for all } z \in B,
\tag{35}
$$

$$
|\varphi(z,y)| \leq C\delta(z)^{1-n} \text{ and } |\nabla_z \varphi(z,y)| \leq C\delta(z)^{-n}
\tag{36}
$$

everywhere, and

$$
\varphi(z,y) = 0 \text{ unless } y \in D(z) = \partial B \cap B(z, 2\delta(z)).
\tag{37}
$$

Such a function φ is easy to construct, for instance because partitions of unity in z allow us to restrict to z in a ball of radius ρ whose distance to ∂B is about ρ. Set

$$
v(z) = \int_{\partial B} \varphi(z,y) f(y)\, dH^{n-1}(y)
\tag{38}
$$

for $z \in B$. Note that v is smooth in B and

$$
\begin{aligned}
|\nabla v(z)| &= \left| \int_{\partial B} \nabla_z \varphi(z, y) f(y) \, dH^{n-1}(y) \right| \\
&= \left| \int_{D(z)} \nabla_z \varphi(z, y) [f(y) - m_{D(z)} f] \, dH^{n-1}(y) \right| \\
&\leq C\delta(z)^{-n} \int_{D(z)} |f(y) - m_{D(z)} f| \, dH^{n-1}(y) \\
&\leq C\delta(z)^{1-n} \int_{D(z)} |\nabla f(y)| \, dH^{n-1}(y),
\end{aligned}
\tag{39}
$$

where we used (35) to remove a constant $m_{D(z)} f$, and then (36) and Poincaré's inequality (12.12) in $D(z)$. We want to evaluate the L^2-norm of the restriction h_δ of $\nabla v(z)$ to the sphere $\partial B(0, 1-\delta)$. Note that (39) says that $|\nabla v(z)| \leq C \int_{\partial B} |\nabla f(y)|$ when $\delta(z) \leq 1/2$, so we do not need to worry about $\delta \leq 1/2$.

Because of (39), h_δ is dominated by the convolution of $|\nabla f|$ with δ^{1-n} times the characteristic function of a ball of radius 2δ in ∂B. The fact that we work on two slightly different spheres here (instead of just $\mathbb{R}^{n-1}$) does not matter; we leave the details. Now we want to use the fact that the convolution maps $L^p \times L^r$ into L^2, if we choose r so that $\frac{1}{p} + \frac{1}{r} = \frac{1}{2} + 1$. The L^r-norm of the convolving function is at most $C\delta^{1-n+\frac{n-1}{r}}$. Also, $1 - n + \frac{n-1}{r} = (1-n)(1 - \frac{1}{r}) = -(n-1)(\frac{1}{p} - \frac{1}{2}) = -(n-1)\frac{2-p}{2p}$, so that

$$
\|h_\delta\|_2^2 \leq C\delta^{-(n-1)\frac{2-p}{p}} \left\{ \int_{\partial B} |\nabla f|^p \right\}^{2/p}.
\tag{40}
$$

To get $\int_B |\nabla v(z)|^2$, we still need to integrate (40) with respect to δ, which is possible because the exponent is more than -1 when $p > \frac{2(n-1)}{n}$. This proves (33).

We still need to check that the radial limit of v is equal to f almost-everywhere on ∂B. Let us first verify that if we set

$$
T_r f(z) = \int_{\partial B} \varphi(rz, y) f(y) dH^{n-1}(y)
\tag{41}
$$

for $z \in \partial B$ and $0 < r < 1$, then

$$
\lim_{r \to 1^-} T_r f = f \quad \text{in } L^1(\partial B)
\tag{42}
$$

for $f \in L^1(\partial B)$. This is a quite standard argument about approximations of identity. Note that (42) holds when f is continuous, by (35)–(37). The general case follows because continuous functions are dense in $L^1(\partial B)$, and the operators T_r are uniformly bounded on $L^1(\partial B)$, by (36) and (37).

Thus (42) holds for our function $f \in W^{1,p}(\partial B)$. Since we also know that v has radial limits almost-everywhere (by Proposition 9.5, (13.5), and Proposition 11.4 (for the radial change of variable)), these limits must coincide with f. This completes our proof of Lemma 32. $\square$

23 Carleson measures

Traditionally Carleson measures are measures on $\mathbb{R}_+^{n+1} = \mathbb{R}^n \times (0, +\infty)$ such that the measure of a "Carleson box" $B(x,r) \times (0,r]$ is less than Cr^n. (Thus the homogeneity is n-dimensional.) Carleson measures show up naturally in a variety of situations in classical analysis, and it looks as if merely invoking the name is already a powerful tool. Actually, we shall not do much more than this in this section.

In the present context the role of $\mathbb{R}_+^{n+1}$ will be played by the set of balls $B(x,r)$ centered on K (our reduced minimizer, almost-minimizer, or quasiminimizer) and contained in Ω. This is a natural generalization; points $(x,t) \in \mathbb{R}_+^{n+1}$ are often seen as representing a scale t and a location x.

The idea of studying functions on $\mathbb{R}^n$ by looking at related functions on $\mathbb{R}_+^{n+1}$ is now very standard; think about harmonic extensions, square functions, Littlewood–Paley theory for instance. The idea of applying the same sort of techniques to study the regularity of (Ahlfors-regular) sets K by means of functions defined on the set of balls $K \times (0, +\infty)$ is much more recent, but quite powerful. You can already see it in [Jo1] and [Jo2], where the P. Jones numbers $\beta(x,t)$ show up, as well as some Carleson measure condition. Incidentally, the $\beta(x,t)$ are the same numbers that we shall use extensively in Part E to measure flatness, see (41.2). This point of view of studying sets in terms of functions defined on the set of balls becomes really explicit in [Se3], and then in [DaSe1,3] where one systematically works on $K \times (0, +\infty)$ or a set of "dyadic cubes", and one does stopping time arguments and analogues of Littlewood–Paley theory. An extremely useful tool there is the transposition of the "corona construction" of L. Carleson to this context (an idea of S. Semmes). See for instance [Ga], Chapter VIII.5, for a description of the corona construction in its original context.

In this section we do not intend to do anything fancy, just apply the definition of a Carleson measure. It is interesting to note that, even that way, the situation will improve and allow the smooth unfolding of proofs in the next sections.

Let us now state our hypotheses for this section. Let $\Omega \subset \mathbb{R}^n$ be open, and $K \subset \Omega$ be closed in Ω. Set

$$\Delta = \big\{ (x,r) \in K \times (0, +\infty) \, ; \, B(x,r) \subset \Omega \big\} \tag{1}$$

and, for $r_0 > 0$,

$$\Delta(r_0) = \big\{ (x,r) \in \Delta \, ; \, 0 < r \leq r_0 \big\}. \tag{2}$$

We think of Δ and Δ_0 as sets of balls centered on K.

We shall assume that K is locally Ahlfors-regular of dimension $n-1$, and more precisely that there are constants $C > 0$ and $r_0 \in (0, +\infty]$ such that

$$C^{-1} r^{n-1} \leq H^{n-1}(K \cap B(x,r)) \leq C r^{n-1} \text{ for } (x,r) \in \Delta(r_0). \tag{3}$$

We are also given a function $u \in W^{1,2}_{\text{loc}}(\Omega \setminus K)$ such that

$$\int_{B(x,r) \setminus K} |\nabla u|^2 \leq C r^{n-1} \text{ for } (x,r) \in \Delta(r_0). \tag{4}$$

Of course we think about Mumford-Shah minimizers and quasiminimizers, as in Remark 10 below. We want to estimate the quantities

$$\omega_p(x,t) = t^{1-\frac{2n}{p}} \left\{ \int_{B(x,t) \setminus K} |\nabla u|^p \right\}^{2/p} \tag{5}$$

for $1 \leq p \leq 2$ and $(x,t) \in \Delta(r_0)$. The exponents are chosen so that the $\omega_p(x,t)$ are dimensionless numbers (which will give a nicer statement for Theorem 8 below), but of course we are interested in

$$\int_{B(x,t) \setminus K} |\nabla u|^p = t^{n-\frac{p}{2}} \omega_p(x,t)^{p/2}. \tag{6}$$

Note that $\omega_2(x,t) \leq C$ for $(x,t) \in \Delta(r_0)$, by (4), and also

$$\omega_p(x,t) \leq t^{1-\frac{2n}{p}} |B(x,t)|^{\frac{2}{p}-1} \int_{B(x,t) \setminus K} |\nabla u|^2 = c \omega_2(x,t) \leq C \tag{7}$$

for $(x,t) \in \Delta(r_0)$ and $1 \leq p \leq 2$, by (5), Hölder and (4). As we shall see now, much better estimates hold on average when $p < 2$.

Theorem 8 [DaSe4]. *Let $K \subset \Omega$ and $u \in W^{1,2}_{\text{loc}}(\Omega \setminus K)$ satisfy the conditions (3) and (4). Then for $p \in [1,2)$ we can find $C_p \geq 0$ such that*

$$\int_{y \in K \cap B(x,r)} \int_{0 < t < r} \omega_p(y,t) \frac{dH^{n-1}(y)dt}{t} \leq C_p r^{n-1} \tag{9}$$

for $x \in K$ and $r > 0$ such that $(x,3r) \in \Delta(r_0)$.

As usual the constant C_p in (9) can be chosen to depend only on p, n, and the constant C in (3) and (4). It blows up when p tends to 2. Our choice of constant 3 is of course not optimal; anything larger than 1 would do. Let us make a few remarks before we prove the theorem.

Remark 10. Theorem 8 applies in particular when (u, K) is a reduced minimizer of the Mumford-Shah functional, or a reduced local almost-minimizer in Ω, or even more generally when $(u, K) \in TRLQ(\Omega)$, as in Standard Assumption 18.14. Our assumptions (3) and (4) then come from Theorem 18.16 and the trivial estimate in Lemma 18.19. [Also see Theorem 18.1 and (18.8) for the simpler case of reduced local almost-minimizers, and Proposition 7.8 for Mumford-Shah minimizers.]

Remark 11. We can see $\frac{dH^{n-1}(y)dt}{t}$ as the analogue on $\Delta(r_0) \subset K \times (0, +\infty)$ of the invariant measure $\frac{dxdt}{t}$ on R_+^{n+1}. Note that it is locally infinite:

$$\int_{K \cap B(x,r)} \int_{0 < t < r} \frac{dH^{n-1}(y)dt}{t} = +\infty \qquad (12)$$

for all $(x, r) \in \Delta(r_0)$, because $H^{n-1}(K \cap B(x, r)) > 0$ and $\int_0^1 \frac{dt}{t}$ diverges. Thus (9) says that $\omega_p(y, t)$ is small very often. We shall see a slightly weaker but very useful manifestation of this in Corollary 38.

We can restate Theorem 8 in terms of Carleson measures.

Definition 13. *A Carleson measure on* $\Delta(r_0)$ *is a (positive Borel) measure* μ *on* $\Delta(r_0)$ *such that*

$$\mu\big([K \cap B(x,r)] \times (0,r)\big) \leq Cr^{n-1} \text{ for } (x, 3r) \in \Delta(r_0). \qquad (14)$$

Here again we demanded that $B(x, 3r) \subset \Omega$ to avoid all potential problems near the boundary; this is not the important part of the definition.

Theorem 8 says that when (3) and (4) hold and $1 \leq p < 2$,

$$\omega_p(y, t) \frac{dH^{n-1}(y)dt}{t} \text{ is a Carleson measure on } \Delta(r_0). \qquad (15)$$

Remark 16. Our proof of Theorem 8 will not use any relation that could exist between u and K (other than (4)). To make this even more clear, let us call

$$g(z) = |\nabla u(z)| \text{ for } z \in \Omega \setminus K, \qquad (17)$$

record that

$$\int_{B(x,r) \setminus K} g(z)^2 \leq Cr^{n-1} \text{ for } (x, r) \in \Delta(r_0), \qquad (18)$$

and forget about u. The proof will only use (18), Hölder, Fubini, and the local Ahlfors-regularity property (3).

We want to apply Hölder, so let us choose two exponents. Define σ by

$$\frac{1}{\sigma} + \frac{p}{2} = 1; \qquad (19)$$

then $2 \leq \sigma < +\infty$ because $1 \leq p < 2$. Also choose b such that

$$0 < b < \frac{1}{\sigma}. \qquad (20)$$

Set $d(z) = \mathrm{dist}(z, K)$ for $z \in \Omega \setminus K$, write

$$g(z)^p = \left\{ g(z)^2 \left(\frac{d(z)}{t} \right)^{\frac{2b}{p}} \right\}^{\frac{p}{2}} \left\{ \frac{d(z)}{t} \right\}^{-b}, \tag{21}$$

and then apply Hölder to the integral in (5), as suggested by (21) and (19). We get that

$$\omega_p(y,t) = t^{1-\frac{2n}{p}} \left\{ \int_{B(y,t) \setminus K} g(z)^p \right\}^{2/p} \leq t^{1-\frac{2n}{p}} I(y,t) J(y,t)^{\frac{2}{p\sigma}}, \tag{22}$$

where

$$I(y,t) = \int_{B(y,t) \setminus K} g(z)^2 \left(\frac{d(z)}{t} \right)^{2b/p} \tag{23}$$

and

$$J(y,t) = \int_{B(y,t) \setminus K} \left(\frac{d(z)}{t} \right)^{-b\sigma}. \tag{24}$$

The following lemma will be useful to estimate $J(y,t)$. It relies on the local Ahlfors-regularity of K.

Lemma 25. *There is a constant C such that*

$$\left| \{ z \in B(y,t) \,;\, d(z) \leq \rho t \} \right| \leq C \rho t^n \tag{26}$$

for all (y,t) such that $(y, 2t) \in \Delta(r_0)$ and all $\rho \in (0,1]$.

It is enough to prove (26) when $\rho < \frac{1}{4}$, because the other case is trivial. Call H the set of (26); we want to cover H with a limited number of balls of radius $3\rho t$. Let Z denote a maximal subset of $K \cap \overline{B}\left(y, \frac{3t}{2}\right)$ whose points lie at mutual distances $\geq 2\rho t$. Let us check that

$$H \subset \bigcup_{z \in Z} \overline{B}(z, 3\rho t). \tag{27}$$

Let $z_0 \in H$ be given, and let $z_1 \in K$ be such that $|z_1 - z_0| = \mathrm{dist}(z_0, K) = d(z_0) \leq \rho t$. Such a point exists, because K is closed in Ω and $B(z_0, 2d(z_0)) \subset B(y, 2t) \subset \Omega$ (recall that $\rho < \frac{1}{4}$). Also, $|z_1 - y| \leq t + 2\rho t < \frac{3t}{2}$ (because $\rho < \frac{1}{4}$), so $z_1 \in B\left(y, \frac{3t}{2}\right)$ and we can find $z \in Z$ such that $|z - z_1| \leq 2\rho t$ (otherwise, we could add z_1 to Z and Z would not be maximal). Thus $|z - z_0| \leq 3\rho t$, and (27) follows.

By definition, the balls $B(z, \rho t)$, $z \in Z$, are disjoint and contained in $B(y, 2t)$. Hence

$$\sum_{z \in Z} H^{n-1}(K \cap B(z, \rho t)) \leq H^{n-1}(K \cap B(y, 2t)) \leq C t^{n-1} \tag{28}$$

by (3) and because $(y, 2t) \in \Delta(r_0)$. But the lower bound in (3) says that

$$H^{n-1}(K \cap B(z, \rho t)) \geq C^{-1} \rho^{n-1} t^{n-1} \tag{29}$$

for $z \in Z$, because $(z, \rho t) \in \Delta(r_0)$ as well. Hence Z has at most $C \rho^{1-n}$ elements, and $|H| \leq C \rho t^n$ by (27). This proves Lemma 25. $\qquad\square$

We may return to the integral $J(y, t)$. Assume that $y \in K \cap B(x, r)$ and $t \leq r$ for some pair (x, r) such that $(x, 3r) \in \Delta(r_0)$, as in Theorem 8 or Definition 13. Then $(y, 2r) \in \Delta(r_0)$ and we can apply Lemma 25. Cut the domain of integration $B(y, t) \setminus K$ into regions B_k, $k \geq 0$, where $d(z) \sim 2^{-k} t$. Then

$$\int_{B_k} \left(\frac{d(z)}{t} \right)^{-b\sigma} \leq C 2^{kb\sigma} |B_k| \leq C 2^{kb\sigma} 2^{-k} t^n \tag{30}$$

by Lemma 25, and

$$J(y, t) = \sum_k \int_{B_k} \left(\frac{d(z)}{t} \right)^{-b\sigma} \leq C t^n \tag{31}$$

by (24) and because $b\sigma < 1$ (by (20)). Now (22) yields

$$\omega_p(y, t) \leq C t^{1 - \frac{2n}{p}} I(y, t) t^{\frac{2n}{p\sigma}} = C t^{1-n} I(y, t) \tag{32}$$

because $1 - \frac{2n}{p} + \frac{2n}{p\sigma} = 1 - \frac{2n}{p} \frac{p}{2} = 1 - n$ by (19).

Next we want to substitute this in (9) and use Fubini. Call A the left-hand side of (9). Then, by (32) and (23),

$$A \leq \iiint_D t^{-n} g(z)^2 \left(\frac{d(z)}{t} \right)^{\frac{2b}{p}} dz \, dH^{n-1}(y) \, dt, \tag{33}$$

where we integrate on

$$D = \{(y, z, t); y \in K \cap B(x, r), z \in B(y, t) \setminus K, \text{ and } 0 < t < r\}. \tag{34}$$

Note that $z \in B(x, 2r) \setminus K$, $y \in K \cap B(z, t)$, and $0 < d(z) \leq t$ on our domain of integration. (The last one comes from the fact that $d(z) \leq |z - y|$.) Let us integrate in y first. Since $H^{n-1}(K \cap B(z, t)) \leq C t^{n-1}$ (by (3) and because $B(z, t) \subset B(x, 3r) \subset \Omega$), we get that

$$A \leq C \int_{z \in B(x, 2r) \setminus K} \int_{0 < d(z) \leq t \leq r} t^{-1} g(z)^2 \left(\frac{d(z)}{t} \right)^{\frac{2b}{p}} dt \, dz. \tag{35}$$

Note that

$$\int_{d(z)}^{+\infty} \left(\frac{d(z)}{t} \right)^{\frac{2b}{p}} \frac{dt}{t} = \int_1^{+\infty} \frac{dt}{t^{1 + \frac{2b}{p}}} < +\infty \tag{36}$$

by a change of variable and because $b > 0$, so that

$$A \leq C \int_{z \in B(x, 2r) \setminus K} g(z)^2 \, dz \leq C r^{n-1}, \tag{37}$$

by (18). This completes our proof of Theorem 8. $\qquad\qquad\square$

The following weaker version of Theorem 8 will be sufficient for most purposes (and it has a nice geometric flavor too).

Corollary 38. *Let $K \subset \Omega$ and $u \in W^{1,2}_{\mathrm{loc}}(\Omega \setminus K)$ satisfy (3) and (4). For each choice of $p \in [1,2)$ and each $\varepsilon > 0$ we can find a constant $C_5 = C_5(n,p,\varepsilon,C)$ such that for every $(x,r) \in \Delta(r_0)$ we can find $y \in K \cap B\left(x, \frac{r}{3}\right)$ and $0 < t < \frac{r}{3}$ such that*

$$t \geq C_5^{-1} r \tag{39}$$

and

$$\omega_p(y,t) \leq \varepsilon. \tag{40}$$

Corollary 38 also applies to almost- and quasiminimizers, as in Remark 10. We shall see in the next sections that it is very easy to use.

To prove the corollary, set

$$\mathcal{A} = \left[K \cap B\left(x, \frac{r}{3}\right) \right] \times \left[C_5^{-1} r, \frac{r}{3} \right] \subset \Delta(r_0). \tag{41}$$

Note that

$$\iint_{\mathcal{A}} \omega_p(y,t) \frac{dH^{n-1}(y)dt}{t} \leq C_p r^{n-1}, \tag{42}$$

by Theorem 8 (applied to the pair $\left(x, \frac{r}{3}\right)$). If we cannot find (y,t) as in the lemma, then $\omega_p(y,t) > \varepsilon$ on $\mathcal{A}$, and

$$\iint_{\mathcal{A}} \omega_p(y,t) \frac{dH^{n-1}(y)dt}{t} \geq \varepsilon \iint_{\mathcal{A}} \frac{dH^{n-1}(y)dt}{t} \tag{43}$$

$$= \varepsilon H^{n-1}\left(K \cap B\left(x, \frac{r}{3}\right) \right) \int_{C_5^{-1} r}^{r/3} \frac{dt}{t}$$

$$\geq C^{-1} \varepsilon r^{n-1} \operatorname{Log}\left(\frac{C_5}{3}\right)$$

by (41) and (3). We choose C_5 so large that this contradicts (42); Corollary 38 follows.   □

24   The property of projections $(n = 2)$

In this section Ω is an open subset of $\mathbb{R}^2$ and (u,K) is a (topologically) reduced or coral (topological) local quasiminimizer in Ω (in short, $(u,K) \in TRLQ(\Omega)$), as in Standard Assumption 18.14. As usual, this includes reduced Mumford-Shah minimizers and almost-minimizers. We want to prove that K has the "property of projections" introduced by F. Dibos and G. Koepfler [DiKo].

For each $\theta \in \mathbb{R}$, denote by L_θ the line (through the origin) with direction $e^{i\theta}$. Also call π_θ the orthogonal projection onto L_θ, and set $L_\theta^\perp = L_{\theta + \frac{\pi}{2}}$ and $\pi_\theta^\perp = \pi_{\theta + \frac{\pi}{2}}$.

Theorem 1 [DiKo]. *Let $\Omega \subset \mathbb{R}^2$ and $(u, K) \in TRLQ(\Omega)$ be given, and suppose that the constant $a \geq 0$ in (7.20) is small enough (depending on the other quasi-minimality constant M). Then there is a constant $C_6 > 0$ such that*

$$H^1\big(\pi_\theta(K \cap B(x,r))\big) + H^1\big(\pi_\theta^\perp(K \cap B(x,r))\big) \geq C_6^{-1}r \tag{2}$$

for all $x \in K$, all $r > 0$ such that $B(x,r) \subset \Omega$ and $r \leq r_0$, and all $\theta \in \mathbb{R}$.

Here r_0 is as in the definition of quasiminimizers, and the constant C_6 depends only on M. Thus, if we are dealing with almost-minimizers, or Mumford-Shah minimizers, we can take C_6 to depend on nothing (but r needs to be small enough).

See [Di] for an expanded version of the original proof. Our proof here will be based on [Lé1]; see also [DaSe5]. Before we do it, a few comments on the notion of rectifiability and the property of projections are in order.

Definition 3. *Let $0 < d \leq n$ be integers, and let $K \subset \mathbb{R}^n$ be a Borel set such that $H^d(K)$ is sigma-finite. We say that K is <u>rectifiable</u> if we can find a countable family $\{\Gamma_m\}$ of d-dimensional C^1 surfaces in $\mathbb{R}^n$, and a Borel set E such that $H^d(E) = 0$, so that*

$$K \subset E \cup \Big(\bigcup_m \Gamma_m\Big). \tag{4}$$

The terminology is not completely standard. Important sources like [Fe] would say "countably rectifiable" here, and Besicovitch would even have used the word "regular" (again!). We would have obtained an equivalent definition if instead of allowing only C^1 surfaces Γ_m we had used images under rotations of d-dimensional Lipschitz graphs, or even images of $\mathbb{R}^d$ by Lipschiz mappings. For this reason, we do not need to be too specific about what we mean by C^1 surfaces here (embedded or not, etc.), because all reasonable choices would yield equivalent definitions. For this and related issues we refer the reader to [Fa], [Fe], [Mat2], or [Morg1].

Rectifiability comes with an opposite twin notion.

Definition 5. *Let $0 \leq d \leq n$ be integers and $K \subset \mathbb{R}^n$ be a Borel set with sigma-finite H^d-measure. We say that K is <u>unrectifiable</u> (Besicovitch would have said irregular) if*

$$H^d(K \cap \Gamma) = 0 \text{ for every } C^1 \text{ surface } \Gamma \text{ of dimension } d \text{ in } \mathbb{R}^n. \tag{6}$$

Because of Definition 3, this is the same as requiring that $H^d(K \cap \Gamma) = 0$ for all rectifiable sets Γ. It is fairly easy to show that every Borel set K such that $H^d(K)$ is sigma-finite can be written as the disjoint union

$$K = K_{\text{rect}} \cup K_{\text{unrec}} \tag{7}$$

of a rectifiable piece K_{rect} and an unrectifiable piece K_{unrec}. The decomposition is unique, modulo sets of H^d-measure zero.

Rectifiable and unrectifiable sets have interesting and opposite behaviors in many respects. For instance, rectifiable sets have densities almost-everywhere, i.e.,

$$\lim_{r \to 0} r^{-d} H^d(K \cap B(x, r)) \tag{8}$$

exists for H^d-almost every $x \in K$ when K is rectifiable, and is even equal to the constant that we get by taking $K = \mathbb{R}^d$. This is not too hard to show, because we can reduce to C^1 surfaces by localization arguments that use analogues of the Lebesgue density theorem. On the opposite, if K is unrectifiable then for H^d-almost every $x \in K$ the limit in (8) does not exist. This is a really difficult theorem, especially in dimensions larger than 1 (where it is due to D. Preiss [Pr]).

Also, rectifiable sets have measure-theoretic (approximate) tangent d-planes almost-everywhere, while unrectifiable sets admit such tangents almost nowhere. Let us not define approximate tangents for the moment (this is done in Exercise 41.21 below); the main difference with usual tangent d-planes is that we want to allow additional sets of measure zero, and even sets of vanishing density at the given point of K, that would be far from the tangent d-plane. Let us also say that when K is Ahlfors-regular of dimension d, approximate tangents are automatically tangent d-planes in the usual sense. See Exercise 41.23.

Let us finally talk about the behavior with respect to projections. Let $\mathcal{G}(d)$ denote the Grassmann manifold of all d-planes in $\mathbb{R}^n$ through the origin. We are not interested in the full structure of $\mathcal{G}(d)$, just the fact that there is a reasonable measure on it (the rotation invariant probability measure). For each $V \in \mathcal{G}(d)$, denote by π_V the orthogonal projection from $\mathbb{R}^n$ to V. If K is rectifiable and $H^d(K) > 0$, then

$$H^d(\pi_V(K)) > 0 \text{ for almost-every } V \in \mathcal{G}(d). \tag{9}$$

This is not too hard to check, again because the main case is when K is a subset of a C^1 surface. When K is unrectifiable,

$$H^d(\pi_V(K)) = 0 \text{ for almost-every } V \in \mathcal{G}(d). \tag{10}$$

This is a hard theorem, due to A. Besicovitch when $d = 1$ and $n = 2$ and to H. Federer in the general case. It still has some mysterious aspects; for instance it is apparently very hard to come up with uniform estimates that generalize (10). [See Open question 33.]

Let us return to the property of projections in Theorem 1, and first check that it implies that K is rectifiable.

Decompose K into a rectifiable and an unrectifiable part, as in (7). Because of (10),

$$H^1(\pi_\theta(K_{\text{unrec}})) = 0 \tag{11}$$

for almost-every $\theta \in \mathbb{R}$. If $x \in K$, $0 < r \leq r_0$, and $B(x, r) \subset \Omega$, we may apply (2) to a $\theta \in \mathbb{R}$ such that (11) holds for θ and for $\theta + \frac{\pi}{2}$. We get that

$$
\begin{aligned}
C_6^{-1} r &\leq H^1\big(\pi_\theta(K \cap B(x, r))\big) + H^1\big(\pi_\theta^\perp(K \cap B(x, r))\big) \qquad (12)\\
&= H^1\big(\pi_\theta(K_{\text{rect}} \cap B(x, r))\big) + H^1\big(\pi_\theta^\perp(K_{\text{rect}} \cap B(x, r))\big)\\
&\leq 2H^1\big(K_{\text{rect}} \cap B(x, r)\big)
\end{aligned}
$$

because π is 1-Lipschitz and by Remark 2.11. To complete our proof of rectifiability we need to know that

$$
\limsup_{r \to 0} r^{-1} H^1(K_{\text{rect}} \cap B(x, r)) = 0 \qquad (13)
$$

for H^1-almost all $x \in K_{\text{unrec}}$. This is a standard density result; it only uses the fact that K_{rect} does not meet K_{unrec}, and that both sets have sigma-finite H^1-measures. The proof uses a standard covering argument, but we shall not do it here because the present argument is not central. See Theorem 6.2 on p. 89 of [Mat2].

From (12) and (13) we deduce that $H^1(K_{\text{unrec}}) = 0$, i.e., that K is rectifiable. This is not a very good motivation for Theorem 1, though. First, the initial proof of existence for minimizers of the Mumford-Shah functional (with functions of Special Bounded Variation) also gives the rectifiability of K. Also, the theorem of Besicovitch on projection is, I think, heavier than some of the arguments of the next sections that give better results (like uniform rectifiability). Note however that (2) is more precise and quantitative than mere rectifiability. When A. Dibos and G. Koepfler proved Theorem 1, they hoped that it would automatically imply better quantitative rectifiability results, but this approach failed so far. It is conceivable (and in my opinion, quite probable) that any Ahlfors-regular set K that satisfies the property of projections described in Theorem 1 is locally uniformly rectifiable (as in Section 27), but this seems quite hard to prove. For us it will be much more convenient to use the minimality (or quasiminimality) of the pair (u, K) again.

Proof of Theorem 1. Let $(u, K) \in TRLQ(\Omega)$ and (x, r) be as in the statement. With the notations of Section 23 (and in particular (23.2)), $(x, r) \in \Delta(r_0)$. Choose any $p \in (1, 2)$; for instance, $p = \frac{3}{2}$ will do. Also let $\varepsilon > 0$ be small, to be chosen later. We may apply Corollary 23.38 to the pair (x, r), because the assumptions (23.3) and (23.4) follow from Theorem 18.16 and Lemma 18.19. (See Remark 23.10.) We get a new pair (y, t), with

$$
y \in K \cap B\left(x, \frac{r}{3}\right), \qquad (14)
$$

$$
C_5^{-1} r \leq t \leq \frac{r}{3}, \qquad (15)
$$

and

$$
\omega_p(y, t) \leq \varepsilon. \qquad (16)
$$

The constant C_5 depends on ε, but this is all right.

Lemma 17. *If ε and a are small enough (depending on M), then*

$$H^1\big(\pi_\theta(K \cap B(y,t))\big) + H^1\big(\pi_\theta^\perp(K \cap B(y,t))\big) \geq \frac{t}{8} \tag{18}$$

for all $\theta \in \mathbb{R}$.

Let us first say why Theorem 1 follows from this lemma. The point is simply that $K \cap B(x,r)$ contains $K \cap B(y,t)$ (by (14) and (15)), so (18) yields (2) with $C_6 = 8C_5$ (by (15)). See how nice it is to have the Carleson measure result here: we just pay the relatively low price of multiplying C_6 by C_5, and in exchange we can assume that $\omega_p(y,t)$ is as small as we want.

We shall prove the lemma by contradiction, so let us assume that (18) fails for some $\theta \in \mathbb{R}$. Set $I = \left[\frac{t}{4}, \frac{t}{2}\right]$ and, for each $\rho \in I$, denote by Q_ρ the square centered at y, with sidelength 2ρ, and with sides parallel to L_θ and $L_\theta^\perp$. See Figure 1. Set

$$X = \{\rho \in I;\, \partial Q_\rho \cap K = \emptyset\}. \tag{19}$$

We want to check that

$$H^1(X) \geq \frac{t}{8}. \tag{20}$$

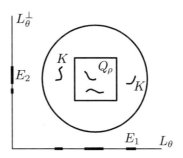

Figure 1

Let us assume that $y = 0$ to simplify the notation, and set

$$E_1 = \pi_\theta(K \cap B(y,t)) \quad \text{and} \quad E_2 = \pi_\theta^\perp(K \cap B(y,t)). \tag{21}$$

If $\rho \in I \setminus X$, then one of the four sides of ∂Q_ρ meets $K \cap B(y,t)$ (because $Q_\rho \subset B(y,t)$). Hence $\rho \in E_1$, or $-\rho \in E_1$, or $\rho \in E_2$, or $-\rho \in E_2$. In other words

$$I \setminus X \subset \varphi(E_1) \cup \varphi(E_2), \tag{22}$$

where we set $\varphi(\xi) = |\xi|$, and hence

$$H^1(I \setminus X) \leq H^1(E_1) + H^1(E_2) \leq \frac{t}{8}, \tag{23}$$

by Remark 2.11 and because we assumed that (18) fails. Obviously (20) follows from this.

Next observe that

$$\int_I \left\{ \int_{\partial Q_\rho \setminus K} |\nabla u|^p \, dH^1 \right\} d\rho \leq \int_{B(y,t) \setminus K} |\nabla u|^p = t^{2-\frac{p}{2}} \, \omega_p(y,t)^{p/2} \leq \varepsilon^{p/2} \, t^{2-\frac{p}{2}}$$

(24)

by (23.6) and (16). By (20) and Chebyshev, we can find $\rho \in X$ such that

$$\int_{\partial Q_\rho} |\nabla u|^p \, dH^1 \leq 10 \varepsilon^{p/2} t^{1-\frac{p}{2}}.$$

(25)

[Recall that $\partial Q_\rho \cap K = \emptyset$ for $\rho \in X$.] We can also require that in addition,

$$u|_{\partial Q_\rho} \in W^{1,p}(\partial Q_\rho)$$

(26)

and the derivative of $u|_{\partial Q_\rho}$ is given by restriction of ∇u, because these additional properties are satisfied for almost every $\rho \in X$. See Corollary 9.13, and also note that X is open in I. Finally recall from Lemma 18.22 that u is (equal almost-everywhere to) a continuous functions on $\Omega \setminus K$, and in particular near ∂Q_ρ.

So let $\rho \in X$ satisfy the properties above. We want to use Lemma 22.16 to find a continuous function u_1 on $\overline{Q}_\rho$ such that $u_1 = u$ on ∂Q_ρ, $u_1 \in W^{1,2}(Q_\rho)$, and

$$\int_{Q_\rho} |\nabla u_1|^2 \leq C \rho^{2-\frac{2}{p}} \left\{ \int_{\partial Q_\rho} \left| \frac{\partial u}{\partial \tau} \right|^p \right\}^{2/p} \leq C \varepsilon t$$

(27)

(by (22.17), (25), (26), and because $\rho \leq t$). Of course we cannot apply Lemma 22.16 directly because Q_ρ is not a disk, but this is not a serious problem. We can find a bilipschitz mapping $\varphi : \overline{Q}_\rho \to \overline{B}(0, \rho)$ which is smooth (or piecewise smooth, to make the construction easier) on Q_ρ, and we can take $u_1 = v_1 \circ \varphi$, where v_1 is the harmonic extension on $\overline{B}(0, \rho)$ of $(u|_{\partial Q_\rho}) \circ \varphi^{-1}$.

Let us use our new function u_1 to construct a competitor for (u, K). Set

$$K_1 = K \setminus Q_\rho,$$

(28)

and extend u_1 to $\Omega \setminus K_1$ by taking $u_1(z) = u(z)$ on $\Omega \setminus [Q_\rho \cup K]$. It is clear that K_1 is closed in Ω (because K is), and $u_1 \in W^{1,2}_{\text{loc}}(\Omega \setminus K_1)$ because of Welding Lemma 14.4. (The argument is essentially the same as for (20.30), so we do not elaborate.) Thus $(u_1, K_1) \in \mathcal{A}$ (see (7.1)), and it is a competitor for (u, K) in $B(y, t)$ (as in Definition 7.2) because we did not modify anything out of $\overline{Q}_\rho \subset B(y, t)$. It is even a topological competitor (as in Definition 7.15), by the proof of Lemma 20.10 (applied with $B = Q_\rho$, and where the analogue of (20.11) holds because $\partial Q_\rho \subset \Omega \setminus K$ by (19)). Since $B(y, t) \subset\subset B(x, r) \subset \Omega$ by assumption, we can apply our hypothesis that $(u, K) \in TRLQ(\Omega)$ and use (7.20).

With the notation of (7.18) and (7.19),

$$\widetilde{E} = \int_{\Omega \cap B(y,t) \setminus K_1} |\nabla u_1|^2 \leq E + \int_{Q_\rho} |\nabla u_1|^2 \leq E + C\varepsilon t, \tag{29}$$

hence

$$\delta E =: \operatorname{Max}\{(\widetilde{E} - E), M(\widetilde{E} - E)\} \leq C M \varepsilon t, \tag{30}$$

and (7.20) says that

$$H^1(K \cap Q_\rho) = H^1(K \setminus K_1) \leq M H^1(K_1 \setminus K) + \delta E + at \leq C M \varepsilon t + at. \tag{31}$$

But we also know from Theorem 18.16 (and (14), (15), and our hypotheses that $B(x,r) \subset \Omega$ and $r \leq r_0$) that

$$H^1(K \cap Q_\rho) \geq H^1\left(K \cap B\left(y, \frac{t}{4}\right)\right) \geq C^{-1} t. \tag{32}$$

(See also the definition of Q_ρ before (19)). Thus, if we choose ε and a small enough, we get the desired contradiction. This completes our proof of Lemma 17 and Theorem 1. $\qquad\square$

Open question 33. Let $K \subset \mathbb{R}^2$ be an Ahlfors-regular set (of dimension 1). Suppose that for all $x \in K$ and $r > 0$,

$$H^1(\pi_\theta(K \cap B(x,r))) \geq C^{-1} r \tag{34}$$

for all $\theta \in \mathbb{R}$ (and where π_θ denotes the orthogonal projection onto a line L_θ of direction $e^{i\theta}$). Can we prove that K is uniformly rectifiable (i.e., that it is contained in an Ahlfors-regular curve, see Section 27)? If the answer is yes we can ask analogous questions in larger dimensions and codimensions, or just require that (34) holds for $\theta \in D(x,r)$, where $D(x,r) \subset [0, 2\pi]$ has measure $\geq C^{-1}$, or just ask similar questions with a single ball (and without assuming Ahlfors-regularity).

Exercise 35. (Could we reduce to rectifiable sets?)

1. Let K be a closed unrectifiable set in a domain Ω, with $H^{n-1}(K) < +\infty$. Show that K is removable for $W^{1,p} \cap L^1_{\text{loc}}(\Omega)$, as in Exercise 10.22. Thus K cannot be the singular set in a reduced Mumford-Shah minimizer, for instance.

2. Find a compact set $K \subset \mathbb{R}^2$ such that the rectifiable part K_{rect} of K is dense in K, but $H^1(K_{rect}) < H^1(K)$; for instance, you can manage so that the unrectifiable part is the Cantor set of Exercise 18.26. Also find a function $u \in W^{1,2}(\mathbb{R}^2 \setminus K)$ that does not lie in $W^{1,2}(\mathbb{R}^2 \setminus \widetilde{K})$ for any closed set $\widetilde{K}$ strictly contained in K.

Thus it seems hard to reduce immediately our study of quasiminimizers (or just Mumford-Shah minimizers) to pairs (u, K) where K is rectifiable. Theorem 1 says that K is rectifiable when (u, K) is a quasiminimizer, but observe that unless we prove the existence of minimizers for the Mumford-Shah functional (for instance), we do not get that the infimum of the functional on $\mathcal{A}$ is the same as its infimum on the smaller class $\mathcal{A}' = \{(u, K) \in \mathcal{A}; K \text{ is rectifiable}\}$.

25   The concentration property $(n = 2)$

In this section we keep the same hypotheses as in the previous one, i.e., that $\Omega \subset \mathbb{R}^2$ is open and $(u, K) \in TRLQ(\Omega)$, as in Standard Assumption 18.14, and we prove that K has the "concentration property" introduced in [DMS]. (Maybe we should say "uniform concentration property" instead, but here everything tends to be uniform anyway.) This property may look a little strange at first, but it will be very useful when we try to prove the convergence of Hausdorff measures for sequences of minimizers. See Section 4 for a presentation of the problem, and Sections 35 and 38–40 for the relevant theorems.

Theorem 1 [DMS]. *Let $\Omega \subset \mathbb{R}^2$ and $(u, K) \in TRLQ(\Omega)$ be given (as in Standard Assumption 18.14), and denote by r_0, M, a the usual quasiminimality constants (as in Definition 7.21). If a is small enough (depending on M), then for each $\varepsilon > 0$ we can find $C_7 = C_7(\varepsilon, M)$ such that if $x \in K$ and $0 < r \leq r_0$ are such that $B(x, r) \subset \Omega$, then there exists a pair (y, t) such that*

$$y \in K \cap B\left(x, \frac{r}{3}\right), \tag{2}$$

$$C_7^{-1} r \leq t \leq \frac{r}{3}, \tag{3}$$

and

$$H^1(K \cap B(y, t)) \geq 2(1 - \varepsilon) t. \tag{4}$$

Thus for each disk $B(x, r) \subset \Omega$ centered on K and with $r \leq r_0$, we can find another disk $B(y, t)$ centered on K, contained in $B(x, r)$, not too small compared to $B(x, r)$, and in which the set K is almost "optimally concentrated". Here optimal concentration corresponds to what we would have if K were a line; it also fits with expected densities at points where K is smooth. This is the reason why the property described in Theorem 1 is called a "concentration property". Note that our examples of sequences $\{K_k\}$ of compact subsets of the line which converge to a limit K_∞ and for which $H^1(K_\infty) > \limsup_{k \to +\infty} H^1(K_k)$ (see after (4.4)) do not satisfy the concentration property in any uniform way. This is of course coherent with the results of Section 35.

Remark 5. There is an analogue of Theorem 1 when $n \geq 3$ and $(u, K) \in TRLQ(\Omega)$ (with a small enough, depending on M and n). The statement is similar; we just have to replace (4) with

$$H^{n-1}(K \cap B(y, t)) \geq (1 - \varepsilon) H^{n-1}(\mathbb{R}^{n-1} \cap B(0, t)). \tag{6}$$

This was proved in [MaSo1] (for Mumford-Shah minimizers, extended in [MaSo4] to almost-minimizers; probably the argument works for quasiminimizers too), as a consequence of the bisection property of [So2]. There is also an independent proof, based on the uniform rectifiability of K, and which will be rapidly described in Remark 75.17.

Very roughly, the idea is that since K is (locally) uniformly rectifiable, we can find pairs (y, t) such that $\omega_p(y, t) \leq \varepsilon$, but also $K \cap B(y, t)$ stays εt-close to some hyperplane P. It is not too difficult to show that in such circumstances something like (6) must hold (with a larger value of ε). One can even show that if π denotes the orthogonal projection onto P, then

$$H^{n-1}\big([P \cap B(y, t)] \setminus \pi(K \cap B(y, t))\big) \leq \widetilde{\varepsilon}\, t^{n-1}, \tag{7}$$

with an $\widetilde{\varepsilon}$ that goes to 0 with ε. The proof is similar to other arguments in [DaSe6], or to the proof of Propositions 42.10 and 44.1 below.

Proof of Theorem 1. Let $(u, K) \in TRLQ(\Omega)$, choose $p \in (1, 2)$, like $p = \frac{3}{2}$, and let $\varepsilon > 0$ be given. Let $\varepsilon_1 > 0$ be very small, depending on ε and M, and to be chosen later. Then let (x, r) be as in the statement. We want to apply Corollary 23.38 to the pair (x, r) and the constant ε_1. The assumptions (23.3) and (23.4) are satisfied, by Theorem 18.16 and Lemma 18.19 (see also Remark 23.10). Thus Corollary 23.38 applies, and gives a new pair (y, t) such that (2) and (3) hold (with $C_7 = C_5$), and in addition

$$\omega_p(y, t) \leq \varepsilon_1. \tag{8}$$

The constant C_5 depends only on ε_1, p if we want to take $p \neq \frac{3}{2}$, and the constants in Theorem 18.16 and Lemma 18.19. These constants depend only on M (if a is small enough, as always), so eventually $C_7 = C_5$ will only depend on M and ε (through ε_1).

Thus it is enough to prove that our pair (y, t) satisfies (4) if a is small enough (depending on M) and ε_1 is small enough (depending on M and ε). As usual, we shall proceed by contradiction, suppose that (4) fails, and try to reach a contradiction by constructing an appropriate competitor for (u, K) in $B(y, t)$. Set

$$X_1 = \big\{\rho \in (0, t)\, ; \, \partial B(y, \rho) \cap K \text{ has at least two points}\big\}. \tag{9}$$

We claim that

$$H^1(X_1) \leq \frac{1}{2} H^1(K \cap B(y, t)) \leq (1 - \varepsilon)t \tag{10}$$

(if (4) fails). This should not surprise the reader, since the radial projection $\pi :$ $z \to |z - y|$ is 1-Lipschitz and at least 2-to-1 from $K \cap \pi^{-1}(X_1)$ onto X_1, but a small verification is needed. Let us not do it now, and refer to Lemma 26.1 below. With the notations of that lemma, (10) holds because $2H^1(X_1) \leq \int_{X_1} N(s)\, ds \leq H^1(K \cap B(y, t))$, by (26.2).

Next set $X = \big(\frac{\varepsilon t}{2}, t\big) \setminus X_1$. Then

$$H^1(X) \geq t - \frac{\varepsilon t}{2} - H^1(X_1) \geq \frac{\varepsilon t}{2}, \tag{11}$$

by (10). Note that

$$\int_X \int_{\partial B(y, \rho) \setminus K} |\nabla u|^p \, dH^1 \, d\rho \leq \int_{B(y, t) \setminus K} |\nabla u|^p = t^{2 - \frac{p}{2}} \omega_p(y, t)^{p/2} \leq \varepsilon_1^{p/2} t^{2 - \frac{p}{2}} \tag{12}$$

by (23.6) and (8). Thus we can use (11) and Chebyshev to find a set of positive measure of radii $\rho \in X$ such that

$$\int_{\partial B(y,\rho) \setminus K} |\nabla u|^p \, dH^1 \leq \frac{4}{\varepsilon} \varepsilon_1^{p/2} t^{1-\frac{p}{2}}. \tag{13}$$

Let us check that we can even choose ρ so that in addition

$$u|_{\partial B(y,\rho) \setminus K} \in W^{1,p}(\partial B(y,\rho) \setminus K), \tag{14}$$

and the tangential derivative of $u|_{\partial B(y,\rho) \setminus K}$ is given by the restriction of ∇u to $\partial B(y,\rho) \setminus K$.

To see this, let us cover $B(y,t) \setminus K$ by countably many little pseudorectangles

$$R_k = \{ \rho e^{i\theta}; \rho_k^- < \rho < \rho_k^+ \text{ and } \theta_k^- < \theta < \theta_k^+ \} \tag{15}$$

that do not meet K. We can change variables and use polar coordinates (as in Proposition 11.4); this transforms $u|_{R_k}$ into a function v_k on the rectangle $(\rho_k^-, \rho_k^+) \times (\theta_k^-, \theta_k^+)$. By Corollary 9.13, v_k is absolutely continuous on $\{\rho\} \times (\theta_k^-, \theta_k^+)$ for almost-every $\rho \in (\rho_k^-, \rho_k^+)$. Call $E_k \subset (\rho_k^-, \rho_k^+)$ the corresponding exceptional set of zero length. Thus for $\rho \in (\rho_k^-, \rho_k^+) \setminus E_k$, u is absolutely continuous on the arc of circle $R_k \cap \partial B(y,\rho)$. If we take $\rho \in X \setminus \left(\underset{k}{\cup} E_k \right)$, then (14) holds. The additional constraint that the tangential derivative be given by the restriction of ∇u can be taken care of similarly (or at the same time).

Also recall from Lemma 18.22 that u is continuous on $\Omega \setminus K$, so we shall not have difficulties with radial limits almost-everywhere. Incidentally, even if we only knew that $u \in W^{1,2}(B(x,r) \setminus K)$, we would be able to prove that for almost-every ρ, $u(\rho e^{i\theta}) = \lim_{r \to \rho} u(r e^{i\theta})$ for a.e. θ anyway.

Choose $\rho \in X$ such that (13) and (14) hold. Set $B = B(y,\rho)$, and let us construct a competitor for (u,K) in B.

Let us assume that $K \cap \partial B$ is not empty, and call ξ the point of intersection. (There is only one, because $\rho \in X = \left(\frac{\varepsilon t}{2}, t \right) \setminus X_1$ and by (9).) If $K \cap \partial B = \emptyset$, we can always pick $\xi \in \partial B$ at random and follow the argument below, but the most reasonable thing to do would be to replace $u|_B$ with the harmonic extension of $u|_{\partial B}$, and conclude as in Section 24.

Let Z be the closed arc of ∂B centered at ξ and with length

$$H^1(Z) = \frac{1}{2M} H^1(K \cap B) \geq C^{-1} \rho, \tag{16}$$

where the inequality comes from Theorem 18.16 (the local Ahlfors-regularity result) and as usual C may depend on M. See Figure 1. If $H^1(K \cap B) > 4\pi M \rho$, we can take $Z = \partial B$, but don't worry, this will not happen.

Let v denote the continuous function on ∂B that coincides with u on $\partial B \setminus Z$ and is linear on Z. Then $v \in W^{1,p}(\partial B)$ (by (14) and because $K \cap \partial B \subset Z$), and

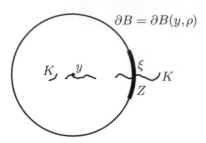

$$\partial B = \partial B(y,\rho)$$

Figure 1

the size of its constant tangential derivative on Z is at most

$$H^1(Z)^{-1}\int_{\partial B\setminus Z}|\nabla u|dH^1 \le H^1(Z)^{-1}H^1(\partial B)^{1-\frac{1}{p}}\Big\{\int_{\partial B\setminus K}|\nabla u|^p dH^1\Big\}^{1/p}, \quad (17)$$

by the constraint after (14) and Hölder's inequality. Then

$$\begin{aligned}
\int_{\partial B}\Big|\frac{\partial v}{\partial\tau}\Big|^p &= \int_{\partial B\setminus Z}\Big|\frac{\partial u}{\partial\tau}\Big|^p + \int_{Z}\Big|\frac{\partial v}{\partial\tau}\Big|^p \\
&\le \int_{\partial B\setminus Z}|\nabla u|^p + H^1(Z)^{1-p}H^1(\partial B)^{p-1}\int_{\partial B\setminus K}|\nabla u|^p \qquad (18)\\
&\le C\int_{\partial B\setminus K}|\nabla u|^p \le C\varepsilon^{-1}\varepsilon_1^{p/2}t^{1-\frac{p}{2}}
\end{aligned}$$

by (17), (16), and (13).

Next call u_1 the harmonic extension of v to B. That is, u_1 is continuous on $\overline{B}$, coincides with v on ∂B, and is harmonic on B. By Lemma 22.16,

$$u_1 \in W^{1,2}(B) \qquad (19)$$

and

$$\int_B|\nabla u_1|^2 \le C\rho^{2-\frac{2}{p}}\Big\{\int_{\partial B}\Big|\frac{\partial v}{\partial\tau}\Big|^p\Big\}^{2/p} \le C\rho^{2-\frac{2}{p}}\varepsilon^{-2/p}\varepsilon_1 t^{\frac{2}{p}-1} \le C\varepsilon^{-2/p}\varepsilon_1 t, \quad (20)$$

by (18) and because $\rho \le t$. Also define u_1 on $\Omega\setminus(\overline{B}\cup K)$ by keeping $u_1 = u$ there. Finally set

$$K_1 = (K\setminus B)\cup Z. \qquad (21)$$

First, K_1 is closed in Ω, because B is open and Z is closed. Next,

$$u_1 \in W^{1,2}_{\text{loc}}(\Omega\setminus(\overline{B}\cup K)) \qquad (22)$$

because the same thing holds for u (and $(u,K)\in\mathcal{A}$; see (7.1)). We even have that

$$u_1 \in W^{1,2}\big(B(y,t)\setminus(\overline{B}\cup K)\big), \qquad (23)$$

for instance by Lemma 18.19 (the trivial estimate). Thus $u_1 \in W^{1,2}(B(y,t) \setminus [K_1 \cup \partial B])$, by (23) and (19). Since the radial limits of u_1 on $\partial B \setminus Z$ from both sides coincide almost-everywhere (by definition of v and u_1, and because u is continuous), we can apply Corollary 14.28. We get that $u_1 \in W^{1,2}(B(y,t) \setminus K_1)$. Then $u_1 \in W^{1,2}_{\text{loc}}(\Omega \setminus K_1)$, by (22), and so $(u_1, K_1) \in \mathcal{A}$. Obviously it is a competitor for (u, K) in B (see Definition 7.2). It is even a topological competitor, by Lemma 20.10 and because $\partial B \setminus K_1 = \partial B \setminus Z$ is connected and contained in $\Omega \setminus K$.

Altogether (u_1, K_1) is a topological competitor for (u, K) in B and, since $\overline{B} \subset B(y,t) \subset \Omega$ by definitions, we can apply the quasiminimality condition (7.20) (see Definition 7.21). Let $E, \widetilde{E}$, and δE be as in (7.18) and (7.19). Then $\widetilde{E} = \int_B |\nabla u_1|^2$,

$$\delta E = \text{Max}\{(\widetilde{E} - E), M(\widetilde{E} - E)\} \leq M\widetilde{E} \leq C_M \varepsilon^{-2/p} \varepsilon_1 t, \qquad (24)$$

where we exceptionally show the dependence on M, and by (20). Now (7.20) says that

$$H^1(K \setminus K_1) \leq M H^1(K_1 \setminus K) + \delta E + a\rho, \qquad (25)$$

and since

$$H^1(K \setminus K_1) - M H^1(K_1 \setminus K) = H^1(K \cap B) - M H^1(Z) = \frac{1}{2} H^1(K \cap B) \geq \frac{\rho}{C_M'} \qquad (26)$$

by (16), we get that

$$\frac{\rho}{C_M'} \leq \delta E + a\rho \leq C_M \varepsilon^{-\frac{2}{p}} \varepsilon_1 t + a\rho \leq 2C_M \varepsilon^{-1-\frac{2}{p}} \varepsilon_1 \rho + a\rho, \qquad (27)$$

where the last inequality holds because $\rho \geq \varepsilon t/2$ (by definition of X). If a is small enough (precisely, $a \leq 1/(2C_M')$), and if we choose ε_1 small enough (depending on M and ε, but this is allowed), we get the desired contradiction. [Recall the general strategy exposed just above (9).] Our proof of Theorem 1 is complete, modulo Lemma 26.1 below. $\qquad \square$

Remark 28. Theorem 1 is still true for quasiminimizers with linear distortion (see Definition 7.52), with almost the same proof. The same remark holds for the results of the last few sections, and also for uniform rectifiability in the next sections. In all these cases, we only construct one competitor, at the scale where we want the desired property. This competitor is constructed with a harmonic extension, but we do not even need to pay attention and use a change of variables, because all energies are small anyway. Also see Remark 75.18 for the analogous properties in higher dimensions.

It should also be noted that we have a little more than the concentration property: if (u, K) satisfies the hypotheses of Theorem 1 or Remark 5 (or even with linear distortion), and if A is (the linear map associated to) an invertible $n \times n$ matrix, then $A(K)$ satisfies the conclusions of Theorem 1 and Remark 5 (with respect to the domain $A(\Omega)$), uniformly in A provided that $||A||$ and $||A^{-1}||$ stay bounded. The proof stays the same, except that we may have to work with ellipses instead of circles. This may be useful if we try to compute $H^{n-1}(A(K))$ and take limits, as in Part D.

26   Average number of points on a circle

The following lemma is not too surprising, but since we intend to use it a few times we shall try to prove it reasonably carefully.

Lemma 1. *Let $B = B(y,t)$ be a disk in $\mathbb{R}^2$ and $K \subset \mathbb{R}^2$ be a closed set such that $H^1(K \cap B) < +\infty$. For each $s \in (0,t)$ call $N(s) \in [0,+\infty]$ the number of points in $K \cap \partial B(y,s)$. Then N is Borel-measurable on $(0,t)$ and*

$$\int_0^t N(s)ds \leq H^1(K \cap B). \tag{2}$$

Of course most of our restrictive hypotheses are just here for convenience.

To prove the lemma it will help to discretize a little. Let $\tau > 0$ be small, and define a nondecreasing sequence of functions $\theta_n : (0,t) \to [0,2\pi]$ as follows. If $y + s \in K$, take $\theta_1(s) = 0$. Otherwise, set

$$\theta_1(s) = \sup \left\{ \theta \in [0,2\pi]; y + se^{i\xi} \notin K \text{ for } 0 \leq \xi \leq \theta \right\}. \tag{3}$$

Now suppose that $\theta_n(s)$ has already been defined. Set

$$\theta_{n+1}(s) = 2\pi \text{ if } \theta_n(s) \geq 2\pi - \tau. \tag{4}$$

Otherwise, set

$$\theta_{n+1}(s) = \theta_n(s) + \tau \text{ if } y + se^{i(\theta_n(s)+\tau)} \in K, \tag{5}$$

and, in the remaining case,

$$\theta_{n+1}(s) = \sup \left\{ \theta \in [\theta_n(s) + \tau, 2\pi]; y + se^{i\xi} \notin K \text{ for } \theta_n(s) + \tau \leq \xi \leq \theta \right\}. \tag{6}$$

In other words, we start from the point $y + s$ and turn along $\partial B(y,s)$ until we get a first point in K. This point is $y + se^{i\theta_1(s)}$. Then we continue our trip along $\partial B(y,s)$, and consider the first point of K after a grace period of length τs. We get our second point $y + se^{i\theta_2(s)} \in K$. We continue like this until $\theta_n(s) \geq 2\pi - \tau$, or there is no point of $K \cap \partial B(y,s)$ before we return to $y + s$ for the first time. When this happens, we set $\theta_{n+1}(s) = 2\pi$, and even $\theta_m(s) = 2\pi$ for all $m \geq n + 1$.

Call $N_\tau(s) + 1$ the smallest integer n such that $\theta_n(s) = 2\pi$; this will be our discrete approximation to the function N. By definitions,

$$y + se^{i\theta_n(s)} \in K \text{ when } 1 \leq n \leq N_\tau(s) \tag{7}$$

because K is closed,

$$\theta_{n+1}(s) \geq \theta_n(s) + \tau \quad \text{when} \quad 1 \leq n \leq N_\tau(s) - 1, \tag{8}$$

and in particular $N_\tau(s) \leq 2\pi\tau^{-1}$. Note also that

$$N_\tau(s) \leq N(s) \quad \text{and} \quad N(s) = \lim_{\tau \to 0} N_\tau(s). \tag{9}$$

Let us prove (by induction on n) that θ_n is lower semicontinuous, i.e., that

$$\theta_n(s) \leq \liminf_{k \to +\infty} \theta_n(s_k) \quad \text{when} \quad s = \lim_{k \to +\infty} s_k. \tag{10}$$

Let $\{s_k\}$ converge to s, and set $\theta_n^* = \liminf_{k \to +\infty} \theta_n(s_k)$. If $\theta_n^* = 2\pi$, there is nothing to prove. So let us assume that $\theta_n^* < 2\pi$. We may also assume that $\theta_n^* = \lim_{k \to +\infty} \theta_n(s_k)$, because anyway we may restrict our attention to a subsequence for which the $\liminf$ in (10) is a limit. Then $\theta_n(s_k) < 2\pi$ for k large enough, and hence $y + s_k\, e^{i\theta_n(s_k)} \in K$ by (7). Since K is closed,

$$y + s e^{i\theta_n^*} \in K. \tag{11}$$

If $n = 1$, then $\theta_1(s) \leq \theta_1^*$ by definition of $\theta_1(s)$. If $n \geq 2$, note that $\theta_n(s_k) \geq \theta_{n-1}(s_k) + \tau$ for k large, because $\theta_n(s_k) < 2\pi$ and by (8). Then

$$\theta_n^* = \lim_{k \to +\infty} \theta_n(s_k) \geq \liminf_{k \to +\infty} \theta_{n-1}(s_k) + \tau \geq \theta_{n-1}(s) + \tau, \tag{12}$$

by induction hypothesis. In this case also $\theta_n(s) \leq \theta_n^*$, by (11) and (12). This proves (10) in all cases.

For each integer $m \geq 0$, set

$$E_m = \{s \in (0,1)\,;\, N_\tau(s) = m\}. \tag{13}$$

Thus $E_0 = \{s\,;\, \theta_1(s) = 2\pi\}$ and for $m \geq 1$, $E_m = \{s\,;\, \theta_m(s) < 2\pi$ but $\theta_{m+1}(s) = 2\pi\}$. Since the functions θ_n are lower semicontinuous, each E_m is a Borel set, and then N_τ is a Borel function. Set

$$K_{m,n} = \{y + s e^{i\theta_n(s)}\,;\, s \in E_m\} \tag{14}$$

for $m \geq 1$ and $1 \leq n \leq m$. Each $K_{m,n}$ is a Borel subset of $B(y,t)$ (because E_m and θ_n are Borel-measurable). Also, the $K_{m,n}$ are all disjoint, because the sets E_m are disjoint and by (8) in particular. Moreover, if $\pi : z \to |y - z|$ denotes the radial projection, then

$$\pi(K_{m,n}) = E_m \quad \text{for} \quad 1 \leq n \leq m. \tag{15}$$

Finally note that $K_{m,n} \subset K$, by (7). Altogether,

$$\int_0^1 N_\tau(s)ds = \sum_{m=1}^\infty m\, H^1(E_m) = \sum_{m=1}^\infty \sum_{n=1}^m H^1(\pi(K_{m,n})) \tag{16}$$

$$\leq \sum_{m=1}^\infty \sum_{n=1}^m H^1(K_{m,n}) \leq H^1(K \cap B(y,t))$$

by (13), (15), and Remark 2.11.

Now we can take limits. Select a sequence $\{\tau_\ell\}$ that tends to 0^+, and note that $N(s) = \lim_{\ell \to +\infty} N_{\tau_\ell}(s)$ pointwise, by (9). Then N is Borel-measurable because each N_τ is, and (2) follows from (16) and Fatou's lemma. This proves Lemma 1. $\square$

We shall need later a similar (but slightly simpler) result.

Lemma 17. *Let $I \subset \mathbb{R}$ be a compact interval and $g : I \to \mathbb{R}$ be a 1-Lipschitz mapping. For $s \in \mathbb{R}$, denote by $N(s)$ the number of points $x \in I$ such that $g(x) = s$. Then N is Borel-measurable and*

$$\int_{\mathbb{R}} N(s)ds \leq H^1(I). \tag{18}$$

Write $I = [a, b]$. For each $\tau > 0$, define a nondecreasing sequence of functions $h_n : \mathbb{R} \to [a - \tau, b + \tau]$, as follows. Start with $h_0(s) = a - \tau$ for $s \in \mathbb{R}$. Then assume that $h_n(s)$ has already been defined. If we can find $x \in [a, b]$ such that $x \geq h_n(s) + \tau$ and $g(x) = s$, set

$$h_{n+1}(s) = \inf \{x \in [a, b] \,;\, x \geq h_n(s) + \tau \text{ and } g(x) = s\}. \tag{19}$$

Otherwise set $h_{n+1}(s) = b + \tau$. This defines Borel-measurable functions h_n, $n \geq 0$. Note that

$$g(h_{n+1}(s)) = s \text{ and } h_{n+1}(s) \geq h_n(s) + \tau \text{ whenever } h_{n+1}(s) \in [a, b]. \tag{20}$$

Next call $N_\tau(s)$ the number of integers n such that $h_n(s) \in [a, b]$. Thus N_τ also is Borel-measurable, and $N_\tau(s) \leq N(s)$ for $s \in \mathbb{R}$, by (20). Moreover,

$$N(s) = \lim_{\tau \to 0^+} N_\tau(s), \tag{21}$$

essentially by construction. Set

$$E_n = \{x \in [a, b] \,;\, x = h_n(g(x))\} \tag{22}$$

for $n \geq 1$. These sets are Borel-measurable and disjoint (by (20)). Also,

$$g(E_n) = \{s \in \mathbb{R} \,;\, h_n(s) \in [a, b]\} = \{s \in \mathbb{R} \,;\, N_\tau(s) \geq n\}, \tag{23}$$

by (22) and (20), and then by definition of $N_\tau(s)$ (and because $\{h_n\}$ is nondecreasing). Hence, for $s \in \mathbb{R}$, $N_\tau(s)$ is the number of integers $n \geq 1$ such that $s \in g(E_n)$. Consequently,

$$\int_{\mathbb{R}} N_\tau(s) ds = \int_{\mathbb{R}} \Big\{ \sum_{n \geq 1} \mathbf{1}_{g(E_n)}(s) \Big\} ds = \sum_{n \geq 1} H^1(g(E_n)) \leq \sum_{n \geq 1} H^1(E_n) \leq H^1(I) \tag{24}$$

by Remark 2.11, because g is 1-Lipschitz, and because the sets E_n are disjoint. Now (18) follows from (24), (21), and Fatou's lemma. This proves Lemma 17. $\square$

27 Uniform rectifiability when $n = 2$

In the next few sections we want to show that K is locally uniformly rectifiable when $(u, K) \in TRLQ(\Omega)$ (as usual) and the dimension is $n = 2$. Let us first give a precise statement, and discuss uniform rectifiability later.

Theorem 1 [DaSe4]. *Let $\Omega \subset \mathbb{R}^2$ and $(u, K) \in TRLQ(\Omega)$ be given, as in Standard Assumption 18.14. Denote by r_0, M, a the quasiminimality constants (as in Definition 7.21). If a is small enough (depending only on M), then for all $x \in K$ and $0 < r \leq r_0$ such that $B(x, 2r) \subset \Omega$,*

$$K \cap B(x, r) \text{ is contained in an Ahlfors-regular curve with constant } \leq C_8, \tag{2}$$

where the constant C_8 depends only on M.

Definition 3. *An Ahlfors-regular curve with constant $\leq C$ is a set of the form $\Gamma = z(I)$, where $I \subset \mathbb{R}$ is a closed interval (not reduced to a point) and $z : I \to \mathbb{R}^n$ is a Lipschitz function such that*

$$|z(x) - z(y)| \leq |x - y| \text{ for } x, y \in I \tag{4}$$

and

$$H^1(\{x \in I \, ; \, z(x) \in B(y, r)\}) \leq C \, r \tag{5}$$

for all $y \in \mathbb{R}^n$ and $r > 0$.

In this definition, we intentionally allow both compact and unbounded intervals I; however for the curves in (2) we can always take I compact, and even $H^1(I) \leq C r$.

Remark 6. As usual, Theorem 1 applies to reduced minimizers of the Mumford-Shah functional, or even reduced or coral local almost-minimizers (topological or not). Then $M = 1$ and r_0 has to be chosen small enough, depending on $||g||_\infty$ or the gauge function h. See Proposition 7.8 and Remark 7.22. There are other examples (see Example 7.23 and Proposition 7.27).

Also, Theorem 1 extends to quasiminimizers with linear distortion. See Remarks 20.48 and 25.28 for some of the steps; the rest of the proof is almost the same.

Remark 7. Theorem 1 does not give a good description of K near $\partial\Omega$. This is normal, because we only assume that (u, K) is a local quasiminimizer. If (u, K) is a reduced quasiminimizer in Ω (not merely local), then we can say a bit more. For instance, if $\partial\Omega$ is an Ahlfors-regular curve and a is small enough (depending on M, as usual), (2) holds for all $x \in K$ and $0 < r \leq r_0$, even if $B(x, r)$ crosses $\partial\Omega$. This is in fact fairly easy to deduce from Theorem 1 and the fact that $H^1(K \cap B(x, r)) \leq Cr$ for all choices of $x \in \mathbb{R}^2$ and $r > 0$. This last fact is easy to check (the proof of Lemma 18.19 still works with very minor modifications), and then one can apply the argument given in Section 7 of [DaSe4]. See also Remark 32.5 below. Of course, since this is so easy to obtain, there is a good reason: even when $x \in \partial\Omega$, (2) does not really give more information on the boundary behavior of K than Theorem 1. For instance, it does not give a lower bound on $H^1(K \cap B(x, r))$. See Part I, and in particular Section 79, for more information on the boundary behavior of quasiminimizers and almost-minimizers.

Remark 8. Theorem 1 can be generalized to higher dimensions [DaSe6], but we shall not give a complete proof here. See Sections 73 and 74 for a few additional details.

Next we want to make a few comments on Ahlfors-regular curves, and then on uniform rectifiability of one-dimensional sets. Not all these comments are needed to understand Mumford-Shah minimizers, but the author cannot resist being a little loquacious here.

Typical examples of (Ahlfors-)regular curves are graphs of Lipschitz functions (from $\mathbb{R}$ to $\mathbb{R}^{n-1}$), or chord-arc curves, but regular curves can be a little more complicated than this. They are allowed to cross themselves to some limited extent, for instance. In terms of smoothness, they are not very different from Lipschitz graphs, and so the difference with C^1-curves is not enormous either.

There is no big difference between regular curves parameterized by compact, or unbounded intervals I. If $z : I \to \mathbb{R}^n$ is as in Definition 3 (we shall call this a _regular parameterization_), then the restriction of z to any (nontrivial) closed interval $J \subset I$ is also a regular parameterization, with a constant C which is at least as good. Conversely, if $I \neq \mathbb{R}$, then we can extend z into a regular parameterization defined on $\mathbb{R}$ and with constant $\leq C + 2$. This is easy to do: one can continue z on each component of $\mathbb{R} \setminus I$ by the parameterization at constant speed 1 of a half-line emanating from the corresponding endpoint of $z(I)$.

Also, if $z : [a, +\infty) \to \mathbb{R}^n$ (for instance) is a regular parameterization, then

$$\lim_{x \to +\infty} |z(x)| = +\infty. \tag{9}$$

This is an easy consequence of the following slightly more general fact: if $z : I \to \mathbb{R}^n$ is a regular parameterization and $B(y, r)$ is any ball in $\mathbb{R}^n$, then

$$\{ x \in I \,;\, z(x) \in B(y, r)\} \text{ is contained in a union of } \leq 8C \text{ intervals of length } r,$$
(10)

where C is the same constant as in (5).

Indeed, let $x_1, \ldots, x_k$ be a collection of points of $z^{-1}(B(y, r)) \subset I$ that lie at mutual distances $\geq \frac{r}{2}$. The intervals $J_j = I \cap \left(x_j - \frac{r}{4}, x_j + \frac{r}{4} \right)$, $1 \leq j \leq k$, are disjoint and contained in $z^{-1}(B(y, 2r))$, by (4) and because $z(x_j) \in B(y, r)$. If $H^1(I) \leq r$, then (10) holds trivially. Otherwise, $H^1(J_j) \geq \frac{r}{4}$ for each j, and hence

$$k \leq \sum_{j=1}^{k} \frac{4}{r} H^1(J_j) = \frac{4}{r} H^1\left(\bigcup_j J_j \right) \leq \frac{4}{r} H^1\left(z^{-1}(B(y, 2r)) \right) \leq 8C, \qquad (11)$$

by (5). So we can choose $x_1, \ldots, x_k$ as above but so that k is maximal, and we still get that $k \leq 8C$. But then $z^{-1}(B(y, r))$ is contained in the union of the intervals $\left[x_j - \frac{r}{2}, x_j + \frac{r}{2} \right]$, by maximality (if not, we can add a new point x_{k+1} in the collection). This proves (10). Of course the constant 8 is not optimal.

Note that (10) gives a bound on how many intervals are needed to cover $z^{-1}(B(y, r))$, but not on how far they are, and indeed z may go very far before it decides to return to $B(y, r)$. Similarly, (9) does not come with bounds on how fast $|z(x)|$ tends to $+\infty$.

Another consequence of (10) is that for $y \in \mathbb{R}^n$, $z^{-1}(y)$ never has more than $8C$ points. So z is not so far from being injective.

Remark 12. If $\Gamma = z(I)$ is a regular curve, then it is also an Ahlfors-regular set. More precisely,

$$r \leq H^1(\Gamma \cap B(x, r)) \leq Cr \text{ for } x \in \Gamma \text{ and } 0 < r \leq \operatorname{diam} \Gamma, \qquad (13)$$

with the same constant C as in (5).

The first inequality is a simple consequence of the connectedness of Γ. To see this, first note that

$$H^1(\Gamma) \geq \operatorname{diam} \Gamma. \qquad (14)$$

Indeed, if y, z are distinct points of Γ and π denotes the orthogonal projection onto the line through y and z, then $\pi(\Gamma)$ contains the segment $[y, z]$ because it is connected and contains both endpoints. Hence $H^1(\Gamma) \geq H^1(\pi(\Gamma)) \geq |y - z|$ by Remark 2.11, and (14) follows by choosing y, z as far from each other as possible.

Now let $x \in \Gamma$ and $0 < r \leq \operatorname{diam} \Gamma$ be given. If $\Gamma \subset B(x, r)$, then $H^1(\Gamma \cap B(x, r)) = H^1(\Gamma) \geq \operatorname{diam} \Gamma \geq r$ and we are happy. Otherwise, each circle $\partial B(x, \rho)$, $0 < \rho < r$, meets Γ, because otherwise it would separate Γ into two open nonempty subsets (one that contains x and another one that contains $K \setminus B(x, r)$). In this case we set $\pi(z) = |z - x|$ and note that

$$H^1(\Gamma \cap B(x, r)) \geq H^1(\pi(\Gamma \cap B(x, r)) \geq H^1((0, r)) = r, \qquad (15)$$

as needed.

The second inequality in (13) is even simpler, because $\Gamma \cap B(x, r)$ is the image under the 1-Lipschitz mapping z of the same set $E = z^{-1}(B(x, r))$ as in (5). Then $H^1(\Gamma \cap B(x, r)) \leq H^1(E) \leq Cr$.

Thus every regular curve is a connected Ahlfors-regular set. The converse is not exactly true; for instance the union of the two axes in $\mathbb{R}^2$ is not a regular curve, because the parameterization z would have to go through the origin infinitely many times, and this is not allowed by (10). However,

$$\text{every connected Ahlfors-regular set of dimension 1 is} \tag{16}$$
$$\text{contained in a regular curve.}$$

We shall only prove this when the regular set is bounded (see Theorem 31.5). The proof in the other case is not much more difficult, but we shall not need it. See Exercise 22, though.

Now we want to say a few words about uniform rectifiability. For one-dimensional sets, the simplest definition is probably the following.

Definition 17. *Let $E \subset \mathbb{R}^n$ be an Ahlfors-regular set of dimension 1 (see Definition 18.9). We say that E is uniformly rectifiable when E is contained in some Ahlfors-regular curve (see Definition 3).*

This looks a little silly, but part of the point of uniform rectifiability is that it has many equivalent definitions. We shall see a few of them soon, but the interested reader should consult [DaSe3] for a very long list with all sorts of funny names. Definition 17 is among the strongest in appearance.

The name uniform rectifiability makes sense: if E is contained in a regular curve, it is rectifiable, but we have a much more precise and quantitative control than if it was merely contained in a countable collection of C^1, or Lipschitz curves.

Uniform rectifiability exists also for Ahlfors-regular sets of integer dimensions d, $d > 1$, but some definitions have to be modified. For instance, in Definition 17, we would need to replace regular curves with an appropriate class of d-dimensional "surfaces" (called ω-regular surfaces, because the parameterizations are not always Lipschitz, but their derivative is controlled by a weight ω). See Section 73 for other definitions. For lots of information on uniform rectifiability, see [DaSe3].

The notion of uniform rectifiability shows up naturally in other contexts (at least two). The first, and most convincing example so far, is in connection with singular integral operators. The typical problem that we would consider in this respect is the following. Let $K(x, y)$ be a d-dimensional Calderón–Zygmund kernel on $\mathbb{R}^n$. We do not wish to say what this means here, but the main example is $K(x, y) = k(x - y)$, where $k : \mathbb{R}^n \setminus \{0\}$ is a smooth (away from 0) real- or complex-valued function which is homogeneous of degree $-d$ and odd. The most celebrated example is the Cauchy kernel $\frac{1}{x-y}$ in the complex plane (and with $d = 1$). The problem is to characterize the Ahlfors-regular measures μ of dimension d in $\mathbb{R}^n$ (as in Lemma 18.11 and (18.12)) for which a given kernel $K(x, y)$, or all kernels in a class, define operators that are bounded on $L^2(\mu)$. Since the integrals do not always converge (due to the singularity of K), we mean by this that

$$||T_\varepsilon f||_{L^2(\mu)} \le C ||f||_{L^2(\mu)},\tag{18}$$

with bounds that do not depend on ε, and where

$$T_\varepsilon f(x) = \int_{|y-x|>\varepsilon} K(x,y)\, f(y)\, d\mu(y).\tag{19}$$

This sort of problems started to be very popular with the question of L^2-boundedness for the Cauchy kernel on Lipschitz graphs (eventually settled in [CMM]), and kept us busy for quite some time. See for instance [Da1,2,3], [Mat1], [DaSe1,3], [Se1,2,3,4] (a partial list). Uniform rectifiability emerged from this, probably because we wanted to give more and more importance to the geometrical aspects, and it really seems to be the right notion here. A very striking result of P. Mattila, M. Melnikov, and J. Verdera [MatMV] says that for Ahlfors-regular measures μ of dimension 1 in the complex plane, the Cauchy kernel $K(x,y) = \frac{1}{x-y}$ defines a bounded operator on $L^2(\mu)$ if and only if the support of μ is uniformly rectifiable. Actually, the lower bound in Ahlfors-regularity is not vital here, and there is even a similar characterization of L^2-boundedness of the Cauchy integral operator for measures μ that are not necessarily Ahlfors-regular. This is a result of X. Tolsa [To1], and almost simultaneously by F. Nazarov, S. Treil, and A. Volberg [NTV]. Let us mention that singular integrals, and to a perhaps lesser extent uniform rectifiability, play a central role in the recent solution by Tolsa [To2] of the Painlevé problem and his proof of the semi-additivity of analytic capacity.

Uniform rectifiability also shows up when you study things like "quasiminimal sets". These are sets E such that, when you perturb E in a ball (say), you cannot reduce $H^d(E)$ by large factors, compared with what you add (for instance). Definitions would be similar to Definition 7.21, but there are different classes of perturbations to consider. See for instance [Al], [DaSe8,9], or [Ri2,3].

The reader may find more information on uniform rectifiability in the survey [Da9], and the references therein.

In spite of this long piece of shameless advertisement, the reader should not think that uniform rectifiability is just the right notion for Mumford-Shah minimizers. First, we know that the right notion is C^1-curves (see Section 6), and it is possible that uniform rectifiability only shows up here because there was prior work on it that could be used easily. For quasiminimizers (as in Example 7.22 and Proposition 7.26), the situation is less clear because they do not have C^1 pieces in a uniform way, so uniform rectifiability is probably closer to being optimal.

Remark 20. Theorem 1 says that K is locally uniformly rectifiable. We can say a little more: K contains big pieces of Lipschitz graphs locally. This means that for x, r as in the statement of Theorem 1, we can find a Lipschitz graph Γ with constant $\leq C_9 = C_9(M)$, such that

$$H^1(K \cap \Gamma \cap B(x, r)) \geq C_9^{-1} r. \tag{21}$$

(By Lipschitz graph we mean the graph of a Lipschitz function $A : \mathbb{R} \to \mathbb{R}$, or the image of such a graph by a rotation.)

This property is a little stronger than local uniform rectifiability; as we shall see, local uniform rectifiability amounts to saying that K contains big pieces of connected sets locally (instead of Lipschitz graphs). The fact that K contains big pieces of Lipschitz graphs locally is a (not completely trivial!) consequence of Theorem 1 and the property of projections (Theorem 24.1). See [DaSe4], p. 303 for details. One can also prove it directly; see Remark 29.93 and Theorem 29.98 below. If we only think about Mumford-Shah minimizers and almost-minimizers with sufficiently rapidly decreasing gauge functions h, there is perhaps not too much point to this, because we shall even get big connected pieces of C^1 arcs.

Our proof of Theorem 1 will be composed of two main steps. First we shall prove that K contains big pieces of connected sets locally. We shall do this in Section 29, after a rapid description of the co-area theorem in Section 28. For the second step we shall forget that (u, K) is a quasiminimizer and deduce the local uniform rectifiability of K from the local existence of big pieces of connected sets. This will be done in Section 31.

Exercise 22. Check our claim that the union of the two axes in $\mathbb{R}^2$ is not a regular curve. Construct a regular curve that contains it. The general case of (16) is a little like that.

28   The co-area formula

We shall use the co-area theorem in the next section; since we like it we want to record its statement in a separate section, to make it easier to find.

Theorem 1. *Let* $\Omega \subset \mathbb{R}^n$ *be open and* $f : \Omega \to \mathbb{R}$ *be Lipschitz. Then*

$$|\nabla f| \, dx = \int_{\mathbb{R}} H^{n-1}|_{f^{-1}(t)} dt. \tag{2}$$

We do not intend to prove this here; see for instance [Fe], p. 248 or 249. Let us just comment on the statement a little.

The identity (2) is an equality between measures on Ω. In more explicit terms, it means that if $A \subset \Omega$ is a Borel set, then

$$\int_A |\nabla f| \, dx = \int_{\mathbb{R}} H^{n-1}(A \cap f^{-1}(t)) \, dt; \tag{3}$$

the fact that $H^{n-1}(A \cap f^{-1}(t))$ is a measurable function of t is part of the statement. Here is another (apparently a little stronger) way of saying this: if $g : \Omega \to [0, +\infty)$ is a nonnegative Borel function, then

$$\int_{\Omega} g(x) |\nabla f(x)| \, dx = \int_{\mathbb{R}} \left\{ \int_{f^{-1}(t)} g(x) \, dH^{n-1}(x) \right\} dt. \tag{4}$$

It is easy to deduce (4) from (3), because we can write g as the (pointwise) limit of a nondecreasing sequence of linear combinations of characteristic functions, and then apply Beppo–Levi to both sides of (4).

There is a similar result for Lipschitz mappings $f : \mathbb{R}^n \to \mathbb{R}^m$, $m < r$; then (3) is replaced with

$$\int_A |Jf| = \int_{\mathbb{R}^m} H^{n-m}(A \cap f^{-1}(t)) \, dt, \tag{5}$$

for an appropriate definition of Jabobian $|Jf|$.

Note that Federer states this with $\Omega = \mathbb{R}^n$, but there is no serious difference: if $f : \Omega \to \mathbb{R}$ is Lipschitz, we can always extend f to a Lipschitz function on $\mathbb{R}^n$, apply the co-area theorem, and then restrict to sets $A \subset \Omega$ in (3). Similarly, it is often possible to apply the theorem with initial functions f that are not exactly Lipschitz. For instance, it could be that f is only locally Lipschitz, but we can apply the theorem with smaller domains where f is Lipschitz, and then go to the limit. If f is not regular enough, we can also try to replace it with smooth approximations of f and see what we get. But one needs to be careful about what happens to sets of measure zero.

We shall only use the inequality $\geq$, which happens to be the easiest to prove. (See [Se5] for an argument.) The reason why the proof is a little unpleasant is that we need to take care of sets of measure zero, and sets where ∇f vanishes. If f is smooth and ∇f never vanishes, then the rapid sketch that follows gives a proof.

First note that the problem is local, and that it is enough to check (3) on characteristic functions of small boxes. In a small box where ∇f is nonzero and essentially constant, the level sets are essentially parallel surfaces, and the distance between $f^{-1}(t)$ and $f^{-1}(t + \delta t)$ is close to $|\delta t| |\nabla f|^{-1}$. If we want to get the Lebesgue measure of the region between these surfaces, we have to multiply by $H^{n-1}(f^{-1}(t))$. Thus the contribution of this region to the left-hand side of (3) looks like $|\delta t| H^{n-1}(f^{-1}(t))$, just like for the right-hand side. (See Figure 1.)

Of course this is not a true proof, but it should convince the reader that the statement is natural.

Let us say a few words about how we shall use Theorem 1. Here is a typical, perhaps oversimplified, situation that may occur. We are given a ball B, a closed set $K \subset B$, and a smooth function f on $B \setminus K$. For some reason, we know that $\int_{B \setminus K} |\nabla f| \leq \varepsilon$, and at the same time, there are two balls B_1, B_2 contained in $B \setminus K$, such that $f = 0$ on B_1 and $f = 1$ on B_2. (See Figure 2.)

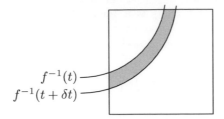

Figure 1.   The area (volume if $n > 2$) of the shaded region (in the box and between the level lines) is about $|\nabla f|^{-1} |\delta t|$ times the length (surface if $n > 2$) of the level set $f^{-1}(t)$ in the box.

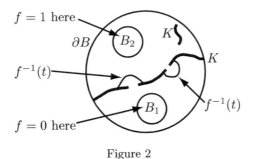

Figure 2

We want to say that since f varies very slowly and yet takes fairly different values on the B_i, K must almost separate B_1 from B_2 in B. The co-area formula allows you to justify this in a quite precise way. Indeed, Let us apply (3) to f in $\Omega = B \setminus K$. We get that $\int_0^1 H^{n-1}(f^{-1}(t)) \, dt \leq \varepsilon$. By Chebyshev, we can find $t \in (0,1)$ such that the level set $\Gamma_t = f^{-1}(t)$ has Hausdorff measure at most ε.

However, $K \cup \Gamma_t$ separates B_1 from B_2 in B, because if γ is a path in B that goes from B_1 to B_2, and if γ does not meet K, then f is defined and continuous on γ, hence takes all values between 0 and 1, and in particular t. This means that γ meets Γ_t, as promised.

29 K contains big pieces of connected sets locally

In this section we prove the following first step towards Theorem 27.1.

Theorem 1.   *Let $\Omega \subset \mathbb{R}^2$ and $(u,K) \in TRLQ(\Omega)$ be given (as in Standard Assumption 18.14). If a is small enough (depending only on M), then for all $x \in K$ and $0 < r \leq r_0$ such that $B(x,r) \subset \Omega$, we can find a compact connected set Γ such that*

$$H^1(\Gamma) \leq C_{10} r \tag{2}$$

$$and \quad H^1(K \cap \Gamma \cap B(x,r)) \geq C_{10}^{-1} r. \tag{3}$$

As usual, C_{10} depends only on M.

Recall that r_0, M, and a are the quasiminimality constants for (u, K).

Our proof will give a little more than Theorem 1; with a little bit more work, we can prove that K contains big pieces of Lipschitz graphs locally, as in Remark 27.20. See Remark 93 and Theorem 98 below.

The proof of Theorem 1 that we give here is slightly different from the original one in [DaSe4], even though it also relies on the co-area theorem. It is based on (the beginning of) the argument used in [DaSe6] to prove that in dimensions $n \geq 3$, K is locally uniformly rectifiable.

We start the argument in the usual way. Choose $p \in (1, 2)$, like $p = \frac{3}{2}$, and let $\varepsilon > 0$ be small. We shall choose ε near the end of the argument, depending only on M. By Theorem 18.16 (on local Ahlfors-regularity) and Lemma 18.19 (the trivial energy estimate), the assumptions (23.3) and (23.4) are satisfied, and Corollary 23.38 says that we can find

$$y \in K \cap B\left(x, \frac{r}{3}\right) \text{ and } 0 < t < \frac{r}{3} \tag{4}$$

such that

$$t \geq C_5^{-1} r \tag{5}$$

and

$$\omega_p(y, t) \leq \varepsilon. \tag{6}$$

Here C_5 depends on M (through the constants in (23.3) and (23.4)), and on our future choice of ε, but this will be all right: we shall not let ε depend on C_5. As usual, we want to work on $B(y, t)$, and the connected set that we shall get will also work on $B(x, r)$ (with a worse constant).

Lemma 7. *We can choose a radius $t_1 \in \left(\frac{t}{2}, t\right)$ such that, if we set $B = B(y, t_1)$,*

$$\partial B \cap K \text{ has at most } C_{11} \text{ points,} \tag{8}$$

$$\int_{\partial B \setminus K} |\nabla u|^p \, dH^1 \leq 4\varepsilon^{p/2} t^{1-\frac{p}{2}}, \tag{9}$$

$$u|_{\partial B \setminus K} \in W^{1,p}(\partial B \setminus K) \text{ and its derivative} \atop \text{is given by the restriction of } \nabla u. \tag{10}$$

Call $N(s)$ the number of points in $\partial B(y, s) \setminus K$. Then

$$\int_{t/2}^{t} N(s) ds \leq H^1(K \cap B(y, t)) \leq Ct, \tag{11}$$

by Lemma 26.1 and the trivial estimate (18.20). Thus most choices of $t_1 \in \left(\frac{t}{2}, t\right)$ satisfy (8) if C_{11} is large enough, depending on M. Similarly,

$$\int_{t/2}^{t} \left\{ \int_{\partial B(y,s) \setminus K} |\nabla u|^p \, dH^1 \right\} ds \leq \int_{B(y,t) \setminus K} |\nabla u|^p$$
$$\leq t^{2-\frac{p}{2}} \omega_p(y, t)^{p/2} \leq t^{2-\frac{p}{2}} \varepsilon^{p/2}, \tag{12}$$

by (23.6) and (6). Hence many choices of t_1 satisfy (9) as well (by Chebyshev). As for (10), it is satisfied for almost all choices of t_1. This follows from Corollary 9.13 on absolute continuity on almost all lines, a change of variables (to get polar coordinates), and a small localisation argument (to reduce to the case of rectangles). The details are the same as for the verification of (20.17) in Lemma 20.21, or the proof (25.14), so we don't repeat them. Lemma 7 follows.   □

Let t_1 and $B = B(y, t_1)$ be as in Lemma 7. (To be honest, there will be a slightly annoying special case where we shall need to replace (9) with a stronger condition, but let us not worry now). Set

$$\ell = (6C_{11}M)^{-1} H^1(K \cap B) \geq \frac{t}{C}, \tag{13}$$

where C_{11} is as in (8) and the last inequality comes from Theorem 18.16. Also set

$$Z = \{ z \in \partial B \, ; \mathrm{dist}(z, \partial B \cap K) \leq \ell \}. \tag{14}$$

Then

$$H^1(Z) \leq 3C_{11}\ell \leq \frac{1}{2M} H^1(K \cap B), \tag{15}$$

by (8) and (13). The most interesting case will be when

$$\partial B \setminus Z \text{ is contained in a single connected component of } \Omega \setminus K. \tag{16}$$

So let us first assume that (16) holds.

Lemma 17. *If ε and a are small enough (depending on M), we can find $z_1, z_2 \in \partial B \setminus Z$ such that*

$$|u(z_1) - u(z_2)| \geq C_{12}^{-1} t^{1/2}, \tag{18}$$

where C_{12} depends only on M.

Suppose not. Note that

$$\text{the length of each connected component of } Z \text{ is at least } 2\ell, \tag{19}$$

because (14) says that Z is a finite union of intervals of length $\geq 2\ell$.

Let v denote the continuous function on ∂B that coincides with u on $\partial B \setminus Z$ and is linear on each component of Z. This makes sense, and also v is absolutely continuous on ∂B, because $u|_{\partial B \setminus Z} \in W^{1,p}(\partial B \setminus Z)$ by (10). Denote by $\frac{\partial v}{\partial \tau}$ the derivative of v on ∂B. Then

$$\left| \frac{\partial v}{\partial \tau} \right| \leq (2\ell)^{-1} C_{12}^{-1} t^{1/2} \leq C C_{12}^{-1} t^{-1/2} \quad \text{on } Z, \tag{20}$$

by (19), because we assumed that (18) fails for all choices of $z_1, z_2 \in \partial B \setminus Z$, and by (13). Here C depends on M, but not on C_{12}. Now

$$\int_{\partial B} \left| \frac{\partial v}{\partial \tau} \right|^p dH^1 = \int_{\partial B \setminus Z} \left| \frac{\partial u}{\partial \tau} \right|^p + \int_Z \left| \frac{\partial v}{\partial \tau} \right|^p \leq 4\varepsilon^{p/2} t^{1-\frac{p}{2}} + C C_{12}^{-p} t^{-p/2} H^1(Z)$$

$$\leq 4\varepsilon^{p/2} t^{1-\frac{p}{2}} + C C_{12}^{-p} t^{1-\frac{p}{2}}$$

(21)

by (10), (9), (20), and because $H^1(Z) \leq 2\pi t$ trivially.

As usual we want to use this to construct a good competitor. Denote by u_1 the harmonic extension of v to B. Thus u_1 is continuous on $\overline{B}$, harmonic in B, and equal to v on ∂B. Then

$$u_1 \in W^{1,2}(B)$$

(22)

and

$$\int_B |\nabla u_1|^2 \leq C t^{2-\frac{2}{p}} \left\{ \int_{\partial B} \left| \frac{\partial v}{\partial \tau} \right|^p \right\}^{2/p} \leq C(\varepsilon + C_{12}^{-2}) t,$$

(23)

by Lemma 22.16 and (21).

Set $u_1(z) = u(z)$ for $z \in \Omega \setminus (K \cup \overline{B})$. This way, u_1 is defined on $\Omega \setminus K_1$, where

$$K_1 = (K \setminus B) \cup Z.$$

(24)

We claim that

$$(u_1, K_1) \text{ is a topological competitor for } (u, K) \text{ in } B.$$

(25)

First, K_1 is closed in Ω (because K and Z are closed and B is open). Next

$$u_1 \in W^{1,2}_{\text{loc}}(\Omega \setminus K_1).$$

(26)

There is no problem in B (by (22)) or in $\Omega \setminus (\overline{B} \cup K)$ (because $u \in W^{1,2}_{\text{loc}}(\Omega \setminus K)$), so we just need to control u_1 near $\partial B \setminus Z$. There we can use the welding Corollary 14.28. Indeed recall from Lemma 18.22 that u is continuous on $\Omega \setminus K$; since u_1 is continuous on $\overline{B}$ and coincides with u on $\partial B \setminus Z$, u_1 is even continuous across $\partial B \setminus Z$; then (22) and the fact that $u \in W^{1,2}(B(y,t) \setminus K)$, (by the trivial energy estimate (18.20)) say that $u_1 \in W^{1,2}(B(y,t) \setminus [K_1 \cup \partial B])$. The details are the same as in Section 25, near (25.23); see also the proof of (20.30) for a similar argument.

Thus $(u_1, K_1) \in \mathcal{A}$ (see the definition (7.1)), and it is a competitor for (u, K) in B because we did not modify anything outside of $\overline{B}$ (see Definition 7.2). Now (25) follows from Lemma 20.10 and our assumption (16). Note that if we just want a competitor (not necessarily topological), (16) will not be needed.

Since $\overline{B} \subset B(y,t) \subset B(x,r) \subset \Omega$ by construction (see Lemma 7, (4) and the statement of Theorem 1), (25) allows us to apply the quasiminimization condition

(7.20). (See Definition 7.21.) With the notation of (7.18) and (7.19), $E \geq 0$ and $\tilde{E} = \int_B |\nabla u_1|^2$, so that

$$\delta E = \text{Max}\{(\tilde{E} - E), M(\tilde{E} - E)\} \leq CM(\varepsilon + C_{12}^{-2})t, \tag{27}$$

by (7.19) and (23). Recall that (7.20) says that

$$H^1(K \setminus K_1) \leq MH^1(K_1 \setminus K) + \delta E + at. \tag{28}$$

Since $K \setminus K_1 = K \cap B$ and

$$H^1(K_1 \setminus K) = H^1(Z) \leq \frac{1}{2M} H^1(K \cap B) \tag{29}$$

by (24) and (15), we get that

$$\frac{1}{2} H^1(K \cap B) \leq CM(\varepsilon + C_{12}^{-2})t + at. \tag{30}$$

If ε, C_{12}^{-1}, and a are small enough (depending on M), (30) contradicts the local Ahlfors-regularity from Theorem 18.16. This contradiction completes our proof of Lemma 17. $\qquad\square$

Remark 31. If (u, K) is a (reduced) local quasiminimizer (not topological), then we do not need (16) to prove Lemma 17.

Let us continue with our assumption (16), and let $z_1, z_2 \in \partial B \setminus Z$ satisfy (18). For $i = 1, 2$, call I_i the connected component of z_i in $\partial B \setminus Z$ and I_i'' the component of $\partial B \setminus K$ that contains z_i and I_i. Note that for $z \in I_i''$,

$$|u(z) - u(z_i)| \leq \int_{I_i''} \left|\frac{\partial u}{\partial \tau}\right| \leq \int_{\partial B \setminus K} |\nabla u| \leq (2\pi t)^{1-\frac{1}{p}} \left\{ \int_{\partial B \setminus K} |\nabla u|^p \right\}^{1/p}$$
$$\leq C\varepsilon^{1/2} t^{1/2} \leq \frac{1}{4} C_{12}^{-1} t^{1/2} \tag{32}$$

by (10), Hölder, (9), and if ε is small enough compared to C_{12}^{-1}. Set

$$L_i = u(I_i'') \tag{33}$$

for $i = 1, 2$. Thus L_i is an interval (because u is continuous on $\partial B \setminus K$), and

$$\text{dist}(L_1, L_2) \geq \frac{1}{2} C_{12}^{-1} t^{1/2} \tag{34}$$

by (18) and (32).

Thus we have a function u on $B \setminus K$ with a very small derivative on average (because $\omega_p(y, t) \leq \varepsilon$), but which takes fairly different values on I_1'' and I_2'' (by

(34)). This should mean that K almost separates I_1'' from I_2'' in B. Let us try to make this more precise.

Call I_i', $i = 1, 2$, the open arc of ∂B with the same center as I_i, but length $H^1(I_i') = H^1(I_i) + \ell$. Note that $\mathrm{dist}(I_i, K) \geq \ell$, by (14). Hence $I_i' \subset I_i''$, and there is even a security interval of length $\frac{\ell}{2}$ on both side. Of course, I_1'' and I_2'' are disjoint, by (34). Call J_1 and J_2 the two connected components of $\partial B \setminus (I_1' \cup I_2')$. By the discussion above,

$$I_1, \ I_2, \ J_1, \ \text{and} \ J_2 \ \text{have lengths} \ \geq \ell. \tag{35}$$

Lemma 36. *There is a compact set $K^* \subset \overline{B} \setminus (I_1' \cup I_2')$ such that*

$$H^1(K^* \cap B \setminus K) \leq C\varepsilon^{1/2} t, \tag{37}$$

K^* *contains* $\partial B \setminus (I_1' \cup I_2')$, *and* K^* *separates* I_1' *from* I_2' *in* $\overline{B}$.

This last condition means that I_1' and I_2' lie in different connected components of $\overline{B} \setminus K^*$.

Lemma 36 will be much easier to prove when

$$u \ \text{is continuous on} \ \overline{B} \setminus K \ \text{and} \ C^1 \ \text{on} \ B \setminus K. \tag{38}$$

The point is that we would like to apply the co-area theorem to u in $B \setminus K$, but if (38) does not hold we may not be able to do that directly. In this bad case we will need the following lemma.

Lemma 39. *There is a C^1 function f on $B \setminus K$ such that*

$$\int_{B \setminus K} |\nabla f| \leq C\varepsilon^{1/2} t^{3/2} \tag{40}$$

and, for each choice of $i \in \{1, 2\}$ and $z \in I_i'$,

$$\limsup_{\substack{\xi \to z \\ \xi \in B}} \ \mathrm{dist}(f(\xi), L_i) \leq \frac{1}{10} C_{12}^{-1} t^{1/2}. \tag{41}$$

When (38) holds, we can take $f = u$. Indeed (41) holds because $I_i' \subset I_i'' \subset \partial B \setminus K$ and u is continuous on $\overline{B} \setminus K$, and (40) holds because

$$\int_{B \setminus K} |\nabla u| \leq (\pi t^2)^{1 - \frac{1}{p}} \left\{ \int_{B \setminus K} |\nabla u|^p \right\}^{1/p} = (\pi t^2)^{1 - \frac{1}{p}} \left\{ t^{2 - \frac{p}{2}} \omega_p(y, t)^{p/2} \right\}^{1/p}$$
$$\leq C t^{3/2} \omega_p(y, t)^{1/2} \leq C t^{3/2} \varepsilon^{1/2}, \tag{42}$$

by (23.6) and (6).

Note that (38) holds when (u, K) is a minimizer of the Mumford-Shah functional, because we know that $u \in C^1(\Omega \setminus K)$ in that case. (See Corollary 3.19.) Since we decided to prove our results also for quasiminimizers, we shall have to pay the price and prove Lemma 39 even when (38) does not hold. But the proof is a little painful and not very interesting, so we shall do it at the end of this section. (This is the place where we shall need to choose t_1 in Lemma 7 a little more carefully.)

So let us see first how to deduce Theorem 1 from Lemma 39. Set

$$A_\ell = \{z \in B \setminus K; f(z) = \ell\} \tag{43}$$

for $\ell \in \mathbb{R}$. We would like to apply the co-area theorem in Section 28 to f and the domain $\Omega' = B \setminus K$, but we cannot really do this directly because f may not be (globally) Lipschitz on Ω'. However, if Ω_k is open and relatively compact in $B \setminus K$, then f is Lipschitz on Ω_k, and we can apply (28.3) in Ω_k with $A = \Omega_k$. We get that

$$\int_{\mathbb{R}} H^1(A_\ell \cap \Omega_k) d\ell = \int_{\Omega_k} |\nabla f(z)| dz \leq C \varepsilon^{1/2} t^{3/2}, \tag{44}$$

by (40). We can apply this to an increasing sequence of open sets Ω_k that converges to $B \setminus K$, and we get that

$$\int_{\mathbb{R}} H^1(A_\ell) d\ell \leq C \varepsilon^{1/2} t^{3/2}. \tag{45}$$

Because of (34), we can find an interval $L \subset \mathbb{R}$ of length $\frac{1}{4} C_{12}^{-1} t^{1/2}$ that lies between L_1 and L_2 and for which

$$\text{dist}(L, L_1 \cup L_2) \geq \frac{1}{8} C_{12}^{-1} t^{1/2}. \tag{46}$$

By (45) and Chebyshev, we can choose $\ell \in L$ such that

$$H^1(A_\ell) \leq C C_{12} \varepsilon^{1/2} t. \tag{47}$$

Set

$$K^* = (K \cap B) \cup [\partial B \setminus (I_1' \cup I_2')] \cup A_\ell. \tag{48}$$

Obviously, (37) holds because of (47). Next we check that K^* is compact. Let $\{z_k\}$ be a sequence in K^* that converges to some limit z; we want to show that $z \in K^*$. If $z \in B$, there is no problem because K is closed in B and A_ℓ is closed in $B \setminus K$. If $z \in \partial B \setminus (I_1' \cup I_2')$, then $z \in K^*$ by (48). We are left with the case when $z \in I_i'$ for some i. In this case, $z_k \in A_\ell$ for k large, because otherwise z would lie in $K \cup (\partial B \setminus (I_1' \cup I_2'))$, which is closed in Ω. In particular, $z_k \in B \setminus K$ (because $A_\ell \subset B \setminus K$) and $f(z_k) = \ell \in L$, which lies at distance $\geq \frac{1}{8} C_{12}^{-1} t^{1/2}$ from L_i (by (46)). This is not compatible with (41), and so our last case is impossible. Hence K^* is compact.

We still need to check that K^* separates I_1' from I_2' in $\overline{B}$. Let us proceed by contradiction and assume that we have a path γ in $\overline{B} \setminus K^*$ that starts on I_1' and ends on I_2'. The last point of $\gamma \cap \overline{I}_1'$ when we run along γ cannot lie on $\partial B \setminus (I_1' \cup I_2') \subset K^*$, because γ does not meet K^*. It cannot lie on I_2' either, because $\mathrm{dist}(I_1', I_2') > 0$ by the discussion before (35). So it lies on I_1'. Hence we can assume that the only point of $\gamma \cap I_1'$ is the initial extremity a_1 of γ. (Otherwise, remove the beginning of γ.) Similarly, we can assume that the only point of $\gamma \cap I_2'$ is the final extremity a_2 of γ. The rest of γ lies in $B \setminus K$, and so f is defined and continuous on it. Hence

$$f(\gamma \setminus \{a_1, a_2\}) \text{ is an interval.} \tag{49}$$

Moreover, (41) says that

$$\limsup_{\substack{\xi \to a_i \\ \xi \in \gamma}} \ \mathrm{dist}(f(\xi), L_i) \leq \frac{1}{10} C_{12}^{-1} t^{1/2}, \tag{50}$$

so this interval gets close to L_1 and to L_2. Recall that L lies between L_1 and L_2, and at distance $\geq \frac{1}{8} C_{12}^{-1} t^{1/2}$ from them. (See (46).) Then $f(\gamma \setminus \{a_1, a_2\})$ contains L, and in particular γ meets $A_\ell = f^{-1}(\ell)$ (because $\ell \in L$). This is a contradiction: γ was not supposed to meet K^*. The conclusion is that K^* separates I_1' from I_2' in $\overline{B}$, and this completes our proof of Lemma 36 (modulo Lemma 39 when (38) does not hold). □

Recall from just before (35) that J_1 and J_2 are the two connected components of $\partial B \setminus (I_1' \cup I_2')$.

Lemma 51. *The two arcs J_1 and J_2 lie in a same connected component of K^*.*

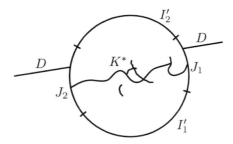

Figure 1

We want to deduce this from the separation property in Lemma 36. (See Figure 1.) This will be easier to do with some help from planar topology; see Exercise 100 below for a more direct approach, though.

Let D be a line through J_1 and J_2, and set $\Gamma_0 = K^* \cup (D \setminus B)$. Let us check that

$$\Gamma_0 \text{ separates } I_1' \text{ from } I_2' \text{ in } \mathbb{R}^2. \tag{52}$$

Indeed, any path that goes from I_1' to I_2' without meeting $J_1 \cup J_2$ contains a path γ from I_1' to I_2' such that $\gamma \cap (I_1' \cup I_2')$ is only composed of the two extremities of γ (see the argument just above (49)). The rest of γ is either contained in B or in $\mathbb{R}^2 \setminus \overline{B}$, because it is connected and does not meet ∂B. In the first case it must meet K^*, because K^* separates I_1' from I_2' in $\overline{B}$. In the second case, it meets $D \setminus B$, because I_1' and I_2' lie on opposite sides of D. This proves (52).

Pick two points $z_1 \in I_1'$ and $z_2 \in I_2'$. By (52), Γ_0 is a closed subset of $\mathbb{R}^2$ which separates z_1 from z_2 (in $\mathbb{R}^2$). Then there is a connected component of Γ_0 that separates z_1 from z_2. See for instance Theorem 14.3 on p. 123 of [Ne].

Call Γ_1 this component. Obviously Γ_1 contains the two pieces of $D \setminus B$, because otherwise it would not separate. Then it contains J_1 and J_2 as well, because they touch $D \setminus B$. Set $\Gamma = \Gamma_1 \cap \overline{B}$. Then

$$J_1 \cup J_2 \subset \Gamma \subset K^* \qquad (53)$$

by construction. Also,

$$\Gamma \text{ is connected.} \qquad (54)$$

This is easy to check. If we had a decomposition of Γ into two disjoint closed sets, we could get a similar decomposition of Γ_1 by adding each half-line of $D \setminus B$ to the closed set that contains its extremity. Now Lemma 51 follows from (53) and (54). $\qquad \square$

We claim that Γ satisfies all the requirements in Theorem 1. Indeed, Γ is compact, connected, and (2) holds because

$$H^1(\Gamma) \leq H^1(K^*) \leq H^1(K \cap B) + H^1(\partial B) + H^1(A_\ell) \leq Ct \leq Cr \qquad (55)$$

by (53), (48), Theorem 18.16 or Lemma 18.19, (47), and (4). As for (3), note that $B = B(y, t_1) \subset B(y, t) \subset B(x, r)$ (see Lemma 7 and (4)), and so

$$H^1(K \cap \Gamma \cap B(x, r)) \geq H^1(K \cap \Gamma \cap B) \geq H^1(\Gamma \cap B) - H^1(\Gamma \cap B \setminus K) \qquad (56)$$

$$\geq \text{dist}(J_1, J_2) - H^1(K^* \cap B \setminus K) \geq \frac{\ell}{2} - C\varepsilon^{1/2} t$$

because Γ connects J_1 to J_2 in $B \cup J_1 \cup J_2$, and by (53), (35), and (37). Recall from (13) that $\ell \geq \frac{t}{C}$; then the right-hand side of (56) is at least $\frac{t}{3C}$ if ε is chosen small enough, and (3) holds with $C_{10} = 3CC_5$, by (5). (Notice that we kept our promise not to let ε depend on C_5.)

This takes care of our main case when (16) holds. When (16) does not hold, i.e., when $\partial B \setminus Z$ is not contained in a single component of $\Omega \setminus K$, we can find two connected components I_1 and I_2 of $\partial B \setminus Z$ that lie in different components of $\Omega \setminus K$. This is because $\partial B \setminus Z \subset \Omega \setminus K$, by (14). In particular, $K \cap \overline{B}$ separates I_1 from I_2 in $\overline{B}$. Call I_i'', $i = 1, 2$, the component of I_i in $\partial B \setminus K$; then $K \cap \overline{B}$ also separates I_1'' from I_2'' in $\overline{B}$.

We can define I_1' and I_2' as before (see a little above (35)), i.e., let I_i' be the arc of ∂B with the same center as I_i, but length $H^1(I_i) + \ell$. Then (35) still holds, where J_1 and J_2 still denote the two components of $\partial B \setminus (I_1' \cup I_2')$. Now Lemma 36 holds trivially, with $K^* = [K \cap B] \cup [\partial B \setminus (I_1' \cup I_2')]$, because this set separates I_1'' from I_2'' in $\overline{B}$. The rest of the argument from there on (that is, Lemma 51 and the final verification) is the same. In other words, we only used (16) and Lemma 17 to find components I_1 and I_2 of $\partial B \setminus Z$ for which Lemma 36 holds, but if (16) does not hold we get Lemma 36 directly.

This completes our proof of Theorem 1, modulo Lemma 39 that we still need to check (at least when (38) fails). $\square$

Proof of Lemma 39. The main point will be the construction of a reasonable C^1 approximation of u in $B \setminus K$, which is very standard but takes some time. We want to use the standard construction with Whitney cubes (except that the cubes will not show up here). We start with a partition of unity.

Set $d(z) = \text{dist}(z, K \cup \partial B)$ for $z \in B \setminus K$. For $k \in \mathbb{Z}$, set

$$A_k = \left\{ z \in B \setminus K ; 2^{-k} \leq d(z) \leq 2^{-k+1} \right\} \qquad (57)$$

and then select a maximal collection $\{x_i\}_{i \in I(k)}$ of points $x_i \in A_k$, with the property that

$$|x_i - x_j| \geq 2^{-k-10} \quad \text{for } i, j \in I(k), \ i \neq j. \qquad (58)$$

We can restrict to k large enough, because for k small A_k and $I(k)$ are empty. Set $B_i = B(x_i, 2^{-k-8})$ and choose a function $\widetilde{\varphi}_i \in C_c^\infty(B_i)$ such that $0 \leq \widetilde{\varphi}_i \leq 1$ everywhere,

$$\widetilde{\varphi}_i(z) = 1 \text{ on } \frac{1}{2} B_i = B(x_i, 2^{-k-9}), \qquad (59)$$

and

$$|\nabla \widetilde{\varphi}_i(z)| \leq C \, d(z)^{-1} \text{ everywhere} . \qquad (60)$$

The disks $\frac{1}{2} B_i$, $i \in I(k)$, cover A_k (by maximality of $\{x_i\}_{i \in I(k)}$). Hence for each $z \in B \setminus K$, there is an $i \in I = \bigcup_k I(k)$ such that $\widetilde{\varphi}_i(z) = 1$. In particular,

$$\widehat{\varphi}(z) = \sum_{i \in I} \widetilde{\varphi}_i(z) \geq 1 \text{ on } B \setminus K. \qquad (61)$$

Also note that for each $z \in B \setminus K$,

$$B\left(z, \frac{d(z)}{3}\right) \text{ meets less than } C \text{ disks } B_i, \ i \in I,$$

$$\text{and all these disks are contained in } B\left(z, \frac{d(z)}{2}\right). \qquad (62)$$

This follows easily from (58) and the other definitions. Because of all this, the functions

$$\varphi_i(z) = \widehat{\varphi}(z)^{-1} \widetilde{\varphi}_i(z) \qquad (63)$$

are all C^∞ on $B \setminus K$, and satisfy the same sort of estimates

$$|\nabla\varphi_i(z)| \leq C d(z)^{-1} \tag{64}$$

as the $\widetilde\varphi_i$ (see (60)). Also,

$$\sum_{i \in I} \varphi_i(z) = 1 \text{ on } B \setminus K. \tag{65}$$

Now set $m_i = \frac{1}{|B_i|} \int_{B_i} u(z)\, dz$ for $i \in I$, and then

$$f(z) = \sum_{i \in I} m_i \varphi_i(z) \text{ on } B \setminus K. \tag{66}$$

Clearly f is C^1 on $B \setminus K$. Next we want to check (40). For each $z \in B \setminus K$, set

$$B(z) = B\left(z, \frac{d(z)}{2}\right) \text{ and } m(z) = \frac{1}{|B(z)|} \int_{B(z)} u. \tag{67}$$

Then

$$\nabla f(z) = \sum_{i \in I} m_i \nabla\varphi_i(z) = \sum_{i \in I} [m_i - m(z)]\nabla\varphi_i(z) \tag{68}$$

because $\sum_{i \in I} \nabla\varphi_i(z) = 0$ on $B \setminus Z$, by (65). To estimate $m_i - m(z)$, note that $B_i \subset B(z)$ when $\nabla\varphi_i(z) \neq 0$, by (62) and because $\varphi_i \in C_c^\infty(B_i)$. Then we can apply the Poincaré inequality (Proposition 12.11) to $u \in W^{1,1}(B(z))$ and get that

$$|m_i - m(z)| \leq \frac{1}{|B_i|} \int_{B_i} |u - m(z)| \leq C d(z)^{-2} \int_{B(z)} |u - m(z)|$$

$$\leq C d(z)^{-1} \int_{B(z)} |\nabla u|, \tag{69}$$

because the radius of B_i is comparable to $d(z)$ by construction, and by (12.12). Hence

$$|\nabla f(z)| \leq \sum_{i \in I} |m_i - m(z)|\, |\nabla\varphi_i(z)| \leq C d(z)^{-2} \int_{B(z)} |\nabla u| \tag{70}$$

by (68), (69), (64), and because the sum has less than C terms (by (62)). Next

$$\int_{B \setminus K} |\nabla f| \leq C \int_{z \in B \setminus K} d(z)^{-2} \int_{w \in B(z)} |\nabla u(w)|. \tag{71}$$

When $w \in B(z)$, we have that $d(w) \geq \frac{1}{2} d(z)$, and then $z \in B\left(w, \frac{d(z)}{2}\right) \subset B(w, d(w))$. Also, $d(z) \geq \frac{2}{3} d(w)$. Hence

$$\int_{B \setminus K} |\nabla f| \leq C \int_{B \setminus K} |\nabla u(w)| d(w)^{-2} \int_{(B \setminus K) \cap B(w, d(w))} dz\, dw \tag{72}$$

$$\leq C \int_{B \setminus K} |\nabla u| \leq C\, t^{3/2}\, \varepsilon^{1/2}$$

by Fubini and (42). This proves (40).

We still need to check (41). Note that

$$|f(z) - m(z)| = \left|\sum_{i \in I} m_i \varphi_i(z) - m(z)\right| \leq \sum_{i \in I} |m_i - m(z)| \varphi_i(z) \qquad (73)$$

$$\leq C \sum_{i \in I} \varphi_i(z) \, d(z)^{-1} \int_{B(z)} |\nabla u| = C \, d(z)^{-1} \int_{B(z)} |\nabla u|$$

for $z \in B \setminus K$, by (66), (65), because (69) holds for all the disks B_i that contain z, and by (65) again.

Next we want to compare $m(z)$ with the values of u on ∂B. Let $i \in \{1, 2\}$ and $z_0 \in I'_i$ be given, and set

$$\delta(z_0) = \text{dist}(z_0, K \cup (\partial B \setminus I'_i)) > 0. \qquad (74)$$

(Recall that I'_i is an open arc and does not meet K.) We shall only consider points $z \in B$ such that

$$|z - z_0| \leq 10^{-2} \delta(z_0). \qquad (75)$$

Note that

$$\text{dist}(z, K \cup (\partial B \setminus I'_i)) \geq \delta(z_0) - |z - z_0| \geq 99|z - z_0| \geq 99 \, d(z) \qquad (76)$$

when (75) holds. Then

$$d(z) = \text{dist}(z, \partial B) \qquad (77)$$

because K is much further. Recall from Lemma 7 that $B = B(y, t_1)$, and write $z = y + Re^{i\theta}$. Thus $R = t_1 - d(z)$, by (77). Set

$$R(z) = \left\{y + \rho e^{is} \, ; \, R - d(z) < \rho < t_1 \text{ and } t_1|s - \theta| < d(z)\right\}$$

$$= \left\{y + \rho e^{is} \, ; \, t_1 - 2d(z) < \rho < t_1 \text{ and } t_1|s - \theta| < d(z)\right\} \qquad (78)$$

[see Figure 2]. Note that diam $R(z) \leq 10 d(z)$, hence

$$R(z) \subset B \setminus K, \qquad (79)$$

by (76). Also, if $I(z) = \{y + t_1 e^{is} \, ; \, t_1|s - \theta| < d(z)\}$ denotes the common boundary of $R(z)$ with B, then

$$I(z) \subset I'_i, \qquad (80)$$

by (76) again.

Call $\ell(z) = H^1(I(z))^{-1} \int_{I(z)} u(\xi) dH^1(\xi)$ the mean value of u on $I(z)$. We claim that

$$|m(z) - \ell(z)| \leq C \, d(z)^{-1} \int_{R(z)} |\nabla u|. \qquad (81)$$

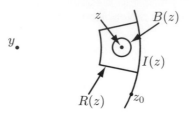

Figure 2

The point is that $u \in W^{1,1}(R(z))$ (by (79)) and $B(z)$ lies somewhere in the middle of $R(z)$, so that (81) is a consequence of the proof of Lemma 13.20 (or its immediate corollary (13.22)) and (11). We cannot apply (13.22) directly here, because $R(z)$ is not exactly a rectangle, but of course the proof of (13.22) applies also in $R(z)$, with only minor modifications. A second difficulty is that in Lemma 13.20 and (13.22) we let B be any ball in the analogue V of $R(z)$, but did not write carefully how the constants depend on B and V. However the proof gives (81), which is also easy to trust by homogeneity. We assume here that the reader would not appreciate being given the details.

Note that $u(\xi) \in L_i =: u(I_i'')$ for $\xi \in I(z)$, because $I(z) \subset I_i' \subset I_i''$. (See (80), (33), and the definitions before (32).) Then the average $\ell(z)$ lies in L_i, because L_i is convex, and

$$\operatorname{dist}(f(z), L_i) \leq |f(z) - \ell(z)| \leq |f(z) - m(z)| + |m(z) - \ell(z)| \leq C\,d(z)^{-1} \int_{R(z)} |\nabla u| \tag{82}$$

by (73) and (81), and because $B(z) \subset R(z)$. By Hölder,

$$d(z)^{-1} \int_{R(z)} |\nabla u| \leq d(z)^{-1} |R(z)|^{1 - \frac{1}{p}} \left\{ \int_{R(z)} |\nabla u|^p \right\}^{1/p}$$

$$\leq C\,d(z)^{1 - \frac{2}{p}} \left\{ \int_{R(z)} |\nabla u|^p \right\}^{1/p}. \tag{83}$$

To estimate the right-hand side of (83), we want to revise our choice of radius t_1 in Lemma 7. Set

$$h(s) = \int_{\partial B(y,s) \setminus K} |\nabla u|^p \, dH^1 \tag{84}$$

for $s \in (0, t)$, and $h(s) = 0$ on $\mathbb{R} \setminus (0, t)$. Then

$$\int_0^t h(s)\,ds = \int_{B(y,t) \setminus K} |\nabla u|^p \leq t^{2 - \frac{p}{2}} \varepsilon^{p/2} \tag{85}$$

by (the proof of) (12). Call h^* the (centered) Hardy–Littlewood maximal function of h. Thus

$$h^*(s) = \sup_{\varepsilon > 0} \frac{1}{2\varepsilon} \int_{s - \varepsilon}^{s + \varepsilon} h(u)\,du. \tag{86}$$

By the Hardy–Littlewood maximal theorem (see for instance [St], p. 5, or Exercise 33.38), h^* lies in weak-L^1, and more precisely there is an absolute constant C such that

$$|\{s \in \mathbb{R}; h^*(s) > \lambda\}| \leq C\lambda^{-1}||h||_1 \leq C\lambda^{-1}\varepsilon^{p/2}t^{2-\frac{p}{2}} \qquad (87)$$

for all $\lambda > 0$. Take $\lambda = 100C\varepsilon^{p/2}t^{1-\frac{p}{2}}$; then

$$\left\{s \in \left(\frac{t}{2}, t\right); h^*(s) > \lambda\right\} \leq \frac{t}{100}, \qquad (88)$$

and by the proof of Lemma 7 we can choose $t_1 \in \left(\frac{t}{2}, t\right)$ so that

$$h^*(t_1) \leq \lambda = 100C\varepsilon^{p/2}t^{1-\frac{p}{2}}, \qquad (89)$$

in addition to the other requirements in Lemma 7.

We may now return to the integral in (83). Note that

$$\int_{R(z)} |\nabla u|^p \leq \int_{t_1-2d(z)}^{t_1} h(s)ds \leq 4d(z)h^*(t_1) \leq 400Cd(z)\varepsilon^{p/2}t^{1-\frac{p}{2}}, \qquad (90)$$

by the second part of (78), (84), and then (86) and (89). Then

$$\mathrm{dist}(f(z), L_i) \leq Cd(z)^{1-\frac{2}{p}}\left\{\int_{R(z)}|\nabla u|^p\right\}^{1/p}$$

$$\leq Cd(z)^{1-\frac{2}{p}}d(z)^{1/p}\varepsilon^{1/2}t^{\frac{1}{p}-\frac{1}{2}} = C\varepsilon^{1/2}d(z)^{1-\frac{1}{p}}t^{\frac{1}{p}-\frac{1}{2}}, \qquad (91)$$

by (82), (83), and (90). Now (91) holds as soon as (75) is satisfied, and since the right-hand side of (91) tends to 0 when z tends to z_0 (because then $d(z) = \mathrm{dist}(z, \partial B)$ tends to 0), we get that

$$\limsup_{\substack{z \to z_0 \\ z \in B}} \mathrm{dist}(f(z), L_i) = 0. \qquad (92)$$

This is actually stronger than (41).

Our proof of Lemma 39 is now complete; Theorem 1 follows as well.   □

Remark 93. With our proof of Theorem 1 and a small amount of additional work, we can prove directly that K contains big pieces of Lipschitz graphs locally (as in Remark 27.20, but we shall give a precise statement soon). To see this, let us first review the main lines of our proof of Theorem 1.

We started from a disk $B(x, r) \subset \Omega$, as in the statement of the theorem, and used Corollary 23.38 (on Carleson measures) to show that for each small enough $\varepsilon > 0$, we can find a new disk $B(y, t) \subset B(x, r)$ such that $t \geq C_5(\varepsilon)r$ and $\omega_p(y, t) \leq \varepsilon$. Then we constructed a new disk $B = B(y, t_1)$, with $t_1 \subset \left(\frac{t}{2}, t\right)$, and a compact connected set $\Gamma \subset \overline{B}$, with the following two main properties. First,

$$H^1(\Gamma \cap B \setminus K) \leq C\varepsilon^{1/2}t, \qquad (94)$$

by (53) and (37). And there are two disjoint arcs J_1, J_2, of ∂B such that $\text{dist}(J_1, J_2) \geq C^{-1}t$, $\Gamma \subset B \cup J_1 \cup J_2$, and Γ contains $J_1 \cup J_2$. (See (35) and (13), the definition of J_1 and J_2 just before (35), (53), the statement of Lemma 36, and also the discussion of the case where (16) fails just before the proof of Lemma 39.)

Pick points $y_1 \in J_1$ and $y_2 \in J_2$. By Proposition 30.14 below, we can find a simple curve γ_1 that goes from y_1 to y_2 and is supported in Γ. Remove the part of γ_1 before we leave J_1 for the last time, and then remove the part after we hit J_2 for the first time. We get a new simple arc γ in Γ which connects J_1 to J_2 and lies in B except for its extremities.

Thus we could replace the words "connected set Γ" with "simple curve Γ" in our statement of Theorem 1. We can do a bit more than this, thanks to the "Rising Sun Lemma" of F. Riesz. Note that $\text{diam}(\gamma) \geq \text{dist}(J_1, J_2) \geq C^{-1}t$, and also

$$\text{length}(\gamma) \leq H^1(K \cap B) + C\varepsilon^{1/2}t \leq Ct \tag{95}$$

by (94) and Lemma 18.19. Then there is a Lipschitz graph G with constant $\leq C'$ such that

$$H^1(G \cap \gamma) \geq \frac{t}{C'}. \tag{96}$$

See [Da1], Proposition 8, p. 168. The proof is not hard, but giving it may lead us a little too far from our main subject. Recall that for us a Lipschitz graph is the image under a rotation of the graph of some Lipschitz function $A : \mathbb{R} \to \mathbb{R}$.

It is important here that C' depends on M, but not on ε. Thus we can choose ε so small that

$$H^1(K \cap G \cap B) \geq H^1(G \cap \gamma) - H^1(\gamma \setminus K) \geq \frac{t}{C'} - H^1(\Gamma \cap B \setminus K) \geq \frac{t}{2C'}, \tag{97}$$

by (96) and (94). Altogether we have proved the following slight improvement of Theorem 1.

Theorem 98. *Let $\Omega \subset \mathbb{R}^2$ and $(u, K) \in TRLQ(\Omega)$ be given. If a is small enough (depending on M), then for all $x \in K$ and $0 < r \leq r_0$ such that $B(x, r) \subset \Omega$, we can find a C_{13}-Lipschitz graph G such that*

$$H^1(K \cap G \cap B(x, r)) \geq C_{13}^{-1}r. \tag{99}$$

As usual, C_{13} depends only on M. We can take $C_{13} = 4C_5(\varepsilon)C'$, where $C_5(\varepsilon)$ comes from Corollary 23.38 and ε was chosen just above (97) (depending on C').

Exercise 100. Here we want to propose a more direct approach to Lemma 51, without the separation result from [Ne]. Let $B \subset \mathbb{R}^2$ be a disk, $J_1, J_2 \subset \partial B$ two disjoint closed arcs of ∂B, and let $K^* \subset B \cup J_1 \cup J_2$ be a compact set which contains $J_1 \cup J_2$ and separates the two components of $\partial B \setminus (J_1 \cup J_2)$ in $\overline{B}$. We want to find a connected set $\Gamma \subset K^*$ that contains J_1 and J_2.

Fix a small $\varepsilon > 0$. Set $J_j^\varepsilon = \{x \in B\,;\, \text{dist}(x, J_j) < \varepsilon\}$ for $j = 1, 2$. Call $K_1 = K^* \setminus (J_1^\varepsilon \cup J_2^\varepsilon)$ and $\varepsilon_1 = \text{dist}(K_1, \partial B) > 0$.

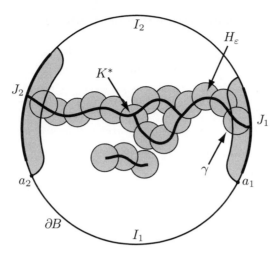

Figure 3

1. Find a covering of K_1 with a finite set of disks B_i, $i \in I$, with the following properties: $B_i = B(x_i, r_i)$ is centered on K_1, $r_i \in \left(\frac{\varepsilon_1}{4}, \frac{\varepsilon_1}{2}\right)$, the boundaries ∂B_i, $i \in I$, never meet each other or one of the ∂J_j^ε tangentially, and no point of $\overline{B}$ ever lies in three ∂B_i (or more), or two ∂B_i and a ∂J_j^ε. See Figure 3 for a rough picture.

2. Set $H_\varepsilon = \overline{J}_1^\varepsilon \cup \overline{J}_2^\varepsilon \cup \left(\bigcup_{i \in I} \overline{B}_i \right)$. Show that $\partial H_\varepsilon \cap \partial B = (\partial J_1^\varepsilon \cap \partial B) \cup (\partial J_2^\varepsilon \cap \partial B)$.

3. Check that $\partial H_\varepsilon \cap K^*$ reduces to $J_1 \cup J_2$.

4. Show that each point of ∂H_ε has a neighborhood in which ∂H_ε is a simple curve, composed of one or two arcs of circles.

5. Call I_1 and I_2 the two components of $\partial B \setminus (J_1 \cup J_2)$, and denote by a_j the point of I_1 that lies at distance ε from J_j. Note that $a_1 \in \partial H_\varepsilon$, and there is an arc γ of ∂H_ε that leaves ∂B at the point a_1. Show that we can continue the path γ in ∂H_ε, and follow it until we hit ∂B again for the first time. (Use Question 4.) Also show that the first point of ∂B that we hit is a_2 (use the hypothesis!).

6. Call H_ε' the connected component of J_1 in H_ε. Check that $J_2 \subset H_\varepsilon'$.

7. Construct a decreasing sequence of connected sets H_{ε_k}' that contain $J_1 \cup J_2$ and lie within ε_k of K^*. Conclude.

30   Simple arcs in connected sets with finite H^1-measure

This section contains well-known results on the existence of rectifiable arcs in connected sets with finite H^1-measure. See for instance [Fa].

Proposition 1. *Let $\Gamma \subset \mathbb{R}^n$ be a compact connected set such that $L = H^1(\Gamma) < +\infty$. Then there is a C_n-Lipschitz surjective mapping $f : [0, L] \to \Gamma$.*

Corollary 2. *The set Γ in Proposition 1 is arcwise connected and rectifiable.*

The corollary is immediate. See in particular Definition 24.3 and the comments that follow.

To prove the proposition we want to use a nice simple lemma on finite graphs. For us a graph G will be given by a set S of vertices, together with a set E of unordered pairs $(x, y) \in S^2$, with $x \neq y$. We see each pair $(x, y) \in E$ as an edge with extremities x and y; then G is the union of all these edges, with the topology that we all imagine. If S is finite, we say that G is finite (and E also is finite).

Lemma 3. *Let G be a finite, connected graph, and let $x_0 \in S$ be a vertex of G. We can find a path in G that starts from x_0, runs along each edge of G exactly twice (once in each direction), and ends up at x_0.*

Here a path merely means a succession of vertices: we do not care about how we run along each edge of the path. The lemma will be easily proved by induction. Note that it is trivial when G has 0 or 1 edge.

Now assume that G has at least two edges. Let $A = (x_0, x_1)$ be an edge that touches x_0. Then denote by G^* the graph G (with the same vertices), minus the edge A.

If G^* is connected, we can apply the induction hypothesis and get a path γ that starts at x_0, runs along each edge of G^* exactly twice, and ends up at x_0. We can complete γ by adding a back-and-forth excursion along A either at the beginning or at the end of γ.

If G^* is not connected, it can be decomposed into two connected pieces G_0 and G_1, with $x_0 \in G_0$ (say). We can use the induction hypothesis to find a path γ_0 in G_0 as in the statement. (Possibly γ_0 is trivial, if $G_0 = \{x_0\}$.) Similarly, $x_1 \in G_1$ (because the missing edge must connect G_0 to G_1), and we can apply the induction hypothesis to get a path γ_1 that starts from x_1, runs along each edge of G_1 twice, and goes back to x_1. To get the desired path on G, we leave from x_0, run along A to x_1, follow γ_1, return to x_0 along A, and finish with the path γ_0. Lemma 3 follows.   □

Now let Γ be a connected set with finite measure $L = H^1(\Gamma)$, as in the statement of Proposition 1. Let $\varepsilon > 0$ be given (smaller than $\frac{1}{2}\operatorname{diam}\Gamma$, which we can assume to be positive). We want to approximate Γ by a graph. Let X_ε denote a maximal subset of Γ with the property that

$$\operatorname{dist}(x, y) \geq 2\varepsilon \quad \text{for} \ x, y \in X_\varepsilon, \ x \neq y. \tag{4}$$

Of course X_ε is finite, because Γ is compact. Also

$$\text{dist}(x, X_\varepsilon) < 2\varepsilon \text{ for all } x \in \Gamma, \tag{5}$$

by maximality. Denote by G_ε the (abstract) graph whose vertices are the points of X_ε and whose edges are the pairs of points $x, y \in X_\varepsilon$ such that $x \neq y$ and $|x - y| \leq 4\varepsilon$. Let us check that G_ε is connected.

Suppose not. Then there is a decomposition of X_ε into two disjoint, nonempty subsets Y and Z, that are not connected to each other in G_ε. This means that we cannot find $y \in Y$ and $z \in Z$ such that $|y - z| \leq 4\varepsilon$. In other words,

$$\text{dist}(Y, Z) > 4\varepsilon. \tag{6}$$

(Recall that our sets are finite.) Set

$$\widehat{Y} = \bigcup_{y \in Y} \overline{B}(y, 2\varepsilon) \text{ and } \widehat{Z} = \bigcup_{z \in Z} \overline{B}(z, 2\varepsilon). \tag{7}$$

These are closed sets, and they are disjoint by (6). Also, (5) says that $\Gamma \subset \widehat{Y} \cup \widehat{Z}$, because $X_\varepsilon = Y \cup Z$. Then $\Gamma = (\Gamma \cap \widehat{Y}) \cup (\Gamma \cap \widehat{Z})$ is a partition of Γ into closed subsets. These sets are not empty, because they contain Y and Z respectively; hence we get a contradiction with the connectedness of Γ. This proves that G_ε is connected.

Lemma 3 gives a path in G_ε that goes through each edge of G_ε exactly twice. If we associate to each edge (x, y) the corresponding interval $[x, y]$ in $\mathbb{R}^n$, we get a path γ in $\mathbb{R}^n$. Call $N(\varepsilon)$ the number of edges in G_ε. Since $H^1([x, y]) \leq 4\varepsilon$ for each edge (x, y) in the graph, we get that

$$\text{length}(\gamma) \leq 8\varepsilon\, N(\varepsilon). \tag{8}$$

Also, $N(\varepsilon) \leq C_n N_1(\varepsilon)$, where $N_1(\varepsilon)$ is the number of elements in X_ε. Indeed, there are at most C_n edges that leave from a given point $x \in X_\varepsilon$, because $\overline{B}(x, 4\varepsilon)$ contains at most C_n points that lie at mutual distances $\geq 2\varepsilon$, as in (4). To estimate $N_1(\varepsilon)$, note that all the balls $B(x, \varepsilon)$, $x \in X_\varepsilon$, are disjoint (by (4)). Let us check that

$$H^1(\Gamma \cap B(x, \varepsilon)) \geq \varepsilon \text{ for } x \in X_\varepsilon. \tag{9}$$

If we can do this, then automatically

$$\text{length}(\gamma) \leq 8\varepsilon\, N(\varepsilon) \leq 8C_n\, \varepsilon\, N_1(\varepsilon) \leq 8C_n \sum_{x \in X_\ell} \varepsilon$$
$$\leq 8C_n \sum_{x \in X_\ell} H^1(\Gamma \cap B(x, \varepsilon)) \leq 8C_n\, H^1(\Gamma) = 8C_n\, L, \tag{10}$$

by (8) and the discussion above.

To prove (9), recall that we took $\varepsilon < \frac{1}{2}\operatorname{diam}\Gamma$, so Γ is not contained in $B(x,\varepsilon)$. Set $\pi(z) = |z - x|$ for $z \in \Gamma$. Then $\pi(\Gamma)$ is not contained in $(0,\varepsilon)$, and since it is connected (because Γ is connected) and contains 0 (because $x \in \Gamma$), $\pi(\Gamma)$ contains $(0,\varepsilon)$. Now

$$\varepsilon = H^1\big(\pi(\Gamma \cap B(x,\varepsilon))\big) \leq H^1(\Gamma \cap B(x,\varepsilon)), \tag{11}$$

by Remark 2.11. This proves (9) and (10).

Because of (10), we can find an $8C_n$-Lipschitz parameterization $f_\varepsilon : [0,L] \to \mathbb{R}^n$ of our path γ. Notice that

$$\operatorname{dist}(f_\varepsilon(t),\Gamma) \leq 2\varepsilon \ \text{ for } \ t \in [0,L], \tag{12}$$

by construction. (For each edge (x,y) of G_ε, $[x,y]$ stays within 2ε of its endpoints (that lie in Γ).) Also,

$$\operatorname{dist}(x, f_\varepsilon([0,L])) \leq \operatorname{dist}(x, X_\varepsilon) \leq 2\varepsilon \ \text{ for } \ x \in \Gamma, \tag{13}$$

because $f_\varepsilon([0,L]) \supset X_\varepsilon$ and by (5).

For each sequence $\{\varepsilon_k\}$ that tends to 0, we can extract a subsequence (which we shall still denote by $\{\varepsilon_k\}$) for which $\{f_{\varepsilon_k}\}$ converges uniformly. (See Exercise 23 if you have a doubt.) Then the limit f is the surjective mapping that we wanted. It is still $8C_n$-Lipschitz, like all the f_{ε_k}, and $f([0,\ell]) \subset \Gamma$ by (12) and because Γ is closed. To prove that f is also onto, let $z \in \Gamma$ be given. For each n, there exists $t_n \in [0,L]$ such that $|f_{\varepsilon_n}(t_n) - z| \leq 2\varepsilon_n$ (by (13)). We can extract a subsequence so that (after extraction) t_n tends to a limit t. Then $|f(t) - z| \leq |f(t) - f_{\varepsilon_n}(t)| + |f_{\varepsilon_n}(t) - f_{\varepsilon_n}(t_n)| + |f_{\varepsilon_n}(t_n) - z| \leq |f(t) - f_{\varepsilon_n}(t)| + 8C_n|t - t_n| + 2\varepsilon_n$, which tends to zero, so $f(t) = z$ (as desired). Thus f has all the desired properties, and Proposition 1 follows. $\qquad\square$

In some arguments it is more pleasant to work with simple arcs; they are also fairly easy to get in the present situation, as in the following improvement of Corollary 2.

Proposition 14. *Let $\Gamma \subset \mathbb{R}^n$ be a compact connected set such that $H^1(\Gamma) < +\infty$. Then for each choice of $x_0, y_0 \in \Gamma$, with $y_0 \neq x_0$, we can find an injective Lipschitz mapping $f : [0,1] \to \Gamma$ such that $f(0) = x_0$ and $f(1) = y_0$.*

A reasonable option would be to take an arc in Γ from x_0 to y_0 and then remove the loops in it until it becomes simple (see for instance [Fa]). Since we like to minimize things here, let us try to find f directly, essentially by minimizing the length of the corresponding arc. Let Γ, x_0, y_0 be given, as in the statement, and set

$$M = \inf \Big\{ m \,;\ \text{there is an } m\text{-Lipschitz function } f : [0,1] \to \Gamma \\ \text{such that } f(0) = x_0 \text{ and } f(1) = y_0 \Big\}. \tag{15}$$

We know from Theorem 1 that $M < +\infty$: we can even find Lipschitz mappings from $[0,1]$ onto Γ. Let $\{f_k\}$ be a minimizing sequence. That is, $f_k : [0,1] \to \Gamma$ is m_k-Lipschitz, $f_k(0) = x_0$, $f_k(1) = y_0$, and $\{m_k\}$ tends to M. As before, we can extract a subsequence, which we shall still denote by $\{f_k\}$, that converges uniformly on $[0,1]$ to some limit f. Then f is M-Lipschitz, $f(0) = x_0$, $f(1) = y_0$, and $f([0,1]) \subset \Gamma$ (because Γ is compact).

Suppose f is not injective. We can find $0 \le t_1 < t_2 \le 1$ such that $f(t_1) = f(t_2)$. Then we can remove the needless loop between t_1 and t_2, reparameterize our arc $[0,1]$, and get an $(1 - t_2 + t_1)M$-Lipschitz mapping $\widetilde{f}$ with the usual properties. Namely, we can take

$$
\begin{cases}
\widetilde{f}(t) = f((1 - t_2 + t_1)\, t) & \text{for} \quad 0 \le t \le \dfrac{t_1}{1 - t_2 + t_1}, \\[2mm]
\widetilde{f}(t) = f\Big((1 - t_2 + t_1)\, t + (t_2 - t_1)\Big) & \text{for} \quad \dfrac{t_1}{1 - t_2 + t_1} \le t \le 1.
\end{cases}
\tag{16}
$$

The existence of $\widetilde{f}$ contradicts (15), so f is injective and satisfies all the required properties. Proposition 14 follows. $\qquad\square$

We shall need later the following extension of Proposition 14 to closed connected sets with locally finite Hausdorff measure.

Proposition 17. *Let $\Gamma \subset \mathbb{R}^n$ be a closed connected set such that $H^1(\Gamma \cap B(0,R)) < +\infty$ for every $R > 0$. Then for each choice of $x, y \in \Gamma$, with $y \ne x$, we can find an injective Lipschitz mapping $f : [0,1] \to \Gamma$ such that $f(0) = x$ and $f(1) = y$.*

Let us first observe that for each radius $R > 0$ we can find $R' \in [R/2, R]$ such that $\Gamma \cap \partial B(0, R')$ is finite. This comes directly from Lemma 26.1 (which says that the integral over $R' < R$ of the number of points in $\Gamma \cap \partial B(0, R')$ is at most $H^1(\Gamma \cap B(0,R)) < +\infty$) and Chebyshev. Because of this, we can choose an increasing sequence of radii R_k, with $R_k \ge 2^k$, such that $\Gamma \cap \partial B(0, R_k)$ is finite for each k. Let us also assume, without loss of generality, that x and y lie in $B(0, R_0)$.

Set $\Gamma_k = \Gamma \cap \overline{B}(0, R_k)$, call $A_k(x)$ the connected component of x_0 in Γ_k, and similarly for $A_k(y)$. Note that $A_k(x)$ and $A_k(y)$ are compact, because their closures are connected anyway. If $A_k(x) = A_k(y)$ for some k, then we can apply Proposition 14 to $\Gamma' = A_k(x)$, and get the desired Lipschitz mapping f. So let us assume that $A_k(x) \ne A_k(y)$ for all k and try to reach a contradiction. It will be useful to know that for each k,

$$
\Gamma'_k = \Gamma_k \cup \partial B(0, R_k) \text{ is connected.}
\tag{18}
$$

Indeed suppose that $\Gamma'_k = E \cup F$, where E and F are disjoint closed subsets of Γ'_k. Since $\partial B(0, R_k)$ is connected, it is contained in E or F; let us assume for definiteness that $\partial B(0, R_k) \subset F$. Then set $F' = [F \cap \Gamma] \cup [\Gamma \setminus B(0, R_k)]$. Notice that F' is closed (because F and Γ arc). Also, E does not meet $\partial B(0, R_k)$ because $\partial B(0, R_k) \subset F$, nor $\Gamma \setminus \overline{B}(0, R_k)$ because $E \subset \Gamma_k$, so E and F' are disjoint. They are both contained in Γ, and their union is Γ. Since Γ is connected, E or F' is empty. Now F' is not empty, because otherwise Γ would be contained in $B(0, R_k)$, Γ_k would be equal to Γ, and $A_k(x)$ would not be different from $A_k(y)$ (because Γ is connected). Altogether, E is empty, and (18) holds.

Fix $k \geq 0$, and call $\mathcal{B}_k$ the set of connected components of Γ_k. Because of (18) and Proposition 1, every point of Γ_k can be connected to $\partial B(0, R_k)$ by a path in Γ_k. So each component of Γ_k contains at least one point of $\Gamma \cap \partial B(0, R_k)$, and hence $\mathcal{B}_k$ is finite.

For each $l > k$, set $A_{k,l}(x) = A_l(x) \cap \Gamma_k$, where $A_l(x)$ still denotes the component of x in Γ_l. Then

$$A_{k,l}(x) \text{ is a union of components } A \in \mathcal{B}_k, \tag{19}$$

because any $A \in \mathcal{B}_k$ that meets $A_l(x)$ is contained in it. Next, $A_{k,l}(x)$ is a non-decreasing function of l, because $A_l(x)$ is. Then (19) says that it is stationary (because $\mathcal{B}_k$ is finite). Set

$$A_{k,\infty}(x) = \bigcup_l A_{k,l}(x); \tag{20}$$

thus $A_{k,\infty}(x)$ coincides with $A_{k,l}(x)$ for l large, and (19) says that it is a union of components $A \in \mathcal{B}_k$. In particular,

$$A_{k,\infty}(x) \text{ and } \Gamma_k \setminus A_{k,\infty}(x) \text{ are closed} \tag{21}$$

(because $\mathcal{B}_k$ is finite). Moreover, if $k < k' < l$, $A_{k,l}(x) = A_l(x) \cap \Gamma_k = A_{k',l}(x) \cap \Gamma_k$, so $A_{k,\infty}(x) = A_{k',\infty}(x) \cap \Gamma_k$. Set $A_\infty(x) = \bigcup_k A_{k,\infty}(x)$. Then

$$A_{k,\infty}(x) = A_\infty(x) \cap \Gamma_k \text{ for all } k, \tag{22}$$

and (21) says that $A_\infty(x)$ and its complement in Γ are closed. Since Γ is connected and $A_\infty(x)$ is not empty, we get that $A_\infty(x) = \Gamma$, (22) says that $A_{k,\infty}(x) = \Gamma_k$, hence $A_{k,l}(x) = \Gamma_k$ for l large, and in particular $y \in A_l(x)$ for some l. This contradicts our assumption that $A_l(x) \neq A_l(y)$ and completes our proof of Proposition 17. $\qquad \square$

Exercise 23 (Montel). Let E be a separable metric space (i.e., with a dense sequence $\{x_i\}$) and let $\{f_k\}$ be a sequence of functions from E to a compact metric space. Suppose that there is a function $\omega : [0, 1] \to [0, +\infty]$ such that $\lim_{t \to 0^+} \omega(t) = 0$ and

$$\text{dist}(f_k(x), f_k(y)) \leq \omega(t) \text{ when } \text{dist}(x, y) \leq t. \tag{24}$$

(It is important that ω does not depend on k.)

1. Extract a subsequence such that $\{f_{k_\ell}(x_i)\}$ converges for each i.

2. Show that $\{f_{k_\ell}(x)\}$ converges for every $x \in E$, and that the limit f also satisfies (24).

3. Show that the convergence is uniform on every compact subset of E.

31   Big pieces of connected sets and uniform rectifiability

In this section we prove that Ahlfors-regular sets of dimension 1 in $\mathbb{R}^n$ that contain big pieces of connected sets are uniformly rectifiable (i.e., contained in a regular curve). In the next section we shall see how to use this to prove Theorem 27.1 on the local uniform rectifiability of K when $(u, K) \in TRLQ(\Omega)$.

Recall from Definition 18.9 that an Ahlfors-regular set of dimension 1 is a closed set $E \subset \mathbb{R}^n$ such that

$$C_0^{-1}r \leq H^1(E \cap B(x, r)) \leq C_0 r \tag{1}$$

for $x \in E$, $0 < r < \mathrm{diam}(E)$, and some constant $C_0 \geq 1$. We also require $\mathrm{diam}(E) > 0$ to avoid trivialities.

As we said in Definition 27.17, a uniformly rectifiable set (of dimension 1) is an Ahlfors-regular set $E \subset \mathbb{R}^n$ (of dimension 1) which is contained in some Ahlfors-regular curve. See Definition 27.3 concerning regular curves, but recall that they are Ahlfors-regular connected sets.

Definition 2. *Let $E \subset \mathbb{R}^n$ be Ahlfors-regular (of dimension 1). We say that E contains big pieces of connected sets if there is a constant $C_1 \geq 1$ such that for all choices of $x \in E$ and $0 < r < \mathrm{diam}(E)$ we can find a compact connected set Γ such that*

$$H^1(\Gamma) \leq C_1 r \tag{3}$$

and

$$H^1(E \cap \Gamma \cap B(x, r)) \geq C_1^{-1} r. \tag{4}$$

We have seen a local version of this briefly in Remark 27.20, and in Theorem 29.1. The main result of this section is the following.

Theorem 5. *If the one-dimensional Ahlfors-regular set $E \subset \mathbb{R}^n$ contains big pieces of connected sets, then it is uniformly rectifiable.*

The converse to Theorem 5 is true, but not very interesting. See Exercise 97.

A good reference for Theorem 5 and similar results is [DaSe3]. The proof given here comes from [DaSe6]. It is slightly different, because in the second part of the argument we shall use an idea of Morel and Solimini ([MoSo2], Proposition 16.25 on page 206) to find a regular curve by minimization, instead of constructing it by hand.

Our proof of Theorem 5 splits naturally into two parts. In the first one, we show that for each $x_0 \in \Gamma$ and $0 < r_0 < \mathrm{diam}(E)$, $E \cap B(x_0, r_0)$ is contained in a not too large connected set $\Gamma = \Gamma(x_0, r_0)$. In the second part, we show that all these connected sets can be organized to give a single regular curve.

Proposition 6. *If $E \subset \mathbb{R}^n$ is an Ahlfors-regular set of dimension 1 that contains big pieces of connected sets, then there is a constant $C_2 > 0$ such that, for each $x_0 \in E$ and $r_0 > 0$, we can find a compact connected set $\Gamma = \Gamma(x_0, r_0)$ such that*

$$E \cap B(x_0, r_0) \subset \Gamma \subset \overline{B}(x_0, r_0), \tag{7}$$

and

$$H^1(\Gamma) \le C_2 r_0. \tag{8}$$

This will be our first part. The general idea is to take the little pieces of connected sets given by Definition 2 and glue them to each other to get a single connected set Γ. As often, all the strength of Definition 2 comes from the fact that although we do not require so much on a given ball $B(x, r)$, we have a similar requirement on *each* such ball, so that we can always complete our information on $B(x, r)$ by looking at smaller balls. This is the same sort of reason that makes the John and Nirenberg theorem on BMO functions work so nicely.

So let $x_0 \in E$ and $r_0 > 0$ be given. As far as Proposition 6 is concerned, we may as well assume that $r_0 < \frac{3}{2} \operatorname{diam}(E)$ if E is bounded, because otherwise the result on $B(x_0, \frac{3}{2} \operatorname{diam}(E))$ would be stronger anyway. We want to construct a nondecreasing sequence of compact connected sets Γ_j that contain more and more of the set

$$E_0 = E \cap B(x_0, r_0). \tag{9}$$

Let Γ_0 be a compact connected set such that

$$H^1(\Gamma_0) \le C_1 r_0 \tag{10}$$

and

$$H^1(E \cap \Gamma_0 \cap B(x_0, r_0)) \ge (2C_1)^{-1} r_0. \tag{11}$$

Such a set exists, by Definition 2 (applied to the pair $\left(x_0, \frac{r_0}{2}\right)$ to get a radius smaller than $\operatorname{diam}(E)$). We can even modify Γ_0 (without changing (10) and (11)), so that $\Gamma_0 \subset \overline{B}(x_0, r_0)$. Indeed, if $\pi : \mathbb{R}^n \to \overline{B}(x_0, r_0)$ is defined by

$$\begin{cases} \pi(z) = z & \text{for} \quad z \in \overline{B}(x_0, r_0), \\ \pi(z) = x_0 + r_0 \dfrac{z - x_0}{|z - x_0|} & \text{for} \quad z \in \mathbb{R}^n \setminus \overline{B}(x_0, r_0), \end{cases} \tag{12}$$

it is easy to see that π is 1-Lipschitz, and so $\pi(\Gamma_0)$ still satisfies (10). Of course $\pi(\Gamma_0)$ is still connected, (11) stays the same, and $\pi(\Gamma_0) \subset \overline{B}(x_0, r_0)$.

We want to continue our construction by induction, so let us assume that we already have a nondecreasing (finite) sequence of compact connected sets $\Gamma_0, \ldots, \Gamma_j$ for some $j \ge 0$, and let us define Γ_{j+1}. Set

$$U_j = E_0 \setminus \Gamma_j \tag{13}$$

and, for each $x \in U_j$,

$$\delta_j(x) = \operatorname{dist}(x, \Gamma_j) \text{ and } B_j(x) = B(x, \delta_j(x)). \tag{14}$$

Note that

$$\delta_j(x) \le 2r_0 < 3\operatorname{diam}(E) \tag{15}$$

because $x \in B(x_0, r_0)$, (11) says that $\Gamma_0 \cap B(x_0, r_0) \ne \emptyset$ (and hence $\Gamma_j \cap B(x_0, r_0) \ne \emptyset$ too), and by our assumption above (9).

By a simple covering argument (see the first pages of [St], or Lemma 33.1 below), we can find a finite or countable set $A_j \subset U_j$ such that

$$\text{the balls } \overline{B}_j(x), \ x \in A_j, \ \text{are disjoint} \tag{16}$$

and

$$U_j \subset \bigcup_{x \in A_j} \overline{B}(x, 5\delta_j(x)). \tag{17}$$

In particular,

$$H^1(U_j) \le \sum_{x \in A_j} H^1(E \cap \overline{B}(x, 5\delta_j(x))) \le C \sum_{x \in A_j} \delta_j(x), \tag{18}$$

by (1) (i.e., the Ahlfors-regularity of E). By (15), we can apply Definition 2 to each $x \in A_j$, with the radius $\frac{1}{4}\delta_j(x)$. We get a compact connected set $\gamma_j(x)$ such that

$$H^1(\gamma_j(x)) \le \frac{1}{4} C_1 \delta_j(x) \tag{19}$$

and

$$H^1(E \cap \gamma_j(x) \cap B_j(x)) \ge \frac{1}{4} C_1^{-1} \delta_j(x). \tag{20}$$

Also, by the same argument as for Γ_0, we can always replace $\gamma_j(x)$ with its image by a contraction onto $\overline{B}(x, \frac{1}{4}\delta_j(x))$ (like π in (12)), and so we may assume that $\gamma_j(x) \subset B_j(x)$.

Call $z_j(x)$ a point of $\Gamma_j \cap \partial B_j(x)$. Such a point exists, by (14). Add to $\gamma_j(x)$ a line segment that connects it to $z_j(x)$. We get a new compact connected set $\alpha_j(x)$ such that

$$H^1(\alpha_j(x)) \le C\delta_j(x), \tag{21}$$
$$H^1(E \cap \alpha_j(x)) \ge (4C_1)^{-1}\delta_j(x), \tag{22}$$

and

$$\alpha_j(x) \subset B_j(x), \text{ except for the single point } z_j(x) \in \Gamma_j \cap \partial B_j(x). \tag{23}$$

Notice that

$$\text{the sets } \alpha_j(x), \ x \in A_j, \ \text{are disjoint,} \tag{24}$$

by (23) and (16). Also, if we set

$$V_j = E \cap B(x_0, 3r_0) \setminus \Gamma_j, \tag{25}$$

then

$$E \cap \alpha_j(x) \subset V_j \cup \{z_j(x)\}. \tag{26}$$

This comes from (23) and (14), plus (15) and the fact that $x \in A_j \subset U_j \subset E_0 \subset B(x_0, r_0)$ for the part about $B(x_0, 3r_0)$. Now

$$\sum_{x \in A_j} \delta_j(x) \leq 4C_1 \sum_{x \in A_j} H^1(E \cap \alpha_j(x)) \leq 4C_1 H^1(V_j) < +\infty \tag{27}$$

by (22), (24), (26), and the Ahlfors-regularity property (1).

Since $\sum_{x \in A_j} \delta_j(x) < +\infty$, we can find a finite set $A'_j \subset A_j$ such that

$$\sum_{x \in A'_j} \delta_j(x) \geq \frac{1}{2} \sum_{x \in A_j} \delta_j(x). \tag{28}$$

Set

$$\Gamma_{j+1} = \Gamma_j \cup \left(\bigcup_{x \in A'_j} \alpha_j(x) \right). \tag{29}$$

By construction, Γ_{j+1} is still compact (because A'_j is finite) and connected (because each $\alpha_j(x)$ is connected and touches the connected set Γ_j).

Our construction of sets Γ_j is now finished. We want to take

$$\Gamma^* = \left\{ \bigcup_{j \geq 0} \Gamma_j \right\}^- \tag{30}$$

as our first approximation of Γ, but let us first see why the process has a chance of converging. Note that for $x \in A'_j$, $\alpha_j(x)$ is contained in Γ_{j+1} (by (29)). Hence it does not meet V_{j+1} (see (25)), and so

$$E \cap \alpha_j(x) \subset (V_j \setminus V_{j+1}) \cup \{z_j(x)\}, \tag{31}$$

by (26). Since all the $\alpha_j(x)$ are disjoint by (24), we get that

$$H^1(V_j \setminus V_{j+1}) \geq \sum_{x \in A'_j} H^1(E \cap \alpha_j(x)) \geq (4C_1)^{-1} \sum_{x \in A'_j} \delta_j(x), \tag{32}$$

by (22). Note that

$$\sum_{j \geq 0} H^1(V_j \setminus V_{j+1}) \leq H^1(V_0) \leq H^1(E \cap B(x_0, 3r_0)) \leq Cr_0 \tag{33}$$

by (25) and (1). Hence

$$\sum_{j\geq 0}\sum_{x\in A'_j}\delta_j(x)\leq Cr_0 < +\infty, \tag{34}$$

by (32) and (33). Now we can verify that

$$E_0 =: E\cap B(x_0, r_0) \text{ is contained in } \Gamma^*, \tag{35}$$

where Γ^* is defined by (30).

Let $z\in E_0$ be given. If $z\in\Gamma_j$ for some $j\geq 0$, there is nothing to prove (because $\Gamma_j\subset\Gamma^*$). So we can assume that $z\in U_j = E_0\setminus\Gamma_j$ for each j. (See the definition (13).) Then $z\in\overline{B}(x, 5\delta_j(x))$ for some $x\in A_j$, by (17), and so

$$\operatorname{dist}(z, \Gamma^*)\leq\operatorname{dist}(z, \Gamma_j)\leq|z-x|+\delta_j(x)\leq 6\delta_j(x), \tag{36}$$

by (14). Note also that

$$\delta_j(x)\leq\sum_{y\in A_j}\delta_j(y)\leq 2\sum_{y\in A'_j}\delta_j(y) \tag{37}$$

for $x\in A_j$, by (28). The right-hand side of (37) tends to 0 when j tends to $+\infty$, by (34). Hence $\operatorname{dist}(z, \Gamma^*) = 0$, by (36) and (37), and $z\in\Gamma^*$ (which is closed by definition). This proves (35).

Next we want to check that

$$H^1(\Gamma^*)\leq Cr_0. \tag{38}$$

Set $\Gamma_0^* = \bigcup_{j\geq 0}\Gamma_j$. We already know that

$$H^1(\Gamma_0^*)\leq H^1(\Gamma_0)+\sum_{j\geq 0}\sum_{x\in A'_j}H^1(\alpha_j(x))\leq Cr_0 + C\sum_{j\geq 0}\sum_{x\in A'_j}\delta_j(x)\leq Cr_0 \tag{39}$$

by (29), (10), (21), and (34). So it is enough to control $\Gamma^*\setminus\Gamma_0^*$.

Let $z\in\Gamma^*\setminus\Gamma_0^*$ be given, and let $\{z_\ell\}$ be a sequence in Γ_0^* which converges to z. If infinitely many z_ℓ lie in a same Γ_j, then $z\in\Gamma_j$ as well. This is impossible, since $z\notin\Gamma_0^*$. Thus for ℓ large enough, $z_\ell\in\alpha_{j(\ell)}(x_\ell)$ for some index $j(\ell)$ that tends to $+\infty$ and some $x_\ell\in A_{j(\ell)}$. By construction, $x_\ell\in A_{j(\ell)}\subset U_{j(\ell)}\subset E_0$ (see (13) and the discussion that follows it). Since $\alpha_{j(\ell)}(x_\ell)\subset\overline{B}_{j(\ell)}(x_\ell)$ by (23), we get that

$$\operatorname{dist}(z_\ell, E_0)\leq\operatorname{dist}(z_\ell, x_\ell)\leq\delta_{j(\ell)}(x_\ell)\leq 2\sum_{x\in A'_{j(\ell)}}\delta_{j(\ell)}(x), \tag{40}$$

by (37). The right-hand side of (40) tends to 0 by (34), hence $\operatorname{dist}(z_\ell, E_0)$ tends to 0 and $z\in E\cap\overline{B}(x_0, r_0)$. Thus $\Gamma^*\setminus\Gamma_0^*\subset E\cap\overline{B}(x_0, r_0)$, and

$$H^1(\Gamma^*)\leq H^1(\Gamma_0^*)+H^1(E\cap\overline{B}(x_0, r_0))\leq Cr_0 \tag{41}$$

by (39) and (1).

Our compact set Γ^* is connected, because each Γ_j is. Let us rapidly check this. Let $\Gamma^* = F_1 \cup F_2$ be a decomposition of Γ^* into disjoint closed subsets. Then for each j, $\Gamma_j = (\Gamma_j \cap F_1) \cup (\Gamma_j \cap F_2)$ is a similar decomposition of Γ_j. Since Γ_j is connected, $\Gamma_j \cap F_1 = \emptyset$ or $\Gamma_j \cap F_2 = \emptyset$. If $\Gamma_j \cap F_1 = \emptyset$ for all j, then $F_1 = \Gamma^* \cap F_1$ is also empty, because all the Γ_j lie in F_2, which lies at positive distance from F_1. If $\Gamma_{j_0} \cap F_1 \neq \emptyset$ for some j_0, then $\Gamma_j \cap F_1 \neq \emptyset$ for all $j \geq j_0$, and hence $\Gamma_j \cap F_2 = \emptyset$ for all $j \geq j_0$. In this case F_2 is empty (by the same argument as above). Our partition of Γ^* was trivial, and Γ^* is connected.

Thus Γ^* has all the properties required in Proposition 6 (see in particular (35) and (41)), except perhaps that it is not contained in $\overline{B}(x_0, r_0)$. This is easy to fix; we can replace Γ^* with $\Gamma = \pi(\Gamma^*)$, where π is the Lipschitz contraction onto $\overline{B}(x_0, r_0)$ defined by (12). This proves Proposition 6. $\square$

Before we move to the second part of our proof of Theorem 5, let us give a small improvement of Proposition 6.

Corollary 42. *Let $E \subset \mathbb{R}^n$ be Ahlfors-regular of dimension 1 and contain big pieces of connected sets. Then there exists a constant $C_3 > 0$ such that for $x \in E$ and $r > 0$, we can find a 1-Lipschitz function $f : [0, C_3 r] \to \mathbb{R}^n$ such that*

$$E \cap B(x, r) \subset f([0, C_3 r]) \subset \overline{B}(x, r). \tag{43}$$

Indeed, let (x, r) be as in the statement. Proposition 6 gives a compact connected set Γ such that $E \cap B(x, r) \subset \Gamma \subset \overline{B}(x, r)$ and $H^1(\Gamma) \leq C_2 r$. Then Proposition 30.1 says that we can find a C_n-Lipschitz function $g : [0, H^1(\Gamma)] \to \Gamma$ which is onto. Set

$$f(t) = g(C_3^{-1} H^1(\Gamma) t) \text{ for } t \in [0, C_3 r]. \tag{44}$$

It is easy to see that f is 1-Lipschitz if $C_3 \geq C_2 C_n$, and that it satisfies (43). The corollary follows. $\square$

Our proof of the remaining part of Theorem 5 will be slightly easier when

$$\text{diam}(E) < +\infty, \tag{45}$$

because it will be possible to cover E in only one step. Since this is also the only case that will be needed for Theorem 27.1, we shall first prove the theorem when (45) holds, and then merely indicate how to deal with the case when E is not bounded. So let us first assume that E is bounded and set

$$M = \inf \Big\{ m > 0 \,;\, \text{there is an } m\text{-Lipschitz function} $$
$$f : [0,1] \to \mathbb{R}^n \text{ such that } E \subset f([0,1]) \Big\}. \tag{46}$$

Note that $M < +\infty$, by Corollary 42 and (45). Also, there is an M-Lipschitz function $f_0 : [0,1] \to \mathbb{R}^n$ such that $E \subset f_0([0,1])$. We can obtain it as the limit of some subsequence of any minimizing sequence $\{f_k\}$ for (46). The argument is the same as in Section 30, just after (30.15), so we omit the easy proof.

Set $I = [0, M]$ and $f(t) = f_0(\frac{t}{M})$ for $t \in I$. Then

$$f \text{ is 1-Lipschitz and } E \subset f(I). \tag{47}$$

We want to prove that f is a regular parameterization, like z in Definition 27.3, so we need to check the analogue of (27.5), i.e., that

$$H^1(\{t \in I ; f(t) \in B(x, r)\}) \leq C_4 r \tag{48}$$

for all $x \in \mathbb{R}^n$ and $r > 0$, and some constant C_4 that depends only on n and the constants C_0, C_1 from (1) and Definition 2. If we can do this, then $f(I)$ will be the desired Ahlfors-regular curve that contains E, and Theorem 5 (when E is bounded) will follow.

Let us proceed by contradiction and assume that we can find (x, r) such that (48) fails. We want to prove the following.

Lemma 49. *If C_4 is large enough, we can find an interval $\widetilde{I}$ and a function $\widetilde{f} : \widetilde{I} \to \mathbb{R}^n$ such that*

$$H^1(\widetilde{I}) < H^1(I) = M, \tag{50}$$

$$\widetilde{f} \text{ is 1-Lipschitz }, \tag{51}$$

and

$$\widetilde{f}(\widetilde{I}) \text{ contains } E. \tag{52}$$

Theorem 5 for bounded sets E will immediately follow from the lemma, because the existence of $\widetilde{f_0}(t) = \widetilde{f}(H^1(\widetilde{I}) t)$ contradicts the definition (46) of M.

So we want to prove the lemma. Let us first check that we can find a radius $\rho \in (2r, 3r)$ such that

$$Z = \{t \in I ; f(t) \in \partial B(x, \rho)\} \text{ is finite.} \tag{53}$$

Set $g(t) = |f(t) - x|$. This is a 1-Lipschitz function defined on I. Call $N(\rho)$ the number of points $t \in I$ such that $g(t) = \rho$; then

$$\int_{\mathbb{R}} N(\rho) \, d\rho \leq H^1(I) < +\infty, \tag{54}$$

by Lemma 26.17, and in particular $N(\rho) < +\infty$ for almost every $\rho > 0$. Any such ρ will satisfy (53).

Choose $\rho \in (2r, 3r)$ such that (53) holds. Denote by T_0 the set of connected components J of $I \setminus Z$ such that $f(J)$ meets $B(x, r)$. If $J \in T_0$, $f(J)$ does not meet $\partial B(x, \rho)$ (because J does not meet Z), and so it stays in $B(x, \rho)$. Let us record this:

$$f(J) \subset B(x, \rho) \text{ for } J \in T_0. \tag{55}$$

Let us also check that

$$H^1(J) \geq r \text{ for } J \in T_0. \tag{56}$$

This is clear if $J = I$, because $H^1(I) \geq C_4 r$ since (48) fails. Otherwise, at least one of the endpoints of J lies in Z. Then $f(\overline{J})$ is a connected set that touches $B(x, r)$ and $\partial B(x, \rho)$, hence $H^1(\overline{J}) \geq H^1(f(\overline{J})) \geq \rho - r \geq r$ because f is 1-Lipschitz and by Remark 2.11. Set

$$X = \bigcup_{J \in T_0} J. \tag{57}$$

Note that X contains $f^{-1}(B(x, r))$, by definition of T_0, and hence

$$H^1(X) \geq C_4 r \tag{58}$$

because we are currently assuming that (48) fails (to get a contradiction).

Our intention is to remove the set X and the restriction of f to X; we shall need to add something to cover $E \cap f(X) \subset E \cap B(x, \rho)$ (by (55)). If $E \cap B(x, \rho)$ is empty we do not need to do this, but we can also take $J_0 = \{0\}$ in the argument below. So let us assume that $E \cap B(x, \rho)$ contains some point y, and apply Corollary 42 to the pair $(y, 2\rho)$. Set $J_0 = [0, 2C_3 \rho]$; thus

$$H^1(J_0) \leq 6\,C_3\,r, \tag{59}$$

and Corollary 42 says that there is a 1-Lipschitz function $\widetilde{f} : J_0 \to \mathbb{R}^n$ such that

$$E \cap B(x, \rho) \subset E \cap B(y, 2\rho) \subset \widetilde{f}(J_0) \subset \overline{B}(y, 2\rho) \subset \overline{B}(x, 3\rho). \tag{60}$$

Now we need to do some dirty work to rearrange the various pieces of $f|_{I \setminus X}$ and connect them to $\widetilde{f}|_{J_0}$ as well. Call $J_1, \ldots, J_N$ the connected components of $I \setminus X$. We shall use the convention that J_1 is the first interval on the left of $I \setminus X$ and J_N is the last one (because they may play a special role), but otherwise we may choose the middle ones in some other order if we find it convenient later. Also, we shall implicitly assume in our notations that $N \geq 2$; the reader may easily check that the argument below also works when $N \leq 1$, with simplifications.

We need a last notation. For $0 \leq j \leq N$, call t_j^- the initial extremity of J_j and t_j^+ its final endpoint. Also set

$$z_j^{\pm} = f(t_j^{\pm}) \text{ for } j \geq 1, \tag{61}$$

and

$$z_0^{\pm} = \widetilde{f}(t_0^{\pm}). \tag{62}$$

Let us first construct our new function $\widetilde{f}$ when

$$t_1^- \in Z, \tag{63}$$

and hence
$$z_1^- = f(t_1^-) \in \partial B(x, \rho). \tag{64}$$
The domain of definition $\widetilde{I}$ of $\widetilde{f}$ will be a compact interval, obtained as the juxtaposition
$$\widetilde{I} = J_0 \cup L_1 \cup J_1' \cup L_2 \cup J_2' \cup \ldots \cup L_N \cup J_N' \tag{65}$$
of contiguous intervals. Here J_0 is still the interval $[0, 2C_3\rho]$ that we got from Corollary 42. The intervals J_j', $1 \leq j \leq N$, are translations
$$J_j' = J_j + \tau_j \tag{66}$$
of the J_j. Finally, L_j, $1 \leq j \leq N$, is a compact interval of length
$$H^1(L_j) = |z_j^- - z_{j-1}^+|. \tag{67}$$
(We shall see why fairly soon.) With all this information, plus the fact that each interval must start at the final extremity of the previous one on the list, all our intervals are uniquely determined.

Now we can define $\widetilde{f}$ on $\widetilde{I}$. We keep the same definition of $\widetilde{f}$ on J_0 as before. For $1 \leq j \leq N$, set
$$\widetilde{f}(t) = f(t - \tau_j) \text{ for } t \in J_j', \tag{68}$$
with τ_j as in (66). We still need to define $\widetilde{f}$ on the L_j, $1 \leq j \leq N$. Write $L_j = [s_j^-, s_j^+]$. If we want $\widetilde{f}$ to be continuous, we have no choice on the value of $\widetilde{f}(s_j^\pm)$. For instance, we must have $\widetilde{f}(s_j^+) = f(t_j^-) = z_j^-$ by (68), (66), and (61). Similarly, for $j \geq 2$, $\widetilde{f}(s_j^-) = f(t_{j-1}^+) = z_{j-1}^+$. Also, $\widetilde{f}(s_1^-) = \widetilde{f}(t_0^+) = z_0^+$, this time by (62). So we decide to take
$$\widetilde{f}(s_j^+) = z_j^-, \quad \widetilde{f}(s_j^-) = z_{j-1}^+, \tag{69}$$
and require that $\widetilde{f}$ be linear on L_j.

This completes our definition of $\widetilde{f}$ when (63) holds. Note that
$$\widetilde{f} \text{ is 1-Lipschitz,} \tag{70}$$
because it is continuous on $\widetilde{I}$, and the various pieces that compose it are 1-Lipschitz. (Concerning $\widetilde{f}|_{L_j}$, this comes from (67) and (69).) Also,
$$\widetilde{f}(\widetilde{I}) \text{ contains } E. \tag{71}$$
Indeed $f(I)$ contains E by (47), and we only removed $f(X)$ from the image (at most). But $f(X) \subset B(x, \rho)$ by (57) and (55), and what we may have lost because of $f(X)$ is contained in $\widetilde{f}(J_0)$, by (60). Let us also record that
$$H^1(\widetilde{I}) = \sum_{j=0}^{N} H^1(J_j) + \sum_{j=1}^{N} H^1(L_j) = H^1(J_0) + H^1(I \setminus X) + \sum_{j=1}^{N} |z_j^- - z_{j-1}^+|$$
$$\leq Cr + H^1(I \setminus X) + \sum_{j=2}^{N} |z_j^- - z_{j-1}^+|, \tag{72}$$

by (65), (66), (67), the fact that $I \setminus X$ is the disjoint union of the intervals J_j, $1 \leq j \leq N$, (59), and the fact that $|z_1^- - z_0^+| \leq Cr$ because $z_1^- \in \partial B(x, \rho)$ (by (64)) and $z_0^+ = \widetilde{f}(t_0^+)$ lies in $\overline{B}(x, 3\rho)$ (by (62) and (60)).

Before we continue to estimate $H^1(\widetilde{I})$, let us construct $\widetilde{f}$ in our second case when (63) fails, i.e., when $t_1^- \notin Z$. We claim that $t_1^- = 0$ (the initial end of I) in this case. Let us even check that for $1 \leq j \leq N$, $t_j^- \in Z$ unless $t_j^- = 0$. Our claim will follow from this, and so will the fact that

$$t_j^- \in Z \text{ when } j \geq 2, \tag{73}$$

which will be needed later. So let $j \geq 1$ be such that $t_j^- \neq 0$. Then t_j^- is the final endpoint of one of the intervals J, $J \in T_0$, that compose X. If $t_j^- < M$, then $t_j^- \in Z$ by definition of T_0. If $t_j^- = M$, then $J_j = \{t_j^-\} = \{M\}$ (because J_j is not empty), but then also t_j^- should lie in Z (because otherwise J_j would be larger); incidentally, this case is impossible because $J_j \in T_0$ does not meet Z. Thus $t_j^- \in Z$ in all cases, and (73) and our claim follows.

Thus $t_1^- = 0$. Note that X is not empty or reduced to a point (by (58)), and hence $t_1^+ < M$. Then $t_1^+ \in Z$ (by the same argument as for (73)), and so

$$z_1^+ = f(t_1^+) \in \partial B(x, \rho), \tag{74}$$

by (61) and (53). This time it is better to intercalate the interval J_0 between J_1' and J_2'; except for this the construction will be very similar. We set

$$\widetilde{I} = J_1' \cup L_1 \cup J_0 \cup L_2 \cup J_2' \cup \ldots \cup L_N \cup J_N', \tag{75}$$

with intervals J_j', $1 \leq j \leq N$, that are still translations of the J_j (as in (66)). We take

$$H^1(L_1) = |z_0^- - z_1^+|, \quad H^1(L_2) = |z_2^- - z_0^+|, \tag{76}$$

and keep $H^1(L_j) = |z_j^- - z_{j-1}^+|$ for $j > 2$, as in (67).

As before, we keep the same definition of $\widetilde{f}$ on J_0, define $\widetilde{f}$ on J_j', $1 \leq j \leq N$, by (68), and interpolate linearly on L_j. Thus, if $L_j = [s_j^-, s_j^+]$, we still have that

$$\widetilde{f}(s_j^+) = z_j^- \text{ for } j \geq 2 \quad \text{and} \quad \widetilde{f}(s_j^-) = z_{j-1}^+ \text{ for } j > 2, \tag{77}$$

but now

$$\widetilde{f}(s_1^-) = z_1^+, \quad \widetilde{f}(s_1^+) = z_0^-, \quad \text{and} \quad \widetilde{f}(s_2^-) = z_0^+. \tag{78}$$

Our new choices of lengths in (76) ensure that $\widetilde{f}$ is 1-Lipschitz on $\widetilde{I}$, as before. We still have that $\widetilde{f}(\widetilde{I})$ contains E, as in (71). As for the analogue of (72), first note that

$$\sum_{j=1}^{N} H^1(L_j) = |z_0^- - z_1^+| + |z_2^- - z_0^+| + \sum_{j \geq 3} H^1(L_j) \leq Cr + \sum_{j \geq 3} |z_j^- - z_{j-1}^+| \tag{79}$$

by (76), because $z_0^-, z_0^+ \in \tilde{f}(J_0) \subset \overline{B}(x, 3\rho)$ (by (62) and (60)), because z_1^+ and z_2^- lie on $\partial B(x, \rho)$ (by (74), (73), (61), and (53)), and by (67). Then

$$H^1(\tilde{I}) = \sum_{j=0}^{N} H^1(J_j) + \sum_{j=1}^{N} H^1(L_j) = H^1(J_0) + H^1(I \setminus X) + \sum_{j=1}^{N} H^1(L_j) \quad (80)$$

$$\leq Cr + H^1(I \setminus X) + \sum_{j \geq 3} |z_j^- - z_{j-1}^+|$$

by (75), the fact that $I \setminus X$ is the disjoint union of the intervals J_j, $1 \leq j \leq N$, (59), and (79).

So far we were able to construct a Lipschitz function $\tilde{f}$ with all the requirements of Lemma 49, except for the estimate (50) on $H^1(\tilde{I})$. Instead we merely have proved that

$$H^1(\tilde{I}) \leq Cr + H^1(I \setminus X) + \sum_{j \geq 2} |z_j^- - z_{j-1}^+|, \quad (81)$$

by (72) or (80). Note that

$$z_j^- \text{ and } z_{j-1}^+ \text{ lie on } \partial B(x, \rho) \quad (82)$$

for $j \geq 2$. For z_j^-, this follows directly from (73), because $z_j^- = f(t_j^-) \in \partial B(x, \rho)$ when $z_j^- \in Z$, by definitions. For z_{j-1}^+, the same argument works because $z_{j-1}^+ < z_j^+ \leq M$.

Let us first assume that $N \leq C_5$, where C_5 will be chosen soon. Then

$$H^1(\tilde{I}) \leq Cr + H^1(I) - H^1(X) + 2C_5\rho \leq Cr + H^1(I) - C_4r + 6C_5r < H^1(I) \quad (83)$$

by (81), (82), (58), and if C_4 is large enough (depending on C_5 in particular). Then (50) holds, and we are happy in this case.

We are left with the case when $N \geq C_5$. In this case, we have that

$$H^1(X) \geq (N-1)r, \quad (84)$$

by (57), (56), and because T_0 has at least $N-1$ elements, by definition of $J_1, \ldots, J_N$ as components of $I \setminus X$. Thus

$$H^1(\tilde{I}) \leq H^1(I) - Nr + Cr + \sum_{j \geq 2} |z_j^- - z_{j-1}^+|, \quad (85)$$

by (81) and (84).

Now we need to proceed carefully, if we do not want to kill the $-Nr$ with a bad estimate for the last sum. Let us first observe that in the construction above, we may as well have reversed the flow of f on some of the inside intervals J_j,

$2 \leq j \leq N - 1$. The only effect of this is to exchange the two points $z_j^{\pm}$. Also recall that we may choose the order in which the intervals J_j, $2 \leq j \leq N - 1$, are chosen (and labelled). We want to do all this in a way that will make the last sum in (85) significantly smaller than Nr.

Let us decompose $\partial B(x, r)$ into less than $C(n)$ disjoint boxes $R_1, \ldots, R_S$ of diameters at most $\frac{r}{2}$. Call Y the set of points $z_j^{\pm}$, $2 \leq j \leq N - 1$. Here is how we want to choose the labels for the J_j and their orientations. Start with the point z_1^+. It lies in some box R_s. If R_s contains some point of Y, choose J_2 and its orientation so that $z_2^- \in R_s$. Otherwise, choose at random. This choice determines a new point z_2^+ (the other end of $f(J_2)$). If $Y \setminus \{z_2^-, z_2^+\}$ has a point in the same box as z_2^+, we choose J_3 and its orientation so that z_3^- lies in this box. Otherwise, we choose them at random. This choice determines z_3^+. We continue the argument in the same way: if $J_2, \ldots, J_j$ have been chosen, we try to choose J_{j+1} and its orientation so that z_{j+1}^- lies in the same box as z_j^+, and if this is impossible we choose at random. Eventually there will be no points left in Y to choose from, because all the intervals were selected, and we can stop.

Note that we do not get unlucky more than S times (once per box): if for some j we cannot choose z_{j+1}^- in the same box as z_j^+ (because the box is now empty), then all the later points z_i^+, $i > j$, will land out of that box. In other words, each box can only make us miserable once.

When we are lucky, $|z_{j+1}^- - z_j^+| \leq \frac{r}{2}$ (because our boxes have a small diameter). Otherwise, $|z_{j+1}^- - z_j^+| \leq 2\rho \leq 4r$, by (82). This is also the case for $j = N - 1$. Thus

$$\sum_{j \geq 2} |z_j^- - z_{j-1}^+| \leq \frac{Nr}{2} + (S + 1) 4r \leq \frac{Nr}{2} + Cr, \qquad (86)$$

and (85) says that

$$H^1(\widetilde{I}) \leq H^1(I) - \frac{Nr}{2} + Cr < H^1(I) \qquad (87)$$

because $N \geq C_5$, and if C_5 is chosen large enough.

This proves (50) in our last case. Lemma 49 follows and, with it, Theorem 5 in the case when E is bounded.   □

Now let $E \subset \mathbb{R}^n$ be unbounded and satisfy the hypotheses of Theorem 5, and let $B = B(x, r)$ be any ball in $\mathbb{R}^n$ (not necessarily centered on E). We want to repeat most of the argument above with $E \cap B$. Set

$$M = \inf \Big\{ m > 0 \, ; \text{ there is an } m\text{-Lipschitz function } f : [0, 1] \to \mathbb{R}^n$$
$$\text{such that } E \cap B \subset f([0, 1]) \Big\}. \qquad (88)$$

(Compare with (46)). Then $M < +\infty$, by Corollary 42 (applied to some point $x_0 \in E \cap B$ and the radius $2r$; if there is no such x_0, then $M = 0$). By the same argument as for (47), we can find a 1-Lipschitz function $f : [0, M] \to \mathbb{R}^n$ such that $E \cap B \subset f([0, M])$. The same argument as above shows that f is an Ahlfors-regular parameterization (like the function z in Definition 27.3), with constants that do not depend on x or r. Note that

$$f([0, M]) \subset \overline{B}, \tag{89}$$

because otherwise we could make $f([0, M])$ shorter by composing with a contraction $\pi : \mathbb{R}^n \to \overline{B}(x, r)$ onto $\overline{B}$, like the one defined by (12). Also

$$M \leq Cr ; \tag{90}$$

we can either get this directly from (88) and Corollary 42, or else from (89) and the fact that f is a regular parameterization.

We want to apply this to a sequence of balls $B_j = B(x_j, r_j)$. Start with $B_0 = B(0, 1)$, and then choose x_j so that $|x_j| = 2^{j\varepsilon}$ and $r_j = |x_j|/2$ (we do not need to take $x_j \in E$). If ε is chosen small enough, depending on n, we can choose the points x_j so that the B_j cover $\mathbb{R}^n$. For instance, when $n = 2$, we can take $x_j = (2^{j\varepsilon}\cos(j/100), 2^{j\varepsilon}\sin(j/100))$.

Let $f_j : [0, M_j] \to \overline{B}_j$ denote the 1-Lipschitz mappings given by the discussion above. We can extend f_j into a mapping $f_j' : [0, M_j'] \to \mathbb{R}^n$, with the following properties:

$$f_j'(M_j') = f_{j+1}(0), \tag{91}$$

$$|f_j'(t)| \geq 2^{j\varepsilon}/4 \text{ for } 0 \leq t \leq M_j' \text{ and } j \geq 1, \tag{92}$$

$$f_j' : [0, M_j'] \to \mathbb{R}^n \text{ is an Ahlfors-regular parameterization} \tag{93}$$

with a constant that does not depend on j, and

$$C^{-1} 2^{j\varepsilon} \leq M_j' \leq C 2^{j\varepsilon} \text{ for } j \geq 0. \tag{94}$$

Next we put together all these mappings to get a single one $f : [0, +\infty) \to \mathbb{R}^n$. Set $M_j'' = \sum\limits_{\ell=0}^{j} M_\ell'$ for $j \geq 0$ and take

$$f(t) = f_j'(t + M_{j-1}'') \text{ for } M_{j-1}'' \leq t < M_j''. \tag{95}$$

Then f is continuous because of (91), 1-Lipschitz because each f_j' is 1-Lipschitz, and also

$$C^{-1}t \leq |f(t)| \leq Ct \text{ for } t > 1, \tag{96}$$

by (94), (92), and also because $|f_j'(t)| \leq C 2^{j\varepsilon}$ for $j \geq 0$ and $0 \leq t \leq M_j'$.

The Ahlfors-regularity of f (i.e., the analogue of (27.5) for f) is fairly easy to deduce from (96), (94), and the Ahlfors-regularity of each piece. (Note that it is enough to verify (27.5) for balls centered at 0 and balls that stay away from the origin.) We leave the verification as an exercise. This completes our proof of Lemma 49 and Theorem 5 in the remaining case when E is not bounded. $\square$

Exercise 97. Show that if $E \subset \mathbb{R}^n$ is Ahlfors-regular of dimension 1 and contained in an Ahlfors-regular curve (see Definition 27.3), it contains big pieces of connected sets (as in Definition 2). Hint: use (27.10).

32   Application to quasiminimizers and a temporary conclusion

Proof of Theorem 27.1. Recall that this is the result of local uniform rectifiability of K when $(u, K) \in TRLQ(\Omega)$. We want to deduce it from Theorems 29.1 and 31.5; this will just involve a localization argument.

So let $\Omega \subset \mathbb{R}^2$ and a reduced or coral quasiminimizer $(u, K) \in TRLQ(\Omega)$ be given, as in Theorem 27.1. Also let $x \in K$ and $0 < r \leq r_0$ be such that $B(x, 2r) \subset \Omega$. We want to find an Ahlfors-regular curve that contains $K \cap B(x, r)$, as in (27.2).

Set $B = B(x, r)$ and $E = (K \cap B) \cup \partial B$. We want to apply Theorem 31.5 to E, so let us check the hypotheses. Let us first verify the Ahlfors-regularity, i.e., that

$$C^{-1}t \leq H^1(E \cap B(y, t)) \leq Ct \tag{1}$$

for $y \in E$ and $0 < t < \operatorname{diam}(E) = 2r$.

Obviously, it is enough to prove (1) for $0 < t < r$. In this case,

$$H^1(E \cap B(y, t)) \leq H^1(\partial B \cap B(y, t)) + H^1(K \cap B(y, t)) \leq 2\pi t + Ct, \tag{2}$$

by Lemma 18.19. Conversely, if $B\left(y, \frac{t}{2}\right)$ meets ∂B, then

$$H^1(E \cap B(y, t)) \geq H^1(\partial B \cap B(y, t)) \geq t. \tag{3}$$

Otherwise, $B\left(y, \frac{t}{2}\right) \subset B$ and

$$H^1(E \cap B(y, t)) \geq H^1\left(K \cap B\left(y, \frac{t}{2}\right)\right) \geq C^{-1}t, \tag{4}$$

by Theorem 18.16. This proves (1).

We also need to check that E contains big pieces of connected sets, as in Definition 31.2. This means that for $y \in E$ and $0 < t < \operatorname{diam}(E)$, we can find a compact connected set Γ such that $H^1(\Gamma) \leq Ct$ and $H^1(E \cap \Gamma \cap B(y, t)) \geq C^{-1}t$. When $B\left(y, \frac{t}{2}\right)$ meets ∂B, we can take $\Gamma = \partial B \cap B(y, t)$ and use (3). When $B\left(y, \frac{t}{2}\right) \subset B$, we note that $E \cap B\left(y, \frac{t}{2}\right) = K \cap B\left(y, \frac{t}{2}\right)$, and we can take the compact connected set Γ provided by Theorem 29.1 (applied to the pair $\left(y, \frac{t}{2}\right)$). Thus E contains big pieces of connected sets. Theorem 31.5 says that E is uniformly rectifiable, i.e., contained in an Ahlfors-regular curve.

As usual, the uniform rectifiability constant, i.e., the Ahlfors-regularity constant for the regular curve that contains E (and hence $K \cap B(x, r)$) depends only on M, through the constants in Theorem 18.16, Lemma 18.19, Theorem 29.1, and what becomes of them in Theorem 31.5.

This completes our proof of Theorem 27.1. $\square$

Remark 5. Let (u, K) be a reduced quasiminimizer in $\Omega \subset \mathbb{R}^2$. (See Definition 7.21 and 8.1.) Assume that a is small enough, depending on M. (The condition is the same as in Theorems 18.16 and 29.1.) Finally assume that $\partial\Omega$ is a uniformly rectifiable Ahlfors-regular set of dimension 1. Then $K \cup \partial\Omega$ is uniformly rectifiable, and in particular K is contained in an Ahlfors-regular curve (with a constant that depends only on M).

The proof is the same as above, with B and ∂B replaced with Ω and $\partial\Omega$. We only use the fact that (u, K) is a full quasiminimizer (not merely local) to get the upper bound for Ahlfors-regularity. This last is proved just like in Lemma 18.19.

The statement is not as nice as it may seem. For instance, it does not even say that $H^1(K \cap B(x, r)) \geq C^{-1} r$ when $B(x, r)$ is a small ball centered on K near $\partial\Omega$; so maybe we are just using $\partial\Omega$ to hide the defects of K. See Part I, and in particular Section 80, for a somewhat more serious study of K near the boundary.

Note that this remark applies to minimizers of the Mumford-Shah functional, or the slightly more general minimizers of Example 7.23 and Proposition 7.27.

Concluding Remark 6. As far as quasiminimizers are concerned, Theorem 27.1 and similar results are probably not far from optimal. On the other hand, for almost-minimizers (and in particular Mumford-Shah minimizers), uniform rectifiability is not the optimal notion, and we shall see better regularity results later. It is possible that the main reason for the intrusion of uniform regularity in connection with the Mumford-Shah functional was some of the authors' background. Hopefully it played a useful role anyway.

In the next part we shall forget a little about our quest for regularity properties of K, and concentrate more on the behavior of minimizers under limits. This will make it a little easier for us to return in Part E to regularity properties, probably with a clearer view of some issues. Anyway we shall try to make the two parts as independent from each other as we can.

D. Limits of Almost-Minimizers

In this part we want to prove that, thanks to the concentration property of [DMS] (see Section 25 and Remark 75.17), the Hausdorff measure H^{n-1} becomes lower semicontinuous when we restrict to sets K for which (u, K) is a reduced almost-minimizer. This is the main ingredient for proving that limits of almost-minimizers are also almost-minimizers (see Theorem 38.3 in particular).

This part is largely independent from Part E, which can (almost) be read first.

33   Vitali covering lemmas

We already used a very simple covering lemma in Section 31 (Lemma 1 below), but in the next section we shall need a more elaborate version of it, to prove the lower semicontinuity of the Hausdorff measure on sequences of sets that satisfy uniform concentration properties. We start with the easiest and most classical lemma.

Lemma 1. *Let $\{B_i\}_{i \in I}$ be a collection of balls in $\mathbb{R}^n$ (open or closed, we do not care). Suppose that $\sup_{i \in I} \operatorname{diam} B_i < +\infty$. Then there is a finite or countable subset I_0 of I such that*

$$\text{the } B_i \text{ , } i \in I_0, \text{ are disjoint;} \tag{2}$$

$$\bigcup_{i \in I} B_i \subset \bigcup_{i \in I_0} 5B_i. \tag{3}$$

As usual, $5B_i$ denotes the ball with the same center as B_i and 5 times the radius. As the proof will show, we could replace 5 with any number larger than 3, but 5 is the traditional constant to use. Lemma 1 is extremely useful and simple, and it has the nice feature of extending to many metric spaces. See also Remark 6.

To prove the lemma, set $M_0 = \sup\{\operatorname{diam} B_i \; ; \; i \in I\} < \infty$ and $M_k = 2^{-k}M_0$ for $k \geq 1$. Decompose I into the disjoint sets $J_k = \{i \in I \; ; \; M_{k+1} < \operatorname{diam} B_i \leq M_k\}$, $k \geq 0$.

Start with a maximal subset $J(0)$ of J_0 such that the balls B_i, $i \in I$, are disjoint. Then define the sets $J(k)$, $k \geq 1$, by induction: choose for $J(k)$ a maximal subset of J_k such that the B_i, $i \in J(k)$, are disjoint from each other and from the B_i, $i \in J(0) \cup \cdots \cup J(k-1)$. By construction,

$$I(k) = J(0) \cup \cdots \cup J(k) \text{ is a maximal subset of } J_0 \cup \cdots \cup J_k \qquad (4)$$

$$\text{such that the } B_i, \ i \in I(k), \text{ are disjoint.}$$

Finally take $I_0 = \bigcup_{k \geq 0} I(k) = \bigcup_{k \geq 0} J(k)$. Obviously (2) holds. Also, each $J(k)$ is at most countable, because the B_i, $i \in J(k)$, are disjoint and each contains lots of rationals. So it is enough to check (3). Let $j \in I$ be given, and let us show that $B_j \subset 5B_i$ for some $i \in I_0$. Let k be such that $j \in J_k$. Because of (4), B_j meets some B_i, $i \in I(k)$. Note that $\operatorname{diam} B_i > \frac{1}{2} \operatorname{diam} B_j$ by definition of J_k; then $B_j \subset 5B_i$, as needed. Lemma 1 follows. $\qquad \square$

Remark 5. The requirement that $\sup\{\operatorname{diam} B_i \ ; \ i \in I\} < \infty$ is needed. Think about the intervals $[0, i]$ in $\mathbb{R}$.

Remark 6. The B_i, $i \in I$, in Lemma 1 do not need to be balls. They can be cubes, or just any objects such that for each $i \in I$ we can find a ball D_i such that $D_i \subset B_i \subset CD_i$, with a constant C that does not depend on i. Of course 5 needs to be replaced with a constant that depends on C, but otherwise the proof stays the same.

For our second covering lemma and its application in the next section, we shall follow rather closely the presentation from [MoSo2], even though we shall state the results with a little less generality. Recall from Definition 2.6 that

$$H^d(A) = \lim_{\tau \to 0^+} H^d_\tau(A) \qquad (7)$$

for all Borel sets A, where

$$H^d_\tau(A) = c_d \inf \left\{ \sum_{i=1}^{\infty} (\operatorname{diam} A_i)^d \ ; \ A \subset \bigcup_{i=1}^{\infty} A_i \ \text{ and } \ \operatorname{diam} A_i \leq \tau \ \text{ for all } i \right\}, \qquad (8)$$

and where the normalizing constant c_d is chosen so that H^d coincides with the Lebesgue measure on $\mathbb{R}^d$.

Definition 9. *A* *Vitali covering* *of $A \subset \mathbb{R}^n$ is a collection $\{B_i\}_{i \in I}$ of sets with positive diameters, such that for all choices of $x \in A$ and $\tau > 0$ we can find $i \in I$ such that $x \in B_i$ and $\operatorname{diam} B_i < \tau$.*

We shall need a slightly more general notion.

Definition 10. *Let $\{B_i\}_{i \in I}$ be a collection of sets of positive diameters. We say that $\{B_i\}_{i \in I}$ is a* *Vitali family* *for the set $A \subset \mathbb{R}^n$ if there is a constant $C \geq 1$ such that $\{^C B_i\}_{i \in I}$ is a Vitali covering of A. Here we set*

$$^C B_i = \left\{ x \in \mathbb{R}^n \ ; \ \operatorname{dist}(x, B_i) \leq C \operatorname{diam} B_i \right\}. \qquad (11)$$

Morel and Solimini call this an approximate Vitali covering, but we decided to change the name because in some cases the B_i themselves do not cover much. It can even happen that each B_i is disjoint from A.

Theorem 12 [DMS], [MoSo2]. *Let $A \subset \mathbb{R}^n$ be a Borel set and $\{B_i\}_{i \in I}$ a Vitali family for A. For each choice of $\lambda < H^d(A)$ and $\tau > 0$, we can find an at most countable set $I_0 \subset I$ such that*

$$\text{the sets } \{B_i\}_{i \in I_0} \text{ are disjoint }; \tag{13}$$

$$\operatorname{diam} B_i < \tau \ \text{ for } \ i \in I_0 ; \tag{14}$$

$$c_d \sum_{i \in I_0} (\operatorname{diam} B_i)^d \geq \lambda. \tag{15}$$

Let us just give two comments before we start the proof. To get something interesting from Theorem 12, we shall need to be careful about our choice of family $\{B_i\}$. That is, if we start with sets B_i that are too thin, then (13) may be too easy to get, and then (15) will not give much information. Because of this, it is not stupid to think of the B_i as being roughly balls.

Also, Theorem 12 is only a part of Theorem 7.14 on p. 83 of [MoSo2], which in particular gives additional information when $H^d(A) < +\infty$. Also see Remark 37.

The proof will use the following interesting lemma.

Lemma 16. *Let $A \subset \mathbb{R}^n$ be a Borel set such that $H^d(A) < +\infty$. For each $\varepsilon > 0$ we can find $\tau > 0$ such that if $\{B_i\}_{i \in I}$ is a finite or countable collection of sets such that $\operatorname{diam} B_i < \tau$ for all i, then*

$$H^d\left(A \cap \left(\bigcup_{i \in I} B_i\right)\right) \leq c_d \sum_{i \in I} (\operatorname{diam} B_i)^d + \varepsilon. \tag{17}$$

This is quite cute. It says that high concentrations of A somewhere (like in $\bigcup_i B_i$) are excluded. As we shall see, this is because otherwise we could complete the covering of A to get a better estimate on $H^d(A)$.

Before we really start with the proof of Lemma 16, note that we can always assume that the B_i are closed, because replacing them with their closure increases the left-hand side of (17) without changing the right-hand side.

Let $\{B_i\}$ be given, with $\operatorname{diam} B_i < \tau$ for all i, and choose a covering of $A \setminus \bigcup_i B_i$ by sets D_j, $j \in J$, with J at most countable, $\operatorname{diam} D_j \leq \tau$ for all j, and

$$c_d \sum_{j \in J} (\operatorname{diam} D_j)^d < H^d\left(A \setminus \bigcup_i B_i\right) + \varepsilon/2. \tag{18}$$

When we group this with the B_i, $i \in I$, we get a covering of A by sets of diameters at most τ, and so

$$H_\tau^d(A) \leq c_d \sum_{i \in I} (\operatorname{diam} B_i)^d + c_d \sum_{j \in J} (\operatorname{diam} D_j)^d \tag{19}$$

$$< c_d \sum_{i \in I} (\operatorname{diam} B_i)^d + H^d(A \setminus \bigcup_i B_i) + \varepsilon/2 .$$

If τ is small enough, $H_\tau^d(A) \geq H^d(A) - \varepsilon/2$. Also, $H^d(A) = H^d(A \cap (\cup_i B_i)) + H^d(A \setminus \cup_i B_i)$ by additivity of Hausdorff measure on Borel sets, and now (17) follows from (19). This proves Lemma 16. $\qquad \square$

Now return to the proof of Theorem 12. We can always assume that the B_i are closed, because replacing B_i with $\overline{B_i}$ preserves the fact that $\{B_i\}$ is a Vitali family for A, and makes (13)–(15) (just a tiny bit) harder to check. We start with the most interesting case when

$$H^d(A) < +\infty . \tag{20}$$

We want to define a sequence of finite sets $J(k) \subset I$, and then take $I_0 = \bigcup_k J(k)$. Take $J(0) = \emptyset$. Next assume that the $J(l)$, $l \leq k$, were chosen, and define $J(k+1)$. Set $I(k) = J(0) \cup \cdots \cup J(k)$, and then

$$V_k = \mathbb{R}^n \setminus \bigcup_{i \in I(k)} B_i . \tag{21}$$

This is an open set, because $I(k)$ is finite and the B_i are closed. Set

$$Z_k = \{j \in I \; ; \; B_j \text{ does not meet any } B_i, i \in I(k)\} . \tag{22}$$

We claim that

$$\{B_j\}_{j \in Z_k} \text{ is a Vitali family for } A \cap V_k , \tag{23}$$

with the same constant C as for the initial family $\{B_i\}_{i \in I}$. Indeed, let $x \in A \cap V_k$ be given. We can find sets B_j, $j \in I$, with arbitrarily small diameters and such that $x \in {}^C B_j$. When $\operatorname{diam}{}^C B_j$ is small enough, B_j does not meet any B_i, $i \in I(k)$, by (21) and because V_k is open. In other words, $j \in Z_k$. This proves our claim (23).

Now let $\tau_k < \tau$ be small, to be chosen soon. Set $Z_k' = \{i \in Z_k \; ; \; \operatorname{diam} B_i \leq \tau_k\}$. The sets ${}^C B_i$, $i \in Z_k'$, cover $A \cap V_k$, by (23). Call D_i the smallest ball that contains ${}^C B_i$, and set $\widehat{B_i} = 5D_i$. Apply Lemma 1 to the balls D_i, $i \in Z_k'$. This gives a set $Z_k^* \subset Z_k' \subset Z_k$ such that

$$\text{the sets } {}^C B_i, i \in Z_k^*, \text{ are disjoint,} \tag{24}$$

and

$$A \cap V_k \subset \bigcup_{i \in Z_k^*} \widehat{B_i} \, . \tag{25}$$

Also note that

$$\operatorname{diam} B_i \leq \tau_k \quad \text{and} \quad B_i \subset V_k \text{ for } i \in Z_k^*, \tag{26}$$

by definition of Z_k'.

If τ_k is small enough, then

$$H^d(A \cap V_k) \leq 2H_{10C\tau_k}^d(A \cap V_k) \leq 2c_d \sum_{i \in Z_k^*} (\operatorname{diam} \widehat{B_i})^d \leq C' \sum_{i \in Z_k^*} (\operatorname{diam} B_i)^d, \tag{27}$$

by (25). Note that (27) still holds (trivially) when $A \cap V_k = \emptyset$. We choose τ_k so small that (27) holds, but also smaller than the constant τ that we get when we apply Lemma 16 to the set $A \cap V_k$ and with the constant $\varepsilon = 2^{-k-1}(H^d(A) - \lambda) > 0$.

Now choose $J(k+1) \subset Z_k^*$ finite, but so large that

$$\sum_{i \in J(k+1)} (\operatorname{diam} B_i)^d \geq \frac{1}{2} \sum_{i \in Z_k^*} (\operatorname{diam} B_i)^d \geq \frac{1}{2C'} H^d(A \cap V_k) \tag{28}$$

if $\sum_{i \in Z_k^*} (\operatorname{diam} B_i)^d < +\infty$, and otherwise

$$\sum_{i \in J(k+1)} (\operatorname{diam} B_i)^d \geq 1. \tag{29}$$

By Lemma 16 and our choice of τ_k,

$$H^d(A \cap V_k \cap (\bigcup_{i \in J(k+1)} B_i)) \leq c_d \sum_{i \in J(k+1)} (\operatorname{diam} B_i)^d + 2^{-k-1}(H^d(A) - \lambda). \tag{30}$$

Note that we already completed our definition of $I(k)$ by induction. We still need to check that $I_0 = \bigcup_{k \geq 0} I(k)$ has the desired properties. Since (13) follows from (21), (24), and (26), while (14) holds because of (26), we just need to check (15). We may as well assume that

$$\sum_k \sum_{i \in J(k)} (\operatorname{diam} B_i)^d < +\infty, \tag{31}$$

because otherwise (15) is trivial. Then $\sum_{i \in J(k)} (\operatorname{diam} B_i)^d$ tends to 0. This prevents (29) from happening infinitely many times, and then (28) says that $H^d(A \cap V_k)$ tends to 0.

Recall from (21) that $V_k = \mathbb{R}^n \setminus \bigcup_{I(k)} B_i$, so

$$H^d(A \setminus \bigcup_{I_0} B_i) = 0 \, . \tag{32}$$

Thus the B_i almost cover A, and since they are disjoint,

$$H^d(A) = \sum_{i \in I_0} H^d(A \cap B_i) \,. \tag{33}$$

For $k \geq 0$,

$$\sum_{i \in J(k+1)} H^d(A \cap B_i) = H^d(A \cap V_k \cap (\bigcup_{i \in J(k+1)} B_i)) \tag{34}$$

$$\leq c_d \sum_{i \in J(k+1)} (\operatorname{diam} B_i)^d + 2^{-k-1}(H^d(A) - \lambda)$$

because the B_i are disjoint and contained in V_k, and by (30). We can sum this over $k \geq 0$ and deduce from (33) that

$$H^d(A) \leq c_d \sum_{i \in I_0} (\operatorname{diam} B_i)^d + H^d(A) - \lambda \,. \tag{35}$$

[Note that $I(0) = \emptyset$ does not give a contribution to (33).] This is just (15). Our proof is thus complete in the special case when $H^d(A) < +\infty$, as in (20).

Now assume that $H^d(A) = +\infty$. Let M be large (compared to λ), to be chosen soon, and let $\rho < \tau$ be so small that $H^d_\rho(A) \geq M$. Since the ^CB_i, $i \in I$, form a Vitali covering of A, we can find $I_1 \subset I$ such that the ^CB_i, $i \in I_1$, cover A, and all the D_i (the smallest balls that contain the ^CB_i) have diameters at most $\rho/5$. Then Lemma 1 allows us to choose $I_0 \subset I_1$ such that the D_i, $i \in I_0$, are disjoint, but the $5D_i$ cover A. Since $H^d_\rho(A) \geq M$, we get that

$$\sum_{i \in I_0} (\operatorname{diam} B_i)^d \geq C^{-1} \sum_{i \in I_0} (\operatorname{diam}(5D_i))^d \geq C^{-1} H^d_\rho(A) \geq C^{-1} M \geq \lambda \tag{36}$$

if M is chosen large enough. This proves Theorem 12 in our last case when $H^d(A) = +\infty$. $\qquad\square$

Remark 37. In the special case when $H^d(A) < +\infty$, we have the additional property that $\sum_{i \in I_0} (\operatorname{diam} B_i)^d = +\infty$, or else (31) holds and $H^d(A \setminus \bigcup_{i \in I_0} B_i) = 0$, as in (32).

Exercise 38. Let μ be a finite positive measure on $\mathbb{R}^n$, and define the (centered) Hardy–Littlewood maximal function μ^* by

$$\mu^*(x) = \sup_{r>0} \left\{ \frac{1}{r^n} \mu(B(x,r)) \right\}. \tag{39}$$

Show that

$$|\{x \in \mathbb{R}^n \,;\, \mu^*(x) > \lambda\}| \leq C_n \, \mu(\mathbb{R}^n) \lambda^{-1} \tag{40}$$

for all $\lambda > 0$. Thus, $\mu^* \in L^1_{\text{weak}}(\mathbb{R}^n)$. Note that this applies in particular to $d\mu = f(x)dx$, with $f \in L^1(\mathbb{R}^n, dx)$.

Hint: cover $\Omega_\lambda = \{x \in \mathbb{R}^n ; \mu^*(x) > \lambda\}$ by balls $B(x, r)$ such that $\mu(B(x, r)) > \lambda r^n$, and apply Lemma 1.

Exercise 41 (Sard). Let $\Omega \subset \mathbb{R}^d$ be open, and $f : \Omega \to \mathbb{R}^n$ be of class C^1. Set $Z = \{x \in \Omega ; Df(x) \text{ is not injective}\}$. We want to show that

$$H^d(f(Z)) = 0. \tag{42}$$

1. Let $x \in Z$ be given. Show that there is a constant C_x such that for each $\varepsilon > 0$ we can find r_0 such that for $r < r_0$, $f(B(x, r))$ can be covered by less than $C_x \varepsilon^{1-d}$ balls of radius εr. Hint: show it first with f replaced with its differential $Df(x)$ at x.

2. Show that for each choice of $x \in Z$, $\tau > 0$, and $\eta \in (0, 1)$, we can find $r_x > 0$ such that $r_x < \tau$, $B(x, r_x) \subset \Omega$, and $f(B(x, 5r_x))$ can be covered by finitely many balls D_j, with $\sum_j (\text{diam } D_j)^d < \eta r_x^d$.

3. Pick $R > 0$ and cover $Z \cap B(0, R)$ by balls $B(x, r_x)$ as above. Then apply Lemma 1. Show that this yields

$$H^d_\tau(f(Z \cap B(0, R))) \leq C\eta|\Omega \cap B(0, R)|. \tag{43}$$

Conclude.

34 Local Hausdorff convergence of sets

The main goal of this section is to give a natural (and of course very classical) definition of convergence for closed subsets of an open set $\Omega \subset \mathbb{R}^n$. The main point will be to make sure that we can extract convergent subsequences from any given sequence of sets in Ω. Note that for $\Omega = \mathbb{R}^n$, say, the usual Hausdorff distance between sets is too large for this; what we need is something like convergence for the Hausdorff metric on every compact subset.

So let $\Omega \subset \mathbb{R}^n$ be a given open set, and let $\{H_m\}_{m \geq 0}$ be an exhaustion of Ω by compact subsets. This means that each H_m is compact, $H_m \subset \text{interior}(H_{m+1})$ for $m \geq 0$, and $\Omega = \bigcup_m H_m$. When $\Omega = \mathbb{R}^n$, a good choice is $H_m = \overline{B}(0, 2^m)$. We set

$$d_m(A, B) = \sup\{\text{dist}(x, B) ; x \in A \cap H_m\} + \sup\{\text{dist}(y, A) ; y \in B \cap H_m\} \tag{1}$$

for $A, B \subset \Omega$ and $m \geq 0$. We use the convention that $\text{dist}(x, B) = +\infty$ when $B = \emptyset$, but $\sup\{\text{dist}(x, B) ; x \in A \cap H_m\} = 0$ when $A \cap H_m = \emptyset$. Thus $d_m(A, B) = 0$ when $A \cap H_m = B \cap H_m = \emptyset$, but $d_m(A, \emptyset) = +\infty$ if $A \cap H_m \neq \emptyset$.

We shall find it more convenient here to use $d_m(A, B)$, rather than the Hausdorff distance between $A \cap H_m$ and $B \cap H_m$. Recall that the Hausdorff distance between A and B is

$$d_{\mathcal{H}}(A, B) = \sup_{x \in A} \operatorname{dist}(x, B) + \sup_{y \in B} \operatorname{dist}(y, A). \tag{2}$$

Here it could happen that $d_m(A, B) << d_{\mathcal{H}}(A \cap H_m, B \cap H_m)$, for instance because points of $A \cap H_m$ can be close to B, even if B does not meet H_m. Thus our functions d_m are a little less sensitive to boundary effects. Also, they have the nice feature that

$$d_m(A, B) \le d_l(A, B) \quad \text{for } 0 \le m \le l, \tag{3}$$

because $H_m \subset H_l$.

Definition 4. *Let* $\{A_k\}$ *be a sequence of subsets of* Ω, *and* $B \subset \Omega$. *We say that* $\{A_k\}$ *converges to* B *if*

$$\lim_{k \to +\infty} d_m(A_k, B) = 0 \quad \text{for all } m \ge 0. \tag{5}$$

This is the notion of convergence that we shall systematically use in the following sections. A few comments may be useful here. The reader may be worried because we did not restrict to closed sets. This is not really needed, but anyway $\{A_k\}$ converges to B if and only if the sequence $\{\overline{A}_k\}$ of closures in Ω converges to B. Also, convergence to B is equivalent to convergence to $\overline{B}$, and if we want to have uniqueness of the limit, we can require that it be closed. [See Exercise 15.] In the later sections, we shall always apply the definition to sequences of closed sets, and take a closed limit.

It is fairly easy to check that our notion of convergence does not depend on the choice of exhaustion $\{H_m\}$. See Exercise 17.

We now come to the main point of this section.

Proposition 6. *For each sequence* $\{A_k\}$ *of subsets of* Ω, *we can find a subsequence of* $\{A_k\}$ *that converges to some closed subset of* Ω.

This is a very standard exercise on complete boundedness, Cauchy sequences, and the diagonal process. We nonetheless give the proof for the convenience of the reader. Let us say that $\{A_k\}$ is a Cauchy sequence if

$$\lim_{k,l \to +\infty} d_m(A_k, A_l) = 0 \quad \text{for every } m \ge 0. \tag{7}$$

It is easy to see that if $\{A_k\}$ converges to some $B \subset \Omega$, it is a Cauchy sequence.

Lemma 8. *Every Cauchy sequence has a limit.*

Let $\{A_k\}$ be a Cauchy sequence, and set

$$B = \{x \in \Omega \, ; \, \lim_{k \to +\infty} \text{dist}(x, A_k) = 0\}. \tag{9}$$

Note that B is closed. We want to check that $\{A_k\}$ converges to B. Let $m \geq 0$ be given. First note that the functions $f_k = \text{dist}(\cdot, A_k)$ are 1-Lipschitz, so they are equicontinuous on H_m. Since they converge to 0 on B, the convergence is uniform on $B \cap H_m$. [See Exercise 22.] Hence

$$\lim_{k \to +\infty} \left\{ \sup\{\text{dist}(x, A_k) \, ; \, x \in B \cap H_m\} \right\} = 0. \tag{10}$$

Note that

$$B = \{x \in \Omega \, ; \, \liminf_{k \to +\infty} \text{dist}(x, A_k) = 0\}, \tag{11}$$

because $\{A_k\}$ is a Cauchy sequence. Indeed, if x lies in this last set and ε is small, we can find arbitrarily large values of k for which A_k meets $B(x, \varepsilon)$. Then all A_l, l large enough, meet $B(x, 2\varepsilon)$, because $d_m(A_k, A_l) \leq \varepsilon$, where we choose m so large that $B(x, \varepsilon) \subset H_m$. The other inclusion is trivial. Next we want to check that

$$\lim_{k \to +\infty} \left\{ \sup\{\text{dist}(x, B) \, ; \, x \in A_k \cap H_m\} \right\} = 0 \tag{12}$$

for all $m \geq 0$. Let ε be given, so small that $B(x, 2\varepsilon) \subset H_{m+1}$ for $x \in H_m$. Let k_0 be such that $d_{m+1}(A_k, A_l) \leq \varepsilon$ for $k, l \geq k_0$. If $k \geq k_0$ and $x \in A_k \cap H_m$, then $\overline{B}(x, \varepsilon)$ meets A_l for all $l \geq k$. Thus $\overline{B}(x, \varepsilon)$ contains points of adherence of the sequence $\{A_l\}$. These points lie in B, by (11). This proves (12), and the lemma follows because of (10). □

Remark 13. If $\{A_k\}$ converges to B and B is closed in Ω, then B is given by both formulae (9) and (11). This follows from our proof of Lemma 8 and the uniqueness of the limit (given that it is closed). See Exercise 15 for this last point.

Now we can prove Proposition 6. Let $\{A_k\}$ be any sequence of sets in Ω. It is enough to show that we can extract a Cauchy subsequence.

For each $m \geq 0$ and $N \geq 0$, cover H_m by finitely many balls $B_{m,N,s}$, $s \in S(m, N)$, that are centered on H_m, contained in H_{m+1}, and with radii smaller than 2^{-N}. Then define functions $g_{m,N,s}$ on $\mathbb{N}$ by $g_{m,N,s}(k) = 1$ if $B_{m,N,s}$ meets A_k, and $g_{m,N,s}(k) = 0$ otherwise. We can use the diagonal process to extract a subsequence $\{k_i\}$ such that every $\{g_{m,N,s}(k_i)\}_i$ has a limit $L_{m,N,s}$ at infinity. Let us check that $\{A_{k_i}\}$ converges.

Let m and N be given, and let i_0 be so large that $g_{m,N,s}(k_i) = L_{m,N,s}$ for $i \geq i_0$ and $s \in S(m, N)$. Let $i, j \geq i_0$ and $x \in H_m \cap A_{k_i}$ be given. Then x lies in some $B_{m,N,s}$, and $g_{m,N,s}(k_i) = 1$ by definition. Then $g_{m,N,s}(k_j) = 1$ also, and $B_{m,N,s}$ meets A_{k_j}. Hence $\text{dist}(x, A_{k_j}) \leq 2^{-N+1}$. The same reasoning can be done with $x \in A_{k_j}$, and altogether $d_m(A_{k_i}, A_{k_j}) \leq 2^{-N+1}$.

Since this can be done for all m and N, $\{A_{k_i}\}$ is a Cauchy sequence, as desired. Proposition 6 follows. □

Exercise 14. Check that the sequence $\{A_k\}$ converges to $\emptyset$ if and only if for every compact set $H \subset \Omega$, we can find k_0 such that $A_k \cap H = \emptyset$ for $k \geq k_0$.

Exercise 15. Check that if A and A' have the same closure in Ω, and similarly B and B' have the same closure, then

$$d_m(A, B) \leq d_{m+1}(A', B') \tag{16}$$

for all $m \geq 0$. Use this to check the affirmations after Definition 4. Also check the uniqueness of closed limits, and our assertion that if $\{A_k\}$ converges, then it is a Cauchy sequence.

Exercise 17. Check that the notion of convergence in Definition 3 does not depend on our choice of exhaustion $\{H_m\}$. Use an analogue of (3).

Exercise 18. Let $\{A_k\}$ be a sequence of subsets of Ω and $B \subset \Omega$. Check that if $\{A_k \cap H\}$ converges to $B \cap H$ (for the Hausdorff distance $d_{\mathcal{H}}$) for every compact set $H \subset \Omega$, then $\{A_k\}$ converges to B. The converse is not true; give a counterexample where H is a closed disk, say.

Exercise 19. Suppose that the sequence $\{A_k\}$ converges to B in Ω, and let $\Omega' \subset \Omega$ be open. Show that $\{A_k \cap \Omega'\}$ converges to $B \cap \Omega'$ in Ω'.

Exercise 20. Let $\{A_k\}$ be a sequence of subsets of Ω. Denote by $\overline{A}_k$ the closure of A_k in $\mathbb{R}^n$ (or equivalently in $\overline{\Omega}$).

1. Suppose that $\{\overline{A}_k\}$ converges to B for the Hausdorff distance $d_{\mathcal{H}}$ in $\overline{\Omega}$. Show that $\{A_k\}$ converges to $B \cap \Omega$ (as in Definition 4).

Suppose from now on that Ω is bounded.

2. Show that there exists a subsequence of $\{\overline{A}_k\}$ that converges for $d_{\mathcal{H}}$ to some set B^*. Thus we can get Proposition 6 in a slightly easier way when Ω is bounded, assuming that we already know that the set of closed subsets of a compact set, with the distance $d_{\mathcal{H}}$, is compact.

Suppose in addition that $\{A_k\}$ itself converges to some closed set $B \subset \Omega$ (as in Definition 4).

3. Show that from each subsequence of $\{\overline{A}_k\}$ we can extract a subsequence that converges in $d_{\mathcal{H}}$ to some (closed) $B^* \subset \overline{\Omega}$, and that $B^* \cap \Omega = B$.

4. We cannot say much more. Take $\Omega = B(0, 1)$ and give examples where $\{\overline{A}_k\}$ does not converge, or converges to sets strictly larger than $\overline{B}$.

Exercise 21. Let $\{A_k\}$ be a sequence of closed subsets of Ω, and assume that the A_k are locally Ahlfors-regular in Ω, with uniform estimates. Suppose that $\{A_k\}$ converges to the closed set $A \subset \Omega$. Show that A is locally Ahlfors-regular in Ω. Hint 1: Lemma 18.11 or Exercise 18.25. Hint 2: this will be done in Lemma 38.8.

Exercise 22. Let $H \subset \mathbb{R}^n$ be compact, and let $\{f_k\}$ be a sequence of functions defined on H. Suppose that the f_k are uniformly bounded (i.e., $|f_k(x)| \leq M$ for some M that does not depend on k), and equicontinuous on H. This last means that for each $\varepsilon > 0$ we can find $\eta > 0$ (independent of k) such that $|f_k(x) - f_k(y)| \le \varepsilon$ as soon as $|x - y| \le \eta$.

1. Show that there is a subsequence of $\{f_k\}$ that converges uniformly on H. [You may use a dense sequence $\{x_m\}$ in H, and reduce to the case when each $\{f_k(x_m)\}_{k\ge 0}$ converges.]

2. Show that if the f_k are uniformly bounded and equicontinuous on H, and if $\{f_k\}$ converges to some function, then the convergence is uniform.

35   Uniform concentration and lower semicontinuity of Hausdorff measure

We are now ready to see why Dal Maso, Morel, and Solimini cared so much about the uniform concentration property of Section 25.

Let us set notation for the next theorem. We are given an open set $\Omega \subset \mathbb{R}^n$, and a sequence $\{E_k\}_{k\ge 0}$ of relatively closed subset of Ω. We assume that

$$\lim_{k\to+\infty} E_k = E \tag{1}$$

for some set $E \subset \Omega$, which we naturally assume to be closed in Ω. Here we use the notion of convergence introduced in the last section. That is, we take an exhaustion of Ω by compact subsets H_m and define the corresponding pseudodistances d_m as in (34.1); then (1) means that $d_m(E_k, E)$ tends to 0 for each $m \ge 0$. See Definition 34.4.

We want to give a sufficient condition under which

$$H^d(E) \le \liminf_{k\to+\infty} H^d(E_k), \tag{2}$$

where $0 < d < n$ is a given integer. Note that something is needed, (2) is not true in general. See the simple counterexample just after (4.4) (i.e., dotted lines). Our main hypothesis is the following. For each $\varepsilon > 0$, we assume that we can find a constant $C_\varepsilon \ge 1$ such that the following property $\mathcal{H}(\varepsilon, C_\varepsilon)$ holds.

$\mathcal{H}(\varepsilon, C_\varepsilon)$: for each $x \in E$, we can find $r(x) > 0$ with the following property. Let $0 < r \le r(x)$ be given. Then for k large enough we can find a ball $B(y_k, \rho_k) \subset \Omega \cap B(x, r)$ such that $\rho_k \ge C_\varepsilon^{-1} r$ and

$$H^d(E_k \cap B(y_k, \rho_k)) \ge (1 - \varepsilon)\,\omega_d\,\rho_k^d. \tag{3}$$

Here ω_d denotes the Lebesgue (or H^d-) measure of the unit ball in $\mathbb{R}^d$.

Theorem 4 [DMS], [MoSo2]. *If $\{E_k\}_{k\ge 0}$ is a sequence of (relatively) closed subsets of Ω, E is a closed subset of Ω, and if for every $\varepsilon > 0$ we can find C_ε such that $\mathcal{H}(\varepsilon, C_\varepsilon)$ holds, then we have the semicontinuity property (2).*

The reader may be surprised that we did not include (1) in the hypotheses. It is not really needed; the main point is that because of $\mathcal{H}(\varepsilon, C_\varepsilon)$, every $x \in E$ is a limit of some sequence of points $x_k \in E_k \cap B(y_k, \rho_k)$. The fact that points of the E_k lie close to E is not needed for (2). But (1) will always hold when we apply Theorem 4 anyway.

Our proof will essentially follow [MoSo2]. Before we start for good, let us notice that we can reduce to the case when

$$L = \lim_{k \to +\infty} H^d(E_k) \quad \text{exists.} \tag{5}$$

Indeed we can always find a subsequence $\{E_{k_j}\}$ for which $H^d(E_{k_j})$ converges to the right-hand side of (2). This subsequence still satisfies the hypotheses of Theorem 4, and if we prove (2) for it, we also get it for $\{E_k\}$ itself.

So we assume (5) and try to prove that $H^d(E) \leq L$. By the same argument as above, we shall always be able to replace $\{E_k\}$ with any subsequence of our choice (because the hypotheses are still satisfied and the conclusion is the same). We shall use this possibility a few times.

Let ε be given, and let C_ε be such that $\mathcal{H}(\varepsilon, C_\varepsilon)$ holds. For each $x \in E$, $\mathcal{H}(\varepsilon, C_\varepsilon)$ gives us a radius $r(x)$. For each integer N, set $r_N(x) = \text{Min}\{2^{-N}, r(x)\}$ and $B_N(x) = B(x, r_N(x))$. We can cover E by countably many balls $B_N(x)$, because each $E \cap H_m$, H_m compact, can be covered by finitely many such balls. Now take all the balls that we get this way (when N varies). This gives a quite large, but still countable, family $\{B_i\}_{i \in I}$ of balls centered on E. By construction,

$$\{B_i\}_{i \in I} \quad \text{is a Vitali covering of } E \tag{6}$$

(see Definition 33.9).

Let $i \in I$ be given. Write $B_i = B(x_i, r_i)$, with $x_i \in E$ and $0 < r_i \leq r(x_i)$. We know from $\mathcal{H}(\varepsilon, C_\varepsilon)$ that for k large enough we can find a ball $D_{i,k} = B(y_{i,k}, \rho_{i,k})$ such that

$$C_\varepsilon^{-1} r_i \leq \rho_{i,k}, \qquad D_{i,k} \subset \Omega \cap B_i, \tag{7}$$

and

$$H^d(E_k \cap D_{i,k}) \geq (1 - \varepsilon)\omega_d \rho_{i,k}^d. \tag{8}$$

Since we can replace $\{E_k\}$ with any subsequence and I is at most countable, we can assume that for each $i \in I$, the sequence $\{y_{i,k}\}$ converges to some limit y_i, and similarly $\{\rho_{i,k}\}$ converges to some $\rho_i \in [C_\varepsilon^{-1} r_i, r_i]$.

Set $D_i = B(y_i, (1 + \varepsilon)\rho_i)$. Note that $D_{i,k} \subset D_i$ for k large enough. Hence

$$H^d(E_k \cap D_i) \geq H^d(E_k \cap D_{i,k}) \geq (1 - \varepsilon)\omega_d \rho_{i,k}^d \geq (1 - \varepsilon)^2 \omega_d \rho_i^d \tag{9}$$

for k large enough.

Note that $B_i \subset 3C_\varepsilon D_i$ by (7), hence (6) says that $\{D_i\}_{i \in I}$ is a Vitali family for E (see Definition 33.10). Now we want to apply Theorem 33.12. Pick any $\lambda < H^d(E)$ and $\tau > 0$. We get a set $I_0 \subset I$ such that

$$\text{the } D_i, i \in I_0, \text{ are disjoint,} \tag{10}$$

$$\text{diam } D_i < \tau \text{ for } i \in I_0, \tag{11}$$

and

$$c_d \sum_{i \in I_0} (\text{diam } D_i)^d \geq \lambda. \tag{12}$$

Let us choose a finite set $I_1 \subset I_0$ such that

$$\lambda \leq (1 + \varepsilon)\, c_d \sum_{i \in I_1} (\text{diam } D_i)^d = (1 + \varepsilon)^{d+1} 2^d c_d \sum_{i \in I_1} \rho_i^d . \tag{13}$$

Recall that c_d is the constant in the definition of H^d (see Definition 2.6). We need to know that $2^d c_d \leq \omega_d$, where as before $\omega_d = H^d(B)$ and B is the unit ball in $\mathbb{R}^d$. This is a nasty little verification, which amounts to checking that in the definition of $H^d(B)$, covering B with little balls is asymptotically optimal. See [Fe], 2.10.35 (and compare with 2.10.2 and 2.7.16 (1)) or look at Exercises 19 and 21.

For k large enough, (9) holds for all $i \in I_1$, and then

$$\lambda \leq (1 + \varepsilon)^{d+1} \omega_d \sum_{i \in I_1} \rho_i^d \leq (1 - \varepsilon)^{-2}(1 + \varepsilon)^{d+1} \sum_{i \in I_1} H^d(E_k \cap D_i)$$
$$\leq (1 - \varepsilon)^{-2}(1 + \varepsilon)^{d+1} H^d(E_k), \tag{14}$$

by (13), (9), and (10). We can let k tend to $+\infty$; we get that $\lambda \leq (1 - \varepsilon)^{-2}(1 + \varepsilon)^{d+1} L$, where $L = \lim_{k \to +\infty} H^d(E_k)$ as in (5). Now λ was any number strictly smaller than $H^d(E)$, so $H^d(E) \leq (1 - \varepsilon)^{-2}(1 + \varepsilon)^{d+1} L$. Finally, this holds for every choice of $\varepsilon > 0$, and (2) follows. This completes our proof of Theorem 4. $\qquad \square$

Theorem 4 has the following simple consequence (essentially, Golab's theorem).

Corollary 15. *Let $\{E_k\}$ be a sequence of closed sets in $\Omega \subset \mathbb{R}^n$. Suppose that $\{E_k\}$ converges to some closed subset E of Ω, and that there is an integer N such that each E_k has at most N connected components. Then*

$$H^1(E) \leq \liminf_{k \to +\infty} H^1(E_k). \tag{16}$$

Golab's theorem corresponds to $N = 1$. As we shall see, the general case follows rather easily. As far as the author knows, Corollary 15 is not much simpler to prove than Theorem 4.

Corollary 15 is one more illustration of the fact that for sets with finite H^1-measure, connectedness is a quite strong regularity property. Once again compare with the counterexample just after (4.4).

We start the proof of Corollary 15 by reducing to the case when $N = 1$. Note that any subsequence of our initial sequence $\{E_k\}$ satisfies the same hypotheses as $\{E_k\}$ itself. We can use this to reduce to the case when $L = \lim_{k \to +\infty} H^1(E_k)$ exists, as for (5). [Take a subsequence for which $H^1(E_k)$ converges to the right-hand side of (16).]

Let $N(k)$ denote the number of connected components of E_k. A second sequence extraction allows us to suppose that $N(k)$ is constant. Call N' this constant value. Denote by $E_k^1, E_k^2, \ldots, E_k^{N'}$ the components of E_k (any choice of order will do). We can use Proposition 34.6 to extract a new subsequence for which each sequence $\{E_k^l\}_{k \geq 0}$ converges to some (closed) limit E^l. If Corollary 15 holds for $N = 1$, we get that

$$H^1(E^l) \leq \liminf_{k \to +\infty} H^1(E_k^l). \tag{17}$$

Now $E = \bigcup_{l=1}^{N'} E^l$ (because one can check from the definition of convergence that $E_k = E_k^1 \cup E_k^2 \cdots \cup E_k^{N'}$ converges to $E^1 \cup E^2 \cdots \cup E^{N'}$, and by uniqueness of limits). Then

$$\begin{aligned} H^1(E) &\leq \sum_{l=1}^{N'} H^1(E^l) \leq \sum_{l=1}^{N'} \liminf_{k \to +\infty} H^1(E_k^l) \\ &\leq \liminf_{k \to +\infty} \Big\{ \sum_{l=1}^{N'} H^1(E_k^l) \Big\} = \liminf_{k \to +\infty} \big\{ H^1(E_k) \big\}, \end{aligned} \tag{18}$$

in particular because the E_k^l, $1 \leq l \leq N'$, are disjoint.

Thus it is enough to prove Corollary 15 when $N = 1$. So let us assume that each E_k is connected.

If E is empty or reduced to one point, then $H^1(E) = 0$ and there is nothing to prove. So we may assume that $\delta = \operatorname{diam}(E) > 0$. Let us check that for each $\varepsilon > 0$, $\{E_k\}$ satisfies $\mathcal{H}(\varepsilon, C_\varepsilon)$ with $C_\varepsilon = 3$. Let $x \in E$ be given, and take $r(x) = \operatorname{Min}\{\delta/3, \operatorname{dist}(x, \mathbb{R}^n \setminus \Omega)\}$. Then let $r \leq r(x)$ be given. Choose $x_k \in E_k$, $k \geq 0$, so that $\{x_k\}$ converges to x. Also choose $z \in E$, with $|z - x| \geq 4\delta/10$, and then $z_k \in E_k$ such that $\{z_k\}$ converges to z. Thus $|z_k - x_k| \geq \delta/3$ for k large enough.

As usual, we may as well assume that $L = \lim_{k \to +\infty} H^1(E_k)$ exists. If $L = +\infty$, there is nothing to prove. Otherwise, $H^1(E_k) < +\infty$ for k large enough. This allows us to apply Proposition 30.14 and find a simple arc $\gamma_k \subset E_k$ that goes from x_k to z_k. Then let y_k be a point of $\gamma_k \cap \partial B(x_k, r/2)$. Such a point exists, because γ_k is connected and z_k lies outside of $B(x_k, r) \subset B(x_k, r(x))$. [Recall that $r(x) \leq \delta/3$.]

Each of the two disjoint arcs that compose $\gamma_k \setminus \{y_k\}$ goes from y_k to a point that lies outside of $B_k = B(y_k, r/3)$. Thus $H^1(E_k \cap B_k) \geq H^1(\gamma_k \cap B_k) \geq 2r/3$. That is, B_k satisfies (3) with $\varepsilon = 0$. Also, $B_k \subset \Omega \cap B(x, r)$ for k large enough. This completes the verification of $\mathcal{H}(\varepsilon, C_\varepsilon)$, Theorem 4 applies, and we get (16). Corollary 15 follows. □

Exercise 19 (the isodiametric inequality). We want to check that

$$|A| \leq \omega_d 2^{-d} (\operatorname{diam} A)^d \tag{20}$$

for all Borel sets $A \subset \mathbb{R}^d$. Here $|A|$ is the Lebesgue measure of A, and ω_d is the Lebesgue measure of the unit ball in $\mathbb{R}^d$. Notice that (20) is an equality when A is a ball.

We shall follow [Fe], 2.10.30, and use Steiner symmetrizations.

Let $A \subset \mathbb{R}^d$ be a Borel set and H a hyperplane through the origin. The Steiner symmetrization of A with respect to H is the set $S_H(A)$ such that, for each line L perpendicular to H, $S_H(A) \cap L$ is empty if $A \cap L$ is empty, and is the closed interval of L centered on $H \cap L$ and with length $H^1(A \cap L)$ otherwise.

1. Check that S_H is measurable and $|S_H(A)| = |A|$.

2. Let E, F be Borel sets in $\mathbb{R}$ and I, J intervals centered at 0 and such that $H^1(I) = H^1(E)$ and $H^1(J) = H^1(F)$. Show that

$$\sup\{|x - y| \,;\, x \in I \text{ and } y \in J\} \leq \sup\{|x - y| \,;\, x \in E \text{ and } y \in F\}.$$

3. Show that $\operatorname{diam} S_H(A) \leq \operatorname{diam} A$.

4. Set $\lambda = \inf\{\operatorname{diam} A \,;\, A \in \mathbb{R}^d \text{ is a Borel set and } |A| = \omega_d\}$, and

$$\Sigma = \{A \subset \mathbb{R}^d \,;\, A \text{ is closed}, |A| = \omega_d, \text{ and } \operatorname{diam} A = \lambda\}.$$

Show that Σ is not empty. [Hint: use Fatou.]

5. Now set $\alpha = \inf\{r > 0 \,;\, \text{we can find } A \in \Sigma \text{ such that } A \subset \overline{B}(0, r)\}$. Show that $\Sigma_1 = \{A \in \Sigma \,;\, A \subset \overline{B}(0, \alpha)\}$ is not empty.

6. Show that $S_H(A) \in \Sigma_1$ when $S \in \Sigma_1$ and H is a hyperplane through the origin.

7. Let $A \in \Sigma_1$ be given, and suppose that $z \in \partial B(0, \alpha)$ and $\varepsilon > 0$ are such that $\partial B(0, \alpha) \cap B(z, \varepsilon)$ does not meet A. Let H be a a hyperplane through the origin and call w the symmetric image of z with respect to H. Show that $S_H(A)$ does not meet $\partial B(0, \alpha) \cap [B(z, \varepsilon) \cup B(w, \varepsilon)]$.

8. Show that $\partial B(0, \alpha) \subset A$ for $A \in \Sigma_1$. Then show that $A = \overline{B}(0, \alpha)$, hence $\lambda = 2$ and (21) holds.

Exercise 21. We want to use Exercise 19 to show that $2^d c_d = \omega_d$. We keep the same notation.

1. Let $\{A_i\}_{i \in I}$ be a covering of $B(0,1)$ in $\mathbb{R}^d$ by countably many sets. Use (20) to show that $\sum (\operatorname{diam} B_i)^d \geq 2^d$.

2. Show that $\omega_d \geq 2^d c_d$. [Recall that c_d was chosen so that $H^d(B(0,1)) = \omega_d$.]

3. Let $\tau > 0$ be given. Show that we can find disjoint balls $B_i \subset B$, of diameters less than τ, and such that $|B(0,1) \setminus \bigcup_i B_i| = 0$. You may use Theorem 33.12 and Remark 33.37.

4. Show that for each $\varepsilon > 0$ we can cover $B(0,1) \setminus \bigcup_i B_i$ by (countably many) sets D_j of diameters less than τ and for which $\sum_j (\operatorname{diam} D_j)^d \leq \varepsilon$.

5. Show that $H^d(B(0,1)) \leq c_d \sum (\operatorname{diam} B_i)^d + c_d \varepsilon \leq c_d 2^d \omega_d^{-1} H^d(B(0,1)) + c_d \varepsilon$. Conclude.

36   A little more on the existence of minimizers

In this section we return briefly to the issue of existence of minimizers for the Mumford-Shah functional on a domain Ω. We want to say how it is possible to get these minimizers as limits of suitable sequences of nearly minimizing pairs, using in particular the lower semicontinuity result of the previous section. Our description here will be a little more precise than what we did in Section 4, but we shall not give a complete proof. Also, we shall restrict to dimension 2, as often in this book. The technique that we describe in this section also works in higher dimensions; see [MaSo3].

We should remind the reader that the standard, and somewhat simpler way to prove existence for this sort of functional is to use its weak description in terms of SBV functions, and then the compactness theorem from [Am]. This is well described in [AFP3]; here we take the slightly silly attitude of trying to see what we can get without using SBV.

The idea of trying to use the uniform concentration property from Section 25 to get existence results for minimizers comes from [DMS]. It is also described in [MoSo2], but in both cases the authors do not go all the way to a complete proof without SBV, due to technical complications. See the comments near the end of this section.

Let us start our description. Let Ω be a domain in the plane, $g \in L^\infty(\Omega)$, and set

$$J(u,K) = \int_{\Omega \setminus K} |u - g|^2 + \int_{\Omega \setminus K} |\nabla u|^2 + H^1(K), \qquad (1)$$

where the competitors lie in the class $\mathcal{U}$ of pairs (u, K) for which K is a relatively closed set in Ω such that $H^1(K) < \infty$, and $u \in W^{1,2}(\Omega \setminus K)$ is defined on $\Omega \setminus K$ and has one derivative in L^2 there. See Section 2. It is customary to assume that

Ω is bounded, but the main point is to find competitors for which $J(u, K) < \infty$. However, to make the proof below work more easily, we shall assume that $\partial\Omega$ has finitely many connected components, and $H^1(\partial\Omega) < \infty$.

Our first goal will be to minimize J in the smaller class

$$\mathcal{U}_N = \{(u, K) \in \mathcal{U} \, ; K \cup \partial\Omega \text{ has at most } N \text{ connected components}\}, \quad (2)$$

where $N \geq 1$ is a given integer. So let $(u_k, K_k) \in \mathcal{U}_N$ be given, with

$$\lim_{k \to +\infty} J(u_k, K_k) = m_N, \quad (3)$$

where we set

$$m_N = \inf \{J(u, K); (u, K) \in \mathcal{U}_N\}. \quad (4)$$

We may assume that $m_N < +\infty$ and that u_k minimizes $J(u_k, K_k)$ with the given K_k, i.e., that

$$J(u_k, K_k) = \inf \{J(v, K_k) \, ; v \in W^{1,2}(\Omega \setminus K_k)\}. \quad (5)$$

This is possible, because Proposition 3.3 says that for any given K_k, the infimum in (5) is reached.

We want to extract from $\{(u_k, K_k)\}$ a convergent subsequence, and then show that the limit minimizes J on $\mathcal{U}_N$. We extract a first subsequence (which we shall still denote $\{(u_k, K_k)\}$ to keep the notation reasonably pleasant), so that the sets $K_k \cup \partial\Omega$ converge (in $\mathbb{R}^n$) to some closed set K^*. Here we can use convergence for the Hausdorff metric on compact sets, or equivalently the notion of convergence from Section 34. The existence of the desired subsequence follows from Proposition 34.6, say.

Set $K = K^* \setminus \partial\Omega \subset \Omega$. Because of Corollary 35.15,

$$H^1(K) = H^1(K^*) - H^1(\partial\Omega) \leq \liminf_{k \to +\infty} H^1(K_k \cup \partial\Omega) - H^1(\partial\Omega) = \liminf_{k \to +\infty} H^1(K_k). \quad (6)$$

We shall need to know that $K^* = K \cup \partial\Omega$ has at most N connected components. First note that we can find a subsequence for which each $K_k \cup \partial\Omega$ has the same number of components, and moreover each of these converges to some limit (see the first lines of the proof of Corollary 35.15). For the separate existence of limits for the components, we can also proceed as follows. Each component K_k^l of $K_k \cup \partial\Omega$ has a finite length $L_k^l \leq H^1(K_k \cup \partial\Omega) \leq m_N + H^1(\partial\Omega) + 1 < +\infty$, by (1) and (3), and if k is large enough. Then Proposition 30.1 says that we can find a CL_k^l-Lipschitz mapping $f_{k,l}$ from $[0, 1]$ onto K_k^l. Since $L_k^l \leq m_N + H^1(\partial\Omega) + 1$ for k large, we can extract a subsequence so that each $\{f_{k,l}\}_k$ converges. Then the sets $\{E_{k,l}\}_k$ also converge. Also, each limit set is connected, since it is a Lipschitz image of $[0, 1]$. Thus K^* has at most N components.

The reader may wonder why we considered the number of components of $K_k \cup \partial\Omega$, rather than K_k itself, especially since this is what cost us the unnatural assumption that $H^1(\partial\Omega) < +\infty$. The point is that we may have a sequence of connected sets K_k, with some parts that tend to the boundary, and for which the intersection of the limit with Ω is no longer connected. There are ways to change the definition of $\mathcal{U}_N$ so that this issue does not arise, and yet we don't need to assume that $H^1(\partial\Omega) < +\infty$. But we do not wish to elaborate here.

Next we want to make a subsequence of $\{u_k\}$ converge. For this we need some little, but uniform amount of regularity for the functions u_k. Here we shall try to use fairly little information (namely, the L^∞-bound (7) and later the fact that $u_k \in W^{1,2}(\Omega \setminus K_k)$), but we could go a little faster by using stronger properties of the u_k. Observe that

$$||u||_\infty \leq ||g||_\infty, \tag{7}$$

by uniqueness of the minimizing function in (5) and a comparison with a truncation of u_k. See the proof of (3.20) for details.

Let $D \subset\subset \Omega \setminus K$ be any relatively compact subset of $\Omega \setminus K$. By (7), the functions u_k define bounded linear forms on $\mathcal{C}_c(D)$, the set of continuous functions with compact support in D (with the sup norm), with uniform bounds. Hence, modulo extraction of a new subsequence, we can assume that these linear forms converge weakly. That is, there is a measure $\mu = \mu_D$ on D such that $\lim_{k\to+\infty} \int_D f(x)u_k(x)dx = \int_D f(x)d\mu(x)$ for every f in (a dense class of) $\mathcal{C}_c(D)$. The measure μ defines a distribution on D, and its derivative is given by

$$\langle \partial_i\mu, f \rangle = - \int_D \partial_i f(x)d\mu(x) = - \lim_{k\to+\infty} \int_D \partial_i f(x)u_k(x)dx \tag{8}$$

for $f \in \mathcal{C}_c^\infty(D)$ and $i = 1, 2$.

Recall that D is relatively compact in $\Omega \setminus K$, and hence lies at positive distance from $K^* = K \cup \partial\Omega$. Since K^* is the limit of the $K_k \cup \partial\Omega$, D does not meet $K_k \cup \partial\Omega$ for k large enough. Then $\int_D \partial_i f(x)u_k(x)dx = - \int_D f(x)\partial_i u_k(x)dx$, and

$$\langle \partial_i\mu, f \rangle = \lim_{k\to+\infty} \int_D f(x)\partial_i u_k(x)dx. \tag{9}$$

In particular,

$$|\langle \partial_i\mu, f \rangle| \leq \liminf_{k\to+\infty} ||f||_2\, ||\partial_i u_k||_{L^2(D)}$$
$$\leq ||f||_2 \liminf_{k\to+\infty} J(u_k, K_k)^{1/2} \leq ||f||_2\, m_N^{1/2} < +\infty. \tag{10}$$

The Riesz representation theorem says that $\partial_i\mu \in L^2(D)$, and so $\mu \in W^{1,2}(D)$. [Note that $\mu \in L^1_{\text{loc}}(D)$; it is even bounded, because

$$\int_D f(x)d\mu(x) \leq ||g||_\infty \int_D |f(x)|dx \text{ for } f \in \mathcal{C}_c(D), \text{ by (7).}]$$

Next

$$||\nabla\mu||_{L^2(D)} = \sup\left\{\left|\int_D \nabla\mu \cdot G\right|; G \in \mathcal{C}_c^\infty(D, \mathbb{R}^2), ||G||_2 \leq 1\right\}. \tag{11}$$

Since

$$\left| \int_D \nabla \mu \cdot G \right| = \left| \int_D \mu \operatorname{div} G \right| = \left| \lim_{k \to +\infty} \int_D u_k \operatorname{div} G \right|$$

$$= \left| \lim_{k \to \infty} \int_D \nabla u_k \cdot G \right| \leq \liminf_{k \to +\infty} ||\nabla u_k||_{L^2(D)} ||G||_2$$

(12)

for all $G \in C_c^\infty(D, \mathbb{R}^2)$, (11) says that

$$||\nabla \mu||_{L^2(D)} \leq \liminf_{k \to +\infty} ||\nabla u_k||_{L^2(D)} .$$

(13)

This is what we get when we consider a single relatively compact open set $D \subset\subset \Omega$. By the diagonal process, we can extract a subsequence so that the computations above work for every $D \subset\subset \Omega$. Then there is a function $u \in W^{1,2}(\Omega \setminus K)$, that coincides on each D with the corresponding function μ_D, and such that

$$\int_{\Omega \setminus K} u(x) f(x) dx = \lim_{k \to +\infty} \int_{\Omega \setminus K} u_k(x) f(x) dx \quad \text{for every } f \in C_c(\Omega \setminus K).$$

(14)

Moreover,

$$\int_{\Omega \setminus K} |\nabla u|^2 \leq \liminf_{k \to +\infty} \int_{\Omega \setminus K_k} |\nabla u_k|^2 ,$$

(15)

by (13) and because for each choice of $D \subset\subset \Omega$, the right-hand side of (13) is smaller than or equal to the right-hand side of (15).

The same reasoning, but a little simpler because we don't have to integrate by parts, yields

$$\int_{\Omega \setminus K} |u - g|^2 \leq \liminf_{k \to +\infty} \int_{\Omega \setminus K_k} |u_k - g|^2 .$$

(16)

Altogether we constructed a pair $(u, K) \in \mathcal{U}_N$ such that

$$J(u, K) = \int_\Omega |u - g|^2 + \int_\Omega |\nabla u|^2 + H^1(K)$$

$$\leq \liminf_{k \to +\infty} \int_{\Omega \setminus K_k} |\nabla u_k|^2 + \liminf_{k \to +\infty} \int_{\Omega \setminus K_k} |u_k - g|^2 + \liminf_{k \to +\infty} H^1(K_k) \quad (17)$$

$$\leq \liminf_{k \to +\infty} J(u_k, K_k) = m_N ,$$

by (1), (15), (16), (6), and (3).

So we can find a minimizer (u_N, K_N) for the restriction of J to $\mathcal{U}_N$, and this for each $N \geq 1$. Now we want to repeat the argument and show that some subsequence of $\{(u_N, K_N)\}$ converges to a pair (u, K) that minimizes J on $\mathcal{U}$.

We may assume that $\inf\{J(u, K) ; (u, K) \in \mathcal{U}\} < +\infty$ because otherwise there won't be minimizers. Note that this is the case if Ω is bounded. Let us also assume that $m_N < +\infty$ for N large. One could also prove this (this is somewhat easier than (21) below), but let us not bother.

We can proceed as above with the sequence $\{(u_N, K_N)\}$, up to the point where we applied Corollary 35.15 to prove (6). Here we have no reason to believe that the sets $K_N \cup \partial\Omega$ have less than M components for some fixed M, so we need to find something else.

We want to get additional information from the fact that (u_N, K_N) minimizes J on $\mathcal{U}_N$. First, we can assume that none of the components of K_N is reduced to one point. Indeed, if x is an isolated point of K_N, u_N has a removable singularity at x, by Proposition 10.1. Then we can simply remove x from K_N, keep the same function u_N, and get an equivalent competitor with one less component.

Claim 18. The sets K_N are locally Ahlfors-regular (as in Theorem 18.1), have the property of projections (as in Theorem 24.1), and more importantly have the concentration property described in Theorem 25.1. All these properties hold with constants that do not depend on N.

The proof is the same as in Sections 18–25. Let us try to say why rapidly. All the estimates that we proved in these sections were obtained by comparing our initial minimizer (or quasiminimizer) (u, K) with other competitors $(\widetilde{u}, \widetilde{K})$. Our point now is that for all the competitors $(\widetilde{u}, \widetilde{K})$ that we used, $\widetilde{K} \cup \partial\Omega$ never has more connected components than $K \cup \partial\Omega$ itself.

For instance, to get the trivial estimates (18.20), we replaced $K \cap \overline{B}(x, r)$ with $\partial B(x, r)$. This does not increase the number of components as soon as $B(x, r) \subset \Omega$ and $K \cap B(x, r)$ is not empty, which is always the case when we use (18.20). For the local Ahlfors-regularity, we removed $K \cap B(x, r)$ from K for some ball $B(x, r) \subset \Omega$ such that $\partial B(x, r)$ does not meet K, and this is also all right. The stories for Theorems 24.1 and 25.1 are similar.

We are now ready to continue the argument above. We have a subsequence of $\{(u_N, K_N)\}$ that converges to some limit (u, K). Because of Claim 18, this subsequence satisfies the hypotheses of Theorem 35.4. [Take $r(x) = \mathrm{Min}(r_0, \mathrm{dist}(x, \mathbb{R}^2 \setminus \Omega))$ and compare $\mathcal{H}(\varepsilon, C_\varepsilon)$ in Section 35 with the conclusion of Theorem 25.1]. Hence

$$H^1(K) \leq \liminf_{N \to +\infty} H^1(K_N). \tag{19}$$

The analogues of (15) and (16) hold with the same proofs, and finally

$$J(u, K) \leq \lim_{N \to +\infty} J(u_N, K_N) = \lim_{N \to +\infty} m_N, \tag{20}$$

as in (17), because (u_N, K_N) minimizes J on $\mathcal{U}_N$.

Recall also that $\{m_N\}$ is nonincreasing. Since we want to show that (u, K) minimizes J on $\mathcal{U}$, it is enough to check that

$$\inf \{J(u, K); (u, K) \in \mathcal{U}\} = \lim_{N \to +\infty} m_N. \tag{21}$$

In other words, for each $\varepsilon > 0$, we have to find some pair $(v, G) \in \mathcal{U}$ such that $J(v, G) \leq \inf\{J(u, K); (u, K) \in \mathcal{U}\} + \varepsilon$, and for which $G \cup \partial\Omega$ has a finite number of connected components.

To do this, the idea is to start from a pair $(v_0, G_0) \in \mathcal{U}$ such that $J(v_0, G_0) - \inf\{J(u, K); (u, K) \in \mathcal{U}\} << \varepsilon$, and modify it slightly. As far as the author knows, (21) is not proved like this in [DMS], where the authors prefer to use the existence result from the SBV approach, nor in [MoSo2], where they avoid the issue. There is a (hopefully) complete argument in [BoDa], Sections 6–8, but which is somewhat more painful than needed because a slightly more complicated version of the Mumford-Shah functional is studied there. Recall that a full proof of existence (in any dimension) is given in [MaSo3]; Probably it contains a proof of (21), but I did not check.

Let us also mention that F. Dibos and E. Séré [DiSé] showed that in dimension 2, the minimum is approached by competitors for which K is composed of arcs of circles.

No matter how, the argument cannot be too simple. The difficulty with it is that we need to show that many of the good properties of minimizers (like local Ahlfors-regularity, or rectifiability) hold on very large parts of the set G_0. This needs the same sort of arguments as in Sections 20–24, but one needs to be more careful because the little piece of G_0 where the Ahlfors-regularity fails, say, could play nasty tricks on you when you want to deal with the property of projections, for instance. In the case of minimizers, this issue simply did not arise because the little piece was empty. In particular, it would seem that the use of some covering lemma is really unavoidable here. We leave the details and refer the reader to [BoDa] or [MaSo3] for a full proof.

Once we have (21), our proof of existence of minimizers for the Mumford-Shah functional J in (1) is complete. $\qquad\square$

37   Limits of admissible pairs

In the next few sections, we want to consider limits of almost-minimizers. One of the main uses for this will come from the following simple idea of A. Bonnet [Bo]. Consider a single Mumford-Shah minimizer (u, K) in a domain Ω, pick any point $x \in \Omega$, and look at the pairs (u_t, K_t), t small, where $K_t = t^{-1}(K - x)$ and $u_t(y) = t^{-1/2}u(x + ty)$. We get minimizers for analogues of the Mumford-Shah functional in the larger domains $\Omega_t = t^{-1}(\Omega - x)$, but where the term $\int_{\Omega_t \setminus K_t} |u - g|^2$ is multiplied by t^2. If we can take a limit of these minimizers (we shall call this a blow-up limit of (u, K)), we should get a minimizer for a simpler functional without the $|u-g|^2$-term. Then we can hope to get a good description of all the tangent objects to (u, K), which in turn will give valuable local information on (u, K) itself. This idea looks so natural now (especially since the $|u - g|^2$-term has always been known to be a minor perturbation term) that it is surprising that we did not have it before! We shall return to this in Section 54.

So we want to take limits of almost-minimizers, and then show that they are themselves almost-minimizers (or better in some cases). First we need to define a notion of limits that will allow us to find enough convergent subsequences.

We shall work in the following setting. We are given a sequence $\{\Omega_k\}$ of open sets in $\mathbb{R}^n$, and an open set $\Omega \subset \mathbb{R}^n$. We assume that

$$\text{for each compact set } H \subset \Omega, H \subset \Omega_k \text{ for } k \text{ large enough.} \tag{1}$$

Thus we do not really assume that $\{\Omega_k\}$ converges to Ω; we allow the Ω_k to be much larger. This does not matter at this point, because we do not want yet to study the behavior of almost-minimizers near the boundary.

We are also given a sequence $\{(u_k, K_k)\}$ of admissible pairs. That is, we assume that for each k,

$$(u_k, K_k) \in \mathcal{A}(\Omega_k) = \{(u, K) \,;\, K \text{ is relatively}$$
$$\text{closed in } \Omega_k \text{ and } u_k \in W^{1,2}_{\text{loc}}(\Omega_k \setminus K)\}, \tag{2}$$

and we even require that for each ball B such that $\overline{B} \subset \Omega$, there is a constant $C(B)$ such that

$$\int_B |\nabla u_k|^2 \leq C(B) \text{ for } k \text{ large enough.} \tag{3}$$

Our goal for this section is to define a notion of convergence for the pairs (u_k, K_k) so that we can extract convergent subsequences, and then show that the limits are admissible pairs in Ω. We shall only look at minimizing properties of the limits in later sections. Let us first define the convergence of sets.

Definition 4. *Let* $\{K_k\}$ *be a sequence of closed sets in the* Ω_k, *and let* K *be a (relatively) closed subset of* Ω. *We say that* $\{K_k\}$ *converges to* K *if for each compact set* $H \subset \Omega$, $\lim_{k \to +\infty} d_H(K_k, K) = 0$, *where we set*

$$d_H(K_k, K) = \sup\{\text{dist}(x, K) \,;\, x \in K_k \cap H\} + \sup\{\text{dist}(y, K_k) \,;\, y \in K \cap H\}. \tag{5}$$

We keep the same convention as for (34.1) for the cases when $K_k \cap H$ or $K \cap H$ is empty. We decided to give a new definition here, because in Section 34 we worked on a fixed domain. However the difference is not great, and of course our two definitions coincide when $\Omega_k = \Omega$ for all k. In particular we could have taken any exhaustion of Ω by compact subsets H_m, and required only that $\lim_{k \to +\infty} d_{H_m}(K_k, K) = 0$ for every m, and this would have given an equivalent definition. [See Exercise 34.17.] Another minor difference is that now we want to restrict to closed sets. This is of course not a serious problem; see the comments after Definition 34.4. Also note that our definition of convergence depends on Ω: for instance taking Ω smaller makes convergence easier to get.

Now let us assume that $\{K_k\}$ converges to K, and define the convergence of $\{u_k\}$. In the situations that we have in mind, we shall have no control on the L^∞-norms of the functions u_k, so in effect we shall only make the gradients converge.

Definition 6. *Let* $\{(u_k, K_k)\}$ *be as above. We say that* $\{(u_k, K_k)\}$ *converges to the pair* (u, K) *if* $\{K_k\}$ *converges to* K *(as in Definition 4),* $u \in L^1_{\text{loc}}(\Omega \setminus K)$, *and* $\{u_k\}$ *converges to* u *in the following way. For each connected component* $\mathcal{O}$ *of* $\Omega \setminus K$, *we can find a sequence* $\{a_k\}$ *of numbers, so that*

$$\{u_k - a_k\} \text{ converges to } u \text{ in } L^1(H) \text{ for every compact set } H \subset \mathcal{O}. \qquad (7)$$

Here we chose a fairly weak notion of convergence for the functions u_k, but in many applications we shall be able to assume much more. In Proposition 25 below we prove the uniform convergence of $\{u_k - a_k\}$ on compact subsets of $\mathcal{O}$ under quite mild assumptions, but for sequences of local minimizers, for instance, it would even be easy to show that the gradients ∇u_k converge to ∇u uniformly on every compact subset of Ω, because the u_k are harmonic. Note that in Definition 6, u is only known up to an additive constant on each component of Ω. This is all right; in effect we are really interested in the (quite weak) convergence of the ∇u_k to ∇u.

Proposition 8. *Let* $\{\Omega_k\}$ *be a sequence of open sets, and assume that* (1) *holds for some open set* Ω. *Also let* $\{(u_k, K_k)\}$ *be a sequence of admissible pairs, as in* (2), *and assume that* (3) *holds. Then we can find a subsequence* $\{(u_{k_j}, K_{k_j})\}$ *that converges to some* (u, K) *(as in Definition 6).*

See Propositions 18 and 25 below for more details about the function u, and the next section for estimates on $H^d(K)$ in some situations. We shall also worry about the minimality properties of (u, K) in the next sections.

The proof of Proposition 8 is quite standard. First we need to know that we can extract a subsequence (which we shall still denote by $\{(u_k, K_k)\}$ to save notation), so that $\{K_k\}$ converges to some relatively closed set $K \subset \Omega$. This is Proposition 34.6, or rather it follows directly from its proof.

Now we want to take care of $\{u_k\}$. Call Δ the set of balls $B = B(x, r)$ such that x has rational coordinates, r is rational, and $3B = B(x, 3r) \subset \Omega \setminus K$. We want to replace $\{u_k\}$ with a subsequence for which

$$\{u_k - m_B u_k\} \text{ converges in } L^1(B) \qquad (9)$$

for every $B \in \Delta$, where $m_B u_k$ denotes the mean value of u_k on B.

Since Δ is countable, it will be enough to get (9) for any single $B \in \Delta$, and then extract a diagonal subsequence. So let $B = B(x, r) \in \Delta$ be given. Note that $2B \subset \Omega_k$ for k large, by (1). Since we can extract as many subsequences as we like, we may as well assume that $2B \subset \Omega_k$ for all k, and that $\int_{2B} |\nabla u_k|^2 \leq C(2B)$ for all k, where $C(2B)$ comes from (3).

Lemma 10. *Let* $A > 0$ *be given, and call* E *the set of functions* $u \in W^{1,1}(2B)$ *such that* $\int_B u = 0$ *and* $\int_{2B} |\nabla u| \leq A$. *Then* E *is relatively compact in* $L^1(B)$.

Once we have the lemma, we can use it to extract subsequences that satisfy (9), because all the $u_k - m_B u_k$ lie in E, with $A \leq \{|2B|\, C(2B)\}^{1/2}$, by (3) and Cauchy–Schwarz.

So let us prove Lemma 10. Note that $E \subset L^1(B)$, by Proposition 12.11. Since $L^1(B)$ is complete, it is enough to check that E is completely bounded. [See Exercise 32.] That is, we just need to show that for each $\varepsilon > 0$, E can be covered by finitely many balls of radius ε in $L^1(B)$.

Let $\varepsilon > 0$ be given, and let $\eta > 0$ be very small, to be chosen soon. Cover B with balls $B_i = B(x_i, \eta r)$, $i \in I$, with centers $x_i \in B$ and at mutual distances larger than $\eta r / 3$. Recall from Proposition 12.11 that

$$\int_{2B} |u - m_{2B} u| \leq Cr \int_{2B} |\nabla u| \leq C' A \qquad (11)$$

for $u \in E$, where we do not care about the dependence of C' (or the various constants C in the next few lines) on r. In particular, (11) says that $|m_{2B} u| \leq C$, since $m_B u = 0$ by definition of E. Then $\int_{2B} |u| \leq C$ for $u \in E$, by (11) again, and

$$|m_{B_i} u| \leq C(\eta) \text{ for } i \in I. \qquad (12)$$

Set $J = [-C(\eta), C(\eta)]$, choose a finite set $S \subset J$ which is η-dense in J, and then denote by Ξ the finite set of all functions $h : I \to S$. For each $h \in \Xi$, set

$$E(h) = \{u \in E \,;\, |m_{B_i} u - h(i)| \leq \eta \text{ for all } i \in I\}. \qquad (13)$$

Then $E \subset \bigcup_{h \in \Xi} E(h)$ by construction, hence it is enough to show that each $E(h)$ is contained in a ball of radius ε in $L^1(B)$.

So let u_1, $u_2 \in E(h)$ be given, and set $v = u_1 - u_2$. Proposition 12.11 says that

$$\int_{B_i} |v| \leq |m_{B_i} v|\, |B_i| + \int_{B_i} |v - m_{B_i} v| \leq 2\eta |B_i| + C_n \eta r \int_{B_i} |\nabla v|. \qquad (14)$$

Then we sum over $i \in I$ and use the fact that the B_i cover B, have bounded overlap, and are contained in $2B$, to get that

$$\|v\|_{L^1(B)} = \int_B |v| \leq \sum_{i \in I} \int_{B_i} |v| \leq C\eta |B| + C\eta r \int_{2B} |\nabla v| \leq C\eta < \varepsilon \qquad (15)$$

if η is small enough. This proves Lemma 10. $\square$

So we may assume that (9) holds for all $B \in \Delta$, and we want to show that $\{u_k\}$ converges to some limit u as in Definition 6. We can verify this independently on each connected component of Ω. So let a component $\mathcal{O}$ of $\Omega \backslash K$ be given. Choose a ball $D = D_{\mathcal{O}} \in \Delta$, with $3D \subset \mathcal{O}$, and set $a_k = m_D u_k$, at least for k large (so that $2D \subset \Omega_k$, say). We already know from (9) that $\{u_k - a_k\}$ converges in $L^1(D)$, but we want to establish convergence in $L^1(B)$ for every ball $B \in \Delta(\mathcal{O}) = \{B \in \Delta \,; 3B \subset \mathcal{O}\}$.

Let $B \in \Delta(\mathcal{O})$ be given. Choose a path $\gamma \subset \mathcal{O}$ from the center of D to the center of B. Then set $\tau = \frac{1}{10} \operatorname{Min} \big\{ \operatorname{diam} D, \operatorname{diam} B, \operatorname{dist}(\gamma, \mathbb{R}^n \setminus \mathcal{O}) \big\}$. Also choose a chain of points $x_j \in \gamma$, $0 \le j \le L$, such that x_0 is the center of D, x_L is the center of B, and $|x_j - x_{j-1}| \le \tau$ for $1 \le j \le L$. Finally set $B_0 = D$, $B_L = B$, and $B_j = B(x_j, 3\tau)$ for $0 < j < L$. Note that all the B_j lie in Δ, by definition of τ. By (9),

$$\{u_k - m_{B_j} u_k\} \text{ converges in } L^1(B_j). \tag{16}$$

For $1 \le j \le L$, the set $V_j = B_j \cap B_{j-1}$ has positive measure. Hence (16) (for j and $j - 1$, and integrated on V_j) says that $\{m_{B_j} u_k - m_{B_{j-1}} u_k\}$ converges. When we add this up, we get that

$$\{m_B u_k - m_D u_k\} = \{m_{B_L} u_k - m_{B_0} u_k\} \quad \text{converges.} \tag{17}$$

Since $m_D u_k = a_k$, (17) and (9) imply the convergence of $\{u_k - a_k\}$ in $L^1(B)$.

This holds for every $B \in \Delta(\mathcal{O})$, and hence (7) holds for every compact set $H \subset \mathcal{O}$ (cover H with finitely many balls in $\Delta(\mathcal{O})$). Thus $\{u_k\}$ converges, as in Definition 6. Proposition 8 follows. $\qquad \square$

Next we want to say a little more on u when $\{(u_k, K_k)\}$ converges to (u, K).

Proposition 18. *Let $\{\Omega_k\}$ and Ω be as above, and in particular assume that (1) holds. Let $\{(u_k, K_k)\}$ be a sequence of admissible pairs such that (2) and (3) hold. Assume in addition that $\{(u_k, K_k)\}$ converges to (u, K) (as in Definition 6). Then $u \in W^{1,2}_{\mathrm{loc}}(\Omega \setminus K)$, and also*

$$\int_{V \setminus K} |\nabla u|^p \le \liminf_{k \to +\infty} \int_{[V \cap \Omega_k] \setminus K_k} |\nabla u_k|^p \tag{19}$$

for every open set $V \subset \Omega$ and $1 \le p \le 2$.

We start with the easier case when V is connected and relatively compact in Ω. Thus $V \subset \Omega_k$ for k large enough, by (1), and (7) says that there is a sequence $\{a_k\}$ such that

$$\{u_k - a_k\} \text{ converges to } u \text{ in } L^1(V). \tag{20}$$

Note also that

$$\int_V |\nabla u_k|^2 \le C \tag{21}$$

for k large enough (so that $V \subset \Omega_k$ in particular), and where C may depend on V, but not on k. This is an easy consequence of (3): just cover V with finitely many small balls contained in Ω.

For each vector-valued test-function $\varphi \in \mathcal{C}_c^\infty(V, \mathbb{R}^n)$,

$$\langle \nabla u, \varphi \rangle = -\langle u, \operatorname{div} \varphi \rangle = -\int_V u \operatorname{div} \varphi$$

$$= -\lim_{k \to +\infty} \int_V (u_k - a_k) \operatorname{div} \varphi = \lim_{k \to +\infty} \int_V \nabla u_k \cdot \varphi \qquad (22)$$

by (20) and because $\nabla a_k = 0$. In particular,

$$|\langle \nabla u, \varphi \rangle| = \lim_{k \to +\infty} \left| \int_V \nabla u_k \cdot \varphi \right| \leq ||\varphi||_{L^2(V)} \liminf_{k \to +\infty} ||\nabla u_k||_{L^2(V)}. \qquad (23)$$

Hence $\nabla u \in L^2(V)$ (by (21) and the Riesz representation theorem), and also (19) holds for $p = 2$. Then the proof of (23) says that

$$|\langle \nabla u, \varphi \rangle| \leq ||\varphi||_{L^q(V)} \liminf_{k \to +\infty} ||\nabla u_k||_{L^p(V)}$$

for $\varphi \in \mathcal{C}_c^\infty(V, \mathbb{R}^n)$ and where q is the dual exponent, which gives (19) when $p < 2$.

This takes care of the case when V is connected and relatively compact in Ω. In general, we can write V as an increasing union of open sets W_ℓ which have a finite number of components and are relatively compact in Ω. Then

$$\int_{W_\ell \setminus K} |\nabla u|^p \leq \liminf_{k \to +\infty} \int_{W_\ell \setminus K_k} |\nabla u_k|^p \leq \liminf_{k \to +\infty} \int_{[V \cap \Omega_k] \setminus K_k} |\nabla u_k|^p \qquad (24)$$

for each ℓ, and (19) follows. This proves Proposition 18. $\qquad \square$

As was explained earlier, the convergence of the u_k is often better than suggested by Definition 6. Here is a simple example if this.

Proposition 25. *Let $\{\Omega_k\}$ and Ω be as above, and in particular assume that (1) holds. Let $\{(u_k, K_k)\}$ be a sequence of admissible pairs such that (2) holds, and assume that there is a constant C such that*

$$\int_{B(x,r) \setminus K_k} |\nabla u_k|^2 \leq C r^{n-1} \quad \text{when } B(x,r) \subset \Omega_k, \ r \leq 1, \text{ and } k \text{ is large enough.}$$

$$(26)$$

Finally assume that $\{(u_k, K_k)\}$ converges to (u, K) (as in Definition 6). Then for every connected component $\mathcal{O}$ of Ω and every compact set $H \subset \mathcal{O}$,

$$\{u_k - a_k\} \text{ converges to } u \text{ uniformly on } H. \qquad (27)$$

Here $\{a_k\}$ is the sequence associated to $\mathcal{O}$ in Definition 6.

Our hypothesis (26) could be weakened considerably, but it is natural because it will be satisfied if, for k large, (u_k, K_k) is a local quasiminimizer in Ω_k (topological or not), with constants r_0, M, and a that do not depend on k. See Lemma 18.19 for the estimates and Standard Assumption 18.14 for the assumptions. The conclusion also could be made stronger; as we shall see, the reason why we have uniform convergence is because we have uniform Hölder estimates on the u_k (away from K_k).

So let $\mathcal{O}$ be a connected component of Ω and H be a compact subset of $\mathcal{O}$. Let $\varepsilon > 0$ be given (as in the definition of uniform convergence). Cover H with a finite collection of balls B_l, so that each B_l has a radius $r_l < \tau$, where $\tau > 0$ will be chosen soon, and $2B_l \subset \Omega \setminus K$. It is enough to show that for each l,

$$|u_k(x) - a_k - u(x)| \le \varepsilon \quad \text{for } x \in B_l \tag{28}$$

as soon as k is large enough. So we fix l and observe that for k large, B_l is contained in $\Omega_k \setminus K_k$ (by (1) and Definition 4), so that if $\tau < 1$, (26) says that $u_k \in W^{1,2}(B_l)$ and

$$\int_{B(x,r)} |\nabla u_k|^2 \le C r^{n-1} \quad \text{whenever } B(x,r) \subset B_l. \tag{29}$$

Then Lemma 18.22 says that

$$|u_k(x) - u_k(y)| \le C|x - y|^{1/2} \quad \text{for } x, y \in B_l. \tag{30}$$

To be precise, Lemma 18.22 was stated under the assumptions that $(u_k, K_k) \in TRLQ(\Omega_k)$ and $B_l \subset \Omega_k \setminus K_k$, but these assumptions were only used to get (29) (through Lemma 18.19), and then we only used the Poincaré inequality in balls contained in B_l. Of course it is important here that the constant in (30) does not depend on k or our choice of balls B_l, but only on the constant in (26).

We now choose τ so small that since $|x - y| < 2r_l \le 2\tau$ for $x, y \in B_l$, (30) yields $|u_k(x) - u_k(y)| \le \varepsilon/3$ for $x, y \in B_l$ (and k large enough). The same choice of τ also gives that $|u(x) - u(y)| \le \varepsilon/3$ for $x, y \in B_l$. The simplest way to check this is probably to notice that (29) holds for u, by Proposition 18, so we can use the same proof as for u_k. But we could also manage without Proposition 18, because we know that the $u_k - a_k$ satisfy the desired estimate uniformly on k, and u is the limit of the $u_k - a_k$ in $L^1(B_l)$ (so that we could extract a subsequence that converges almost-everywhere). If we denote by m_k the mean value of $u_k - a_k - u$ on B_l, our estimates say that

$$|u_k(x) - a_k - u(x) - m_k| \le \frac{1}{|B_l|} \int_{B_l} |u_k(x) - a_k - u(x) - u_k(y) + a_k + u(y)| \, dy \le 2\varepsilon/3. \tag{31}$$

But m_k tends to 0 because $\{(u_k, K_k)\}$ converges to (u, K) (and by (7)), so (28) follows from (31). This completes our proof of Proposition 25. $\qquad\square$

Exercise 32. Complete the proof of Lemma 10. That is, deduce from the complete boundedness of E its relative compactness in $L^1(B)$. You may decide to show the sequential compactness first, i.e., that from each sequence in E we can extract a subsequence that converges in $L^1(B)$. If in trouble, see a book on general topology.

38   Limits of local almost-minimizers are local topological almost-minimizers

In this section we want to show that limits of coral local almost-minimizers (topological or not) are coral local topological almost-minimizers. We assume, as in Section 37, that $\{\Omega_k\}$ is a sequence of open sets in $\mathbb{R}^n$, and that Ω is an open set such that (37.1) holds. We are also given a sequence of local almost-minimizers (u_k, K_k), with the following uniform estimates. Let h be a gauge function. That is, h is nondecreasing and $h(r)$ tend to 0 when r tends to 0^+, as in (7.5). We assume that for each k,

$$(u_k, K_k) \text{ is a coral local topological almost-minimizer in } \Omega_k, \text{ with gauge function } h. \tag{1}$$

See Definitions 7.17 and 8.24.

Remark 2. The reader should not worry about the word "coral". Our assumption (1) is satisfied if (u_k, K_k) is a reduced local almost-minimizer, or a topologically reduced local topological almost-minimizer in Ω_k, with gauge function h. [See Definitions 7.15, 7.17, 8.1, and 8.7.] This comes from Remark 8.30 (with an even simpler proof in the nontopological case), but you can also prove it the long way, by saying that K_k is locally Ahlfors-regular (by Theorem 18.16), hence trivially coral.

Theorem 3. *Let $\{\Omega_k\}$ be a sequence of open sets in $\mathbb{R}^n$, Ω an open set such that (37.1) holds, and $\{(u_k, K_k)\}$ a sequence of local almost-minimizers in the Ω_k that satisfy (1) for some (fixed) gauge function h. Suppose in addition that $\{(u_k, K_k)\}$ converges to some pair (u, K), as in Definition 37.6. Then*

$$(u, K) \text{ is a coral local topological almost-minimizer in } \Omega \text{ with gauge function } h^+, \tag{4}$$

where we set $h^+(r) = \lim_{\rho \to r+} h(\rho)$ for $r > 0$.

Remarks 5. a) Theorem 3 is inspired of a result in [Bo]. The main differences are that we increased the ambient dimension and adapted the proof to almost-minimizers and topological almost-minimizers, but the general idea is the same. The generalization to higher dimensions and almost-minimizers was also done in [MaSo4] (before this and independently).

b) We shall see in the next section an analogue of Theorem 3 for limits of almost-minimizers (no longer local) near the boundary.

c) Even if the (u_k, K_k) are (plain) almost-minimizers, we only get that (u, K) is a topological almost-minimizer. For instance, (u_k, K_k) could be obtained from a single Mumford-Shah minimizer (v, G) by blow-ups at a point near which G is a nice C^1 curve. Then we can take $\Omega = \mathbb{R}^2$, K is a straight line, and u is constant on each of the two components of $\mathbb{R}^2 \setminus K$. Our definition of convergence does not pay any attention to the values of these two constants, and it is easy to see that if they are equal, (u, K) is not a local almost-minimizer, because we could just remove any piece of K. This is not merely a trick. First, there is no reasonable way to choose the two constants. But even if we did choose them different, (u, K) would still not be a local minimizer in the whole plane (as one would perhaps have expected). The point is that we could remove $K \cap B(0, R)$ for enormous values of R and modify u in $B(0, 2R)$ to ameliorate the jump, and we could do this with a loss of energy much smaller than $2R$.

If we only want to take limits of Mumford-Shah minimizers in a domain, for instance, then we should not use the definition of convergence of Section 37 (and we could get plain local almost-minimizers). The difference is that we have some control on the size of the functions u_k themselves, so we can take a subsequence for which the u_k converge, and do things as suggested in Section 36. It is perhaps a little unfortunate that we do not include a full statement on the convergence of Mumford-Shah minimizers in a domain, but the reader may find such statements in other sources (such as [AFP3] or [MoSo2]), and also the proof below works, with some simplifications. See Exercise 107.

The point of our definition of convergence (and the reason why we only get topological almost-minimizers) is that we do not want to assume any control on the size of the functions u_k, but only on their gradients. Our typical example is blow-up sequences; see more about these in Section 40.

d) As we shall see in Remark 63, the pair (u, K) is not necessarily topologically reduced, even if the (u_k, K_k) were. This is not too sad, since the results of Part C and E apply to coral almost-minimizers. Also, there are circumstances where we can say that (u, K) is topologically reduced. See Remark 64.

e) Theorem 3 is easy to localize, since we may always restrict to a smaller open set Ω without losing (37.1). Recall that the restriction of a local topological almost-minimizer to a smaller domain is a local topological almost-minimizer with the same gauge function; see Exercise 7.58.

f) Theorem 3 extends to local topological almost-minimizers with linear distortion (as defined in Section 7.c); see Remark 65 below.

We would have liked to have a similar statement for quasiminimizers, but there are difficulties with this. The point is that the quasiminimality condition (7.20) does not seem to go to the limit well. See Remark 70.

On the other hand, we do not really need our gauge function h to tend to 0 at 0 for Theorem 3. That is, there is a constant $\varepsilon_0 > 0$ such that Theorem 3 still holds under the weaker assumption that $h(r) \leq \varepsilon_0$ for r small enough. Indeed the only places where we shall (apparently) use the fact that $h(r)$ tends to 0 are in Lemma 6 (to get the lower semicontinuity of Hausdorff measure) and Lemma 8 (for the local Ahlfors-regularity). In both places we just need $h(r)$ to be small enough for r small.

g) In Theorem 3 we assume that (1) holds with a fixed gauge function, but in applications, (1) will often be satisfied with gauge functions h_k that get better, and then we shall also get better results in (4). For instance, if $\{(u_k, K_k)\}$ is a blow-up sequence for an almost-minimizer, then (u, K) will be a local topological minimizer in $\mathbb{R}^n$ (i.e., with a gauge function equal to 0). See Proposition 40.9.

h) Our statement of Theorem 3 is far from optimal. For instance, it could be that each (u_k, K_k) minimizes a functional like the one in (7.26) (i.e., something like the Mumford-Shah functional, but with slowly varying coefficients). Then the proof below will show that the limit (u, K) (locally topologically) minimizes the same functional, but our statement says much less. We shall try to ameliorate this slightly and give a more general statement in Remark 73 and Corollary 78 below, but anyhow we may end up referring to the proof (rather than the statement) later.

One of the main points of the proof will be the following semicontinuity result.

Lemma 6. *Keep the notation and hypotheses of Theorem 3. Then*

$$H^{n-1}(K \cap V) \leq \liminf_{k \to +\infty} H^{n-1}(K_k \cap V) \tag{7}$$

for every open set $V \subset \Omega$.

Let $V \subset \Omega$ be open. We want to apply Theorem 35.4 to the sequence $\{K_k \cap V\}$ in the open set V. The sequence converges to $K \cap V$ (compare Definitions 37.4 and 34.4, and maybe see Exercise 34.19).

Let $a > 0$ be small, to be chosen very soon, and choose $r_0 > 0$ so small that $h(r_0) < a$. Then Remark 7.22 says that $(u_k, K_k) \in TRLQ(\Omega_k)$, with constants r_0, $M = 1$, and a. If a is small enough, we can apply the analogue of Theorem 25.1 in dimension n (see Remark 25.5) and get the concentration property for the K_k.

Set $r(x) = \frac{1}{2} \operatorname{Min}\{r_0, \operatorname{dist}(x, \mathbb{R}^n \setminus V)\}$ for each $x \in K \cap V$. Now let $\varepsilon > 0$ and $0 < r < r(x)$ be given, and let x_k be a point of K_k that lies as close to x as possible. Recall that $B(x, 2r) \subset \Omega$ and $\{x_k\}$ converges to x (because $\{K_k\}$ converges to K). Thus for k large enough, $B(x_k, r) \subset \Omega_k$, by (37.1), and we can apply Theorem 25.1 to the pair (x_k, r). We get a point $y_k \in K_k \cap B(x_k, r/3)$ and a radius ρ_k such that $C_7^{-1} r \leq \rho_k \leq r/3$ and $H^{n-1}(K_k \cap B(y_k, \rho_k)) \geq (1 - \varepsilon) H^{n-1}(\mathbb{R}^{n-1} \cap B(0, \rho_k)) = (1-\varepsilon)\omega_{n-1}\rho_k^{n-1}$, as in (25.2), (25.3), and (25.6). In other words, (35.3) holds. Since $B(y_k, t_k) \subset V \cap B(x, r)$, we have just checked that $\mathcal{H}(\varepsilon, C_\varepsilon)$ holds, with $C_\varepsilon = C_7$. Then Theorem 35.4 applies, (7) holds, and Lemma 6 follows. As announced in Remark 5.f, we do not need h to tend to 0, but just $h(r)$ to be small enough for r small. $\qquad\square$

Let us record another fact before we get to the main part of the proof.

Lemma 8. *There is a constant $C \geq 1$ such that*

$$C^{-1}r^{n-1} \leq H^{n-1}(K \cap B(x,r)) \leq Cr^{n-1} \tag{9}$$

for $x \in K$ and $0 < r < 1$ such that $B(x,2r) \subset \Omega$.

The constant C depends on n and the gauge function h. The main point of the proof will be that the sets K_k verify (9) with uniform estimates, by Theorem 18.1 or 18.16. So let x, r be as in the lemma. We claim that for each small $\delta > 0$,

we can cover $K \cap B(x,r)$ with less than $C\delta^{1-n}$ balls of radius δr, (10)

where C depends only on the local regularity constants for the K_k. Indeed let x, r, and $\delta < 1/2$ be given. Pick a maximal set $Y \subset K \cap B(x,r)$ whose points lie at mutual distances at least $\delta r/2$. Of course Y is finite, and for k large enough each $B(y, \delta r/4)$ contains a ball of radius $\delta r/5$ centered on K_k and whose double is contained in Ω_k. Then $H^{n-1}(K_k \cap B(y, \delta r/4)) \geq C^{-1}(\delta r/5)^{n-1}$ for $y \in Y$, by the uniform local Ahlfors-regularity of the K_k. All these sets are disjoint, and their total H^{n-1}-measure is less than Cr^{n-1}, by local Ahlfors-regularity again. So Y has at most $C\delta^{1-n}$ elements. Our claim (10) follows, since the balls $B(y, \delta r)$, $y \in Y$, cover $K \cap B(x,r)$.

The second inequality in (9) is an easy consequence of (10). Indeed, (10) implies that for δ small, $H^{n-1}_{2\delta r}(K \cap B(x,r)) \leq C\delta^{1-n}(\delta r)^{n-1}$, where $H^{n-1}_{2\delta r}$ is as in (2.7) and (2.8). The upper bound in (9) follows by letting δ tend to 0.

We shall not need the first inequality, so we only sketch its proof. The simplest is probably to consider the restrictions μ_k of H^{n-1} to the K_k. Note that $\mu_k(B(x,3r/2)) \leq Cr^{n-1}$ (by local Ahlfors-regularity of K_k again), so we can extract a subsequence for which the restriction of μ_k to $B(x,3r/2)$ converges weakly to some measure μ. Call K^* the support of μ in $B(x,3r/2)$. If $z \in K^*$, then any ball D centered at z meets K_k for k large, because $\mu(D/2) > 0$. Thus $z \in K$ (because $\{K_k\}$ converges to K). Conversely, if $z \in K \cap B(x,3r/2)$ and $t < r/10$, then for k large we can find $z_k \in K_k \cap B(z,t/2)$, and since $B(z_k, t/2) \subset \Omega_k$ for k large (by (37.1)), the local Ahlfors-regularity of K_k says that $\mu_k(B(z,t)) \geq \mu_k(B(z_k,t/2)) \geq C^{-1}t^{n-1}$. Thus $\mu(B(z,t) \geq C^{-1}t^{n-1}$ for $t < r/10$, and in particular $z \in K^*$. We also have that $\mu(B(z,t)) \leq Ct^{n-1}$, by the uniform local Ahlfors-regularity of the μ_k.

Altogether $K^* = K \cap B(x,3r/2)$, and $C^{-1}t^{n-1} \leq \mu(B(z,t)) \leq Ct^{n-1}$ for $z \in K \cap B(x,3r/2)$ and $t < r/2$. Now the proof of Lemma 18.11 (or Exercise 18.25) says that on $B(x,r)$, μ is equivalent to the restriction of H^{n-1} to K. Lemma 8 follows. □

The general principle for our proof of Theorem 3 is simple. We consider a topological competitor (u^*, K^*) for (u,K), as in Definitions 7.2 and 7.15, and use it to construct interesting competitors for the (u_k, K_k). Then we use the almost-minimality of (u_k, K_k) and get valuable estimates on (u^*, K^*).

So let (u^*, K^*) be a topological competitor for (u, K) in some ball $B_1 = B(x, r_1)$ such that $\overline{B}_1 \subset \Omega$. Let us already choose $r_2 > r_1$, pretty close to r_1, such that $\overline{B}(x, r_2) \subset \Omega$.

We need to choose rather carefully a radius $r \in (r_1, r_2)$ on which we shall do the main surgery of the construction. We shall use a small parameter τ, and construct competitors that depend on τ. At the end of the argument, τ will tend to zero, but for the moment it is fixed.

Cover $K' = K \cap \overline{B}(x, r_2)$ with balls $B(y, \tau r_1)$, $y \in Y$. Choose the centers y, $y \in Y$, on K', and at mutual distances larger than $\tau r_1/2$. Then Y has at most $C\tau^{1-n}$ points, by the proof of (10). Here and in the rest of the argument, constants C will be allowed to depend on x, r_1, r_2, and the distance from $B(x, r_2)$ to $\mathbb{R}^n \setminus \Omega$, but not on τ. We shall keep some dependence on r_1 in the estimates below, but this is just to keep track of the homogeneity and reassure the reader, and we shall not need the information. Consider

$$\Sigma_1 = \bigcup_{y \in Y} B(y, \tau r_1) \quad \text{and} \quad \Sigma_2 = \bigcup_{y \in Y} \overline{B}(y, 2\tau r_1). \tag{11}$$

Note that

$$K \cap \overline{B}(x, r_2) \subset \Sigma_1 \quad \text{and} \quad \text{dist}(y, \Sigma_1) \geq \tau r_1 \text{ for } y \in \overline{B}(x, r_2) \setminus \Sigma_2. \tag{12}$$

Our first conditions on r will be that $r_1 < r < (r_1 + r_2)/2$ and

$$H^{n-1}(Z) \leq C\tau r_1^{n-1}, \text{ where we set } Z = \partial B \cap \Sigma_2 \text{ and } B = B(x, r). \tag{13}$$

Note that

$$\int_{r_1 < r < (r_1+r_2)/2} H^{n-1}(\partial B(x, r) \cap \Sigma_2)) \, dr \tag{14}$$

$$\leq \sum_{y \in Y} \int_{r_1 < r < (r_1+r_2)/2} H^{n-1}(\partial B(x, r) \cap \overline{B}(y, 2\tau r_1) \, dr \leq C(\tau r_1)^n \, (\sharp Y) \leq C\tau r_1^n,$$

by Fubini. If we choose the constant C in (13) large enough (depending on $(r_2 - r_1)/r_1$ as well), half the radii $r \in (r_1, (r_1 + r_2)/2)$ will satisfy (13).

Before we state our second condition on r, we need to extract a subsequence or two. Note that $\overline{B}(x, r_2) \setminus \Sigma_1$ is a compact subset of $\Omega \setminus K$ (by (12)). Hence we can find a finite collection $\mathcal{O}_\ell$, $\ell \in L$, of connected components of $\Omega \setminus K$ such that

$$\overline{B}(x, r_2) \setminus \Sigma_1 \subset \bigcup_{\ell \in L} \mathcal{O}_\ell. \tag{15}$$

[Cover $\overline{B}(x, r_2) \setminus \Sigma_1$ with finitely many open balls contained in $\Omega \setminus K$.] The convergence of $\{u_k\}$ to u (as in Definition 37.6) tells us that for each $\ell \in L$, there is a sequence $\{a_{\ell,k}\}$ such that

$$\{u_k - a_{\ell,k}\} \text{ converges to } u \text{ in } L^1(\mathcal{O}_\ell \cap \overline{B}(x, r_2) \setminus \Sigma_1). \tag{16}$$

Since we may replace $\{(u_k, K_k)\}$ with any subsequence, we can assume that

$$\overline{B}(x, r_2) \subset \Omega_k \text{ and } \overline{B}(x, r_2) \setminus \Sigma_1 \subset \Omega_k \setminus K_k \tag{17}$$

for all k (and not just for k large enough). This is by (37.1), because $\overline{B}(x, r_2) \setminus \Sigma_1$ is a compact set in Ω that does not meet K (by (12)), and by the convergence of $\{K_k\}$ to K.

Let $\{\varepsilon_m\}$ be a sequence of extremely small numbers, to be chosen near the end of the argument. It may depend on τ in particular. Because of (16), we can replace $\{(u_k, K_k)\}$ with a subsequence for which

$$\sum_{\ell \in L} \int_{\mathcal{O}_\ell \cap \overline{B}(x,r_2) \setminus \Sigma_1} |u - u_k + a_{\ell,k}| \leq \varepsilon_k. \tag{18}$$

Let us choose a representative for each u_k (that is, defined everywhere). Our second condition on r is that for each k,

$$\sum_{\ell \in L} \int_{\mathcal{O}_\ell \cap \partial B \setminus \Sigma_1} |u - u_k + a_{\ell,k}| \leq C2^k \varepsilon_k. \tag{19}$$

If C is large enough (and we still omit to keep track of the dependence on r_1 and r_2), most radii in $(r_1, (r_1 + r_2)/2)$ satisfy (19). [We paid an extra 2^k to get all the conditions simultaneously.] For our third condition, note that $\overline{B}(x, r_2) \subset \Omega_k$ for all k (by (17)). Then

$$\int_{B(x,r_2) \setminus K_k} |\nabla u_k|^2 \leq Cr_2^{n-1}, \tag{20}$$

by Lemma 18.19. We want r to be such that

$$\int_{\partial B \setminus K_k} |\nabla u_k|^2 \leq C2^k \tag{21}$$

for all k, which is again easy to get if C in (21) is chosen large enough (depending on r_1 and r_2). We also make a similar requirement for u itself, i.e., that

$$\int_{\partial B \setminus K} |\nabla u|^2 \leq C. \tag{22}$$

This is possible, again by Fubini and Chebyshev, and because $\int_{\partial B(x,r_2) \setminus K} |\nabla u|^2 \leq Cr_2^{n-1}$, by (20) and the lower semicontinuity inequality (37.19).

Finally note that for almost every r and every k, the restriction of u_k to $\partial B \setminus K_k$ lies in $W_{\text{loc}}^{1,2}(\partial B \setminus K_k)$, its derivative in the tangential directions can be computed almost-everywhere on $\partial B \setminus K_k$ in terms of the restriction of ∇u_k (for which we choose a representative in advance), and $u_k(y)$ is the radial limit of u_k almost-everywhere on $\partial B \setminus K_k$. See Proposition 9.5 (for the absolute continuity on hyperplanes), Lemma 13.9 (for the radial limits), Proposition 11.4 (to allow the distortion from planes to spheres), and Lemma 18.22 for the continuity of u_k. The same things also hold for u itself (instead of u_k), with similar proofs as for the u_k.

We choose $r \in (r_1, (r_1 + r_2)/2)$ with all the properties above.

Now we want to use (u^*, K^*) to construct a competitor (u_k^*, K_k^*) for (u_k, K_k) in B. We take

$$K_k^* = [K^* \cap B] \cup Z \cup [K_k \setminus \overline{B}], \tag{23}$$

with $Z = \partial B \cap \Sigma_2$ as in (13) (see Figure 1). If we want to get a competitor in B, we have to keep

$$u_k^*(y) = u_k(y) \text{ for } y \in \Omega_k \setminus [K_k \cup \overline{B}]. \tag{24}$$

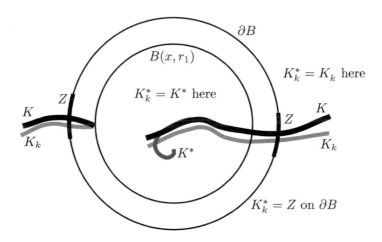

Figure 1

The values of u_k^* on $\partial B \setminus Z$ do not really matter (because u_k^* just needs to be defined almost-everywhere), but it is most reasonable to keep the formula (24) there, because this is the radial limit almost-everywhere on $\partial B \setminus Z$ of u_k^* and u_k, with access from outside.

We still need to define u_k^* in $B \setminus K^*$, and we shall do it separately on each connected component. So let V be a connected component of $B \setminus K^*$. Set $\partial(V) = \partial V \cap \partial B \setminus \Sigma_1$. If $\partial(V)$ is empty, we take $u_k^*(y) = 0$ on V. This will not cause trouble, because $\partial V \cap \partial B \setminus Z = \partial V \cap \partial B \setminus \Sigma_2 \subset \partial(V)$. Otherwise, first note that $\partial(V) \subset B(x, r_2) \setminus \Sigma_1 \subset \bigcup_{\ell \in L} \mathcal{O}_\ell$, by (15). We claim that

$$\partial(V) \text{ is contained in } \mathcal{O}_\ell \text{ for some } \ell \in L. \tag{25}$$

Let y_1, $y_2 \in \partial(V)$ be given. Let ℓ_1, ℓ_2 be such that $y_j \in \mathcal{O}_{\ell_j}$ for $j = 1, 2$. Choose $z_j \in V$ so close to y_j that $z_j \in \mathcal{O}_{\ell_j}$ and also $z_j \notin \overline{B}(x, r_1)$. Since both z_j lie in the same component V of $B \setminus K^*$, K^* does not separate them from each other in Ω. Since (u^*, K^*) is a topological competitor for (u, K) in $B(x, r_1)$ and the z_j lie out of $\overline{B}(x, r_1) \cup K$, K does not separate them from each other either. In other words, z_1 and z_2 lie in the same component of $\Omega \setminus K$, and $\ell_1 = \ell_2$. This proves (25).

The following lemma will help us find an extension to B of the restriction of $u_k - u$ to $\partial(V)$.

Lemma 26. *For each $\varepsilon > 0$, there is a constant $C(\varepsilon, \tau, r_1)$ such that if U is an open subset of ∂B and $h \subset W^{1,2}(U) \cap L^1(U)$, we can find a function $v \in W^{1,2}(B)$ such that*

$$\lim_{\rho \to 1^-} v(x + \rho(y - x)) = h(y) \text{ for almost-every } y \in U^\tau, \tag{27}$$

where we set

$$U^\tau = \{y \in U \,;\, \mathrm{dist}(y, \partial B \setminus U) \geq \tau r_1\}, \tag{28}$$

and also such that

$$\int_B |\nabla v|^2 \leq \varepsilon \int_U |\nabla h|^2 + C(\varepsilon, \tau, r_1) \left\{ \int_U |h| \right\}^2. \tag{29}$$

The verification is fairly easy, but we leave it for the end of this section, and in the mean time continue with our proof of Theorem 3. Let $\ell \in L$ be given, and apply the lemma to the function $h = u - u_k + a_{\ell,k}$ and the open set $U = \partial B \cap \mathcal{O}_\ell \setminus \overline{\Sigma}_1$. Note that $h \in L^1(U)$, with a norm less than $C2^k \varepsilon_k$, by (19), and $h \in W^{1,2}(U)$, with

$$\int_U |\nabla h|^2 \leq C2^k, \tag{30}$$

by (12), (17), (21) and (22). We take $\varepsilon = 2^{-2k}$ in Lemma 26; this gives a function $v_\ell \in W^{1,2}(B)$ such that

$$\int_B |\nabla v_\ell|^2 \leq C2^{-k} + C(k, \tau, r_1) \, ||h||_1 \leq C2^{-k} + C(k, \tau, r_1) \, \varepsilon_k^2. \tag{31}$$

Note that $\partial B \cap \mathcal{O}_\ell \setminus Z = \partial B \cap \mathcal{O}_\ell \setminus \Sigma_2 \subset U^\tau$, by (13) and (12). Hence

$$\lim_{\rho \to 1^-} v_\ell(x + \rho(y - x)) = u(y) - u_k(y) + a_{\ell,k} \text{ for almost-every } y \in \partial B \cap \mathcal{O}_\ell \setminus Z. \tag{32}$$

We are now ready to define u_k^* on $B \setminus K_k^* = B \setminus K^*$. For each component V of $B \setminus K_k^*$, let $\ell \in L$ be such that $\partial(V) \subset \mathcal{O}_\ell$, as in (25), and set

$$u_k^* = u^* - v_\ell + a_{\ell,k} \text{ on } V. \tag{33}$$

This complete our definition of (u_k^*, K_k^*).

Lemma 34. *The pair* (u_k^*, K_k^*) *is a topological competitor for* (u_k, K_k) *in* B.

First, K_k^* is closed, by (23), and because Z is closed (by (11) and (13)) and contains $\partial B \cap [K^* \cup K_k] = \partial B \cap [K \cup K_k]$ by (12), (17), and because (u^*, K^*) is a competitor for (u, K) in $B(x, r_1) \subset\subset B$. Next,

$$u_k^* \in W_{\text{loc}}^{1,2}(\Omega_k \setminus K_k^*). \tag{35}$$

Outside of $\overline{B}$, this comes from (24) and the fact that $u_k \in W_{\text{loc}}^{1,2}(\Omega_k \setminus K_k)$. Inside B, this comes from (33), and the facts that $K_k^* \cap B = K^* \cap B$, $u^* \in W_{\text{loc}}^{1,2}(\Omega \setminus K^*)$, and each v_ℓ lies in $W^{1,2}(B)$. We do not need to worry because u_k^* has different definitions on different components V of $B \setminus K_k^*$, by definition of $W_{\text{loc}}^{1,2}$.

We still have to check that for each $y \in \partial B \setminus Z$, $u_k^* \in W^{1,2}(D)$ for some small disk D centered at y. Note that y lies at some positive distance from Σ_2, hence if D is small enough, $D \setminus \partial B$ has two pieces, one contained in $\Omega_k \setminus K_k$ and the other one contained in some V as above. The function u_k^* has radial limits almost-everywhere on $\partial B \cap D$ from both sides, and these limits coincide by (32), (33), the fact that u and u^* coincide near ∂B, and our final conditions on r (see about six lines below (22)). This proves (35).

We just checked that (u_k^*, K_k^*) is a competitor for (u_k, K_k) in B (compare with Definition 7.2), but for Lemma 34 we still need to check the topological condition (7.16), i.e., that

$$\begin{aligned}&\text{if } y, z \in \Omega_k \setminus [K_k \cup \overline{B}] \text{ are separated by } K_k \text{ in } \Omega_k,\\&\qquad\text{then } K_k^* \text{ also separates them in } \Omega_k.\end{aligned} \tag{36}$$

Let us first check that for k large enough,

$$\begin{aligned}&\text{if } y, z \in \partial B \setminus Z \text{ are not separated by } K \text{ in } \Omega,\\&\qquad\text{then } K_k \text{ does not separate them in } \Omega_k.\end{aligned} \tag{37}$$

Let $y, z \in \partial B \setminus Z$ be given, and suppose that K does not separate them in Ω. Then there is an arc $\gamma \subset \Omega \setminus K$ that goes from y to z. Now $\gamma \subset \Omega_k \setminus K_k$ for k large enough, because γ is compact, by (37.1), and because (1) says that $\{K_k\}$ converges to K as in Definition 37.4. Thus K_k does not separate y from z for k large. This is almost (37), except that we have to be careful not let the notion of "k large" depend on y and z. Note that all the points of $\partial B \setminus Z$ lie at distance $\geq \tau r_1$ from Σ_1, by (12). Thus for each point $y \in \partial B \setminus Z$, $\partial B \cap B(y, \tau r_1)$ is contained in a single component of $\partial B \setminus \Sigma_1$. Then $\partial B \setminus Z$ only meets a finite number of components of $\partial B \setminus \Sigma_1$. Choose a point $y_i \in W_i \cap [\partial B \setminus Z]$ for each such component W_i of $\partial B \setminus \Sigma_1$, and then apply the argument above to the finite number of pairs (y_i, y_j) thus obtained. We get that for k large enough, if K does not separate y_i from y_j in Ω, then K_k does not separate them from each other in Ω_k. The same thing holds when we replace y_i with any other point of its component in $\partial B \setminus \Sigma_1$, because it is contained in $[\Omega \setminus K] \cap [\Omega_k \setminus K_k]$ by (12) and (17). Similarly we can replace y_j with any other point of its component in $\partial B \setminus \Sigma_1$. Thus we can reach all pairs (y, z) in $(\partial B \setminus Z)^2$, and (37) follows.

Now we turn to (36). Let y, $z \in \Omega_k \setminus [K_k \cup \overline{B}]$ be given, assume that K_k separates them in Ω_k but K_k^* does not, and let us try to find a contradiction. There is a curve $\gamma \subset \Omega_k \setminus K_k^*$ that goes from y to z. Since K_k^* and K_k coincide outside of $\overline{B}$, γ meets ∂B. Let y_1 be the first point of $\gamma \cap \partial B$ (when we go from y to z along γ). Then $y_1 \in \partial B \setminus Z$ (because $Z \subset K_k^*$, by (23)), and K_k still separates y_1 from z (because the arc of γ between y and y_1 lies in $\Omega_k \setminus [K_k^* \cup \overline{B}] = \Omega_k \setminus [K_k \cup \overline{B}]$ by (23)). Similarly, we can replace z with the last point z_1 of $\gamma \cap \partial B$, and K_k still separates y_1 from z_1 in Ω_k. See Figure 2.

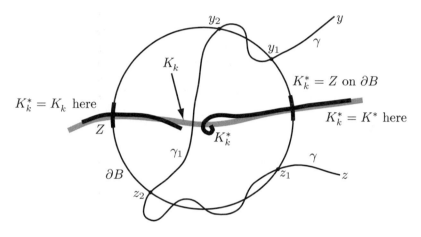

Figure 2

Call ∂_1 the set of points of $\gamma \cap \partial B \setminus Z$ that do not lie in the same connected component of $\Omega_k \setminus K_k$ as y_1. Note that $z_1 \in \partial_1$. Also, all the points of $B(y_1, \tau r_1)$ lie in the same component of $\Omega_k \setminus K_k$ as y_1, by (12), (17), and because $y_1 \in \partial B \setminus Z = \partial B \setminus \Sigma_2$. Thus $\operatorname{dist}(\partial_1, y_1) \geq \tau r_1$, and in particular there is a first point $z_2 \in \partial_1$ when you start from y_1 and run along γ. Note that the points of γ just before z_2 lie in the same component of $\Omega_k \setminus K_k$ as z_2. This is because $z_2 \in \partial B \setminus Z$ (since γ does not meet $Z \subset K_k^*$), and $\partial B \setminus Z$ lies at positive distance from K_k, by (12) and (17) again. Hence there is a small portion of γ, just before we hit z_2, which does not meet $\partial B \setminus Z$ (otherwise, z_2 would not be minimal). Then there is a last point $y_2 \in \gamma \cap \partial B \setminus Z$ before we hit z_2. (Note that y_1 is a candidate, so the set is not empty.) Now y_2 lies in the same component of $\Omega_k \setminus K_k$ as z_1, because otherwise y_2 would be a point of ∂_1 strictly before z_2.

Call γ_1 the part of γ between y_2 and z_2. It does not meet $\partial B \setminus Z$ by definition of y_2, hence it does not meet ∂B at all (because γ does not meet Z). It meets K_k because y_2 and z_2 lie in different components of $\Omega_k \setminus K_k$, but it does not meet K_k^* because γ does not meet K_k^*. Thus it cannot be contained in the complement of

$\overline{B}$, where K_k and K_k^* coincide by (23). Since it does not meet ∂B, it is contained in B. Hence $\gamma_1 \subset B \setminus K_k^* = B \setminus K^*$, by (23), and so K^* does not separate y_2 from z_2 in $\overline{B}$. Since these points lie in $\partial B \setminus K$ and (u^*, K^*) is a topological competitor for (u, K) in $B(x, r_1) \subset\subset B$, K does not separate y_2 from z_2 in Ω (see (7.16)). Then we use (37) and see that K_k does not separate y_2 from z_2 in Ω_k. This is not true: y_2 and z_2 lie in different components of $\Omega_k \setminus K_k$ by definition. This is the desired contradiction; (36) and Lemma 34 follow. $\square$

We may now start with the accounting. For each $\ell \in L$, denote by $V(\ell)$ the union of the components V of $B \setminus K^*$ such that $\partial(V) \subset \mathcal{O}_\ell$. Then (33) says that $\nabla u_k^* = \nabla u^* - \nabla v_\ell$ on $V(\ell)$. Set

$$A = \left\{ \int_{B \setminus K^*} |\nabla u^*|^2 \right\}^{1/2} \text{ and } \beta_k = \left\{ \sum_{\ell \in L} \int_{V(\ell)} |\nabla v_\ell|^2 \right\}^{1/2}. \tag{38}$$

Then

$$\int_{B \setminus K^*} |\nabla u_k^*|^2 \leq A^2 + 2\beta_k A + \beta_k^2 \tag{39}$$

by Cauchy–Schwarz, and hence

$$\int_{B \setminus K^*} |\nabla u_k^*|^2 \leq A^2 + C2^{-k} + C(k, \tau, r_1) \varepsilon_k^2 \tag{40}$$

by (31), and with constants that depend also on A and the total number of components $\mathcal{O}_\ell$, $\ell \in L$. We may now choose the sequence $\{\varepsilon_k\}$ so small that the right-hand side of (40) is less than $C2^{-k}$. [Notice that it is all right to let $\{\varepsilon_k\}$ depend on L, because L was known when we chose the subsequence that satisfies (18) and (19).] Then

$$\limsup_{k \to +\infty} \int_{B \setminus K^*} |\nabla u_k^*|^2 \leq A^2 = \int_{B \setminus K^*} |\nabla u^*|^2. \tag{41}$$

Consider the quantities

$$E = \int_{B(x,r_1) \setminus K} |\nabla u|^2 \text{ and } E^* = \int_{B(x,r_1) \setminus K^*} |\nabla u^*|^2, \tag{42}$$

and their analogues

$$E_k = \int_{B \setminus K_k} |\nabla u_k|^2 \text{ and } E_k^* = \int_{B \setminus K_k^*} |\nabla u_k^*|^2 \tag{43}$$

at the level k. Recall that u and u^* coincide almost-everywhere on $B \setminus B(x, r_1)$, so

$$E - E^* = \int_{B \setminus K} |\nabla u|^2 - \int_{B \setminus K^*} |\nabla u^*|^2 \leq \liminf_{k \to +\infty} \int_{B \setminus K_k} |\nabla u_k|^2 - \int_{B \setminus K^*} |\nabla u^*|^2$$

$$\leq \liminf_{k \to +\infty} \int_{B \setminus K_k} |\nabla u_k|^2 - \limsup_{k \to +\infty} \int_{B \setminus K^*} |\nabla u_k^*|^2 \leq \liminf_{k \to +\infty} \left[E_k - E_k^* \right] \tag{44}$$

by Proposition 37.18, (41), and (43). Then set

$$
\begin{aligned}
\Delta L &= H^{n-1}(K \cap \overline{B}(x,r_1)) - H^{n-1}(K^* \cap \overline{B}(x,r_1)) \\
&= H^{n-1}(K \cap B) - H^{n-1}(K^* \cap B) \\
&\le \liminf_{k \to +\infty} H^{n-1}(K_k \cap B) - H^{n-1}(K^* \cap B) \\
&\le \liminf_{k \to +\infty} \left\{ H^{n-1}(K_k \cap \overline{B}) - H^{n-1}(K^* \cap B) \right\} \\
&\le \liminf_{k \to +\infty} \left\{ H^{n-1}(K_k \cap \overline{B}) - H^{n-1}(K_k^* \cap \overline{B}) + H^{n-1}(Z) \right\} \\
&\le \liminf_{k \to +\infty} \left\{ H^{n-1}(K_k \cap \overline{B}) - H^{n-1}(K_k^* \cap \overline{B}) + C\tau r_1^{n-1} \right\}
\end{aligned}
\tag{45}
$$

because K and K^* coincide outside of $\overline{B}(x,r_1)$ (since (u^*, K^*) is a competitor for (u, K) in $B(x, r_1)$), by Lemma 6, (23), and (13). Set

$$
J_{x,r_1}(u, K) = \int_{\Omega \cap B(x,r_1) \setminus K} |\nabla u|^2 + H^{n-1}(K \cap \overline{B}(x,r_1)),
\tag{46}
$$

and define $J_{x,r_1}(u^*, K^*)$ similarly. Then

$$
\begin{aligned}
J_{x,r_1}(u, K) - J_{x,r_1}(u^*, K^*) &= [E - E^*] + \Delta L \\
&\le \liminf_{k \to +\infty} [E_k - E_k^*] + \liminf_{k \to +\infty} \left\{ H^{n-1}(K_k \cap \overline{B}) - H^{n-1}(K_k^* \cap \overline{B}) \right\} + C\tau r_1^{n-1} \\
&\le \liminf_{k \to +\infty} \left\{ J_{x,r}(u_k, K_k) - J_{x,r}(u_k^*, K_k^*) \right\} + C\tau r_1^{n-1} \le h(r) r_1^{n-1} + C\tau r_1^{n-1}
\end{aligned}
\tag{47}
$$

by (42), (45), (44), (43), a definition like (46), and (7.7), which we can apply because (u_k^*, K_k^*) is a topological competitor for (u_k, K_k) in B (by Lemma 34).

Notice that the left-hand side of (47) does not depend on our construction. Hence we can take r_2 as close as we want to r_1 and then τ as small as we want. Thus the left-hand side is no larger than $h^+(r_1) r_1^{n-1}$, where $h^+(r_1) = \lim_{r \to r_1^+} h(r)$ is as in the statement of Theorem 3. This proves the analogue of (7.7) for (u, K) and (u^*, K^*), with $h^+(r_1)$. So (u, K) is an almost-minimizer with gauge function h^+. It is coral, because it is locally Ahlfors-regular, by Lemma 8. Our proof of Theorem 3 is complete, modulo Lemma 26 which will be proved at the end of the section.   □

The following corollary of the proof will be useful when we deal with limits of minimizers. We put it before other general comments on Theorem 3 to make sure that the reader does not have too much time to forget the proof.

Corollary 48. *Keep the notation and hypotheses of Theorem 3, and suppose that* $\overline{B}(x, r) \subset \Omega$. *Then*

$$
\limsup_{k \to +\infty} H^{n-1}(K_k \cap B(x, r)) \le H^{n-1}(K \cap \overline{B}(x, r)) + h^+(r) r^{n-1}
\tag{49}
$$

and

$$\limsup_{k \to +\infty} \int_{B(x,r) \setminus K_k} |\nabla u_k|^2 \leq \int_{B(x,r) \setminus K} |\nabla u|^2 + h^+(r) r^{n-1}. \qquad (50)$$

Note that this goes in the opposite direction to the semicontinuity results in Lemma 6 and Proposition 37.18. The general idea is that if the inequality in one of these two estimates were very strict, we could use (u, K) to construct a very good competitor for the (u_k, K_k).

We start with the proof of (49), but of course the argument for (50) will be similar. Before we start for good, let us extract a first subsequence such that $H^{n-1}(K_k \cap B(x,r))$ (for the subsequence) is equal to the $\limsup$ in (49). It is enough to prove (49) for this subsequence, or even for any subsequence of this subsequence. In particular, let us already replace our subsequence with another one for which $\int_{B(x,r) \setminus K_k} |\nabla u_k|^2$ has a limit too (perhaps not equal to our initial $\limsup$ in (50), but we don't care). This does not cost much, and will simplify a little the discussion below.

Also, let us now call r_1 the radius in the present corollary, to get closer to the notation of the theorem.

Let us rapidly review some parts of the argument above. We started from any topological competitor (u^*, K^*) for (u, K) in a ball $B_1 = B(x, r_1)$. Then we chose $r_2 > r_1$, quite close to r_1, and selected a radius $r \in (r_1, r_2)$. Then we constructed, for each choice of a small parameter $\tau > 0$ and k in an extracted subsequence, a topological competitor (u_k^*, K_k^*) to (u_k, K_k) in $B = B(x, r)$ (see Lemma 34). Then we used our hypothesis that (u_k, K_k) is a topological quasiminimizer, and the lower semicontinuity of H^{n-1} and energy integrals to derive the desired inequalities. This is the chain of inequalities that leads to (47). If any of our two lower semicontinuity inequalities is strict, we can get an even better estimate. To make this precise, first extract a new subsequence such that $\int_{B(x,r) \setminus K_k} |\nabla u_k|^2$ and $H^{n-1}(K_k \cap B(x,r))$ have limits (to simplify the notation) and set

$$\delta_E(x,r) = \lim_{k \to +\infty} \int_{B(x,r) \setminus K_k} |\nabla u_k|^2 - \int_{B(x,r) \setminus K} |\nabla u|^2 \qquad (51)$$

and

$$\delta_L(x,r) = \lim_{k \to +\infty} H^{n-1}(K_k \cap B(x,r)) - H^{n-1}(K \cap B(x,r)). \qquad (52)$$

Both quantities are nonnegative, by Proposition 37.18 and Lemma 6. The proof of (44) also yields

$$E - E^* \leq \liminf_{k \to +\infty} [E_k - E_k^*] - \delta_E(x,r) \qquad (53)$$

(just change the first inequality). Similarly,

$$\Delta L \leq \liminf_{k \to +\infty} \left\{ H^{n-1}(K_k \cap \overline{B}) - H^{n-1}(K_k^* \cap \overline{B}) + C\tau r_1^{n-1} \right\} - \delta_L(x,r), \qquad (54)$$

by the proof of (45). [Again keep track of what we win at the first inequality.] Then (47) becomes

$$J_{x,r_1}(u, K) - J_{x,r_1}(u^*, K^*) \leq h(r)r^{n-1} + C\tau r_1^{n-1} - \delta_E(x,r) - \delta_L(x,r). \quad (55)$$

We may apply all this with $(u^*, K^*) = (u, K)$, which is obviously a topological competitor for itself in any ball $B(x, r_1)$ such that $\overline{B}(x, r_1)$. Then the left-hand side of (55) vanishes, and we get that

$$\delta_E(x,r) + \delta_L(x,r) \leq h(r)r^{n-1} + C\tau r_1^{n-1}. \quad (56)$$

This is not bad, but since we started with $B(x, r_1)$, we would prefer estimates with $\delta_E(x, r_1) + \delta_L(x, r_1)$. Set $A = B(x, r) \setminus \overline{B}(x, r_1)$; then

$$H^{n-1}(K \cap B(x,r)) = H^{n-1}(K \cap \overline{B}(x,r_1)) + H^{n-1}(K \cap A) \quad (57)$$
$$\leq H^{n-1}(K \cap \overline{B}(x,r_1)) + \liminf_{k \to +\infty} H^{n-1}(K_k \cap A)$$
$$= H^{n-1}(K \cap \partial B(x,r_1)) + \lim_{k \to +\infty} H^{n-1}(K_k \cap B(x,r_1))$$
$$- \delta_L(x,r_1) + \liminf_{k \to +\infty} H^{n-1}(K_k \cap A),$$

by Lemma 6, because A is open, and by the analogue of the definition (52) for r_1. Since

$$\lim_{k \to +\infty} H^{n-1}(K_k \cap B(x,r_1)) + \liminf_{k \to +\infty} H^{n-1}(K_k \cap A) \leq \liminf_{k \to +\infty} H^{n-1}(K_k \cap B(x,r)), \quad (58)$$

we get that

$$H^{n-1}(K \cap B(x,r)) \leq H^{n-1}(K \cap \partial B(x,r_1)) - \delta_L(x,r_1) + \liminf_{k \to +\infty} H^{n-1}(K_k \cap B(x,r)), \quad (59)$$

and then

$$-\delta_L(x,r) = H^{n-1}(K \cap B(x,r)) - \lim_{k \to +\infty} H^{n-1}(K_k \cap B(x,r)) \quad (60)$$
$$\leq H^{n-1}(K \cap \partial B(x,r_1)) - \delta_L(x,r_1),$$

by (52). Hence (56) says that

$$\delta_L(x,r_1) \leq H^{n-1}(K \cap \partial B(x,r_1)) + h(r)r^{n-1} + C\tau r_1^{n-1}. \quad (61)$$

This is what we get when we fix $r_2 > r_1$, take τ small, and apply the construction of Theorem 3. We may now let τ tend to zero, then let r_2 tend to r_1^+, and (61) yields

$$\lim_{k \to +\infty} H^{n-1}(K_k \cap B(x,r_1)) = H^{n-1}(K \cap B(x,r_1)) + \delta_L(x,r_1) \quad (62)$$
$$\leq H^{n-1}(K \cap \overline{B}(x,r_1)) + h^+(r)r^{n-1},$$

by (52). This proves (49).

For (50), we can proceed with the integrals exactly as we did with the Hausdorff measures, except that we use Proposition 37.18 instead of Lemma 6, and we do not need to worry about the contribution of $\partial B(x, r_1)$. This completes our proof of Corollary 48. $\square$

Remark 63. Unfortunately, the almost-minimizer (u, K) in Theorem 3 is not always topologically reduced. Here is a simple example. Consider the set $K = \partial B(0, 1) \subset \mathbb{R}^2$, and let u be a locally constant function on $\mathbb{R}^2 \setminus K$. Consider the gauge function h defined by $h(r) = \text{Min}(10r, 2\pi)$, say. Then (u, K) is a local topological almost-minimizer on $\mathbb{R}^2$, with gauge function h. Indeed consider a topological competitor for (u, K) in some ball $B = B(x, r)$. If $r \leq 1/2$, B does not contain a whole component of $\mathbb{R}^2 \setminus K$ and the best that we can do is replace $K \cap B$ with a line segment with the same endpoints. We win less than $10r^2$ this way. In all cases we cannot win more than 2π (by removing the whole circle).

Suppose in addition that the constant values of u on the two components of $\mathbb{R}^2 \setminus K$ are different. Then (u, K) is topologically reduced (as in Definition 8.7), and even reduced (as in Definition 8.1).

Now consider the constant sequence equal to (u, K). With our definitions, it converges to any pair (v, K), provided that we take v constant on each component of $\mathbb{R}^2 \setminus K$. In particular, we can take $v = 0$, and for this choice (v, K) is not topologically reduced. The reader may argue that this is because we did not choose a notion of convergence that is sufficiently selective, but this is not the case. We can take a sequence of functions u_n that are locally constant, with different values for the constants (so that each (u_n, K) is a reduced local topological almost-minimizer), but for which $\{u_n\}$ converges to $v = 0$. Thus the problem is rather that the notion of reduced pair is very unstable.

Let us also say a few words about strong local topological almost-minimizers (as in Definition 8.21). If K is the unit circle and u locally constant, as before, (u, K) is still a strong local topological almost-minimizer, as long as $h(1) > 2\pi$ and we choose the constant values of u on the two components of $\mathbb{R}^2 \setminus K$ sufficiently far apart. [This affirmation would need to be checked, but it is easy to believe.] But with our definition of limits, the limit may still be given by the same K and $u = 0$. Then it is no longer a strong local topological almost-minimizer. Of course this is due to the fact that strong local topological almost-minimizers are not stable under adding constants to u in a component of $\mathbb{R}^2 \setminus K$. It could be that this problem is relatively easy to fix, but it seems just as simple to content ourselves with coral almost-minimizers.

Remark 64. Yet there are cases when we can say that (u, K) in Theorem 3 is reduced. If our pairs (u_k, K_k) are local almost-minimizers (topological or not) with gauge functions h_k that tend to 0, then Theorem 3 says that (u, K) is a local almost-minimizer with gauge function $h = 0$. [See for instance the first lines of the proof of Proposition 40.9, or the argument below (60.6), for details about this.] In this case, (u, K) is easily seen to be reduced (because it is locally Ahlfors-regular, and any topological competitor $(\widetilde{u}, \widetilde{K})$ with a strictly smaller $\widetilde{K}$ and an extension $\widetilde{u}$ of u would do strictly better on the comparison ball. See Remark 8.31 for more details.

Remark 65. Theorem 3 also works with local topological almost-minimizers with linear distortion (as defined near Definition 7.45). That is, if we replace (1) with the assumption that the pairs (u_k, K_k) are coral local topological almost-minimizers with linear distortion, with fixed constant L in (7.46) and gauge function h, then we get that (u, K) is a coral local topological almost-minimizer with linear distortion at most L, with the same gauge function h^+ as before.

Fortunately, most of the proof stay the same, but a few modifications are needed. First, we need to know that

$$H^{n-1}(A(V \cap K)) \leq \liminf_{K \to +\infty} H^{n-1}(A(V \cap K_k)) \tag{66}$$

for all bijective affine mappings A and all open sets $V \subset \Omega$. The proof is the same as for Lemma 6, except that we now use the uniform concentration property for the sets $A(K_k)$ (with A fixed). See Remarks 25.28 and 75.18.

We start the argument as before. Given a competitor (u^*, K^*) for (u, K), we construct the same competitors (u_k^*, K_k^*), and only change the estimates at the end of the argument. We know from Lemma 34 that (u_k^*, K_k^*) is a topological competitor for (u_k, K_k) in $B = B(x, r)$. The definition of almost-minimizers with linear distortion gives two matrices $A_{1,k} = A_{1,k}(x, r)$ and $A_{2,k} = A_{2,k}(x, r)$ such that (7.46) and (7.47) hold, and we can replace our initial sequence with a sub-sequence, so that $\{A_{1,k}\}$ and $\{A_{2,k}\}$ converge to invertible matrices A_1 and A_2. Note that A_1 and A_2 still satisfy (7.46), with the same L. The proof of (41) yields

$$\limsup_{k \to +\infty} \int_{B \setminus K^*} |A_2 \nabla u_k^*|^2 \leq \int_{B \setminus K^*} |A_2 \nabla u^*|^2. \tag{67}$$

Thus, if we set

$$J_{x,r_1}(u, K) = \int_{\Omega \cap B(x,r_1) \setminus K} |A_2 \nabla u|^2 + H^{n-1}(A_1(K \cap \overline{B}(x, r_1))) \tag{68}$$

as in (46), and similarly for $J_{x,r_1}(u^*, K^*)$, (66) and (67) allow us to prove that

$$J_{x,r_1}(u, K) - J_{x,r_1}(u^*, K^*) \leq h(r)r^{n-1} + C\tau r_1^{n-1} \tag{69}$$

as we did for (47). We may conclude as before (i.e., by letting r tend to a given r_1 and τ tend to 0 in the construction). We should just be a little careful, because we do not want our matrices A_1 and A_2 to depend on the specific competitor (u^*, K^*). So, even before we consider (u^*, K^*), we select a first sequence of radii $s > r_1$ that tends to r_1. For each choice of s and k, we have matrices $A_{1,k,s}$ and $A_{2,k,s}$, and we preventively extract a subsequence so that they converge to matrices $A_{1,s}$ and $A_{1,s}$. Then we replace our sequence of radii s with a subsequence for which the $A_{j,s}$ converge to limits A_j, and these are the matrices A_1 and A_2 that we associate to (u, K) as in Definition 7.45. For the verification, we do as above, except that when we apply Lemma 34, we just use the fact that (u_k^*, K_k^*) is a topological competitor for (u_k, K_k) in $B(x, s)$, where s is the smallest element of our sequence such that $r \leq s$. Then we use the matrices $A_{1,k,s}$ and $A_{2,k,s}$ instead of $A_{1,k}$ and $A_{2,k}$ before. This leads to the same estimates, except that we need to replace $h(r)$ with $h(s)$ in (69). The local topological almost-minimality with linear distortion of (u, K) follows, as announced. $\qquad\square$

Remark 70. The analogue of Theorem 3 for quasiminimizers seems harder to get. The point is that we should be able to get (7.20) for (u, K) and a competitor (u^*, K^*) from the same thing with the (u_k, K_k) and the (u_k^*, K_k^*) that we construct. More precisely, we get an estimate like

$$H^{n-1}(K_k \setminus K_k^*) \leq M H^{n-1}(K_k^* \setminus K_k) + \delta E_k + ar^{n-1}, \tag{71}$$

and we wish to get something like

$$H^{n-1}(K \setminus K^*) \leq M H^{n-1}(K^* \setminus K) + \delta E + ar^{n-1}. \tag{72}$$

The situations where (72) will be hard to get are typically when K^* is almost contained in K, because then the right-hand side of (72) is small. However, if we don't pay attention and use the same construction as in Theorem 3, the set K_k^*, which coincides with K^* in B (by (23)), could have a big piece that lies outside of K_k (recall that $K_k \cap B$ may be almost disjoint from $K \cap B$, even if it is very close to it). If this is the case, the right-hand side of (71) is pretty large, (71) does not mean much, and (72) will not be easy to deduce from (71).

What a nice argument should do is try to construct another competitor (u_k^*, K_k^*) for which K_k^* is almost contained in K_k (and not K), and then use the quasiminimality of (u_k, K_k) with that competitor. This may be hard to do, and it is not too clear that there is a point. The reader may consult [Da8], where an argument of this type is done with quasiminimal sets; the author does not know whether something similar works here.

Remark 73. There are many ways in which Theorem 3 can be generalized. Let us try to indicate a typical direction in which this can be done, without being too specific. Let $\{\Omega_k\}$, Ω, $\{(u_k, K_k)\}$ be as usual. In particular, assume that (37.1) holds, that $(u_k, K_k) \in \mathcal{A}(\Omega_k)$ (as in (37.2)), and also that $\{(u_k, K_k)\}$ converges to some (u, K) as in Definition 37.6. We want to replace the almost-minimality assumptions (1) or (2) with slightly more general ones. For instance, we give ourselves bounded positive functions f_k and g_k on each Ω_k and a continuous gauge function h, and we assume that

$$\int_{B(x,r)\setminus K_k} |\nabla u_k(y)|^2 f_k(y)dy + \int_{K_k \cap \overline{B}(x,r)} g_k(y)dH^{n-1}(y)$$

$$\leq \int_{B(x,r)\setminus K_k^*} |\nabla u_k^*(y)|^2 f_k(y)dy + \int_{K_k^* \cap \overline{B}(x,r)} g_k(y)dH^{n-1}(y) + h(r)r^{n-1} \quad (74)$$

whenever $\overline{B}(x,r) \subset \Omega_k$, and (u_k^*, K_k^*) is a competitor for (u_k, K_k) in $B(x,r)$ (plain or topological, there are two possible statements). We also assume that the pair (u_k, K_k) is reduced, as in Definition 8.1 (in the plain case) or 8.7 (in the topological case), or that it is just coral, as in Definition 8.24.

Finally, we suppose that $\{f_k\}$ and $\{g_k\}$ converge to some limits f and g on Ω; to simplify the discussion, let us assume that

$$\{f_k\} \text{ and } \{g_k\} \text{ converge uniformly on compact subsets of } \Omega \quad (75)$$

and that

$$f \text{ and } g \text{ are lower semicontinuous.} \quad (76)$$

This last means that $\{y \in \Omega; f(y) > \lambda\}$ is open for every λ, and similarly for g. We would like to know whether

$$\int_{B(x,r)\setminus K} |\nabla u(y)|^2 f(y)dy + \int_{K \cap \overline{B}(x,r)} g(y)dH^{n-1}(y) \quad (77)$$

$$\leq \int_{B(x,r)\setminus K^*} |\nabla u^*(y)|^2 f(y)dy + \int_{K^* \cap \overline{B}(x,r)} g(y)dH^{n-1}(y) + h(r)r^{n-1}$$

when $\overline{B}(x,r) \subset \Omega$ and (u^*, K^*) is a topological competitor for (u, K) in $B(x,r)$. Let us first give a statement; later on we shall try to convince the reader that the hypotheses are reasonable and not too hard to check.

Corollary 78. *Let us suppose, in addition to the assumptions above, that for each ball $B = B(x,r)$ such that $\overline{B} \subset \Omega$, there is a constant $C(B) \geq 0$ such that*

$$H^{n-1}(B \cap K_k) + \int_{B\setminus K_k} |\nabla u_k|^2 \leq C(B) \text{ for } k \text{ large enough} \quad (79)$$

and a function $\eta_B(\delta)$ such that $\lim_{\delta \to 0+} \eta_B(\delta) = 0$ and

$$K \cap B \text{ can be covered by less than } \eta_B(\delta)\,\delta^{-n} \text{ balls of radius } \delta r \quad (80)$$

for every $\delta < 1$. Let us also assume that for each relatively compact open set $V \subset\subset \Omega$,

$$H^{n-1}(K \cap V) \leq \liminf_{k \to +\infty} H^{n-1}(K_k \cap V). \quad (81)$$

Then (77) holds for all balls $B(x,r)$ such that $\overline{B}(x,r) \subset \Omega$ and all topological competitors (u^*, K^*) for (u, K) in $B(x,r)$.

Corollary 78 will be a simple consequence of our proof of Theorem 3; we essentially listed the properties that we used. In particular, (79) implies (37.3), which is needed for Proposition 37.18. Similarly, (81) is a substitute for Lemma 6, and (80) replaces (10), which is the only part of Lemma 8 that we used. The construction of the competitor (u_k^*, K_k^*) is the same as before, and we only need to be careful about the last estimates (41)–(47). We now set

$$E = \int_{B(x,r_1)\setminus K} |\nabla u(y)|^2 f(y) dy, \qquad E^* = \int_{B(x,r_1)\setminus K^*} |\nabla u^*(y)|^2 f(y) dy, \qquad (82)$$

$$E_k = \int_{B\setminus K_k} |\nabla u_k(y)|^2 f_k(y) dy, \text{ and } E_k^* = \int_{B\setminus K_k^*} |\nabla u_k^*(y)|^2 f_k(y) dy. \qquad (83)$$

We still have that

$$E - E^* = \int_{B\setminus K} |\nabla u(y)|^2 f(y) dy - \int_{B\setminus K^*} |\nabla u^*(y)|^2 f(y) dy \qquad (84)$$

because the two pairs coincide outside of $B(x,r_1)$. Then we need to check that

$$\int_{B\setminus K} |\nabla u(y)|^2 f(y) dy \le \liminf_{k\to+\infty} \int_{B\setminus K_k} |\nabla u_k(y)|^2 f(y) dy. \qquad (85)$$

We know from Proposition 37.18 that

$$\int_{V\setminus K} |\nabla u|^2 \le \liminf_{k\to+\infty} \int_{V\setminus K_k} |\nabla u_k|^2 \qquad (86)$$

for any open set $V \subset B$. Hence (85) would hold immediately if f was a linear combination with positive coefficients of characteristic functions of open sets. This is where we use the lower semicontinuity of f. For each integer m and $y \in \Omega$, denote by $h_m(y)$ the integer such that $h_m(y)-1 < 2^m f(y) \le h_m(y)$. Then h_m is the sum of the characteristic functions of the open sets $\{y : 2^m f(y) > j\}, j \ge 0$. Note that the sum is finite because f is bounded on B. [Recall that $B \subset\subset \Omega$, each f_k is bounded, and use (75).] Thus (85) holds for each function $[h_m - 1]^+ = \text{Max}(h_m - 1, 0)$, and so

$$\int_{B\setminus K} |\nabla u(y)|^2\, 2^{-m}[h_m(y) - 1]^+\, dy \le \liminf_{k\to+\infty} \int_{B\setminus K_k} |\nabla u_k(y)|^2\, 2^{-m}[h_m(y) - 1]^+\, dy$$

$$\le \liminf_{k\to+\infty} \int_{B\setminus K_k} |\nabla u_k(y)|^2 f(y)\, dy, \qquad (87)$$

because $2^{-m}[h_m(y) - 1]^+ \le f(y)$ everywhere. Now (85) follows from (87), because $\int_{B\setminus K} |\nabla u(y)|^2 < +\infty$ (by (79) and (86)) and the $2^{-m} h_m$ converge uniformly to f. Then

$$E - E^* \le \liminf_{k\to+\infty} \int_{B\setminus K_k} |\nabla u_k(y)|^2 f(y) dy - \int_{B\setminus K^*} |\nabla u^*(y)|^2 f(y) dy, \qquad (88)$$

by (84) and (85). We still have that

$$\limsup_{k\to+\infty} \int_{B\setminus K^*} |\nabla u_k^*(y)|^2 f(y) dy \le \int_{B\setminus K^*} |\nabla u^*(y)|^2 f(y) dy \qquad (89)$$

by the same argument as for (41). Then

$$E - E^* \le \liminf_{k\to+\infty} \int_{B\setminus K_k} |\nabla u_k(y)|^2 f(y) dy - \limsup_{k\to+\infty} \int_{B\setminus K^*} |\nabla u_k^*(y)|^2 f(y) dy$$

$$\le \liminf_{k\to+\infty} \left\{ \int_{B\setminus K_k} |\nabla u_k(y)|^2 f_k(y) dy - \int_{B\setminus K^*} |\nabla u_k^*(y)|^2 f_k(y) dy \right\} \qquad (90)$$

$$= \liminf_{k\to+\infty} [E_k - E_k^*]$$

by (88), (89), the uniform convergence of $\{f_k\}$ to f on B, the uniform bound (79), a corresponding uniform bound for the $\int_{B\setminus K^*} |\nabla u_k^*(y)|^2$ that comes from (89), and our definitions (83). This is the analogue of (44) above. Next we need to estimate

$$\Delta L = \int_{K\cap \overline{B}(x,r_1)} g(y) \, dH^{n-1}(y) - \int_{K^*\cap \overline{B}(x,r_1)} g(y) \, dH^{n-1}(y)$$

$$= \int_{K\cap B} g(y) \, dH^{n-1}(y) - \int_{K^*\cap B} g(y) \, dH^{n-1}(y) \qquad (91)$$

(because K and K^* coincide outside of $\overline{B}(x,r_1)$). Note that

$$\int_{K\cap B} g(y) \, dH^{n-1}(y) \le \liminf_{k\to+\infty} \int_{K_k\cap B} g(y) \, dH^{n-1}(y), \qquad (92)$$

because of (81). The argument uses the lower semicontinuity of g and the uniform bounds from (79), and it is the same as for (85). Then

$$\Delta L \le \liminf_{k\to+\infty} \int_{K_k\cap \overline{B}} g(y) \, dH^{n-1}(y) - \int_{K^*\cap B} g(y) \, dH^{n-1}(y) \qquad (93)$$

$$\le \liminf_{k\to+\infty} \left\{ \int_{K_k\cap \overline{B}} g(y) \, dH^{n-1}(y) - \int_{K_k^*\cap \overline{B}} g(y) \, dH^{n-1}(y) + CH^{n-1}(Z) \right\}$$

$$\le \liminf_{k\to+\infty} \left\{ \int_{K_k\cap \overline{B}} g(y) \, dH^{n-1}(y) - \int_{K_k^*\cap \overline{B}} g(y) \, dH^{n-1}(y) \right\} + C(B)\, \tau\, r_1^{n-1}$$

$$\le \liminf_{k\to+\infty} \left\{ \int_{K_k\cap \overline{B}} g_k(y) \, dH^{n-1}(y) - \int_{K_k^*\cap \overline{B}} g_k(y) \, dH^{n-1}(y) \right\} + C(B)\, \tau\, r_1^{n-1}$$

because $B \subset \overline{B}$, by (91) and (92), by (23), because g is bounded on B, by (13), and finally because the g_k converge to g uniformly on B and we have uniform bounds on $H^{n-1}(K_k \cap \overline{B})$ and $H^{n-1}(K_k^* \cap \overline{B})$. The first ones come from (79), and then we use the fact that $H^{n-1}(K_k^* \cap \overline{B}) \le H^{n-1}(K_k \cap \overline{B}) + C(B)\,\tau\, r_1^{n-1}$. Now (93) is the analogue of (45), and we may conclude as for (47). This completes our proof of Corollary 78. □

Let us say rapidly why the hypotheses in Corollary 78 should often be easy to check. There should never be serious difficulty with the verification of (79). For instance, if we assume that for each compact set $H \subset \Omega$, there is a constant $M_H \geq 1$ such that

$$1 \leq f_k(y) \leq M_H \text{ and } 1 \leq g_k(y) \leq M_H \text{ when } y \in \Omega_k, \qquad (94)$$

then (79) follows by the usual truncation argument (add a sphere ∂B to K_k, remove $K_k \cap B$, replace u_k with a constant in B, and compare).

The main difficulty a priori is with (80) and (81). For (80), it seems that the most natural way to get it is to prove that the sets K_k are locally Ahlfors-regular in the corresponding domains Ω_k, with estimates that do not depend on k (but may even diverge badly when one gets close to $\partial\Omega$), and then use the proof of (10) above. This would give (80), with $\eta_B(\delta) = C(B)\,\delta$.

For instance, it would be enough to require that for each relatively compact open set $\Omega' \subset\subset \Omega$ there are constants M and r_0 such that for k large,

$$(u_k, K_k) \text{ is a (topologically) reduced local}$$
$$\text{(topological) } (r_0, M, a)\text{-quasiminimizer in } \Omega', \qquad (95)$$

where the constant $a > 0$ is chosen so small (depending only on n) that we can apply Theorem 18.16 and hence get the local Ahlfors-regularity of K_k. [With the notation of Standard Assumption 18.14, (95) means that $(u_k, K_k) \in TRLQ(\Omega')$, with constants (r_0, M, a).] Of course it is important that the constants r_0 and M do not depend on k.

Note that (95) could be fairly easy to check, much easier than the conclusion of Corollary 78. Even getting (1), with a fixed gauge function, would often be easier. The point is that the gauge function with which we would prove (80) could be much larger than the one we use in (74) and (77).

For instance, a typical use of Corollary 78 would be to show that if (74) holds with functions h that tend to 0, then (77) holds with $h = 0$. This should not prevent us from using less precise information like (95), or even (1) with a fairly large $\widetilde{h}$, just to get (80) and (81).

The situation for (81) is similar. The easiest way seems to use the concentration property. Here again, if we assume (95) or (1), we can apply Remark 25.5 or Theorem 25.1 to get constants C_ε for which the $\mathcal{H}(\varepsilon, C_\varepsilon)$ hold, and (81) follows from Theorem 35.4 as in the proof of Lemma 6.

Remarks 96. As before, the limit (u, K) in Corollary 78 is not necessarily topologically reduced, but it will be coral if the K_k are (uniformly in k) locally Ahlfors-regular.

Corollary 78 should still hold when we replace the energy term in (74) with integrals like $\int_{B(x,r)\setminus K_k} |F_k(y)\nabla u_k(y)|^2\,dy$, where the F_k are continuous, matrix-valued functions and the F_k are uniformly bounded and invertible, say, and converge uniformly to a matrix-valued limit F. One could even try to distort the

H^{n-1}-term in a similar way, but the formula is harder to write down. The arguments should be the same as for Corollary 78, with the modifications suggested in Remark 65.

Our proof of Corollary 48 should also go through in the context of Remark 73 and Corollary 78, with minor modifications.

The reader should be aware that we did not check these two extensions seriously.

It looks quite believable that there is a proof of Theorem 3 and its variants that uses the SBV setting, but I don't know of any such proof in the literature.

Proof of Lemma 26. Let $B = B(x, r)$ (with $r \leq r_1$), $U \subset \partial B$, $h \in W^{1,2}(U) \cap L^1(U)$, and ε be as in the statement. Since we do not care about the way the constant $C(\varepsilon, \tau, r_1)$ will depend on r_1, we may assume that B is the unit ball. It will be convenient to set

$$\alpha = \left\{ \int_U |\nabla h|^2 \right\}^{1/2} \quad \text{and} \quad \beta = \int_U |h|. \tag{97}$$

Our first task will be to define a function g on ∂B that coincides with h in a neighborhood of U^τ (defined in (28)). Set $U^* = \{ y \in \partial B \, ; \, \text{dist}(y, \partial B \setminus U) \geq \tau/2 \}$. Thus $\text{dist}(U^\tau, \partial B \setminus U^*) \geq \tau/2$, by definition of U^τ and the triangle inequality. We can find a smooth function φ on ∂B such that $\varphi(y) = 1$ in a neighborhood of U^τ, $\varphi(y) = 0$ outside of U^*, $0 \leq \varphi(y) \leq 1$ everywhere, and $|\nabla \varphi| \leq 10\tau^{-1}$ on ∂B. We take

$$g(y) = \varphi(y) \, h(y) \tag{98}$$

on U and $g(y) = 0$ on $\partial B \setminus U$. There will be no matching problem, since $\varphi(y) = 0$ on a neighborhood of $\partial B \setminus U$. We shall need to know that $h \in L^2(U^*)$ and estimate its norm. Choose an exponent $p > 2$, smaller than the Sobolev exponent $\frac{2n}{n-2}$ if $n > 2$. We want to estimate $\|h\|_{L^p(U^*)}$ first. Let D be any ball in ∂B of the form $D = \partial B \cap B(z, \tau/2)$, with a center $z \in U^*$. Note that $h \in W^{1,2}(D)$ because $D \subset U$. Proposition 12.23 (plus a smooth change of variables) tells us that $h \in L^p(D)$ and

$$\|h - m_D h\|_{L^p(D)} \leq C(\tau) \left\{ \int_D |\nabla h|^2 \right\}^{1/2} \leq C(\tau)\alpha. \tag{99}$$

Here $m_D h$ is the mean value of h on D (maybe computed against some smooth weight coming from the change of variable above, but this does not matter). Note that $|m_D h| \leq C(\tau)\beta$ trivially. We can cover U^* with less than $C\tau^{1-n}$ balls D as above, and apply (99) to each of them. This yields

$$\|h\|_{L^p(U^*)} \leq C(\tau)(\alpha + \beta). \tag{100}$$

Now we write $\frac{1}{2} = \frac{s}{p} + (1 - s)$ for some $s \in (0, 1)$ and use Hölder and (100) to get that

$$\|h\|_{L^2(U^*)} \leq \|h\|_p^s \, \|h\|_1^{1-s} \leq C(\tau)(\alpha + \beta)^s \beta^{1-s}. \tag{101}$$

We may now return to (98) and compute derivatives. Note that $g \in W^{1,2}(\partial B)$ and $\nabla g = \varphi \nabla h + h \nabla \varphi$, because $h \in W^{1,2}(U)$, $\varphi = 0$ out of U^*, and by definition. [See Exercise 9.15 concerning the product of $h \in W^{1,2}$ with a smooth function.] Then

$$\left\{ \int_{\partial B} |\nabla g|^2 \right\}^{1/2} \leq \left\{ \int_U |\nabla h|^2 \right\}^{1/2} + C\tau^{-1} \|h\|_{L^2(U^*)} \tag{102}$$

$$\leq \alpha + C(\tau)(\alpha + \beta)^s \beta^{1-s}.$$

Now we want to extend g to B. Let us *not* use the harmonic extension and set

$$v(ty) = \psi(t)g(y) \text{ for } y \in \partial B \text{ and } 0 \leq t < 1, \tag{103}$$

where ψ is a smooth function on $[0,1]$ that we want to choose now. Let η be a small positive parameter, to be chosen soon, and take ψ supported on $[1 - \eta, 1]$, with $\psi(t) = 1$ in a neighborhood of 1, and $0 \leq \psi(t) \leq 1$ and $|\psi'(t)| \leq C\eta^{-1}$ everywhere. Obviously, v has a radial limit equal to $g(y)$ everywhere on ∂B, and this limit is $h(y)$ on a neighborhood of U, by (98). This takes care of (27).

It is easy to check that $v \in W^{1,2}(B)$ and $|\nabla v(ty)|^2 \leq |\psi(t)|^2 |\nabla g(y)|^2 + |\psi'(t)|^2 |g(y)|^2$ almost-everywhere; you may use Lemmas 11.1 and 11.2, for instance, or directly the definitions, a polar change of coordinates, and Fubini. [See Exercise 9.16.] Then

$$\int_B |\nabla v(ty)|^2 \leq \int_B |\psi(t)|^2 |\nabla g(y)|^2 + \int_B |\psi'(t)|^2 |g(y)|^2$$

$$\leq \eta \int_{\partial B} |\nabla g(y)|^2 + C\eta^{-2} \int_{\partial B} |g(y)|^2 \tag{104}$$

$$\leq \eta \{ \alpha + C(\tau)(\alpha + \beta)^s \beta^{1-s} \}^2 + C(\tau) \eta^{-2} (\alpha + \beta)^{2s} \beta^{2-2s}$$

$$\leq 2\eta \alpha^2 + C(\tau, \eta) (\alpha + \beta)^{2s} \beta^{2-2s} \leq 3\eta \alpha^2 + C(\tau, \eta, s) \beta^2,$$

by (102), (101), and a brutal estimate. So we may take $\eta = \varepsilon/3$, and (29) follows from (104). This proves Lemma 26. $\square$

Exercise 105. Complete the proof of Lemma 8.

Exercise 106. Theorem 3 gives another (more complicated) proof of the fact that when (u, K) is a (topologically reduced) almost-minimizer on Ω and we add a constant to u on a component of $\Omega \setminus K$, we get another almost-minimizer with the same (continuous) gauge function. [See Exercise 7.57.] Check that the ingredients in the proof of Exercise 7.57 are present in the proof of Theorem 3.

Exercise 107. Let $\Omega \subset \mathbb{R}^n$ be given, suppose that $\{f_k\}$, $\{g_k\}$, and $\{h_k\}$ are sequences of bounded functions on Ω, and that there is a constant $M \geq 1$ such that

$$M^{-1} \leq f_k(x) \leq M, \; M^{-1} \leq g_k(x) \leq M,$$

$$\text{and } |h_k(x)| \leq M \text{ for } x \in \Omega \text{ and } k \geq 0. \tag{108}$$

Assume that $\{h_k\}$ converges to some function h almost-everywhere on Ω (or weakly if you wish), and that $\{f_k\}$ converges to f and $\{g_k\}$ converges to g, in both cases uniformly on compact subsets of Ω. Also assume that f and g are lower semicontinuous. Finally, let $(u_k, K_k) \in \mathcal{A}$ be a reduced (or coral) minimizer of

$$J_k(u, K) = \int_K f_k(x) \, dH^{n-1}(x) + \int_{\Omega \backslash K} |\nabla u(x)|^2 g_k(x) \, dx + \int_{\Omega \backslash K} |u(x) - h_k(x)|^2 dx.$$
(109)

Suppose that the sets K_k converge (as in Section 34) to a limit K (a closed set in Ω), and that the functions u_k converge to a limit u (say, weakly in $L^2(\Omega)$). Show that (u, K) minimizes the functional J defined as in (109), but with f_k, g_k, h_k replaced with f, g, h. Hint: extract a subsequence so that the ∇u_k converge weakly, and copy the arguments in Sections 37 and 38. This may take a while, but there should be no bad surprise. Some things get easier, too, because you won't have to glue competitors (like in most of this section) or worry about the topological condition. Also, convergence is easier to deal with.

Exercise 110. Give the correct statement (and proof) of an extension of Corollary 58 in the context of Remark 73 and Corollary 78.

39   Limits of almost-minimizers up to the boundary

In this section we want to extend Theorem 38.3 to limits of almost-minimizers all the way up to the boundary of Ω. As we shall see, this is mostly a matter of checking that the proof goes through. For the results of this section, the reader may also consult [MaSo4] for an earlier and only slightly different approach. The limiting properties of energy and surface are possibly better treated there too.

So far we only considered the local behavior of almost- and quasiminimizers, and we shall continue after this section, but it is reasonable to consider this extension before we forget entirely about Theorem 38.3. Later on, we shall return to the boundary behavior of minimizers and almost-minimizers, and Theorems 10 and 34 below may be useful in such a context.

Let us first recall our definition of almost-minimizers and topological almost-minimizers. We are given an open set Ω and a gauge function $h : (0, +\infty) \to [0, +\infty)$ such that

$$h \text{ is nondecreasing and } \lim_{r \to 0+} h(r) = 0. \tag{1}$$

[Actually, we shall not really need that $\lim_{r \to 0+} h(r) = 0$, but $h(r)$ should be small enough for r small.]

An <u>almost-minimizer</u> in Ω is an admissible pair $(u, K) \in \mathcal{A} = \mathcal{A}(\Omega)$ (see (7.1)) such that

$$
\int_{\Omega \cap B(x,r) \setminus K} |\nabla u|^2 + H^{n-1}(K \cap \overline{B}(x,r))
$$
$$
\leq \int_{\Omega \cap B(x,r) \setminus \widetilde{K}} |\nabla \widetilde{u}|^2 + H^{n-1}(\widetilde{K} \cap \overline{B}(x,r)) + h(r) r^{n-1} \tag{2}
$$

for all balls $B = B(x,r)$ and all competitors $(\widetilde{u}, \widetilde{K})$ for (u, K) in B. [See Definition 7.6.] The definition of competitor for (u, K) in B is the same as before (see Definition 7.2), but we also allow balls B that are not contained in Ω.

A <u>topological almost-minimizer</u> is a pair $(u, K) \in \mathcal{A}$ such that (2) holds for all balls $B(x,r)$ (that meet Ω, say) and all topological competitors $(\widetilde{u}, \widetilde{K})$ for (u, K) in B. We keep the same definition of a topological competitor for (u, K) in B as in the local case: it is a competitor for (u, K) in B such that

if $y, z \in \Omega \setminus [K \cup \overline{B}(x,r)]$ are separated by K in Ω,

then $\widetilde{K}$ also separates them in Ω. $\tag{3}$

See Definitions 7.15 and 7.17. One may consider different definitions; as we said just after Definition 7.17, we chose this one mainly because it looks simplest. Of course, our main examples of almost-minimizers are minimizers of the Mumford-Shah or similar functionals, as in Proposition 7.8. And probably our main examples of topological almost-minimizers will be blow-up limits of almost-minimizers. See Theorem 10.

So the only difference in the definitions with the local (topological) almost-minimizers studied so far is that we also allow balls B that touch or cross the boundary $\partial\Omega$. In particular, almost-minimizers (respectively, topological almost-minimizers) are local almost-minimizers (respectively, local topological almost-minimizers). But this time we expect some amount of regularity near the boundary.

With almost-minimizers, we can keep the same notion of reduction as before (see Definition 8.1), and there is no special difficulty because the notion is local. For topological almost-minimizers, we may probably define topological reduction as in Definition 8.7, but since we do not want to check the existence of equivalent topologically reduced topological almost-minimizer, we shall not bother to do so. Instead we shall consider coral almost-minimizers, as in Definition 8.24. There is no additional difficulty with this, since the notion is local, and the main point is still to get local Ahlfors-regularity.

Remark 4. Suppose we already know that (u, K) is a local almost-minimizer in Ω (respectively, a topological one), with gauge function h, and we want to show that it is an almost-minimizer in Ω (respectively, a topological one), with gauge function $h(2r)$. Then it is enough to prove (2) for all balls B centered on $\partial\Omega$

and all competitors for (u, K) in B (respectively, all topological competitors for (u, K) in B).

Indeed, we already know (2) when $B \subset\subset \Omega$, so it is enough to consider balls B such that $\overline{B}$ meets $\partial\Omega$. Now let $(\widetilde{u}, \widetilde{K})$ be a (topological) competitor for (u, K) in B, and let B' be a ball centered on $\partial\Omega$, with a radius at most twice the radius of B, and that contains B. It is easy to see that $(\widetilde{u}, \widetilde{K})$ is also a (topological) competitor for (u, K) in B', so (2) holds for B', hence also for B with the larger $h(2r)$. $\square$

For this section we need to consider a sequence of open sets Ω_k that really converges to Ω. So we keep our assumption (37.1) that

$$\text{for each compact set } H \subset \Omega, \ H \subset \Omega_k \text{ for } k \text{ large enough,} \qquad (5)$$

but we shall also require Ω to be sufficiently large. We shall have two different statements, Theorems 10 and 34. In the first one, we shall simply require that

$$\Omega_k \subset \Omega \ \text{ for } k \text{ large enough,} \qquad (6)$$

which will give a simpler proof. Our convergence assumption in the second one will be more natural, but there will be a slightly stronger regularity assumption on $\partial\Omega$.

Now let $\{(u_k, K_k)\}$ be a sequence of almost-minimizers. More precisely, assume that

$$(u_k, K_k) \text{ is a reduced or coral almost-minimizer in } \Omega_k, \text{ with gauge function } h, \qquad (7)$$

or merely that

$$(u_k, K_k) \text{ is a coral topological almost-minimizer in } \Omega_k, \text{ with gauge function } h, \qquad (8)$$

where it is important to keep the same function h for all k. We want to know that if $\{(u_k, K_k)\}$ converges to a pair (u, K), then (u, K) is a topological almost-minimizer in Ω.

Here is our first (very weak) assumption on $\partial\Omega$; for each ball $B(x, r)$ centered on $\partial\Omega$, there is a function $\eta_B(\tau)$, defined on $(0, 1)$ and that tends to 0 with τ, such that for every $\tau \in (0, 1)$,

$$\partial\Omega \cap B(x, r) \text{ can be covered with fewer than } \eta_B(\tau)\tau^{-n} \text{ balls of radius } \tau r. \qquad (9)$$

For instance, if $\partial\Omega$ is contained in an Ahlfors-regular set of any dimension $d < n$, (9) holds with $\eta(\tau) = C\tau^{n-d}$.

Theorem 10. *Let Ω, h, $\{\Omega_k\}$, and $\{(u_k, K_k)\}$ be as above. That is, assume that (1), (5), (6), (7) or (8), and (9) hold. Finally assume that $\{(u_k, K_k)\}$ converges to the pair (u, K), as in Definition 37.6. Then (u, K) is a coral topological almost-minimizer in Ω, with gauge function $h(4r)$.*

Here again, we do not get more than a topological almost-minimizer, even if the (u_k, K_k) were plain almost-minimizers. If we want a plain almost-minimizer, we should use a different notion of convergence (and get some control on the sizes of the u_k). See Remark 38.5.c.

As we said above, we do not really need our gauge function h to tend to 0 at 0, but it needs to have a small enough limit at 0.

Theorem 10 can be generalized in the spirit of Remark 38.73 and Corollary 38.78 (i.e., to almost-minimizers of slightly different functionals), and it also holds with almost-minimizers with linear distortion; see Remark 43 below.

The proof of Theorem 10 will follow the same lines as for Theorem 38.3 above. We are given a topological competitor (u^*, K^*) for (u, K) in a ball $B(x, r_1)$ and we want to prove (the analogue of) (2) with $h(4r_1)$. We already know about the case when $B(x, r_1) \subset\subset \Omega$, by Theorem 38.3. So it is enough to restrict to the case when $x \in \partial\Omega$, and prove (2) with $h(2r_1)$ (see Remark 4).

As before, we use a small parameter τ. We first apply (9) to the ball $B(x, 2r_1)$ to cover $B(x, 2r_1) \cap \partial\Omega$ with less than $\eta_{B(x,2r_1)}(\tau)\tau^{-n}$ open balls B_j of radius $2\tau r_1$. Since we shall not care about how $\eta_{B(x,2r_1)}(\tau)$ depends on x and r_1, let us just call it $\eta(\tau)$. Set $\Sigma_4 = \bigcup_j 4B_j$. Thus $|\Sigma_4| \leq C \eta(\tau) r_1^n$, so we can choose a first radius $r_2 \in (r_1, \frac{3}{2}r_1)$ such that

$$H^{n-1}(\Sigma_4 \cap \partial B(x, r_2)) \leq C \eta(\tau) r_1^{n-1}. \tag{11}$$

The argument is the same as for (38.13) and (38.14). We want to choose $r \in (r_2, r_2 + \tau r_1)$, but, as before, we shall need to be careful. Note that if we set

$$\Sigma_3 = \bigcup_j 3\overline{B}_j, \quad B = B(x, r), \quad \text{and} \quad Z_0 = \partial B \cap \Sigma_3, \tag{12}$$

then for any radius $r \in (r_2, r_2 + \tau r_1)$, the radial projection of Z_0 on $\partial B(x, r_2)$ will be contained in $\Sigma_4 \cap \partial B(x, r_2)$, and consequently

$$H^{n-1}(Z_0) \leq C\eta(\tau) r_1^{n-1}. \tag{13}$$

Next set

$$\Sigma_2 = \bigcup_j 2B_j \quad \text{and} \quad H = \Omega \cap \overline{B}(x, 7r_1/4) \setminus \Sigma_2. \tag{14}$$

Note that H is a compact subset of Ω, because the B_j cover $B(x, 2r_1) \cap \partial\Omega$. Then $K \cap H$ also is compact, and our proof even gives that

$$\text{dist}(K \cap H, \partial\Omega) \geq \tau r_1. \tag{15}$$

Let τ_1 be another small parameter, much smaller than τ, and cover $K \cap H$ with open balls of radius $\tau_1 r_1$. Because of (15), and by the uniform local Ahlfors-regularity of the K_k in the Ω_k, we can do this with fewer than $C(\tau)\tau_1^{1-n}$ such balls D_i, $i \in I$. The argument is the same as for (38.10). We shall use the set

$$Z = Z_0 \cup \Big\{ \partial B \cap \Big[\bigcup_{i \in I} 2\overline{D}_i \Big] \Big\}, \tag{16}$$

and we want to select r so that

$$H^{n-1}(Z) \leq H^{n-1}(Z_0) + C(\tau)\, \tau_1\, r_1^{n-1} \leq C\, \eta(\tau)\, r_1^{n-1} + C(\tau)\, \tau_1\, r_1^{n-1}. \tag{17}$$

This is not a problem, the same argument as for (38.13) and (38.14) shows that most radii $r \in (r_2, r_2 + \tau r_1)$ satisfy (17) if we take $C(\tau)$ large enough. But let us not choose r yet.

First we want to have an analogue of (38.19). Set

$$H_1 = H \setminus \Big\{ \bigcup_{i \in I} 2D_i \Big\}. \tag{18}$$

This is a compact subset of Ω (by the comment after (14)), and it does not meet K (by definition of the D_i). Then it is contained in a finite number of connected components $\mathcal{O}_\ell$, $\ell \in L$, of $\Omega \setminus K$. For each ℓ, there is a sequence $\{a_{\ell,k}\}$ such that

$$\{u_k - a_{\ell,k}\} \text{ converges to } u \text{ in } L^1(\mathcal{O}_\ell \cap H_1), \tag{19}$$

because $\{u_k\}$ converges to u as in Definition 37.6. We can extract a subsequence so that

$$H_1 \subset \Omega_k \setminus K_k \quad \text{for all } k \tag{20}$$

(and not just for k large) and

$$\sum_{\ell \in L} \int_{\mathcal{O}_\ell \cap H_1} |u - u_k + a_{\ell,k}| \leq \varepsilon_k, \tag{21}$$

where the values of the small numbers ε_k will be chosen near the end of the argument. Because of (21), most radii $r \in (r_2, r_2 + \tau r_1)$ are such that

$$\sum_{\ell \in L} \int_{\partial B \cap \mathcal{O}_\ell \cap H_1} |u - u_k + a_{\ell,k}| \leq C(\tau)\, 2^k\, \varepsilon_k \tag{22}$$

for each k. This is our analogue of (38.19). We still have some latitude to choose r. We use it to ensure that

$$\int_{\Omega_k \cap \partial B \setminus K_k} |\nabla u_k|^2 \leq C(\tau)\, 2^k\, r_1^{n-1} \quad \text{and} \quad \int_{\Omega \cap \partial B \setminus K} |\nabla u|^2 \leq C(\tau)\, r_1^{n-1}. \tag{23}$$

These can be obtained as (38.20) and (38.21), because $\int_{\Omega_k \cap B(x,2r) \setminus K_k} |\nabla u_k|^2 \leq C r_1^{n-1}$ (by the proof of Lemma 18.19; see the remark just after the statement) and (for the second part) by Proposition 37.18. As usual, we choose r with all the properties above, and also such that $u \in W_{\text{loc}}^{1,2}(\Omega \cap \partial B \setminus K)$, with a derivative given almost everywhere by the restriction of ∇u, such that u coincides almost-everywhere on $\Omega \cap \partial B \setminus K$ with its radial limits (from both sides), and such that we have similar properties for the u_k on $\Omega_k \cap \partial B \setminus K_k$. See a little below (38.22) for a longer discussion. This completes our choice of r; now Z_0 and Z are chosen too.

We may now define our competitor (u_k^*, K_k^*). We set

$$K_k^* = [K^* \cap B \cap \Omega_k] \cup [Z \cap \Omega_k] \cup [K_k \setminus \overline{B}], \qquad (24)$$

$$u_k^*(y) = u_k(y) \text{ for } y \in \Omega_k \setminus [K_k \cup B \cup Z], \qquad (25)$$

and try to use Lemma 38.26 to define u_k^* on $\Omega_k \cap B \setminus K^*$. This is where our hypothesis that $\Omega_k \subset \Omega$ for k large will be useful. Indeed it will be easier to define u_k^* on $\Omega \cap B \setminus K^*$, because u^* is already defined there. We proceed as in Section 38, component by component. Let V be a component of $\Omega \cap B \setminus K^*$. If ∂V does not meet $\partial B \setminus Z$, we can take $u_k^*(y) = 0$ on V. Otherwise, consider $\partial(V) = \partial V \cap \partial B \setminus Z$. Note first that $\Omega \cap \partial B \setminus Z_0 \subset H$ by (12) and (14), then $\Omega \cap \partial B \setminus Z \subset H_1$ by (16) and (18), so it is contained in the same finite union of components $\mathcal{O}_\ell$, $\ell \in L$, of $\Omega \setminus K$ that we considered in (19)–(22). We claim that

$$\partial(V) \text{ is contained in a single } \mathcal{O}_\ell, \ \ell \in L. \qquad (26)$$

The proof is the same as for (38.25). The starting point is that since $\Omega \cap \partial B \setminus Z \subset H_1$ for some compact set $H_1 \subset \Omega \setminus K$, it must lie in $\Omega \setminus K$, and at positive distance from $\partial \Omega \cup K$. [The proof even says that the distance is at least $\tau_1 r_1$, but we don't need this.] Then we use the fact that (u^*, K^*) is a topological competitor for (u, K) in $B(x, r_1)$, as before.

For each $\ell \in L$, we can define u_k^* (simultaneously) on all the components V such that $\partial(V) \subset \mathcal{O}_\ell$. We set $h_\ell = u - u_k + a_{\ell,k}$ on $\partial B \cap \mathcal{O}_\ell \setminus Z$, and choose an extension v_ℓ of h_ℓ to B, as in Lemma 38.26. Then we set

$$u_k^* = u^* - v_\ell + a_{\ell,k} \text{ on } V \qquad (27)$$

when $\partial(V) \subset \mathcal{O}_\ell$. This completes the definition of (u_k^*, K_k^*). We need to check that

$$(u_k^*, K_k^*) \text{ is a topological competitor for } (u_k, K_k) \text{ in } B. \qquad (28)$$

First K_k^* is closed in Ω_k. This is easy, and the proof is the same as for Lemma 38.34. Also, $u_k^* \in W_{\text{loc}}^{1,2}(\Omega_k \setminus K_k^*)$; it even has an extension in $W_{\text{loc}}^{1,2}([\Omega_k \setminus \overline{B}] \cup [\Omega \cap \overline{B} \setminus (K^* \cup Z)])$. Here again the argument is the same as for (38.35). It is also true that for k large enough,

$$\text{if } y, z \in \partial B \cap \Omega \setminus Z \text{ are not separated by } K \text{ in } \Omega, \qquad (29)$$
$$\text{then } K_k \text{ does not separate them in } \Omega_k,$$

with the same proof as for (38.37). We want to show that for k large,

$$\text{if } y, z \in \Omega_k \setminus [K_k \cup \overline{B}] \text{ are separated by } K_k \text{ in } \Omega_k, \qquad (30)$$
$$\text{then } K_k^* \text{ also separates them in } \Omega_k.$$

So we let $y, z \in \Omega_k \setminus [K_k \cup \overline{B}]$ be given and assume that K_k separates them in Ω_k, but not K_k^*. We use a curve $\gamma \subset \Omega_k \setminus K_k^*$ that goes from y to z. By the same sort of arguments as for (38.36), we can find a subarc γ_1 of γ that is contained in B, except for its endpoints y_2 and z_2 that lie on $\partial B \cap \Omega \setminus Z$, and such that these two points lie in different components of $\Omega_k \setminus K_k$ (just like our original points y and z). Then (29) says that

$$K \text{ separates } y_2 \text{ from } z_2 \text{ in } \Omega. \tag{31}$$

Since $\gamma_1 \subset \gamma \subset \Omega_k \setminus K_k^*$ and $\gamma_1 \subset B$ (except for y_2 and z_2), and since $\Omega_k \setminus K_k^* \subset \Omega \setminus K^*$ in B (by (6) and (24)), we see that K^* does not separate y_2 from z_2 in $\Omega \cap \overline{B}$, and even less in Ω. But (u^*, K^*) is a topological competitor for (u, K) in $B(x, r_1) \subset\subset B$, so K does not separate y_2 from z_2 in Ω either, a contradiction with (31). This proves (30), and (28) follows.

The accounting is the same as in the proof of Theorem 38.3, between (38.38) and (38.47), except that the right-hand side of (38.47) is now less than $h(r)r^{n-1} + C\eta(\tau)r_1^{n-1} + C(\tau)\tau_1 r_1^{n-1}$ (coming from (17)). We take τ very small, and τ_1 even smaller (depending on τ), and we get (2) for (u, K) (in $B(x, r_1)$, with $h(r_1)$ replaced by $h(2r_1) + \varepsilon$, and with ε as small as we want). This completes our proof of almost-minimality. We already know from the proof Theorem 38.3 that K is locally Ahlfors-regular, hence (u, K) is coral, so Theorem 10 follows. □

Now we want an analogue of Theorem 10 without the somewhat unnatural assumption (6). Instead we shall assume that for each compact set $H \subset \mathbb{R}^n$,

$$\lim_{k \to +\infty} \left\{ \sup \left\{ \operatorname{dist}(y, \Omega) \, ; \, y \in \Omega_k \cap H \right\} \right\} = 0. \tag{32}$$

But we shall also need an additional regularity hypothesis on Ω to make the proof work smoothly. We shall assume that $\partial \Omega$ is locally Lipschitz, and more precisely that there is a small radius $r_0 > 0$ such that for $0 < r < 2r_0$ and $x \in \partial\Omega$, we can find an orthonormal set of coordinates where

$$\partial\Omega \cap B(x, r) \text{ is the graph of some } 10^{-1}\text{-Lipschitz function,}$$
$$\text{with } \Omega \cap B(x, r) \text{ lying on one side only of } \partial\Omega \cap B(x, r). \tag{33}$$

See Remark 43 for a slightly weaker variant of (33).

Theorem 34. *Let $\{\Omega_k\}$ be a sequence of open sets, and assume that it converges to some open set Ω, as in (5) and (32), and that Ω satisfies the Lipschitz condition (33). Also let $\{(u_k, K_k)\}$ be a sequence of almost-minimizers in the Ω_k, and more precisely assume that (7) or (8) holds for some gauge function h (as in (1)). Finally assume that $\{(u_k, K_k)\}$ converges to the pair (u, K), as in Definition 37.6. Then (u, K) is a coral topological almost-minimizer in Ω, with a gauge function $\tilde{h}$ such that $\tilde{h}(r) = h(4r)$ for $r < r_0$.*

Fortunately, most of the proof of Theorem 10 goes through. We can easily reduce to the case of a topological competitor (u^*, K^*) for (u, K) in a ball $B(x, r_1)$ centered on $\partial\Omega$, and such that $r_1 < r_0$. Indeed, our statement allows us to take $\widetilde{h}(r)$ very large for $r \geq r_0$, so that there will be nothing to prove in that case. Note that when $r_1 \leq r_0$, (9) for $B(x, 2r_1)$ follows from (33), so we can start the argument as before. The only point where we need to be careful is when we try to define the function u_k^* on $\Omega_k \cap B \setminus K_k^*$. We cannot use the formula (27) directly, because it only defines a function $\widetilde{u}_k^*$ on $\Omega \cap B \setminus K^*$, and we no longer know that $\Omega_k \subset \Omega$. So we need to modify $\widetilde{u}_k^*$ slightly, and for this we shall use a deformation of B.

By (33) and because $r_1 \leq r_0$, $\Gamma = \partial\Omega \cap B(x, 2r_1)$ is the graph of some 10^{-1}-Lipschitz function, with $\Omega \cap B(x, 2r_1)$ on one side only. Note that, by (12) and (16), Z contains all the points of ∂B that lie at distance less than τr_1 from Γ. We claim that for k large enough, there is a C^1-diffeomorphism Φ_k of B, with the following properties:

$$\Phi_k(y) = y \text{ in a neighborhood of } \partial B \setminus Z, \tag{35}$$

$$\Phi_k(\Omega \cap B) \text{ contains } \Omega_k \cap B, \tag{36}$$

and

$$\|D\Phi_k - Id\|_\infty \leq C(r_1, \tau)\, d_k, \tag{37}$$

where $D\Phi_k$ denotes the differential of Φ_k, and $d_k = \sup\{\, \mathrm{dist}(y, \Omega)\,;\, y \in \Omega_k \cap \overline{B}\}$, which tends to zero by (32). See Figure 1. Our mapping Φ_k is easy to construct, the point is merely to move Γ up or down a little bit, so that it gets past all the points of $\Omega_k \cap B$. Maybe we should point out that since we have a good Lipschitz description of Ω in $B(x, 2r_1)$, which is substantially larger than B, we know that every point of $\Omega_k \cap B$ lies within $2d_k$ of $\Omega \cap B$ (and not merely Ω). We leave the other details of the construction of Φ_k. Now we set

$$K_k^* = [\Phi_k(K^* \cap B) \cap \Omega_k] \cup [Z \cap \Omega_k] \cup [K_k \setminus \overline{B}] \tag{38}$$

instead of (24), and

$$u_k^*(y) = \widetilde{u}_k^*(\Phi_k^{-1}(y)) \text{ for } y \in \Omega_k \cap B \setminus K_k^*, \tag{39}$$

which makes sense because $\Phi_k^{-1}(y) \in \Omega \cap B \setminus K$, by (36) and (38). Note that $u_k^* \in W^{1,2}(\Omega_k \cap B \setminus K_k^*)$ because $\widetilde{u}_k^* \in W^{1,2}(\Omega \cap B \setminus K^*)$, and in addition

$$\int_{\Omega_k \cap B \setminus K_k^*} |\nabla u_k^*|^2 \leq [1 + C(r_1, \tau)\, d_k] \int_{\Omega \cap B \setminus K^*} |\nabla \widetilde{u}_k^*|^2 \tag{40}$$

by (37) and a change of variables. The integral $\int_{\Omega \cap B \setminus K^*} |\nabla \widetilde{u}_k^*|^2$ can be estimated as for Theorems 38.3 and 10, and the contribution of the extra term with $C(r_1, \tau)\, d_k$ will become negligible when we let k tend to $+\infty$, by (32) (and our estimates on $\int_{\Omega \cap B \setminus K^*} |\nabla \widetilde{u}_k^*|^2$). Similarly,

$$H^{n-1}(K_k^* \cap B) \leq [1 + C(r_1, \tau)\, d_k]\, H^{n-1}(K^* \cap B) \tag{41}$$

by (37) and (38), and the extra term $C(r_1, \tau) \, d_k H^{n-1}(K^* \cap B)$ will not cause any harm either. At this point we can use our new function u_k^* as we did before, and the proof of Theorem 10" goes through. Note in particular that (35) allows a smooth transition between the definitions of u_k^* inside and outside B. This completes our proof of Theorem 34. $\square$

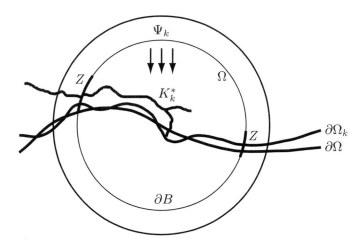

Figure 1

Remark 42. We can generalize Theorem 34 a little bit further. First, we can replace (33) with the assumption that $\Omega \cap B(x, r)$ is a cone centered at x; this will be used in Section 79, and the proof above works almost without modification (that is, Φ_k is easy to construct). Also, we can replace 10^{-1} in (33) with any positive constant λ, but then we need to adapt the geometry of the construction a little bit. For instance, we can do the argument with balls replaced with boxes centered at x, with sides parallel to the axes and height $(1 + \lambda)$ times their other sides. This way $\partial \Omega$ does not leave the boxes through the top or bottom, and we can find Φ_k as above.

Remark 43. We can generalize Theorems 10 and 34 as we did for Theorem 38.3 in Remark 38.73 and Corollary 38.78. These modifications are orthogonal to what we did in this section, and (the author thinks that) they work the same way here. The same sort of comment applies to Corollary 38.48.

Finally, we can replace "almost-minimizers" with "almost-minimizers with linear distortion $\leq L$" in the statements of Theorems 10 and 34, provided that we keep L fixed. The necessary modifications of the proof are the same as in Remark 38.65.

40   Blow-up limits

In this section we mostly make more explicit some of the results of Section 38 in the special case of limits of a local almost-minimizer under blow-up sequences. We consider a fixed domain $\Omega \subset \mathbb{R}^n$, and a local almost-minimizer (u, K) in Ω. We also give ourselves a sequence $\{y_k\}$ in Ω and a sequence $\{t_k\}$ of positive numbers. We consider the domains

$$\Omega_k = \frac{1}{t_k}(\Omega - y_k) \tag{1}$$

and the pairs (u_k, K_k) defined by

$$K_k = \frac{1}{t_k}(K - y_k), \tag{2}$$

$$u_k(x) = \frac{1}{\sqrt{t_k}}\, u(y_k + t_k x) \ \text{ for } x \in \Omega_k \setminus K_k. \tag{3}$$

Note that (3) makes sense, because $y_k + t_k x \in \Omega \setminus K$ when $x \in \Omega_k \setminus K_k$. Obviously, $(u_k, K_k) \in \mathcal{A}(\Omega_k)$ just because $(u, K) \in \mathcal{A}(\Omega)$; see (2.1) or (7.1) for the definition. Similarly, (u_k, K_k) is reduced (respectively, topologically reduced, or coral) as soon as (u, K) itself is reduced (respectively, topologically reduced, or coral). The main point of the normalization in (3) is that if we assume that

(u, K) is a local almost-minimizer in Ω, with gauge function $h(r)$,   (4)

as in Definition 7.14, then

(u_k, K_k) is a local almost-minimizer in Ω_k, with gauge function $h(t_k r)$,   (5)

and similarly for local topological almost-minimizers (as in Definition 7.17). Let us rapidly check this. Given a competitor $(\widetilde{u}_k, \widetilde{K}_k)$ to (u_k, K_k) in a ball $B = B(x, r) \subset\subset \Omega_k$, we can construct a competitor $(\widetilde{u}, \widetilde{K})$ to (u, K) by the inverse transformation to the one in (2) and (3). That is, we set $\widetilde{K} = y_k + t_k \widetilde{K}_k$ and $\widetilde{u}(z) = \sqrt{t_k}\, u(t_k^{-1}(z - y_k))$ for $z \in \Omega \setminus \widetilde{K}$. We get a competitor for (u, K) in $B^* = y_k + t_k B$, and we can apply (7.7) to it. Note that if we set

$$\Delta L = H^{n-1}(K \cap \overline{B}^*) - H^{n-1}(\widetilde{K} \cap \overline{B}^*) \ \text{ and}$$
$$\Delta_k L = H^{n-1}(K_k \cap \overline{B}) - H^{n-1}(\widetilde{K}_k \cap \overline{B}), \tag{6}$$

then $\Delta L = t_k^{n-1} \Delta_k L$. Similarly, $\Delta E = t_k^{n-1} \Delta_k E$, where

$$\Delta E = \int_{B^* \setminus K} |\nabla u|^2 - \int_{B^* \setminus \widetilde{K}} |\nabla \widetilde{u}|^2 \ \text{ and}$$

$$\Delta_k E = \int_{B \setminus K_k} |\nabla u_k|^2 - \int_{B \setminus \widetilde{K}_k} |\nabla \widetilde{u}_k|^2. \tag{7}$$

[We get a contribution of t_k^n from the Jacobian, and t_k^{-1} comes from the square of the size of the derivative.] Now (7.7) says that $\Delta E + \Delta L \leq h(t_k r)[t_k r]^{n-1}$, which yields $\Delta_k E + \Delta_k L \leq h(t_k r) r^{n-1}$, as needed for (5). The story for topological almost-minimizers is the same.

We shall assume here that

$$\{t_k\} \text{ tends to } 0 \quad \text{and} \quad \lim_{k \to +\infty} t_k^{-1} \operatorname{dist}(y_k, \mathbb{R}^n \setminus \Omega) = +\infty. \tag{8}$$

These conditions ensure that Ω_k converges to $\mathbb{R}^n$ (i.e., that every compact set in $\mathbb{R}^n$ is contained in Ω_k for k large). When we talk about blow-ups, we usually mean that (in addition to the first part of (8)) we take all y_k equal to some point $y \in \Omega$. Here we decided to let y_k depend on k (but not get too close to the boundary) because this does not cause trouble. Also, we do not need to take y_k in K, even though the most interesting blow-up sequences will be defined that way.

For each sequence $\{(u_k, K_k)\}$ as above, we can extract a subsequence that converges. This comes from Proposition 37.8; also see Definition 37.6 for the convergence.

We want to check that when $\{(u_k, K_k)\}$ converges, the limit is a "reduced global minimizer" in $\mathbb{R}^n$. By definition, a <u>reduced global minimizer</u> in $\mathbb{R}^n$ will be a topologically reduced local topological almost-minimizer in $\mathbb{R}^n$, with gauge function $h = 0$. It would perhaps be more logical to call this a reduced local minimizer in $\mathbb{R}^n$, but bad habits were already taken. We shall say more about global minimizer in Section 54 and the ones after.

Proposition 9. *Let (u, K) be a reduced or coral local almost-minimizer in Ω, or a topologically reduced or coral topological local almost-minimizer in Ω. Let $\{y_k\}$ be a sequence in Ω, and $\{t_k\}$ a sequence of positive numbers. Assume that (8) holds, and that $\{(u_k, K_k)\}$ converges to some limit (v, G) (as in Definition 37.6). Then (v, G) is a reduced global minimizer in $\mathbb{R}^n$.*

Fix $\ell \geq 0$, and apply Theorem 38.3 to the sequence $\{(u_k, K_k)\}$, $k \geq \ell$. The domains Ω_k converge to $\mathbb{R}^n$ as in (37.1), and we can take the common gauge function $h_\ell(r) = \sup\{h(t_k r); k \geq \ell\}$. Note that $h_\ell(r) = h(A_\ell r)$, where $A_\ell = \sup\{t_k; k \geq \ell\}$, because h is nondecreasing and t_k tends to 0. Thus h_ℓ is a gauge function.

Theorem 38.3 says that (v, G) is a coral topological local almost-minimizer in $\mathbb{R}^n$, with gauge function h_ℓ. This holds for any ℓ, and

$$\lim_{\ell \to +\infty} h_\ell(r) = \lim_{\ell \to +\infty} h(A_\ell r) = 0$$

because A_ℓ tends to zero and h is a gauge function, so we can also take the gauge function $\tilde{h} = 0$. In other words, (v, G) is a coral topological local minimizer in $\mathbb{R}^n$.

To complete the proof of Proposition 9, we just need to check that (v, G) is topologically reduced too. This follows from Remark 8.31. [The idea is that if v

has an extension $\widetilde{v}$ to $\mathbb{R}^n \setminus \widetilde{G}$, with $\widetilde{G} \subset G$, and for which $(\widetilde{v}, \widetilde{G})$ is a topological competitor for (v, G), then $H^{n-1}(G \setminus \widetilde{G}) = 0$ (because (v, G) is a topological local minimizer and the contributions of the gradients are the same), hence $\widetilde{G} = G$ because G is coral.] Proposition 9 follows. $\qquad \square$

Proposition 9 can be generalized slightly. Suppose for instance that (u, K) is reduced and satisfies the following variant of almost-minimality:

$$\int_{B(x,r)\setminus K} |\nabla u(z)|^2 f(z)dz + \int_{K \cap \overline{B}(x,r)} g(z)dH^{n-1}(z) \tag{10}$$

$$\leq \int_{B(x,r)\setminus K^*} |\nabla u^*(z)|^2 f(z)dz + \int_{K^* \cap \overline{B}(x,r)} g(z)dH^{n-1}(z) + h(r)r^{n-1}$$

for all balls $B = B(x, r) \subset\subset \Omega$ and all (topological) competitors (u^*, K^*) for (u, K) in B.

Let us assume that f and g are continuous and positive, that we took a sequence $\{y_k\}$ that converges to some point $y \in \Omega$ and, as before, that $\{(u_k, K_k)\}$ converges to some limit (v, G). We want to apply Corollary 38.78 to the sequence $\{(u_k, K_k)\}$ and the domains Ω_k. The condition (38.74) is satisfied, with $f_k(z) = f(y_k + t_k z)$ and $g_k(z) = g(y_k + t_k z)$. These functions converge to the constants $f(y)$ and $g(y)$, uniformly on compact subsets of $\mathbb{R}^n$. The hypothesis (38.79) holds because locally the functions f_k and g_k are uniformly bounded from above and below (see near (38.94)). As for (38.80) and (38.81), we can get them because they follow from (38.95), for instance. [See the comments after (38.95).]

So we apply Corollary 38.78, and we get that

$$f(y) \int_{B\setminus G} |\nabla u(z)|^2 + g(y)H^{n-1}(G \cap \overline{B})$$

$$\leq f(y) \int_{B\setminus G^*} |\nabla u^*(z)|^2 + g(y)H^{n-1}(G^* \cap \overline{B}) \tag{11}$$

for all topological competitors (u^*, G^*) for (v, G) in a ball B. In other words, (v, G) is almost a global minimizer in $\mathbb{R}^n$, except for the fact that we multiplied the two terms of the functional by different constants $f(y)$ and $g(y)$. It is also reduced, because it is coral and by the same argument as above (i.e., the proof of Remark 8.31 applies). Thus, if we set $\lambda = (f(y)/g(y))^{1/2}$, the pair $(\lambda v, G)$ is a reduced global minimizer in $\mathbb{R}^n$.

Remark 12. Blow-up limits of local almost-minimizers with linear distortion $\leq L$ (topological or not, and accordingly reduced or coral) are reduced global minimizers of a variant of our usual functional in $\mathbb{R}^n$.

Indeed let (v, G) be such a blow-up limit. The proof of Proposition 9, together with Remark 38.65, shows that (v, G) is a reduced topological almost-minimizer in $\mathbb{R}^n$, with linear distortion $\leq L$ and gauge function 0. This means that for each $R > 0$ we can find matrices A_1 and A_2 such that

$$H^{n-1}(A_1(G \cap \overline{B}(0,M))) + \int_{B(0,M)\setminus G} |A_2 \nabla v|^2$$

$$\leq H^{n-1}(A_1(\widetilde{G} \cap \overline{B}(0,M))) + \int_{B(0,M)\setminus \widetilde{G}} |A_2 \nabla \widetilde{v}|^2 \qquad (13)$$

when $(\widetilde{v}, \widetilde{G})$ is a topological competitor for (v, G) in $B(0, M)$ (see Definition 7.45).

We can even find fixed matrices A_1 and A_2 that work for all topological competitors. Indeed, if A_1 and A_2 work for all topological competitors in $B(0, M)$, it trivially works for $M' < M$ too. Then we take for (A_1, A_2) any limit of pairs (A_1, A_2) that work for radii M that tend to $+\infty$, and for this pair (13) holds for every topological competitor for (v, G).

It is not clear to the author whether the study of this slightly more general type of global minimizers reduces to the study of the standard ones (with $A_1 = A_2 = I$). Certainly we can compose with A_1 and reduce to the case when $A_1 = I$. We may also assume that A_2 is diagonal with positive terms, because all we care about is the quadratic form $v \to |A_2 v|^2$. But the author does not know what the local minimizers of $H^1(K) + \int_{\mathbb{R}^2 \setminus K} \left\{ \frac{\partial u}{\partial x} \right\}^2 + a \left\{ \frac{\partial u}{\partial y} \right\}^2$ in $\mathbb{R}^2$ should be.

Remark 14. There are also versions of Proposition 9 where we consider a (not merely local) almost-minimizer (u, K) on some domain Ω, take blow-up limits at a point $y \in \partial\Omega$, and show that the limit is a topological minimizer on an appropriate limit domain (like a half-space). We shall need to do this in Section 79.

E. Pieces of C^1 Curves for Almost-Minimizers

This part is mostly devoted to the following result ([Da5], [AFP1]): if (u, K) is a reduced local (plain or topological) almost-minimizer in $\Omega \subset \mathbb{R}^2$, with sufficiently small gauge function (any positive power of the radius will do), then for each disk $B(x, r)$ centered on K and contained in Ω we can find $y \in B(x, r/2)$ and $t \in (r/C, r/2)$ such that $K \cap B(y, t)$ is a nice C^1 curve through $B(y, t)$. The constant C depends on r and the gauge function, not on (u, K) or $B(x, r)$. See Theorem 51.4 for a more precise statement.

Let us try to give a very rapid preview of the general organization of a proof. The reader may also have a look at the beginning of Section 53 (which deals with an extension of the main part of the argument to situations where K looks like a propeller) for a slightly more detailed and technical account of how it works.

There are actually two parts in this result. The main one consists in finding a combination of initial conditions in a disk $B(x, 2r)$ that imply that $K \cap B(x, r)$ is a C^1 curve through $B(x, r)$. There are slightly different possible combinations that work (depending on which proof we choose), but usually one assumes that $K \cap B(x, 2r)$ is pretty flat, and that $r^{-1} \int_{B(x,2r) \setminus K} |\nabla u|^2$, or some L^p version of it, is pretty small. The difficult part is to find a quantity $\psi(x, 2r)$ (that depends on our combination of initial conditions on $B(x, 2r)$) for which we can show that $\psi(x, r) \leq \psi(x, 2r)$ (and then iterate). The central part of the argument that will be presented here dates from 1998, but was not written anywhere so far. As in [Da5], the leading quantity is $\psi(x, r) = \omega_2(x, r) = r^{-1} \int_{B(x,2r) \setminus K} |\nabla u|^2$, so our main estimate will show that under suitable smallness conditions of other quantities, $\omega_2(x, r) < \omega_2(x, 2r)$. This estimate will be based on an argument of A. Bonnet [Bo], that proved decay of the energy in the slightly different context of global minimizers in the plane, with a connected singular set K. See Sections 47 (for the Bonnet estimate) and 48 (for our main decay estimate).

We should already mention here that this argument does not work in higher dimensions, but the argument of [AFP1] does. It uses a leading quantity $\psi(x, 2r)$ that is a combination of energy (as before) and tilt. We shall return to this in Section 75.

Besides the central decay estimate, we shall need a few auxiliary results, whose general effect is to show that $\omega_2(x, r)$ controls many other quantities (that we need to keep small if we want to apply our main decay estimate). We shall try to isolate many of these in little separate sections, because they may often be used for different purposes. Also, some of them work the same way in higher dimensions, and we want to record that.

Once we have conditions on $B(x, 2r)$ that imply that $K \cap B(x, r)$ is a C^1 curve, we still need to show that we can find sufficiently large disks $B(x, 2r)$ like this in a given $B(x_0, r_0)$. The best way to do this seems to be to use our Carleson measure estimates from Section 23 on an L^p version of $\omega_2(x, r)$, possibly in connection with similar Carleson measure estimates on the flatness of K that come from its uniform rectifiability (as in Theorem 27.1), to find a good starting disk $B(x, 2r)$ (by Chebyshev). This is what we do in the next section. Note also that this part of the argument extends to higher dimensions; see [Ri1] and Section 75, or [MaSo2] for a different proof.

Along the way, we shall find a few different sets of conditions that force $K \cap B(x, r)$ to be a nice C^1 curve through $B(x, r)$, and simple consequences of these results. We give a short list here, to help the reader find specific results that she would be looking for. A basic case when we can say that $K \cap B$ is a chord-arc curve is given in Theorem 50.1. Then the gauge function h just needs to be small. If we assume that $h(r)$ tends to 0 with r, we get asymptotically straight chord-arc curves; see Corollary 50.25. If $h(t) = Ct^\alpha$, we get $C^{1,\gamma}$ curves; see Corollary 50.33. Also see Corollary 50.51 for an intermediate condition that gives C^1 curves.

Recall that Theorem 51.4 will tell us that each disk $B \subset\subset \Omega$ centered on K contains a disk with comparable size centered on K and where K is a nice C^1 curve. In particular, K is C^1 except on a set of dimension < 1 (Theorem 51.20).

We shall see later weaker sufficient conditions on the restriction of K and u to a disk that imply that K is a C^1 curve in a smaller disk (say, when $h(t) = Ct^\alpha$). See Corollary 51.17 (where we only assume flatness and small L^p-energy), Corollary 52.25 (where we only assume flatness and a large jump), and Proposition 70.2 (where we only assume flatness and the absence of large holes in K).

Also see Theorems 53.4 and 53.31 for analogues of the previous results when K looks a lot like a propeller (a union of three half-lines emanating from a point and making $120°$ angles).

As a consequence of these results and a blow-up argument, Proposition 60.1 will tell us that if in a disk the energy of u (or an L^p variant, $p > 1$) is small enough, then in a smaller disk K is a C^1 curve or spider (a C^1 version of a propeller).

Many of the lemmas in this part are not really needed for a fast proof of the main theorem, but we hope that taking a more leisurely route will be a little nicer.

41 Uniform rectifiability, the Peter Jones numbers $\beta(x, r)$, and why K is flat in many places

Standard Assumption 1. In the next few sections, Ω will be an open set in $\mathbb{R}^n$ and (u, K) will be a local, plain or topological, almost-minimizer in Ω, with gauge function h. See Definition 7.14 or 7.17 for local almost-minimizers, and (7.5) for the definition of a gauge function. Actually, $h(r)$ does not need to tend to 0 when r tends to 0, but we shall always have to assume that it is sufficiently small. Our results will extend to almost-minimizers with linear distortion (see Section 7.c), but as before we shall only do rapid comments inside remarks to this effect. The first such remark is at the end of Section 45; up until there only minor modifications are needed.

In addition, (u, K) will be assumed to be coral (see Definition 8.24). Recall that this is the case when (u, K) is reduced, or topologically reduced if it is a topological almost-minimizer. [See Definitions 8.1 and 8.7, and Remark 8.30 for the implication.]

To simplify the exposition, we shall write this assumption as

$$(u, K) \in TRLAM, \text{ or } (u, K) \in TRLAM(\Omega, h)$$

if we want to be explicit.

In this section we just want to use simple Carleson measure estimates to find lots of balls where K is fairly flat and u has low energy. These balls may be used later as starting points for the C^1-theorems.

To measure flatness, we shall use the following numbers. For $x \in K$ and $r > 0$, set

$$\beta_K(x, r) = \inf_P \left\{ \sup \left\{ r^{-1} \operatorname{dist}(y, P) \, ; \, y \in K \cap \overline{B}(x, r) \right\} \right\}, \tag{2}$$

where the infimum is taken over all hyperplanes P through x. Note that $\beta_K(x, r)$ is a dimensionless number, thanks to the normalization by r^{-1}. Also, $\beta_K(x, r) \leq 1$ in all cases, and the idea is that if it is much smaller, then K is very flat in $\overline{B}(x, r)$. The notation comes from [Jo1], which is the first place (to our knowledge) where these sorts of numbers were used to study systematically the quantitative rectifiability properties of some sets. In [Jo1], the main point was to estimate the $\beta_K^2(x, r)$ associated to an Ahlfors-regular curve in the plane, show that they satisfy a Carleson measure estimate, and use this to prove the L^2-boundedness of the corresponding Cauchy integral operator. The $\beta_K(x, r)$ also show up in [Jo2], where they are used to characterize the compact sets E in the plane for which there is a rectifiable curve with finite length that contains them. This is a little like a travelling salesman problem with infinitely many points, but where we only want to know whether the salesman can visit all his clients in finite time. Also compare with the definition of uniform rectifiability in Definition 27.17, where we require that E be contained in an Ahlfors-regular curve.

Let us first describe the situation for one-dimensional sets.

Theorem 3 [Jo1], [DaSe1], [Ok]. *Let Γ be an Ahlfors-regular curve in $\mathbb{R}^n$. Then the numbers $\beta_\Gamma(x, r)$ associated to Γ as in (2) satisfy the following Carleson condition:*

$$\int_{y \in \Gamma \cap B(x, r)} \int_{0 < t < r} \beta_\Gamma^2(y, t) \, dH^1(y) \frac{dt}{t} \leq Cr \tag{4}$$

for all $x \in \Gamma$ and $r > 0$. The constant C depends only on the Ahlfors-regularity constant for Γ.

See Definition 27.3 for Ahlfors-regular curves. We added the references [DaSe1] and [Ok] because the initial results of P. Jones only worked in ambient dimension 2; the generalization for Ahlfors-regular sets (even of larger dimensions) was done in [DaSe1], and for general one-dimensional subsets of $\mathbb{R}^n$, the extension is due to K. Okikiolu [Ok]; see the next paragraph. Also see [DaSe3], and in particular Theorem 1.57 on page 22, for more information on this and connected issues.

Unfortunately, Theorem 3 is a little too complicated and too far from our main interests to allow a proof in this book. So we have to refer the reader to [DaSe1]. Note that Theorem 3 looks a little like a quantitative form of the fact that Lipschitz functions have derivatives almost-everywhere (so that their graphs have tangents almost-everywhere). It is a little harder than just a result on approximation of Lipschitz functions by affine functions because Γ may come back onto itself (and have multiple points). Part of the proof consists in showing that this does not happen too often.

The reader should not be worried by the Carleson measure condition (4). Since Ahlfors-regularity gives a size condition for each ball centered on Γ, it is very natural that in (4) we also get a condition for each ball. The choice of measures and the constant Cr are natural in view of the invariance of our problem with respect to dilations. Finally note that there is even a characterization in terms of the $\beta_\Gamma^2(y, t)$ of sets that are contained in a rectifiable curve. See [Jo2], generalized in [Ok] for larger ambient dimensions. [The proof in [Jo2] uses conformal mappings!]

If K is a subset of an Ahlfors-regular curve, it satisfies (4) as well, because $\beta_K(x, r)$ is a nondecreasing function of K and in addition we integrate on a smaller set. Because of this, we have the following direct consequence of Theorem 27.1 and Theorem 3. [See also Remark 7.22].

Corollary 5. *There are absolute constants C and ε_0 such that if $n = 2$ and $(u, K) \in TRLAM(\Omega, h)$, then*

$$\int_{y \in K \cap B(x,r)} \int_{0 < t < r} \beta_K^2(y, t) \, dH^1(y) \frac{dt}{t} \leq Cr \qquad (6)$$

for all $x \in K$ and $r > 0$ such that $B(x, 3r) \subset \Omega$ and $h(3r) \leq \varepsilon_0$.

The situation in higher dimensions is similar, even though the proofs are more complicated. We still have that if $(u, K) \in TRLAM(\Omega, h)$, then K is locally uniformly rectifiable [DaSe6]. See Part H for a little more detail, including a definition of uniform rectifiability. Then there is an analogue of Theorem 3 for higher-dimensional uniformly rectifiable set, but we need to replace $\beta_K(y, t)$ with variants $\beta_{K,q}(y, t)$ constructed with L^q-norms, as follows:

$$\beta_{K,q}(x, r) = \inf_P \left\{ r^{1-n} \int_{y \in K \cap B(x,r)} \{r^{-1} \operatorname{dist}(y, P)\}^q \, dH^{n-1}(y) \right\}^{1/q}, \qquad (7)$$

where we still take the infimum over all hyperplanes P through x. Note that we are essentially taking an average in the integral, at least when $0 < r < 1$ and $B(x,r) \subset \Omega$, because $H^{n-1}(K \cap B(x,r))$ is comparable to r^{n-1} by local Ahlfors-regularity. [See Theorem 18.1 (or Theorem 18.16 for a topological almost-minimizer).] Thus $\beta_{K,q}(x,r) \leq C\beta_K(x,r) \leq C$, by Hölder. The definition (7) can be modified to work on (locally Ahlfors-regular) sets of any integer dimension d: simply consider d-planes P and replace the normalization factor r^{1-n} with r^{-d}.

Theorem 8 [DaSe1]. *If K is a uniformly rectifiable set of dimension d in $\mathbb{R}^n$, then for each $q \in [1, \frac{2d}{d-2})$ there is a constant C_q such that*

$$\int_{y \in K \cap B(x,r)} \int_{0 < t < r} \beta_{K,q}^2(y,t)\, dH^d(y)\frac{dt}{t} \leq C_q r^d \qquad (9)$$

for $x \in K$ and $r > 0$.

See also Theorem 4.1.3 p. 213 of [DaSe3] for an alternative (but not much shorter) proof. Here again the Carleson condition (9) is natural. The converse is also true. In fact, if K is Ahlfors-regular of dimension d and (9) holds for any $q \in [1, \frac{2d}{d-2})$, then K is uniformly rectifiable.

When we apply Theorem 8 to a uniformly rectifiable set of codimension 1 that contains $K \cap B(x,r)$, for instance given by Theorem 27.1 or Theorem 74.1, we get the following analogue of Corollary 5.

Corollary 10. *For each choice of $n \geq 2$ and $1 \leq q < \frac{2n-2}{n-3}$, there are constants $\varepsilon(n)$ and $C(n,q)$ such that if $(u, K) \in TRLAM(\Omega, h)$ for some open set $\Omega \subset \mathbb{R}^n$, then*

$$\int_{y \in K \cap B(x,r)} \int_{0 < t < r} \beta_{K,q}^2(y,t)\, dH^{n-1}(y)\frac{dt}{t} \leq C(n,q)\, r^{n-1} \qquad (11)$$

for all $x \in K$ and $r > 0$ such that $B(x, 3r) \subset \Omega$ and $h(3r) \leq \varepsilon(n)$.

This is not entirely satisfactory yet, because we are really interested in the $\beta_K(x,r)$. Fortunately,

$$\beta_K(x,r) \leq C(q)\,\beta_{K,q}(x,2r)^{\frac{q}{n+q-1}} \qquad (12)$$

for $x \in K$ and $0 < r < 1$ such that $B(x, 2r) \subset \Omega$, and with a constant $C(q)$ that depends also on n and the local Ahlfors-regularity constant for K. This is (1.73) on p. 27 of [DaSe3], but we shall check it anyway. But first note that we also control the local Ahlfors-regularity constant here, with bounds that depend only on n, at least if $h(3r)$ is small enough.

To prove (12), let P realize the infimum in the definition of $\beta_{K,q}(x, 2r)$, and let $z \in K \cap \overline{B}(x, r)$ be as far as possible from P. Set $D = \text{dist}(z, P) \leq r$. Then

$$
\begin{aligned}
\beta_{K,q}(x, 2r)^q &= (2r)^{1-n} \int_{y \in K \cap B(x, 2r)} \{(2r)^{-1} \text{dist}(y, P)\}^q \, dH^{n-1}(y) \\
&\geq (2r)^{1-n} \int_{y \in K \cap B(z, D/2)} \{(2r)^{-1} \text{dist}(y, P)\}^q \, dH^{n-1}(y) \qquad (13) \\
&\geq (2r)^{1-n} \left(\frac{D}{4r}\right)^q H^{n-1}(B(z, D/2)) \geq C^{-1} r^{1-n-q} D^{q+n-1},
\end{aligned}
$$

by Ahlfors-regularity; (12) follows because $\beta_K(x, r) \leq D/r$. Thus (11) implies that

$$
\int_{y \in K \cap B(x, r)} \int_{0 < t < r} \beta_K(y, t)^{\frac{2(n+q-1)}{q}} \, dH^{n-1}(y) \frac{dt}{t} \leq C_q \, r^{n-1}. \qquad (14)
$$

We lost some information here, but this will not matter. When $n = 2$, we could even take $q = +\infty$ and get the exponent 2, as in Corollary 5. Otherwise, the simplest is to take $q = 2$, which works for all n, and get the exponent $n + 1$.

In terms of Definition 23.13, we just obtained that

$$
\beta_K^{n+1}(y, t) \, dH^{n-1}(y) \frac{dt}{t} \text{ is a Carleson measure on } \Delta(r_0) \qquad (15)
$$

if $h(3r_0) \leq \varepsilon(n)$, and with a Carleson constant that depends only on n. (See (23.1) and (23.2) for the definition of $\Delta(r_0)$.)

Now recall from (23.15) or Theorem 23.8 that if $h(3r_0)$ is small enough (depending on n), then

$$
\omega_p(y, t) \, dH^{n-1}(y) \frac{dt}{t} \text{ is a Carleson measure on } \Delta(r_0), \qquad (16)
$$

for each exponent $p \in [1, 2)$, where ω_p is as in (23.5). As usual, the Carleson constant in (16) depends only on n and p. What we really wanted is the following easy consequence of (15) and (16).

Proposition 17. *For each choice of n, $\tau > 0$, and $p \in [1, 2)$, we can find constants $\varepsilon(n)$ and $C(n, p, \tau)$ such that if $(u, K) \in TRLAM(\Omega, h)$, $x \in K$, $r > 0$, $B(x, r) \subset \Omega$, and $h(r) \leq \varepsilon(n)$, we can find $y \in K \cap B(x, r/2)$ and $t \in (\frac{r}{C(n,p,\tau)}, \frac{r}{2})$ such that*

$$
\beta_K(y, t) + \omega_p(y, t) < \tau. \qquad (18)
$$

The proof is the same as for Corollary 23.38. $\square$

Remark 19. We shall see later another proof of Proposition 17, at least in dimension 2, where our use of uniform rectifiability is replaced with a limiting argument, plus the knowledge of all global minimizers in $\mathbb{R}^2$ for which u is constant. See Lemma 60.5 and Remark 60.8.

Remark 20. Proposition 17 also holds for (r_0, M, a)-quasiminimizers (that is, when $(u, K) \in TRLQ(\Omega)$, as in Standard Assumption 18.14), provided that we take a small enough (depending on n) and $r \leq r_0$. Then C depends on a, M, and n. The proof is the same.

Exercise 21. [About approximate and true tangent planes.] Let $E \subset \mathbb{R}^n$ be such that $H^d(E) < +\infty$ (sigma-finite would be enough) for some integer d, and let $x \in E$ be such that the d-dimensional upper density of E at x, namely $D_E(x) = \limsup_{r \to 0} r^{-d} H^d(E \cap B(x, r))$, is positive. Also let P be a d-dimensional affine plane through x. We say that P is an approximate tangent plane to E at x when

$$\lim_{r \to 0} r^{-d} H^d(E \cap P_{\varepsilon r} \cap B(x, r)) = 0 \tag{22}$$

for every $\varepsilon > 0$, where $P_{\varepsilon r} = \{y \in \mathbb{R}^n \,;\, \mathrm{dist}(y, P) \leq \varepsilon r\}$ is an εr neighborhood of P.

1. Verify that if P is an approximate tangent plane to E at x, and F is another measurable set such that $D_F(x) = 0$, then P is also an approximate tangent plane to $E \cup F$ at x.

2. Show that if E is rectifiable (i.e., if it is contained in a countable union of C^1 surfaces of dimension d, plus a set of H^d-measure zero), then it has an approximate tangent plane at x for H^d-almost every $x \in E$. You may need to use Theorem 6.2 in [Mat2].

Note that the converse (i.e., the fact that the existence of approximate tangent planes H^d-almost everywhere characterizes rectifiability) is true, but much harder. See [Mat2], in particular Theorem 15.19, for this and all connected questions.

From now on, suppose that E is locally Ahlfors-regular of dimension d. It will even be enough to assume that for each each $x \in E$, we can find $r > 0$ and $C > 0$ such that $C^{-1} t^d \leq H^d(E \cap B(y, t)) \leq C t^d$ for $y \in E \cap B(x, r)$ and $0 < t < r$.

3. Show that if P is an approximate tangent plane to E at x, then P is also a true tangent plane. That is, for each $\varepsilon > 0$, we can find $r_0 > 0$ such that $E \cap B(x, r) \subset P_{\varepsilon r}$ for $r < r_0$. Hint. Proceed as for (12): if $E \cap B(x, r)$ is not contained in $P_{\varepsilon r}$, then $(2r)^{-d} H^d((E \cap P_{\varepsilon r/2} \cap B(x, 2r)) \geq C^{-1} \varepsilon^d$. Which is incompatible with (22) if r is small enough.

4. Show that the tangent plane P is unique (if it exists).

Exercise 23. Let E be an Ahlfors-regular set of dimension d in $\mathbb{R}^n$.

1. Show that $\lim_{r \to 0} \beta_E(x, r) = 0$ if E has a tangent plane at x.

2. Show that the converse is not true, even when $d = 1$ and $n = 2$.

3. Suppose now that E is uniformly rectifiable, so that (9) holds. Let $q \in [1, \frac{2d}{d-2})$ be given. Show that for H^d-almost all $y \in E$, the integral $\int_0^1 \beta_{K,q}(y,t)^2 \frac{dt}{t}$ converges.

4. Show that the convergence at y of the integral above, even with $d = 1$, $n = 2$, and $q = +\infty$, is not enough to imply the existence of a tangent. [If we removed the power 2, the convergence of the integral would give a tangent, but on the other hand Theorem 8 would not hold anymore.] Nonetheless, uniformly rectifiable sets have tangent d-planes almost-everywhere, because they are rectifiable and by Exercise 21.

42 Lower bounds on the jump of u when K is flat and ∇u is small

We continue to suppose that $(u, K) \in TRLAM(\Omega, h)$, as in Assumption 41.1, and we want to prove estimates on the normalized jump of u when we cross a ball where K is fairly flat and ∇u is fairly small.

Some extra notation will be useful. It will be convenient to restrict to balls such that

$$\overline{B}(x,r) \subset \Omega \quad \text{and} \quad \beta_K(x,r) < 1/2. \tag{1}$$

For each such ball $B(x,r)$, choose a hyperplane $P(x,r)$ through x such that

$$\text{dist}(y, P(x,r)) \leq r\beta_K(x,r) \quad \text{for } y \in K \cap \overline{B}(x,r). \tag{2}$$

Denote by $z_{\pm}(x,r)$ the two points of $B(x,r)$ that lie at distance $3r/4$ from $P(x,r)$ and whose orthogonal projections on $P(x,r)$ are equal to x (see Figure.1).

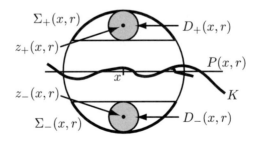

Figure 1

Then set

$$D_{\pm} = D_{\pm}(x,r) = B(z_{\pm}(x,r), r/4). \tag{3}$$

Note that $D_{\pm} \subset B(x,r) \setminus K \subset \Omega \setminus K$, by (1). Our first normalized jump is

$$J(x,r) = r^{-1/2} |m_{D_+} u - m_{D_-} u|, \tag{4}$$

where as usual the normalization by $r^{-1/2}$ will make $J(x,r)$ dimensionless, and $m_{D_\pm} u = |D_\pm|^{-1} \int_{D_\pm} u$ denotes the mean value of u on $D_\pm$.

We shall also need variants of $w_p(x,r)$ and $J(x,r)$ on the circle. First set

$$w_p^*(x,r) = r^{1-\frac{2n}{p}+\frac{2}{p}} \left\{ \int_{\partial B(x,r)\setminus K} |\nabla u|^p \right\}^{2/p} \tag{5}$$

for $B(x,r) \subset \Omega$ [compare with (23.5)].

Convention 6. We shall set $w_p^*(x,r) = +\infty$ when the restriction of u to $\partial B(x,r)\setminus K$ does not lie in $W^{1,p}(\partial B(x,r)\setminus K)$, or its derivative on $\partial B(x,r)\setminus K$ is not given by the restriction of ∇u there, or when it is not equal, H^{n-1}-everywhere on $\partial B(x,r)\setminus K$, to the radial limit of u. Note that even with this convention, $w_p^*(x,r) < +\infty$ for almost-every $r < \operatorname{dist}(x, \mathbb{R}^n\setminus\Omega)$ when $x \in \Omega$. This comes from Propositions 9.5 and 11.4 (on the absolute continuity of u on almost all hyperplanes, and to transform spheres into planes locally), and the beginning of Section 13. [Also see the proof of Corollary 14.28.]

For the analogue of $J(x,r)$ on the circle, assume that (1) holds and call $\Sigma_+(x,r)$ and $\Sigma_-(x,r)$ the two components of $\{y \in \partial B(x,r); \operatorname{dist}(y, P(x,r)) > r/2\}$, with $\Sigma_+(x,r)$ on the same side of $P(x,r)$ as D_+. Then set

$$J^*(x,r) = r^{-1/2} \left| m_{\Sigma_+(x,r)} u - m_{\Sigma_-(x,r)} u \right| \tag{7}$$

when in addition $w_p^*(x,r) < +\infty$. There is not much difference between J and J^*. We claim that

$$|J^*(x,r) - J(x,r)| \leq C w_1(x,r)^{1/2} \leq C w_p(x,r)^{1/2} \tag{8}$$

when (1) holds and $w_p^*(x,r) < +\infty$. In fact,

$$|m_{\Sigma_\pm(x,r)} u - m_{D_\pm(x,r)} u| \leq C r^{1-n} \int_{y\in B(x,r); \operatorname{dist}(y,P(x,r))>r/2} |\nabla u|$$
$$= C r^{1/2} w_1(x,r)^{1/2}, \tag{9}$$

by Convention 6, the proof of Lemma 13.20, and the definition (23.5) of w_1. There is a minor difference between the geometry here and in that lemma because $B(x,r)$ is not the product of a ball with an interval, but this is easy to fix with a change of variables; also, Lemma 13.20 does not come with an expression for the constant $C(B,V)$, but the reader may easily check that we did not mess up with the homogeneity in (9). Our claim (8) follows from (9) and Hölder.

Proposition 10. *For each $p \in (\frac{2(n-1)}{n}, 2]$ there is a constant $\tau_0 = \tau_0(n,p)$ such that*

$$J^*(x,r) \geq \tau_0 \tag{11}$$

when $x \in K$ and $\overline{B}(x,r) \subset \Omega$ are such that

$$\beta_K(x,r) + \omega_p^*(x,r) + h(r) < \tau_0 \tag{12}$$

and, in the special case when (u, K) is a topological almost-minimizer,

$$\Sigma_+(x,r) \text{ and } \Sigma_-(x,r) \text{ lie in the same component of } \Omega \setminus K. \tag{13}$$

Remark 14. Proposition 10 stays true for (r_0, M, a)-quasiminimizers (i.e., when $(u, K) \in TRLQ(\Omega)$, as in Standard Assumption 18.14), provided that the quasi-minimality constant a is small enough (depending on n) and we restrict to $r \leq r_0$. Then τ_0 depends on a, M, n, and p. The proof is the same.

The proof of Proposition 10 will probably remind the reader of what we did for the property of projections and the concentration property in Sections 24 and 25. Let $B = B(x,r)$ be as in the statement, and set

$$Z = \{y \in \partial B \, ; \text{dist}(y, P(x,r)) \leq 3\tau r\}, \tag{15}$$

where $\tau \geq \tau_0$ will be chosen soon. Note that

$$U = \{y \in \partial B \, ; \text{dist}(y, P(x,r)) > \tau r\} \subset \Omega \setminus K, \tag{16}$$

by (2) and (12). We claim that if (11) fails, then we can find a function $f \in W^{1,p}(\partial B)$ that coincides with u on $\partial B \setminus Z$ and for which

$$r^{1 - \frac{2n}{p} + \frac{2}{p}} \left\{ \int_{\partial B} |\nabla f|^p \right\}^{2/p} \leq C(\tau)\tau_0. \tag{17}$$

The argument that follows is almost the same as for Lemma 38.26, proved near the end of Section 38, and to which we refer for additional details. We may as well assume that $m_{\Sigma_+(x,r)} u = 0$, because adding a constant to u and f does not change the estimates. Then $|m_{\Sigma_-(x,r)} u| \leq \tau_0 r^{1/2}$ because (11) fails. We take $f(y) = \varphi(y)u(y)$ on ∂B, with a smooth function φ supported on $U_1 = \{y \in \partial B \, ; \text{dist}(y, P(x,r)) > 2\tau r\}$, equal to 1 on a neighborhood of $\partial B \setminus Z$, and such that $|\nabla \varphi| \leq C\tau^{-1} r^{-1}$ everywhere. Then $f \in W^{1,p}(\partial B)$, and

$$|\nabla f| \leq |\varphi \nabla u| + |u \nabla \varphi|. \tag{18}$$

The contribution to (17) of the first term is less than τ_0, by (12) and (5). For the second one, we cover U_1 by less than $C(\tau)$ balls B_j of radius τr centered on U_1, and set $D_j = B_j \cap \partial B$. Then $D_j \subset U \subset \partial B \setminus K$, and

$$\int_{D_j} |u - m_{D_j} u|^p dH^{n-1} \leq C(\tau) \, r^p \int_{D_j} |\nabla u|^p dH^{n-1} \tag{19}$$

$$\leq C(\tau) \, r^{\frac{p}{2} + n - 1} \omega_p^*(x,r)^{p/2} \leq C(\tau) \, r^{\frac{p}{2} + n - 1} \tau_0^{p/2},$$

by our proof of Proposition 12.23 (or Exercise 12.29 with $r = p$), (5), and (12). Also, if D_j lies in the same component of $\partial B \setminus P(x, r)$ as $\Sigma_\pm(x, r)$, then

$$|D_j|\,|m_{D_j}u - m_{\Sigma_\pm(x,r)}u|^p \leq C(\tau)\,r^p \int_U |\nabla u|^p \leq C(\tau)\,r^{\frac{p}{2}+n-1}\tau_0^{p/2}, \qquad (20)$$

for the same reasons. Since $|D_j|\,|m_{\Sigma_\pm(x,r)}u|^p \leq |D_j|\tau_0^p r^{p/2} \leq C r^{\frac{p}{2}+n-1}\tau_0^p$, (19) and (20) yield

$$\int_{D_j} |u|^p dH^{n-1} \leq C(\tau)\,r^{\frac{p}{2}+n-1}\tau_0^{p/2}. \qquad (21)$$

[The powers of τ_0 look wrong because the homogeneity in (12) is not perfect.] We may now sum over j and get that

$$\int_{U_1} |u|^p \leq \sum_j \int_{D_j} |u|^p \leq C(\tau)\,r^{\frac{p}{2}+n-1}\tau_0^{p/2}. \qquad (22)$$

Thus

$$\int_{\partial B} |u\nabla\varphi|^p \leq C(\tau)\,r^{-p} \int_{U_1} |u|^p \leq C(\tau)\,r^{-\frac{p}{2}+n-1}\tau_0^{p/2}, \qquad (23)$$

and the contribution of the second half of (18) to (17) is also less than $C(\tau)\tau_0$; (17) follows.

Now there is a function $v \in W^{1,2}(B)$ such that

$$\int_B |\nabla v|^2 \leq C(\tau)\,\tau_0\,r^{n-1}, \qquad (24)$$

and whose radial limits on ∂B coincide with f almost-everywhere. When $n = 2$, this follows from (17) and Lemma 22.16. When $n > 2$, we use Lemma 22.32 instead, and this is where we need p to be larger than $\frac{2(n-1)}{n}$.

We may now introduce a competitor (v, G). Set $G = [K \setminus B] \cup Z$, and extend the function v of (24) to $\Omega \setminus G$ by taking $v = u$ out of B. Then $v \in W^{1,2}(\Omega \setminus G)$, by Corollary 14.28 and Convention 6. Hence (v, G) lies in $\mathcal{A}(\Omega)$, and it is a competitor for (u, K) in B. If (u, K) is a topological almost-minimizer, then (v, G) is a topological competitor, by (13), (16), and Lemma 20.10. Note that

$$H^{n-1}(G \cap \overline{B}) + \int_B |\nabla v|^2 \leq H^{n-1}(Z) + C(\tau)\,\tau_0\,r^{n-1} \leq C\tau r^{n-1} + C(\tau)\,\tau_0\,r^{n-1}, \qquad (25)$$

by (24) and (15), while $H^{n-1}(K \cap B) \geq C^{-1}r^{n-1}$ by local Ahlfors-regularity. This gives a contradiction to the almost-minimality of (u, K) (i.e., (7.7)) if τ, and then τ_0, are chosen small enough. [Recall that $h(r) \leq \tau_0$, by (12).] We also get a contradiction if (u, K) is a local quasiminimizer, a is small enough, and τ and τ_0 are small enough, depending also on M. This proves Proposition 10 and Remark 14. $\qquad \square$

43 Normalized jumps often tend to get larger

In this section we just want to check that when we fix $x \in K$ and consider decreasing sequences of radii r, the normalized jumps $J(x,r)$ easily get very large, especially once they reach a large enough value. We keep the same notation and assumptions as in the previous section, and in particular we assume that $(u, K) \in TRLAM(\Omega, h)$, as in Standard Assumption 41.1. [But $(u, K) \in TRLQ(\Omega)$, as in Standard Assumption 18.14, would do as well.] Our main ingredient will be the following easy estimate.

Lemma 1. Let $x \in K$ and r, r_1 be given, with $0 < r_1 \leq r \leq 2r_1$. Assume that $\overline{B}(x,r) \subset \Omega$ and $\beta_K(x,r) \leq 10^{-1}$. Then

$$\left| \left(\frac{r_1}{r} \right)^{1/2} J(x, r_1) - J(x, r) \right| \leq C\omega_1(x, r)^{1/2} \leq C\omega_p(x, r)^{1/2} \leq C(1 + h(r))^{1/2} \quad (2)$$

for $1 \leq p \leq 2$. Here C depends only on n.

Recall from (42.4) that $r^{1/2} J(x, r) = \left| m_{D_+(x,r)} u - m_{D_-(x,r)} u \right|$. Set

$$V = \{ y \in B(x,r) \, ; \, \operatorname{dist}(y, P(x,r)) > \beta_K(x,r) \, r \}, \quad (3)$$

where $P(x,r)$ is as in (42.2), and denote by $V_\pm$ the component of V that contains $D_\pm(x,r)$. Then

$$\left| m_{D_\pm(x,r)} u - m_{V_\pm} u \right| \leq Cr^{1-n} \int_{V_\pm} |\nabla u|, \quad (4)$$

because $V_\pm \subset \Omega \setminus K$, and by the proof of Proposition 12.11. Next we want to compare these averages to the mean value of u on the $D_\pm(x, r_1)$. Note that $\beta_K(x, r_1) \leq 1/5$ because $\beta_K(x,r) \leq 10^{-1}$, so (42.1) holds for the pair (x, r_1) and we can define $P(x, r_1)$ and the disks $D_\pm(x, r_1)$. Modulo exchanging the names of these disks, we may assume that the center $z_+(x, r_1)$ of $D_+(x, r_1)$ lies in the closure of the component of $B(x,r) \setminus P(x,r)$ that contains V_+ (just because the union of this closure and its analogue for V_- contains $B(x,r)$). Then $z_-(x, r_1)$ lies in the closure of the component of $B(x,r) \setminus P(x,r)$ that contains V_-, by symmetry with respect to x. And we can find a ball $B_\pm$ of radius $r/10$ such that $B_\pm \subset D_\pm(x, r_1) \cap V_\pm$. Since neither $D_\pm(x, r_1)$ nor $V_\pm$ meet K, we get that

$$\left| m_{V_\pm} u - m_{D_\pm(x,r_1)} u \right| \leq \left| m_{V_\pm} u - m_{B_\pm} u \right| + \left| m_{B_\pm} u - m_{D_\pm(x,r_1)} u \right| \quad (5)$$

$$\leq Cr^{1-n} \int_{B(x,r) \setminus K} |\nabla u|,$$

by the same argument as for (4), i.e., by Proposition 12.11. Then

$$\left| m_{D_\pm(x,r)} u - m_{D_\pm(x,r_1)} u \right| \leq Cr^{1-n} \int_{B(x,r) \setminus K} |\nabla u| \leq Cr^{1/2} \omega_1(x, r)^{1/2} \quad (6)$$

$$\leq Cr^{1/2} \omega_p(x, r)^{1/2} \leq Cr^{1/2} \omega_2(x, r)^{1/2} \leq C(1 + h(r))^{1/2} r^{1/2},$$

by (4), (5), (23.6), Hölder, and the trivial energy bound (18.8) (note that its proof also works for topological almost-minimizers). Now (6) and the definition (42.4) immediately give an estimate on $r^{1/2}J(x,r) - r_1^{1/2}J(x,r_1)$ from which (2) follows.

$\square$

Here are two immediate consequences of Lemma 1, to be used later.

Lemma 7. *Let $x \in K$ and $r > 0$ be such that $\overline{B}(x,r) \subset \Omega$. Let $r_1 \in (0,r)$ be such that*

$$\beta_K(x,t) \leq 10^{-1} \quad for \ r_1 \leq t \leq r. \tag{8}$$

Then

$$J(x,r_1) \geq \left(\frac{r}{r_1}\right)^{1/2}[J(x,r) - C'], \tag{9}$$

where $C' = (1 + h(r))C$ for some constant C that depends only on n.

Of course (9) is interesting only when $J(x,r) > C(1 + h(r))$.

When $r_1 \leq r \leq 2r_1$, (9) follows directly from (2). Otherwise we can use repeatedly the special case when $r = 2r_1$, and we get that

$$J(x,r_1) \geq \sqrt{2}J(x,2r_1) - C'\sqrt{2} \geq \sqrt{2}^2 J(x,4r_1) - C'\sqrt{2}^2 - C'\sqrt{2} \tag{10}$$
$$\geq \cdots \geq \sqrt{2}^k J(x,2^k r_1) - C'\sqrt{2}^k \left(1 + \sqrt{2}^{-1} + \sqrt{2}^{-2} + \cdots\right)$$
$$\geq \sqrt{2}^k \left[J(x,2^k r_1) - \frac{\sqrt{2}C'}{1 - \sqrt{2}}\right],$$

where we stop at the largest integer k for which $2^k r_1 \leq r$. Now (9) follows from (10) and the special case when $r \leq 2r_1$.

$\square$

Lemma 11. *For each choice of constants $C_0 > 0$ and $\tau > 0$, we can find $\eta > 0$ and $\varepsilon > 0$ such that if $x \in K$ and $r > 0$ are such that $\overline{B}(x,r) \subset \Omega$, and if*

$$\beta_K(x,t) \leq 10^{-1} \quad for \ \eta r \leq t \leq r, \quad J(x,r) \geq \tau, \ and \ \omega_1(x,r) \leq \varepsilon, \tag{12}$$

then $J(x,\eta r) \geq C_0$.

We take $\eta = 2^{-k}$, with k so large that $\sqrt{2}^k \tau \geq 2C_0$. For $0 \leq j \leq k-1$, (2) says that

$$\sqrt{2}^j J(x,2^j \eta r) \geq \sqrt{2}^{j+1}J(x,2^{j+1}\eta r) - C\sqrt{2}^j \omega_1(x,2^{j+1}\eta r)^{1/2}. \tag{13}$$

Note that $\omega_1(x,2^{j+1}\eta r) \leq (2^{j+1}\eta)^{1-2n}\omega_1(x,r)$, just because we integrate on a smaller set (see (23.5)). When we sum (13) over j, $0 \leq j < k$, we get that

$$J(x,\eta r) \geq \sqrt{2}^k J(x,r) - C\eta^{\frac{1}{2}-n}\omega_1(x,r)^{1/2} \geq \sqrt{2}^k \tau - C\eta^{\frac{1}{2}-n}\varepsilon^{1/2} \geq C_0, \tag{14}$$

by definition of k and if ε is small enough (depending also on η). Lemma 11 follows.

$\square$

The typical way to apply the lemmas will be the following. We start with a first ball $B(x,r)$ such that $\beta_K(x,r) + \omega_p(x,r) + h(r)$ is very small. Such balls are easy to find, by Proposition 41.17. Then we can find a new radius $r_1 \in [r/2, r]$ that satisfies the condition (42.12) of Proposition 42.10 (by Chebyshev). If (u, K) is a topological almost-minimizer, let us assume (42.13) as well. Then we get that $J^*(x, r_1)$ is not too small. Because of (42.8), $J(x, r_1)$ is not too small either (if $\omega_p(x,r)$ is small enough). So we may apply Lemma 11 and get that $J(x, \eta r_1) \geq C_0$. If we choose C_0 large enough, we can then apply Lemma 7, and get that $J(x, r')$ is even larger for $r' \leq \eta r_1$. This stays true as long as $\beta_K(x,t) \leq 10^{-1}$ for $r' \leq t \leq r$. Of course, our hypothesis that $\beta_K(x,r) + \omega_p(x,r) + h(r)$ is very small gives this for some time, but if we want to continue the estimate all the way to $r' = 0$, we shall need additional information (for instance, with self-improving estimates).

At this point, we merely wish to convince the reader that in our future proof of self-improving estimates, $J(x,r)$ will be an easy quantity to control.

44 On the size of the holes in K

In this section we want to show that if K is reasonably flat in $B(x,r)$ and the normalized jump $J(x,r)$ is large, then it is possible to add a correspondingly small set to K so that the union separates the two balls $D_\pm(x,r)$ in $B(x,r)$. The proof will be a simple application of the co-area formula (see Section 28). This will be easier to use when $n = 2$, because our separating set in $B(x,r)$ will contain a curve, but for the moment we do not need to worry.

We still assume in this section that $(u, K) \in TRLAM(\Omega, h)$, as in Standard Assumption 41.1 (but as before $(u, K) \in TRLQ(\Omega)$, as in Standard Assumption 18.14, would do as well), and that $B(x,r)$ is a ball centered on K and such that $\overline{B}(x,r) \subset \Omega$ and $\beta_K(x,r) < 1/2$, as in (42.1). Thus we can define a good hyperplane $P(x,r)$ and balls $D_\pm = D_\pm(x,r)$, as in (42.2) and (42.3).

Proposition 1. *We can find a compact set $F(x,r) \subset \overline{B}(x,r)$ for which*

$$K \cap \overline{B}(x,r) \subset F(x,r) \subset \{y \in \overline{B}(x,r) \,; \mathrm{dist}(y, P(x,r)) \leq r\beta_K(x,r)\}; \quad (2)$$

$$F(x,r) \quad \text{separates } D_+(x,r) \text{ from } D_-(x,r) \text{ in } \overline{B}(x,r); \quad (3)$$

$$H^{n-1}(F(x,r) \cap B(x,r) \setminus K) \leq Cr^{n-1}\omega_1(x,r)^{1/2} J(x,r)^{-1} \quad (4)$$

$$\leq Cr^{n-1}\omega_p(x,r)^{1/2} J(x,r)^{-1}$$

for $p > 1$.

As usual, (3) means that the balls $D_\pm(x,r)$ lie in different connected components of $\overline{B}(x,r) \setminus F(x,r)$. The constants C in (4) depend only on n. There would be no serious difficulty with proving the analogue of (4) for $H^{n-1}(F(x,r) \setminus K)$ (see Remark 21), but this will not be needed.

The idea of the proof is that if $J(x,r)$ is large and $\omega_1(x,r)$ is small, there are only very few paths in $B \setminus K$ that go from D_+ to D_-. In other words, K should almost disconnect D_+ from D_- in $\overline{B}(x,r)$. To fill the holes, we would like to use some level set of u, whose size would be controlled by the co-area formula. See Figure 1.

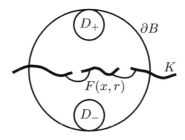

Figure 1

The co-area formula will be easier to apply if we first replace u on $V = B(x,r) \setminus K$ with a slightly different function. The following notation will be useful. For $0 < \lambda < 1$, define a strip around $P(x,r)$ by

$$S(\lambda) = \{y \in \mathbb{R}^n ; \operatorname{dist}(y, P(x,r)) \le \lambda r\}, \tag{5}$$

and then call $\Sigma_\pm(\lambda)$ the connected component of $B(x,r) \setminus S(\lambda)$ that meets $D_\pm$.

Let us first find a function $v \in W^{1,1}(V)$ such that

$$v(y) = m_{D_\pm(x,r)}u \quad \text{for } y \in \Sigma_\pm(2/3) \tag{6}$$

and

$$\int_V |\nabla v| \le C \int_V |\nabla u| . \tag{7}$$

This is easy to do. We start with two smooth functions $\varphi_\pm$ on $B(x,r)$ such that $0 \le \varphi_\pm(y) \le 1$ everywhere, $\varphi_\pm(y) = 1$ on $\Sigma_\pm(2/3)$, $\varphi_\pm(y) = 0$ out of $\Sigma_\pm(1/2)$, and $|\nabla \varphi_\pm(y)| \le Cr^{-1}$ everywhere. Then we set $\varphi(y) = 1 - \varphi_+(y) - \varphi_-(y)$ and

$$v(y) = \varphi(y)\, u(y) + \varphi_+(y)\, m_{D_+(x,r)}u + \varphi_-(y)\, m_{D_-(x,r)}u \quad \text{for } y \in V. \tag{8}$$

We have (6) trivially. Concerning (7), note that

$$\nabla v(y) = \varphi(y)\, \nabla u(y) - \mathbf{1}_{\Sigma_+(1/2)}(y)\, \nabla\varphi_+(y)\, [u(y) - m_{D_+(x,r)}u] \tag{9}$$
$$- \mathbf{1}_{\Sigma_-(1/2)}(y)\nabla\varphi_-(y)\, [u(y) - m_{D_-(x,r)}u].$$

Since $\Sigma_\pm(1/2) \subset B(x,r) \setminus K$, we can apply (the proof of) Proposition 12.11 and get that

$$\int_{\Sigma_\pm(1/2)} |\nabla\varphi_\pm(y)|\, |u(y) - m_{D_\pm(x,r)}u|\, dy \tag{10}$$
$$\le Cr^{-1} \int_{\Sigma_\pm(1/2)} |u(y) - m_{D_\pm(x,r)}u|\, dy \le C \int_{\Sigma_\pm(1/2)} |\nabla u(y)|\, dy ,$$

and then (7) follows from (8) and (9).

Now we want to replace v with a smooth function w on V with essentially the same properties. More precisely, we want to keep

$$w(y) = m_{D_\pm(x,r)} u \ \text{ for } y \in \Sigma_\pm(3/4) \tag{11}$$

and

$$\int_V |\nabla w| \leq C \int_V |\nabla u|, \tag{12}$$

as in (6) and (7). Our construction is very standard, it is almost the same as for the Whitney extension theorem with Whitney cubes, but let us sketch it anyway. For each $z \in V$, set $B(z) = B(z, 10^{-3} \operatorname{dist}(z, \partial V))$, and then choose a maximal set $Z \subset V$ such that the $B(z)$, $z \in Z$, are disjoint. Note that if $y \in V$, then $B(y)$ meets some $B(z)$, $z \in Z$, and hence $y \in 4B(z)$ (because $|y - z|$ is much smaller than $\operatorname{dist}(y, \partial V)$). Thus the $4B(z)$, $z \in Z$, cover V.

For each $z \in Z$, choose a smooth function φ_z supported in $5B(z)$, such that $\varphi_z(y) = 1$ for $y \in 4B(z)$, $0 \leq \varphi_z(y) \leq 1$ and $|\nabla \varphi_z(y) \leq C \operatorname{dist}(z, \partial V)^{-1}$ everywhere. Set $\Phi(y) = \sum_{z \in Z} \varphi_z(y)$ on V. Note that $\Phi(y) \geq 1$ because the $4B(z)$ cover V, and the sum near a given point y has less than C terms because the $B(z)$ are disjoint and all the $B(z)$ for which $y \in 5B(z)$ have radii close to $10^{-3} \operatorname{dist}(y, \partial V)$.

Next set $\varphi_z^*(y) = \varphi_z(y)/\Phi(y)$; note that $\sum_{z \in Z} \varphi_z^*(y) = 1$ on V. Finally set $m_z = m_{B(z)} v$ for $z \in Z$ and $w(y) = \sum_{z \in Z} m_z \varphi_z^*(y)$ for $y \in V$.

If $y \in \Sigma_\pm(3/4)$ then $m_z = m_{D_\pm(x,r)} u$ for all the $z \in Z$ such that $5B(z)$ contains y (by (6)), hence $w(y) = m_{D_\pm(x,r)} u$ (because $\sum_{z \in Z} \varphi_z^*(y) = 1$). Thus (11) holds. Also,

$$\nabla w(y) = \sum_{z \in Z} m_z \nabla \varphi_z^*(y) = \sum_{z \in Z} [m_z - m(y)] \, [\nabla \varphi_z^*(y)], \tag{13}$$

where we denote by $m(y)$ the mean value of v on $B(y) = B(y, 10^{-1} \operatorname{dist}(y, \partial V))$, and because $\sum_{z \in Z} \nabla \varphi_z^*(y) = 0$. The sum at a given point has less than C terms, and each of them is less than

$$C \operatorname{dist}(y, \partial V)^{-1} |m_z - m(y)| \leq C \operatorname{dist}(y, \partial V)^{-n} \int_{B(y)} |\nabla v|,$$

by Proposition 12.11 and because all the balls $5B(z)$ that contain y are contained in $B(y) \subset V$. Thus $|\nabla w(y)| \leq C \operatorname{dist}(y, \partial V)^{-n} \int_{B(y)} |\nabla v|$, and to get (12) we just need to integrate on V, apply Fubini, and use (7).

Let us apply the co-area formula (28.3) to w on V. We get that

$$\int_{\mathbb{R}} H^{n-1}(\Gamma_t) dt = \int_V |\nabla w| \leq C \int_V |\nabla u|, \tag{14}$$

where we call $\Gamma_t = \{y \in V ; w(y) = t\}$ the level sets of w in V. We may assume that $J(x,r) > 0$, because otherwise we can take $F(x,r) = [K \cup P(x,r)] \cap \overline{B}(x,r)$

and (2)–(4) hold trivially. By (14) and Chebyshev, we can find $t \in \mathbb{R}$ such that

$$t \text{ lies strictly between } m_{D_+(x,r)} u \text{ and } m_{D_-(x,r)} u \tag{15}$$

and

$$H^{n-1}(\Gamma_t) \le C |m_{D_+(x,r)} u - m_{D_-(x,r)} u|^{-1} \int_V |\nabla u| \tag{16}$$

$$= C\, r^{-1/2}\, J(x,r)^{-1} \int_V |\nabla u| = C r^{n-1} J(x,r)^{-1}\, \omega_1(x,r)^{1/2},$$

where we also used the definitions (42.4) and (23.5).

Set $F = \Gamma_t \cup [K \cap B(x,r)] \subset B(x,r)$. This set is closed in $B(x,r)$, because Γ_t is closed in $V = B(x,r) \setminus K$ and K is closed. The main point of adding a level set is that

$$F \text{ separates } \Sigma_+(3/4) \text{ from } \Sigma_-(3/4) \text{ in } B(x,r). \tag{17}$$

Indeed if γ is a path in $B(x,r)$ that goes from $\Sigma_+(3/4)$ to $\Sigma_-(3/4)$ and if γ does not meet K, then $\gamma \subset V$, hence w is defined and continuous on γ. Then there is a point $y \in \gamma$ such that $w(y) = t$, by (11) and (15). Thus $y \in \Gamma_t \subset F$, and (17) follows.

It could be that F is not contained in the strip $S = S(\beta_K(x,r))$ (see (5)), but this is easy to fix. We can replace F with $F' = \pi(F)$, where π denotes an appropriate contraction of $S(3/4) \cap B(x,r)$ onto $S \cap B(x,r)$. More precisely, choose π so that

$$\pi(y) = y \text{ on } S \cap B(x,r), \tag{18}$$

$$\pi : S(3/4) \cap B(x,r) \to S \cap B(x,r) \text{ is a limit of homeomorphisms}$$
$$\pi_k : S(3/4) \cap B(x,r) \to S(\beta_K(x,r) + 2^{-k}) \cap B(x,r), \tag{19}$$

and π is Lipschitz with constant less than 10. Then

$$H^{n-1}(F' \setminus K) \le 10^{n-1} H^{n-1}(F \setminus K) \le C r^{n-1} \omega_1(x,r)^{1/2} J(x,r)^{-1}, \tag{20}$$

because π does not move any point of K (by (18) and because $K \cap B(x,r) \subset S$ by (42.2)), and then by (16). Also, F' still separates $\Sigma_+(3/4)$ from $\Sigma_-(3/4)$ in $B(x,r)$, by (17) and (19). Indeed, if $\gamma \subset B(x,r)$ is a curve from $\Sigma_+(3/4)$ to $\Sigma_-(3/4)$, it must meet all the $\pi_k(F)$, hence also F'.

Set $F(x,r) = F' \cup [S \cap \partial B(x,r)]$. It is compact because F' is closed in $B(x,r)$ and contained in S. Next (2) holds by definition (and because π does not move $K \cap B(x,r)$), and (3) holds because F' is contained in S and separates $\Sigma_+(3/4)$ from $\Sigma_-(3/4)$ in $B(x,r)$. Finally, (4) follows from (20) and Hölder (for the second part). This completes the proof of Proposition 1. $\qquad \square$

Remark 21. If we want to find $F(x,r)$ such that in addition

$$H^{n-1}(F(x,r) \setminus K) \leq Cr^{n-1}\,\omega_1(x,r)^{1/2}\,J(x,r)^{-1}, \tag{22}$$

we can consider the topological sphere B^* obtained by gluing together two copies of $B(x,r)$ along $\partial B(x,r)$. Then u can be extended by symmetry (with respect to $\partial B(x,r)$); we get a function $u^* \in W^{1,1}(B^* \setminus K^*)$, where K^* is the union of $K \cap \overline{B}(x,r)$ and its symmetric copy. The same argument as above applies to u^* and gives a set $F^*(x,r) \subset B^*$ that separates top from bottom and almost entirely lies in K^*, and then $F^*(x,r) \cap \overline{B}(x,r)$ satisfies (2), (3), and (22).

Remark 23. Proposition 1 is true trivially when $\beta_K(x,r) < 1/2$ and K separates $D_+(x,r)$ from $D_-(x,r)$ in $\overline{B}(x,r)$. We simply take $F(x,r) = K \cap \overline{B}(x,r)$, and (4) even holds with a right-hand side equal to 0, regardless of the values of $J(x,r)$ and $\omega_1(x,r)$.

Exercise 24. Complete the details of the construction of w, and in particular the proof of (12). Why can we not use the proof of Lemma 11.1 directly to construct w?

45 $\omega_p^*(x,r)$ (sometimes with flatness) controls the surface and energy inside

Here we give two results that say that $H^{n-1}(K \cap B(x,r))$ is not much larger than $H^{n-1}(P \cap B(x,r))$, where P is any hyperplane through x. In the first one, we shall assume that for some $p \in (\frac{2(n-1)}{n}, 2]$, $\omega_p^*(x,r)$ and $\beta_K(x,r)$ are small, and we shall get that $H^{n-1}(K \cap B(x,r)) - H^{n-1}(P \cap B(x,r))$ is correspondingly small. In the second one, we shall assume that $n = 2$ and $K \cap \partial B$ consists of two points, not too close to each other, and we shall get an estimate in terms of $\omega_p^*(x,r)$ alone. The proofs will be comparable, and also of the same vein as in Section 42. In both cases we shall control the energy $\int_{B(x,r)\setminus K} |\nabla u|^2$ at the same time, but this is not the most important in general.

For both results we assume that $(u,K) \in TRLAM(\Omega,h)$, as in Standard Assumption 41.1, and this time quasiminimizers would not do. See Remark 12.

Proposition 1. *There is a constant C that depends only on n and, for each $p \in (\frac{2(n-1)}{n}, 2]$, a constant C_p that depends also on p, so that if $x \in K$ and $\overline{B}(x,r) \subset \Omega$, then*

$$H^{n-1}(K \cap \overline{B}(x,r)) + \int_{B(x,r)\setminus K} |\nabla u|^2 \tag{2}$$

$$\leq H^{n-1}(P(x,r) \cap B(x,r)) + C_p\,r^{n-1}\,\omega_p^*(x,r) + Cr^{n-1}\,\beta_K(x,r) + r^{n-1}\,h(r).$$

Of course $P(x,r)$ in (2) could be replaced with any other hyperplane through x. Recall that $\omega_p^*(x,r)$ is the normalized (square of) L^p norm of ∇u on $\partial B(x,r)$ defined in (42.5). Maybe we should observe that Proposition 1 will not be so easy to use in the core of our C^1 theorem. The difficulty comes from the term with $\beta_K(x,r)$, which is usually pretty large in our arguments. In this respect, Proposition 9 below will be better (but it only works in dimension 2). On the other hand, Proposition 1 will be useful as a preparation lemma, because it still gives some modestly good estimate on $H^{n-1}(K \cap \overline{B}(x,r))$, and transforms a good estimate on $\omega_p^*(x,r)$ (or even $\omega_p(x,r)$, if we may replace r with a comparable radius, and by Chebyshev), with $p < 2$, into an energy estimate (with $p = 2$).

For the proof we may as well assume that $\beta_K(x,r) < 1/10$, say, because otherwise we may use the trivial estimate (18.8), i.e., that $H^{n-1}(K \cap \overline{B}(x,r)) + \int_{B(x,r) \setminus K} |\nabla u|^2 \leq Cr^{n-1} + h(r)r^{n-1}$. [This estimate is still valid, with the same proof, for topological almost-minimizers.] We want to construct a competitor (v, G) for (u, K) in $B(x,r)$, with

$$G = [P(x,r) \cap B(x,r)] \cup Z \cup [K \setminus B(x,r)], \qquad (3)$$

where $Z = \{y \in \partial B(x,r) ; \operatorname{dist}(y, P(x,r)) \leq r\beta_K(x,r)\}$. Note that

$$H^{n-1}(G \cap \overline{B}(x,r)) \leq H^{n-1}(P(x,r) \cap B(x,r)) + Cr^{n-1} \beta_K(x,r). \qquad (4)$$

Now we need to define the function v. Denote by $S_{\pm}$ the two components of $\partial B(x,r) \setminus Z$ and by $V_{\pm}$ the corresponding components of $B(x,r) \setminus G = B(x,r) \setminus P(x,r)$. We may of course assume that $\omega_p^*(x,r) < \infty$, because otherwise (2) is trivial. Then $u \in W^{1,p}(S_{\pm})$, by Convention 42.6 and the definition of $P(x,r)$ in (42.2). Let us fix a sign $\pm$ for a moment, and extend to the whole sphere the restriction of u to $S_{\pm}$. If $S_{\pm}$ was a true half-sphere, we would do that by symmetry with respect to $P(x,r)$. This is not the case here, but $\beta_K(x,r) < 1/10$, hence there is a smooth homeomorphism φ of $\partial B(x,r)$, with bilipschitz constant reasonably close to 1, and which transforms $S_{\pm}$ into one of the two true half-spheres of $\partial B(x,r)$ bounded by $\mathcal{C} = \partial B(x,r) \cap P(x,r)$. We take the restriction of $u \circ \varphi^{-1}$ to this half-sphere, extend it by symmetry, and compose with φ to get an extension $u_{\pm}$ of the restriction of u to $S_{\pm}$. We claim that $u_{\pm} \in W^{1,p}(\partial B(x,r))$, with

$$\int_{\partial B(x,r)} |\nabla_\tau u_{\pm}|^p \, dH^{n-1} \leq C \int_{S_{\pm}} |\nabla_\tau u|^p \, dH^{n-1} \leq C \int_{S_{\pm}} |\nabla u|^p \, dH^{n-1}, \qquad (5)$$

where we denote by $\nabla_\tau u_{\pm}$ the gradient of $u_{\pm}$ on the sphere. Indeed, it is enough to check that our symmetric extension $u_{\pm} \circ \varphi^{-1}$ lies in $W^{1,p}(\partial B(x,r))$, and that the analogue for it of the first inequality holds. It is clear that the extension $u_{\perp} \circ \varphi^{-1}$ lies in $W^{1,p}(\partial B(x,r) \setminus \mathcal{C})$, with the appropriate estimate. [Recall that the great circle $\mathcal{C} = \partial B(x,r) \cap P(x,r)$ is the boundary of $\varphi(S_{\pm})$.] Now we use the analogue of Corollary 14.28 on $\partial B(x,r)$, where the two radial limits on $\mathcal{C}$ coincide by symmetry, and get the result.

Next we apply Lemma 22.32 to get an extension $v_\pm$ of $u_\pm$ to $B(x,r)$ such that

$$\int_{B(x,r)} |\nabla v_\pm|^2 \le C(n,p)\, r^{n-\frac{2n}{p}+\frac{2}{p}} \left\{ \int_{\partial B(x,r)} |\nabla_\tau u_\pm|^p \right\}^{2/p} \tag{6}$$

$$\le C(n,p)\, r^{n-1} \omega_p^*(x,r),$$

by (5) and the definition (42.5). As usual, by extension we mean that the radial limits of $v_\pm$ coincide with $u_\pm$ almost-everywhere on $\partial B(x,r)$.

We take $v(y) = v_\pm(y)$ on $V_\pm$, and keep $v(y) = u(y)$ out of $B(x,r)$. Then $v \in W^{1,2}(B(x,r) \setminus G) \cap W^{1,2}_{\mathrm{loc}}(\Omega \setminus [\overline{B}(x,r) \cup K])$, and it has radial limits on $\partial B(x,r) \setminus Z$ from both sides that coincide (by Convention 42.6 again). Thus $v \in W^{1,2}_{\mathrm{loc}}(\Omega \setminus G)$, by Corollary 14.28, and (v,G) is a competitor for (u,K) in $B(x,r)$. It is even a topological competitor, because $K \cap \partial B(x,r) \subset G$ and $G \cap \overline{B}(x,r)$ separates the two components of $\partial B(x,r) \setminus G$ in $\overline{B}(x,r)$; see Lemma 8 below for details. The almost-minimality condition (7.7) says that

$$\int_{B(x,r) \setminus K} |\nabla u|^2 + H^{n-1}(K \cap \overline{B}(x,r)) \tag{7}$$

$$\le \int_{B(x,r) \setminus G} |\nabla v|^2 + H^{n-1}(G \cap \overline{B}(x,r)) + r^{n-1}\, h(r).$$

Now (2) follows from (7), (6), and (4), which proves Proposition 1 modulo Lemma 8. $\qquad\square$

Lemma 8. *Suppose $(u,K) \in \mathcal{A}$ (with $\mathcal{A}$ as in (2.1)), B is a ball such that $\overline{B} \subset \Omega$, and (v,G) is a competitor for (u,K) in B. Also assume that $K \cap \partial B \subset G$ and that whenever I, J are two connected components of $\partial B \setminus G$ that lie in different components of $\Omega \setminus K$, $G \cap \overline{B}$ separates I from J in $\overline{B}$. Then (v,G) is a topological competitor for (u,K) in B.*

Note that in the statement, any component of $\partial B \setminus G$ (like I and J) is automatically contained in $\Omega \setminus K$, since $K \cap \partial B \subset G$. To prove the lemma, we just need to check (7.16). We shall proceed by contradiction, so let us assume that we can find y, $z \in \Omega \setminus [K \cup \overline{B}]$ such that K separates them in Ω, but G does not.

Let γ be a path in $\Omega \setminus G$ from y to z. By contradiction assumption, γ meets K, and since $K = G$ out of $\overline{B}$, this must happen in $\overline{B}$. So γ meets $\overline{B}$. See Figure 1.

Call y_1 the first point of ∂B that we meet when we go from y to z along γ. Note that γ does not meet K between y and y_1, again because $K = G$ out of $\overline{B}$. Also, $y_1 \in \partial B \setminus G \subset \partial B \setminus K$. Thus y_1 lies in the connected component of $\Omega \setminus K$ that contains y (call it $V(y)$).

Similarly let z_1 be the first point of ∂B when we go backwards from z to y along γ. Then $z_1 \in \partial B \cap V(z)$, where $V(z)$ is the component of $\Omega \setminus K$ that contains z. Note that $V(z) \ne V(y)$ by contradiction assumption.

Next call y_2 the first point of the closed set $\partial B \setminus V(y)$ when we follow γ from y_1 to z_1. Such a point exists, because $z_1 \in \partial B \setminus V(y)$, and because γ stays in $V(y)$ for some time after leaving from y_1. Note that y_2 lies on $\partial B \setminus K$ because $\partial B \cap K \subset \partial B \cap G$ and γ does not meet G.

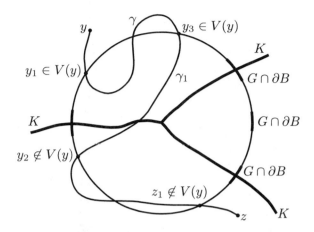

Figure 1

Now start from y_2 and run backwards along γ. We stay out of ∂B for some time, because if y_2 was the limit of a sequence of points of $\gamma \cap \partial B$ before y_2, then some of these points would be in the same component of $\Omega \setminus K$ as y_2, in contradiction with the definition of y_2. Then we can talk about the first point of ∂B when we start from y_2 and run backwards along γ. Call this point y_3. Note that y_3 lies between y_1 and y_2 (just because $y_1 \in \partial B$).

Call γ_1 the arc of γ between y_3 and y_2. Then $\gamma_1 \setminus \{y_3, y_2\}$ does not meet ∂B, by definition of y_3. On the other hand, $y_2 \notin V(x)$ and $y_3 \in V(x)$, by definition of y_2. So $\gamma_1 \setminus \{y_3, y_2\}$ meets K, and this can only happen inside $\overline{B}$, because $K = G$ out of $\overline{B}$. Thus $\gamma_1 \setminus \{y_3, y_2\} \subset B$ (because it is connected, meets $\overline{B}$, but does not meet ∂B). Since y_2 and y_3 lie on $\partial B \setminus G$ (because γ does not meet G) and belong to different components of $\Omega \setminus K$, our assumption says that $G \cap \overline{B}$ separates them in $\overline{B}$. This contradicts the fact that $\gamma_1 \subset \overline{B}$ and $\gamma_1 \subset \gamma$ does not meet G. Lemma 8 follows. $\square$

The next proposition only works in the plane, but it does not have the slightly unpleasant boundary term $\beta_K(x,r)$. We shall see in the next section that the hypothesis (10) is fairly easy to get, for instance by averaging the conclusion of Proposition 1.

Proposition 9 ($n = 2$ only). *Let $1 < p \leq 2$ be given, and let us still assume that $(u, K) \in TRLAM(\Omega, h)$, as in Standard Assumption 41.1. Let $x \in K$ and $r > 0$ be such that $\overline{B}(x,r) \subset \Omega$ and*

$$K \cap \partial B(x,r) \text{ consists of two points } a \text{ and } b, \text{ with } |b-a| \geq r. \qquad (10)$$

Then

$$H^1(K \cap \overline{B}(x,r)) + \int_{B(x,r)\setminus K} |\nabla u|^2 \leq |b - a| + C_p\, r\, \omega_p^*(x,r) + r\, h(r). \qquad (11)$$

The proof follows the same lines as for Proposition 1, except that we take $G = [a,b] \cup [K \setminus B(x,r)]$. Then $\partial B(x,r) \setminus G$ still has two components $S_\pm$, namely, the two arcs of $\partial B(x,r) \setminus \{a,b\}$. These arcs are neither too long nor too short, and so we can extend the restriction of u to each of these arcs to the whole circle, with the same estimate (5) as before. We get functions $u_\pm$ on $\partial B(x,r)$, which we can in turn extend to $B(x,r)$. (This time, we may use Lemma 22.16.) We get functions $v_\pm$ which satisfy the analogue of (6), and which we can use as before. This gives a topological competitor (v,G) for (u,K) in $B(x,r)$, and the almost-minimality condition (7) now yields (11), because $H^1(G \cap \overline{B}(x,r)) = |b - a|$. Proposition 9 follows. $\qquad\qquad\qquad\qquad\qquad\qquad\qquad\qquad\qquad\qquad\qquad\qquad\qquad\quad \square$

Remark 12. The results of this section are typical ones for which quasiminimality will not really help. For instance, if in Proposition 1 we assume that (u,K) is a reduced local (r_0, M, a)-quasiminimizer in Ω, plain or topological (as in Standard Assumption 18.14), then instead of (2) we get that

$$H^{n-1}(K \cap \overline{B}(x,r)) + \int_{B(x,r)\setminus K} |\nabla u|^2 \qquad\qquad\qquad\qquad\qquad (13)$$

$$\leq M H^{n-1}(P(x,r) \cap B(x,r)) + M C_p\, r^{n-1}\, \omega_p^*(x,r) + C r^{n-1}\, \beta_K(x,r) + a\, r^{n-1}.$$

This will not be too useful, unless M is quite close to 1 (but then we are very close to the world of almost-minimizers). The problem comes from the fact that in the argument above, we replace a piece of K with the whole hyperdisk $P(x,r) \cap B(x,r)$, instead of just adding a small piece as we often do. For this we may end up paying a lot, and this is logical. Indeed (u,K) could be a minimizer for some functional with a term $\int_K b(y)\, dH^{n-1}(y)$, where b, even though bounded from above and below, will happen to be significantly larger on $P(x,r)$ than on K. Then it does not make sense to replace $K \cap B(x,r)$ with $P(x,r) \cap B(x,r)$, even though the last one could be somewhat smaller in terms of Hausdorff measure alone. For the same reason, we cannot expect K to have large pieces of C^1 curves or surfaces when (u,K) is merely a quasiminimizer, because already any Lipschitz graph minimizes a functional like $\int_K b(y)\, dH^{n-1}(y)$. (Take $b = 1$ on the graph, and $b = M$ outside.)

On the other hand, we do not need the gauge function h to tend to 0 at 0; it may even be very large (but then (2) and (11) may be essentially useless).

Remark 14. So far all the results of this part apply to almost-minimizers with linear distortion, with the same proofs. This is still the case for this section, but the statements need to be modified slightly. Recall from Definition 7.45 that if $(u,K) \in TRLAM$, but with linear distortion $\leq L$, then we associate two invertible matrices A_1 and A_2 to each pair (x,r), and almost-minimality is computed after linear distortion of the H^{n-1}-term by A_1 and of the energy term by A_2. Propositions 1 and 9 stay true in this context, but we have to replace (2) with

$$H^{n-1}(A_1(K \cap \overline{B}(x,r))) + \int_{B(x,r) \setminus K} |A_2 \nabla u|^2 \tag{15}$$

$$\leq H^{n-1}(A_1(P(x,r) \cap B(x,r))) + C_p \, r^{n-1} \, \omega_p^*(x,r) + C r^{n-1} \, \beta_K(x,r) + r^{n-1} \, h(r),$$

where this time $H^{n-1}(A_1(P(x,r) \cap B(x,r)))$ could depend on the orientation of $P(x,r)$. Similarly, (11) should be replaced with

$$H^1(A_1(K \cap \overline{B}(x,r))) + \int_{B(x,r) \setminus K} |A_2 \nabla u|^2 \leq H^1(A_1([a,b])) + C_p \, r \, \omega_p^*(x,r) + r \, h(r).$$
$$\tag{16}$$

As usual, the various constants get larger, depending on our bound L for the A_j and their inverse.

46   Simple consequences when $n = 2$, and the existence of good disks

Here we continue to assume that $(u, K) \in TRLAM(\Omega, h)$, as in Standard Assumption 41.1, but we restrict our attention to the two-dimensional case and draw some easy conclusions from the last few sections. Let $x \in K$ and $r > 0$ be given, with $\overline{B}(x,r) \subset \Omega$. Let us also assume that $\beta_K(x,r) < 1/2$, as in (42.1), but implicit stronger assumptions will come soon.

In Section 44 we constructed a compact set $F(x,r)$ that splits $\overline{B}(x,r)$ and for which

$$H^1(F(x,r) \cap B(x,r) \setminus K) \leq C \, r \, \omega_1(x,r)^{1/2} J(x,r)^{-1}, \tag{1}$$

where $J(x,r)$ is the normalized jump from (42.4). See Proposition 44.1. We also know from Proposition 45.1 that for $p > 1$,

$$H^1(K \cap \overline{B}(x,r)) \leq 2r + C_p \, r \, \omega_p^*(x,r) + C r \, \beta_K(x,r) + r \, h(r). \tag{2}$$

Also, recall from (44.2) that $F(x,r) \subset \{y \in \overline{B}(x,r) \, ; \, \mathrm{dist}(y, P(x,r)) \leq r \beta_K(x,r)\}$, so that $H^1(F(x,r) \cap \partial B(x,r)) \leq C r \beta_K(x,r)$. Thus (1) and (2) yield

$$H^1(F(x,r)) \leq 2r + C r \, \beta_K(x,r) + C \, r \, \omega_1(x,r)^{1/2} J(x,r)^{-1} + C_p \, r \, \omega_p^*(x,r) + r \, h(r). \tag{3}$$

We want to use (3) to control the following good set of radii:

$$\mathcal{G}(x,r) = \{t \in (0,r) \, ; \quad F(x,r) \cap \partial B(x,t) = K \cap \partial B(x,t) \tag{4}$$

and this set contains exactly two points a and b, with $|b - a| \geq t\}.$

When $t \in \mathcal{G}(x,r)$, we shall often say that $B(x,t)$ is a good disk. The next lemma is a simple consequence of (2) and (3). It says that when the error terms in the right-hand side of (3) are small, most $t \in (0,r)$ lie in $\mathcal{G}(x,r)$.

Lemma 5. *With the notation above,*

$$H^1([0,r] \setminus \mathcal{G}(x,r)) \leq Cr\,\beta_K(x,r) + C\,r\,\omega_1(x,r)^{1/2}J(x,r)^{-1} \tag{6}$$
$$+ C_p\,r\,\omega_p^*(x,r) + r\,h(r).$$

Let us check this. Recall from (44.2) that

$$K \cap \overline{B}(x,r) \subset F(x,r) \subset S = \{y \in \overline{B}(x,r)\,;\,\mathrm{dist}(y,P(x,r)) \leq r\beta_K(x,r)\}. \tag{7}$$

Because of the term $Cr\,\beta_K(x,r)$ in (6), we may assume that $\beta_K(x,r) < 10^{-1}$ and that $t \geq 10\beta_K(x,r)\,r$. Then $B(x,t)$ widely sticks out of S. Call $\partial_\pm$ the two components of $\partial B(x,t) \setminus S$. Then (44.3) and (7) say that they lie in different components of $\overline{B}(x,r) \setminus F(x,r)$. This implies that each of the two intervals that compose $\partial B(x,t) \setminus \partial_+ \cup \partial_-$ meets $F(x,r)$. Thus

$$F(x,r) \cap \partial B(x,t) \text{ contains at least two points } a,b, \text{ with } |b-a| \geq t, \tag{8}$$

where the estimate on $|b-a|$ comes from (7) and the fact that $t \geq 10\beta_K(x,r)\,r$. Thus

$$[10\beta_K(x,r)r,r) \setminus \mathcal{G}(x,r) \subset \mathcal{B}_1 \cup \mathcal{B}_2, \tag{9}$$

where

$$\mathcal{B}_1 = \{t \in (10\beta_K(x,r)r,r)\,;\,F(x,r) \cap \partial B(x,t) \neq K \cap \partial B(x,t)\}, \tag{10}$$

and

$$\mathcal{B}_2 = \{t \in (10\beta_K(x,r)r,r)\,;\,K \cap \partial B(x,t) \text{ has more than two points}\}. \tag{11}$$

First note that $\mathcal{B}_1$ is the image, under the 1-Lipschitz radial projection $y \to |y-x|$, of $F(x,r) \cap B(x,r) \setminus K$. Hence

$$H^1(\mathcal{B}_1) \leq H^1(F(x,r) \cap B(x,r) \setminus K) \leq C\,r\,\omega_1(x,r)^{1/2}J(x,r)^{-1}, \tag{12}$$

by Remark 2.11 and (1). Next we want to apply Lemma 26.1 to the set $K_0 = K \cap B(x,r)$. Of course K_0 is not closed, but this is extremely easy to fix, for instance by restricting to smaller closed balls and taking a limit. We get that the number $N(t)$ of points in $\partial B(x,t) \cap K$ is a Borel-measurable function of $t < r$, and that

$$\int_0^r N(t)dt \leq H^1(K_0) \leq 2r + C_p\,r\,\omega_p^*(x,r) + Cr\,\beta_K(x,r) + r\,h(r), \tag{13}$$

where the last inequality comes from (2). Set $\mathcal{G} = (10\beta_K(x,r)r,r) \setminus \mathcal{B}_1 \cup \mathcal{B}_2$. Then $N(t) = 2$ for $t \in \mathcal{G}$, by (8), and

$$\int_0^r N(t)dt \geq \int_{\mathcal{G}} N(t)dt + \int_{\mathcal{B}_2} N(t)dt \geq 2H^1(\mathcal{G}) + 3H^1(\mathcal{B}_2)$$
$$\geq 2[r - 10\beta_K(x,r)r - H^1(\mathcal{B}_1) - H^1(\mathcal{B}_2)] + 3H^1(\mathcal{B}_2) \tag{14}$$
$$= 2r - 2H^1(\mathcal{B}_1) + H^1(\mathcal{B}_2) - 20\beta_K(x,r)r,$$

and (13) says that

$$
\begin{aligned}
H^1(\mathcal{B}_2) &\le \int_0^r N(t)dt - 2r + 2H^1(\mathcal{B}_1) + 20\beta_K(x,r)r \\
&\le C_p \, r \, \omega_p^*(x,r) + Cr \, \beta_K(x,r) + r \, h(r) + 2H^1(\mathcal{B}_1).
\end{aligned}
\tag{15}
$$

Then we use (12), and (6) follows from (9). This proves Lemma 5. □

The following property of good disks will be used in Section 48 for the main decay estimate.

Lemma 16. *Let $t \in \mathcal{G}(x,r)$ be such that $t \ge 10\beta_K(x,r)r$, and write $F(x,r) \cap \partial B(x,t) = \{a,b\}$. Then a and b lie in the same connected component of $F(x,r) \cap \overline{B}(x,t)$.*

This is a simple consequence of a standard result in plane topology. Call I_1 and I_2 the two intervals that compose $\partial B(x,t) \setminus F(x,r)$. Since $t \ge 10\beta_K(x,r) \, r$, I_1 and I_2 get out of $S = \{y \, ; \mathrm{dist}(y, P(x,r)) \le r\beta_K(x,r)\}$, and on different sides of S. Then (44.2) and (44.3) say that they lie in different components of $\overline{B}(x,r) \setminus F(x,r)$. Then we can apply Theorem 14.3 on p. 123 of [Ne] as we did near Lemma 29.51. That is, we call D the line through a and b (the two points of $F(x,r) \cap \partial B(x,t)$), and consider $F^* = [F(x,r) \cap \overline{B}(x,t)] \cup [D \setminus \overline{B}(x,t)]$. Then F^* separates I_1 from I_2 in the plane (see the proof of (29.52) for details). Pick a point z_i in each I_i. The aforementioned theorem from [Ne] says that there is a component Γ of F^* that separates z_1 from z_2. Then $\Gamma \setminus [D \setminus \overline{B}(x,t)]$ is a connected subset of $F(x,r) \cap \overline{B}(x,t)$ that contains a and b. See the end of the proof of Lemma 29.51 for details. We could also use Exercise 29.100 for a direct argument. □

Remark 17. The results of this section stay true when K separates $D_+(x,r)$ from $D_-(x,r)$ in $\overline{B}(x,r)$, and we can even replace $J(x,r)^{-1}$ with 0 (i.e., omit the terms $\omega_1(x,r)^{1/2} J(x,r)^{-1}$ in the upper bounds). This is because we take $F(x,r) = K \cap \overline{B}(x,r)$ in that case. See Remark 44.23, and (42.3) for the definition of $D_\pm(x,r)$.

Remark 18. For almost-minimizers with linear distortion, we shall need a different notion of good disks (because the distortion of energy will force us to do linear changes of variables), and also we need an additional argument to get the good disks, because we only know from (45.15) or (45.16) that the length of $K \cap B(x,r)$ is almost minimal after a linear change of variable. Fortunately, the notion does not depend too much on the change of variable.

Lemma 19. *For each choice of $L \ge 1$ and $\delta > 0$, we can find $\varepsilon > 0$ such that, if B is an invertible 2×2 matrix such that $||B|| \le L$ and $||B^{-1}|| \le L$, $I = [a,b]$ is a line segment in the plane, and $F \subset \mathbb{R}^2$ is a compact connected set that contains a and b and for which $H^1(F) \le (1+\varepsilon)H^1(I)$, then $H^1(B(F)) \le (1+\delta)H^1(B(I))$.*

Indeed, let F be as in the lemma. Proposition 30.14 says that it contains a simple arc γ that goes from a to b. Since $H^1(\gamma) \geq H^1(I)$,

$$H^1(B(F \setminus \gamma)) \leq LH^1(F \setminus \gamma) \leq L\varepsilon H^1(I), \qquad (20)$$

so we just need to estimate $H^1(B(\gamma))$. Let $z : J \to \mathbb{R}^2$ be a parameterization of γ by arclength (see Exercise 57.31 for details). Thus $H^1(J) = H^1(\gamma) \leq H^1(F) \leq (1 + \varepsilon)H^1(I)$ by assumption. Set $v = (b - a)/|b - a|$ and call π the orthogonal projection on the vector line through v. Then

$$H^1(I) = |b - a| = \left| \int_J z'(t)dt \right| = \left| \int_J \pi(z'(t))dt \right| \leq \int_J |\pi(z'(t))|dt, \qquad (21)$$

and hence $\int_J \left[1 - |\pi(z'(t))|\right] dt \leq H^1(J) - H^1(I) \leq \varepsilon H^1(I)$.

Notice that $\operatorname{dist}(z'(t), \pm v) \leq 2 \left[1 - |\pi(z'(t))|\right]^{1/2}$ almost-everywhere (because $|z'(t)| = 1$), so

$$\int_J \operatorname{dist}(z'(t), \pm v) \, dt \leq 2 \int_J \left[1 - |\pi(z'(t))|\right]^{1/2} dt$$

$$\leq 2\{H^1(J)\}^{1/2} \left\{ \int_J \left[1 - |\pi(z'(t))|\right] dt \right\}^{1/2} \leq 3\varepsilon^{1/2} H^1(I), \qquad (22)$$

by Cauchy–Schwarz. We apply B and get that

$$H^1(B(\gamma)) \leq \int_J |B(z'(t))|dt \leq H^1(J)\,|B(v)| + L \int_J \operatorname{dist}(z'(t), \pm v) \, dt \qquad (23)$$

$$\leq (1 + \varepsilon)\,H^1(I)\,|B(v)| + 3L\varepsilon^{1/2} H^1(I) = (1 + \varepsilon)\,H^1(B(I)) + 3L\varepsilon^{1/2} H^1(I).$$

We may now combine this with (20) and get that

$$H^1(B(F)) \leq (1 + \varepsilon)H^1(B(I)) + 4L\varepsilon^{1/2} H^1(I) \leq (1 + \delta)H^1(B(I)) \qquad (24)$$

if ε is small enough. Lemma 19 follows. $\square$

Let us discuss almost-minimizers with linear distortion now. Assume that $(u, K) \in TRLAM(\Omega, h)$ as before, but with linear distortion at most L, and let (x, r) be as in the beginning of this section. Associated with (x, r) (as in Definition 7.45) are two invertible matrices A_1 and A_2. Call A_3 the inverse of the transpose of A_2. The convenient definition of good radius (to be used later) is a radius $t \leq r/L$ such that

$$A_3(F(x, r)) \cap \partial B(A_3(x), t) = A_3(K) \cap \partial B(A_3(x), t),$$
$$\text{and this set has exactly two points} \qquad (25)$$

(compare with (4)). Still call $\mathcal{G}(x, r)$ this set of good radii; we claim that

$$H^1([0, r/L] \setminus \mathcal{G}(x, r)) \leq \delta r, \qquad (26)$$

with δ as small as we want, provided that $\beta_K(x,r)$, $\omega_1(x,r)^{1/2} J(x,r)^{-1}$, $\omega_p^*(x,r)$, and $h(r)$ are small enough, depending on δ and L.

The proof is the same as before; we get that the length of $A_3(F(x,r))$ is almost minimal from (45.15), (1), the fact that $\beta_K(x,r)$ is small, and Lemma 19.

Even though (26) is less precise than (6), it will be enough. Note also that the analogue of Lemma 16 holds: if t is a good radius, the two points of $A_3(F(x,r)) \cap \partial B(A_3(x),t)$ lie in the same component of $A_3(F(x,r)) \cap \overline{B}(A_3(x),t)$.

47 Bonnet's lower bound on the derivative of energy

The main ingredient in our proof of the C^1-theorem (Theorem 51.4 below and the related results in Sections 50 and 51) is a decay estimate on the normalized energy $r^{-1} \int_{B(x,r)} |\nabla u|^2$ when r tends to zero. Here we shall obtain this decay with a variant of A. Bonnet's argument from [Bo], which he used to prove that the normalized energy is nondecreasing when (u, K) is a global minimizer in the plane and K is connected.

For this section we can forget momentarily about almost-minimizers, and consider the following setting. Let Ω be an open subset of the unit disk in the plane. We shall assume that

$$\Omega = B(0,1) \setminus K, \text{ where } K \text{ is a closed subset of } B(0,1) \text{ such that } H^1(K) < +\infty. \tag{1}$$

Note that we do not require any regularity for Ω (or K), although our condition (8) may force the existence of large rectifiable pieces of K if the set $\mathcal{B}$ below is large.

Also let $u \in W_{loc}^{1,2}(\Omega)$ be such that

$$E(r) = \int_{\Omega \cap B(0,r)} |\nabla u|^2 < +\infty \quad \text{for } 0 < r < 1. \tag{2}$$

We assume that u is a local minimizer of energy, in the sense that

$$\int_{\Omega \cap B(0,r)} |\nabla u|^2 \leq \int_{\Omega \cap B(0,r)} |\nabla v|^2 \tag{3}$$

whenever $0 < r < 1$ and $v \in W_{loc}^{1,2}(\Omega)$ coincides with u almost-everywhere on $\Omega \setminus B(0,r)$. Of course we can think about the case when (u, K) is a local minimizer of $H^1(K) + \int_{B(0,1) \setminus K} |\nabla u|^2$ in $B(0,1)$, and $\Omega = B(0,1) \setminus K$. We want to find lower bounds on the derivative of $E(r)$. [See Proposition 7 for a statement.] We start with an easy observation.

Lemma 4. *For almost-every $r \in (0,1)$, the derivative $E'(r)$ exists and is given by*

$$E'(r) = \int_{\Omega \cap \partial B(0,r)} |\nabla u|^2. \tag{5}$$

Moreover, E is absolutely continuous on $(0,1)$, i.e.,

$$E(r) - E(s) = \int_s^r E'(t)dt \quad for\ 0 < s < r < 1. \tag{6}$$

This is easy. We may define $E'(r)$ by (5), and then (6) and the fact that $E'(t) < +\infty$ for almost-every $t \in (0,1)$ are obtained by writing $E(r)$ in polar coordinates and applying Fubini. Then (6) says that E is absolutely continuous and that $E'(r)$ is really its derivative almost-everywhere. See Proposition 9.1 for details. $\qquad\square$

Proposition 7. *Let Ω be open, assume that (1) holds for some $K \subset B(0,1)$, and let $u \in W^{1,2}_{loc}(\Omega)$ satisfy (2) and (3). Call $\mathcal{B}$ the set of radii $r \in (0,1)$ such that*

$$there\ is\ a\ compact\ connected\ set\ H\ such\ that\ K \cap \partial B(0,r) \subset H \subset K. \tag{8}$$

Then

$$E(r) \leq r\alpha(r)E'(r) \quad for\ almost\ every\ r \in \mathcal{B}, \tag{9}$$

where we denote by $2\pi r\alpha(r)$ the length of the longest connected component of $\Omega \cap \partial B(0,r)$.

We decided not to take any chances with (8); we could have tried to ask for less, for instance that all the points of $K \cap \partial B(0,r) = \partial B(0,r) \setminus \Omega$ lie in a same connected component of $K = B(0,1) \setminus \Omega$ (not necessarily a compact one), but in practice (8) will be as easy to check anyway. To prove the proposition, we shall need to know that

$$E(r) = \int_{\Omega \cap \partial B(0,r)} u \frac{\partial u}{\partial r} dH^1 \quad for\ almost\ every\ r \in (0,1). \tag{10}$$

Normally this relies on an integration by parts, which we shall do now without proper justification. The complete proof will be done at the end of the section, but you may also consult [MoSo2] for a justification of the integration by parts, at least for the case when K is rectifiable (as in our main application). Note first that u is harmonic on Ω, by Lemma 15.17, so there is no problem with the definition of the radial derivative $\frac{\partial u}{\partial r}$. Also, u satisfies the Neumann condition $\frac{\partial u}{\partial n} = 0$ on K (see Remark 15.21, but here again let us not justify for the moment). Call $\Omega_r = \Omega \cap B(0,r)$. Then

$$E(r) = \int_{\Omega_r} \nabla u \cdot \nabla u = -\int_{\Omega_r} u\,\Delta u + \int_{\partial\Omega_r} u \frac{\partial u}{\partial n} = \int_{\partial\Omega_r \setminus K} u \frac{\partial u}{\partial n}, \tag{11}$$

by Green's theorem, and (10) follows because $\partial \Omega_r \setminus K = \Omega \cap \partial B(0, r)$, and the normal direction there is radial.

Let $r \in \mathcal{B}$ be given. Decompose the open subset of the circle $\Omega \cap \partial B(0, r)$ into disjoint open arcs I_j, $j \in J$. Because of (1), and by Lemma 26.1, we know that $K \cap \partial B(0, r)$ is finite for almost all $r < 1$, so we may assume that J is finite, but this will not really help here. We consider $\partial B(0, r)$ itself as an arc, to accommodate the case when $\partial B(0, r) \subset \Omega$. Also, we are not disturbed by the case when $\Omega \cap \partial B(0, r)$ is empty, we just take $J = \emptyset$. This case is not too interesting anyway, because we can check directly that $E(r) = 0$ when this happens. Anyway, (10) is written as

$$E(r) = \sum_{j \in J} \int_{I_j} u \frac{\partial u}{\partial r} \, dH^1. \tag{12}$$

We want to modify this expression, and for this it will be very useful to know that for almost every choice of $r \in \mathcal{B}$,

$$\int_{I_j} \frac{\partial u}{\partial r} \, dH^1 = 0 \ \text{ for all } j \in J. \tag{13}$$

This is the place where we shall use (8). Fix $r \in \mathcal{B}$ and $j \in J$. Call $U_-(j)$ the connected component of $\Omega \setminus I_j$ that contains the points of $\Omega \cap B(0, r)$ that lie very close to I_j. Similarly denote by $U_+(j)$ the connected component of $\Omega \setminus I_j$ that contains the points of $\Omega \setminus \overline{B}(0, r)$ near I_j. See Figure 1.

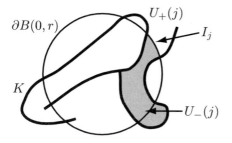

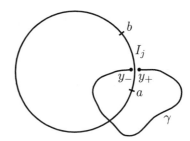

Figure 1.   Figure 2

Let us first check that $U_+(j) \neq U_-(j)$. If I_j is the whole circle $\partial B(0, r)$, or even the circle minus a point, this is trivial because $U_-(j) \subset B(0, r)$. So we may assume that I_j has two endpoints, which we call a and b. Choose two points $y_\pm$ in Ω, very close to the middle of I_j and to each other, and on different sides of I_j. If $U_+(j) = U_-(j)$, there is a simple path $\gamma \subset \Omega \setminus I_j$ that connects y_- to y_+. (See Figure 2.) We may assume that γ does not meet the interval (y_-, y_+). If y_- and y_+ were correctly chosen, we get a simple loop γ^* in Ω by adding (y_-, y_+) to γ. Now γ^* separates the extremities a and b, for instance because when we go from a to b along I_j, we cross γ^* exactly once, and there the index with respect to γ^* changes by 1. This contradicts (8), which says that K connects a to b. So $U_+(j) \neq U_-(j)$. Next we want to check that

$$\overline{U}_+(j) \subset B(0,1) \quad \text{or} \quad \overline{U}_-(j) \subset B(0,1). \tag{14}$$

[The reader may want to verify first that this is the case for all the arcs I_j in Figure 1.] As before, we can assume that I_j has two distinct extremities a and b, because otherwise $U_-(j) \subset B(0,r)$. Pick points $y_\pm \in U_\pm(j)$, very close to each other, so that in particular the segment $[y_-, y_+]$ is contained in Ω and crosses I_j neatly. Also choose $s > r$ so close to 1 that the compact connected set H promised by (8) lies in $B(0,s)$. If (14) fails, then for each choice of sign $\pm$ we can find a simple curve $\gamma_\pm \subset U_\pm(j) \cap B(0,s)$ that goes from $y_\pm$ to some point $z_\pm \in U_\pm(j) \cap \partial B(0,s)$. Then set $\gamma = \gamma_- \cup [y_-, y_+] \cup \gamma_+$. We can complete γ by some arc of $\partial B(0,s)$ from z_+ to z_-, not necessarily contained in Ω, to get a loop γ^*. The index relative to that loop changes exactly by 1 when we go from a to b along I_j, so the indices of a and b are different, and γ^* separates a from b. On the other hand, $H \subset K \cap B(0,s)$ (by (8) and our choice of s), and $\gamma^* \subset \Omega \cup \partial B(0,s)$, so γ^* cannot meet H. This gives the desired contradiction, because H is connected and contains a and b (by (8)). So we have (14).

To prove (13) we shall proceed as for (10). That is, we shall first give a proof that relies on a perhaps convincing, but not fully justified integration by parts, and then the complete verification will be done at the end of this section.

We first use (14) to choose $U = U_-(j)$ or $U_+(j)$, so that $\overline{U} \subset B(0,1)$. Then

$$0 = \int_U \Delta u = \int_{\partial U} \frac{\partial u}{\partial n} \, dH^1 = \int_{\partial U \setminus K} \frac{\partial u}{\partial n} \, dH^1 \tag{15}$$

because u is harmonic in $U \subset \Omega$, by Green, and because of the Neumann condition $\frac{\partial u}{\partial n}$ on $K \cap B(0,1)$. Now $\partial U \setminus K = I_j$, because $\overline{U} \subset B(0,1)$. Hence $\int_{\partial U \setminus K} \frac{\partial u}{\partial n} \, dH^1 = \pm \int_{I_j} \frac{\partial u}{\partial r} \, dH^1$, because I_j is reached from only one side (by definition of $U_\pm(j)$ and because $U_+(j) \neq U_-(j)$). This completes our soft proof (13). See the end of the section for more painful verifications.

We may now return to the expression of $E(r)$ in (12). For each $j \in J$, denote by m_j the mean value of u on I_j. By (2), we may restrict to radii r such that $\int_{\partial B(0,r)} |\nabla u|^2 < +\infty$, and then m_j is well defined. Next

$$\int_{I_j} u \frac{\partial u}{\partial r} \, dH^1 = \int_{I_j} [u - m_j] \frac{\partial u}{\partial r} \, dH^1$$

$$\leq \left\{ \int_{I_j} [u - m_j]^2 \, dH^1 \right\}^{1/2} \left\{ \int_{I_j} \left(\frac{\partial u}{\partial r} \right)^2 dH^1 \right\}^{1/2} \tag{16}$$

by (13) and Cauchy–Schwarz. Let $\lambda > 0$ be given, to be chosen later. From the general inequality $ab \leq \frac{1}{2}[\lambda^{-1}a^2 + \lambda b^2]$ we deduce that

$$\int_{I_j} u \frac{\partial u}{\partial r} \, dH^1 \leq \frac{1}{2\lambda} \left\{ \int_{I_j} [u - m_j]^2 \, dH^1 \right\} + \frac{\lambda}{2} \left\{ \int_{I_j} \left(\frac{\partial u}{\partial r} \right)^2 \, dH^1 \right\}. \qquad (17)$$

Now we use the following simple lemma of Wirtinger, which we shall also prove at the end of the section.

Lemma 18. *If $I \subset \mathbb{R}$ is a bounded interval and $f \in W^{1,2}(I)$, then $f \in L^2(I)$ and*

$$\int_I [f - m_I(f)]^2 \leq \left(\frac{|I|}{\pi} \right)^2 \int_I |f'|^2, \qquad (19)$$

where we denote by $m_I(f)$ the mean value of f on I.

Even though u is defined on an arc of circle, we can apply Lemma 18 to it. We just need to replace u' with the tangential derivative $\frac{\partial u}{\partial \tau}$. In the special case when $I_j = \partial B(0, r)$, we pick any point $\xi \in \partial B(0, r)$ at random, and apply the lemma to the restriction of u to $\partial B(0, r) \setminus \{\xi\}$. This amounts to forgetting that u is also continuous across ξ, but will give a good enough estimate anyway. [We could get a twice better constant in this case, if we used the continuity across ξ.] So we get that

$$\int_{I_j} [u - m_j]^2 \, dH^1 \leq \left(\frac{H^1(I_j)}{\pi} \right)^2 \int_{I_j} \left(\frac{\partial u}{\partial \tau} \right)^2 \, dH^1 \leq (2r\alpha(r))^2 \int_{I_j} \left(\frac{\partial u}{\partial \tau} \right)^2 \, dH^1, \qquad (20)$$

because $H^1(I_j) \leq 2\pi r\alpha(r)$ by definition.

Now we choose $\lambda = 2r\alpha(r)$; then $\frac{1}{2\lambda}(2r\alpha(r))^2 = \frac{\lambda}{2} = r\alpha(r)$, and

$$\int_{I_j} u \frac{\partial u}{\partial r} \, dH^1 \leq r\alpha(r) \int_{I_j} \left(\frac{\partial u}{\partial \tau} \right)^2 \, dH^1 + r\alpha(r) \int_{I_j} \left(\frac{\partial u}{\partial r} \right)^2 \, dH^1$$

$$= r\alpha(r) \int_{I_j} |\nabla u|^2 \, dH^1, \qquad (21)$$

by (17) and (20). Finally, we sum over $j \in J$ and get that

$$E(r) = \sum_{j \in J} \int_{I_j} u \frac{\partial u}{\partial r} \, dH^1 \leq r\alpha(r) \sum_{j \in J} \int_{I_j} |\nabla u|^2 \, dH^1 = r\alpha(r) E'(r), \qquad (22)$$

by (12), (5), and because $\Omega \cap \partial B(0, r)$ is the disjoint union of the I_j. This completes our proof of Proposition 7, modulo the verifications of (10), (13), and Lemma 18. $\qquad \square$

Proof of (10). The main point here will be to get (10) directly, without integrating by parts. Note that our Neumann condition $\frac{\partial u}{\partial n} = 0$ on $\partial\Omega \cap B(0,1)$ is also the result of an integration by parts (and duality). We shall simply shunt the two integrations by parts, and work with competitors for u. First observe that

$$u \text{ is bounded on } \Omega \cap B(0,r) \text{ for every } r \in (0,1). \tag{23}$$

Indeed (1) says that $H^1(K)$ is finite, and then Lemma 26.1 tells us that $K \cap \partial B(0,s)$ is finite for almost all $s < 1$. Because of (2), we can find $s \in (r,1)$ such that $K \cap \partial B(0,s)$ is finite and $\int_{\partial B(0,s)\backslash K} |\nabla u|^2 < +\infty$. Then u is bounded on each of the (finitely many) arcs that compose $\Omega \cap \partial B(0,s)$, and it is easy to see that u is also bounded on $B(0,s)$. To be honest, we should say that there could be connected components of Ω that are compactly contained in $B(0,s)$, and we can only say that u is constant on those. However Proposition 7 does not change when you replace u with 0 in these components, and so we may assume that we did. See Section 15 for details. Also, the reader should not worry (in the future) about the absolute continuity of u on almost all circles, or things like this, because here u is harmonic on Ω.

With (23) out of the way, let $r \in (0,1)$ be given, and consider competitors of the form $u_\lambda = u + \lambda\varphi u$, where φ is a smooth function with compact support in $B(0,r)$. Note that u_λ is smooth in Ω (because u is harmonic), and $\int_{\Omega \cap B(0,r)} |\nabla u_\lambda|^2 < +\infty$ by (2) and (23). Also, u_λ coincides with u out of $B(0,r)$, so $\int_{\Omega \cap B(0,r)} |\nabla u|^2 \leq \int_{\Omega \cap B(0,r)} |\nabla u_\lambda|^2$ by (3). Since this holds for all small λ, we get the scalar product condition

$$\int_{\Omega \cap B(0,r)} \nabla u \cdot \nabla(\varphi u) = 0, \tag{24}$$

as in Proposition 3.16 or Lemma 15.17. Since $\nabla(\varphi u) = \varphi \nabla u + u \nabla\varphi$, we get that

$$\int_{\Omega \cap B(0,r)} \varphi |\nabla u|^2 = -\int_{\Omega \cap B(0,r)} u \nabla u \cdot \nabla\varphi. \tag{25}$$

Naturally we want to apply this to radial functions φ. For all k such that $2^{-k} < r$, set $\varphi_k(x) = h_k(|x|)$, where h_k is a smooth function such that

$$h_k(t) = 1 \text{ for } t \leq r-2^{-k}, \ h_k(t) = 0 \text{ for } t \geq r, \text{ and } 0 \leq -h_k'(t) \leq 2^{k+1} \text{ everywhere.} \tag{26}$$

Then (25) says that

$$\int_{\Omega \cap B(0,r)} h_k(|x|) |\nabla u|^2 = \int_{\Omega \cap B(0,r)} u(x) f_{k,r}(x) \frac{\partial u}{\partial r}(x) \, dx, \tag{27}$$

where we set $f_{k,r}(x) = -h_k'(|x|)$. Note that

$$f_{k,r} \text{ is radial and supported on } \overline{B}(0,r) \backslash B(0, r-2^{-k}), \tag{28}$$

$$0 \leq f_{k,r}(x) \leq 2^{k+1} \text{ everywhere, and } \int_{r-2^{-k}}^r f_{k,r}(t)\,dt = 1. \tag{29}$$

Lemma 30. *Let us assume that for each* $r \in (0,1)$, *we chose functions* $f_{k,r}$, $2^{-k} < r$, *with the properties* (28) *and* (29). *Then for each* $s \in (0,1)$ *and each function* $w \in L^1(B(0,s))$, *we can find a subsequence* $\{k_l\}$ *such that for almost-every* $r \in (0,s)$,

$$\lim_{l \to +\infty} \int_{B(0,r)} f_{k_l,r}(x)\, w(x)\, dx = \int_{\partial B(0,r)} w(x)\, dH^1(x). \tag{31}$$

Before we prove the lemma, let us see how we can deduce (10) from it. Note that $w(x) = \mathbf{1}_\Omega(x) u(x) \frac{\partial u}{\partial r}(x)$ lies in $L^1(B(0,s))$ for all $s < 1$, by (2) and (23). For each $s < 1$, we can find a subsequence $\{k_l\}$ and a set of full measure in $(0,s)$ on which (31) holds. Let us check that the identity in (10) holds for all these radii. When we restrict to the subsequence $\{k_l\}$, (31) says that the right-hand side of (27) converges to $\int_{\partial B(0,r)} w(x)\, dH^1(x)$, which is the right-hand side of (10). On the other hand, the left-hand side converges to $E(r)$, by (26), and (10) follows.

So it is enough to prove the lemma, which is a standard fact on approximations of the identity. Incidentally, we have been a little lazy here. In fact, (31) holds without even taking a subsequence, but a slightly more complicated proof would be needed (with a maximal function argument). Here we just want to show some convergence in L^1, and then extract a subsequence to get convergence almost-everywhere (with respect to the radii). For the sake of definiteness, let us also define $f_{k,r}$ when $2^{-k} \geq r$, and let us simply take $f_{k,r} = 0$ then. Set

$$A_k(w,r) = \int_{B(0,r)} f_{k,r}(x)\, w(x)\, dx$$
$$= \int_{\partial B(0,r)} \int_{r-2^{-k}}^{r} f_{k,r}(ty/r)\, w(ty/r) \frac{t\,dt}{r}\, dH^1(y), \tag{32}$$

where the last expression comes from polar coordinates and a dilatation (set $x = ty/r$),

$$L(w,r) = \int_{\partial B(0,r)} w(x)\, dH^1(x), \quad \text{and} \quad \delta_k(w,r) = |L(w,r) - A_k(w,r)| \tag{33}$$

for $w \in L^1(B(0,s))$, $k \geq 0$, and $r < s$.

We start with an estimate of $\delta_k(w,r)$ when w is smooth. First write

$$\delta_k(w,r) = \left| \int_{\partial B(0,r)} \left\{ w(x) - \int_{r-2^{-k}}^{r} f_{k,r}(tx/r)\, w(tx/r) \frac{t\,dt}{r} \right\} dH^1(x) \right|. \tag{34}$$

We need to know that $a = \int_{r-2^{-k}}^{r} f_{k,r}(tx/r) \frac{t\,dt}{r}$ is often close to 1.

When $2^{-k} < r$, $\displaystyle\int_{r-2^{-k}}^{r} f_{k,r}(tx/r)dt = \int_{r-2^{-k}}^{r} f_{k,r}(t)dt = 1$ because $|x| = r$, $f_{k,r}$ is radial (by (28)), and by (29). We still need to multiply by t/r, but since $r - 2^{-k} \leq t \leq r$ on the domain of integration, we get that $1 - \frac{2^{-k}}{r} \leq a \leq 1$.

When $2^{-k} \geq r$, we just need to know that $f_{k,r} = 0$, hence $a = 0$. Now

$$\left| w(x) - \int_{r-2^{-k}}^{r} f_{k,r}(tx/r)\, w(tx/r)\, \frac{tdt}{r} \right|$$

$$\leq (1-a)|w(x)| + \int_{r-2^{-k}}^{r} f_{k,r}(tx/r)\, |w(tx/r) - w(x)|\, \frac{tdt}{r}$$

$$\leq (1-a)|w(x)| + a\, 2^{-k}||\nabla w||_\infty \leq \frac{2^{-k}}{r}|w(x)| + 2^{-k}||\nabla w||_\infty. \qquad (35)$$

We may now return to (34), integrate over $\partial B(0,r)$, and get that

$$\delta_k(w,r) \leq 2\pi\, 2^{-k}\{||w||_\infty + r||\nabla w||_\infty\}. \qquad (36)$$

This would be more than enough to settle the case when w is smooth, but of course the point is to use (36) to help with the general case. So let $w \in L^1(B(0,s))$ be given. For each small $\varepsilon > 0$, choose a smooth function w_ε such that $||w - w_\varepsilon|| \leq \varepsilon$. Here and below, $||\cdot||$ denotes the norm in $L^1(B(0,s))$. First,

$$\int_0^s |L(w,r) - L(w_\varepsilon,r)|dr \leq \int_0^s \int_{\partial B(0,r)} |w(x) - w_\varepsilon(x)|\, dH^1(x)dr = ||w - w_\varepsilon|| \leq \varepsilon. \qquad (37)$$

Similarly,

$$\int_0^s |A_k(w,r) - A_k(w_\varepsilon,r)|dr \leq \int_0^s \int_{B(0,r)} f_{k,r}(x)\, |w(x) - w_\varepsilon(x)|\, dx\, dr \qquad (38)$$

$$\leq \int_{x \in B(0,s)} \int_{|x| \leq r \leq |x|+2^{-k}} 2^{k+1}|w(x) - w_\varepsilon(x)|\, dr\, dx \leq 2||w - w_\varepsilon|| \leq 2\varepsilon,$$

by Fubini, (28), and (29). Finally,

$$\int_0^s \delta_k(w,r)\, dr \leq \int_0^s \{\delta_k(w_\varepsilon,r) + |L(w,r) - L(w_\varepsilon,r)| + |A_k(w,r) - A_k(w_\varepsilon,r)|\}\, dr$$

$$\leq C\, 2^{-k}\{||w_\varepsilon||_\infty + s||\nabla w_\varepsilon||_\infty\} + 3\varepsilon = 2^{-k}C(w_\varepsilon) + 3\varepsilon \qquad (39)$$

by (36), (37), and (38). If k is large enough (depending on ε through w_ε, but this is all right), this is less than 4ε. Thus we proved that

$$\lim_{k \to +\infty} \int_0^s \delta_k(w,r)\, dr = 0. \qquad (40)$$

We may now select our subsequence $\{k_l\}$ so that $\int_0^s \delta_{k_l}(w, r)\, dr \leq 4^{-l}$ for all l. Set

$$Z(l) = \left\{ r \in (0, s)\, ; \delta_{k_l}(w, r) \geq 2^{-l} \right\}. \tag{41}$$

Then $|Z(l)| \leq 2^{-l}$ by Chebyshev. Next denote by $Z^*(m)$ the union of all the $Z(l)$, $l \geq m$. We still have that $|Z^*(m)| \leq 2^{-m+1}$. The intersection of all the $Z^*(m)$ has measure zero, and so almost-every $r \in (0, s)$ lies outside of some $Z^*(m)$. For such an r, we have that $\delta_{k_l}(w, r) < 2^{-l}$ for l large enough, and in particular $A_{k_l}(w, r)$ tends to $L(w, r)$ (by (33)). Then (31) holds, by (32). So we finally proved that almost all $r \in (0, s)$ satisfy (31). This completes our proof of Lemma 30 and, as was said earlier, (10) follows. $\qquad\square$

Remark 42. For the proof of (10) we did not use our assumption (1) that $H^1(K) < +\infty$, except to get the boundedness of u on the $B(0, r)$ in a brutal way (and there much less is needed). The situation is different for the proof of (13), where the author was not able to make an argument like the one above work correctly, mostly because the geometric situation with the domains $U_\pm(j)$, and in particular the way they depend on r, is a little too complicated. So we shall use (1) in the proof of (13). Of course (1) is not really a regularity condition on K, but as we shall see, it implies some regularity near the circles $\partial B(0, r)$, $r \in \mathcal{B}$.

Proof of (13). For each $r \in \mathcal{B}$, select a compact connected set $H_r \subset K$ that contains $K \cap \partial B(0, r)$. There are a few cases that are easier to treat. First consider

$$\mathcal{B}_1 = \{ r \in (0, 1)\, ; K \cap \partial B(0, r) \text{ has at most one point} \}, \tag{43}$$

which is obviously contained in $\mathcal{B}$. Let us check that

$$\text{(13) holds for almost-every } r \in \mathcal{B}_1. \tag{44}$$

Let us try the same sort of argument as for (10): consider competitors for u of the form $u_\lambda = u + \lambda\varphi$, where φ is a smooth radial function supported on $B(0, r)$ (in fact, the same one as above). Since (3) says that $\int_{\Omega \cap B(0,r)} |\nabla u|^2 \leq \int_{\Omega \cap B(0,r)} |\nabla u_\lambda|^2$ for all small λ, we get the scalar product condition

$$0 = \int_{\Omega \cap B(0,r)} \nabla u \cdot \nabla \varphi = \int_{\Omega \cap B(0,r)} \frac{\partial u}{\partial r} \frac{\partial \varphi}{\partial r}, \tag{45}$$

because φ is radial. When we apply this with the same choice of functions $\varphi_k(x) = h_k(|x|)$ as above, we get that $\int_{\Omega \cap B(0,r)} f_{k,r}(x) \frac{\partial u}{\partial r}(x)\, dx = 0$, where the $f_{k,r}$ are as in (27) and Lemma 30. Then we can apply Lemma 30, because $\mathbf{1}_\Omega \frac{\partial u}{\partial r} \in L^1(B(0, s))$ for all $s < 1$, and we get that for almost all $r < 1$, $\lim_{l \to +\infty} \int_{B(0,r)} f_{k_l,r}(x) \frac{\partial u}{\partial r}(x)\, dx = \int_{\partial B(0,r)} \frac{\partial u}{\partial r}(x)\, dH^1(x)$ for some subsequence $\{k_l\}$. Thus $\int_{\partial B(0,r)} \frac{\partial u}{\partial r}\, dH^1 = 0$ for almost all $r \in (0, 1)$. This proves (44), because $\partial B(0, r)$ has only one component I when $r \in \mathcal{B}_1$, and $I = \partial B(0, r)$, perhaps minus a point.

We are left with $\mathcal{B} \backslash \mathcal{B}_1$. We claim that there is an at most countable collection $\{K_k\}$ of compact connected subsets of K such that whenever $r \in \mathcal{B} \backslash \mathcal{B}_1$, $K \cap \partial B(0,r)$ is contained in some K_k.

Indeed consider a sequence of radii $s_\ell < 1$ that tends to 1. Then call $\mathcal{R}(\ell)$ the set of connected components of $K \cap \overline{B}(0, s_\ell)$ that are not reduced to one point. If $H \in \mathcal{R}(\ell)$, then H is compact (because anyway $\overline{H}$ is connected) and $H^1(H) > 0$. Hence each $\mathcal{R}(\ell)$ is at most countable (by (1)), and we can define $\{K_k\}$ as the collection of all the sets H that lie in some $\mathcal{R}(\ell)$. This works, because if $r \in \mathcal{B} \backslash \mathcal{B}_1$, then $K \cap \partial B(0,r)$ has at least two points, hence (8) says that for ℓ large enough it is contained in some $H \in \mathcal{R}(\ell)$.

For each of the compact connected sets $K_k \subset K$ in our collection, there is a compact interval L_k and a Lipschitz mapping $z_k : L_k \to \mathbb{R}^2$ such that $K_k = z_k(L_k)$. See Proposition 30.1. For each k, call W_k the set of points $t \in L_k$ that are an endpoint of L_k, or at which z_k is not differentiable, or for which $z_k'(t)$ exists, but its radial part $\langle z_k(t), z_k'(t) \rangle$ vanishes. Also call π the radial projection, defined by $\pi(x) = |x|$. Then $H^1(\pi(W_k)) = 0$, because z_k is Lipschitz (hence differentiable almost everywhere), and by Sard's Theorem. (See Exercise 33.41.) Thus

$$H^1(\mathcal{B}_2) = 0, \quad \text{where } \mathcal{B}_2 = \{r \in \mathcal{B} \backslash \mathcal{B}_1 ; r \in \pi(W_k) \text{ for some } k\}. \qquad (46)$$

We are left with $\mathcal{B}_3 = \mathcal{B} \backslash [\mathcal{B}_1 \cup \mathcal{B}_2]$. Let $r \in \mathcal{B}_3$ be given, and let I be one of the components of $\partial B(0,r) \backslash K$. We want to show that $\int_I \frac{\partial u}{\partial r} dH^1 = 0$. (Compare with (13).) Let k be such that $K \cap \partial B(0,r) \subset K_k$. Call a_1, a_2 the endpoints of I; note that $a_1 \neq a_2$ because $r \notin \mathcal{B}_1$. Each a_i, $i = 1, 2$, lies in $K \cap \partial B(0,r) \subset K_k$, so we can find $t_i \in L_k$ such that $a_i = z_k(t_i)$. Also, t_i is an interior point of L_k, z_k has a derivative at t_i, and $\langle a_i, z_k'(t_i) \rangle \neq 0$. Choose a small interval $T_i \subset L_k$ centered at t_i, and set $\gamma_i = z_k(T_i)$. If T_i is small enough, γ_i is contained in a cone of the form $\Gamma_i = \{x \in \mathbb{R}^2 ; |x - a_i| \leq M|\pi(x) - \pi(a_i)|\}$. See Figure 3. We shall not care whether each γ_i is simple, for instance, but the main point is that there is a small interval $S = [s_0, s_1]$ centered at r, such that each circle $\partial B(0,s)$, $s \in S$, crosses each γ_i somewhere in Γ_i.

Denote by U_- the connected component of $\Omega \backslash I = B(0,1) \backslash [K \cup I]$ that contains the points of $B(0,r)$ near I, and by U_+ the component of $\Omega \backslash I$ that contains the points of $B(0,1) \backslash \overline{B}(0,r)$ near I. We know from (14) that $\overline{U}_- \subset B(0,1)$ or $\overline{U}_+ \subset B(0,1)$. Let us assume that this is the case for U_-; the argument for U_+ would be similar.

As usual, we want to compare u with competitors $u_\lambda = u + \lambda \psi$, but this time we define ψ slightly differently. Set $A_0 = \pi^{-1}([s_0, r])$, and call A the connected component of I in $A_0 \backslash [\gamma_1 \cup \gamma_2]$. We use a radial function φ such that $\varphi(x) = 0$ in a neighborhood of $\{|x| \leq s_0\}$, and $\varphi(x) = 1$ in a neighborhood of $\{|x| \geq r\}$, and then we take $\psi(x) = \mathbf{1}_{U_-}(x) - \mathbf{1}_A(x)\varphi(x)$.

Let us first check that this is a smooth function on Ω. The most obvious place where ψ may have discontinuities is the interval I. Let $x_0 \in I$, be given, let D be

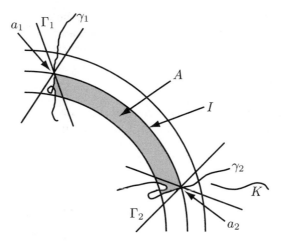

Figure 3

a small ball centered at x_0, and consider $x \in D$. If $x \in \overline{B}(0, r)$ (and D is small enough), then $x \in U_- \cap A$, $\varphi(x) = 1$, and so $\psi(x) = 0$. If $x \in D \setminus \overline{B}(0, r)$, then $\mathbf{1}_A(x) = 0$ and $x \in U_+$, which is disjoint from U_- (see between (13) and (14)), so $\psi(x) = 0$. Thus ψ is smooth across I.

Recall that U_- is one of the components of $\Omega \setminus I$ (see the definition after (13)), so $\mathbf{1}_{U_-}$ is smooth on $\Omega \setminus I$. We can also expect singularities of $\mathbf{1}_A$ on ∂A. This includes the curves γ_i, but these curves are contained in K, so we don't care about discontinuities of ψ there. Since we already took care of I, we are left with $\partial A \cap \partial B(0, s_0)$. [In particular, I is the only piece of $\partial A \cap \partial B(0, r)$.] But any jump of $\mathbf{1}_A$ on $\partial B(0, s_0)$ would be killed by $\varphi(x)$, which vanishes for $|x| \leq s_0$. Altogether, ψ is smooth on Ω.

Next, ψ has compact support in $B(0, 1)$, because $\overline{U}_- \subset B(0, 1)$. Thus u_λ is a competitor for u for every λ, which leads to the scalar product condition

$$0 = \int_\Omega \nabla u \cdot \nabla \psi = -\int_{A \cap \Omega} \nabla u \cdot \nabla \varphi = -\int_{A \cap \Omega} \frac{\partial u}{\partial r} \frac{\partial \varphi}{\partial r}. \tag{47}$$

We can take $\varphi(x) = 1 - h_k(|x|)$, where h_k is as in (26), and we get that

$$\int_{A \cap \Omega} f_{k,r}(x) \frac{\partial u}{\partial r}(x) \, dx = 0, \tag{48}$$

where $f_{k,r}(x) = -h'_k(|x|)$ as before. We want to prove that for almost-every choice of r as before (and all choices of I),

$$\lim_{k \to +\infty} \int_{A \cap \Omega} f_{k,r}(x) \frac{\partial u}{\partial r}(x) \, dx = \int_I \frac{\partial u}{\partial r} \, dH^1. \tag{49}$$

Once we prove (49), (13) will follow from (48) and we will be happy.

Unfortunately, it seems that Lemma 30 is hard to apply directly, because A depends on r. So we shall do things more dirtily. This time we shall not avoid using a maximal function. Consider $g(r) = \int_{\Omega \cap \partial B(0,r)} |\nabla u|^2 dH^1$. Then fix $s \in (0,1)$, and define a Hardy–Littlewood maximal function $g^{*,s}$ by

$$g^{*,s}(r) = \sup_{\varepsilon > 0} \left\{ \frac{1}{2\varepsilon} \int_{(0,s) \cap [r-\varepsilon, r+\varepsilon]} g(t)dt \right\}. \tag{50}$$

Since $g \in L^1(0,s)$ by (2), the Hardy–Littlewood maximal theorem says that $g^{*,s}$ lies in $L^1_{\text{weak}}(0,s)$, and in particular $g^{*,s}(r) < +\infty$ for almost-every $r < s$. See [St] or Exercise 33.38. We shall restrict our attention to the radii r such that

$$g^{*,s}(r) < +\infty \quad \text{for some } s \in (r,1). \tag{51}$$

Obviously it is enough to prove (49) when $\int_I |\nabla u|^2 dH^1 < +\infty$ and (51) holds.

Let I' be any compact subinterval of I, and set $\Gamma(I') = \{x \in B(0,1); x/|x| \in I'\}$. Then

$$\lim_{k \to +\infty} \int_{\Gamma(I') \cap \Omega} f_{k,r}(x) \frac{\partial u}{\partial r}(x)\, dx = \int_{I'} \frac{\partial u}{\partial r} dH^1. \tag{52}$$

The argument is the same as for (36), i.e., the proof of (31) when w is smooth, and the point is that u is harmonic in Ω, which contains a small neighborhood of I'.

For the remaining part, first recall from (26) that $f_{k,r}$ is supported on the annulus $R_k = \{x \in B(0,r)\,; |x| \geq r - 2^{-k}\}$. For k large enough,

$$\Gamma(I') \cap \Omega \cap R_k = \Gamma(I') \cap R_k \subset A \cap R_k \tag{53}$$

(because Ω contains a neighborhood of I' and by definition of A, see Figure 3), and also

$$|A \cap R_k \setminus [\Gamma(I') \cap \Omega \cap R_k]| \leq 2^{-k}[H^1(I \setminus I') + 2M2^{-k}], \tag{54}$$

where M is as in the definition of the cones Γ_i, just above Figure 3. Here the point is that each of our curves γ_i, $i = 1, 2$, is contained in the corresponding Γ_i and reaches all the way to $\partial B(0, s_0)$, so that A does not escape from the region bounded by $\partial B(0, r)$, $\partial B(0, s_0)$, and the (exterior part of) the Γ_i. Now

$$\left| \int_{A \cap \Omega} f_{k,r}(x) \frac{\partial u}{\partial r}(x)\, dx - \int_{\Gamma(I') \cap \Omega} f_{k,r}(x) \frac{\partial u}{\partial r}(x)\, dx \right|$$

$$\leq 2^{k+1} |A \cap R_k \setminus [\Gamma(I') \cap \Omega \cap R_k]|^{1/2} \left\{ \int_{R_k \cap \Omega} |\nabla u|^2 \right\}^{1/2}$$

$$\leq 2^{k+1} \left\{ 2^{-k}[H^1(I \setminus I') + 2M2^{-k}] \right\}^{1/2} \left\{ \int_{r-2^{-k}}^r g(t)dt \right\}^{1/2} \tag{55}$$

$$\leq 2^{k+1} \left\{ 2^{-k}[H^1(I \setminus I') + 2M2^{-k}] \right\}^{1/2} \left\{ 2^{-k+1} g^{*,s}(r) \right\}^{1/2}$$

$$\leq C \left[H^1(I \setminus I') + 2^{-k} \right]^{1/2},$$

by (29), Cauchy–Schwarz, (53), (54), (50), and (51), and with a constant C that depends on r and M (but this does not matter). We also have the simpler estimate

$$\left| \int_{I \setminus I'} \frac{\partial u}{\partial r} \, dH^1 \right| \le H^1 (I \setminus I')^{1/2} \left\{ \int_I |\nabla u|^2 dH^1 \right\}^{1/2}. \tag{56}$$

We can make the right-hand sides of (55) and (56) as small as we want by taking I' close to I and k large enough. Then $\delta_k = \int_{A \cap \Omega} f_{k,r}(x) \frac{\partial u}{\partial r}(x) \, dx - \int_I \frac{\partial u}{\partial r} \, dH^1$ also can be made as small as we wish, because of (52). In other words, δ_k tends to 0, and (49) holds. We just proved (49) for almost-every $r \in \mathcal{B}_3$. As was noted above, (13) follows. This was our last remaining case, so our verification of (13) almost-everywhere is complete. □

Proof of Lemma 18. First we should observe that the lemma stays the same if we replace I with λI and $f(x)$ with $f(x/\lambda)$ for some $\lambda > 0$; both sides of (19) are simply multiplied by λ. Because of this, we can assume that $I = (0, \pi)$. So let $f \in W^{1,2}(0, \pi)$ be given. We know from Proposition 9.1 that

$$f(x) = \int_0^x f'(t)dt + C \quad \text{for } x \in (0, \pi). \tag{57}$$

In particular, f has a continuous extension to $[0, \pi]$. Let us also extend f to $[-\pi, \pi]$ by symmetry, i.e., set $f(-x) = f(x)$ for $0 \le x \le \pi$. This gives a continuous function, with a continuous 2π-periodic extension (i.e., $f(-\pi) = f(\pi)$). So

$$f(x) = \sum_{k=0}^{\infty} a_k \cos kx, \tag{58}$$

because f is even, and then

$$\int_I |f(x) - m_I f|^2 \, dx = \frac{1}{2} \int_{[-\pi, \pi]} |f(x) - m_I f|^2 \, dx = \frac{\pi}{2} \sum_{k=1}^{\infty} |a_k|^2. \tag{59}$$

Also, f is absolutely continuous on $[-\pi, \pi]$, and its derivative is the odd extension of f'. This is easily deduced from (57) and the analogous formula for $x < 0$. Then $f'(x) = -\sum_{k=1}^{\infty} k \, a_k \sin kx$ in $L^2(-\pi, \pi)$, and

$$\int_I |f'(x)|^2 \, dx = \frac{1}{2} \int_{[-\pi, \pi]} |f'(x)|^2 \, dx = \frac{\pi}{2} \sum_{k=1}^{\infty} k^2 |a_k|^2 \ge \int_I |f(x) - m_I f|^2 \, dx. \tag{60}$$

This is (19), because in the present case $|I| = \pi$. Lemma 18 follows. This also completes our detailed proof of Proposition 7. □

Remark 61. We shall need to know the cases when (19) is an identity. When $I = (0, \pi)$, this forces equality in (60), and so all the coefficients a_k, $k > 1$, must vanish. In other words, there are coefficients A and B such that $f(x) = A\cos x + B$ on $(0, \pi)$. In the general case when $I = (a, b)$, we get that (19) is an identity if and only if we can find A, B such that

$$f(x) = A\cos\frac{\pi(x - a)}{b - a} + B \quad \text{for } x \in (a, b). \tag{62}$$

48 The main estimate on the decay of the normalized energy $\omega_2(x, r)$ ($n = 2$)

We continue to assume that $(u, K) \in TRLAM(\Omega, h)$, as in Standard Assumption 41.1.

The main difficulty when we try to prove results like Theorems 50.1 or 51.4 below is to find collections of estimates on a ball $B(x, r)$ centered on K that will imply the same (or slightly better) estimates on $B(x, r/2)$, say. Here we shall concentrate on one main quantity, the normalized energy $\omega_2(x, r) = r^{-1}\int_{B(x,r)\setminus K}|\nabla u|^2$, and show that under suitable circumstances, it will decrease when r gets smaller.

Proposition 1. *For each choice of $\gamma \in (0, 1)$, we can find constants $a < 1/4$, $\tau > 0$, and $C_0 \geq 1$ such that if $(u, K) \in TRLAM(\Omega, h)$, and $x \in K$ and $r > 0$ are such that $\overline{B}(x, r) \subset \Omega$ and*

$$\beta_K(x, r) + \omega_2(x, r) + J(x, r)^{-1} + h(r) < \tau, \tag{2}$$

then

$$\omega_2(x, ar) \leq \gamma\omega_2(x, r) + C_0\,\omega_2(x, r)^{1/2}\,J(x, r)^{-1} + C_0\,h(r). \tag{3}$$

See (23.5), (41.2), and (42.4) for the definitions. The reader should have in mind that $J(x, r)^{-1}$ is very easy to control, and we control $h(r)$ by definitions, so the last two terms of (3) will not disturb us too much, and (3) will imply that $\omega_2(x, ar) < \omega_2(x, r)$ in the interesting situations. See Section 49 for details.

To prove the proposition, we first replace r with a radius $r_0 \in [r/2, r]$ such that

$$\omega_2^*(x, r_0) := \int_{\partial B(x,r_0)\setminus K}|\nabla u|^2 \leq 2r^{-1}\int_{B(x,r)\setminus K}|\nabla u|^2 = 2\omega_2(x, r), \tag{4}$$

where the first equality comes from the definition (42.5), and the inequality is easy to get by Chebyshev. Note that

$$\beta_K(x, r_0) \leq 2\beta_K(x, r) \leq 2\tau, \tag{5}$$

by definition (see (41.2)) and (2). Let us apply Proposition 44.1 to the pair (x, r_0). We get a compact set $F = F(x, r_0)$ such that $K \cap \overline{B}(x, r_0) \subset F \subset \overline{B}(x, r_0)$,

$$H^1(F \cap B(x, r_0) \setminus K) \leq C r \, w_2(x, r_0)^{1/2} J(x, r_0)^{-1}, \tag{6}$$

and F separates $D_+(x, r_0)$ from $D_-(x, r_0)$ in $\overline{B}(x, r_0)$. [Recall from (42.2) and (42.3) that the $D_\pm(x, r_0)$ are large disks in $B(x, r_0)$ that lie on both sides of the good approximating line $P(x, r_0)$.]

Next we want to apply Lemma 46.5 to find a good disk $B(x, r_1)$. Recall that this means that $r_1 \in \mathcal{G}$, where

$$\mathcal{G} = \mathcal{G}(x, r_0) = \{ t \in (0, r_0) \, ; \, F \cap \partial B(x, t) = K \cap \partial B(x, l) \text{ and } \tag{7}$$
$$\text{this set contains exactly two points } a \text{ and } b, \text{ with } |b - a| \geq t \}$$

is as in (46.4). Now

$$H^1([0, r_0] \setminus \mathcal{G}) \leq C r_0 \beta_K(x, r_0) + C r_0 \omega_1(x, r_0)^{1/2} J(x, r_0)^{-1} + C r \omega_2^*(x, r_0) + r_0 h(r_0)$$
$$\leq C r \beta_K(x, r) + C r \omega_2(x, r)^{1/2} J(x, r_0)^{-1} + C r \omega_2(x, r) + r h(r)$$
$$\leq C \tau r \tag{8}$$

by Lemma 46.5, (5), (23.5) and Cauchy–Schwarz, (4), and (2). For the second inequality, we also used the fact that $\left| (r/r_0)^{1/2} J(x, r_0) - J(x, r) \right|$ is very small (by Lemma 43.1 and (2)) and $J(x, r) \geq \tau^{-1}$, so $J(x, r) \leq 2 J(x, r_0)$.

If τ is small enough, $H^1([0, r_0] \setminus \mathcal{G}) \leq r/8$, and we can choose $r_1 \in (r/4, r_0) \cap \mathcal{G}$. Let us also demand that $\omega_2^*(x, r_1) := \int_{\partial B(x, r_1) \setminus K} |\nabla u|^2 \leq C r^{-1} \int_{B(x, r) \setminus K} |\nabla u|^2 = C w_2(x, r)$. [See the definition (42.5) again, but here we also want the extra conditions from Convention 42.6 concerning the absolute continuity on $\partial B(x, r_1) \setminus K$ and the radial limits.]

We want to construct a competitor (v, G) for (u, K) in $B(x, r_1)$. We take

$$G = K \cup [F \cap B(x, r_1)] ; \tag{9}$$

notice that G is closed because $F \cap \partial B(x, t) = K \cap \partial B(x, t)$, by (7). Then choose v on $B(x, r_1) \setminus G = B(x, r_1) \setminus F$ so that its boundary values on the two arcs of $\partial B(x, r_1) \setminus K$ coincide with (those of) u, and its energy $E(v) = \int_{B(x, r_1) \setminus G} |\nabla v|^2$ is minimal under this constraint. See Section 15 for details. Then we keep $v = u$ out of $B(x, r_1)$. This gives a function in $W^{1,2}_{loc}(\Omega \setminus G)$, by the usual combination of Corollary 14.28 and Convention 42.6. Thus (v, G) is a competitor for (u, K) in $B(x, r_1)$, and it is even a topological one because G contains K. So we may apply (7.7), and

$$\int_{B(x, r_1) \setminus K} |\nabla u|^2 + H^1(K \cap \overline{B}(x, r_1)) \leq E(v) + H^1(G \cap \overline{B}(x, r_1)) + r_1 h(r_1). \tag{10}$$

Set $E(u) = \int_{B(x,r_1)\setminus K} |\nabla u|^2$. Then

$$
\begin{aligned}
E(u) - E(v) &\leq H^1([F \cap B(x,r_1)] \setminus K) + r_1 \, h(r_1) \\
&\leq C \, r \, \omega_2(x,r_0)^{1/2} \, J(x,r_0)^{-1} + r \, h(r) \\
&\leq C \, r \, \omega_2(x,r)^{1/2} \, J(x,r)^{-1} + r \, h(r),
\end{aligned}
\tag{11}
$$

by (10), (9), (6), and the fact that $J(x,r_0)^{-1} \leq 2J(x,r)^{-1}$ if τ is small enough, which we still get from Lemma 43.1. We want to deduce from this that

$$
\int_{B(x,r_1)\setminus G} |\nabla(u-v)|^2 \leq C r \, \omega_2(x,r)^{1/2} J(x,r)^{-1} + rh(r).
\tag{12}
$$

This will be useful later, because the results of Section 47 will apply to v and show that its energy in $B(x,ar)$ is small. First notice that u is a competitor in the definition of v. That is, it lies in $W^{1,2}(B(x,r_1) \setminus K) \subset W^{1,2}(B(x,r_1) \setminus G)$ and has the same boundary values on $\partial B(x,r_1) \setminus K$. This is also the case of $v + \lambda(u - v)$ for every $\lambda \in \mathbb{R}$. Then $E(v) \leq \int_{B(x,r_1)\setminus G} |\nabla[v + \lambda(u-v)]|^2$ for all λ. Hence the derivative of the right-hand side at $\lambda = 0$ vanishes, and $\nabla(u-v)$ is orthogonal to ∇v in $L^2(B(x,r_1) \setminus G)$. Thus $E(u) = E(v) + \int_{B(x,r_1)\setminus G} |\nabla(u-v)|^2$ by Pythagoras, and (12) is the same as (11).

Now we want to apply the results of Section 47 to v on $\Omega' = B(x,r_1) \setminus G$. Of course the difference between $B(x,r_1)$ and $B(0,1)$ (as in Section 47) will not matter. The assumptions (47.1)–(47.3) are satisfied because $H^1(G \cap B(x,r_1)) < +\infty$ and by definition of v, and so we can apply Lemma 47.4 and Proposition 47.7.

Set $V(t) = \int_{B(x,t)\setminus G} |\nabla v|^2$ for $0 < t < r_1$. Lemma 47.4 tells us that $V'(t)$ exists almost-everywhere, and that

$$
V(t) = V(s) + \int_s^t V'(\rho)\, d\rho \quad \text{for } 0 < s < t < r_1.
\tag{13}
$$

Then Proposition 47.7 says that

$$
V(\rho) \leq \rho \, \alpha(\rho) \, V'(\rho) \text{ for almost every } \rho \in \mathcal{B},
\tag{14}
$$

where $2\pi\alpha(\rho)\rho$ denotes the length of the longest connected component of $\partial B(x,\rho)\setminus G$ and $\mathcal{B}$ is the nice set of radii $\rho \in (0,r_1)$ for which the analogue of (47.8) holds.

Now let $\rho \in [ar,r_1) \cap \mathcal{G}$ be given. Here a is as in the statement of Proposition 1; it will be chosen soon, much smaller than $1/4$ (so that $ar < r_1$). The other constants C_0 and τ will be chosen later, depending on a. In particular, we can take $\tau < a/20$, so that $10\beta_K(x,r_0)\,r_0 \leq ar$ (by (5)). Then Lemma 46.16 tells us that the two points of $G \cap \partial B(x,\rho) = F \cap \partial B(x,\rho)$ lie in a same component of $G \cap \overline{B}(x,\rho)$. In particular, ρ satisfies (47.8) and lies in $\mathcal{B}$. Moreover, if τ is small enough (again depending on a), then $\alpha(\rho) \leq \frac{11}{20}$, because the two points of

$G \cap \partial B(x, \rho)$ lie on K (by (7) and (9)), and all the points of $K \cap B(x, r_0)$ lie at distance at most $r_0 \beta_K(x, r_0) \leq 2r_0\tau$ from the line $P(x, r_0)$ through x (by (42.2) and (5)). Note that the constraint that $|b - a| \geq t$ in (7) eliminates the case when the two points of $K \cap \partial B(x, \rho)$ are close to each other. Altogether, (14) says that

$$V(\rho) \leq \frac{11}{20} \rho V'(\rho) \text{ for almost every } \rho \in [ar, r_1) \cap \mathcal{G}. \tag{15}$$

Next we want to plug (15) back into (13), get a differential inequality on most of $[ar, r_1)$, and eventually a good estimate on $V(ar)/V(r_1)$. It will be easier to deal with the function $W(t) = \text{Log}(V(t))$. Let us assume that $V(ar) > 0$; otherwise (20) below will be automatically satisfied. Then W is defined on $[ar, r_1)$. It has a derivative at all the points of $[ar, r_1)$ where $V'(t)$ exists (i.e., almost-everywhere), and $W'(t) = V'(t)/V(t)$. Moreover, we also have the analogue of (13) for W, i.e.,

$$W(t) = W(s) + \int_s^t W'(\rho) \, d\rho = W(s) + \int_s^t \frac{V'(\rho)}{V(\rho)} \, d\rho \text{ for } ar \leq s < t < r_1. \tag{16}$$

Indeed, (13) and Proposition 9.1 say that $V \in W^{1,1}(ar, r_1)$, then Lemma 11.7 says that $W \in W^{1,1}(ar, r_1)$, and (16) for $s > ar$ follows from Proposition 9.1. The case of $s = ar$ is obtained by continuity. Now

$$W(r_1) - W(ar) = \int_{ar}^{r_1} \frac{V'(\rho)}{V(\rho)} \, d\rho \geq \int_{[ar,r_1)\cap\mathcal{G}} \frac{V'(\rho)}{V(\rho)} \, d\rho \geq \frac{20}{11} \int_{[ar,r_1)\cap\mathcal{G}} \frac{d\rho}{\rho} \tag{17}$$

by (16), because $W'(\rho) \geq 0$ almost-everywhere, and by (15).

Recall from (8) that $H^1([0, r_0] \setminus \mathcal{G}) \leq C\tau r$. Then

$$\int_{[ar,r_1)\cap\mathcal{G}} \frac{d\rho}{\rho} \geq \int_{[ar+C\tau r, r_1)} \frac{d\rho}{\rho} \geq \text{Log}\left(\frac{r_1}{(a+C\tau)r}\right)$$
$$\geq \text{Log}\left(\frac{r_1}{2ar}\right) \geq \text{Log}\left(\frac{1}{8a}\right) \tag{18}$$

if $a < 1/8$ and $\tau \leq a/C$, say. Recall that $W(t) = \text{Log}(V(t))$, so that

$$\frac{V(r_1)}{V(ar)} = \exp(W(r_1) - W(ar)) \geq \exp\left\{\frac{20}{11} \text{Log}\left(\frac{1}{8a}\right)\right\} = (8a)^{-\frac{20}{11}}. \tag{19}$$

But $V(t) = \int_{B(x,t)\setminus G} |\nabla v|^2$, so

$$\int_{B(x,ar)\setminus G} |\nabla v|^2 = V(ar) \leq (8a)^{\frac{20}{11}} V(r_1) = (8a)^{\frac{20}{11}} \int_{B(x,r_1)\setminus G} |\nabla v|^2. \tag{20}$$

Finally,

$$\int_{B(x,r_1)\setminus G} |\nabla v|^2 \leq \int_{B(x,r_1)\setminus K} |\nabla u|^2 \leq r w_2(x, r) \tag{21}$$

by definition of v and because $r_1 \leq r$, so that

$$\int_{B(x,ar)\setminus G} |\nabla v|^2 \leq (8a)^{\frac{20}{11}} \, r \, \omega_2(x,r). \tag{22}$$

We may now combine (22) with (12) and get that

$$\omega_2(x,ar) = (ar)^{-1} \int_{B(x,ar)\setminus K} |\nabla u|^2 \tag{23}$$

$$\leq 2(ar)^{-1} \int_{B(x,ar)\setminus G} |\nabla v|^2 + 2(ar)^{-1} \int_{B(x,ar)\setminus G} |\nabla(u-v)|^2$$

$$\leq 128 \, a^{\frac{9}{11}} \omega_2(x,r) + Ca^{-1} \omega_2(x,r)^{1/2} \, J(x,r)^{-1} + 2a^{-1}h(r).$$

If a is small enough, $128 \, a^{\frac{9}{11}} < \gamma$, and (23) implies (3). This completes our proof of Proposition 1.   □

Remark 24. It is easy to see from the proof that when γ is small, we can take $a = \gamma^s$ for any given $s > 1$. Of course this forces us to take very large constants C and τ^{-1}. If we continued the proof of our main theorems on these grounds, we would eventually get estimates like $\omega_2(x,t) \leq Ct^\alpha$ for any $\alpha < 1$ (and provided that $h(t)$ decreases fast enough). However, if we want optimal decay results, it is more reasonable to prove first that K is a nice curve, and then go for more precise estimates without the extra difficulties that we have now because we need to control holes in K, and so on.

Remark 25. Most of the arguments of this section would work well with quasi-minimizers (instead of our almost-minimizers), except that we used the results of Sections 45 and 46 to get lots of good disks, and these results used almost-minimality in a rather important way. See Remark 45.12.

    However, so far we never used the fact that the gauge function h tends to zero; we just need it to be small enough (as in (2)). In other words, our arguments still work for (r_0, M, a')-quasiminimizers, provided that we take M sufficiently close to 1 and a' sufficiently small (depending on the constants γ and a from Proposition 1, for instance). But this is not too much the spirit of quasiminimizers, where the idea is rather to take M large.

Remark 26. Proposition 1 stays true when K separates $D_+(x,r)$ from $D_-(x,r)$ in $\overline{B}(x,r)$, and we can even replace $J(x,r)^{-1}$ with 0 (i.e., omit the terms with $J(x,r)^{-1}$) in (2) and (3). This is because all our estimates that use the jumps eventually come from Proposition 44.1, for which this observation holds trivially. See Remarks 44.23 and 46.17. Also note that this separation property is hereditary: if K separates $D_+(x,r)$ from $D_-(x,r)$ in $\overline{B}(x,r)$, if $r_0 \in [r/2, r]$ (as happens near (4), for instance), and if $\beta_K(x,r_0) \leq 10^{-1}$ (so that the disks $D_\pm(x,r_0)$ are defined), then K also separates $D_+(x,r_0)$ from $D_-(x,r_0)$ in $\overline{B}(x,r_0)$.

Remark 27. Proposition 1 also holds when $(u, K) \in TRAM$, with linear distortion $\leq L$. We start the argument as in the standard case: we select radii r_0 and r_1, and then construct a competitor (v, G) for (u, K) in $B(x, r_1)$. This time we have matrices A_1 and A_2 as in Definition 7.45 (for the pair (x, r_1), say), and we choose v so that it minimizes $\int_{B(x,r_1) \backslash G} |A_2 \nabla v|^2$ with the given boundary values on $\partial B(x, r_1) \backslash G$.

Call A_3 the inverse of the transpose of A. By Exercise 7.59 (or an easy computation), $w = v \circ A_3^{-1}$ minimizes the standard energy $\int_{A_3(B(x,r_1) \backslash G)} |\nabla w|^2$ with the given values on $A_3(\partial B(x, r_1) \backslash G)$, and we can apply the results of Section 47 to it.

Recall from (46.25) and (46.26) that we have lots of good radii $t \in [0, r_1/L]$ such that $A_3(G) \cap \partial B(A_3(x), t) = A_3(F) \cap \partial B(A_3(x), t) = A_3(K) \cap \partial B(A_3(x), t)$ (by (9)), and this set has exactly two points. In addition, these two points lie in the same component of $A_3(G) \cap \overline{B}(A_3(x), t)$ (by the analogue of Lemma 46.16).

All this allows us to apply Lemma 47.4 and Proposition 47.7 to w in $A_3(B(x, r_1) \backslash G)$. That is, we set $V(t) = \int_{B(A_3(x), t) \backslash A_3(G)} |\nabla w|^2$, and eventually get that $V(ar)/V(r_1/L) \leq C a^{20/11}$, as in (19). We may then return to v on $B(x, r_1) \backslash G$, and conclude with the same estimates as in the standard case. We lose constants that depend on L in the translation, but this is all right, we can choose a accordingly small. $\qquad\square$

This was the main point where we needed to be careful with linear distortion. Unless the author forgot something important, the rest of this part (except Section 53 on propellers) should apply to almost-minimizers with linear distortion, with only minor modifications in the proofs.

49 Self-improving sets of estimates $(n = 2)$

This is the last section of preparation for our main C^1-theorems. We still assume that $(u, K) \in TRLAM(\Omega, h)$, as in Standard Assumption 41.1, and we use the previous sections to find lists of estimates on u and K in a disk that automatically imply the same ones on smaller disks.

Choose $\gamma = 1/4$, and denote by $a < 1/4$, τ, and C_0 the constants that we get from Proposition 48.1.

Proposition 1. *For each $\varepsilon > 0$ we can find constants $\tau_3 < \tau_2 < \tau_1 < \varepsilon$ such that if $(u, K) \in TRLAM(\Omega, h)$, $x \in K$, and $r > 0$ are such that $\overline{B}(x, r) \subset \Omega$, and if*

$$\beta_K(x, r) \leq \tau_1, \quad \omega_2(x, r) \leq \tau_2, \quad and \quad J(x, r)^{-1} + h(r) \leq \tau_3, \tag{2}$$

then (2) also holds with r replaced by ar.

Often we can say a little more; we shall see this as we proceed with the proof.

Let us first decide to take $\tau_1 \leq \mathrm{Min}(\varepsilon, \tau/3)$. Since $\tau_1 + \tau_2 + \tau_3 < 3\tau_1 \leq \tau$, we can apply Proposition 48.1, and (48.3) says that

$$\omega_2(x, ar) \leq \omega_2(x, r)/4 + C_0\,\omega_2(x, r)^{1/2}\,J(x, r)^{-1} + C_0\,h(r) \tag{3}$$

$$\leq \tau_2/4 + C_0\,\tau_2^{1/2}\,\tau_3 + C_0\,\tau_3 < \tau_2,$$

by (2), and if τ_3 is small enough with respect to τ_2. This takes care of $\omega_2(x, ar)$. This was the most important part of the argument, where we actually proved some decay for one of our quantities (namely, ω_2). The rest is mostly a question of hierarchy between the various quantities (i.e., deciding which ones control the other).

To estimate $J(x, ar)$, first observe that $a < 1/4$, so we can pick a_* such that $a^{1/2} < a_* \leq 1/2$. If τ_1 is small enough, $\beta_K(x, t) \leq 10^{-1}$ for $ar \leq t \leq r$, and Lemma 43.7 says that $J(x, ar) \geq a^{-1/2}[J(x, r) - C] \geq a_*^{-1}J(x, r)$ (if τ_3 is small enough, because $J(x, r) \geq \tau_3^{-1}$). Then

$$J(x, ar)^{-1} \leq a_* J(x, r)^{-1} \leq J(x, r)^{-1}/2, \tag{4}$$

and of course $J(x, ar)^{-1} + h(x, ar) < J(x, r)^{-1} + h(r) \leq \tau_3$, since h is nondecreasing. So the proposition will follow as soon as we check that $\beta_K(x, ar) \leq \tau_1$.

Lemma 5. *Keep the assumptions of Proposition 1. Then*

$$\beta_K(x, r/2) \leq C\omega_2(x, r)^{1/2} + C\omega_2(x, r)^{1/4}J(x, r)^{-1/2} + Ch(r)^{1/2} \leq C\tau_2^{1/2} + C\tau_3^{1/2}. \tag{6}$$

Note that Proposition 1 will follow at once, because $\beta_K(x, ar) \leq \frac{2}{a}\beta_K(x, r/2)$ (by definition, see (41.2)), and because we can take τ_2 and τ_3 as small as we like, depending on τ_1 and a.

The second inequality in (6) is easy. To prove the first one, first use Chebyshev to choose $r_0 \in [3r/4, r]$ such that

$$\omega_2^*(x, r_0) := \int_{\partial B(x, r_0) \setminus K} |\nabla u|^2 \leq 4r^{-1}\int_{B(x, r) \setminus K} |\nabla u|^2 = 4\,\omega_2(x, r). \tag{7}$$

Then apply Proposition 44.1 to the pair (x, r_0). This gives a compact set $F(x, r_0)$ such that in particular

$$K \cap \overline{B}(x, r_0) \subset F(x, r_0) \subset \overline{B}(x, r_0) \tag{8}$$

and

$$H^1(F(x, r_0) \cap B(x, r_0) \setminus K) \leq C\,r_0\,\omega_2(x, r_0)^{1/2}J(x, r_0)^{-1}$$

$$\leq C\,r\,\omega_2(x, r)^{1/2}J(x, r)^{-1}, \tag{9}$$

where the last inequality comes from the definition of w_2 and a simple application of Lemma 43.1 or Lemma 43.7, as above.

If the τ_i are small enough, then $\beta_K(x, r_0)$, $w_2(x, r_0)$, $J(x, r_0)^{-1}$, and $h(r_0)$ are as small as we want, so Lemma 46.5 says that $r^{-1}H^1([0, r_0] \setminus \mathcal{G}(x, r_0)$ is as small as we want, where $\mathcal{G}(x, r_0)$ is the set of good radii defined by (46.4). Here we just want to find a good radius $r_1 \in \mathcal{G}(x, r_0)$ such that $2r/3 < r_1 < r_0$ and $w_2^*(x, r_1) \leq 10 w_2(x, r)$ (which is easy to get, as for (7)).

By definition (see (46.4)), $K \cap \partial B(x, r_1) = F(x, r_0) \cap \partial B(x, r_1)$, and this set has exactly two points. Call these points y_1 and y_2, and note that $|y_1 - y_2| \geq r_1$, again by (46.4). We can apply Proposition 45.9 and get that

$$H^1(K \cap B(x, r_1)) \leq |y_1 - y_2| + C r w_2^*(x, r_1) + r h(r) \leq |y_1 - y_2| + C r w_2(x, r) + r h(r). \tag{10}$$

Then (8) and (9) say that

$$
\begin{aligned}
H^1(F(x, r_0) \cap B(x, r_1)) &\leq H^1(K \cap B(x, r_1)) + C r w_2(x, r)^{1/2} J(x, r)^{-1} \\
&\leq |y_1 - y_2| + C r w_2(x, r) + r h(r) + C r w_2(x, r)^{1/2} J(x, r)^{-1}.
\end{aligned}
\tag{11}
$$

Next $F(x, r_0) \cap \overline{B}(x, r_1)$ separates the two components of $\partial B(x, r_1) \setminus \{y_1, y_2\}$ in $\overline{B}(x, r_1)$, by (44.3). Then there is a connected subset $\Xi \subset F(x, r_0) \cap \overline{B}(x, r_1)$ that still separates. See Theorem 14.3 on p. 123 of [Ne], and the proof of (29.52) for details on how to apply it in a disk. Then Ξ connects y_1 to y_2, so (11) and Pythagoras say that

$$\text{dist}(z, [y_1, y_2]) \leq C r \{w_2(x, r) + h(r) + w_2(x, r)^{1/2} J(x, r)^{-1}\}^{1/2} \quad \text{for } z \in \Xi. \tag{12}$$

Also note that $H^1(\Xi) \geq |y_1 - y_2|$, so

$$
\begin{aligned}
H^1(K \cap B(x, r_1) \setminus \Xi) &\leq H^1(F(x, r_0) \cap B(x, r_1) \setminus \Xi) \\
&\leq H^1(F(x, r_0) \cap B(x, r_1)) - H^1(\Xi) \\
&\leq C r \{w_2(x, r) + h(r) + w_2(x, r)^{1/2} J(x, r)^{-1}\},
\end{aligned}
\tag{13}
$$

by (8) and (11). Finally,

$$\text{dist}(y, \Xi) \leq C r \{w_2(x, r) + h(r) + w_2(x, r)^{1/2} J(x, r)^{-1}\} \quad \text{for } y \in K \cap \overline{B}(x, r/2). \tag{14}$$

This comes from (13), the local Ahlfors-regularity of K near y (see Theorem 18.16), and the fact that $r_1 > 2r/3$: if $d = \text{dist}(y, \Xi)$ is too large, $B(y, \text{Min}(d, r/10))$ is contained in $K \cap B(x, r_1) \setminus \Xi$ and has too much mass for (13).

The first half of (6) follows from (12) and (14). This completes our proof of Lemma 5 and Proposition 1. $\qquad\qquad\qquad\qquad\qquad\qquad\qquad\qquad\qquad\qquad$ $\square$

Proposition 1 does not require h to tend to 0, we just need $h(r)$ to be small enough. Next we really want to assume that h tends to 0 and see how fast $w_2(x, \cdot)$

tends to 0, depending on h. It will be convenient to use the following regularized versions of h. Let $b > 0$ be such that $a^b = 2/3$, where $a < 1/4$ is the same constant as before, which comes from Proposition 48.1 applied with $\gamma = 1/4$. Set

$$\widetilde{h}_r(t) = \sup\left\{ \left(\frac{t}{s}\right)^b h(s)\, ; \, t \leq s \leq r\right\} \tag{15}$$

for $0 < t \leq r$, and $\widetilde{h}_r(t) = h(t)$ for $t \geq r$. Note that $h(t) \leq \widetilde{h}_r(t)$ for all t, because we can take $s = t$ in (15). If h increases sufficiently slowly, for instance, if $h(r) = Cr^\alpha$ for some $\alpha \leq b$, then $\widetilde{h}_r = h$. In general we want to see $\widetilde{h}_r$ as a regularized version of h; see (16) below.

Let us rapidly check that $\widetilde{h}_r$ is a gauge function (i.e., that $\widetilde{h}_r$ is nondecreasing and $\lim_{t\to 0+} h(t) = 0$). Its restriction to $[r, +\infty)$ is clearly nondecreasing, because h is. If $0 < t < t' \leq r$ and $s \in [t, r]$, then either $s \geq t'$ and $\left(\frac{t}{s}\right)^b h(s) \leq \left(\frac{t'}{s}\right)^b h(s) \leq$ $\widetilde{h}_r(t')$ by definition of $\widetilde{h}_r(t')$, or else $s \in [t, t')$ and $\left(\frac{t}{s}\right)^b h(s) \leq h(s) \leq h(t') \leq \widetilde{h}_r(t')$ because h is nondecreasing. So $\widetilde{h}_r$ is nondecreasing. Similarly, if $0 < t < t' < r$ and $s \in [t, r]$, then either $s \geq t'$ and $\left(\frac{t}{s}\right)^b h(s) \leq \left(\frac{t}{t'}\right)^b h(s) \leq \left(\frac{t}{t'}\right)^b h(r)$, or else $s \in [t, t')$ and $\left(\frac{t}{s}\right)^b h(s) \leq h(s) \leq h(t')$. Thus $\widetilde{h}_r(t) \leq \text{Max}\left\{ \left(\frac{t}{t'}\right)^b h(r), h(t')\right\}$. This is enough to show that $\lim_{t\to 0+} h(t) = 0$; hence $\widetilde{h}_r$ is a gauge function.

The main point of the definition (15) is that

$$\widetilde{h}_r(t) \geq \left(\frac{t}{t'}\right)^b \widetilde{h}_r(t') \text{ for } 0 < t < t' \leq r, \tag{16}$$

because if $s \in [t', r]$, then $(\frac{t'}{s})^b h(s) = (\frac{t'}{t})^b \left(\frac{t}{s}\right)^b h(s) \leq (\frac{t'}{t})^b \widetilde{h}_r(t)$. In fact, $\widetilde{h}_r$ was chosen to be the smallest gauge function such that $h \leq \widetilde{h}_r$ and (16) holds. Note that since $a^b = 2/3$, (16) with $t = at'$ (and then a change of names) says that

$$\widetilde{h}_r(at) \geq \frac{2}{3}\, \widetilde{h}_r(t) \text{ for } 0 < t \leq r. \tag{17}$$

Proposition 18. *There exist constants $\tau > 0$ and $C_1 \geq 1$ such that whenever $(u, K) \in TRLAM(\Omega, h)$ (see Standard Assumption 41.1), $x \in K$, $r > 0$, $\overline{B}(x, r) \subset \Omega$, and*

$$\beta_K(x, r) + \omega_2(x, r) + J(x, r)^{-1} + h(r) \leq \tau, \tag{19}$$

we have the following estimates for $0 < t < r$:

$$J(x, t)^{-1} \leq 2\left(\frac{t}{r}\right)^b J(x, r)^{-1}; \tag{20}$$

$$\omega_2(x, t) \leq C_1\left(\frac{t}{r}\right)^b [\omega_2(x, r) + J(x, r)^{-2}] + C_1\widetilde{h}_r(t); \tag{21}$$

$$\beta_K(x, t/2) \leq C_1\left(\frac{t}{r}\right)^{b/2} [\omega_2(x, r) + J(x, r)^{-2}]^{1/2} + C_1\widetilde{h}_r(t)^{1/2}. \tag{22}$$

Recall that $\widetilde{h}_r$ is the gauge function defined in (15), and that b is a positive constant. If $h(t)$ decreases fast enough when t tends to 0^+, we get that $w_2(x, t)$ and $\beta_K(x, t)$ decrease at least like t^b and $t^{b/2}$ respectively. But the exponent b is not optimal. If h decreases very slowly, then $\widetilde{h}_r$ is comparable to h, and governs the decay of $w_2(x, t)$ and $\beta_K(x, t)$.

To prove the proposition, take ε small (to be chosen just below (25)), and apply Proposition 1. This gives constants $\tau_3 < \tau_2 < \tau_1$. Take $\tau = \tau_3$. If (19) holds, then we get that

$$\beta_K(x, a^n r) \leq \tau_1, \ \ w_2(x, a^n r) \leq \tau_2, \ \text{ and } \ J(x, a^n r)^{-1} + h(a^n r) \leq \tau_3 \ \text{ for } n \geq 0, \tag{23}$$

by multiple applications of Proposition 1. We also get that

$$J(x, a^n r)^{-1} \leq \frac{1}{2} J(x, a^{n-1} r)^{-1} \leq \cdots \leq 2^{-n} J(x, r)^{-1}, \tag{24}$$

by (4). If $0 < t < r$, we choose $n \geq 0$ such that $a^{n+1} r < t \leq a^n r$, and we note that

$$J(x, t) \geq \left(\frac{t}{a^n r}\right)^{-1/2} [J(x, a^n r) - C] \geq \frac{1}{2} \left(\frac{t}{a^n r}\right)^{-1/2} J(x, a^n r)$$

$$\geq \frac{1}{2} \left(\frac{t}{a^n r}\right)^{-1/2} 2^n J(x, r) \geq \frac{1}{2} \left(\frac{t}{a^n r}\right)^{-b} 2^n J(x, r) \tag{25}$$

$$\geq \frac{1}{2} \left(\frac{t}{a^n r}\right)^{-b} \left(\frac{a^n r}{r}\right)^{-b} J(x, r) = \frac{1}{2} \left(\frac{t}{r}\right)^{-b} J(x, r)$$

by Lemma 43.7, if ε is small enough (so that $J(x, a^n r) \geq J(x, r) \geq 2C$), by (24), because $b < 1/2$ and $a^{-b} = 3/2 < 2$, and by definition of n. This proves (20).

For the proof of (21) and (22), it will be convenient to set

$$A = w_2(x, r) + C_2 J(x, r)^{-2}, \tag{26}$$

where C_2 will be chosen soon. Let us first check that if we prove that

$$w_2(x, a^n r) \leq \left(\frac{2}{3}\right)^n A + C_2 \widetilde{h}_r(a^n r) \ \text{ for } n \geq 0, \tag{27}$$

then (21) will follow. Indeed, if $t < r$ and $n \geq 0$ is such that $a^{n+1} r < t \leq a^n r$, then we will get that

$$w_2(x, t) = t^{-1} \int_{B(x,t) \setminus K} |\nabla u|^2 \leq t^{-1} a^n r \, w_2(x, a^n r) \leq a^{-1} \left[\left(\frac{2}{3}\right)^n A + C_2 \widetilde{h}_r(a^n r)\right]$$

$$\leq a^{-1} \left[\left(\frac{2}{3}\right)^n A + \frac{3}{2} C_2 \widetilde{h}_r(t)\right] = a^{-1} \left[\left(\frac{a^n r}{r}\right)^b A + \frac{3}{2} C_2 \widetilde{h}_r(t)\right]$$

$$\leq 2 a^{-1} \left[\left(\frac{t}{r}\right)^b A + C_2 \widetilde{h}_r(t)\right] \tag{28}$$

by definition of w_2, (27), (17) and the fact that $\widetilde{h}_r$ is nondecreasing (or directly (16)), and because $a^b = 2/3$. Thus (21) follows from (27).

We shall prove (27) by induction on n. The case of $n = 0$ is trivial, so let us assume that (27) holds for n and check it for $n + 1$. First,

$$
\begin{aligned}
w_2(x, a^{n+1}r) &\leq \frac{1}{4} w_2(x, a^n r) + C_0\, w_2(x, a^n r)^{1/2} J(x, a^n r)^{-1} + C_0 h(a^n r) \\
&\leq \frac{1}{2} w_2(x, a^n r) + 2C_0^2 J(x, a^n r)^{-2} + C_0 h(a^n r) \\
&\leq \frac{1}{2} w_2(x, a^n r) + 2C_0^2\, 2^{-2n} J(x, r)^{-2} + C_0 h(a^n r)
\end{aligned}
\tag{29}
$$

by (3) and (24). Next $J(x, r)^{-2} \leq C_2^{-1} A$ by (26), $h(a^n r) \leq \widetilde{h}_r(a^n r)$ trivially (by (15)), so the induction hypothesis yields

$$
\begin{aligned}
w_2(x, a^{n+1}r) &\leq \frac{1}{2} \left(\frac{2}{3}\right)^n A + \frac{1}{2} C_2\, \widetilde{h}_r(a^n r) + 2C_0^2\, 2^{-2n} C_2^{-1} A + C_0 \widetilde{h}_r(a^n r) \\
&\leq \left(\frac{2}{3}\right)^{n+1} A + \left[\frac{1}{2} C_2 + C_0\right] \widetilde{h}_r(a^n r)
\end{aligned}
\tag{30}
$$

if C_2 is large enough (compared to C_0^2). Now $\widetilde{h}_r(a^n r) \leq \frac{3}{2} \widetilde{h}_r(a^{n+1}r)$ by (17), so

$$
\left[\frac{1}{2} C_2 + C_0\right] \widetilde{h}_r(a^n r) \leq \left[\frac{3}{4} C_2 + \frac{3}{2} C_0\right] \widetilde{h}_r(a^{n+1}r) \leq C_2\, \widetilde{h}_r(a^{n+1}r)
\tag{31}
$$

if C_2 is large enough, and (27) holds for $n + 1$. [This was our second and last constraint on C_2.] This completes our proof of (21).

Concerning (22), observe that the pairs $(x, a^n r)$, $n \geq 0$, satisfy the hypotheses of Proposition 1, by (23). So we may apply Lemma 5 and get that

$$
\begin{aligned}
\beta_K(x, a^n r/2) &\leq C w_2(x, a^n r)^{1/2} + C w_2(x, a^n r)^{1/4} J(x, a^n r)^{-1/2} + C h(a^n r)^{1/2} \\
&\leq C w_2(x, a^n r)^{1/2} + C J(x, a^n r)^{-1} + C h(a^n r)^{1/2} \\
&\leq C\left(\frac{2}{3}\right)^{n/2} A^{1/2} + C\, C_2^{1/2}\, \widetilde{h}_r(a^n r)^{1/2} + C 2^{-n} J(x, r)^{-1} + C h(a^n r)^{1/2} \\
&\leq C\left(\frac{2}{3}\right)^{n/2} A^{1/2} + C \widetilde{h}_r(a^n r)^{1/2}
\end{aligned}
\tag{32}
$$

by (27), (24), (26), and (15). Now let $t \leq r$ be given, and choose $n \geq 0$ such that $a^{n+1}r < t \leq a^n r$. Then

$$
\begin{aligned}
\beta_K(x, t/2) &\leq a^{-1} \beta_K(x, a^n r/2) \leq C a^{-1}\left(\frac{2}{3}\right)^{n/2} A^{1/2} + C a^{-1} \widetilde{h}_r(a^n r)^{1/2} \\
&\leq C a^{-1}\left(\frac{t}{r}\right)^{b/2} A^{1/2} + C a^{-1} \widetilde{h}_r(t)^{1/2},
\end{aligned}
\tag{33}
$$

by (32), because $\left(\frac{2}{3}\right)^{n+1} = a^{(n+1)b} \leq \left(\frac{t}{r}\right)^b$, and by (16) or (17). This proves (22), and Proposition 18 follows as well. $\square$

Remark 34. When K separates $D_+(x, r)$ from $D_-(x, r)$ in $\overline{B}(x, r)$, we can replace $J(x, r)^{-1}$ with 0 in the various estimates. This comes from similar observations in the previous sections, as well as the heredity of this separation property. See Remark 48.26, but also Remarks 44.23 and 46.17.

Propositions 1 and 18 give us all the information we need to get the regularity theorems in the next section. What we still have to do is translate our estimates on the $\beta_K(x, t)$, for instance, into simpler geometric information about curves.

50   Three conditions for K to contain a nice curve $(n = 2)$

We are now ready to state the main regularity results of this part. We still restrict to dimension 2 and keep the same hypotheses as in Standard Assumption 41.1. That is, Ω is an open set in the plane, h is a nondecreasing positive function on $(0, +\infty)$, and $(u, K) \in TRLAM(\Omega, h)$ is a coral local almost-minimizer or local topological almost-minimizer in Ω, with gauge function h. See Sections 7 and 8 for definitions. Our first theorem only requires h to be small enough, and not even to tend to zero (even though we required this in the definition of a gauge function). Later on we shall assume more decay on h and get more regularity on the obtained curves. See Corollaries 25, 33, and 51 below.

a. Chord-arc curves

Theorem 1. *For each $\varepsilon > 0$, we can find a constant $\tau(\varepsilon) > 0$ such that if $(u, K) \in TRLAM(\Omega, h)$, $x \in K$, $r > 0$, $\overline{B}(x, r) \subset \Omega$, and*

$$\beta_K(x, r) + \omega_2(x, r) + J(x, r)^{-1} + h(r) \leq \tau(\varepsilon), \tag{2}$$

then we can find $\rho \in (r/2, 2r/3)$ such that $K \cap B(x, \rho)$ is a $(1+\varepsilon)$-chord-arc (open) curve with its endpoints on $\partial B(x, \rho)$.

Definition 3. *A $(1 + \varepsilon)$-chord-arc (open) curve is a simple rectifiable open curve γ (i.e., not a loop) such that, for each choice of points y and z in γ, the length of the arc of γ from y to z is at most $(1 + \varepsilon)|y - z|$.*

Let us prove Theorem 1. Let (u, K), x, and r be as in the statement, and then set $K_0 = K \cap B(x, 2r/3)$. Note that

$$\beta_K(y, r/3) + \omega_2(y, r/3) \leq 6\beta_K(x, r) + 3\omega_2(x, r) \quad \text{for } y \in K_0, \tag{4}$$

by the definitions (41.2) and (23.5). [Note for professionals of β_K: we need $6\beta_K(x, r)$ instead of just 3 because we restricted to lines through x in (41.2).]

Also, $J(y, r/3)^{-1} \leq 2J(x, r)^{-1} \leq 2\tau(\varepsilon)$ by (2) and the proof of Lemma 43.1 (the fact that we slightly move the center of the ball does not matter, especially since here $\beta_K(x, r)$ is as small as we want). Altogether,

$$\beta_K(y, r/3) + \omega_2(y, r/3) + J(y, r/3)^{-1} + h(r/3) \leq 6\tau(\varepsilon) \quad \text{for } y \in K_0, \tag{5}$$

by (2) and because h is nondecreasing. If we choose $\tau(\varepsilon)$ small enough, we can apply Proposition 49.1 to the pairs $(y, r/3)$, $y \in K_0$, and with any small ε' given in advance. We get that

$$\beta_K(y, a^n r/3) + \omega_2(y, a^n r/3) + J(y, a^n r/3)^{-1} \leq 3\varepsilon' \text{ for } y \in K_0 \text{ and } n \geq 0, \quad (6)$$

because we can iterate our application of Proposition 49.1. Here $a < 1$ is still the same constant as in Section 48. Thus (6) implies that

$$\beta_K(y, t) + \omega_2(y, t) + J(y, t)^{-1} \leq C\varepsilon' \text{ for } y \in K_0 \text{ and } 0 < t \leq r/3, \quad (7)$$

with a constant C that depends on a (but we don't mind). This comes again from the definitions (41.2) and (23.5), and Lemma 43.7. Next set

$$\mathcal{G}'(y, t) = \{\rho \in (0, t) \, ; K \cap \partial B(y, \rho) \text{ has two points,}$$
$$\text{at distance} \geq \rho \text{ from each other}\}. \quad (8)$$

We want to use Lemma 46.5 to estimate the size of $[0, t] \backslash \mathcal{G}'(y, t)$. Let us first choose $t' \in (t - 10^{-4}t, t)$ such that $\omega_2^*(y, t') \leq 10^5 \omega_2(y, t) \leq C\varepsilon'$. [Compare the definitions (42.5) and (23.5), and use Chebyshev.] Then $[0, t'] \cap \mathcal{G}'(y, t) = \mathcal{G}'(y, t') \subset \mathcal{G}(y, t')$, where $\mathcal{G}(y, t)$ is defined in (46.4). Thus Lemma 46.5 and (7) imply that

$$H^1([0, t] \backslash \mathcal{G}'(y, t)) \leq 10^{-4}t + H^1([0, t'] \backslash \mathcal{G}'(y, t')) \leq 10^{-3}t \quad (9)$$

for $y \in K_0$ and $0 < t \leq r/3$, if ε' is small enough. Similarly, Lemma 46.5 and (2) yield

$$H^1([0, r] \backslash \mathcal{G}'(x, r)) \leq 10^{-3}r. \quad (10)$$

For $y \in K_0$ and $0 < t \leq r/3$, and also for $y = x$ and $t = r$, call $P(y, t)$ a line through y such that

$$\text{dist}(z, P(y, t)) \leq t\,\beta_K(y, t) \leq C\varepsilon't \text{ for } z \in K \cap \overline{B}(y, t), \quad (11)$$

as in (42.2). Because of (9) and (10), we also have that

$$\text{dist}(z, K) \leq 2 \cdot 10^{-3}t \text{ for } z \in P(y, t) \cap B(y, t) \quad (12)$$

(take $\rho \in \mathcal{G}'(y, t)$ close to $|z - y|$, and note that at least one of the two points of $K \cap \partial B(y, \rho)$ must be very close to z).

Lemma 13. *Choose $\rho_1 \in (62r/100, 63r/100) \cap \mathcal{G}'(x, r)$. [This is possible, by (10).] Call a, b the two points of $K \cap \partial B(x, \rho_1)$. Then there is a simple curve $\gamma \subset K \cap \overline{B}(x, \rho_1)$ that goes from a to b.*

We shall get γ as a subset of γ^*, where γ^* is the limit of a sequence of polygonal lines γ_j which we construct by induction. We start with $\gamma_0 = [a, b]$. Next assume that we already constructed γ_j, $j \geq 0$, and that

$$\gamma_j \text{ is composed of segments of lengths } \leq 2 \cdot 10^{-2j} \, r$$
$$\text{with endpoints in } K \cap B(x, r_j), \tag{14}$$

where we set $r_j = \rho_1 + \sum_{k=1}^{j} 2 \cdot 10^{-2k} \, r < 2r/3$. To get γ_{j+1}, we replace each of the segments L that compose γ_j by a polygonal line with the same endpoints. We choose the intermediate additional points in K, close to L, and so that each of the new segments has a length between $10^{-3} H^1(L)$ and $10^{-2} H^1(L)$. This is possible because of (12), applied with $y = x$ and $t = r$ when $j = 0$, and with for y any of the endpoints of L and $t = H^1(L)$ when $j > 1$. For this last case, we also use the fact that endpoints of L lie in $K \cap B(x, r_j) \subset K_0$. The new polygonal line γ_{j+1} that we get obviously satisfies (14). We had to let r_j increase slightly with j because when j gets large, the directions of the approximating lines $P(y, t)$ may turn, and then we cannot be sure that some γ_j will not go out of $B(x, \rho_1)$.

Now we want to take limits. We parameterize each γ_j with a piecewise affine mapping $f_j : [0, 1] \to \gamma_j$. The affine pieces of f_j are parameterizations (with constant speeds) of the line segments that compose γ_j. We make sure to choose each f_{j+1} so that it coincides with f_j at all the points of $[0,1]$ that correspond to extremities of the line segments that compose γ_j. Then $\|f_{j+1} - f_j\|_\infty \leq 3 \cdot 10^{-2j} \, r$, by (14) and because we chose our additional vertices of γ_{j+1} close to the line segments L. Thus the f_j converge uniformly to some function f. Set $\gamma^* = f([0, 1])$. Then $\gamma^* \subset K_0$, by (14), and it goes from a to b.

Since $\gamma^* \subset K_0$, it has finite length, and Proposition 30.14 says that we can find a simple curve $\gamma \subset \gamma^*$ that goes from a to b. Then $\gamma \subset B(x, \rho_1)$, because $K \cap \partial B(x, \rho_1) = \{a, b\}$ (by definition of ρ_1) and γ meets $B(x, \rho_1)$ by construction. This proves Lemma 13. $\qquad \square$

Choose $\rho \in (r/2, 6r/10) \cap \mathcal{G}'(x, r)$. This is possible, by (10). We want to check that

$$K \cap \overline{B}(x, \rho) = \gamma \cap \overline{B}(x, \rho). \tag{15}$$

We already know that $\gamma \subset K$. To prove the other inclusion, let us proceed by contradiction and assume that we can find $x_1 \in K \cap \overline{B}(x, \rho) \setminus \gamma$. Also choose a point x_2 in $\gamma \cap B(x, r/4)$. This can be done, because γ goes from a to b and stays very close to $P(x, r)$, by (11). By the proof of Lemma 13 (i.e., with sequences of polygonal lines from x_1 to x_2), we can find a simple arc $\gamma' \subset K_0$ that goes from x_1 to x_2. Call x_3 the first point of γ when we run along γ' from x_1 to x_2, and let γ'' denote the arc of γ' between x_1 and x_3. Note that $\gamma'' \setminus \{x_3\} \subset K \setminus \gamma$. Hence it does not meet $\partial B(x, \rho)$, because $K \cap \partial B(x, \rho)$ has only two points (by definition of ρ), and these points lie on γ (because γ is simple and goes from a to b through $B(x, r/4)$). In particular, $x_3 \in K \cap \overline{B}(x, \rho) \subset K_0$.

For t small enough, $K \cap \partial B(x_3, t)$ has at least three points, one on γ'' and two on γ. Thus $\mathcal{G}'(x_3, t) = \emptyset$ for t small, in contradiction with (9). This proves our claim (15).

Note that $K \cap \overline{B}(x, \rho) = \gamma \cap \overline{B}(x, \rho)$ is a simple (open) curve that connects the two points of $K \cap \partial B(x, \rho)$, so our proof of Theorem 1 will be complete once we check that if $\tau(\varepsilon)$ is small enough, then $\gamma \cap \overline{B}(x, \rho)$ is $(1 + \varepsilon)$-chord-arc. Let y_1, $y_2 \in \gamma \cap \overline{B}(x, \rho)$ be given and denote by γ^* the arc of γ between y_1 and y_2. We want to show that

$$H^1(\gamma^*) \leq (1 + \varepsilon)|y_1 - y_2|. \tag{16}$$

Let us start with the case when $|y_1 - y_2| \geq r/100$. In Lemma 13, when we chose $\rho_1 \in (62r/100, 63r/100) \cap \mathcal{G}'(x, r)$, we may also have required that

$$\omega_2^*(x, \rho_1) = \int_{\partial B(x, \rho_1)} |\nabla u|^2 \leq C\omega_2(x, r) \leq C\tau(\varepsilon). \tag{17}$$

Let us assume that we did. Since $\rho_1 \in \mathcal{G}'(x, r)$, we may apply Proposition 45.9 to the pair (x, ρ_1). We get that

$$H^1(\gamma) \leq H^1(K \cap B(x, \rho_1)) \leq |b - a| + C r \omega_2^*(x, \rho_1) + r h(r) \leq |b - a| + C\tau(\varepsilon) r \tag{18}$$

by Lemma 13, Proposition 45.9, (17), and (2). We may as well assume that y_1 lies between a and y_2 on γ (otherwise exchange y_1 and y_2). Then

$$H^1(\gamma) \geq |a - y_1| + H^1(\gamma^*) + |y_2 - b| \geq |a - b| - |y_1 - y_2| + H^1(\gamma^*) \tag{19}$$

because γ is simple and by the triangle inequality. Thus $H^1(\gamma^*) \leq |y_1 - y_2| + C\tau(\varepsilon) r \leq (1 + C'\tau(\varepsilon))|y_1 - y_2|$, by (18), (19), and because $|y_1 - y_2| \geq r/100$; (16) follows if $\tau(\varepsilon)$ is small enough.

When $|y_1 - y_2| < r/100$, we shall use a similar argument, but at a smaller scale. Set $t = 2|y_1 - y_2|$. Note that $y_1 \in \gamma \subset K_0$ and $t < r/3$. By (9) and Chebyshev, we can choose $s \in (t/2, t) \cap \mathcal{G}'(y_1, t)$ so that in addition

$$\omega_2^*(y_1, s) = \int_{\partial B(y_1, s)} |\nabla u|^2 \leq C\omega_2(y_1, t) \leq C\varepsilon', \tag{20}$$

where we used (7) for the last inequality. Call a' and b' the two points of $K \cap \partial B(y_1, s)$. Proposition 45.9 applies and yields

$$H^1(K \cap B(y_1, s)) \leq |b' - a'| + Cs \omega_2^*(y_1, s) + sh(s) \leq |b' - a'| + Cs\varepsilon' + Cs\tau(\varepsilon), \tag{21}$$

by (7) and (2). Note that $B(y_1, s) \subset B(x, \rho_1)$, because $y_1 \in B(x, \rho)$, $\rho \leq 6r/10$, $s \leq t = 2|y_1 - y_2| < r/50$, and $\rho_1 \geq 62r/100$. Then

$$H^1(K \cap B(y_1, s)) \geq H^1(\gamma \cap B(y_1, s)) \geq |a' - y_1| + H^1(\gamma^*) + |y_2 - b'| \tag{22}$$

by Lemma 13, because $\gamma \cap B(y_1, s)$ is a simple arc that goes from a' to b', and if y_1 lies between a' and y_2 (otherwise, just exchange the names of a' and b'). As before, $|a' - y_1| + |y_2 - b'| \geq |b' - a'| - |y_1 - y_2|$ by the triangle inequality, and

$$
\begin{aligned}
H^1(\gamma^*) &\leq H^1(K \cap B(y_1, s)) - |a' - y_1| - |y_2 - b'| \\
&\leq H^1(K \cap B(y_1, s)) - |b' - a'| + |y_1 - y_2| \\
&\leq |y_1 - y_2| + Cs\,\varepsilon' + Cs\,\tau(\varepsilon) \\
&\leq (1 + 2C\varepsilon' + 2C\tau(\varepsilon)) |y_1 - y_2| \leq (1 + \varepsilon) |y_1 - y_2|
\end{aligned}
\tag{23}
$$

by (22), (21), because $s \leq t = 2|y_1 - y_2|$, and if $\tau(\varepsilon)$ (and hence ε' also, see the discussion after (5)) is small enough.

So (16) holds in all cases, and $\gamma \cap \overline{B}(x, \rho)$ is $(1+\varepsilon)$-chord-arc. This completes our proof of Theorem 1. $\qquad\square$

Remark 24. The argument above does not need to rely on the fact that $n = 2$. In fact, as soon as we have the analogue of (11) and (12) for all the pairs (y, t), $y \in K \cap B(x, 2r/3)$ and $t < r/3$ (and with small enough constants, depending on the dimension), then there is a set Σ such that $K \cap B(x, r/2) \subset \Sigma \subset K \cap B(x, r)$ and which is homeomorphic to a ball in $\mathbb{R}^{n-1}$. The homeomorphism is Hölder, as well as its inverse, and we can even make the exponents as close to 1 as we want. This is a fairly old theorem of Reifenberg [Re]. Once we know that $K \cap B(x, r/2)$ is a fairly nice surface (without holes), it becomes somewhat easier to study its regularity. Here we merely took advantage of the fact that the proof is simpler for one-dimensional sets.

b. Asymptotically straight chord-arc curves

Corollary 25. *There is a constant $\tau > 0$ such that if $(u, K) \in TRLAM(\Omega, h)$, this time with a gauge function h that tends to 0 at 0, if $x \in K$, $r > 0$, $\overline{B}(x, r) \subset \Omega$, and*

$$
\beta_K(x, r) + \omega_2(x, r) + J(x, r)^{-1} + h(r) \leq \tau,
\tag{26}
$$

then we can find $\rho \in (r/2, 2r/3)$ such that $\Gamma = K \cap B(x, \rho)$ is a uniformly asymptotically straight chord-arc (open) curve, with its endpoints on $\partial B(x, r)$. By this we mean that Γ is a C-chord-arc for some $C \geq 1$ (which we could even take as close to 1 as we wish, by Theorem 1), and also that the length of the arc of Γ between $y \in \Gamma$ and $z \in \Gamma$ is at most $(1 + \varphi(|y - z|))|y - z|$, where $\varphi(t)$ is a function of $t > 0$ that tends to 0 when t tends to 0.

This will be easy to check. As we shall see, we can take $\varphi(t) = C\tau \left(\frac{t}{r}\right)^b + \widetilde{Ch}_{r/3}(t)$. We need to be a little more careful near the end of the proof of Theorem 1, when we estimate $H^1(\gamma^*)$, where γ^* is the arc of γ between y_1 and y_2, and y_1, y_2 are points of $\gamma \cap B(x, \rho)$ such that $|y_1 - y_2| \leq r/100$. Our proof of (23) actually yields

$$\begin{aligned}
H^1(\gamma^*) &\leq H^1(K \cap B(y_1, s)) - |b' - a'| + |y_1 - y_2| \\
&\leq |y_1 - y_2| + C s \omega_2^*(y_1, s) + sh(s) \\
&\leq |y_1 - y_2| + C s \omega_2(y_1, t) + sh(s) \\
&\leq \{1 + C\omega_2(y_1, t) + Ch(t)\} |y_1 - y_2|,
\end{aligned} \qquad (27)$$

by the first two lines of (23), the first part of (21), (20), and where we still set $t = 2|y_1 - y_2|$. So we just need to show that $\omega_2(y_1, t) \leq \varphi(t)$ for some function φ that tends to zero.

Recall that $y_1 \in \gamma \subset K \cap B(x, \rho_1) \subset K_0$ (because $\rho_1 < 2r/3$, see Lemma 13). If τ in (26) is small enough, the analogue of (5) with τ holds (with the same proof), i.e.,

$$\beta_K(y_1, r/3) + \omega_2(y_1, r/3) + J(y_1, r/3)^{-1} + h(r/3) \leq 6\tau. \qquad (28)$$

Also take τ six times smaller than in Proposition 49.18. Then (49.21) says that

$$\omega_2(y_1, t) \leq C \left(\frac{t}{r}\right)^b [\omega_2(y_1, r/3) + J(y_1, r/3)^{-2}] + C\widetilde{h}_{r/3}(t) \leq C\tau \left(\frac{t}{r}\right)^b + C\widetilde{h}_{r/3}(t), \qquad (29)$$

where $\widetilde{h}_{r/3}$ is the function defined by (49.15), which tends to zero. [We checked that it is a gauge function just after (49.15).]

This gives the desired control on $\omega_2(y_1, t)$. Corollary 25 follows, by (27). $\square$

Remark 30. It is not possible to prove that $K \cap B(x, \rho)$ is a C^1 curve under the hypotheses of Corollary 25, i.e., when h merely tends to zero. Indeed, consider the curve $\Gamma \subset B(0, 1)$ given in polar coordinates by $\theta = \text{Log}(1 + \text{Log}(1/r))$. We leave it to the reader to verify that $K = \Gamma \cup (-\Gamma) \cup \{0\}$ is an asymptotically straight (open) chord-arc curve, even though there is no tangent at the origin. Incidentally, the choice of $\theta = a \text{Log}(1/r)$ would give a logarithmic spiral, which is a typical chord-arc curve (with a constant that tends to 1 when a tends to 0).

Set $u = 0$ in $B(0, 1)$. The pair (u, K) is a topological almost-minimizer in $B(0, 1)$. The point is that K separates $B(0, 1)$ into two connected components, and the competitors of (u, K) must do the same thing. Then the best competitors are precisely obtained by replacing an arc of K by the corresponding chord. The best gauge function h corresponds to the function φ in the definition of asymptotically straight chord-arc curves, and in particular it tends to zero. Of course there is nothing special about our choice of $\theta(r) = \text{Log}(1 + \text{Log}(1/r))$, the main point is that it tends to ∞ sufficiently slowly.

c. Curves of class $C^{1,\gamma}$

We shall now (at last) worry about the case when $h(t) = Ct^\alpha$ for some $\alpha > 0$, or at least decreases sufficiently fast. Some definitions will be useful.

Definition 31. *A <u>Lipschitz graph</u> with constant $\leq M$ in $\mathbb{R}^2$ is the graph of some function $A : I \to \mathbb{R}$ such that $|A(x) - A(y)| \leq M|x - y|$ for $x, y \in I$ (and where $I \subset \mathbb{R}$ is any interval), or the image of such a graph by a rotation of the plane.*

Definition 32. *Let Γ be a simple curve of class C^1. For each $x \in \Gamma$, denote by $\xi(x)$ the direction of the tangent to Γ at x. Thus $\xi(x)$ is a unit vector, defined modulo multiplication by ± 1. Let $\gamma \in (0, 1)$ be given. We say that Γ is a $C^{1,\gamma}$ curve with constant $\leq M$ when $|\xi(x) - \xi(y)| \leq M|x - y|^\gamma$ for $x, y \in \Gamma$.*

We start with a simple statement with $C^{1,\gamma}$ curves. In the next subsection we shall give weaker conditions that yield C^1 curves.

Corollary 33. *Suppose that $(u, K) \in TRLAM(\Omega, h)$, with $h(t) = C_0 t^\alpha$, $\alpha > 0$. Set $\gamma = \frac{1}{2} \operatorname{Min}\{\alpha, b\}$, where b is the positive constant from Proposition 49.18. For each $\varepsilon > 0$, we can find a constant $\tau_0 > 0$ such that if $x \in K$, $r > 0$, $\overline{B}(x, r) \subset \Omega$, and*

$$\beta_K(x, r) + \omega_2(x, r) + J(x, r)^{-1} + h(r) \leq \tau_0, \tag{34}$$

then $\Gamma = K \cap B(x, r/2)$ is a Lipschitz graph with constant $\leq \varepsilon$, as well as a $C^{1,\gamma}$ curve with constant $\leq \varepsilon$. In addition, the two endpoints of Γ lie on $\partial B(x, r/2)$. The constant $\tau_0 > 0$ depends on ε, α, C_0, but not on Ω, (u, K), and of course x or r.

Remarks 35. This is probably the result of this part that we shall use the most. We called it corollary to insist on the fact that most of the information was already contained in Theorem 1. Some slightly more general variants (with the same gauge function) will be given in the next sections. See for instance Corollaries 51.17 and 52.25, and Proposition 70.2. Also note that we do not need any upper bound on $J(x, r)$ if K separates $D_+(x, r)$ from $D_-(x, r)$ in $\overline{B}(x, r)$; see Remark 62.

Our choice of γ is not at all optimal. To be more specific, the fact that γ looks like $\alpha/2$ for α small looks reasonable (to the author), but the limitation by $b/2$ is not. For optimal results, one should redo the comparison arguments quietly, now that we know that $K \cap B(x, r/2)$ is a nice curve. A typical good estimate could also involve a bootstrap on regularity, with estimates for the decay of $\omega_2(x, t)$ that use the current known amount of regularity of K. But we do not intend to do this here.

For the record, let us recall that when $h(r) \leq Cr$ (which includes Mumford-Shah minimizers), $K \cap B(x, r/2)$ is a $C^{1,1}$ curve (the derivative is Lipschitz). See [Bo] or [AFP3].

Proof of Corollary 33. For the moment, let us not use the precise form of the function h. If τ_0 is small enough, we can still apply (the proof of) (5) as in Theorem 1 and for (28), and we get that

$$\beta_K(y, r/3) + \omega_2(y, r/3) + J(y, r/3)^{-1} + h(r/3) \leq 6\tau_0$$
$$\text{for } y \in K_0 = K \cap B(x, 2r/3). \tag{36}$$

Then Proposition 49.18 tells us that

$$
\begin{aligned}
\beta_K(y, t/2) &\leq C \left(\frac{t}{r}\right)^{b/2} [\omega_2(y_1, r/3) + J(y_1, r/3)^{-2}]^{1/2} + C\, \widetilde{h}_{r/3}(t)^{1/2} \\
&\leq C\tau_0^{1/2} \left(\frac{t}{r}\right)^{b/2} + C\, \widetilde{h}_{r/3}(t)^{1/2} = Ca(t/2)
\end{aligned}
\tag{37}
$$

for $y \in K_0$ and $t < r$, where we set

$$
a(t) = \tau_0^{1/2} \left(\frac{t}{r}\right)^{b/2} + \widetilde{h}_{r/3}(2t)^{1/2}
\tag{38}
$$

to simplify our notation.

Next we want to find tangent lines to K and show that they vary slowly. Let us still call $P(y, t)$, for $y \in K_0$ and $0 < t < r/6$, a line through y such that

$$
\text{dist}\,(z, P(y, t)) \leq t\,\beta_K(y, t) \leq Cta(t) \quad \text{for } z \in K \cap \overline{B}(y, t),
\tag{39}
$$

as in (42.2) and (11). Also denote by $T(y, t) \in \mathbb{S}^1/\{\pm 1\}$ a unit tangent vector to $P(y, t)$, defined modulo multiplication by ± 1. We claim that

$$
\text{dist}\,(T(y, t), T(y', t')) \leq Ca(t)
\tag{40}
$$

when $y, y' \in K_0$, $|y - y'| \leq t/4$, and $t/2 \leq t' \leq t \leq r/6$. [The precise choice of a distance on $\mathbb{S}^1/\{\pm 1\}$ does not matter, but let us take $\text{dist}(v, w) = \text{Min}(|v - w|, |v + w|)$.]

First assume that $|y' - y| \geq t/100$. Note that $y' \in K \cap B(y, t)$, so (39) says that it lies within $Cta(t)$ of $P(y, t)$. Also recall that $P(y, t)$ goes through y by definition. Thus, if $v \in \mathbb{S}^1/\{\pm 1\}$ denotes the direction of $z - y$, we get that

$$
\text{dist}\,(T(y, t), v) = |y' - y|^{-1}\,\text{dist}(y', P(y, t)) \leq Ca(t).
\tag{41}
$$

Similarly, $y \in K \cap B(y', t')$, so $\text{dist}\,(y, P(y', t')) \leq Ct'a(t') \leq Cta(t)$ (by (38) and because $\widetilde{h}_{r/3}$ is nondecreasing) and $\text{dist}\,(T(y', t'), v) \leq Ca(t)$. This proves (40) when $|y' - y| \geq t/100$.

If $|y' - y| < t/100$, the simplest is to compare with a third point. Let $z \in K_0 \cap B(y, t) \cap B(y', t')$ such that $|z - y| \geq t/50$. Such a point exists, for instance because $B(y, t) \cap B(y', t')$ contains $B(y, t/4)$ and because all the points of $P(y, t) \cap B(y, t)$ lie within $2 \cdot 10^{-3}t$ of K, by (12). [Actually, we do not really need (12) here, the local Ahlfors-regularity of K would be enough to give a point $z \in K \cap B(y, t) \cap B(y', t')$ reasonably far from y and y'.] With a little more work, we could choose z so that $z \in K_0 = K \cap B(x, 2r/3)$, but the simplest is to notice that our proofs also give (37) and (39) when z lies barely out of $B(x, 2r/3)$. Now

$$
\begin{aligned}
\text{dist}\,(T(y, t), T(y', t')) &\leq \text{dist}\,(T(y, t), T(z, t)) + \text{dist}\,(T(z, t), T(y', t')) \\
&\leq Ca(t),
\end{aligned}
\tag{42}
$$

by our first case and because $|y' - z| \geq t/100$. This proves (40).

Now set $a^*(t) = \sum_{k \geq 0} a(2^{-k}t)$; we shall see soon that the series converges and that

$$\lim_{t \to 0} a^*(t) = 0, \tag{43}$$

but for the moment let us admit this and finish the argument. Because of (40) (applied with $y' = y$ and a few radii in a geometric sequence),

$$\text{dist}\,(T(y,t), T(y,t')) \leq C a^*(t) \ \text{ for } 0 < t' \leq t \tag{44}$$

when $y \in K_0$ and $0 < t \leq r/6$. So (43) says that we can define

$$T(y) = \lim_{t \to 0^+} T(y,t).$$

Moreover,

$$\text{dist}\,(T(y,t), T(y)) \leq C a^*(t) \ \text{ for } y \in K_0 \text{ and } 0 < t \leq r/6, \tag{45}$$

by (44). Call $P(y)$ the line through y with direction $T(y)$. Then

$$\text{dist}\,(z, P(y)) \leq C t a^*(t) \ \text{ for } z \in K \cap \overline{B}(y,t), \tag{46}$$

by (39) and (45) (and because $P(y)$ and $P(y,t)$ both go through y).

Since this is the case for all small t, (43) says that $P(y)$ is a tangent line to K at y. The direction of $P(y)$ depends on y nicely, because if y, $y' \in K_0$ are such that $|y' - y| \leq r/24$, then

$$\begin{aligned}
\text{dist}\,(T(y), T(y')) &\leq \text{dist}\,(T(y), T(y, 4|y'-y|)) \\
&\quad + \text{dist}\,(T(y, 4|y'-y|), T(y', 4|y'-y|)) \\
&\quad + \text{dist}(T(y', 4|y'-y|), T(y')) \\
&\leq C a^*(4|y'-y|), \tag{47}
\end{aligned}$$

by (40) and (45).

It is time to compute a^* when $h(t) = C_0 t^\alpha$. For the sake of Corollary 33, we can assume that $\alpha \leq b$ (because we took $\gamma = \frac{1}{2} \text{Min}\{\alpha, b\}$). Then (49.15) leads to

$$\begin{aligned}
\widetilde{h}_r(t) &= \sup\{t^b s^{-b} h(s)\,;\, t \leq s \leq r\} \\
&= C_0 \sup\{t^b s^{-b} s^\alpha\,;\, t \leq s \leq r\} = C_0 t^\alpha = h(t), \tag{48}
\end{aligned}$$

because $s^{\alpha-b}$ is maximal when $s = t$. Now (38) says that

$$a(t) = \tau_0^{1/2} \left(\frac{t}{r}\right)^{b/2} + \widetilde{h}_{r/3}(2t)^{1/2} = C_1 t^{b/2} + C_0 t^{\alpha/2} \tag{49}$$

with $C_1 = \tau_0^{1/2} r^{-b/2}$, and then it is easy to see that $a^*(t) \leq C a(t)$.

Thus (47) says that $\text{dist}\,(T(y), T(y')) \leq C|y'-y|^{\alpha/2}$. Hence the curve γ that we constructed in Theorem 1 (and which contains $K \cap B(x, r/2)$) is of class $C^{1,\gamma}$. [Recall that $\gamma = \alpha/2$ in the present case.] If in addition τ_0 and $h(r)$ are small enough, γ is $C^{1,\gamma}$ with constant $\leq c$, and also Lipschitz with constant $\leq \varepsilon$. This proves Corollary 33. $\qquad\square$

d. Curves of class C^1

We end this section with a result about C^1 curves. This time, instead of asking that $h(t) = C_0 t^\alpha$, we shall only require that

$$\int_0^1 h(t)^{1/2} \frac{dt}{t} < +\infty, \tag{50}$$

and apply the same computations as above.

Corollary 51. *For each $\varepsilon > 0$, we can find a constant $\tau_1 > 0$ such that if $(u, K) \in TRLAM(\Omega, h)$, $x \in K$, $r > 0$, $\overline{B}(x, r) \subset \Omega$, and*

$$\beta_K(x, r) + \omega_2(x, r) + J(x, r)^{-1} + \int_0^{2r} h(t)^{1/2} \frac{dt}{t} \le \tau_1, \tag{52}$$

then $\Gamma = K \cap B(x, r/2)$ is a Lipschitz graph with constant $\le \varepsilon$, as well as a C^1 curve. In addition, the two endpoints of Γ lie on $\partial B(x, r/2)$.

As the reader already guesses, the main point of the argument will be to control the right-hand side of (47). First observe that since h is nondecreasing, (52) implies that

$$\sum_{k \ge 0} h(2^{-k} r)^{1/2} \le \mathrm{Log}(2)^{-1} \int_0^{2r} h(t)^{1/2} \frac{dt}{t} \le \mathrm{Log}(2)^{-1} \tau_1. \tag{53}$$

In particular, if τ_1 is small enough, we can apply Theorem 1 to find $\rho \in (r/2, 2r/3)$ such that $\Gamma_0 = K \cap B(x, \rho)$ is a $(1 + \varepsilon)$-chord-arc (open) curve, with its endpoints on $\partial B(x, \rho)$. We can also start the proof of Corollary 33, with τ_0 as small as we want, and go as far as (47), provided that we prove (43).

So let $t \le r/6$ be given, and let us estimate

$$\widetilde{h}_{r/3}(2t) = \sup \left\{ s^{-b}(2t)^b h(s) \, ; \, 2t \le s \le r/3 \right\}. \tag{54}$$

Let k be such that $2^{-k-1} r < t \le 2^{-k} r$ and, for $2t \le s \le r/3$, let l be such that $2^{-l-1} r < s \le 2^{-l} r$. Then $1 \le l \le k$ and $h(s) \le h(2^{-l} r)$. Since $s^{-b}(2t)^b \le 8^b 2^{-(k-l)b}$, we get that

$$\widetilde{h}_{r/3}(2t) \le 8^b \sup_{1 \le l \le k} 2^{-(k-l)b} h(2^{-l} r). \tag{55}$$

Now (38) says that $a(t) \le \tau_0^{1/2} \left(\frac{t}{r} \right)^{b/2} + 8^{b/2} \sup_{1 \le l \le k} \left\{ 2^{-(k-l)b} h(2^{-l} r) \right\}^{1/2}$, and

$$a^*(t) = \sum_{j \ge 0} a(2^{-j} t) = C \tau_0^{1/2} \left(\frac{t}{r} \right)^{b/2} + 8^{b/2} \sum_{j \ge 0} \sup_{1 \le l \le j+k} 2^{-(k+j-l)b/2} h(2^{-l} r)^{1/2}$$

$$\le C \tau_0^{1/2} \left(\frac{t}{r} \right)^{b/2} + C \sum_{j \ge 0} \sum_{1 \le l \le j+k} 2^{-(k+j-l)b/2} h(2^{-l} r)^{1/2}$$

$$\le C \tau_0^{1/2} \left(\frac{t}{r} \right)^{b/2} + C \sum_{l \ge 1} h(2^{-l} r)^{1/2} \sum_{j \ge \mathrm{Max}(0, l-k)} 2^{-(k+j-l)b/2}. \tag{56}$$

The series in j converges. When $l \geq k$, its sum is less than C; when $l < k$, the sum is less than $C2^{-(k-l)b/2}$. Altogether,

$$a^*(t) \leq C\tau_0^{1/2} \left(\frac{t}{r}\right)^{b/2} + C\sum_{l \geq k} h(2^{-l}r)^{1/2} + C\sum_{1 \leq l < k} 2^{-(k-l)b/2} h(2^{-l}r)^{1/2}. \quad (57)$$

All this holds for $t \leq r/6$, with k defined below (54). Since $a^*(t)$ is a nondecreasing function of t by definition,

$$a^*(t) \leq a^*(r/6) \leq C\tau_0^{1/2} + C\tau_1 \quad \text{for } t \leq r/6 \quad (58)$$

by (57) and (52) or (53). Also, when t tends to 0, $\tau_0^{1/2} \left(\frac{t}{r}\right)^{b/2}$ tends to 0 trivially, and $\sum_{l \geq k} h(2^{-l}r)^{1/2}$ tends to 0 by (53) and because k tends to $+\infty$. The last sum in (57) is

$$\sum_{1 \leq l < k} 2^{-(k-l)b/2} h(2^{-l}r)^{1/2} \leq \sum_{l < k/2} 2^{-(k-l)b/2} h(r)^{1/2} + \sum_{l > k/2} 2^{-(k-l)b/2} h(2^{-k/2}r)^{1/2}$$

$$\leq C2^{-kb/4} h(r)^{1/2} + Ch(2^{-k/2}r)^{1/2}, \quad (59)$$

which also tends to 0. Thus $\lim_{t \to 0} a^*(t) = 0$ as in (43), we can continue the argument as in the proof of Corollary 33, and eventually (47) says that

$$\text{dist}\,(T(y), T(y')) \leq Ca^*(4|y' - y|) \quad (60)$$

when $y, y' \in K_0 = K \cap B(x, 2r/3)$ are such that $|y' - y| \leq r/24$.

Recall that $\Gamma_0 = K \cap B(x, \rho)$ is a $(1+\varepsilon)$-chord-arc curve, with its extremities on $\partial B(x, \rho)$. Also, $\Gamma_0 \subset K_0 = K \cap B(x, 2r/3)$, because $\rho \leq 2r/3$. Then (60) says that Γ_0 is of class C^1, and we even get some estimates on the modulus of continuity of the tangent direction $T(y)$.

Also, even though (60) and (58) only give an estimate on $\text{dist}\,(T(y), T(y'))$ when $|y' - y| \leq r/24$, we still get that $\text{dist}\,(T(y), T(y')) \leq C\tau_0^{1/2} + C\tau_1$ for y, $y' \in \Gamma_0 = K \cap B(x, \rho)$, by adding a few intermediate points if needed. Recall that we can make τ_0 as small as we wish, so we get that $\text{dist}\,(T(y), T(y')) \leq \varepsilon/2$ if τ_1 is small enough. Then $K \cap B(x, r/2)$ is a Lipschitz graph with constant $\leq \varepsilon$, and the two endpoints of Γ lie on $\partial B(x, r/2)$. This completes our proof of Corollary 51. $\qquad \square$

Remark 61. The integral conditions (50) and (52) are essentially optimal if one wants to get C^1 curves, as one can check by adapting the example of Remark 30 (i.e., taking curves of equation $\theta = f(r)$ with functions f that tend very slowly to infinity). We leave the details.

Remark 62. When K separates $D_+(x, r)$ from $D_-(x, r)$ in $\overline{B}(x, r)$, we can replace $J(x, r)^{-1}$ with 0 in the various estimates, including Theorem 1 and Corollaries

25, 33, and 51. This comes from similar observations in the previous sections, in particular Remark 49.34. See also Remarks 44.23, 46.17, and 48.26.

Remark 63. We also can justify Remark 62 by abstract nonsense, as follows. Suppose $(u, K) \in TRLAM(\Omega, h)$ as usual, and let $x \in K$ and $r > 0$ be such that $\overline{B}(x, r) \subset \Omega$, $\beta_K(x, r) < 10^{-1}$ and K separates $D_+(x, r)$ from $D_-(x, r)$ in $\overline{B}(x, r)$. Notice that $(u, K \cap B(x, r)) \in TRLAM(B(x, r), h)$. If we add a monstrous constant to u in the connected component of $D_+(x, r)$ in $B(x, r) \setminus K$, we still get a coral topological almost-minimizer in $B(x, r)$, with the same gauge function h. [See Exercise 7.57.] Then we can apply the various results of this section to this new pair, maybe with $B(x, r)$ replaced with a slightly smaller $B(x, r')$, and with a jump J that is now as large as we want. This gives the results announced in Remark 62.

51 $K \cap B(x, r)$ is often a nice C^1 curve $(n = 2)$

Let us assume in this section that Ω is an open set in the plane and $(u, K) \in TRLAM(\Omega, h)$, as in Standard Assumption 41.1. This means that it is a coral local almost-minimizer in Ω, topological or not, with gauge function h. We shall also assume that

$$\int_0^1 h(t)^{1/2} \frac{dt}{t} < +\infty, \tag{1}$$

as in (50.50), so as to be able to apply Corollary 50.51 and get C^1 curves. If we assumed that $h(r) = C_0 r^\alpha$ for some $\alpha > 0$, we would be able to use Corollary 50.33 instead, and get more regularity. But this will not really be the point here. The following definition will be useful.

Definition 2. *Let B be a disk centered on K. We shall say that $K \cap B$ is a nice C^1 curve, or even that B is a nice C^1 disk (for K), if $K \cap \overline{B}$ is a simple C^1 curve, with its two endpoints on ∂B, and in addition it is a 10^{-1}-Lipschitz graph. [See Definition 50.31.]*

This definition, in particular with the choice of 10^{-1}, is somewhat arbitrary. That is, all the results below would still hold if we replaced 10^{-1} with a smaller constant, with the same proof. Note that

$$\text{if } K \cap B(x, r) \text{ is a nice } C^1 \text{ curve, then } K \cap B(x, r')$$
$$\text{is a nice } C^1 \text{ curve for } 0 < r' < r, \tag{3}$$

which will often be convenient.

If $h(r) = C_0 r^\alpha$ for some $\alpha > 0$, $K \cap B$ is automatically $C^{1,\gamma}$ for some $\gamma > 0$, and even $C^{1,1}$ if (u, K) is a Mumford-Shah minimizer [Bo] or $h(r) = Cr$ [AFP3] but again this is not the point of this section. Here we merely worry about the existence of many nice C^1 disks for K. We start with the main result of C^1 regularity.

Theorem 4 ($n = 2$). *We can find constants $\varepsilon_0 > 0$ and $C_1 > 2$ such that, if $(u, K) \in TRLAM(\Omega, h)$, $x \in K$, $r > 0$, $\overline{B}(x,r) \subset \Omega$, and $\int_0^{2r} h(t)^{1/2} \frac{dt}{t} < \varepsilon_0$, then we can find $y \in K \cap B(x, r/2)$ and $t \in [C_1^{-1}r, r/2]$ such that $K \cap B(y,t)$ is a nice C^1 curve.*

This first showed up in [Da5], but it was almost immediately obtained by different means and generalized to higher dimensions in [AmPa], [AFP1], and [Ri1]; see also [AFP3] and [MaSo2]. The generalization will be rapidly discussed in Section 75.

The following lemma (and Corollary 17 below) will be helpful. Together they give a nice sufficient condition for the regularity of K near a point. We shall see a variant in Corollary 52.25 and more results of the same type in Sections 53, 60, and 70.

Lemma 5. *For each choice of $p \in (1,2]$ and $\tau > 0$ we can find constants $\eta_1 > 0$ and $\varepsilon_1 > 0$ with the following properties. Suppose as above that $(u, K) \in TRLAM(\Omega, h)$, and let $x \in K$ and $r > 0$ be such that $\overline{B}(x,r) \subset \Omega$. If*

$$\beta_K(x,r) + \omega_p(x,r) + h(r) \leq \varepsilon_1, \tag{6}$$

then $\omega_2(x, \eta_1 r) \leq \tau$. If in addition

$$K \text{ does not separate } D_+(x,r) \text{ from } D_-(x,r) \text{ in } \overline{B}(x,r), \tag{7}$$

then $J(x, \eta_1 r)^{-1} \leq \tau$ as well.

This lemma is still valid in higher dimensions, with trivial modifications. It seems from the proof that one needs to take $p > 2(n-1)/n$, but Exercise 32 says that this is not a serious restriction.

The statement is not optimal. The restriction that $p > 1$ is not needed (see Exercise 32), and there are different ways to replace the upper bound on $\omega_p(x,r)$ by apparently weaker conditions, like lower bounds on the jump of u, or a better (bilateral) control on how close $K \cap B(x,r)$ is to a plane. In both cases, the main point is to prevent K from being close to a half-line. See Corollary 52.25, Proposition 60.1, and Proposition 70.2.

The main point of Lemma 5 is that if $\varepsilon_1 > 0$ and τ are small enough, the conclusions will allow us to apply Corollary 50.51 to the pair $(x, \eta_1 r)$. See Corollary 17.

We start the proof with the most interesting case when (7) holds. We first choose $r_1 \in [r/2, r]$ so that $\omega_p^*(x, r_1) \leq C\omega_p(x,r) \leq C\varepsilon_1$. [Compare (42.5) with (23.5).] Next we want to apply Proposition 42.10 to the pair (x, r_1). The first assumption (42.12) is obviously satisfied if ε_1 is small enough. Also recall from just above (42.7) that $\Sigma_+(x, r_1)$ and $\Sigma_-(x, r_1)$ are the two components of $\{z \in \partial B(x, r_1); \text{dist}(z, P(x, r_1)) > r_1/2\}$. By (7) and the first part of (6), they lie in the same component of $\overline{B}(x,r) \setminus K$. Then they lie in the same component of $\Omega \setminus K$ as well, as required for (42.13). So (42.11) holds, i.e., $J^*(x, r_1) \geq \tau_0 = \tau_0(2, p)$.

If ε is small enough, then $J(x, r_1) \geq \tau_0/2$, because (42.8) says that

$$|J^*(x, r_1) - J(x, r_1)| \leq C\omega_1(x, r_1)^{1/2} \leq C\omega_p(x, r)^{1/2} \leq C\varepsilon^{1/2}. \tag{8}$$

Next we want to apply Lemma 43.11, with the constants $\tau_0/2$ and $C_0 = 2\tau^{-1}$. Call $\eta = \eta(C_0, \tau_0/2)$ the small constant that we get that way. If ε_1 is small enough (depending also on η, but this is all right), the hypothesis (43.12) is satisfied, and we get that $J(x, \eta r_1) \geq C_0 = 2\tau^{-1}$. Then

$$J(x, t) \geq C_0/2 = \tau^{-1} \quad \text{for } \eta r_1/4 \leq t \leq \eta r_1, \tag{9}$$

because Lemma 43.1 (applied twice if needed) says that

$$J(x, t) \geq J(x, \eta r_1) - C\omega_1(x, \eta r_1) \geq C_0 - C\omega_1(x, \eta r_1), \tag{10}$$

and $\omega_1(x, \eta r_1) \leq C\omega_p(x, r_1) \leq C\varepsilon_1 \leq C_0/2$, by the end of (8) and if ε_1 is small enough. Here and in the estimates below, C depends also on η, but this does not matter because we can choose ε_1 last.

Let us choose a last intermediate radius $r_2 \in [\eta r_1/2, \eta r_1]$ such that

$$\omega_p^*(x, r_2) \leq C\omega_p(x, r_1) \leq C\varepsilon_1.$$

Proposition 45.1 says that

$$H^1(K \cap B(x, r_2)) + r_2\, \omega_2(x, r_2) \leq 2r_2 + Cr_2\, \omega_p^*(x, r_2) + Cr_2\, \beta_K(x, r_2) + r_2\, h(r_2)$$
$$\leq 2r_2 + C\varepsilon_1\, r, \tag{11}$$

by (6). On the other hand, Proposition 44.1 gives us a set $F(x, r_2) \subset \overline{B}(x, r_2)$ such that

$$K \cap \overline{B}(x, r_2) \subset F(x, r_2) \subset \{y \in \overline{B}(x, r_2)\,;\, \text{dist}(y, P(x, r_2)) \leq r_2\, \beta_K(x, r_2)\}$$
$$\subset \{y \in \overline{B}(x, r_2)\,;\, \text{dist}(y, P(x, r_2)) \leq C\varepsilon_1\, r_2\}, \tag{12}$$

$$H^1(F(x, r_2) \cap B(x, r_2) \setminus K) \leq Cr_2\, \omega_p(x, r_2)^{1/2} J(x, r_2)^{-1} \leq Cr_2\, \varepsilon_1^{1/2} \tag{13}$$

(by (9) and (6)), and $F(x, r_2)$ separates $D_+(x, r_2)$ from $D_-(x, r_2)$ in $B(x, r_2)$. In particular, $H^1(F(x, r_2) \cap B(x, r_2)) \geq 2r_2 - C\varepsilon_1\, r_2$ because of (12), and then (13) says that

$$H^1(K \cap B(x, r_2)) \geq H^1(F(x, r_2) \cap B(x, r_2)) - Cr_2\, \varepsilon_1^{1/2} \geq 2r_2 - Cr_2\, \varepsilon_1^{1/2}. \tag{14}$$

We compare this to (11) and get that $\omega_2(x, r_2) \leq C\varepsilon_1^{1/2}$. Finally choose $\eta_1 = \eta/4$. Note that $r_1 \leq r \leq 2r_1$, so $\eta r_1/4 \leq \eta_1 r \leq \eta r_1/2$, and (9) says that $J(x, \eta_1 r) \geq \tau^{-1}$. Also, $r_2/4 \leq \eta r_1/4 \leq \eta_1 r \leq \eta r_1/2 \leq r_2$, so $\omega_2(x, \eta_1 r) \leq 4\omega_2(x, r_2) \leq C\varepsilon_1^{1/2} \leq \tau$ if ε_1 is small enough. This proves the lemma in the special case when (7) holds.

We may now assume that K separates $D_+(x, r)$ from $D_-(x, r)$ in $\overline{B}(x, r)$. We need to keep the same constant η_1 as above (because the statement of Lemma 5 does not allow η_1 to depend on (u, K)), even though much larger values would work as well.

As before we choose a first radius $r_1 \in [r/2, r]$ such that

$$\omega_p^*(x, r_1) \le C\omega_p(x, r) \le C\varepsilon,$$

and then apply Proposition 45.1. We get that

$$H^1(K \cap B(x, r_1)) + r_1 \omega_2(x, r_1)$$
$$\le 2r_1 + Cr_1 \omega_p^*(x, r_1) + Cr_1 \beta_K(x, r_1) + r_1 h(r_1) \le 2r_1 + C\varepsilon_1 r. \quad (15)$$

On the other hand, K separates $D_+(x, r)$ from $D_-(x, r)$ in $\overline{B}(x, r)$ and $\beta_K(x, r) \le \varepsilon_1$ (by (6)), so $H^1(K \cap B(x, r_1)) \ge 2r_1 - Cr\varepsilon_1$. Then (15) says that $r_1\omega_2(x, r_1) \le C\varepsilon_1 r$, and

$$\omega_2(x, \eta_1 r) \le \eta_1^{-1}\omega_2(x, r_1) \le C\varepsilon_1 \le \tau \quad (16)$$

if ε_1 is small enough. This completes our proof of Lemma 5. $\qquad\square$

Corollary 17 ($n = 2$). *For each $p \in (1, 2]$, there is a constant $\varepsilon_2 > 0$ such that if $(u, K) \subset TRLAM(\Omega, h)$, $x \in K$, $r > 0$, $\overline{B}(x, r) \subset \Omega$, and $\beta_K(x, r) + \omega_p(x, r) + \int_0^r h(t)^{1/2} \frac{dt}{t} \le \varepsilon_2$, then $K \cap B(x, r/2)$ is a nice C^1 curve (see Definition 2).*

Let us first prove the corollary and then take care of Theorem 4. Let τ_1 be as in Corollary 50.51, but where we take $\varepsilon = 10^{-2}$. Then apply Lemma 5 with $\tau = \tau_1/4$. This gives constants η_1 and ε_1.

Now let (u, K), x, and r be as in the statement, and let $y \in K \cap B(x, r/2)$ be given. If ε_2 is small enough, then we can apply Lemma 5 to the pair $(y, r/2)$, and we get that $\omega_2(y, \eta_1 r/2) \le \tau_1/4$.

By assumption, we also have that $\beta_K(y, \eta_1 r/2) \le 8\eta_1^{-1}\beta_K(x, r) \le 8\eta_1^{-1}\varepsilon_2 \le \tau_1/4$ (if ε_2 is small enough). If in addition K does not separate $D_+(x, r)$ from $D_-(x, r)$ in $\overline{B}(x, r)$, Lemma 5 also says that $J(y, \eta_1 r/2)^{-1} \le \tau_0/4$. In this case, we can apply Corollary 50.51 directly, and we get that $K \cap B(y, \eta_1 r/4)$ is a Lipschitz graph with constant $\le 10^{-2}$, as well as a C^1 curve.

If K separates $D_+(x, r)$ from $D_-(x, r)$ in $\overline{B}(x, r)$, then K also separates $D_+(y, \eta_1 r/2)$ from $D_-(y, \eta_1 r/2)$ in $\overline{B}(y, \eta_1 r/2)$, because $\beta_K(x, r) < \varepsilon_2$ (and if ε_2 is small enough). Then Remark 50.62 says that we can still apply Corollary 50.51, with $J(x, \eta_1 r)^{-1}$ replaced by 0, and in this case also $K \cap B(y, \eta_1 r/4)$ is a Lipschitz graph with constant $\le 10^{-2}$, as well as a C^1 curve.

Set $\Gamma_y = K \cap B(y, \eta_1 r/4)$ for $y \in K \cap B(x, r/2)$. So we know that Γ_y is a 10^{-2}-Lipschitz graph over some line $L(y)$, and that crosses $B(y, \eta_1 r/4)$. Since $\beta_K(x, r) < \varepsilon_2$, all the lines $L(y)$, $y \in K \cap B(x, r/2)$, make angles less than 2.10^{-2}, say, with $P(x, r)$. Thus the Γ_y are also 4.10^{-2}-Lipschitz graphs over $P(x, r)$, and it is easy to see that $K \cap B(x, r/2)$ is also a 4.10^{-2}-Lipschitz graph over $P(x, r)$, and that crosses $B(x, r/2)$. [Use a small connectedness argument to see that it does not end before it meets $\partial B(x, r/2)$.] Since we also know that each Γ_y is C^1, we get the description announced in the statement of the corollary. This proves Corollary 17.   □

We are now ready to prove Theorem 4. Let $x \in K$ and $r > 0$ be as in the statement, and let us apply Proposition 41.17 with any $p \in (1, 2)$ (like $p = 3/2$) and $\tau = \varepsilon_2/2$, where ε_2 comes from Corollary 17. Note that if ε_0 is small enough,

$$h(r) \leq \mathrm{Log}(2)^{-1} \int_0^{2r} h(t)^{1/2} \frac{dt}{t} < \mathrm{Log}(2)^{-1}\varepsilon_0 \leq \varepsilon(n), \tag{18}$$

where $\varepsilon(n)$ is as in Proposition 41.17, because h is nondecreasing and by assumption. We get $y \in K \cap B(x, r/2)$ and $t_1 \in [C(2, p, \tau)^{-1}r, r/2]$ such that $\beta_K(y, t_1) + \omega_p(y, t_1) < \varepsilon_2/2$, and now Corollary 17 (applied to the pair (y, t_1)) tells us that $K \cap B(y, t_1/2)$ is a nice C^1 curve. In other words, we can take $t = t_1/2$ in the conclusion of Theorem 4.

This completes our proof of Theorem 4, with $C_1 = 2C(2, p, \varepsilon_2/2)$.   □

Definition 19. *A C^1 point of K is a point $x \in K$ such that $K \cap B(x, r)$ is a nice C^1 curve for some $r > 0$ (or equivalently for r small enough, see (3)). See Definition 2 for nice C^1 curves.*

Theorem 20 (n=2). *There is an absolute constant $\delta > 0$ such that if $(u, K) \in TRLAM(\Omega, h)$, with any gauge function h such that (1) holds, and if we denote by K^* the set of points of K that are not C^1 points, then the Hausdorff dimension of K^* is at most $1 - \delta$. In particular, $H^1(K^*) = 0$.*

Of course this applies when $h(r) = C_0 r^\alpha$ for some $\alpha > 0$, and in particular to Mumford-Shah minimizers. There is a similar result in dimensions $n > 2$; see Corollary 75.16 or directly [Ri1], or [MaSo2] for an approach with bisections. The main point of the argument is the following immediate consequence of Theorem 4.

Lemma 21. *We can find a constant C_1, that does not depend on the precise gauge function h, and a positive radius r_1, that is allowed to depend on h, such that for all $x \in K$ and $0 < r < r_1$ such that $\overline{B}(x, r) \subset \Omega$, we can find $y \in K \cap B(x, r/2)$ and $t \in [C_1^{-1}r, r/2]$ such that all points of $K \cap B(y, t)$ are C^1 points of K.*

Theorem 20 is an abstract consequence of Lemma 21 (i.e., the fact that K^* is porous in K) and the local Ahlfors-regularity of K. This is done in [DaSe7], Lemma 5.8 on page 24, for instance, but for the convenience of the reader we shall give a slightly different proof here, that does not use generalized dyadic cubes.

First note that it is enough to prove that the Hausdorff dimension of $K \cap B$ is less than $1 - \delta$ when $B = B(x_0, r)$ is a ball centered on K, such that $3B \subset \Omega$, and with radius $r < r_1$, where r_1 is as in Lemma 21. This is because K is a countable union of such sets, and countable unions do not increase the Hausdorff dimension (as will be clear from the end of the proof).

So let such a ball B be given, and set $K_0^* = K^* \cap B$. Also set $\tau = (5C_1)^{-1}$, where $C_1 > 1$ is as in Lemma 21, and consider the successive scales $r(k) = \tau^k r$, $k \geq 0$. For each $k \geq 0$, choose a maximal subset $X(k)$ of K_0^*, with the property that

$$\text{dist}(x,y) \geq r(k) \quad \text{for } x, y \in X(k), x \neq y. \tag{22}$$

By maximality,

$$\text{the } \overline{B}(x, r(k)), \ x \in X(k), \text{ cover } K_0^*. \tag{23}$$

We want to estimate the numbers

$$N(k) = \sharp X(k) \quad \text{(the number of points in } X(k)), \tag{24}$$

because this will give good bounds on the Hausdorff measure of K_0^*.

We can apply Lemma 21 to the pair $(x, r(k)/2)$ for each $x \in X(k)$, and we get a ball B_x centered on $K \cap B(x, r(k)/4)$, with radius $r(k)/(2C_1)$, and which does not meet K^*. Then $\frac{1}{2}B_x \subset W(k)$ where we set

$$W(k) = \{y \in K \, ; \, \frac{r(k)}{4C_1} \leq \text{dist}(y, K_0^*) \leq r(k)\}, \tag{25}$$

where the second inequality comes from the fact that $x \in K_0^*$. Thus

$$H^1(W(k)) \geq \sum_{x \in X(k)} H^1(K \cap \frac{1}{2}B_x) \geq N(k) \, C^{-1} C_1^{-1} r(k), \tag{26}$$

because the balls $\frac{1}{2}B_x$, $x \in X(k)$, are disjoint (by (22)) and K is locally Ahlfors-regular (see Theorem 18.1 or 18.16). Here we use the fact that we get estimates on the Ahlfors-regularity constants that do not depend on h, with the only constraint that $h(r)$ be small enough for the radius of our initial ball B (which we may assume by the same argument as before).

By definition of τ and the $r(k)$, the sets $W(k)$, $k \geq 0$, are disjoint. If $0 \leq k_0 < k$, each point of $W(k)$ is contained in some $B(x, \frac{3}{2}r(k_0))$, $x \in X(k_0)$, by (23) and (25). Hence

$$\sum_{k > k_0} H^1(W(k)) = H^1\left(\bigcup_{k > k_0} W(k) \right) \tag{27}$$

$$\leq H^1\left(\bigcup_{x \in X(k_0)} K \cap B(x, \frac{3}{2}r(k_0)) \right) \leq CN(k_0) \, r(k_0),$$

this time by the easy estimate in the local Ahlfors-regularity of K. Altogether, (26) and (27) yield

$$\sum_{k > k_0} N(k)r(k) < CC_1 \sum_{k > k_0} H^1(W(k)) \leq CC_1 N(k_0) \, r(k_0). \tag{28}$$

Next let k_1 and k be given, with $k < k_1$. For each $x \in X(k)$, there are at most $Cr(k_1)^{-1}r(k)$ points of $X(k_1)$ in $\overline{B}(x, r(k))$. Indeed, for each such point y we have that $H^1(K \cap B(y, r(k_1)/2)) \geq C^{-1}r(k_1)$ by local Ahlfors-regularity, the sets $K \cap B(y, r(k_1)/2)$ are disjoint by (22), and they are all contained in $K \cap B(x, 2r(k))$ (whose measure is at most $Cr(k)$).

Since every point of $X(k_1)$ is contained in some $\overline{B}(x, r(k))$, $x \in X(k)$, (by (23)),

$$N(k_1) \leq Cr(k_1)^{-1}r(k)N(k) \quad \text{for } 0 \leq k < k_1. \tag{29}$$

Of course (29) stays true when $k = k_1$.

Now let k_0 and $k_1 > k_0$ be given, and consider the contribution to the left-hand side of (28) of indices k such that $k_0 < k \leq k_1$. Each term is larger than or equal to $C^{-1}N(k_1)\, r(k_1)$ (by (29)), and so (28) yields

$$(k_1 - k_0)N(k_1)\, r(k_1) \leq CC_1 N(k_0)\, r(k_0). \tag{30}$$

If the integer L is chosen larger than $2CC_1$, (30) says that $N(k_0 + L)\, r(k_0 + L) \leq \frac{1}{2}N(k_0)\, r(k_0)$ for all $k_0 \geq 0$, which leads to $N(jL)\, r(jL) \leq C2^{-j}$ for all $j \geq 0$. Recall that $r(jL) = \tau^{jL}r$ for some $\tau < 1$. Choose $\delta > 0$ so that $\tau^{L\delta} = 1/2$. Then $[r^{-1}r(jL)]^{\delta} = \tau^{jL\delta} = 2^{-j}$ and $N(jL) \leq C\, 2^{-j}r(jL)^{-1} = Cr^{-\delta}r(jL)^{\delta-1}$.

Now (23) says that for each $j \geq 0$ we can cover K_0^* with less than $Cr^{-\delta}r(jL)^{\delta-1}$ balls of radius $r(jL)$. Then $H^{1-\delta}(K_0^*) < +\infty$, directly by Definition 2.6, and in particular the Hausdorff dimension of K_0^* is at most $1 - \delta$. We also get that $H^1(K_0^*) = 0$ trivially. This completes our proof of Theorem 20. $\square$

Remark 31. The results of the last sections are still valid when $(u, K) \in TRLAM$ with linear distortion $\leq L$; as the reader probably noted, what we did in the last three sections was mostly mechanical consequences of the estimates from the previous sections, and in particular the decay estimate for energy (see Remark 48.27).

Exercise 32. Show that Lemma 5 also holds for any $p > 0$. Use Hölder (or the log-convexity of norms), the trivial bound on $\omega_2(x, r)$, and Lemma 5.

52 Jump and flatness control surface and energy

In this section we return to the case when $(u, K) \in TRLAM(\Omega, h)$ (as in Standard Assumption 41.1) for some domain $\Omega \subset \mathbb{R}^n$ (and any $n \geq 2$) and try to get bounds on $H^{n-1}(K \cap B(x, r))$ and the energy $\omega_2(x, \tau r)$ when $K \cap B(x, r)$ is very flat and the normalized jump $J(x, r)$ is very large. Then we shall see how to use this in dimension 2 to get nice C^1 disks. See Corollary 25. We start with an estimate on $H^{n-1}(K \cap B(x, r))$.

Lemma 1. *Suppose that $(u, K) \in TRLAM(\Omega, h)$, $x \in K$, and $\overline{B}(x, r) \subset \Omega$. Then*

$$H^{n-1}(K \cap \overline{B}(x, r)) \leq H^{n-1}(P \cap B(x, r)) + Cr^{n-1} \beta_K(x, r)^{1/2} + r^{n-1} h(r), \quad (2)$$

where P is any $(n-1)$-plane through x.

Set $B = B(x, r)$. It is enough to prove (2) when $\beta_K(x, r) < 10^{-1}$, since otherwise the trivial estimate $H^{n-1}(K \cap B) \leq H^{n-1}(\partial B) + r^{n-1} h(r)$ (coming from (18.8) which, as we have already seen, also holds for topological almost-minimizers) already implies (2). Choose an $(n-1)$-plane $P = P(x, r)$ through x, so that

$$\text{dist}(y, P) \leq r\beta_K(x, r) \quad \text{for } y \in K \cap \overline{B}, \quad (3)$$

as in (42.2). Then set

$$Z = \{y \in \partial B \, ; \text{dist}(y, P) \leq r\beta_K(x, r)^{1/2} \quad \text{and} \quad G = [P \cap B] \cup Z \cup [K \setminus B]. \quad (4)$$

Note that G is closed in Ω, because $K \cap \partial B \subset Z$. We want to find $v \in W^{1,2}_{\text{loc}}(\Omega \setminus G)$ such that (v, G) is a competitor for (u, K) in B. We keep $v = u$ out of B, and in $B \setminus P$ we want to set $v(y) = u(\Phi(y))$, where Φ is a function of class C^1 which is defined on $\mathbb{R}^n \setminus [(P \cap B) \cup Z]$ and equal to the identity outside of B. We choose Φ so that it is a diffeomorphism from $B \setminus P$ onto the smaller domain

$$V = \{y \in B \, ; \text{dist}(y, P) > r\beta_K(x, r)\}, \quad (5)$$

and by definition of Z we can do this so that

$$\|D\Phi - Id\| \leq C\beta_K(x, r)^{1/2} \quad \text{everywhere on } B \setminus P. \quad (6)$$

Then v is well defined, by (3), and it lies in $W^{1,2}_{\text{loc}}(\Omega \setminus G)$ because Φ is smooth. [See Proposition 11.4.] Thus (v, G) is a competitor for (u, K) in B. It is even a topological competitor, by Lemma 45.8. The almost-minimality condition (7.7) yields

$$H^{n-1}(K \cap \overline{B}) + \int_{B \setminus K} |\nabla u|^2 \leq H^{n-1}(G \cap \overline{B}) + \int_{B \setminus G} |\nabla v|^2 + h(r) r^{n-1}. \quad (7)$$

Here $H^{n-1}(G \cap \overline{B}) = H^{n-1}(P \cap B) + H^{n-1}(Z) \leq H^{n-1}(P \cap B) + C\beta_K(x, r)^{1/2} r^{n-1}$ by (4), and

$$\int_{B \setminus G} |\nabla v|^2 \leq (1 + C\beta_K(x, r)^{1/2}) \int_V |\nabla u|^2 \leq \int_{B \setminus K} |\nabla u|^2 + Cr^{n-1} \beta_K(x, r)^{1/2} \quad (8)$$

by a change of variable in the integral, (6), and the trivial estimate (18.8). Lemma 1 follows. $\qquad\square$

Next we want to get good upper bounds on the normalized energy $w_2(x, \tau r)$ in a smaller ball, in terms of $\beta_K(x, r)$ as before, but also the normalized jump $J(x, r)$. Recall that $w_2(x, r) = r^{1-n} \int_{B(x,r) \setminus K} |\nabla u|^2$.

Lemma 9. *Let $(u, K) \in TRLAM(\Omega, h)$, $x \in K$, and $\overline{B}(x, r) \subset \Omega$ be given. Suppose that $\beta_K(x, r) < 1/2$ and $w_p^*(x, r) < +\infty$ for some $p \in (\frac{2(n-1)}{n}, 2]$. Then*

$$w_2(x, \tau r) \leq C\tau + C\tau^{1-n} \left\{ w_1(x, r)^{1/2} J(x, r)^{-1} + \beta_K(x, r)^{1/2} + h(r) \right\} \ for \ 0 < \tau < 1. \tag{10}$$

The constant C depends only on the dimension n.

Here we required that $\beta_K(x, r) < 1/2$ only to be allowed to define $J(x, r)$, and $w_p^*(x, r) < +\infty$ will be used because we want to construct a new competitor by gluing two pieces along $\partial B(x, r)$. See Convention 42.6.

We keep the same notation as in Lemma 1, and define G, Φ, and v as above. We want to replace v by a new function w on $B \setminus G = B \setminus P$. We choose $w \in W^{1,2}(B \setminus P)$ so that its radial limit coincides almost-everywhere on $\partial B \setminus Z$ with u (and hence with its radial limit from outside of B, by Convention 42.6), and so that $E(w) = \int_{B \setminus P} |\nabla w|^2$ is minimal under these constraints. Let us check that such a function w exists. First, the restriction of u to $\partial B \setminus Z \subset \partial B \setminus K$ lies in $W^{1,p}$ because $w_p^*(x, r) < +\infty$, hence Lemma 22.32 says that it has an extension in $W^{1,2}(B \setminus P)$. In other words, the analogue of $\mathcal{F}$ in (15.2) is not empty. [We could also get this directly, since $v \in \mathcal{F}$.] Then Proposition 15.6 applies and gives a minimizing w.

The function w is harmonic on each of the two components of $B \setminus P$ (call them $U_\pm$). When we take the values of w on $U_\pm$ and symmetrize them with respect to P, we get a function $w_\pm$ on B which is harmonic. The argument is the same as in Lemma 17.6. Then

$$\int_B |\nabla w_\pm|^2 \leq 2E(w) \leq 2 \int_{B \setminus G} |\nabla v|^2 \leq C \int_{B \setminus K} |\nabla u|^2 \leq Cr^{n-1} \tag{11}$$

by symmetry, because v is one of the competitors in the definition of w, by (8), and by the trivial estimate (18.8).

Brutal estimates on harmonic functions on B yield $|\nabla w_\pm| \leq Cr^{-1/2}$ on $\frac{1}{2}B$. We can integrate this on $B(x, t) \setminus P$ and get that

$$\int_{B(x,t) \setminus P} |\nabla w|^2 \leq Ct^n r^{-1} \ \text{for } 0 \leq t \leq r/2. \tag{12}$$

Next we want to show that ∇v is close to ∇w, because then (12) will give good upper bounds on energy integrals for v, and then for u. As usual, this will come from a comparison with (u, K).

Set $w(y) = u(y)$ on $\Omega \setminus [K \cup B]$. Note that $w \in W^{1,2}_{\text{loc}}(\Omega \setminus G)$ because the radial limits of u (from outside) and w (from inside) coincide almost-everywhere on $\partial B \setminus Z$ (by definition of w), and by Corollary 14.28 on welding. Then (w, G) is a topological competitor for (u, K) in B, by the same argument as for Lemma 1, and (7.7) yields

$$H^{n-1}(K \cap \overline{B}) + \int_{B \setminus K} |\nabla u|^2 \leq H^{n-1}(G \cap \overline{B}) + \int_{B \setminus G} |\nabla w|^2 + h(r) \, r^{n-1}. \qquad (13)$$

Next we need a lower bound for $H^{n-1}(K \cap \overline{B})$. Denote by $F = F(x, r)$ the compact set provided by Proposition 44.1. Recall that F stays within $r\beta_K(x, r)$ of P (by (44.2)) and separates the two components of $\{y \in \overline{B} ; \operatorname{dist}(y, P) \geq r\beta_K(x, r)\}$ in $\overline{B}$ (by (44.3)). Then

$$H^{n-1}(F \cap B) \geq (1 - \beta_K(x, r))^{n-1} H^{n-1}(P \cap B), \qquad (14)$$

for instance because its orthogonal projection on P contains $P \cap B(x, (1 - \beta_K(x, r))r)$. Since

$$H^{n-1}(F \cap B \setminus K) \leq Cr^{n-1}\omega_1(x, r)^{1/2} J(x, r)^{-1} \qquad (15)$$

by (44.4), we get that

$$
\begin{aligned}
H^{n-1}(K \cap \overline{B}) &\geq H^{n-1}(K \cap B) \geq H^{n-1}(F \cap B) - Cr^{n-1}\omega_1(x, r)^{1/2} J(x, r)^{-1} \\
&\geq H^{n-1}(P \cap B) - Cr^{n-1}\omega_1(x, r)^{1/2} J(x, r)^{-1} - Cr^{n-1}\beta_K(x, r) \\
&\geq H^{n-1}(G \cap \overline{B}) - Cr^{n-1}\omega_1(x, r)^{1/2} J(x, r)^{-1} - Cr^{n-1}\beta_K(x, r)^{1/2},
\end{aligned}
\qquad (16)
$$

where the last inequality comes from (4). Then (13) says that

$$\int_{B \setminus K} |\nabla u|^2 \leq \int_{B \setminus G} |\nabla w|^2 + Cr^{n-1}\{\omega_1(x, r)^{1/2} J(x, r)^{-1} + \beta_K(x, r)^{1/2} + h(r)\}, \qquad (17)$$

and the same estimate holds for $\int_{B \setminus G} |\nabla v|^2$, by (8). Recall that w minimizes $\int_{B \setminus G} |\nabla w|^2$ under some boundary constraints, and v satisfies them. Then $w + \lambda(v - w)$ is a competitor in the definition of w for all real λ, and the derivative of $\int_{B \setminus G} |\nabla[w + \lambda(v - w)]|^2$ at $\lambda = 0$ vanishes. Then $\nabla(v - w)$ is orthogonal to ∇w in $L^2(B \setminus G)$, and Pythagoras says that

$$
\begin{aligned}
\int_{B \setminus G} |\nabla(v - w)|^2 &= \int_{B \setminus G} |\nabla v|^2 - \int_{B \setminus G} |\nabla w|^2 \\
&\leq Cr^{n-1}\{\omega_1(x, r)^{1/2} J(x, r)^{-1} + \beta_K(x, r)^{1/2} + h(r)\},
\end{aligned}
\qquad (18)
$$

by the analogue of (17) for $\int_{B \setminus G} |\nabla v|^2$. Next

$$
\begin{aligned}
\int_{B(x,t) \setminus P} |\nabla v|^2 &\leq 2 \int_{B(x,t) \setminus P} |\nabla w|^2 + 2 \int_{B(x,t) \setminus P} |\nabla(v - w)|^2 \\
&\leq Ct^n r^{-1} + Cr^{n-1}\{\omega_1(x, r)^{1/2} J(x, r)^{-1} + \beta_K(x, r)^{1/2} + h(r)\}
\end{aligned}
\qquad (19)
$$

for $0 \leq t \leq r/2$, by (12).

Recall that $v = u \circ \Phi$ on $B \setminus P$, where $\Phi : B \setminus P \to V$ is a diffeomorphism that satisfies (6) (and V is defined by (5)). Let us assume that Φ was also chosen so that $|\Phi(y) - y| \leq Cr\beta_K(x,r)$ on $B \setminus P$, which is easy to arrange.

Let $\tau \leq 1/2$ be given, and choose $t = \tau r + Cr\beta_K(x,r)$. Then $\Phi^{-1}(V \cap B(x,\tau r)) \subset B(x,t) \setminus P$, and

$$
\begin{aligned}
\int_{V \cap B(x,\tau r)} |\nabla u|^2 &\leq (1 + C\beta_K(x,r)^{1/2}) \int_{\Phi^{-1}(V \cap B(x,\tau r))} |\nabla v|^2 \\
&\leq (1 + C\beta_K(x,r)^{1/2}) \int_{B(x,t) \setminus P} |\nabla v|^2 \qquad\qquad (20) \\
&\leq \int_{B(x,t) \setminus P} |\nabla v|^2 + C\beta_K(x,r)^{1/2} r^{n-1} \\
&\leq C\tau^n r^{n-1} + Cr^{n-1} \{ \omega_1(x,r)^{1/2} J(x,r)^{-1} + \beta_K(x,r)^{1/2} + h(r) \},
\end{aligned}
$$

by (6) and (19).

We also need to control what happens on $W = B \setminus V$. Recall from (8) that

$$
\int_{B \setminus G} |\nabla v|^2 \leq (1 + C\beta_K(x,r)^{1/2}) \int_V |\nabla u|^2 \leq \int_V |\nabla u|^2 + Cr^{n-1}\beta_K(x,r)^{1/2}. \quad (21)
$$

Then (7) says that

$$
\begin{aligned}
\int_{B \setminus K} |\nabla u|^2 &\leq H^{n-1}(G \cap \overline{B}) - H^{n-1}(K \cap \overline{B}) + \int_{B \setminus G} |\nabla v|^2 + h(r)\, r^{n-1} \qquad (22) \\
&\leq Cr^{n-1} \{ \omega_1(x,r)^{1/2} J(x,r)^{-1} + \beta_K(x,r)^{1/2} + h(r) \} + \int_{B \setminus G} |\nabla v|^2 \\
&\leq Cr^{n-1} \{ \omega_1(x,r)^{1/2} J(x,r)^{-1} + \beta_K(x,r)^{1/2} + h(r) \} + \int_V |\nabla u|^2,
\end{aligned}
$$

by (16) and (21). Thus

$$
\int_{B \setminus [K \cup V]} |\nabla u|^2 \leq Cr^{n-1} \{ \omega_1(x,r)^{1/2} J(x,r)^{-1} + \beta_K(x,r)^{1/2} + h(r) \}, \qquad (23)
$$

and now (20) and (23) yield

$$
\begin{aligned}
\omega_2(x,\tau r) = (\tau r)^{1-n} \int_{B(x,\tau) \setminus K} |\nabla u|^2 &\leq (\tau r)^{1-n} \Big\{ \int_{V \cap B(x,\tau)} |\nabla u|^2 + \int_{B \setminus [K \cup V]} |\nabla u|^2 \Big\} \\
&\leq C\tau + C\tau^{1-n} \{ \omega_1(x,r)^{1/2} J(x,r)^{-1} + \beta_K(x,r)^{1/2} + h(r) \}. \qquad (24)
\end{aligned}
$$

This proves (10) when $\tau \leq 1/2$. The other case follows at once from the trivial estimate (18.8). This completes our proof of Lemma 9. $\square$

Corollary 25 ($n = 2$). *There exist a constant $\varepsilon > 0$ such that if Ω is an open set in the plane, h is a nondecreasing positive function, $(u, K) \in TRLAM(\Omega, h)$, $x \in K$, $r > 0$, $\overline{B}(x, r) \subset \Omega$, and*

$$\beta_K(x, r) + J(x, r)^{-1} + \int_0^{2r} h(t)^{1/2} \frac{dt}{t} \leq \varepsilon \qquad (26)$$

then $K \cap B(x, r/2)$ is a nice C^1 curve (see Definition 51.2).

Let τ_1 be as in Corollary 50.51, where we take $\varepsilon = 10^{-2}$. Then let x and r be as in the statement, and let $y \in K \cap B(x, r/2)$ be given. Choose $r' \in (r/4, r/2)$ such that $\omega_2^*(y, r') < +\infty$, and apply Lemma 9 to the pair (y, r'), and with a constant τ so small that $C\tau$ in (10) is less than $\tau_1/5$.

Note that $J(y, r') \geq J(x, r) - C \geq (2\varepsilon)^{-1}$ by the proof of Lemma 43.7, (26), and if ε is small enough. Also, $\omega_1(y, r') \leq C\omega_2(y, r') \leq C$ by (18.8), $\beta_K(y, r') \leq 8\beta_K(x, r) \leq 8\varepsilon$, and $h(r') \leq h(r) \leq \log(2)^{-1} \big(\int_r^{2r} h(t)^{1/2} \frac{dt}{t} \big)^2 \leq C\varepsilon^2$. Altogether, (10) says that $\omega_2(x, \tau r') \leq 2\tau_1/5$ (if ε is small enough).

We are now ready to apply Corollary 50.51. Indeed

$$\beta_K(y, \tau r') \leq 8\tau^{-1} \beta_K(x, r) \leq 8\tau^{-1} \varepsilon \leq \tau_1/5$$

by (26) and

$$J(y, \tau r') \geq J(y, r') - C \geq (3\varepsilon)^{-1} \geq 5\tau_1^{-1}$$

by Lemma 43.7 and (26). And of course

$$\int_0^{2\tau r'} h(r)^{1/2} \frac{dt}{t} \leq \varepsilon \leq \tau_1/5,$$

by (26) again. Altogether, the pair (y, r') satisfies (50.52), and Corollary 50.51 says that $\Gamma_y = K \cap B(y, r'/2)$ is a 10^{-2}-Lipschitz graph as well as a C^1 curve, with its endpoints on $\partial B(y, r'/2)$.

We can easily deduce the fact that $K \cap B(x, r/2)$ is a nice C^1 curve from this description of each Γ_y, $y \in K \cap B(x, r/2)$, and the fact that $\beta_K(x, r) \leq \varepsilon$. The argument is the same as for the conclusion of Corollary 51.17. This completes our proof of Corollary 25. $\qquad\square$

Remark 27. When K separates $D_+(x, r)$ from $D_-(x, r)$ in $\overline{B}(x, r)$, then we can set $J(x, r)^{-1} = 0$ in Lemma 9 and Corollary 25, and still get the same conclusion. See Remarks 44.23, 46.17, 48.26, 49.34, and 50.62.

Remark 28. The estimates in this section also go through when $(u, K) \in TRLAM$ with linear distortion. For Lemma 1 we should just replace (2) with

$$H^{n-1}(A_1(K \cap \overline{B}(x, r))) \leq H^{n-1}(A_1(P \cap B(x, r))) + Cr^{n-1} \beta_K(x, r)^{1/2} + r^{n-1} h(r), \qquad (29)$$

where A_1 comes from (7.47), and this time we need to take $P = P(x, r)$. For Lemma 9, we proceed as before, but minimize energy on $A_3(B(x, r) \setminus P)$, where A_3

is the transpose of the inverse of A_2 in (7.47). Then we need to replace $H^{n-1}(K \cap \overline{B})$ with $H^{n-1}(A_1(K \cap \overline{B}))$ in the Hausdorff measure estimates (13)–(16); the main point is the analogue of (14), which we can get because $A_1(F \cap B(x,r))$ stays very close to $A_1(P \cap B(x,r))$, and separates. This yields

$$\int_{B \setminus K} |A_2 \nabla u|^2 \le \int_{B \setminus G} |A_2 \nabla w|^2 + C r^{n-1} \{ \omega_1(x,r)^{1/2} J(x,r)^{-1} + \beta_K(x,r)^{1/2} + h(r) \},$$

$$(30)$$

as in (17), and the rest of the argument stays the same.

53   Similar results with spiders and propellers ($n = 2$)

The previous results of this part give situations where $K \cap B$ is a nice curve. In this section we worry about how to show that $K \cap B$ is the union of three C^1 curves that meet at a point. The techniques are quite similar to those used before, and for this reason we shall mostly review the arguments without insisting too much on the details. Thus the proof of Theorem 4 below could be used as a review of what we did in the previous sections. We start with definitions for propellers and spiders.

Definition 1. *A propeller is a closed subset of the plane obtained as the union of three (closed) half-lines emanating from a same point x_0 and making $120°$ angles at this point. We shall call x_0 the center of the propeller.*

Definition 2. *A spider is a closed subset S of the plane obtained as the union of three simple curves γ_i, $1 \le i \le 3$, of class C^1 that all start from a same point x_0, make $120°$ angles at x_0, and for which the $\gamma_i \setminus \{x_0\}$ are disjoint. We call the curves γ_i the legs of S, x_0 the center of S, and the three other endpoints of the γ_i the endpoints of S (when S is bounded). When each γ_i is of class C^α, we say that S is of class C^α.*

It is a little unfortunate that, due to the lack of more appropriate animal species, our spiders have only three legs. Our definition allows both bounded and unbounded spiders, but all the examples that we shall see will be bounded. Here is the analogue of Definition 51.2 for spiders, followed by an analogue of Corollary 51.17.

Definition 3. *Let K be a closed subset of some planar domain Ω, and let B be a disk centered on K and contained in Ω. We say that $K \cap B$ is a nice spider (or that B is a spider disk for K) if $K \cap B$ is a spider, with its center inside B and its three endpoints on ∂B, and each of its three legs is a 10^{-1}-Lipschitz graph. We say that $x \in K$ is a spider point of K if $K \cap B(x,r)$ is a nice spider centered at x for some small $r > 0$.*

As before, the choice of 10^{-1} is somewhat arbitrary, and all the results below would work equally well with any smaller constant. Note that if $K \cap B(x,r)$ is a nice spider, then for every $r' < r$, $K \cap B(x,r')$ is either a nice spider or a nice C^1 curve.

Theorem 4 $(n = 2)$. *Let $p \in (1,2]$ be given. There is a positive constant ε such that, if Ω is an open set in the plane, h is a gauge function, $(u, K) \in TRLAM(\Omega, h)$, $x \in K$, $r > 0$, $\overline{B}(x,r) \subset \Omega$, and if in addition $\omega_p(x,r) + \int_0^{2r} h(t)^{1/2} \frac{dt}{t} \leq \varepsilon$ and there is a propeller P centered at x such that*

$$\text{dist}(y, P) \leq \varepsilon r \text{ for } y \in K \cap \overline{B}(x,r), \tag{5}$$

then $K \cap B(x, r/2)$ is a nice spider.

See Remark 41 below concerning attribution, and Remark 40 about connected results in later sections.

The proof will follow the same general route as for Corollary 51.17. We just give the main ingredients, because the details would be too similar to what we did in the previous sections.

a. Preliminary reductions

First we need a lower bound on the jumps of u when we cross the three branches of P (the analogue of $J(x,r)$ below). To define these jumps, consider the three disks $D_j(x,r)$, $1 \leq j \leq 3$, contained in the different components of $B(x,r) \setminus P$, of radius $r/4$, and as far away from P as possible. [You may glance at Figure 1.] Denote by $m_{D_j(x,r)} u$ the mean value of u on $D_j(x,r)$, and then set

$$\delta_{i,j}(x,r) = |m_{D_i(x,r)} u - m_{D_j(x,r)} u| \tag{6}$$

when the disks $D_i(x,r)$ and $D_j(x,r)$ lie in the same connected component of $\overline{B}(x,r) \setminus K$, and $\delta_{i,j}(x,r) = +\infty$ otherwise. Finally set

$$J_s(x,r) = r^{-1/2} \text{Min} \{\delta_{i,j}(x,r) ; 1 \leq i,j \leq 3 \text{ and } i \neq j\}. \tag{7}$$

Lemma 8. *There is an absolute constant $\tau_0 > 0$ such that if (x,r) is as in the statement of Theorem 4, and if ε is small enough, then $J_s(x,r) \geq \tau_0$.*

Let us first prove that if we choose τ_0 small enough,

$$\delta_{i,j}(x,r) \geq 2\tau_0 r^{1/2} \text{ for some } i \neq j. \tag{9}$$

If not, all the $\delta_{i,j}(x,r)$ are given by (6), and

the disks $D_j(x,r)$, $1 \leq j \leq 3$, lie in
the same connected component of $\overline{B}(x,r) \setminus K$. $\qquad(10)$

We reach the desired contradiction in essentially the same way as for Proposition 42.10. That is, we choose a radius $r_1 \in [r/2, r]$ such that $\omega_p^*(x, r_1) \leq C\varepsilon$, and then we construct a competitor (v, G) for (u, K) in $B(x, r_1)$. We set

$$G = Z \cup [K \setminus \overline{B}(x, r_1)], \text{ with } Z = \{y \in \partial B(x, r_1) \, ; \, \text{dist}(y, P) \leq C_0^{-1} r\}, \quad (11)$$

where C_0 is chosen so large that the local Ahlfors-regularity of K yields

$$H^1(Z) \leq H^1(K \cap \overline{B}(x, r_1)) - C_0^{-1} r. \quad (12)$$

Because (9) fails and $\omega_p^*(x, r_1) \leq C\varepsilon$, we can extend the restriction of u to $\partial B(x, r_1) \setminus Z$ and get a function v on the whole $\partial B(x, r_1)$ such that $r^{1-\frac{2}{p}} \left\{ \int_{\partial B(x, r_1) \setminus K} |\nabla u|^p \right\}^{2/p}$ is as small as we want (if τ_0 and ε are small enough). Then we can further extend v to $B(x, r_1)$, with $r^{-1} \int_{B(x, r_1) \setminus K} |\nabla v|^2$ as small as we want. We keep $v = u$ out of $B(x, r_1)$, and get a competitor (v, G). It is even a topological competitor, by (10) and Lemma 20.10. Because of (12) and our estimate on $\int_{B(x, r_1) \setminus K} |\nabla v|^2$, (v, G) does significantly better than (u, K), and we get the desired contradiction if $h(r)$ is small enough.

So (9) holds, and we may as well assume that $\delta_{1,2}(x, r) \geq 2\tau_0 r^{1/2}$. Let us also assume that $J_s(x, r) < \tau_0$ and try to reach a contradiction. Then $\delta_{j,3}(x, r) < \tau_0 r^{1/2}$ for some $j \in \{1, 2\}$, and we may as well assume that $j = 2$. In particular,

$$D_2(x, r) \text{ and } D_3(x, r) \text{ lie in the same connected component of } \overline{B}(x, r) \setminus K. \quad (13)$$

We start with the case when $D_1(x, r)$ and $D_3(x, r)$ lie in the same connected component of $\overline{B}(x, r) \setminus K$. Then (10) holds by (13), and altogether

$$\delta_{1,2}(x, r) \geq 2\tau_0 r^{1/2}, \; \delta_{2,3}(x, r) < \tau_0 r^{1/2}, \text{ and } \delta_{1,3}(x, r) > \tau_0 r^{1/2}, \quad (14)$$

where the last part comes from the triangle inequality.

Let us choose $r_1 \in [\frac{4r}{5}, r]$ so that $\omega_p^*(x, r_1) \leq C\varepsilon$. As before, we want to build a competitor (v, G) for (u, K) in $B(x, r_1)$. As suggested by Figure 1, the lower bounds in (14) and the fact that $\omega_p(x, r) \leq \varepsilon$ say that $K \cap B(x, r)$ almost separates $D_1(x, r)$ from $D_2(x, r)$ and $D_3(x, r)$ in $B(x, r)$. To make this more precise, we can proceed as in Proposition 44.1, apply the co-area formula to a small modification u^* of u in $B(x, r_1) \setminus K$, and get a level set Γ of u^* (corresponding to some value between $m_{D_1}(x, r)$ and the two other $m_{D_j}(x, r)$) such that

$$H^1(\Gamma) \leq C\tau_0^{-1} r^{-1/2} \int_{B(x, r) \setminus K} |\nabla u| \leq C\tau_0^{-1} \varepsilon^{1/2} r, \quad (15)$$

and which separates $D_1(x, r) \cap B(x, r_1)$ from $D_2(x, r) \cap B(x, r_1)$ and $D_3(x, r) \cap B(x, r_1)$ in $B(x, r_1) \setminus K$. If needed, we can replace Γ by a set Γ' with the same properties and which stays very close to P (by (5)), and because we can apply to Γ a

Lipschitz contraction onto some neighborhood of P. Then $H^1([K \cup \Gamma'] \cap B(x, r_1)) \geq \frac{99}{50} r_1$ if ε is small enough, because $[K \cup \Gamma'] \cap B(x, r_1)$ stays close to P and separates $D_1(x, r) \cap B(x, r_1)$ from $D_2(x, r) \cap B(x, r_1)$ and $D_3(x, r) \cap B(x, r_1)$ in $B(x, r_1)$. Then (15) says that

$$H^1(K \cap B(x, r_1)) \geq \frac{98}{50} r_1. \tag{16}$$

Here we could also prove (16) directly, without the co-area formula (but with an argument which is slightly messier to write down). Indeed, because of the lower bounds on the $|m_{D_1(x,r)}u - m_{D_j(x,r)}u|$, $j = 2, 3$ provided by (14), and since $\int_{B(x,r) \setminus K} |\nabla u|$ is very small, we can argue that most of the little line segments that cross P orthogonally on the two legs that separate $D_1(x, r)$ from $D_2(x, r)$ and $D_3(x, r)$ must meet K.

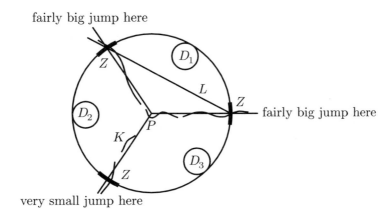

fairly big jump here

very small jump here

Figure 1

Call L the line segment that connects the two points of $P \cap \partial B(x, r_1)$ that are closest to $D_1(x, r)$ (see Figure 1 again), and let Z be the same small neighborhood of P in $\partial B(x, r_1)$ as in (11). Then set $G = [K \setminus B(x, r_1)] \cup Z \cup L$. Note that G is significantly shorter than K, by (16) (and if C_0 in (11) is large enough).

We can use the fact that $|m_{D_2(x,r)}u - m_{D_3(x,r)}u| \leq \tau_0 r^{1/2}$ (by (14) and (6)) and the smallness of $\omega_p(x, r)$ and $\omega_p^*(x, r_1)$ to construct a function v on $B(x, r_1) \setminus L$ that has the same boundary values on $\partial B(x, r_1) \setminus Z$ as u, and for which $r^{-1} \int_{B(x,r_1) \setminus L} |\nabla v|^2$ is as small as we want (if τ_0 and ε are small enough).

We set $v = u$ outside of $B(x, r_1)$, and the pair (v, G) is a topological competitor for (u, K) in $B(x, r_1)$, because (10) holds and by Lemma 20.10. Since (v, G) does significantly better than (u, K), we get a contradiction if $h(r)$ is small enough. This settles our first case when $D_1(x, r)$ and $D_3(x, r)$ lie in the same connected component of $\overline{B}(x, r) \setminus K$.

We may now assume that $D_1(x, r)$ and $D_3(x, r)$ lie in different components of $\overline{B}(x, r) \setminus K$. Then $D_1(x, r)$ and $D_2(x, r)$ also lie in different components, by (13).

Choose r_1 as above, and note that (16) holds, directly because K stays close to P and separates $D_1(x,r) \cap B(x,r_1)$ from $D_2(x,r) \cap B(x,r_1)$ and $D_3(x,r) \cap B(x,r_1)$ in $B(x,r_1)$. We can still define (v,G) as in the previous case, and it is still a topological competitor for (u,K) in $B(x,r_1)$. This time, this follows from Lemma 45.8. More precisely, $K \cap \partial B(x,r_1) \subset G$, and if I, J are two components of $\partial B(x,r_1) \setminus G$ that lie in different components of $\Omega \setminus K$ (that is, if one of them is the one that meets $D_1(x,r)$), by (13)), then G separates I from J in $\overline{B}(x,r_1)$ (because of L).

The same estimates as before give a contradiction. This completes our proof of Lemma 8. $\square$

The next step of the proof consists in going from the small lower bound on the jump $J_s(x,r)$ that we got from Lemma 8 to a much more generous lower bound on a smaller disk $B(x,\eta_1 r)$. This corresponds to what we did in Section 43, and we can essentially use the same arguments. More specifically, we get that if $\eta_1 > 0$ is a small constant, and then ε and r_0 are small enough, depending also on η_1,

$$ J_s(x,\eta_1 r) \geq \frac{1}{2}\,\eta_1^{-1/2}\tau_0. \tag{17} $$

Here we keep the same sort of definition for $J_s(x,t)$, $t \leq r$, as for $J_s(x,r)$ above, provided that $K \cap B(x,t)$ stays within $10^{-2}t$, say, of some propeller $P(x,t)$ centered at x. Note that this is definitely the case for $t \in [\eta_1 r, r]$ if ε is small enough. There is a small difference with Section 43, because we set $\delta_{i,j}(x,r) = +\infty$ when the disks $D_i(x,r)$ and $D_j(x,r)$ lie in different connected components of $\overline{B}(x,r) \setminus K$. Then we can easily check that the corresponding disks in $B(x,\eta_1 r)$ also lie in different components of $\overline{B}(x,\eta_1 r) \setminus K$ (if ε is small enough). This takes care of the new part of (17); the old part goes as in Section 43 (i.e., with the same proofs as for the Poincaré estimates in Section 12).

Next we want to use (17) to control the holes of K in $B(x,\eta_1 r)$, as we did in Proposition 44.1. For $y \in K$ and $t > 0$, denote by $P(y,t)$ a propeller such that $y \in P(y,t)$ and $\sup\{\operatorname{dist}(z,P)\,;\,z \in K \cap \overline{B}(y,t)\}$ is smallest.

Lemma 18. *We can find a compact set $F_s = F_s(x,\eta_1 r) \subset \overline{B}(x,\eta_1 r)$ such that*

$$ K \cap \overline{B}(x,\eta_1 r) \subset F_s \subset \{y \in \overline{B}(x,\eta_1 r)\,;\,\operatorname{dist}(y,P(x,\eta_1 r)) \leq 2\varepsilon r\}, \tag{19} $$

$$ F_s \text{ separates the } D_j(x,\eta_1 r) \text{ from each other in } \overline{B}(x,\eta_1 r), \tag{20} $$

$$ H^1([F_s \cap B(x,\eta_1 r)] \setminus K) \leq C\eta_1 r\,\omega_p(x,\eta_1 r))^{1/2}\,J_s(x,\eta_1 r))^{-1} \leq C(\eta_1)\,r\,\varepsilon^{1/2}. \tag{21} $$

When all the $D_j(x,\eta_1 r)$ lie in the same component of $\overline{B}(x,\eta_1 r) \setminus K$, the argument is the same as in Section 43. That is, call m_j the mean value of u on the three disks $D_j(x,\eta_1 r)$, and assume for definiteness that $m_1 < m_2 < m_3$. We first construct a smooth function v on $B(x,\eta_1 r) \setminus K$ that is equal to m_j on $\frac{1}{2}D_j(x,\eta_1 r)$, and $\int_{B(x,\eta_1 r)\setminus K} |\nabla v| \leq C \int_{B(x,\eta_1 r)\setminus K} |\nabla u|$ (which is as small as we want). Then we choose two numbers λ_1 and λ_2, with $m_1 < \lambda_1 < m_2 < \lambda_2 < m_3$, such that if we set $\Gamma_i = v^{-1}(\lambda_i)$, $H^1(\Gamma_i) \leq C[(\eta_1 r)^{1/2} J_s(x,\eta_1 r)]^{-1} \int_{B(x,\eta_1 r)\setminus K} |\nabla v| \leq C\eta_1 r\, \omega_1(x,\eta_1 r))^{1/2} J_s(x,\eta_1 r))^{-1}$. This is possible, by the co-area theorem, and by definition of $J_s(x,\eta_1 r)$. Then

$$F = [K \cap B(x,\eta_1 r)] \cup \Gamma_1 \cup \Gamma_2 \cup \{y \in \partial B(x,\eta_1 r) ; \mathrm{dist}(y, P(x,\eta_1 r)) \leq 2\varepsilon r\} \quad (22)$$

satisfies (20) and (21), and we can easily arrange (19) without destroying (20) and (21) by applying to F a contraction onto the desired strip around the propeller $P(x,\eta_1 r)$. The details are the same as in Section 44.

When all the $D_j(x,\eta_1 r)$ lie in different components of $\overline{B}(x,\eta_1 r) \setminus K$, we can take $F_s = K \cap \overline{B}(x,\eta_1 r)$.

Finally, assume that two of the $D_j(x,\eta_1 r)$ (say, $D_1(x,\eta_1 r)$ and $D_2(x,\eta_1 r)$) lie in the same component of $\overline{B}(x,\eta_1 r) \setminus K$, but not the third one. Note that (17) says that

$$|m_{D_1(x,\eta_1 r)} u - m_{D_2(x,\eta_1 r)} u| \geq \frac{1}{2} \eta_1^{-1/2} \tau_0\, r^{1/2}. \quad (23)$$

As before, we build a smooth approximation v of u and use the co-area theorem to find λ between $m_{D_1(x,\eta_1 r)} u$ and $m_{D_2(x,\eta_1 r)} u$, such that $\Gamma = v^{-1}(\lambda)$ is short. Define F as in (22), but with $\Gamma_1 \cup \Gamma_2$ replaced by Γ. Then F separates $D_1(x,\eta_1 r)$ from $D_2(x,\eta_1 r)$, by the same argument as before (and in Section 44). We do not need to add anything to separate $D_3(x,\eta_1 r)$ from the other two disks, since K already does this. So we may conclude as before. $\qquad\square$

We shall also need the following analogue of Proposition 45.1.

Lemma 24. *Suppose that $(u, K) \in TRLAM(\Omega, h)$, $x \in K$, $\overline{B}(x,t) \subset \Omega$, $1 < p \leq 2$, and $\omega_p^*(x,t) < +\infty$ (see (42.5) and Convention 42.6). Also suppose that there is a propeller P centered at x such that*

$$\mathrm{dist}(y, P) \leq \beta t \text{ for } y \in K \cap \overline{B}(x,t) \quad (25)$$

for some $\beta > 0$. Then

$$H^1(K \cap \overline{B}(x,t)) + \int_{B(x,t)\setminus K} |\nabla u|^2 \leq 3t + C_p\, t\, \omega_p^*(x,t) + C\beta t + t\, h(t). \quad (26)$$

The case when $\beta \geq 10^{-1}$ is trivial. Otherwise, we compare (u, K) with a competitor (v, G), where G is obtained by replacing $K \cap \overline{B}(x,t)$ with $[P \cap B(x,t)] \cup Z$, and Z is the union of three arcs of $\partial B(x,t)$ of length $3\beta t$ and centered at the points of $P \cap \partial B(x,t)$. We keep $v = u$ outside of $B(x,t)$, and inside we define v separately on each of the three components of $B(x,t) \setminus P$, to be a suitable extension of the restriction of u to the corresponding component of $\partial B(x,t) \setminus Z$. The comparison naturally yields (26); the details are the same as in Section 45. $\square$

When $\beta < 10^{-1}$ and $K \cap \partial B(x,t)$ is composed of exactly three points that lie at mutual distances greater than t, we can improve slightly on (26), as we did in Proposition 45.9. Let P_0 denote the piece of propeller that connects the three points of $K \cap \partial B(x,t)$. That is, P_0 is a spider centered somewhere in $B(x,t)$, composed of three line segments, and whose endpoints are the three points in question. Incidentally, P_0 is also the shortest set in $\overline{B}(x,t)$ that separates the three components of $\partial B(x,t) \setminus K$ from each other.

We can replace $[P \cap B(x,t)] \cup Z$ by P_0 in the argument above, and we get that

$$H^1(K \cap \overline{B}(x,t)) + \int_{B(x,t) \setminus K} |\nabla u|^2 \leq H^1(P_0) + C_p\, t\, \omega_p^*(x,t) + t\, h(t). \qquad (27)$$

Let us return to Theorem 4. We can find $t \in (\eta_1 r/2, \eta_1 r)$ such that

$$\omega_p^*(x,t) = t^{1-\frac{2}{p}} \left\{ \int_{\partial B(x,t) \setminus K} |\nabla u|^p \right\}^{2/p} \leq C(\eta_1 r)^{1-\frac{4}{p}} \left\{ \int_{B(x,r) \setminus K} |\nabla u|^p \right\}^{2/p} \qquad (28)$$

$$\leq C\eta_1^{1-\frac{4}{p}} \omega_p(x,r) \leq C\eta_1^{-3}\varepsilon$$

(by (42.5), Chebyshev, (23.5), and our main hypothesis). The construction of $F_s(x,\eta_1 r)$ also works with the radius t, and when we compare (21) with Lemma 24 and (28), we get that

$$H^1(F_s(x,t) \cap B(x,t)) = H^1([F_s(x,t) \cap B(x,t)] \setminus K) + H^1(K \cap B(x,t))$$

$$\leq C(\eta_1)\, t\, \varepsilon^{1/2} + 3t + C_p\, t\, \omega_p^*(x,t) + C\eta_1^{-1}\varepsilon\, t + t\, h(t) \quad (29)$$

$$\leq 3t + C(\eta_1)\, t\, \varepsilon^{1/2} + t\, h(t) \leq 3t + \tau t,$$

where we can take τ as small as we want, provided that we choose ε small enough (depending on η_1).

Since $F_s(x,t)$ stays very close to P (by (19)) and separates the $D_j(x,t)$ in $\overline{B}(x,t)$ (by (20)), (29) says that most circles $\partial B(x,\rho)$, $0 < \rho < t$, meet $F_s(x,t)$ exactly three times, and at points whose mutual distances are greater than ρ. Let us call "good circles" the circles $\partial B(x,\rho)$ for which, in addition to this, $K \cap \partial B(x,\rho) = F_s(x,t) \cap \partial B(x,\rho)$. Then $\partial B(x,\rho)$ is a good circle for most choices of $\rho \in (0,t)$, by (19) and (21). See Section 46 for additional details.

Also notice that since $F_s(x,t)$ satisfies (19) and (20), $H^1(F_s(x,t) \cap B(x,t)) \geq 3t - \tau t$, with τ as small as we want, and then (21) says that $H^1(K \cap B(x,t)) \geq 3t - 2\tau t$. Thus (26) shows that $\omega_2(x,t) \leq 4\tau$ (that is, $\omega_2(x,t)$ is as small as we want). We can use this estimate to reduce the proof of Theorem 4 to the case when $p = 2$, or just use it as a starting point in the induction that follows.

b. Self-reproducing estimates

At this point we arrived at a disk $B(x,t)$ where we have a good control of all the quantities that we shall need. That is, $K \cap B(x,t)$ stays as close as we want to the propeller P (directly by (5)), the jump $J_s(x,t)$ is as large as we want (if η_1 is chosen accordingly), by (17), and even $\omega_2(x,t)$ is as small as we want, as we just said.

We need to find a way to propagate a set of conditions like this to smaller and smaller scales, and as before we shall base our argument on the fact that we can force a decay on $\omega_2(x, \cdot)$, and at the same time keep some control on the other numbers.

So we want to check that the decay estimates in Section 48 also go through. That is, if $K \cap B(x, t)$ stays very close to some propeller $P(x, t)$, $\omega_2(x, t)$ and $h(t)$ are small enough, and the normalized jump $J_s(x, t)$ is large enough, then

$$\omega_2(x, at) \leq \gamma \omega_2(x, t) + C \omega_2(x, t)^{1/2} J_s(x, t)^{-1} + Ch(t), \tag{30}$$

with constants $0 < a < 1/4$ and $0 < \gamma < 1$, as in (48.3).

The argument does not need to be modified too much. We consider the auxiliary function v defined on the complement of $K \cup F_s(x, t)$, which coincides with u on $\partial B(x, t) \backslash K$ and minimizes the energy under this constraint. The starting point is still that (since a is fixed here), $\partial B(x, \rho)$ is a good circle for most choices of $\rho \in [at, t]$.

For almost all such ρ, we can apply Proposition 47.7 (Bonnet's lower bound on the derivative of energy). Indeed the three points of $F_s(x, t) \cap \partial B(x, \rho)$ lie on the same connected component of $F_s(x, t) \cap \overline{B}(x, \rho)$, because $F_s(x, t)$ separates and by plane topology. The argument is the same as in Lemma 46.16. The lengths of the three components of $\partial B(x, \rho) \setminus F_s(x, t)$ are quite close to $2\pi t/3$, which is even a little smaller than what we had in Section 48 (when there were only two components), and leads to an even slightly faster decay. At any rate, Bonnet's lower bounds lead to an estimate like (30) for v, and finally (30) itself follows because ∇u is very close to ∇v in L^2-norm (by (21), almost-minimality, and Pythagoras).

Once we have (30), we can continue the argument as in Sections 49 and 50. The point is still to use the decay provided by (30) to show that $\omega_2(x, \cdot)$ gets smaller. We easily prove at the same time that $J_s(x, \cdot)$ gets larger, and the reason why the argument works is that the distance from K to propellers can be controlled in terms of these two quantities (so that it never gets so large that we can't apply (30)).

Here the main point is that we have estimates like (27) above (because we can find lots of good circles), which give a very good control on $H^1(K \cap B(x, t))$ compared with the length of the piece of propeller P_0 whose endpoints are the three points of $K \cap \partial B(x, t)$. We also have a very good control on the size of the holes in K, i.e., on how much length must be added to K to get a set $F_s(x, t)$ that separates $B(x, t)$ into three pieces, as in Lemma 18. When we put together these two estimates, we find that indeed $K \cap B(x, t)$ stays quite close to P_0, with estimates in terms of $\omega_2(x, t)$ and $J_s(x, t)$ only. This is not an extremely good control, but it is enough to allow a new application of the argument on a smaller disk, and start an iteration.

There is a little additional technical difficulty because P_0 is not necessarily centered very close to x, but this is not too bad. The most natural way to deal with this is to allow a mixture of the argument presented in this section and what we did in the previous ones, and study together all the situations where $K \cap B(x,t)$ stays close to a propeller (not necessarily centered close to x) or a line. We do not give the details, because they are quite similar to what we did in Sections 49 and 50 (and they would be quite boring). One may always consult [Da5], but there also the arguments tend to be more elliptic the second time around, when it comes to spiders.

Another possibility would be to change the center x at each iteration, so as to get a propeller centered near the center of the disk at each iteration. If we do this, we get a limit point $x_\infty \in B(x, \eta_1 r/10)$, say, and the information that $K \cap B(x_\infty, t)$ stays quite close to a propeller centered at x_∞ for each radius $t \leq \eta_1 r$, say. This looks bad, because we do not seem to control what happens far from x_∞. However we can recover the missing information, because we can apply Corollary 51.17 to balls D contained in $B(x, \eta_1 r)$ (say) and whose radius is somewhat smaller, but comparable to their distance to x_∞.

Altogether, we finally prove that if x and r are as in the statement of Theorem 4, then $K \cap B(x, \eta r)$ is a nice spider. Here η is of the order of magnitude of η_1 in the first part of the argument. If ε is small enough, (5) forces the center of the spider to lie in $B(x, \eta r/10)$, say. Recall that we want to prove that the whole $K \cap B(x, r/2)$ is a nice spider, so we need to control what happens out of $B(x, \eta r/2)$ too. The simplest is to use Corollary 51.17 on lots of little balls centered near $P \cap B(x, r) \setminus B(x, \eta r/2)$ and argue as for the end of Corollary 51.17. This completes our discussion of the proof of Theorem 4. $\square$

Let us also state an analogue of Corollary 52.25 for spiders.

Theorem 31. *There is an absolute positive constant ε with the following property. Let Ω be an open set in the plane, suppose that $(u, K) \in TRLAM(\Omega, h)$ (as in Standard Assumption 41.1, and with some nondecreasing positive function h), and let $x \in K$ and $r > 0$ be such that $\overline{B}(x,r) \subset \Omega$. Suppose in addition that $\int_0^{2r} h(t)^{1/2} \frac{dt}{t} \leq \varepsilon$, that there is a propeller P centered at x such that*

$$\text{dist}(y, P) \leq \varepsilon r \text{ for } y \in K \cap B(x,r), \tag{32}$$

and that $J_s(x,r) \geq \varepsilon^{-1}$, where $J_s(x,r)$ is the normalized smallest jump defined by (6) and (7). Then $K \cap B(x, r/2)$ is a nice spider.

Just as in Section 52, the main point is to show that if the constant η is chosen small enough, $\omega_2(x, \eta r)$ is as small as we wish. Then we still have that $K \cap B(x, \eta r)$ is very close to P, the normalized jump $J_s(x, \eta r)$ is even larger (by the same arguments as in Section 43), and we can apply an iteration scheme as in the second part of Theorem 4 (or Sections 49 and 50). The result is that $K \cap B(x, \eta r/10)$ is a nice spider, but since we can control the rest of $K \cap B(x, r/2)$ directly with Corollary 52.25, this is enough to conclude.

To get the desired bound on $\omega_2(x, \eta r)$, it is enough to prove that for $\tau < 10^{-1}$,

$$\omega_2(x, \tau r) \leq C\tau + C\tau^{-1}\{\omega_1(x,r)^{1/2} J_s(x,r)^{-1} + \varepsilon^{1/2} + h(r)\}, \tag{33}$$

as in Lemma 52.9, and where ε now plays the role of $\beta_K(x,r)$.

The proof of (33) goes just as in Section 52, so let us only review the argument. First we choose a radius $r_1 \in [r/2, r]$ such that $\omega_2^*(x, r_1) \leq C$, and set $B = B(x, r_1)$. Then we define a competitor (v, G) for (u, K) in B. We set $G = [K \setminus B] \cup P \cup Z$, where Z is an $\varepsilon^{1/2}$-neighborhood of P in ∂B. [We take $\varepsilon^{1/2}$ to have enough room for the homeomorphisms Φ below.] As usual, we choose $v \in W^{1,2}(B \setminus P)$ with the same radial limits as u almost-everywhere on $\partial B \setminus Z$, and so that $E = \int_{B \setminus P} |\nabla v|^2$ is as small as possible.

Note that, by the co-area theorem and the proof of Proposition 44.1, we can add a set of H^1-measure less than $Cr\omega_1(x, r)^{1/2} J_s(x, r)^{-1}$ to K and get a set that separates the three main regions in B. The length of this set is at least $3r - Cr\varepsilon^{1/2}$, so $H^1(K \cap B) \geq 3r - Cr\varepsilon^{1/2} - Cr\omega_1(x, r)^{1/2} J_s(x, r)^{-1}$, and

$$H^1(G \cap \overline{B}) \leq 3r + Cr\varepsilon^{1/2} \leq H^1(K \cap B) + Cr\varepsilon^{1/2} + Cr\omega_1(x, r)^{1/2} J_s(x, r)^{-1}. \quad (34)$$

Since (u, K) is an almost-minimizer, we get that

$$\int_{B \setminus K} |\nabla u|^2 \leq E + H^1(G \cap \overline{B}) - H^1(K \cap B) + rh(r)$$
$$\leq E + Cr\varepsilon^{1/2} + Cr\omega_1(x, r)^{1/2} J_s(x, r)^{-1} + rh(r), \quad (35)$$

by (34). On the other hand, we can obtain a competitor v_1 in the definition of v by taking $v_1 = u \circ \Phi$, where ϕ is a diffeomorphism from $B \setminus P$ to the complement in B of an εr-neighborhood of P (so that u is defined on it). This proves that

$$E \leq \int_{B \setminus K} |\nabla u|^2 + C\varepsilon^{1/2} r \quad (36)$$

as in (52.8). Thus u, v, and v_1 almost have the same energy, and by Pythagoras we can say that ∇v_1 is very close to ∇v in L^2. Then we use the fact that ∇v is bounded on $\frac{1}{2}B \setminus P$ (it even vanishes at x), so that $\int_{B(x, \tau r) \setminus P} |\nabla v|^2 \leq C\tau^2 r$. We can deduce from this a similar estimate for v_1 (because ∇v_1 is so close to ∇v), and then (33) follows by looking more closely at the construction of v_1 in terms of v. [We also need to control the part of $\int_{B \setminus K} |\nabla u|^2$ which comes from a thin strip near P, and does not contribute to the energy of v_1.] This completes our description of the proof of Theorem 31. $\qquad \square$

Remark 37. We could also prove (33) with Bonnet's lower bounds on the derivative of energy, as we did in Section 48 (and our proof of Theorem 4). Note that there is a similarity between the two arguments anyway; in both cases we use a lower bound on the decay of some auxiliary function v that is very close to u.

Remark 38. We could make our proof of Theorems 4 and 31 somewhat simpler at places, by using the results of the previous sections (rather than their proofs, as we did). The basic point is that when x and r are as in Theorems 4 and 31, we know from the previous sections that $K \cap B(x, r/2) \setminus B(x, r/100)$ is composed of three nice C^1 curves. Also see the last lines of our proof of Theorem 4.

Remark 39. If $h(r) \leq Cr^\alpha$ for some $\alpha > 0$ and $K \cap B(x,r)$ is a nice spider, it is automatically of class $C^{1+\beta}$ for some $\beta > 0$. This can be proved directly, just like the corresponding result for nice curves.

Also, when we only know that $h(r)$ tends to 0 (or even, is small enough), we still have an analogue of Theorem 50.1 (where we get a chord-arc spider) and Corollary 50.25 (where we get an asymptotically straight chord-arc spider). We leave the definition of chord-arc spiders to the reader's sagacity. See Exercise 43.

Remark 40. We shall see later slight improvements of Theorems 4 and 31. For instance, in Proposition 60.1, we shall get that $K \cap B(x, \eta r)$ is a nice C^1 curve or spider as soon as r is small enough and $\omega_p(x, r)$ is small enough, and in Proposition 70.2 we shall get a similar conclusion when r is small and $K \cap B(x,r)$ is ε-close in Hausdorff distance to $P \cap B(x,r)$, where P is a line or a propeller. A simple consequence is that, if any blow-up limit of K at x is a line or propeller, then for r small enough $K \cap B(x,r)$ is a nice C^1 curve or spider centered at x.

Remark 41. Theorems 4 and 31 are minor variants of results from [Da5]. Similar results were also rediscovered (independently and with slightly different proofs) by F. Maddalena and S. Solimini. Their argument uses Remark 38, a nice averaging argument over directions, and ideas similar to the ones above, but to the author's knowledge it was not published.

Remark 42. Hopefully Theorems 4 and 31 stay true for almost-minimizers with linear distortion, but the author checked this even less than for the other statements. At least, spiders and propellers should be defined differently; the story with the 120° has to be modified, depending on the matrices A_1 in Definition 7.45.

Exercise 43

1. Formulate a definition of chord-arc spiders with constant less than C. [Suggestion: use three chord-arc curves, and add a special condition about angles near the center. When C is large, we just need to require that the union of any two of the curves is a chord-arc curve, but when C is close to 1, I don't see any especially beautiful definition.]

2. Prove that when S is a chord-arc spider with constant $\leq 1+\varepsilon$ and x_1, x_2, x_3 lie on the three legs of S, the length of the part of S that lies between those points (the component of the center in $S \setminus \{x_1, x_2, x_3\}$) is at most $(1+C\varepsilon)L$, where L is the length of the shortest connected set that contains the x_j.

3. Prove that the infinite chord-arc spiders are the images of propellers by bilipschitz mappings. We can most probably even say more: they are images of propellers under bilipschitz mappings of the whole plane. This is a classical result (but not trivial) for chord-arc curves, and the lazy author has to admit that he did not check the statement for spiders. In both cases, if the chord-arc constant is close to 1, we (should) get that the bilipschitz constant is close to 1.

F. Global Mumford-Shah Minimizers in the Plane

Global (Mumford-Shah) minimizers in the plane should be at the center of the study of the Mumford-Shah conjecture (described in Section 6). Indeed, we know from Section 40 that every blow-up limit of a reduced almost-minimizer (which includes any minimizer of the Mumford-Shah functional in a domain) is a global minimizer. Moreover, in dimension 2, we can hope that the situation is sufficiently simple to allow a complete description of all global minimizers; see Conjecture 18 below. And such a description would be enough to prove the Mumford-Shah conjecture (see Section 71).

Our goal for this part is to describe most of the known results on global minimizers in the plane. As we shall see, the main additional ingredients are the fact that limits of global minimizers are global minimizers, and monotonicity arguments based on Bonnet's lower bound in Section 47. We start with a description in $\mathbb{R}^n$, but very rapidly we shall restrict to global minimizers in the plane, in particular because we do not know of any analogue of Section 47 in higher dimensions.

54 Global minimizers in $\mathbb{R}^n$ and a variant of the Mumford-Shah conjecture

a. Definitions

Global minimizers in $\mathbb{R}^n$ are what we have called so far local topological almost-minimizers in $\mathbb{R}^n$, with gauge function $h = 0$. Since we shall use global minimizers a lot and the definition is a little simpler for them, let us recall it. We work in $\mathbb{R}^n$, and the set of admissible pairs is

$$\mathcal{A} = \mathcal{A}(\mathbb{R}^n) = \{(u, K)\,; K \subset \mathbb{R}^n \text{ is closed and } u \in W^{1,2}_{\text{loc}}(\mathbb{R}^n \setminus K)\}. \quad (1)$$

Let a pair $(u, K) \in \mathcal{A}$ be given. A topological competitor for (u, K) is a pair $(v, G) \in \mathcal{A}$ such that for R large enough,

$$G \setminus \overline{B}(0, R) = K \setminus \overline{B}(0, R) \text{ and } v(x) = u(x) \text{ for } x \in \mathbb{R}^n \setminus [K \cup \overline{B}(0, R)], \quad (2)$$

and

$$\text{if } x, y \in \mathbb{R}^n \setminus [K \cup \overline{B}(0, R)] \text{ are separated by } K,$$

$$\text{then they are also separated by } G. \tag{3}$$

As usual, "K separates x from y" means that x and y lie in different connected components of $\mathbb{R}^n \setminus K$. Note that if (2) and (3) hold for some R, they also hold for all $R' > R$.

Definition 4. *A global minimizer is an admissible pair* $(u, K) \in \mathcal{A}$ *such that*

$$\int_{B(0,R)\setminus K} |\nabla u|^2 + H^{n-1}(K \cap \overline{B}(0, R)) \leq \int_{B(0,R)\setminus G} |\nabla v|^2 + H^{n-1}(G \cap \overline{B}(0, R)) \tag{5}$$

for all topological competitors (v, G) *for* (u, K) *and all radii* R *so large that* (2) *and* (3) *hold.*

Remark 6. If (u, K) is a global minimizer, $\int_{B(0,R)\setminus K} |\nabla u|^2 + H^{n-1}(K \cap \overline{B}(0, R)) \leq \omega_{n-1} R^{n-1}$ for every $R > 0$, by (18.8). Also, if we know that $\int_{B(0,R)\setminus K} |\nabla u|^2 + H^{n-1}(K \cap \overline{B}(0, R)) < +\infty$ for every R, then (5) is the same for all R such that (2) and (3) hold, and hence it is enough (for a given pair (v, G)) to check (5) for one such R. Similarly, under this mild additional constraint on (u, K), we can replace $\overline{B}(0, R)$ with $B(0, R)$ in (2), (3) and (5), and get an equivalent definition.

Remark 7. If (u, K) is a global minimizer and Ω is a connected component of $\mathbb{R}^n \setminus K$, and if you replace $u(x)$ with $-u(x)$ or $u(x) + C$ (for some constant C) in Ω, this still gives a global minimizer. This is a special case of Exercise 7.57, but let us prove it anyway.

This is easier to check when Ω is bounded, because every topological competitor for the new pair $(\tilde{u}, K)$ is also a topological competitor for (u, K) (take R so large that $\Omega \subset B(0, R)$). It is very easy to see in this case that $\tilde{u} \in W^{1,2}(\mathbb{R}^n \setminus K)$, and that the integrals in (5) are the same for $\tilde{u}$ as for u. Hence $(\tilde{u}, K)$ is a global minimizer.

When Ω is not bounded, consider any topological competitor $(\tilde{v}, G)$ for $(\tilde{u}, K)$, and let R be so large that the analogue of (2) and (3) hold. We want to define a function v on $\mathbb{R}^n \setminus G$, and we shall do it component by component. If U is a component of $\mathbb{R}^n \setminus G$ that does not meet $\Omega \setminus B(0, R)$, keep $v = \tilde{v}$ on U. Otherwise, set $v(x) = -\tilde{v}(x)$ or $v(x) = \tilde{v}(x) - C$ on U (depending on the way $\tilde{u}$ was defined). It is easy to see that $v \in W^{1,2}_{\text{loc}}(\mathbb{R}^n \setminus G)$, because this is the case for $\tilde{v}$, and we only added a constant or changed the sign of $\tilde{v}$ locally. Also, we claim that v coincides with u out of $B(0, R) \setminus G$. This is clear for the values in U when U does not meet $\Omega \setminus B(0, R)$, because $v = \tilde{v} = \tilde{u} = u$ there. Otherwise, (3) tells us that $U \setminus B(0, R) \subset \Omega$, where $\tilde{u} = -u$ or $\tilde{u} = u + C$. Then $v = u$ on $U \setminus B(0, R)$, as needed. Thus (v, G) is a topological competitor for (u, K). The integrals in (5) are not modified, and so $(\tilde{u}, K)$ is also a global minimizer.

Remark 8. We shall restrict our attention to topologically reduced, or coral global minimizers (as in Definitions 8.7 and 8.24). First observe that the two notions coincide in the present case. If the global minimizer (u, K) is topologically reduced, then Theorem 18.16 says that K is locally Ahlfors-regular, and in particular it is coral. Conversely, if (u, K) is coral, then it is topologically reduced. This is Remark 8.31. The main point is that if $(\widetilde{u}, \widetilde{K})$ is a topological competitor for (u, K) such that $\widetilde{K} \subset K$ and $\widetilde{u}$ is an extension of u, then $H^{n-1}(K \setminus \widetilde{K}) = 0$ because (u, K) is a global minimizer. Then $K \setminus \widetilde{K}$ does not meet the core of K. Since K is assumed to be coral, it is equal to its core, and $\widetilde{K} = K$.

When we restrict to topologically reduced global minimizers, we shall not lose anything important, for the following reason. We claim that for each a global minimizer (u, K) there is a topologically reduced global minimizer $(\widetilde{u}, \widetilde{K})$ such that

$$\widetilde{K} \subset K, \ \widetilde{u} \text{ is an extension of } u, \text{ and } H^{n-1}(K \setminus \widetilde{K}) = 0. \tag{9}$$

Since (u, K) is sufficiently close to $(\widetilde{u}, \widetilde{K})$, it will be natural to restrict our attention to $(\widetilde{u}, \widetilde{K}))$.

To prove the claim, notice that we have just checked that (u, K) satisfies (8.9), so Proposition 8.8 tells us how to find a topologically reduced pair $(\widetilde{u}, \widetilde{K}) \in \mathcal{A}$ such that (9) holds. To show that $(\widetilde{u}, \widetilde{K})$ is also a global minimizer, we shall use Proposition 8.22, even though there would be slightly more rapid arguments here. Note that because $\Omega = \mathbb{R}^n$ and $h = 0$, there is no difference between our global minimizers (i.e., local topological almost-minimizers) and "strong local topological almost-minimizers" (as in Definitions 8.19 and 8.21). The point is that the topological competitors are the same, and we do not care in which ball they are topological competitors, because anyway our gauge function is $h \equiv 0$ and hence (7.7) does not depend on r (large). Then Proposition 8.22 says that $(\widetilde{u}, \widetilde{K})$ is a global minimizer, as needed.

Notation 10. *We shall denote by RGM, or $RGM(\mathbb{R}^n)$, the set of topologically reduced (or equivalently coral, see Remark 8) global minimizers. In this context, we shall often forget the word "topological", which will be considered implicit.*

b. General regularity properties

We already know about some of the properties of the reduced global minimizers, because they hold for all coral almost-minimizers.

If $(u, K) \in RGM$, then K is an (unbounded) Ahlfors-regular set, by Theorem 18.16. Note that this means that there is an absolute constant $C = C(n)$ such that $C^{-1} r^{n-1} \leq H^{n-1}(K \cap B(x, r)) \leq C r^{n-1}$ for all $(u, K) \in RGM$, all $x \in K$, and all $r > 0$, and not just for r small enough. The point is that the usual restrictions on r that come from the distance from x to the boundary of Ω, or the size of $h(r)$, do not exist here. We could also get this information from the invariance under dilations described below.

Among other properties, K uniformly rectifiable, as in Theorem 27.1 when $n = 2$ and Theorem 74.1 in higher dimensions. Thus, in dimension 2, K is contained in an (unbounded) Ahlfors-regular curve, with a uniform bound on the Ahlfors constant.

Here is a more precise regularity result that works almost-everywhere.

Theorem 11 [Da5], [AFP1], [Ri1]. *There is a constant $C = C(n)$ such that for each $(u, K) \in RGM(\mathbb{R}^n)$ and all choices of $x \in K$ and $r > 0$, there is a point $y \in K \cap B(x, r)$ and a radius $t \in (C^{-1}r, r)$ such that $K \cap B(y, t)$ is a nice C^1 surface.*

For more precise statements (and in particular a definition of nice C^1 surfaces), see Theorem 51.4 when $n = 2$, and Theorem 75.2 for higher dimensions. Recall that this also implies that the set of points $x \in K$ such that K is not a nice C^1 surface near x has Hausdorff dimension $d < n - 1$. [See Theorem 51.20 and Corollary 75.16.] Here we say C^1, but once we know that $K \cap B(y, t)$ is a nice C^1 surface, it is fairly easy to get more regularity. In fact, $K \cap B(y, t)$ is even real-analytic: see [LeMo] for the 2-dimensional case and [KLM] when $n > 2$.

The initial references for Theorem 11 are [Da5] (only for $n = 2$) and [AFP1] (independently, and for all n), completed by [Ri1] to get exactly the same quantifiers as in the statement.

Concerning u, we know from the proof of Lemma 18.19 that

$$\int_{B(x,r)} |\nabla u|^2 \leq Cr^{n-1} \quad \text{for every ball } B(x, r), \tag{12}$$

where we can take $C = H^{n-1}(\partial B(0, 1))$. Also, u is harmonic on $\mathbb{R}^n \setminus K$, with Neumann boundary conditions $\frac{\partial u}{\partial n} = 0$ on K, weakly and pointwise wherever this makes sense. See Lemma 15.17 and Remark 15.21.

In addition, we know that the class RGM is stable under translations, dilations, and rotations. That is, if $(u, K) \in RGM$ and if we set $\widetilde{K} = \frac{1}{t}(K - y)$ and $\widetilde{u}(x) = \frac{1}{\sqrt{t}} u(y + tx)$ for some $y \in \mathbb{R}^n$ and $t > 0$, then $(\widetilde{u}, \widetilde{K}) \in RGM$. See (40.5). The statement and proof for rotations would be similar. Also, RGM is stable under limits:

$$\begin{aligned} &\text{if a sequence of reduced global minimizers in } \mathbb{R}^n \\ &\text{converges to a pair } (u, K), \text{ then } (u, K) \in RGM(\mathbb{R}^n). \end{aligned} \tag{13}$$

See Definition 37.6 for the convergence; (13) comes from Theorem 38.3 (applied with $\Omega_k = \Omega = \mathbb{R}^n$ and $h = 0$), plus the fact that (u, K) is coral (by Remark 38.64, or just because it is Ahlfors-regular), and because coral global minimizers are topologically reduced (by Remark 8).

However, we expect much more than this to be true. In dimension 2, we even think that we know all the global minimizers (see Conjecture 18). Even in higher dimensions, it is reasonable to think that there are very few global minimizers. If this is the case, this will provide a good way to get information on minimizers of the Mumford-Shah functional in a domain, because we will already know about the tangent objects. We shall see how this could go in Section 71.

The notion of global minimizer looks a little more complicated than for usual minimizers of functionals, because it corresponds to a sort of Dirichlet problem at infinity, but where the Dirichlet data depend on the candidate-minimizer itself. In addition, most asymptotic behaviors near infinity do not give rise to any minimizer. We should also note that in dimensions larger than 2, it is not known whether the function u is essentially determined by K and the knowledge that $(u, K) \in RGM$ (we have to say "essentially", because we should allow the transformations suggested by Remark 7). See the beginning of Section 63 below for the positive answer in dimension 2.

c. A slightly stronger form of the Mumford-Shah conjecture

Let us now restrict our attention to the plane and describe more precisely what we expect the global minimizers to look like. For higher dimensions, and in particular $n = 3$, see Section 76 (but much less is known).

We start with examples of reduced global minimizers. The first one is when

$$K = \emptyset \quad \text{and} \quad u \text{ is constant on } \mathbb{R}^2, \tag{14}$$

which obviously gives a reduced global minimizer. The second example is when

$$K \text{ is a line and } u \text{ is constant on each of the two components of } \mathbb{R}^2 \setminus K. \tag{15}$$

It is also easy to check that $(u, K) \in RGM$ when (15) holds, but for this the topological condition (3) is needed. The point is that G needs to separate the plane when (v, G) is a topological competitor for (u, K). Incidentally, notice that the values of the two constants in (15) do not matter, as in Remark 7.

If we did not have the topological condition, (15) would not define a global minimizer, even if we chose extremely different constants C_1, C_2 on the two components of $\mathbb{R}^2 \setminus K$. This is easy to check; one could do better than (u, K) by removing $K \cap B(0, R)$ for some very large R, and modifying u on $B(0, 2R)$ to smooth out the jump. A reasonably easy homogeneity argument shows that this would cost less than $C|C_2 - C_1|^2$ in the energy term, much less than the gain of $2R$ in length (see Exercise 20).

Our third example is when

$$K \text{ is a propeller and } u \text{ is constant on each} \atop \text{of the three components of } \mathbb{R}^2 \setminus K. \tag{16}$$

By propeller we mean the union of three (closed) half-lines with a common end-point, and that make $120°$ angles at that point. It is a little less obvious that

$(u, K) \in RGM$ when (16) holds, but it is true. The point is that if x_1, x_2, and x_3 are three points of the unit circle such that the three intervals I_1, I_2, and I_3 that compose $\partial B(0,1) \setminus \{x_1, x_2, x_3\}$ have equal lengths, then every compact subset G of the closed unit ball that separates the intervals I_j from each other in that ball must be such that $H^1(G) \geq 3$. See Lemma 59.1 for a proof.

We shall see in Section 58 that all the reduced global minimizers for which u is locally constant on $\mathbb{R}^2 \setminus K$, or even for which u is locally constant somewhere on $\mathbb{R}^2 \setminus K$, are as in (14), (15), or (16). Notice that since u is harmonic, if it is constant in a neighborhood of some point $x \in \mathbb{R}^2 \setminus K$, it must be constant on the component of x in $\mathbb{R}^2 \setminus K$.

There is a last known example, for which u is not locally constant. This is when we can find an orthonormal basis of $\mathbb{R}^2$ where

$$ K = \{(x,0) \,;\, x \leq 0\} \quad \text{and} \quad u(r\cos\theta, r\sin\theta) = \pm\sqrt{\frac{2}{\pi}}\, r^{1/2} \sin\frac{\theta}{2} + C \qquad (17) $$

for $r > 0$ and $-\pi < \theta < \pi$. Because of Remark 7, the values of the constant C and the constant sign $\pm$ do not matter. The fact that (17) defines a global minimizer is far from trivial, but true [BoDa]. See more about this in Section 62.

Conjecture 18. *All the reduced global minimizers in $\mathbb{R}^2$ are as in (14), (15), (16), or (17).*

It would be very interesting to have an appropriate generalization of this conjecture to higher dimensions, and more particularly in dimension 3. See Section 76 for a first attempt.

It will be convenient to call "exotic minimizers" the reduced global minimizers that are not as in our short list (14)–(17) of examples. Thus the conjecture says that there is no exotic minimizer.

Conjecture 18 is a variant of the Mumford-Shah conjecture on the minimizers of the Mumford-Shah functional in a domain (see Section 6). We shall check in Section 71 that it is a little stronger. The idea is that all the blow-up limits of reduced minimizers of the Mumford-Shah functional are reduced global minimizers (by Propositions 7.8 and 40.9), so if we know that all these blow-ups are as in (14)–(17), this gives enough information on the initial minimizer to conclude.

We don't know if the converse is true, because we don't know whether all reduced global minimizers can be realized as blow-up limits of reduced Mumford-Shah minimizers. We only know that this is the case for the examples (14) (trivially), (15) and (16) (by a calibration argument, see [ABD]). But we do not know whether one can find a domain Ω and a bounded function g on Ω such that some reduced minimizer for the Mumford-Shah functional in (2.13) (and with $a = b = c = 1$) has a blow-up limit as in (17) at a point $y \in \Omega$. And one could also imagine that there are exotic minimizers that cannot be realized as blow-up limits of reduced Mumford-Shah minimizers.

d. A few partial results

There are a few special cases where the conclusion of Conjecture 18 is known to hold. We start with a result of A. Bonnet.

Theorem 19 [Bo]. *If $(u, K) \in RGM(\mathbb{R}^2)$ and K is connected, then (u, K) is as in our short list* (14)–(17).

So there is no exotic minimizer with K connected. This is the main ingredient in the result that says that if (u, K) is a reduced minimizer for the Mumford-Shah functional in a smooth domain Ω and K_0 is an *isolated* connected component of K, then K_0 is a finite union of C^1 arcs [Bo]. The main reason is that then the blow-up limits of (u, K) at points of K_0 are reduced global minimizers for which the singular set is connected. We shall prove Theorem 19 in Section 61; the main ingredient is a monotonicity argument based on the lower bound in Section 47. The fact that isolated components of Mumford-Shah minimizers are finite union of C^1 arcs follows from (the proof) of Proposition 71.2 below.

This theorem is important in many respects. It is the first time one actually proved a full C^1 result on a whole component of K, as well as the first place (in the context of Mumford-Shah minimizers) where blow-up techniques were systematically used. It is also the first appearance of a monotonicity formula in this context. We shall use all these ideas a lot in this part.

There is no exotic minimizer (u, K) for which K is contained in a countable union of straight lines. This is proved in [Lé3]; the proof relies on a magic formula which allows us to compute the square of the complex derivative $\frac{\partial u}{\partial z}$ of u in $\mathbb{R}^2 \backslash K \approx \mathbb{C} \backslash K$, just in terms of (the Beurling transform of the Hausdorff measure on) K. We shall prove the formula in Section 63 and a special case of the theorem in Section 64.

The magic formula also proves the uniqueness of u given K (and assuming that $(u, K) \in RGM$), modulo the trivial transformations from Remark 7. The point is that $(\frac{\partial u}{\partial z})^2$ determines u in each component of $\mathbb{C} \backslash K$, modulo a sign (that comes from taking a square root) and an additive constant. Thus, if we are given a closed set K, we can compute a function u such that (u, K) is the only possible global minimizer (modulo trivial transformations) with the given set K, if there is one. However this does not tell us which sets K will give global minimizers.

If (u, K) is an exotic minimizer, then $\mathbb{R}^2 \backslash K$ is connected [DaLé]. In other words, the only pairs $(u, K) \in RGM$ for which $\mathbb{R}^2 \backslash K$ is not connected are our examples (15) and (16). We shall give an idea of the proof in Section 66. The proof also gives a reasonably precise description of each connected component K_0 of K when $(u, K) \in RGM$: K_0 is either a single point or a locally C^1 chord-arc tree with possible branching with $120°$ angles. See Section 57 for a first description.

Note that since $\mathbb{R}^2 \backslash K$ is connected when (u, K) is an exotic minimizer, the topological condition (3) is void. In other words, exotic minimizers are reduced local almost-minimizers in the plane, with gauge function $h = 0$.

It is also known that if (u, K) is an exotic minimizer, then all its limits under blow-ins (i.e., when $y_k = 0$ and t_k tends to $+\infty$ in Section 40) are exotic. More precisely, if $(u, K) \in RGM$ and one blow-in limit of K is empty, a line, a propeller, or a half-line, then (u, K) itself is of the same type. This is easy for the empty set (because singular sets are Ahlfors-regular). See Proposition 59.10 for the case of lines and propellers, and [BoDa] for the somewhat more complicated case of half-lines. Also see Section 62 for a rapid description of the argument.

A tempting way to try to understand global minimizers is to study their singular singular set, i.e., the set $K^\sharp$ of points of K near which K is neither a C^1 curve nor a C^1 spider. This is also the set of points x for which $\omega_2(x, r) = r^{-1} \int_{B(x,r) \backslash K} |\nabla u|^2$ does not tend to 0 with r. For instance, if (u, K) is an exotic global minimizer, then $K^\sharp$ is infinite, and even unbounded. Also, the singular singular set of a limit of reduced global minimizers is the limit of the singular singular sets. See Section 67 for details.

Surprisingly, the case of global minimizers in a half-plane (which corresponds to blow-ups of Mumford-Shah minimizers in Ω at regular points of the boundary) is easier, and the only possible global minimizers are when K is the empty set, or a half-line that emanates from the boundary and is perpendicular to it. There is also a similar result in cones with aperture $< \frac{3\pi}{2}$. See Section 66.c for a more precise description of the global minimizers in some cones, and Section 79 for its applications to the boundary behavior of Mumford-Shah minimizers in domains.

We shall see in Part G a few other applications of our study of global minimizers to local almost-minimizers in a domain.

Exercise 20. Set $G = \mathbb{R} \backslash [-1, 1] \subset \mathbb{R}^2$, and call $P_\pm = \{(x, y) \in \mathbb{R}^2 \,;\, \pm y > 0\}$.

1. Check that we can find $v \in W^{1,2}(\mathbb{R}^2 \backslash G)$ such that $v(x, y) = 0$ for $(x, y) \in P_- \backslash B(0, 2)$ and $v(x, y) = 1$ for $(x, y) \in P_+ \backslash B(0, 2)$. Set $\alpha = \int_{\mathbb{R}^2 \backslash G} |\nabla v|^2$.

2. Set $v_R(z) = v(z/R)$. Check that $v_R \in W^{1,2}(\mathbb{R}^2 \backslash RG)$, with $\int_{\mathbb{R}^2 \backslash RG} |\nabla v_R|^2 = \alpha$. Compare with our claim just above (16).

55 No bounded connected component in $\mathbb{R}^2 \backslash K$

In the rest of this part, we shall restrict to dimension 2, assume that (u, K) is a reduced global minimizer in the plane (in short, $(u, K) \in RGM(\mathbb{R}^2)$, see Notation 54.10), and try to prove various regularity results for (u, K).

Proposition 1. *If* $(u, K) \in RGM(\mathbb{R}^2)$, *then* $\Omega = \mathbb{R}^2 \backslash K$ *does not have any bounded connected component.*

This result extends to $\mathbb{R}^n$, with essentially the same proof. It is due to Bonnet [Bo], but the reader will see that it is very easy once we have regularity results. Also see Proposition 68.1 for a variant where (u, K) is a local almost-minimizer in a domain.

Let us prove the proposition by contradiction, and assume that Ω_0 is a bounded connected component of $\Omega = \mathbb{R}^2 \setminus K$. First observe that

$$u \text{ is constant on } \Omega_0, \tag{2}$$

because otherwise we could replace u with a constant on Ω_0. This would give a competitor for (u, K) because Ω_0 is bounded and we do not modify K (so (54.2) and (54.3) hold), and then (54.5) would fail because we reduced the energy.

Next we want to find a C^1 point in $\partial\Omega_0$. Recall from Definitions 51.19 and 51.2 that a C^1 point of K is a point $x \in K$ such that for $r > 0$ small enough, $K \cap \overline{B}(x, r)$ is a nice C^1 curve. This means that $K \cap \overline{B}(x, r)$ is a simple C^1 curve, with its two endpoints in $\partial B(x, r)$, and in addition it is a 10^{-1}-Lipschitz graph (after rotation, if needed). By Theorem 51.20,

$$H^1\text{-almost-every point of } K \text{ is a } C^1 \text{ point.} \tag{3}$$

On the other hand, $H^1(\partial\Omega_0) > 0$. This is very easy to see. For instance, pick two small balls B_1 and B_2 with the same radius ρ, and such that $B_1 \subset \Omega_0$ and $B_2 \subset \Omega \setminus \Omega_0$. Call L the line through the centers of B_1 and B_2. Then each line segment parallel to L and that meets B_1 and B_2 must also meet $\partial\Omega_0$. This means that if π is the orthogonal projection onto a line perpendicular to L, then $\pi(\partial\Omega_0)$ contains $\pi(B_1)$, which is a line segment of length 2ρ. Then $H^1(\partial\Omega_0) \geq 2\rho$, by Remark 2.11.

So we can find a C^1 point of K in $\partial\Omega_0$. Call it x. Also let r be such that $K \cap B(x, r)$ is a nice C^1 curve. Set $V = B(x, r) \setminus K$. We claim that

$$\begin{array}{c} \text{one of the two components of } V \text{ lies in } \Omega_0, \\ \text{and the other one in some other component of } \Omega. \end{array} \tag{4}$$

It is clear that one of the components of V lies in Ω_0, because otherwise x would not belong to $\partial\Omega_0$. If both components lay in Ω_0, we could simply remove $K \cap B(x, r)$ from K; this would not create a new jump, because of (2), and we would get a (strictly better) topological competitor for (u, K), because both sides of $\partial B(x, r)$ lie in Ω_0, so (54.3) already holds out of $B(x, r)$. [If not immediately convinced, see Lemma 20.10.] This is impossible because $(u, K) \in RGM$, and our claim follows.

At this point, we should probably mention a minor detail that we did not always address properly. By definition, we know that $K \cap B(x, r)$ is a C^1 curve, but actually

$$K \cap B(x, r) \text{ is a } C^{1+\alpha} \text{ curve for some } \alpha > 0. \tag{5}$$

We do not need to prove this, because our proof of (3) gives us nice C^1 disks where $K \cap B(x, r_0)$ is $C^{1+\alpha}$. However (5) is automatically true as soon as $B(x, r)$ is a nice C^1 disk. See Exercise 11. One can even take $\alpha = 1$ by [Bo], and a further study with the curvature equation shows that $K \cap B(x, r)$ is even C^∞ (because $h \equiv 0$ here). Actually, it is even an analytic curve [LeMo].

Next we want to check that

$$\nabla u \text{ is bounded on } B(x, r/2) \setminus K. \tag{6}$$

In fact, the restriction of u to $V = B(x,r) \setminus K$ is the function $f \in W^{1,2}(V)$ that has boundary values on $\partial B(x,r) \setminus K$ equal to u there, and for which $\int_V |\nabla f|^2$ is minimal, as in Section 15. This comes directly from the local minimality of (u, K) (consider competitors of the form (f, K)). Because of (5) in particular, we can apply Proposition 17.15. We get that the restriction of f to each of the two components of $V \cap B(x,r) = B(x,r) \setminus K$ has a C^1 extension up to the boundary $K \cap B(x,r)$, and of course (6) follows from this.

We are now ready to construct a competitor for (u, K). Let $t < r/2$ be small, to be chosen later, and set $B = B(x,t)$ and $G = K \setminus B(x,t/2)$. Call ∂_0 and ∂_1 the two components of $\partial B \setminus K$, with $\partial_0 \subset \Omega_0$ and ∂_1 contained in some other component of Ω (see (4)). Denote by m_1 the mean value of u on ∂_1. Recall that u is harmonic on Ω, so there is no difficulty with the existence of traces on ∂B, or the existence of tangential derivatives. By (6), we can find a smooth function v on $B \setminus G$ which coincides with u on ∂_1, is equal to m_1 on ∂_0, and for which

$$\int_{B \setminus G} |\nabla v|^2 \leq Ct^2. \tag{7}$$

Here C depends on an upper bound for $|\nabla u|$ in (6), but as long as it does not depend on t, we won't care.

We also set $v = m_1$ on $\Omega_0 \setminus B$ and $v = u$ on $\Omega \setminus [K \cup B \cup \Omega_0]$. Now v is defined on $\mathbb{R}^2 \setminus G$. Let us check that $v \in W^{1,2}_{loc}(\mathbb{R}^2 \setminus G)$. There is no problem in $B \setminus G$, or even near points of $\partial_0 \cup \partial_1$, because v is continuous across $\partial_0 \cup \partial_1$ (and by Corollary 14.28). The other points of $\mathbb{R}^2 \setminus G$ lie in the open sets $\Omega_0 \setminus \overline{B}$ or $\Omega \setminus [K \cup \overline{B} \cup \Omega_0]$, and there is no problem there either. So $(v, G) \in \mathcal{A}$.

Next we claim that (v, G) is a topological competitor for (u, K). Let R be so large that $\Omega_0 \cup B \subset B(0, R)$. Then we did not modify (u, K) out of $B(0, R)$ (i.e., (54.2) holds), and we just need to check the topological condition (54.3). Let y, $z \in \mathbb{R}^2 \setminus [K \cup \overline{B}(0, R)]$ be given, and suppose that G does not separate them. Let $\gamma \subset \mathbb{R}^2 \setminus G$ be a path from y to z. If γ does not meet B, then it does not meet K either (because $K \setminus G \subset B$), K does not separate y from z, and we are happy. If γ meets B, let y_1 denote the first point of ∂B when we run along γ, starting from y. Since the arc of γ between y and y_1 is contained in $\mathbb{R}^2 \setminus [G \cup B] \subset \mathbb{R}^2 \setminus K$, we see that y_1 lies in the same component of $\mathbb{R}^2 \setminus K$ as y. This forces y_1 to lie on ∂_1, because $\partial_0 \subset \Omega_0$, $\Omega_0 \subset B(0, R)$, and $y \notin B(0, R)$. Similarly, call z_1 the first point of ∂B when we run backwards along γ, starting from z. Then z_1 also lies on ∂_1, and in the same component of $\mathbb{R}^2 \setminus K$ as z. Since ∂_1 is connected, K does not separate y from z, (54.3) holds, and (v, G) is a topological competitor for (u, K). But

$$\begin{aligned} H^1(G \cap B(0, R)) &= H^1(K \cap B(0, R)) - H^1(K \cap B(x, r/2)) \\ &\leq H^1(K \cap B(0, R)) - t \end{aligned} \tag{8}$$

because $K \cap B(x, r/2)$ is a simple curve that goes from $\partial B(x, r/2)$ to x and back to $\partial B(x, r/2)$), while

$$\int_{B(0,R)\setminus G} |\nabla v|^2 \leq \int_{B(0,R)\setminus K} |\nabla u|^2 + \int_{B\setminus G} |\nabla v|^2 \leq \int_{B(0,R)\setminus K} |\nabla u|^2 + Ct^2 \quad (9)$$

because out of B, ∇v is either zero or equal to ∇u, and by (7). When we add (8) and (9), and if t is small enough, we get a contradiction with the minimality condition (54.5). This proves the proposition. $\qquad \square$

Remarks 10. In dimension 2, there are stronger results than Proposition 1.

We shall see in Proposition 56.1 an improvement of Proposition 1 (in dimension 2 only): all the components of Ω are reasonably fat, in the sense that they contain disks of arbitrarily large sizes.

Also, if K is not a line or a propeller, $\Omega = \mathbb{R}^2 \setminus K$ is connected, so it cannot have bounded or thin components. [See Section 66.]

In both cases, the proof is somewhat more complicated, and does not extend to higher dimensions.

Given the fact that the idea of the proof is very simple, it is a little sad that we used the C^1 theorem (3). It looks as if it was for convenience, but the truth is that the author does not know any clean proof that only uses basic facts. See Exercise 12 for a proof that "only" uses uniform rectifiability (and only works in dimension 2). This makes a nice review exercise, but it is not even convincingly simple. Exercise 13 gives a slightly easier argument, but that uses Proposition 40.9 on blow-up limits, or Bonnet's decay estimate from Section 47.

Exercise 11. We want to check that (5) holds when $(u, K) \in RGM$ and $B(x, r)$ is a nice C^1 disk.

1. First a brutal proof: Show that for any $y \in K \cap B(x, r)$, we can apply Corollary 52.25 to the pair (y, r) for r small, with $J(y, r)^{-1} = 0$ because of Remark 52.27.

2. For another proof, let $y \in K \cap B(x, r)$ be given, and set $E(t) = \int_{B(y,t)\setminus K} |\nabla u|^2$ for t small. Show that for each constant $a < 2$, there is a $t_0 > 0$ such that $E'(t) \geq aE(t)/r$ for almost-every $t \leq t_0$. [Use Proposition 47.7.]

3. Deduce from this (and Lemma 47.4) that $E(t) \leq C(a, y)t^a$ and $\omega_2(y, t) \leq C(a, y)t^{a-1}$ for t small. Conclude. (Use (the proof of) Corollary 51.17.)

Exercise 12. We want to see how Proposition 1 can be proved without using the full C^1 theorem. But instead we shall use the uniform rectifiability of K. We still assume that Ω_0 is a bounded component of $\Omega = \mathbb{R}^2 \setminus K$.

1. Pick $x \in \partial\Omega_0$ and let $r < \text{diam}(\Omega_0)/2$ be given. Show that there is a component Ω_1 of Ω, different from Ω_0, and which meets both $B(x, r/2)$ and $\partial B(x, r)$. Hint: if not, change the constant values of u on the components of

Ω that meet $B(x, r/2)$, and then notice that you can remove $K \cap B(x, r/2)$ from K.

2. Show that (for x and r as above) $H^1(\partial\Omega_0 \cap B(x, r)) \geq r/10$.

3. Fix $p \in (1, 2)$ ($p = 3/2$ will do). Show that for each $\varepsilon > 0$ we can find a point $y \in \partial\Omega_0 \cap B(x, r)$ and a radius $t \in (0, r)$ such that $\omega_p(y, t) + \beta_K(y, t) \leq \varepsilon$. Hint: use the proof of Corollary 23.38 and Corollary 41.5.

4. Choose $t_1 \in (t/2, t)$ such that $\omega_p^*(y, t_1) \leq C\varepsilon$ (see (42.5) and maybe Convention 42.6). Set $B_1 = B(y, t_1)$, and call Z a $10^{-2}t$-neighborhood of $K \cap \partial B_1$ in ∂B_1. Show that Z (or equivalently $\partial B_1 \setminus Z$) has two components (instead of just one). Hint: to construct a competitor, look at the proof of Proposition 42.10, for instance.

5. Call ∂_1 and ∂_2 the two components of $\partial B_1 \setminus Z$, and assume that they lie in the same component Ω_1 of Ω. Check that $\Omega_1 \neq \Omega_0$. Show that the normalized jump $J^*(y, t_1)$ from (42.7) is not too small. Hint: Proposition 42.10.

6. Continue with the same assumption as in Question 5. Show that $H^1(K \cap B_1) \geq 21t_1/10$. Hint: you do not need the co-area theorem here, just the fact that most "vertical" rays from the component of ∂_1 in $\partial B_1 \setminus K$ to its analogue for ∂_2 meet K at least once (because of the jump), and, almost half of the time, at least twice (because there is a curve in Ω_0 that goes from close to y to ∂B_1).

7. Check that ∂_1 and ∂_2 lie in different components of Ω, and that one of these components is Ω_0.

8. Use again Lemma 22.16 and the construction of Proposition 1 to get the desired contradiction.

9. Why does this proof fail in higher dimensions?

Exercise 13. Yet another approach to Proposition 1. Still consider $(u, K) \in RGM(\mathbb{R}^2)$, and assume that Ω_0 is a bounded component of $\Omega = \mathbb{R}^2 \setminus K$.

1. Show that $\partial\Omega_0$ contains a (nontrivial) simple rectifiable curve γ. [You may use Theorem 14.3 on p. 123 of [Ne] (see Lemma 29.51) to get a connected piece, and then Proposition 30.17.] Call $z : I \to K$ a parameterization of γ by arclength.

2. Show that for almost every point $x \in \gamma$, we can find $t \in I$ such that $x = z(t)$ and the derivative $z'(t)$ exists. Show that γ has a tangent at such a point.

3. Show K also has a tangent line at almost-every point x as above. You will need to know that

$$t^{-1}H^1([K \setminus \gamma] \cap B(x, t)) \text{ tends to zero for } H^1\text{-almost every } x \in \gamma, \quad (14)$$

and also to use the Ahlfors-regularity of K. See Theorem 6.2 in [Mat2] for (14), and Exercise 41.21 for hints on the other part.

4. Now let (v, G) denote a blow-up limit of (u, K) at x. See the beginning of Section 40 for definitions, Proposition 37.8 for existence, and Proposition 40.9 for some extra information that you will need. Show that G is a line (use **2.**) and that v is locally constant (if in trouble, see Lemma 59.6).

5. Show that $\lim_{t \to 0} \omega_2(x, t) = 0$. Hint: (38.50).

6. Adapt the rest of the proof in Exercise 12. Some verifications are easier, because we already know that γ crosses small balls centered at x.

7. Find another way to show that $\lim_{t \to 0} \omega_2(x, t) = 0$, without using blow-ups. Suggestion: here again, use the fact that γ crosses small balls $B(x, t)$, and (14). This will allow you to show that the complement of the set $\mathcal{B}$ in Proposition 47.7 is very small, and then that $\omega_2(x, t)$ decays fast enough (almost as in Exercise 11).

56   No thin connected component in $\mathbb{R}^2 \setminus K$ either; John domains

We continue with $(u, K) \in RGM(\mathbb{R}^2)$ and improve slightly the result from the previous section. This time we want to show that all the components of $\Omega = \mathbb{R}^2 \setminus K$ contain disks of arbitrarily large radii. We shall also rapidly discuss a consequence, the fact that all the components of Ω are John domains with center at infinity.

Proposition 1 [BoDa]. *There is a constant C such that if $(u, K) \in RGM(\mathbb{R}^2)$, Ω_0 is a connected component of $\mathbb{R}^2 \setminus K$, $x \in \Omega_0$, and $r > 0$, then $\Omega_0 \cap B(x, r)$ contains a disk of radius $C^{-1} r$.*

Note that we shall see later (and independently) that Ω is connected when K is not a line or a propeller, and Proposition 1 will be a trivial consequence. We still include the original proof from [BoDa] because it is shorter and uses the following nice application of Carleson measures. Also see Lemma 68.14 for a local version in a domain $\Omega \subset \mathbb{R}^2$.

Lemma 2. *For each $C_0 > 0$ there is a constant $C_1 > 0$ such that if $(u, K) \in RGM(\mathbb{R}^2)$, $x \in K$, $r > 0$, E is a (measurable) subset of $K \cap B(x, r)$, and $H^1(E) \geq C_0^{-1} r$, then we can find $y \in E$ and $t \in (C_1^{-1} r, 10^{-1} r)$ such that $K \cap B(y, t)$ is a nice C^1 curve.*

See Definition 51.2 for nice C^1 curves. Lemma 2 extends to higher dimensions, with the same argument (and Corollary 51.17 replaced with Theorem 75.2 at the end). Fix $p \in (1, 2)$. We start with (41.15) and (41.16), which say

$$[\beta_K^3(y, t) + \omega_p(y, t)] \frac{dH^1(y) dt}{t} \text{ is a Carleson measure on } K \times (0, +\infty). \quad (3)$$

Also see (23.1) and (23.2) for the definitions, and note that the restriction to $t \leq 1$ came from the fact that we had a possibly nonzero gauge function h. Here we could also get the result without constraints on t from the statement of (41.15) and (41.16), by invariance of RGM under dilations. Recall from Definition 23.13 that (3) says that

$$\int_{y \in K \cap B(x,r)} \int_{t \in (0,r)} [\beta_K^3(y,t) + \omega_p(y,t)] \frac{dH^1(y)dt}{t} \leq C_p \, r. \tag{4}$$

Let us check that for each $\varepsilon > 0$ we can find $C_2 = C_2(\varepsilon, p, C_0)$ such that if E is as in the statement of Lemma 2, we can find $y \in E$ and $t \in (C_2^{-1}r, 10^{-1}r)$ such that

$$\beta_K(y,t) + \omega_p(y,t) \leq \varepsilon. \tag{5}$$

Indeed, otherwise the contribution of $E \times (C_2^{-1}r, 10^{-1}r)$ to the left-hand side of (4) would be at least $\varepsilon^3 H^1(E) \int_{C_2^{-1}r}^{10^{-1}r} \frac{dt}{t} \geq \varepsilon^3 C_0^{-1} r \, Log\left(\frac{C_2}{10}\right)$, which is larger than $C_p \, r$ if C_2 is chosen large enough.

So we can find $y \in E$ and $t \in (C_2^{-1}r, 10^{-1}r)$ such that (5) holds. If ε was chosen small enough, Corollary 51.17 tells us that $K \cap B(y, t/2)$ is a nice C^1 curve. [Recall that $h = 0$ here.] Lemma 2 follows, with $C_1 = 2C_2$. $\square$

Let Ω_0, x, and r be as in Proposition 1. If $B(x, r/4) \subset \Omega_0$, there is nothing to prove. So we can assume that $\partial\Omega_0$ meets $B(x, r/4)$. Set $E = \partial\Omega_0 \cap B(x, 3r/4)$, and let us check that

$$H^1(E) \geq r/C. \tag{6}$$

If Ω_0 is the only component of Ω that meets $B(x, r/2)$, then $K \cap B(x, r/2) \subset \partial\Omega_0$ (because Ω is dense), and $H^1(E) \geq H^1(K \cap B(x, r/2)) \geq r/C$, by Ahlfors-regularity of K. Otherwise, $B(x, r/2)$ meets a component Ω_1 of Ω other than Ω_0. Pick $x \in \Omega_1 \cap B(x, r/2)$. Proposition 55.1 says that Ω_1 is unbounded, so there is a curve $\gamma_1 \subset \Omega_1$ that goes from x_1 to $\partial B(x, r)$. Similarly, there is a curve $\gamma_0 \subset \Omega_0$ that goes from $B(x, r/4)$ to $\partial B(x, r)$. For each radius $\rho \in r/2, 3r/4$, the circle $\partial B(x, \rho)$ meets both γ_j, $j = 0, 1$, and so it meets $\partial\Omega_0$. Then the image of $E = \partial\Omega_0 \cap B(x, 3r/4)$ by the radial projection $z \to |z - x|$ contains an interval of length $r/4$, and Remark 2.11 says that $H^1(E) \geq r/4$. Thus (6) holds in all cases.

Lemma 2 gives us a point $y \in E$ and a radius $t \in (r/C_1, r/10)$ such that $K \cap B(y, t)$ is a nice C^1 curve. Both components of $B(y, t) \setminus K$ contain a disk of radius $r/(10C_1)$ and, since $y \in E \subset \partial\Omega_0$, at least one of these components is contained in Ω_0. The corresponding disk is contained in $\Omega_0 \cap B(x, r)$, and Proposition 1 follows.
                    $\square$

Note that our proof does not work in higher dimensions, because the existence of γ_1 and γ_2 does not give any good lower bound on the surface measure of $\partial\Omega_0 \cap B(x, r)$. It would be interesting to know whether Proposition 1 and Proposition 7 below generalize to higher dimensions.

Our next result is a little more precise than Proposition 1. In effect, it says that every connected component Ω_0 of $\Omega = \mathbb{R}^2 \setminus K$ is a John domain centered at infinity. This means that every point of Ω_0 can be connected to infinity by a curve that stays proportionally far from K. The precise statement is as follows.

Proposition 7 [BoDa].   *There is a constant $C \geq 1$ such that if $(u, K) \in RGM(\mathbb{R}^2)$, Ω_0 is a connected component of $\mathbb{R}^2 \setminus K$, and $x_0 \in \Omega_0$, there is a Lipschitz mapping $z : [0, +\infty) \to \Omega_0$ such that $z(0) = x_0$,*

$$z \text{ is 1-Lipschitz, i.e., } |z(s) - z(t)| \leq |s - t| \text{ for } s, t \in [0, +\infty), \tag{8}$$

and

$$\operatorname{dist}(z(t), K) = \operatorname{dist}(z(t), \partial\Omega_0) \geq C^{-1}\{t + \operatorname{dist}(x_0, K)\} \text{ for } t \geq 0. \tag{9}$$

See Proposition 68.16 for a local version in a domain Ω.

We call (the image of) z an escape path from x_0 to infinity. Notice that indeed (9) forces $z(t)$ to go to infinity, almost as fast as t. The union of the balls $\frac{1}{2C} B\big(z(t), t + \operatorname{dist}(x_0, K)\big)$, $t \geq 0$, forms a nice nontangential access region to x_0 in Ω_0. In this respect, the case when x_0 is very close to K (and, at the limit, lies on $\partial\Omega_0$) is perhaps the most interesting.

Proposition 7 is a special case of Lemma 20.1 and Remark 20.5 in [BoDa]. It was used there to give a good control on the variations of the function u and its conjugate harmonic function v, in particular when we approach the boundary K. [See Section 21 in [BoDa].] The main step of the argument is to find a first path γ, as follows.

Lemma 10.   *There is a constant C_3 such that if $(u, K) \in RGM(\mathbb{R}^2)$, Ω_0, and $x_0 \in \Omega_0$ are as above, and if $d_0 = \operatorname{dist}(x_0, K)$, there is a curve γ such that*

$$\gamma \subset \Omega_0 \cap B(x_0, C_3 d_0), \ \operatorname{dist}(\gamma, K) \geq C_3^{-1} d_0, \tag{11}$$

and

$$\gamma \text{ goes from } x_0 \text{ to some point } x_1 \in \Omega_0 \text{ such that } d_1 = \operatorname{dist}(x_1, K) \geq 2 d_0. \tag{12}$$

We shall proceed by contradiction and compactness. First, it is enough to prove the lemma when $x_0 = 0$ and $d_0 = 1$. If the lemma fails, then for each $k \geq 0$, we can find $(u_k, K_k) \in RGM(\mathbb{R}^2)$ and a component $\Omega_{k,0}$ of $\mathbb{R}^2 \setminus K_k$ such that $0 \in \Omega_{k,0}$, $\operatorname{dist}(0, K_k) = 1$, and the lemma fails for $x_0 = 0$ with the constant $C_3 = 2^k$, say.

Next we use Proposition 37.8 to find a subsequence, which we still call $\{(u_k, K_k)\}$ for convenience, that converges to some limit (u, K). Proposition 40.9 tells us that (u, K) is a reduced global minimizer. Note that $\operatorname{dist}(0, K) = 1$ because $\{K_k\}$ converges to K. Call Ω_0 the component of 0 in $\mathbb{R}^2 \setminus K$; Proposition 1 tells us that Ω_0 contains a disk of radius 3. Call γ a path in Ω_0 that goes from 0 to the center of this disk, and set $\delta = \operatorname{dist}(\gamma, K) > 0$. If k is large enough, dist $(\gamma, K_k) \geq \delta/2$ and the endpoint of γ lies at a distance larger than 2 from K_k, because $\{K_k\}$ converges to K. Then γ satisfies (11) and (12) with respect to K_k, and with $C_3 = \text{Max}\{2/\delta, \text{diam}(\gamma)\}$. This gives the desired contradiction with the definition of $\{(u_k, K_k)\}$. Lemma 10 follows.   □

Once we have the curve γ in Lemma 10, it is easy to iterate the construction and get the desired escape curve. Before we do this, let us replace γ with a curve γ_0 with the same endpoints, that satisfies (11) with the constant $2C_3$, and which in addition is smooth and has a total length less than $C_4 d_0$.

Now we continue γ_0 with a curve γ_1 that starts at x_1, ends at a point $x_2 \in \Omega_0$ such that $d_2 = \text{dist}(x_2, K) \geq 2d_1$, stays at distance greater than $(2C_3)^{-1} d_1$ from K, is contained in $B(x_1, 2C_3 d_1)$, and whose total length is at most $C_4 d_1$. And we iterate the construction.

The image of z is the union of all the curves γ_j, $j \geq 0$, that we construct this way. The simplest way to define z is to take the arclength parameterization, i.e., parameterize the γ_j successively and by arclength. Then (8) is automatic, and the verification of (9) is straightforward. Since we do not intend to use Proposition 7 here, and anyway the proof on pages 126–127 of [BoDa] uses almost the same notation and is reasonable, we skip the details. This completes our proof of Proposition 7.   □

Remark 13. If we could extend Proposition 1 to higher dimensions, the analogue of Proposition 7 would follow easily, by the same argument as here.

57   Every nontrivial connected component of K is a chord-arc tree

Suppose $(u, K) \in RGM(\mathbb{R}^2)$ and let K_0 be a connected component of K. We cannot exclude the case when K_0 is reduced to one point, but otherwise we shall prove that K_0 is a chord-arc tree. It is even composed of C^1 arcs that meet with angles of 120°; we shall say a few words about this at the end of this section, but with no proof.

Proposition 1. *If $(u, K) \in RGM(\mathbb{R}^2)$, K_0 is a connected component of K, and K_0 is not reduced to one point, then K_0 is a chord-arc tree. This means that if x and y are two distinct points of K_0, there is a unique simple curve $\gamma_{x,y} \subset K_0$ that goes from x to y, and*

$$\text{length}(\gamma_{x,y}) \leq C|x - y|. \tag{2}$$

The constant C is absolute.

This first showed up in [Bo], maybe with a slightly different statement, and there are a few different proofs. We start with the observation that

$$K \text{ does not contain any (nontrivial) simple loop,} \tag{3}$$

which is a simple consequence of Proposition 55.1: if γ is a nontrivial simple loop in K, then there is at least one connected component of $\mathbb{R}^2 \setminus K$ that is contained in the bounded component of $\mathbb{R}^2 \setminus \gamma$, and this component is bounded too.

Next let x and y be distinct points of K_0. Proposition 30.17 says that there is a simple curve $\gamma_{x,y} \subset K_0$ that goes from x to y. Let us check that it is unique. Suppose we have two simple curves γ_1 and γ_2 in K, that go from x to y. For $i = 1, 2$, call $z_i : I_i \to K$ an injective parameterization of γ_i.

Let us first verify that if $z_1(I_1) \subset z_2(I_2)$, the two arcs γ_i are equivalent. Since I_2 is compact, z_2 is a homeomorphism from I_2 to $z_2(I_2)$. Since $z_1(I_1) \subset z_2(I_2)$, we can set $f_1 = z_2^{-1} \circ z_1$, and it is a continuous injective mapping from I_1 to a subinterval of I_2. But the extremities of that interval are $z_2^{-1}(x)$ and $z_2^{-1}(y)$, which are the extremities of I_2. So f_1 is a homeomorphism from I_1 to I_2, $z_2 \circ f_1 = z_1$, and γ_1 is equivalent to γ_2.

We are left with the case when $z_1(I_1)$ is not contained in $z_2(I_2)$. Pick $w \in z_1(I_1) \setminus z_2(I_2)$, call y_1 the first point of γ_2 that we hit when we go from w to y along γ_1, and let x_1 denote the first point of γ_2 that we hit when we run backwards on γ_1 from w to x. If we put together the arc of γ_1 from x_1 to y_1 and the arc of γ_2 from y_1 back to x_1, we get a simple loop in K, because the γ_i are simple, and our two pieces are essentially disjoint. This loop is nontrivial because $y_1 \neq w$. Its existence contradicts (3), and this completes our verification of uniqueness for $\gamma_{x,y}$. Proposition 1 will thus follow from the next lemma.

Lemma 4. *There is an absolute constant C such that if γ is a simple arc in K, then γ is chord-arc, with constant $\leq C$.*

See Proposition 68.18 for a local version in a domain.

The conclusion means that if x, y lie on γ and $\gamma_{x,y}$ is the arc of γ between x and y, then length $(\gamma_{x,y}) \leq C|x - y|$. Let us also mention here that we used the length of γ, and not the Hausdorff measure of its support, mostly out of affectation. Length is supposed to be simpler, because it can be measured directly from a parameterization (with subdivisions). It is the total variation of an (absolutely continuous) parameterization of γ. However, it is very easy to check that $H^1(\gamma) \leq$ length (γ) for any curve, and a little more difficult to show that $H^1(\gamma) = $ length (γ) when γ is simple. See Exercise 31, or the argument from (63.10) to (63.15), or Lemma 3.2 on p. 29 of [Fa]. If you do not want to use this fact, just replace length with Hausdorff measure everywhere.

As was said earlier, there are a few slightly different proofs of this. See [Bo] for the first one, Exercise 29 for a proof based on Proposition 56.7, and [DaLé] for the only proof so far without any compactness argument. This last one is based on Section 47 and a monotonicity argument.

Here we shall prove the lemma by contradiction and compactness. We shall try to avoid using strong regularity results for minimizers, but the argument would be a tiny bit shorter if we did. So we assume that for each $k \geq 0$, there is a pair $(u_k, K_k) \in RGM(\mathbb{R}^2)$, a simple arc γ_k in K_k, and points x_k, y_k of K_k such that the length of the arc of γ_k between x_k and y_k is larger than $2^k |x_k - y_k|$. We may as well assume that x_k and y_k are the extremities of γ_k, because anyway the arc of γ_k between x_k and y_k is still a simple arc in K_k. By translation and dilation invariance of RGM, we can also assume that $x_k = 0$ and $\mathrm{diam}(\gamma_k) = 1$. Note that

$$\mathrm{length}\,(\gamma_k) = H^1(\gamma_k) \leq H^1(K_k \cap \overline{B}(0,1)) \leq C, \tag{5}$$

by the easy part of Ahlfors-regularity for K_k, so $|y_k| = |x_k - y_k| \leq 2^{-k}\,\mathrm{length}\,(\gamma_k) \leq 2^{-k}C$, and $\{y_k\}$ tends to 0. Let $z_k : [0,1] \to K_k$ denote a Lipschitz parameterization of γ_k. By (5), we can choose z_k with a Lipschitz constant less than C, and then we can extract a subsequence (which we still denote as the initial sequence, as usual) so that $\{z_k\}$ converges to a Lipschitz mapping z. Let us also use Proposition 37.8 to extract our subsequence so that $\{(u_k, K_k)\}$ converges to some limit (u, K). Note that (u, K) is a reduced global minimizer, by Proposition 40.9, and that $z([0,1]) \subset K$, by definition of a limit.

We cannot conclude immediately, because we do not know that z defines a simple loop. This is not so bad, but we want to show that $\gamma = z([0,1])$ bounds some component of $\mathbb{R}^2 \setminus K$. In particular, we should exclude the case where z merely runs along a single arc of curve, once in each direction. So we shall try to find a place where γ and the γ_k should be a little nicer.

First observe that $z'(t)$ exists for almost every $t \in [0,1]$, because z is Lipschitz. [See for instance Proposition 9.1 and Exercise 9.17.] Moreover, if $X = \{t \in [0,1]\,;\, z'(t) = 0\}$, then $H^1(z(X)) = 0$, by Sard's theorem. [See Exercise 33.41, but of course things are slightly simpler in dimension 1.] Set

$$E = \{t \in [0,1]\,;\, |z(t)| \geq 1/2,\ z'(t) \text{ exists, and } z'(t) \neq 0\}. \tag{6}$$

Then $H^1(z(E)) \geq 1/2$, because $\mathrm{diam}(\gamma) = \lim_{k \to +\infty} \mathrm{diam}(\gamma_k) = 1$, and $z(0) = 0$. We claim that for almost every point $w \in z(E)$,

$$K \text{ has a tangent line at } w. \tag{7}$$

We can deduce this from Theorem 51.20 (which says that for almost every point $w \in K$, there is a radius $r > 0$ such that $K \cap B(w,r)$ is a nice C^1 curve), but we could also prove this directly. Let us merely indicate how we could do it, because the reader probably knows anyway, and we already have the other (heavy) proof. First, it is enough to show that $z(E)$ has a tangent line at almost every point. The point is that the upper density $\limsup_{r \to 0} \left[r^{-1} H^1(B(w,r) \cap K \setminus z(E)) \right]$ is equal to 0 for almost every $w \in z(E)$. [See Theorem 6.2 in [Mat2].] For such a point, K has a tangent line at w as soon as $z(E)$ has a tangent at w. This uses the Ahlfors-regularity of K, a little bit as in the proof of (41.12). See Exercise 41.21. Then we also need to check that $z(E)$ has a tangent at almost every point. This is not an immediate consequence of the fact that $z'(t)$ exists and $z'(t) \neq 0$ for $t \in E$, because z is not injective and different pieces of $z(E)$ could interact, but we can get it by the same sort of density argument as above. [See Exercise 30.] And it is a standard fact about rectifiable curves.

Choose $t \in E$ such that (7) holds for $w = z(t)$, and then $r < 10^{-1}$ so small that

$$|z(s) - z(t) - (s-t)z'(t)| \leq 10^{-1}|s-t||z'(t)|$$
$$\text{for all } s \in [0,1] \text{ such that } |s-t||z'(t)| \leq 2r, \tag{8}$$

and also such that $\beta_K(w, 2r) \leq \varepsilon$, where $\varepsilon > 0$ will be chosen soon. [See (41.2) for the definition of β_K.] Our plan for the near future is to get the following description for z. There is a small interval J centered at t such that the curve $z(J)$ crosses $B(w,r)$ reasonably neatly (because of (8)), and then $z([0,1] \setminus J)$ never comes back to $B(w, r/2)$. Later on, we shall prove that $z([0,1])$ bounds some component of $\mathbb{R}^2 \setminus K$ and get our contradiction.

We first try to show a similar behavior for the curves γ_k. Set $w_k = z_k(t)$; then w_k tends to w because $\{z_k\}$ tends to z, and

$$\beta_{K_k}(w_k, r) \leq 3\varepsilon \quad \text{for } k \text{ large enough,} \tag{9}$$

because $\beta_K(w, 2r) \leq \varepsilon$ and K_k tends to K. Lemma 52.1 says that

$$H^1(K_k \cap B(w_k, r)) \leq 2r + C\beta_{K_k}(w_k, r)^{1/2}\, r \leq 2r + C\varepsilon^{1/2}r. \tag{10}$$

Recall that $|w| \geq 1/2$ by definition of E, so $B(w_k, r)$ lies reasonably far from the origin. Since γ_k starts from 0, goes through w_k, and ends at y_k (which is very close to 0), we can define points y_k' and x_k' as follows: y_k' is the first point of $\partial B(w_k, r)$ when we run along γ_k from w_k to y_k, and x_k' is the first point of $\partial B(w_k, r)$ when we run backwards along γ_k from w_k to 0. See Figure 1. Call γ_k' the open arc of γ_k between x_k' and y_k'. Thus $\gamma_k' \subset B(w_k, r)$.

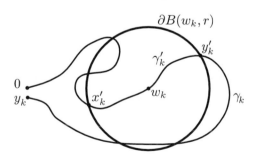

Figure 1

Notice that $H^1(\gamma_k') \geq 2r$ because γ_k is simple and x_k' and y_k' lie on $\partial B(w_k, r)$. Since $\gamma_k' \subset K_k \cap B(w_k, r)$ by definition, (10) says that

$$H^1(K_k \cap B(w_k, r) \setminus \gamma_k') = H^1(K_k \cap B(w_k, r)) - H^1(\gamma_k') \leq C\varepsilon^{1/2}r. \tag{11}$$

Let us verify that

$$\gamma_k \setminus \gamma_k' \text{ does not meet } B(w_k, r/2). \tag{12}$$

Indeed $\gamma_k \setminus \gamma_k'$ is composed of two arcs, each with an extremity that lies outside of $B(w_k, r)$ (namely, at 0 and y_k). If $\gamma_k \setminus \gamma_k'$ meets $B(w_k, r/2)$, one of these arcs must cross the annulus $B(w_k, r) \setminus B(w_k, r/2)$ entirely, and its contribution to the left-hand side of (11) is at least $r/2$ (by Remark 2.11). This contradicts (11) if ε is chosen small enough, and (12) follows.

Let us try to interpret (12) in terms of z. Some control on the parameterizations will be useful. Write $x_k' = z_k(t_k^-)$ and $y_k' = z_k(t_k^+)$. We can extract a new subsequence, so that $\{t_k^\pm\}$ converges to a limit $t^\pm$. Let us check that

$$t^- < t < t^+, \quad \text{and} \quad r/2 \leq |t^\pm - t|\,|z'(t)| \leq 2r. \tag{13}$$

First recall that $w_k = z_k(t)$ lies between x_k' and y_k' on γ_k, so $t_k^- < t < t_k^+$. Since $|w_k - x_k'| = |w_k - y_k'| = r$ and z_k is C-Lipschitz, we even get that $t_k^- \leq t - C^{-1}r$ and $t_k^+ \geq t + C^{-1}r$. The first part of (13) follows. For the second part, let k be so large that $|z(s) - z_k(s)| \leq 10^{-1}r$ for $s \in [0, 1]$. First let us only consider s such that $|s - t|\,|z'(t)| \leq 2r$. Then (8) says that

$$|z_k(s) - z_k(t) - (s - t)z'(t)| \leq 10^{-1}|s - t|\,|z'(t)| + 2 \cdot 10^{-1}r \leq 4 \cdot 10^{-1}r. \tag{14}$$

In particular, we get that

$$|z_k(s) - z_k(t)| \leq |(s - t)z'(t)| + 4 \cdot 10^{-1}r < r \quad \text{when } |s - t|\,|z'(t)| \leq r/2 \tag{15}$$

and

$$|z_k(s) - z_k(t)| \geq |(s - t)z'(t)| - 4 \cdot 10^{-1}r > r \quad \text{when } |s - t|\,|z'(t)| = 2r. \tag{16}$$

This proves that $r/2 < |t_k^\pm - t|\,|z'(t)| < 2r$ (by definition of x_k' and y_k'); (13) follows.

Set $J = [t^-, t^+]$, and let us verify that

$$z([0, 1] \setminus J) \text{ does not meet } B(w, r/2). \tag{17}$$

Let $s \in [0, 1] \setminus J$ be given. By definition, of t^- and t^+, $s \in [0, 1] \setminus [t_k^-, t_k^+]$ for k large, so (12) says that $z_k(s)$ lies outside of $B(w_k, r/2)$. Then $z(s) \notin B(w, r/2)$, because w_k tends to w and $z_k(s)$ tends to $z(s)$. This proves (17).

We have finally arrived at the description of z that was alluded to before. Note that the inequality in (8) holds for all $s \in J$, because of (13). Thus the behavior of z in J is essentially linear, and $z(J)$ crosses $B(w, r)$ (see more details later). In addition, (17) says that $z(s)$ does not return to $B(w, r/2)$ after leaving J. We want to show that

$$\text{at least one of the connected components of } \mathbb{R}^2 \setminus z([0, 1]) \text{ is bounded.} \tag{18}$$

This will give the desired contradiction because if W is a bounded component of $\mathbb{R}^2 \setminus z([0, 1])$ and ξ is any point of $W \setminus K$, then the component of ξ in $\mathbb{R}^2 \setminus K$ is contained in W because $z([0, 1]) \subset K$. This is impossible, by Proposition 55.1.

There are a few ways to check (18). The simplest is perhaps to find two points $\xi^\pm$ of $B(w, r/2) \backslash z([0, 1])$ for which the indices relative to $z([0, 1])$ are different. [See Figure 2.] These points will necessarily lie in different components of $\mathbb{R}^2 \backslash z([0, 1])$, because indices are locally constant on $\mathbb{R}^2 \backslash z([0, 1])$, and then there is at most one unbounded component. Then (18) will follow. We choose ξ^+ and ξ^- in $B(w, r/2)$, at distance $r/4$ from the line L through w and parallel to $z'(t)$, and on different sides of L. By (8), $z(J)$ stays at distance less than $r/8$ from L, so it does not get close to $\xi^\pm$. Moreover the index of $\xi^\pm$ relative to $z([0, 1])$ does not change when we replace $z(J)$ with the line segment $[z(t^-), z(t^+)]$, parameterized with constant speed. This is because we can find a homotopy from (our parameterization of) $z(J)$ to this new curve, such that none of the intermediate images ever get close to $\xi^\pm$.

The segment $[\xi^-, \xi^+]$ crosses $[z(t^-), z(t^+)]$ transversally. Indeed $z(t^\perp) \in \partial B(w, r)$ because $x_k' = z_k(t_k^-)$ and $y_k' = z_k(t_k^+)$ lie on $\partial B(w_k, r)$, w_k tends to w, $t_k^\pm$ tends to $t^\pm$, and the z_k are uniformly Lipschitz and tend to z; in addition $z(t^+)$ and $z(t^-)$ almost lie opposite to each other in $\partial B(w, r)$, for instance by (8). Altogether the index relative to the new curve changes by exactly 1 when one goes from ξ^+ to ξ^- across $[z(t^-), z(t^+)]$, and (18) follows.

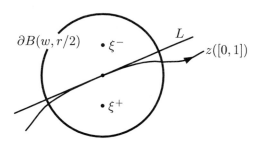

Figure 2

One could also show (18) by finding a nontrivial simple loop in $z([0, 1])$. Let us first sketch the argument when $K \cap B(w, r)$ is a nice C^1 curve (which we may assume, if we are ready to use Theorem 51.20). Since $z(0) = z(1) = 0$, we can find s_1 and $s_2 \in [0, 1]$, with $0 < s_1 < t < s_2 < 1$, such that $z(s_i) \in \partial B(w, r/3)$. Since z is not necessarily injective, there may be a few different choices of s_i like this, but let us take s_1 as small and s_2 as large as possible. Then $G = z([0, 1] \backslash [s_1, s_2])$ stays out of $B(w, r/3)$, by definition of the s_i. Since $z(0) = z(1) = 0$, G is connected, and $H^1(G) < +\infty$ because z is Lipschitz. By Proposition 30.14, we can find a simple arc Γ_1 in G that goes from $z(s_1)$ to $z(s_2)$.

Call Γ_2 the open arc of $K \cap B(w, r/3)$ between $z(s_1)$ and $z(s_2)$; Γ_2 does not meet Γ_1, because Γ_1 stays out of $B(w, r/3)$. Since in addition Γ_1 and Γ_2 are simple, $\Gamma_1 \cup \Gamma_2$ forms a simple loop. This loop is contained in K (and in fact in $z([0, 1])$, because $K \cap B(w, r/3)$ must be contained in $z([0, 1])$, but this does not matter). To show that it is nontrivial, let us just check that $z(s_1) \neq z(s_2)$. The point is that the

s_i lie in J, by (17), and then (8) says that $|z(s_2) - z(s_1)| \geq \frac{9}{10}|s_2 - s_1| |z'(t)| > 0$. Thus K contains a nontrivial simple loop, and we get a contradiction with (3).

There is an analogous argument when we do not assume that $K \cap B(w,r)$ is a nice C^1 curve. We want to define s_1 and s_2 as before, but so that the $z(s_i)$ lie on a circle $\partial B(w,\rho)$, $\rho \in (0, r/3)$, rather than $\partial B(w, r/3)$. We need the extra flexibility because we want to choose ρ such that

$$K \cap \partial B(w, \rho) \text{ is reduced to two points.} \qquad (19)$$

We shall see why we care soon, but let us first say why we can find ρ. Recall that $\beta_K(w, 2r) \leq \varepsilon$, so Lemma 52.1 says that

$$H^1(K \cap B(w,r)) \leq 2r + C\beta_K(w,r)^{1/2} r \leq 2r + C\varepsilon^{1/2}r, \qquad (20)$$

as for (10) above. Also, (8) and the continuity of z say that for $0 < \rho < r$, $\partial B(w,\rho) \cap z(J)$ has at least two different points (almost opposed on $\partial B(w,\rho)$). Thus if $N(\rho)$ denotes the number of points in $K \cap \partial B(w,\rho)$, then $N(\rho) \geq 2$ for $\rho < r$. Since $\int_0^r N(\rho)d\rho \leq H^1(K \cap B(w,r)) \leq 2r + C\varepsilon r$ by Lemma 26.1 and (20), we can easily find $\rho < r/3$ such that $N(\rho) = 2$, as needed.

So we choose $\rho < r/3$ such that (19) holds and define s_1 and s_2 as above. That is, s_1 is the first point of $[0,1]$ such that $z(s_1) \subset \partial B(w,\rho)$ and s_2 is the last one. As before, $G = z([0,1] \setminus [s_1, s_2])$ is connected, so we can find a simple curve $\Gamma_1 \subset G$ that goes from $z(s_1)$ to $z(s_2)$. Note that Γ_1 does not meet $B(w,\rho)$, by definition of s_1 and s_2.

Our condition (19) is useful, because it will allow us to find a simple curve $\Gamma_2 \subset K \cap \overline{B}(w,\rho)$ that goes from $z(s_1)$ to $z(s_2)$. Indeed consider the smallest point $s_2' > t$ such that $z(s_2') \in \partial B(w,\rho)$ and the largest $s_1' < t$ such that $z(s_1') \in \partial B(w,\rho)$. By (17), both s_i' lie in J. By (8), the two points $z(s_i')$ are distinct (they almost lie opposite to each other on $\partial B(w,\rho)$). By (19), one of them is $z(s_1)$ and the other one is $z(s_2)$. Then $z([s_1', s_2'])$ is a connected set that contains $z(s_1)$ and $z(s_2)$, and Proposition 30.14 tells us that there is a simple curve in $z([s_1', s_2'])$ that goes from $z(s_1)$ to $z(s_2)$. This is the curve Γ_2 that we wanted, since $z([s_1', s_2']) \subset K \cap \overline{B}(w,\rho)$.

As before, Γ_1 and Γ_2 are simple and only meet at $z(s_1)$ and $z(s_2)$ (because Γ_1 does not meet $B(w,\rho)$). We put them together and get a nontrivial simple loop in K, as needed.

This completes our proof of Lemma 4 and Proposition 1. $\qquad\qquad\qquad\square$

We end this section with a slightly more precise description of what nontrivial connected components of K look like. The following description comes from Section 10 in [DaLé], to which we refer for proofs and additional details. Also see Lemma 67.3 for a proof that $r_1(x) \geq r_3(x)$.

Let $(u, K) \in RGM(\mathbb{R}^2)$ be given, and let K_0 be a connected component of K, not reduced to one point. Then K_0 is a (possibly infinite) union of C^1 arcs of curves, that can only meet by sets of three and with $120°$ angles. We even have lower bounds on the size of the C^1 arcs, in terms of various functions that measure the distance to the extremities of K_0.

A first way to measure such a distance is to set

$$r_1(x) = \sup\{r > 0\,;\ \text{there is an injective Lipschitz function } f : [-r, r] \to K_0$$
$$\text{such that } f(0) = x \text{ and } |f'(t)| = 1 \text{ almost-everywhere}\}.$$
(21)

Thus $r_1(x)$ measures how far we can travel in K_0 in at least two opposite directions. If you think about a finite tree, this is a reasonable, but a priori not perfect way to define the distance to the extremities. It appears that it may overestimate this distance, for instance if x is the center of a tree with three branches, two long ones and a short one. In this case, $r_1(x)$ measures the size of the second longest branch. Notice that if you move x a tiny bit to the short leg, $r_1(x)$ suddenly becomes smaller. The reader should not worry too much about this, the second part of Proposition 26 below shows that the two shortest legs must have comparable sizes anyway.

A different way to proceed is to define what we mean by extremities of K_0. As we shall see in Section 60, there are two different types of points in K. The points of low energy are the points $x \in K$ such that $\lim_{r \to 0} \left\{ r^{-1} \int_{B(x,r) \setminus K} |\nabla u|^2 = 0 \right\}$. For such a point x, we can show that for r small, $K \cap B(x, r)$ is either a nice C^1 curve or a nice spider (as in Definition 53.3). The other points, or points of high energy, are actually such that $\liminf_{r \to 0} \left\{ r^{-1} \int_{B(x,r) \setminus K} |\nabla u|^2 \right\} > 0$. They form what we shall call the singular singular set $K^\sharp$. We can decide that the extremities of K_0 are the points of $K_0 \cap K^\sharp$, and set

$$r_2(x) = \text{dist}(x, K_0 \cap K^\sharp).$$
(22)

Here we defined $r_2(x)$ in terms of the ambient Euclidean distance, but using the length of the shortest path in K between two points would give an equivalent distance, by Proposition 1. A minor variant is to set

$$r_3(x) = \text{dist}(x, K^\sharp),$$
(23)

which could be a little smaller because of points of $K^\sharp \setminus K_0$. Let us introduce two last radii $r_4^\pm(x)$. We start with the (most important) case when x is a C^1 point of K, which means that for r small, $K \cap B(x, r)$ is a nice C^1 curve. Then we set

$$r_4(x) = \lim_{r \to 0} \Big\{ \inf \{ \text{length}\,(\gamma)\,;\ \gamma \text{ is a curve in } \mathbb{R}^2 \setminus K \text{ that}$$
$$\text{connects the two components of } B(x, r) \setminus K \} \Big\}.$$
(24)

For the other points of K_0, we can set

$$r_4^+(x) = \limsup_{y \to x\,;\, y \in K'} r_4(y) \quad \text{or} \quad r_4^-(x) = \liminf_{y \to x\,;\, y \in K'} r_4(y),$$
(25)

where K' denotes the set of C^1 points of K_0. Note that $r_4^\pm(x) = r_4(x)$ for C^1 points of K.

Proposition 26 [DaLé]. *There is a universal constant $C > 0$ such that*

$$C^{-1}r_1(x) \le r_2(x), r_3(x), r_4^+(x), r_4^-(x) \le Cr_1(x) \tag{27}$$

when $(u, K) \in RGM(\mathbb{R}^2)$, K_0 is a nontrivial connected component, and $x \in K_0$. Moreover, if $r_1(x) > 0$, $K \cap B(x, C^{-1}r_1(x))$ is a nice C^1 curve (as in Definition 51.2) or a nice spider (as in Definition 53.3).

In the proof of Proposition 26, the most important part is the second half (which is proved first). For this, one uses a monotonicity argument based on Bonnet's lower bound in Section 47. This gives a fast decay of the normalized energy $\omega_2(x, r) = r^{-1} \int_{B(x,r) \setminus K} |\nabla u|^2$ when $r < r_1(x)$, and then we just need to know a little about points of low energy. See Section 66.b for a few more details. Also see Lemma 67.3 for a proof that $r_1(x) \ge r_3(x)$.

We can also use the radii $r_j(x)$ to estimate the jump of u: there is a constant $C > 0$ such that

$$C^{-1}r_1(x)^{1/2} \le |\operatorname{Jump}(x)| \le Cr_1(x)^{1/2} \text{ for all } C^1 \text{ points } x \in K, \tag{28}$$

where $\operatorname{Jump}(x)$ denotes the difference between the two limits of u at x (corresponding to accesses from both sides of K), which is defined modulo its sign. See Proposition 10.7 in [DaLé].

The obvious limit of our description of a nontrivial component K_0 is that we don't know whether it can have an infinite number of extremities (or equivalently, whether it can be composed of an infinite number of C^1 branches). We know from Proposition 26 that the extremities of K_0 are points of high energy (i.e., lie in $K^\sharp$), and we shall see in Section 67 that $K^\sharp$ is either empty, reduced to one point, or (perhaps) infinite. If this last case occurs, we don't know either if infinitely many points of $K^\sharp$ can lie on the same component K_0.

Even though the author did not check carefully, he thinks that there is an analogue of our description of K_0 when $(u, K) \in TRLAM(\Omega, h)$, if h decays fast enough. The main ingredients should be Proposition 68.1 for the absence of small loops (this time, big loops are allowed), Proposition 68.18 for the chord-arc condition, and Corollary 66.29, which will give a nice C^1 or spider disk centered at x in situations where two distinct arcs leave from x (i.e., when $r_1(x) > 0$).

Exercise 29. [Another proof of Lemma 4.] Let γ be a simple arc in K, and call x, y the extremities of γ. We want to show that length $(\gamma) \le C|x - y|$.

1. Check that it is enough to show that $\operatorname{diam}(\gamma) \le C|x - y|$.

2. From now on, we assume for simplicity that $\operatorname{diam}(\gamma) \ge 10|x - y|$. Call $\widehat{\gamma}$ the loop obtained by adding the segment $[x, y]$ to γ. Find a point $\xi \in \mathbb{R}^2 \setminus \widehat{\gamma}$

such that $|\xi - x| \geq \text{diam}(\gamma)/2$, and ξ lies in a bounded component of $\mathbb{R}^2 \setminus \widehat{\gamma}$. There are probably many ways to do this; here is a brutal one. Find a point $z \in \gamma$ such that $K \cap B(z, r)$ is a C^1 curve for r small, and then prove that two points of $B(z, r) \setminus K$ that lie very close to z, but on different sides of K, cannot both be in the unbounded component of $\mathbb{R}^2 \setminus \widehat{\gamma}$.

3. Consider the escape path from ξ to infinity given by Proposition 56.7. Show that this path crosses $\widehat{\gamma}$. Call η a point of intersection. Show that $\eta \in [x, y]$, and that $\text{dist}(\eta, K) \geq C^{-1} \text{diam}(\gamma)$. Conclude. [Note about self-contained-ness: you do not really need the parameterization z from Proposition 56.7; the properties of the combination of paths that we constructed in Section 56 are enough. Which is not shocking since the statement of Proposition 56.7 is just a way to code them.]

Exercise 30. [Rectifiable curves have tangent lines almost-everywhere.]
Let $z : [0, 1] \to \mathbb{R}^n$ be a Lipschitz mapping, and call $\Gamma = z([0, 1])$ its image. We want to show that Γ has a tangent line at w for H^1-almost every $w \in \Gamma$.

1. Check that it is enough to verify this for H^1-almost every $w \in z(E)$, where we set $E = \{t \in [0, 1] \,;\, z'(t) \text{ exists and } z'(t) \neq 0\}$. [Notice the similarity with (6).]

2. Let $t \in E$, $w = z(t)$, and pick $r > 0$ such that (8) holds. Call I the interval of length $|z'(t)|^{-1} r$ centered at t. Show that $z(I)$ has a tangent at w.

3. Call $E(\rho) = \{t \in E \,;\, z([t - \rho, t + \rho]) \text{ has a tangent at } z(t)\}$. Show that Γ has an approximate tangent line at almost every $w \in z(E(\rho))$. [See Exercise 41.21 for the definition.] This is the right place to use the fact that when X, Y are disjoint measurable sets with finite H^1 measure, the upper density $\limsup_{r \to 0} \{r^{-1} H^1(Y \cap B(x, r))\}$ vanishes at H^1-every point $x \in X$. [See Theorem 6.2 in [Mat2].]

4. Show that Γ has an approximate tangent line at H^1-almost every point of $z(E)$ and conclude. [You may use Exercise 41.21].

Exercise 31. [Length and Hausdorff measure; see the argument between (63.10) and (63.15) for a different proof.] Let $z : [0, 1] \to \mathbb{R}^n$ be continuous, and define the length of (the arc defined by) z by $L = \sup \left\{ \sum_{i=1}^{m} |z(s_i) - z(s_{i-1})| \right\}$, where the supremum is taken over all choices of subdivisions $0 = s_0 < s_1 < \cdots < s_m = 1$ of $[0, 1]$.

1. Check that if $z' \in L^1([0, 1])$ (as in Section 9), then $L \leq \int_{[0,1]} |z'(t)| dt$.

2. From now on, suppose that $L < +\infty$. For $0 \leq t_1 \leq t_2 \leq 1$, call $L(t_1, t_2)$ the length of the arc defined by the restriction of z to $[t_1, t_2]$. That is, keep the same definition as for L, but with $s_0 = t_1$ and $s_m = t_2$. Show that $L(t_1, t_3) = L(t_1, t_2) + L(t_2, t_3)$ for $0 \leq t_1 \leq t_2 \leq t_3 \leq 1$. Show that $L(0, t)$ is a continuous nondecreasing function of t.

3. Show that $z' \in L^1([0,1])$, and $\int_{[0,1]} |z'(t)|dt \leq L$. One way to do this is to approximate z with piecewise linear maps.

4. [Parameterization by arclength.] Set $m(l) = \sup\{t \in [0,1]\,;\, L(0,t) \leq l\}$ and $f(l) = z(m(l))$ for $0 \leq l \leq L$. Show that m is bijective as soon as z is not constant on any nontrivial interval. Also show that f is Lipschitz, with $|f(l) - f(l')| \leq |l - l'|$, and that $|f'(l)| = 1$ almost-everywhere. [Use Question 1.]

5. Show that $H^1(z([0,1])) \leq L$. You can do this directly from the definitions, or use the previous question and Remark 2.11.

6. Suppose in addition that z is injective. Show that $H^1(z([0,1])) = L$. [Because of Question 4, we may assume that z is 1-Lipschitz, with $|z'(t)| = 1$ almost-everywhere. Then consider the image by z of the Lebesgue measure on $[0,1]$, show that it is absolutely continuous with respect to H^1, and that its density is 1. This is perhaps a little harder than the average exercise in this book, but you may find additional help near (63.10).]

58   The case when u is locally constant somewhere $(n = 2)$

Here we want to show that if $(u, K) \in RMG(\mathbb{R}^2)$ (a reduced global minimizer, as in Notation 54.10) and u is locally constant somewhere, then K is either empty, a line, or a propeller, and u is constant on each component of $\Omega = \mathbb{R}^2 \setminus K$. To the author's knowledge, this first showed up in [BoDa].

The special case when u is assumed to be constant on every component of Ω is very useful and slightly simpler, so we shall give a special argument; see Proposition 10.

Let us first observe that if u is locally constant near a point $x \in \Omega$, then it is constant on the component of x in Ω, because it is harmonic. Our first step will be to show that this component is convex.

Proposition 1. *Suppose $(u, K) \in RMG(\mathbb{R}^2)$, Ω_0 is a connected component of $\Omega = \mathbb{R}^2 \setminus K$, and u is constant on Ω_0. Then Ω_0 is convex.*

This is not shocking. If Ω_0 was smooth and bounded (which is not the case, but let us just imagine), we could try to replace Ω_0 with its convex hull and keep u equal to the same constant there. We would save on length because the boundary of the convex hull is shorter, and we would also save on the size of the gradient of u wherever we modified it. Our proof will be a little more complicated than this, but the idea will be the same.

Remark 2. In higher dimensions, we can still get some information on the sign of the mean curvature of $\partial\Omega_0$, but this information seems much harder to use than convexity.

We shall prove the proposition by contradiction, so let us assume that Ω_0 is not convex. Let $y_1, y_2 \in \Omega_0$ be such that the segment $[y_1, y_2]$ is not contained in Ω_0.

First suppose that there is a little neighborhood V of $[y_1, y_2]$ that does not meet any component of Ω other than Ω_0. [Think about the case when Ω_0 would look like a ball, minus a line segment.] Pick $y \in [y_1, y_2] \setminus \Omega_0 = [y_1, y_2] \cap K$, and choose r so small that $B = B(y, r) \subset\subset V$. Set $\widetilde{K} = K \setminus B$ and keep $\widetilde{u} = u$ on Ω. Also set $\widetilde{u} = \lambda$ on $B \cap K$, where λ is the constant value of u on Ω_0. Since u does not take any values other than λ near $\overline{B}$, $u \in W^{1,2}_{\text{loc}}(\mathbb{R}^2 \setminus \widetilde{K})$ and so $(\widetilde{u}, \widetilde{K})$ is a competitor for (u, K) in B. It is even a topological competitor, by Lemma 20.10. But $\nabla \widetilde{u} = \nabla u$ almost everywhere, and $H^1(K \setminus \widetilde{K}) > 0$ by Ahlfors-regularity of K. This is impossible because $(u, K) \in RMG(\mathbb{R}^2)$.

So every neighborhood of $L = [y_1, y_2]$ meets some other component of Ω. We can replace y_1 or y_2 by a nearby point of Ω_0, so that now L meets a component Ω_1 of Ω other than Ω_0.

Let v be a unit vector orthogonal to L, and set $L_t = L + tv$ for $t \in \mathbb{R}$. We want to replace L with some L_t to get a few additional properties that will simplify our life. First observe that if t is small enough, then the endpoints of L_t still lie in Ω_0 and L_t still meets Ω_1. We want to take t such that in addition

$$K \cap L_t \text{ is finite.} \tag{3}$$

Let us check that this is the case for almost every $t < 1$, say. In fact, if $N(t)$ denotes the number of points in $K \cap L_t$, then $\int_{t=0}^{1} N(t)dt \leq H^1(K \cap B) < +\infty$, where B is a ball that contains all the L_t, $t \leq 1$. This is not exactly Lemma 26.1 (where instead $N(t)$ was the number of points on $\partial B(0, t)$), but the same proof works. We could also deduce this formally from Lemma 26.1, maybe at the expense of losing a factor 2 or 4 in the estimate, as follows. We would first prove a variant of Lemma 26.1, with the disks $B(0, t)$ replaced by squares with sides parallel to L and v. To do this, we would just use a bilipschitz mapping of $\mathbb{R}^2$ that transforms the $B(0, t)$ into such squares. Then we would just have to notice that our numbers $N(t)$ are less than the number of points on the boundaries of such squares (centered reasonably far from L).

It will also be convenient to know that

$$\text{for each } x \in K \cap L_t, \text{ there is an } r > 0 \text{ such} \atop \text{that } K \cap B(x, r) \text{ is a nice } C^1 \text{ curve.} \tag{4}$$

See Definition 51.2 for nice C^1 curves. Theorem 51.20 says that H^1-almost every point of K is a C^1 point, so the bad set of points $x \in K$ that do not satisfy the condition in (4) has zero measure, its orthogonal projection on the direction of v too (by Remark 2.11), and (4) holds for almost every t. Let us finally demand that

$$K \text{ be transverse to } L_t \text{ at } x \text{ for all } x \in K \cap L_t. \tag{5}$$

The notion makes sense because of (4), and (5) holds for almost all choices of t, by Sard. More precisely, we can cover the set of C^1 points of K with a countable collection of C^1 curves Γ_j. Fix j and $x \in \Gamma_j$. There is only one L_t that goes through x, and K is not transverse to L_t at x if and only if the tangent to Γ_j at x is parallel to L. This means that if π denotes the restriction to Γ_j of the orthogonal projection on the line with direction v, then the derivative of π at x is zero. Thus x is a critical point of π, $t \approx \pi(x)$ is a critical value, and Sard's Theorem says that the set of such t has zero measure. [See Exercise 33.41 for Sard's theorem.] Then we take a countable union over j, and we get that (5) holds for almost all t.

Let us choose t small, with all the properties above. Pick $y_3 \in L_t \cap \Omega_1$, and let $I = [a, b]$ be a subinterval of L_t that contains y_3, is contained in the complement of Ω_0, and is maximal with this property. See Figure 1.

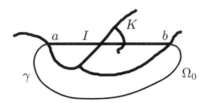

Figure 1

By maximality of I, a and b lie in $\partial \Omega_0 \subset K$, and we can find $r > 0$ such that $K \cap B(a, r)$ is a nice C^1 curve, transverse to L_t at a, and similarly for b. Since a and b lie in $\partial \Omega_0$, we can find a simple curve γ from a to b, which is contained in Ω_0 except for the endpoints a and b. In particular, $\gamma \cap I = \{a, b\}$ and, when we add the interval I to γ, we get a simple loop $\widehat{\gamma}$. Call D the bounded component of $\mathbb{R}^2 \setminus \widehat{\gamma}$. We want to define a competitor $(\widehat{u}, \widehat{K})$. Set

$$\widehat{K} = [K \cup I] \setminus D \tag{6}$$

and

$$\begin{cases} \widehat{u}(x) = u(x) \text{ for } x \in \mathbb{R}^2 \setminus [K \cup \Omega_0 \cup D \cup I], \\ \widehat{u}(x) = \lambda \text{ for } x \in \Omega_0 \cup D, \end{cases} \tag{7}$$

where λ still denotes the constant value of u on Ω_0. We want to check that $(\widehat{u}, \widehat{K})$ is a competitor for (u, K) in B, where B is any ball that contains $\overline{D}$. Obviously $\widehat{K}$ is closed, and we did not modify anything out of $\overline{D}$. In particular, $\widehat{u} \in W^{1,2}_{\text{loc}}(\mathbb{R}^2 \setminus [K \cup \overline{D}])$. Since $\widehat{u}$ is constant on D, we get that $\widehat{u} \in W^{1,2}_{\text{loc}}(\mathbb{R}^2 \setminus [(K \cup \overline{D}) \setminus D])$. So we only need to show that $\widehat{u} \in W^{1,2}_{\text{loc}}$ near points of $\partial D \setminus I \subset \gamma$, since the rest of $[(K \cup \overline{D}) \setminus D])$ lies in $\widehat{K}$. But $\gamma \setminus I$ is contained in Ω_0, and $\widehat{u}$ is constant there. So $\widehat{u} \in W^{1,2}_{\text{loc}}(\mathbb{R}^2 \setminus \widehat{K})$, and $(\widehat{u}, \widehat{K})$ is a competitor for (u, K).

We also need to check the topological condition (54.3). Let B be large enough to contain $\overline{D}$, and let $x, y \in \mathbb{R}^2 \setminus [K \cup B]$ be given. We assume that $\widehat{K}$ does not separate x from y, and we want to check that K does not separate them either. Call Γ a path from x to y in $\mathbb{R}^2 \setminus \widehat{K}$. If Γ does not meet $\overline{D}$, then it does not meet K either, by (6), and we are happy. Otherwise, call x' the first point of $\overline{D}$ that we hit when we run from x to y along Γ. Since the arc of Γ between x and x' lies outside of $\overline{D}$, it does not meet K and x' lies inside the same component of $\mathbb{R}^2 \setminus K$ as x. But x' cannot lie on I, which is contained in $\widehat{K}$, so $x' \in \gamma$, and consequently x' and x both lie inside Ω_0. Similarly, y lies inside Ω_0, and K does not separate x from y.

So $(\widehat{u}, \widehat{K})$ is a topological competitor, and (54.5) holds. Fix a ball B that contains $\overline{D}$. Obviously $\int_{B \setminus \widehat{K}} |\nabla \widehat{u}|^2 \leq \int_{B \setminus K} |\nabla u|^2$, so if we prove that $H^1(\widehat{K} \cap B) < H^1(K \cap B)$ as well, we shall get the desired contradiction with (54.5). Notice that $H^1(\widehat{K} \cap B) = H^1(K \cap B) + H^1(I \setminus K) - H^1(K \cap D) = H^1(K \cap B) + H^1(I) - H^1(K \cap \overline{D})$ by (6), so it will be enough to check that

$$H^1(K \cap \overline{D}) > H^1(I). \tag{8}$$

Let $\xi \in (a, b)$ (the interior of I) be given. Call P_ξ the half-line that starts from I perpendicularly to I and in the direction of D (i.e., downwards in the case of Figure 1). Observe that P_ξ eventually leaves D, because D is bounded. Call ξ' the first point where this happens. Then $\xi' \in \gamma$, because I is impossible. Now $\xi \in \mathbb{R}^2 \setminus \Omega_0$ by definition of I, and $\xi' \in \Omega_0$ by definition of γ. This forces the segment $[\xi, \xi']$ to meet $\partial \Omega_0 \subset K$. Call π the orthogonal projection onto L_t. Then $\pi(K \cap \overline{D})$ contains ξ (because $[\xi, \xi'] \subset \overline{D}$). Altogether, $\pi(K \cap \overline{D})$ contains (a, b) and $H^1(K \cap \overline{D}) \geq H^1(I)$, by Remark 2.11.

We still need to make sure that the inequality in (8) is strict. The simplest is to note that (4) and (5) give us a little bit of extra length near a and b. That is, if r is small enough, $K \cap B(a, r)$ is a nice C^1 curve that is transverse to L_t at a. We can assume that γ also is smooth and transverse to K at a. Then $\widehat{\gamma}$ crosses $K \cap B(a, r)$, and if r is small enough, one of the two parts of $K \cap B(a, r) \setminus \{a\}$ is contained in D (call it K_0). The contribution of K_0 to $\pi(K \cap \overline{D})$ is strictly less than its length (by transversality), which is enough to get the strict inequality. [Note that both behaviors pictured in Figure 2 make us win something in the

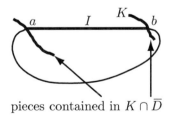

pieces contained in $K \cap \overline{D}$

Figure 2

estimate above.] More precisely, $\pi(K \cap \overline{D} \setminus K_0)$ still contains $(a, b) \setminus \pi(K_0)$ because $\pi(K \cap \overline{D})$ contains (a, b), and so

$$\begin{aligned} H^1(I) &= H^1(I \cap \pi(K_0)) + H^1(I \setminus \pi(K_0)) \\ &< H^1(K_0) + H^1(K \cap \overline{D} \setminus K_0) = H^1(K \cap \overline{D}). \end{aligned} \tag{9}$$

This proves (8), a contradiction ensues, and Proposition 1 follows. $\qquad\square$

Next we characterize the reduced global minimizers for which u is locally constant. See Proposition 19 below for the more general case when u is only locally constant somewhere.

Proposition 10. *Let $(u, K) \in RMG(\mathbb{R}^2)$ be such that u is constant on each connected component of $\mathbb{R}^2 \setminus K$. Then K is either empty, a line, or a propeller.*

Recall that a propeller is a union of three half-lines that start from the same point and make $120°$ angles at that point. As the reader may have guessed, Proposition 10 is nothing else than a characterization of minimal sets of dimension 1 in the plane, where the notion of minimal is with respect to a Dirichlet condition at infinity, and includes the topological constraint (54.3) for the competitors.

There is almost-surely an analogue of Proposition 10 in dimension 3, which comes from a result of Jean Taylor [Ta]. That is, the reduced global minimizers in $\mathbb{R}^3$ for which u is locally constant must come in four types, corresponding to the empty set, planes, unions of three half-planes glued along a line and making angles of $120°$, and unions of six plane sectors bounded by four lines emanating from a center (and that bound four convex regions of space). Here "almost-surely" means that she only states the local result, but the global case seems to follow reasonably easily. See Section 76 for a little more.

The proof below is essentially by hand, except that it uses the convexity from Proposition 1. We start with an easy bound on the number of components, which holds for any reduced global minimizer. But of course much better is known; there is no exotic minimizer with more than one component in $\mathbb{R}^2 \setminus K$ (see Theorem 66.1).

Lemma 11. *If $(u, K) \in RMG(\mathbb{R}^2)$, then $\mathbb{R}^2 \setminus K$ has at most six connected components.*

Suppose not, and let Ω_i, $1 \leq i \leq 7$, denote distinct components of $\mathbb{R}^2 \setminus K$. Let $R > 0$ be such that $B(0, R)$ meets all Ω_i. For each $r > R$, $\partial B(0, r)$ also meets all the Ω_i, because none of them is bounded (by Proposition 55.1). This forces $K \cap \partial B(0, r)$ to have at least seven points (pick a point in each $\Omega_i \cap \partial B(0, r)$, and notice that you have to cross a boundary between two consecutive such points). Now apply Lemma 26.1: if $N(r)$ denotes the number of points in $K \cap \partial B(0, r)$, then for all $R_1 > R$,

$$H^1(K \cap B(0, R_1)) \geq \int_0^{R_1} N(r)dr \geq 7(R_1 - R) > 2\pi R_1 \qquad (12)$$

if R_1 is large enough. This contradicts the trivial estimate that you get by replacing $K \cap B(0, R_1)$ with $\partial B(0, R_1)$. [See (18.8).] Lemma 11 follows. □

Now let $(u, K) \in RGM(\mathbb{R}^2)$ be as in Proposition 10, and call Ω_j, $1 \leq j \leq m$, the connected components of $\mathbb{R}^2 \setminus K$. Since each Ω_j is convex, so is $\overline{\Omega}_j$, and also $\partial_{j,k} = \overline{\Omega}_j \cap \overline{\Omega}_k$ for $j < k$. Since $\partial_{j,k} \subset \partial \Omega_j \subset K$ and K has empty interior, all the points of a given $\partial_{j,k}$ lie on a line. And since it is convex,

$$\text{each } \partial_{j,k} \text{ is a line, a half-line, or a line segment.} \qquad (13)$$

Let us verify that

$$K \text{ is the union of all the } \partial_{j,k}, j < k. \tag{14}$$

Let $x \in K$ be given. Recall that there is only a finite number of components Ω_j in $\mathbb{R}^2 \setminus K$, so if x does not lie on any $\partial_{j,k}$, then there is a little ball B centered on x such that $2B$ only meets one Ω_j. Set $\widetilde{K} = K \setminus B$, and extend u to B by setting $u(y) = \lambda_j$ there, where λ_j is the constant value of u on Ω_j. Then $(u, \widetilde{K})$ is a competitor for (u, K) in B. It satisfies the topological condition (54.3) by Lemma 20.10, and since it is clearly better than (u, K) because $H^1(K \cap B) > 0$, we get a contradiction that proves (14).

Observe that two different $\partial_{j,k}$ never have an intersection larger than a point, because otherwise this intersection would contain an open interval I (by (13)), then (by (13) and (14)) there would be a point $x \in I$ and a small radius r such that $K \cap B(x,r) = I \cap B(x,r)$. But j and k would be determined by the two components of $B(x,r) \setminus K$, a contradiction.

Call N the number of half-lines, plus twice the number of lines. We want to show that $N \leq 3$. Choose $R > 0$ so large that $B(0,R)$ contains all the $\partial_{j,k}$ that are reduced to line segments, all the extremities of the $\partial_{j,k}$ that are half-lines, at least one point of the $\partial_{j,k}$ that are lines, and also all the intersections between different $\partial_{j,k}$. Then $\partial B(0,r) \cap K$ has exactly N points for $r > R$.

So let us assume that $N \geq 4$ and construct a competitor better than (u, K). Fix $r > R$ and call ξ_j, $1 \leq j \leq N$, the points of $K \cap \partial B(0,r)$. Since $N \geq 4$, we can find two consecutive points (call them ξ_1 and ξ_2) such that the length of the arc of $\partial B(0,r)$ between them is no more than $\pi r/2$. We define a piece of propeller F as follows. Pick a point α in the convex hull of $\{0, \xi_1, \xi_2\}$, and set $F = [0, \alpha] \cup [\alpha, \xi_1] \cup [\alpha, \xi_2]$ We choose α so that these three segments make $120°$ angles at α. See Figure 3. Then $H^1(F) \leq br$, where $b < 2$ is a constant that we

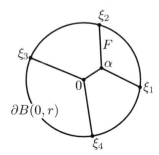

Figure 3

could compute. The point is that since the angle of $[0, \xi_1]$ with $[0, \xi_2]$ is at most $\pi/2$, α lies strictly inside the convex hull, with a definite gain in length compared to $[0, \xi_1] \cup [0, \xi_2]$. Now set

$$\widetilde{K} = [K \setminus B(0,r)] \cup F \cup \left(\bigcup_{j \geq 3} [0, \xi_j] \right). \tag{15}$$

Call I_j, $1 \leq j \leq N$ the connected component of $\partial B(0,r) \setminus K$. The point of choosing $\widetilde{K}$ as in (15) is that

$$\widetilde{K} \text{ separates the various } I_j \text{ from each other in } \overline{B}(0,r). \tag{16}$$

Call λ_j the constant value of u on the connected component of I_j in $\mathbb{R}^2 \setminus K$. We set $\widetilde{u} = \lambda_j$ on the component of I_j in $\overline{B}(0,r) \setminus \widetilde{K}$, and we keep $\widetilde{u} = u$ out of $\overline{B}(0,r) \cup K$. It is easy to see that $\widetilde{u}$ is locally constant on $\mathbb{R}^2 \setminus \widetilde{K}$. Furthermore, $(\widetilde{u}, \widetilde{K})$ is a topological competitor for (u, K), by (16) and Lemma 45.8. Since $(u, K) \in RGM$, (54.5) says that

$$H^1(K \cap B(0,r)) \leq H^1(\widetilde{K} \cap B(0,r)) = H^1(F) + (N-2)r = (N - 2 + b)r. \tag{17}$$

On the other hand, since $K \cap \partial B(0,s)$ has N points for $R < s < r$, Lemma 26.1 says that

$$H^1(K \cap B(0,r)) \geq \int_0^r \sharp(K \cap \partial B(0,s)) \, ds \geq N(r - R), \tag{18}$$

which contradicts (17) if r is large enough (because $b < 2$). Hence $N \leq 3$.

If $N = 0$, then K is bounded, $\mathbb{R}^2 \setminus K$ has only one component, and K must be empty. [Otherwise, simply remove K and get a better competitor.] Similarly, $N = 1$ is impossible, because all the components of $\mathbb{R}^2 \setminus K$ are unbounded, so they meet $\partial B(0,r)$ for r large. Hence there is only one such component, and we could remove any piece of K to get a better competitor. We also have a direct contradiction with (14).

If $N = 2$, then $\mathbb{R}^2 \setminus K$ has at most two components, because they all meet $\partial B(0,r)$ for r large. Then $K = \partial_{1,2}$, by (14). This forces K to be a line, as expected.

Finally assume that $N = 3$. Then there are at least three components, because otherwise there would be only one $\partial_{j,k}$. There are also no more than three components, since each of them meets $\partial B(0,r)$ for r large. And since each component meets $\partial B(0,r)$, each $\partial_{j,k}$ meets $\partial B(0,r)$ too. [If $I_j \subset \Omega_j$ and $I_k \subset \Omega_k$ are consecutive intervals of $\partial B(0,r)$, then their common endpoint lies on $\partial_{j,k}$.] This means that the three $\partial_{j,k}$ are half-lines, and K is their union. The verification that these half-lines have a common endpoint and make $120°$ angles is easy (see Exercise 59.15). This completes our proof of Proposition 10. $\qquad \square$

Proposition 19. *Let $(u, K) \in RMG(\mathbb{R}^2)$, and suppose that u is constant on at least one of the connected components of $\mathbb{R}^2 \setminus K$. Then K is either empty, a line, or a propeller, and u is constant on each component of $\mathbb{R}^2 \setminus K$.*

This was first proved in [BoDa], using Proposition 1 on convexity and a series of blow-in limits. There is another approach, based on the fact that if K is neither empty, nor a line or a propeller, then $\mathbb{R}^2 \setminus K$ is connected. [See Section 66.] Then Proposition 19 reduces to Proposition 10. Here we shall follow the argument from Section 18 of [BoDa].

Before we start with the proof, let us observe that we do not know how to extend Proposition 19 to higher dimensions (i.e., prove that u is locally constant everywhere as soon as it is locally constant somewhere). As we shall see, convexity helps a lot.

To prove Proposition 19, we start from a reduced global minimizer (u, K) in the plane and assume that u is constant on the component Ω_0 of $\mathbb{R}^2 \setminus K$. Proposition 1 says that Ω_0 is convex, and we want to reduce to the simpler situation where Ω_0 is a convex cone by replacing (u, K) with a blow-in limit.

Without loss of generality, we can assume that $0 \in \Omega_0$. Set

$$J = \big\{ \theta \in \mathbb{S}^1 \,;\, \text{there is a sequence } \{y_k\} \text{ in } \Omega_0 \text{ such that}$$
$$|y_k| \text{ tends to } +\infty \text{ and } \frac{y_k}{|y_k|} \text{ tends to } \theta \big\}. \tag{20}$$

Our medium-term purpose is to show that J is an interval in the circle, not too small, and that Ω_0 looks a lot like the cone over J at infinity. First observe that J is not empty, because Ω_0 is unbounded and $\mathbb{S}^1$ is compact. We claim that

for every $\theta \in J$, the half-line $D_\theta = \big\{ r\,\theta \,;\, r \geq 0 \big\}$ is contained in Ω_0. $\tag{21}$

This will be an easy consequence of the convexity of Ω_0. Let $B \subset \Omega_0$ be a little ball centered at the origin, and $r\,\theta \in D_\theta$ be given. Let $\{y_k\}$ be a sequence in Ω_0 such that $\frac{y_k}{|y_k|}$ tends to θ. The direction of the (oriented) line segment $[r\,\theta, y_k]$ tends to θ, hence the line through $r\,\theta$ and y_k goes through B for k large enough. This proves that $r\,\theta$ lies in Ω_0, by convexity, and (21) follows.

Next we want to show that

$$J \text{ contains an interval of length } \delta, \tag{22}$$

where δ is a positive absolute constant. Our argument is a little shorter here than in Lemma 18.13 of [BoDa], but it is based on the same principles.

Recall from Proposition 56.1 that for each $r > 0$, $\Omega_0 \cap B(0, r)$ contains a disk of radius $C^{-1} r$. Then we can find $x \in \Omega_0 \cap B(0, r)$ such that $|x| \geq \frac{r}{2C}$ and $B(x, \frac{r}{2C}) \subset \Omega_0$. We apply this with a sequence of radii r_k that tend to $+\infty$, and we find a sequence of points $x_k \in \Omega_0$. Let us extract a subsequence, so that the $\theta_k = \frac{x_k}{|x_k|}$ converge to some limit $\theta \in \mathbb{S}^1$. We just need to show that J contains the interval of length $\delta = \frac{1}{2C}$ centered at θ.

Let ξ lie in this interval, and let k be so large that $\mathrm{dist}(\xi, \theta_k) \leq \delta$. Set $y_k = |x_k|\,\xi$. Note that $|y_k| = |x_k| \geq \frac{r_k}{2C}$ (which tends to $+\infty$), and that $|y_k - x_k| =$

$|x_k| \operatorname{dist}(\xi, \theta_k) \leq r_k \delta$, so that $y_k \in B(x_k, \frac{r_k}{2C}) \subset \Omega_0$. Then $\xi \in J$, by (20), and (22) follows.

Our next claim is that either $\Omega_0 = \mathbb{R}^2$, or else

$$J \text{ is a closed interval of } \mathbb{S}^1, \text{ with } \delta \leq H^1(J) \leq \pi. \tag{23}$$

First let us show that if I is an interval of $\mathbb{S}^1$ such that $H^1(I) < \pi$, and if its two endpoints lie in J, then $I \subset J$. This is easy. Call θ_1 and θ_2 the endpoints; (21) says that the two half-lines D_{θ_i} lie in Ω_0, and then their convex hull is also contained in Ω_0, by Proposition 1. Since the convex hull contains all the half-lines D_θ, $\theta \in I$, J contains I (by (20)).

We already know that J contains an interval I_0 of length δ. If θ_1 is any point of J, we can find $\theta_2 \in I_0$ at distance (strictly) less than 2 from θ_1, and apply the argument above to the shortest interval I in $\mathbb{S}^1$ with endpoints θ_1 and θ_2. We get that $I \subset J$. Thus all points of J can be connected to I_0 in J, and J is an interval. It is easy to see that it is closed, directly by (20) or with the help of (21). Finally, if $H^1(J) > \pi$, every point $\theta \in \mathbb{S}^1$ lies in an interval of length less than π whose endpoints lie in J, so $\theta \in J$, $J = \mathbb{S}^1$, and $\Omega_0 = \mathbb{R}^2$ (by (21)). This proves that (23) holds if $\Omega_0 \neq \mathbb{R}^2$.

We are now ready for our first blow-in limit. Let $\{r_k\}$ be a sequence that tends to $+\infty$, and define pairs (u_k, K_k) by

$$K_k = \frac{1}{r_k} K \text{ and } u_k(x) = \frac{1}{\sqrt{r_k}} u(r_k x) \text{ for } x \in \mathbb{R}^2 \setminus K_k. \tag{24}$$

By invariance of RGM under dilations, $(u_k, K_k) \in RGM(\mathbb{R}^2)$ for each k. By Proposition 37.8, we can extract a subsequence (which we shall still denote the same way, as usual) so that the (u_k, K_k) converge to some limit (u^1, K^1). By (54.13), $(u^1, K^1) \in RGM(\mathbb{R}^2)$. We want to know a little more about (u^1, K^1). Denote by

$$\mathcal{C}(J) = \{r\theta \,;\, r \geq 0 \text{ and } \theta \in J\} \tag{25}$$

the (closed) cone over J, and by $\mathcal{C}^\circ(J)$ its interior. We want to check that

$$\partial \mathcal{C}(J) \subset K^1 \text{ and } \mathcal{C}^\circ(J) \text{ is one of the components of } \mathbb{R}^2 \setminus K^1. \tag{26}$$

First, $\mathcal{C}(J) \subset \Omega_0$, by (21), and in particular it does not meet K. It does not meet any K_k either (by (24) and because it is a cone), and hence its interior does not meet K^1 (because K^1 is the limit of the K_k). So $\mathcal{C}^\circ(J)$ is contained in a component of $\mathbb{R}^2 \setminus K^1$.

Now let I be any closed interval of $\mathbb{S}^1$, with endpoints $\theta \in \mathbb{S}^1 \setminus J$ and $\theta' \in J$. By definition of J, $\Omega_0 \cap D_\theta$ is bounded. Let ρ be such that $\Omega_0 \cap D_\theta \subset B(0, \rho)$. For each $r > \rho$, $r\theta$ lies out of Ω_0, while $r\theta' \in \Omega_0$ (by (21)). Thus the arc rI meets K for every $r > \rho$. Equivalently, rI meets K_k for $r > \rho/r_k$. Since r_k tends to 0, K_k tends to K^1, and I is closed, rI meets K^1 for every $r > 0$.

If a is one of the endpoints of J and we apply this to very short intervals centered at a, we get that $ra \in K^1$ for all $r > 0$. Hence $\partial \mathcal{C}(J) \subset K^1$, and the rest of (26) follows easily.

Next observe that u^1 is constant on $\mathcal{C}^\circ(J)$. Indeed let B be any ball in $\mathcal{C}^\circ(J)$. Then $r_k B$ is contained in Ω_0 for each k, so u is constant on $r_k B$, u_k is constant on B, and u^1 also is constant on B. See Definition 37.6 for the convergence of $\{u_k\}$, and maybe recall that u^1 is determined on $\mathcal{C}^\circ(J)$ modulo an additive constant.

So our pair (u^1, K^1) is like (u, K), except that now the analogue of Ω_0 is a convex cone. This will be useful, because we shall see that in this case u^1 is also constant on a new component Ω_1 of $\mathbb{R}^2 \setminus K^1$ adjacent to $\mathcal{C}^\circ(J)$. This is our next goal.

We may as well forget about the case when our initial domain Ω_0 was the whole plane, because this case is trivial. Then $\mathcal{C}(J)$ is a cone with aperture $\leq \pi$, and in particular $H^1(\partial \mathcal{C}(J)) > 0$. Let $x \in \partial \mathcal{C}(J) \setminus \{0\}$ be a C^1 point of K^1. We know from Theorem 51.20 that almost every point of $\partial \mathcal{C}(J)$ is like this. [In the present situation, given the fact that the two half-lines that compose $\partial \mathcal{C}(J)$ already lie in K^1, there would be simpler ways to get this.] Let $r > 0$ be so small that $K^1 \cap B(x, r)$ is a nice C^1 curve. Here this just means that $K^1 \cap B(x, r) = \partial \mathcal{C}(J) \cap B(x, r)$, and that this set is a diameter of $B(x, r)$. Call Ω_1 the component of $\mathbb{R}^2 \setminus K^1$ that meets $B(x, r) \setminus \mathcal{C}(J)$. We want to show that

$$u^1 \text{ is constant on } \Omega_1, \tag{27}$$

and to this effect let us first show that

$$\text{the boundary values of } \nabla u^1 \text{ on } \partial \Omega_1 \text{ vanish in a neighborhood of } x. \tag{28}$$

Recall that the restriction of u^1 to the half-disk $V = \Omega_1 \cap B(x, r)$ is the harmonic function on V that minimizes the energy $\int_V |\nabla u^1|^2$, given its values on $\partial V \cap \Omega_1$. Lemma 17.6 tells us that on V, u^1 is the restriction of a harmonic function $\widehat{u}^1$ defined on $B(x, r)$ (and which is symmetric with respect to $K^1 \cap B(x, r)$). Thus the boundary values of ∇u^1 are well defined.

In short, (28) holds because u^1 is already constant on $\mathcal{C}^\circ(J)$, K^1 is straight in $B(x, r)$, and there is a formula that relates the curvature of K^1 to the jump of $|\nabla u^1|$. [See [MuSh2].] Let us verify (28) rapidly, without officially talking about curvature. Let w be a unit vector orthogonal to the direction of $K^1 \cap B(x, r)$ and pointing in the direction of Ω_1 (seen from x). Also let φ be a smooth nonnegative bump function supported on $B(x, r/2)$ and such that $\varphi(y) = 1$ for $y \in B(x, r/4)$. Set $\Phi_t(y) = y + t\varphi(y)w$ for $y \in \mathbb{R}^2$ and $t > 0$. If t is small enough, Φ_t is a diffeomorphism, and $\Phi_t(y) = y$ out of $B(x, r)$.

We want to use Φ_t to define a competitor (v_t, G_t). Set $G_t = \Phi_t(K^1)$, keep $v_t(y) = u^1(y)$ for y out of $B(x, r) \cup K^1$, and also for $y \in V_t = \Phi_t(B(x, r) \cap \Omega_1)$. Notice that $V_t \subset B(x, t) \cap \Omega_1$ by definition of Φ_t, so our definition makes sense. [See Figure 4.] We are left with $\Phi_t(B(x, r) \cap \mathcal{C}^\circ(J))$, where we simply keep for v_t the constant value of u^1 on $\mathcal{C}^\circ(J)$. Note that v_t is defined and smooth on $\mathbb{R}^2 \setminus G_t$,

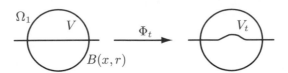

Figure 4

and (v_t, G_t) is a topological competitor for (u^1, K^1) in $B(x,r)$. [The topological condition is easy because Φ_t is a homeomorphism.] Now we can compare (v_t, G_t) with (u^1, K^1). We won the energy

$$\Delta_E = \int_{B(x,r)\backslash K^1} |\nabla u^1|^2 - \int_{B(x,r)\backslash G_t} |\nabla v_t|^2 = \int_{B(x,r)\cap\Omega_1\backslash V_t} |\nabla u^1|^2, \qquad (29)$$

which is at least $C^{-1}t \int_{\partial\Omega_1\cap B(x,r/4)} |\nabla u^1|^2$ for t small, by definition of Φ_t and because ∇u^1 is continuous. On the other hand, we only lost the length

$$\begin{aligned} \Delta_L &= H^1(G_t \cap B(x,r)) - H^1(K^1 \cap B(x,r)) \\ &= H^1(\Phi_t(K^1 \cap B(x,r))) - H^1(K^1 \cap B(x,r)), \end{aligned} \qquad (30)$$

which is less than Ct^2 (by direct computation). If $\int_{\partial\Omega_1\cap B(x,r/4)} |\nabla u^1|^2 > 0$ and t is small enough, we get a contradiction with the minimality of (u^1, K^1). This proves (28).

Recall that the restriction of u^1 to $V = \Omega_1 \cap B(x,r)$ can be extended harmonically to $B(x,r)$. This extension $\widehat{u}^1$ has a gradient that vanishes on a line segment, by (28). Since $\widehat{u}^1$ is harmonic on $B(x,r)$, it is the real part of an analytic function f, and the derivative f' vanishes on the same segment. By analytic continuation, $f' = 0$ and $\widehat{u}^1$ is constant. This proves (27).

So we found a second component Ω_1 (in addition to $\mathcal{C}^\circ(J)$) where u^1 is constant. This component is connected, and we can associate to it an interval J_1 of $\mathbb{S}^1$ as we did for Ω_0. In the case of Ω_0, we found it preferable to take an origin inside Ω to check certain things, but observe that J stays the same if we change the origin in the definition (20), and (u^1, K^1) does not change either.

It is easy to see that J and J_1 have disjoint interiors (by definition (20) of J_1 and because $\mathcal{C}^\circ(J)$ does not meet Ω_1). Now we take a second blow-in sequence $\{(u_k^1, K_k^1)\}$, and extract a subsequence that converges. Call (u^2, K^2) the limit. Then $(u^2, K^2) \in RGM(\mathbb{R}^2)$, K^2 contains $\partial\mathcal{C}(J) \cup \partial\mathcal{C}(J_1)$, and $\mathcal{C}^\circ(J)$ and $\mathcal{C}^\circ(J_1)$ are two components of $\mathbb{R}^2 \setminus K^2$ where u_2 is constant. The proof is the same as for (u^1, K^1).

If $J \cup J_1 = \mathbb{S}^1$, we can stop. We shall see soon how to conclude in this case. Otherwise, we continue. We find a C^1 point $x \in \partial[\mathcal{C}(J) \cup \mathcal{C}(J_1)]$ and a small radius

$r > 0$ such that $K \cap B(x, r)$ is a diameter of $B(x, r)$. One of the components of $B(x, r) \setminus K$ lies in $\mathcal{C}(J) \cup \mathcal{C}(J_1)$, and the other one lies in some third component Ω_2. The proof of (27) shows that u^2 is constant on Ω_2, and we can define a new interval J_2.

We can continue this argument, until we get intervals $J, J_1, \dots, J_m$ that cover $\mathbb{S}^1$. This has to happen eventually, because all our intervals have disjoint interiors, and their length is at least $\delta > 0$, by (22) or (23). So let m be such that $J \cup J_1 \cup \cdots \cup J_m = \mathbb{S}^1$. Then the corresponding next pair (u^{m+1}, K^{m+1}) is simple: K^{m+1} is a finite union of half-lines emanating from the origin, and u^{m+1} is constant on each component of $\mathbb{R}^2 \setminus K^{m+1}$. It is easy to see that K^{m+1} is a line or a propeller (even without using Proposition 10). Which incidentally shows that $m = 1$ or 2.

Now we just need to trace back what we did with blow-ins.

Lemma 31. *Let $(u, K) \in RGM(\mathbb{R}^2)$, and let (v, G) be a blow-in limit of (u, K). This means that (v, G) is the limit of some sequence $\{(u_k, K_k)\}$ defined by (24), and with a sequence $\{r_k\}$ that tends to $+\infty$. Suppose in addition that v is locally constant. If G is a line, then K is a line. If G is a propeller, then K is a propeller. In all cases, u is constant on every component of $\mathbb{R}^2 \setminus K$.*

Lemma 31 is a trivial consequence of Proposition 59.10, where it will also be noted that we do not need to assume that v is locally constant. Of course our proof of Proposition 59.10 will not rely on Proposition 19.

Return to our last pair (u^{m+1}, K^{m+1}). If $m = 1$, we arrived at (u^2, K^2), for which K^2 is a line and u^2 is locally constant. Two applications of Lemma 31 show that the same thing is true with (u, K). Similarly, if $m = 2$, K^3 is a propeller, u^3 is locally constant, and three applications of Lemma 31 give the desired description of (u, K). We got rid of the other case when $\Omega_0 = \mathbb{R}^2$ long ago. This completes our proof of Proposition 19 (modulo Lemma 31). $\qquad\square$

59 Simple facts about lines and propellers; blow-in limits

In this section we want to relax a little and check that lines and propellers actually give global minimizers, and only with locally constant functions u. We shall use this to check that if (u, K) is a global minimizer and one of the blow-in limits of K is a line or a propeller, then K itself is a line or a propeller. The same sort of argument will also give (in Section 70) new sufficient conditions for $K \cap B$ to be a nice C^1 curve or spider when (u, K) is an almost-minimizer.

We start with the verification of minimality, which we only do in the case of propellers because lines are much simpler. In both cases, there are short proofs using calibrations (and that generalize to somewhat more complicated situations). See [ABD].

Lemma 1. *Let $P \subset \mathbb{R}^2$ be a propeller and u be constant on each component of $\mathbb{R}^2 \setminus P$. Then $(u, P) \in RGM(\mathbb{R}^2)$.*

Recall that propellers are unions of three half-lines emanating from a common center and making $120°$ angles at that point. Without loss of generality, we can assume that the center is the origin. Let (v, G) be a topological competitor for (u, P), and let R be so large that (v, G) coincides with (u, P) outside of $B(0, R)$. Call a_i, $1 \leq i \leq 3$, the intersections of P with $\partial B(0, R)$. Also call I_1, I_2, and I_3 the three components of $\partial B(0, R) \setminus P$, with I_i opposite to a_i. The topological condition (54.3) says that the three I_j lie in different components of $\mathbb{R}^2 \setminus G$. Grosso modo, our plan is to show that

$$G \cap \overline{B}(0, R) \text{ contains a connected set that contains the three } a_i, \qquad (2)$$

and then deduce from (2) that

$$H^1(G \cap \overline{B}(0, R)) \geq 3R. \qquad (3)$$

Then the minimality condition (54.5) will follow at once, because $H^1(P \cap \overline{B}(0, R)) + \int_{B(0,R) \setminus P} |\nabla u|^2 = 3R$.

We use the same separation result from plane topology as before. By Theorem 14.3 on p. 123 of [Ne] (or Exercise 29.100), there is a (closed) connected subset G_0 of G that separates I_1 from I_2. In particular, G_0 contains a_3, and at least one of the two points a_1 and a_2, because otherwise there would be an arc of $\partial B(0, R)$ that does not meet G_0 and connects I_1 to I_2. [Recall that $G \cap \partial B(0, R) = \{a_1, a_2, a_3\}$.] Assume for definiteness that G_0 contains a_2 and a_3.

We may as well assume that $H^1(G \cap B(0, R))$ is finite, because otherwise there is nothing to prove. Then Proposition 30.17 tells us that there is a simple arc γ_1 in G_0 that goes from a_2 to a_3. [We used Proposition 30.17 because we wanted to make the discussion short, but Proposition 30.14 and a tiny bit of topology would have been enough.] Notice that $\gamma_1 \subset \overline{B}(0, R)$ because if it wandered on the other side of some a_i, it would eventually have to come back to a_i, and would not be simple. So we have a first path $\gamma_1 \subset G \cap \overline{B}(0, R)$ that goes from a_2 to a_3.

We can also do the argument above with the intervals I_2 and I_3, and we get a second simple arc $\gamma_2 \subset G \cap \overline{B}(0, R)$ that goes from a_1 to a_2 or from a_1 to a_3. Then $\gamma_1 \cup \gamma_2$ is connected and contains the three a_i, so (2) holds. We could also deduce the existence of γ_1 and γ_2 from (2) and Proposition 30.14, but this is not really the point here.

Call z the first point of γ_1 when you run along γ_2 from a_1 in the direction of a_2 or a_3, let γ_2' be the subarc of γ_2 between a_1 and z, and set $\gamma = \gamma_1 \cup \gamma_2'$. Thus γ is composed of three simple arcs that go from z to the three a_i. One of these arcs can be reduced to one point, if z coincides with one of the a_i, but this does not matter. Also, the three arcs are disjoint, except for the common origin z. So

$$H^1(G \cap \overline{B}(0, R)) \geq H^1(\gamma) \geq |z - a_1| + |z - a_2| + |z - a_3|. \qquad (4)$$

To complete the proof of (3) and Lemma 1, we just need to check that if we set $f(z) = |z-a_1|+|z-a_2|+|z-a_3|$, then $f(z) \geq 3R$ for all choices of $z \in \overline{B}(0, R)$. By compactness, there is a point $z \in \overline{B}(0, R)$ where $f(z)$ is minimal. If $z \in B(0, R)$, we can differentiate f at z and get that the angles of the three radii $[z, a_i]$ with each other are equal to $120°$, which forces $z = 0$. [See Exercise 15.] Then we can eliminate the case when $z \in \partial B(0, R)$, for instance by almost the same variational argument. So $z = 0$ and $f(z) = 3R$, as desired. Lemma 1 follows. $\square$

Remark 5. A close look at our proof of Lemma 1 (and in particular the cases of equality) shows that if a_1, a_2, and a_3 are three distinct points in the plane, then there is a unique shortest compact connected set γ that contains the a_i, and γ is either composed of two line segments between a_i's, or else a piece of propeller. It is of course easy to determine which of the two is optimal, depending on whether two intervals $[a_i, a_j]$ make an angle larger than $120°$ or not.

If a_1, a_2, and a_3 are contained in a circle ∂B, γ is also the shortest compact set that separates the three components of $\partial B \setminus \{a_1, a_2, a_3\}$ in $\overline{B}$. See Exercise 16 for a slightly different approach to this sort of estimate, and [Morg3] for a generalization to higher dimensions.

Next we want to check the (essential) uniqueness of u in the case of lines and propellers. We shall see later that the essential uniqueness of u given K is a general result for $(u, K) \in RGM(\mathbb{R}^2)$, because of the magic formula that relates the gradient of u to the Beurling transform of the restriction of H^1 to K. See Section 63. We do not know whether this general uniqueness holds in higher dimensions, though.

Lemma 6. *Let $(u, K) \in RGM(\mathbb{R}^2)$ be given, and suppose that K is the empty set, a line, or a propeller. Then u is constant on each connected component of $\mathbb{R}^2 \setminus K$.*

In the three cases, the main point will be that u is harmonic and

$$\int_{B(0,R)\setminus K} |\nabla u|^2 \leq 2\pi R, \tag{7}$$

which comes from the trivial energy estimate (18.8) (or its proof).

If K is empty, u is harmonic on the plane and Harnack (or direct estimates on the Poisson kernel) says that

$$\sup_{x \in B(0,R/2)} |\nabla u(x)| \leq \frac{C}{R^2} \int_{B(0,R)\setminus K} |\nabla u| \leq \frac{C}{R} \left\{ \int_{B(0,R)\setminus K} |\nabla u|^2 \right\}^{1/2}, \tag{8}$$

which tends to 0 by (7). Thus $|\nabla u|$ is zero, and u is constant.

Now assume that K is a line, and choose a component Ω_1 of $\mathbb{R}^2 \setminus K$. By Lemma 17.6 (applied to any ball centered on K), the restriction of u to Ω_1 has a harmonic extension $\widehat{u}_1$ to the plane, obtained from the restriction of u to Ω_1

by symmetry. This extension still satisfies (7), hence (8) says that its gradient vanishes, and u is constant on Ω_1.

Finally assume that K is a propeller and (without loss of generality) that K is centered at the origin. Let Ω_1 be a component of $\mathbb{R}^2 \setminus K$, identify $\mathbb{R}^2$ with the complex plane, and define ψ on Ω_1 by $\psi(z) = z^{3/2}$. Any continuous determination of the square root will do. Thus ψ is a conformal mapping from Ω_1 to the half-plane $H = \psi(\Omega_1)$. Set $v(y) = u(\psi^{-1}(y))$ for $y \in H$. Then v is harmonic, and

$$\int_{B(0,R)\cap H} |\nabla v(y)|^2 = \int_{B(0,R^{2/3})\cap\Omega_1} |\nabla u|^2 \leq 2\pi R^{2/3}, \tag{9}$$

by the conformal invariance of energy integrals (see Lemma 17.8) and (7).

Moreover, Lemma 17.11 tells us that for each R, v minimizes the energy $\int_{B(0,R)\cap H} |\nabla v(y)|^2$ given its boundary values on $H \cap \partial B(0,R)$. Then Lemma 17.6 says that the restriction of v to $B(0,R) \cap H$ has a harmonic extension to $B(0,R)$, obtained by symmetry. This is true for all R, so v has a symmetric harmonic extension to the whole plane. This extension still satisfies (9), which is analogous to (7), but with an even better decay. Then the analogue of (8) for the extension of v shows that it is constant. Thus u is constant on Ω_1, as needed for Lemma 6. $\square$

Proposition 10. *Let* $(u, K) \in RGM(\mathbb{R}^2)$ *be given, and suppose that one of the blow-in limits of* K *is a line or a propeller. Then* K *itself is a line or a propeller, and* u *is constant on each component of* $\mathbb{R}^2 \setminus K$.

Recall that a blow-in limit of K is a closed set K_∞ obtained as the limit of a sequence $\{K_k\}$, where $K_k = r_k^{-1} K$ and the sequence $\{r_k\}$ tends to $+\infty$. Let us assume that K_∞ is a propeller and show that K is a propeller; the case of a line would be similar.

Set $u_k(x) = r_k^{-1/2} u(r_k x)$ for $x \in \mathbb{R}^2 \setminus K_k$. Thus $(u_k, K_k) \in RGM$, by invariance of RGM under dilations. By Proposition 37.8, we can extract a subsequence (which will be denoted the same way), so that $\{(u_k, K_k)\}$ converges to some limit (u_∞, K_∞). We already know that K_∞ is a propeller, and since (54.13) tells us that $(u_\infty, K_\infty) \in RGM$, Lemma 6 says that u_∞ is locally constant on $\mathbb{R}^2 \setminus K_\infty$. Call x_0 the center of the propeller K_∞, and let $r > 0$ be given. Corollary 38.48 says that

$$\lim_{k \to +\infty} \int_{B(x_0,r)\setminus K_k} |\nabla u_k|^2 = 0. \tag{11}$$

Since for k large, $K_k \cap B(x_0,r)$ is as close as we want to the propeller K_∞, we can apply Theorem 53.4 and get that

$$K_k \cap B(x_0, r/2) \text{ is a nice spider.} \tag{12}$$

In fact, Remark 53.39 even says that $K_k \cap B(x_0, r/2)$ is a spider of class $C^{1+\beta}$ for some $\beta > 0$. Since this is true with uniform estimates, this means that if y_1, y_2 lie in the same branch of the spider $K_k \cap B(x_0, r/2)$, then the angle of the tangent lines to K_k at y_1 and y_2 is at most $C|y_1 - y_2|^\beta$. [See Exercise 17 for an argument that avoids this.] Because of this, if w is any point of $B(x_0, r/3)$, say, and if $\rho < r/10$, there is a propeller P (possibly centered out of $B(w, \rho)$) such that

$$\text{dist}(z, P) \leq C\rho^{1+\beta} \text{ for } z \in K_k \cap B(w, \rho). \tag{13}$$

Now fix $R > 0$ and consider $K \cap B(0, R)$. If we took r large enough ($r > 3|x_0|$ to be precise), then the origin lies in $B(x_0, r/3)$, so (13) holds with $w = 0$. Apply (13) with $\rho = r_k^{-1}R$ (which is less than $r/10$ if k is large enough), and recall that $K = r_k K_k$. Set $P' = r_k P$ and $z' = r_k z$. Then $z' \in K \cap B(0, R)$ when $z \in K_k \cap B(0, \rho)$ and (13) is written

$$r_k^{-1} \text{dist}(z', P') \leq C (r_k^{-1}R)^{1+\beta} \text{ for } z' \in K \cap B(0, R). \tag{14}$$

We may now let k tend to $+\infty$ in (14); even though P' depends on k, we get that $K \cap B(0, R)$ is contained in some propeller. This is true for all R, hence the whole K is contained in a line or a propeller.

Recall that for all k, $K \cap B(r_k x_0, r_k r/2) = r_k \cdot [K_k \cap B(x_0, r/2)]$ is a nice spider (by (12)). Since $B(r_k x_0, r_k r/2)$ contains any $B(0, R)$ for k large, we get that K is a line or a propeller (instead of just being contained in it). This also forces u to be locally constant, by Lemma 6. We may even add that since some blow-in limit of K is a propeller, K cannot be a line (and hence it is a propeller). This completes our proof of Proposition 10 when K_∞ is a propeller; the case of a line is easier. □

Exercise 15. Let a_1, a_2, $a_3 \in \mathbb{R}^2$ be given.

1. Compute the gradient of $|x - a_i|$ at $x \neq a_i$.

2. Show that $|x - a_1| + |x - a_2| + |x - a_3|$ can only be minimal when x is one of the a_i or the center of a propeller that contains the a_i (one in each branch).

Exercise 16. Consider a regular hexagon H of unit sidelength, and call $L_1, \ldots, L_6$ its sides. We are interested in estimates on the length of connected sets that touch the three opposite sides L_2, L_4, and L_6.

1. For $x \in H$, call $d_j(x)$, $1 \leq j \leq 3$, the distance from x to the line that contains L_{2j}. Show that $d_1(x) + d_2(x) + d_3(x) = \frac{3\sqrt{3}}{2}$.

2. Show that if $x \in H$ and Γ is a union of three curves that connect x to each of the three L_{2j}, $j \leq 3$, then $H^1(\Gamma) \geq \frac{3\sqrt{3}}{2}$. When is there equality?

3. Prove the same estimate for a compact connected set E that touches the three L_{2j}, or for a compact set E that separates the three L_{2j-1} from each other in H.

4. Return to Γ from Question 2, and suppose that the three curves that compose it are rectifiable. Call π_j the orthogonal projection on the line through the origin with the same direction as L_{2j}, and define a sort of tilt as follows. If v is a unit vector, set $\tau(v) = \inf_{1 \leq j \leq 3} |\pi_j(v)|$, and then set $T^2(\Gamma) = \int_\Gamma \tau(v(x))^2 dH^1(x)$, where $v(x)$ denotes a unit tangent vector to Γ at x (defined almost-everywhere on Γ). Show that $H^1(\Gamma) \geq \frac{3\sqrt{3}}{2} + C^{-1}T^2(\Gamma)$.

Comment. The author finds it amusing that the lengths of all the pieces of propellers centered near the middle of H and that connect the three L_{2j} are the same. The estimate with the tilt may be useful; the author thinks about an analogue of Theorem 53.4 in dimension 3, where we would be led to considering surfaces contained in a bolt $H \times [0,1]$ and that separate the faces $L_{2j-1} \times [0,1]$, and where it could be interesting to estimate an excess amount of surface (compared to the product of a propeller by $[0,1]$) in terms of an appropriate tilt.

Exercise 17. Another way to end the proof of Proposition 10. We start the argument just after (12).

1. Show that K is arcwise connected.

2. Show that for each $x \in \mathbb{R}^2$, the quantity $f(r) = r^{-1} \int_{B(x,r) \setminus K} |\nabla u|^2$ is non-decreasing. [Hint: Proposition 47.7.]

3. Show that $f(r) = 0$. [Hint: (11).] Conclude.

60   Another regularity result for local almost-minimizers; points of low energy

In this section we use a compactness argument, and the fact that the only global minimizers in the plane for which u is constant correspond to lines and propellers, to derive an additional regularity result for general almost-minimizers in a domain. This is Proposition 1, where we show that a control on the normalized energy $\omega_2(x,r)$ (or a variant $\omega_p(x,r)$, $p > 1$) and $h(r)$ is enough to get a nice C^1 curve or spider.

We shall call "point of low energy" a point $x \in K$ such that $\omega_2(x,r)$ tends to 0 when r tends to 0. Proposition 1 says that if $x \in K$ is a point of low energy, there is a neighborhood of x where K is a nice C^1 curve or spider. We shall see that the other points of K (the points of high energy) are those for which $\liminf_{r \to 0} \omega_2(x,r) > 0$. We shall say a few additional things about them in Section 67, at least in the case of global minimizers in $\mathbb{R}^2$.

So we return to the general context of almost-minimizers in a domain Ω and start with the following improvement of Corollary 51.17 and Theorem 53.4.

Proposition 1. *For each choice of $p \in (1, 2]$, there exist constants $\tau > 0$ and $\eta > 0$ such that, if $\Omega \subset \mathbb{R}^2$, $(u, K) \in TRLAM(\Omega, h)$, $x \in K$, and $r > 0$ are such that*

$$B(x, r) \subset \Omega \quad and \quad \omega_p(x, r) + \int_0^r h(t)^{1/2} \frac{dt}{t} \leq \tau, \tag{2}$$

then $K \cap B(x, \eta r)$ is a nice C^1 curve or a nice C^1 spider.

See Standard Assumption 41.1 for the definition of $TRLAM$, Definitions 51.2 and 53.3 for nice curves and spiders, and (23.5) for $\omega_p(x, r)$.

Remarks 3. The difference with Corollary 51.17 and Theorem 53.4 is that we don't need to know that $K \cap B(x, r)$ stays close to a line or a propeller; we shall get it from the compactness argument.

As usual, the condition on h is satisfied when $h(t) \leq C_0 t^\alpha$ for some $\alpha > 0$ and r is small enough; when $(u, K) \in RGM$, the only condition left in (2) is $\omega_p(x, r) \leq \tau$. And $K \cap B(x, \eta r)$ is actually better than C^1.

To prove the proposition, our plan is to show that K is very close to a line or a propeller, and then apply Corollary 51.17 or Theorem 53.4. To measure (bilateral) closeness to lines or propellers, we set

$$\gamma_K(x, r) = \inf_P \Big\{ \sup \big\{ r^{-1} \operatorname{dist}(y, P) \,;\, y \in K \cap B(x, r) \big\}$$

$$+ \sup \big\{ r^{-1} \operatorname{dist}(z, K) \,;\, y \in P \cap B(x, r) \big\} \Big\} \tag{4}$$

for $x \in K$ and $r > 0$, where the infimum is taken over all the propellers P that contain x. Note that we do not require P to be centered at x, or even on $B(x, r)$, so we can choose propellers that coincide with lines near $B(x, r)$. Thus $\gamma_K(x, r)$ is also small when K is close to a line near $B(x, r)$.

Lemma 5. *For each $\varepsilon_1 > 0$ and each $p \in [1, 2]$, we can find constants $\tau_1 > 0$ and $\eta_1 > 0$ such that if $(u, K) \in TRLAM(\Omega, h)$, $x \in K$, and $r > 0$ are such that $B(x, r) \subset \Omega$ and $\omega_p(x, r) + h(r) \leq \tau_1$, then $\gamma_K(x, \eta_1 r) \leq \varepsilon_1$.*

We shall prove this by contradiction and compactness. So we fix p and ε_1, and we assume that for each $k \geq 0$ we can find a domain Ω_k, a gauge function h_k, a pair $(u_k, K_k) \in TRLAM(\Omega_k, h_k)$, and a ball $B(x_k, r_k)$ centered on K_k, contained in Ω_k, and for which the statement of Lemma 5 fails with $\eta_1 = 2^{-k}$ and $\tau_1 = 2^{-4k}$. Thus

$$r_k^{1 - \frac{4}{p}} \Big\{ \int_{B(x_k, r_k) \setminus K_k} |\nabla u_k|^p \Big\}^{2/p} + h_k(r_k) \leq 2^{-4k} \quad \text{but} \quad \gamma_{K_k}(x_k, 2^{-k} r_k) \geq \varepsilon_1. \tag{6}$$

By translation and dilation invariance, we can assume that $x_k = 0$ and $r_k = 2^k$ for all k. Then the domains Ω_k converge to the plane. Set $h_m^*(r) = \sup_{k \geq m} h_k(r)$. Notice that $h_k(r) \leq 2^{-4k}$ for $r \leq r_k$, by (6), so we can easily check that each $h_m^*(r)$ tends to 0 at 0. Thus each h_m^* is a gauge function. Also, $\lim_{m\to+\infty} h_m^*(r) = 0$ for all r, again by (6).

By Proposition 37.8, we can extract a subsequence $\{k_l\}$, so that $\{(u_{k_l}, K_{k_l})\}$ converges to a pair (u, K). First fix m and apply Theorem 38.3 with $h = h_m^*$. We get that (u, K) is a local topological almost-minimizer in the plane, with gauge function $h_m^{*+}(r) = \lim_{s\to r+} h_m^*(s)$.

Next let m go to infinity; we get that (u, K) is a local topological almost-minimizer with gauge function $h = 0$. It is also coral, because K is locally Ahlfors-regular. Hence it is topologically reduced. [See Remark 38.64 or Remark 54.8.] Altogether, $(u, K) \in RGM(\mathbb{R}^2)$.

We claim that u is locally constant on $\mathbb{R}^2 \setminus K$. Indeed, if B is a ball in the plane,

$$\int_{B\setminus K} |\nabla u|^p \leq \liminf_{l\to+\infty} \int_{B\setminus K_{k_l}} |\nabla u_{k_l}|^p \leq \liminf_{l\to+\infty} \int_{B(0,r_l)\setminus K_{k_l}} |\nabla u_{k_l}|^p = 0 \qquad (7)$$

by Proposition 37.18, because $B \subset B(0, r_k)$ for k large, by (6), and because $r_k = 2^k$.

Thus u is locally constant, and Proposition 58.10 says that K is a line or a propeller. [The empty set is impossible, because $0 \in K$.] And $\gamma_{K_{k_l}}(0, 1)$ tends to zero, because K_{k_l} tends to K. This is a contradiction with (6), because $x_k = 0$ and $r_k = 2^k$. Lemma 5 follows. $\qquad\square$

Remark 8. We can use Lemma 5 as a substitute to Corollary 41.5 (on the fact that $\beta_K(x, r)^2 \frac{dH^1(x)dr}{r}$ is a Carleson measure) to find lots of pairs (y, t) such that $K \cap B(y, t)$ is very flat and $\omega_p(y, t)$ is very small, as in Proposition 41.17 (but only with $n = 2$). The point is still that for any $p < 2$, we can find many pairs (x, r) such that $\omega_p(x, r)$ is very small (by (23.15) or Theorem 23.8). Then we can apply Lemma 5 to any such such pair (x, r) and get a new pair (y, t), with $y \in B(x, \eta_1 r)$ and $t = \eta_1 r/10$, so that $\omega_p(y, t)$ is still very small, and in addition $K \cap B(y, t)$ is as flat as we want. [If $K \cap B(x, \eta_1 t)$ is close to a propeller, we can still find the desired point y away from the center of the propeller.]

Some readers will find this approach pleasant, because it does not use any regularity theory for uniformly rectifiable sets (here, Ahlfors-regular curves). And it makes the present book a little more self-contained. However, the author's point of view is that the regularity results for Ahlfors-regular curves are nice and natural, and it is good to avoid compactness arguments when we can. Incidentally, there is no vicious circle in this new approach. We did not need to use the fact that almost-every point of K is a C^1 point to prove Propositions 55.1, 58.1, and 58.10 (used in the proof of Lemma 5). See the various exercises to that effect.

We may now prove Proposition 1. Let $1 < p \leq 2$, $(u, K) \in TRLAM(\Omega, h)$, $x \in K$, and $r > 0$ be given, and assume that (2) holds. Also let ε_1 be small, to be chosen soon, and let η_1 and τ_1 be given by Lemma 5.

Notice that

$$\omega_p(x, r/2) + h(r/2) \le 2^{\frac{4}{p}-1}\omega_p(x, r) + \log(2)^{-1}\int_0^r h(t)^{1/2} \le C\tau \le \tau_1,$$

by (23.5) and (2), and if τ is small enough. Then Lemma 5 says that $\gamma_K(x, \eta_1 r/2) \le \varepsilon_1$. Thus there is a propeller P such that $\text{dist}(y, P) \le \varepsilon_1\eta_1 r/2$ for $y \in K \cap B(x, \eta_1 r/2)$ (we shall not need the other part of (4)).

Let ε be the same as in Theorem 53.4, and first assume that the center of P lies at distance less than $\varepsilon\eta_1 r/4$ from x. If we take ε_1 smaller than $\varepsilon/2$, (53.5) holds for $B(x, \eta_1 r/2)$. Also, $\omega_p(x, \eta_1 r/2) \le \eta_1^{1-\frac{4}{p}}\omega_p(x, r) \le \varepsilon$ if τ is small enough. Theorem 53.4 tells us that $K \cap B(x, \eta_1 r/4)$ is a nice spider. We still have that $K \cap B(x, \rho)$ is a nice C^1 curve or spider for $\rho \le \eta_1 r/4$, by Definitions 51.2 and 53.3. Thus we get the desired result if we choose $\eta \le \eta_1/4$.

Now suppose that the center of P lies at distance at least $\varepsilon\eta_1 r/4$ from x. Then all the points of $K \cap B(x, \varepsilon\eta_1 r/8)$ lie at distance at most $\varepsilon_1\eta_1 r/2$ from the branch of propeller that gets closest to x. Choose ε_1 so small that $\varepsilon_1\eta_1 r/2 \le \varepsilon' r'$, where $r' = \varepsilon\eta_1 r/8$ and ε' is the constant in Corollary 51.17. If in addition τ is small enough (depending on ε and ε' in particular), we can apply Corollary 51.17, and we get that $K \cap B(x, r'/2) = K \cap B(x, \varepsilon\eta_1 r/16)$ is a nice C^1 curve. This takes care of the second case, if we choose $\eta \le \varepsilon\eta_1/16$. This completes our proof of Proposition 1. $\qquad\square$

Lemma 9. *Assume that* $(u, K) \in TRLAM(\Omega, h)$. *If* $K \cap B(x, r)$ *is a nice* C^1 *curve or spider, then* $\lim_{r \to 0} \omega_2(x, r) = 0$.

First assume that $K \cap B(x, r)$ is a nice C^1 curve for some small $r > 0$. Remark 52.27 allows us to apply Lemma 52.9 to pairs (x, t), $t < r$, with $J(x, t)^{-1} = 0$. This yields

$$\omega_2(x, t) \le C\tau + C\tau^{-1}[\beta_K(x, t)^{1/2} + h(t)]. \tag{10}$$

Then we choose $\tau = [\beta_K(x, t)^{1/2} + h(t)]^{1/2}$, let t tend to 0, and get the desired result.

We are left with the case when $K \cap B(x, r)$ is a nice spider centered at x. Then we simply use (53.33) instead of Lemma 52.9, and get that $\lim_{r \to 0} \omega_2(x, r) = 0$ as before.

We can also prove the lemma by compactness, a little like for Lemma 5. Indeed if $\limsup_{r \to 0} \omega_2(x, r) > 0$, we can find a blow-up sequence $\{(u_k, K_k)\}$, with $K_k = r_k^{-1}[K - x]$, $u_k(y) = r_k^{-1/2}u(x + r_k y)$, $\lim_{k \to +\infty} r_k = 0$, and

$$\liminf_{k \to +\infty} r_k^{-1}\int_{B(x, r_k) \setminus K} |\nabla u|^2 > 0. \tag{11}$$

By Proposition 37.8, we can extract a subsequence that converges to a limit (u_∞, K_∞). By Proposition 40.9, $(u_\infty, K_\infty) \in RGM(\mathbb{R}^2)$. Obviously, K_∞ is a line or a propeller, so Lemma 59.6 tells us that u_∞ is locally constant. Then Corollary 38.48, applied with a gauge function $\tilde{h}(t) = h(r_{k_0} t)$ as small as we want, says that

$$\limsup_{k \to +\infty} r_k^{-1} \int_{B(x,r_k) \setminus K} |\nabla u|^2 = \limsup_{k \to +\infty} \int_{B(0,1) \setminus K_k} |\nabla u_k|^2 = 0, \qquad (12)$$

which contradicts (11). $\qquad \square$

Definition 13. Let $(u, K) \in TRLAM(\Omega, h)$ be given. A point of low energy of K is a point $x \in K$ such that

$$\lim_{r \to 0} \omega_2(x, r) = \lim_{r \to 0} \frac{1}{r} \int_{B(x,r) \setminus K} |\nabla u|^2 = 0. \qquad (14)$$

A point of high energy of K is a point $x \in K$ such that $\limsup_{r \to 0} \omega_2(x, r) > 0$.

Remark 15. The dichotomy is clearer than the definition suggests. Let us assume that $\int_0^{r_0} h(r)^{1/2} \frac{dr}{r} < +\infty$ for some $r_0 > 0$, as in Proposition 1. Let $1 \le p \le 2$ be given (and $p = 2$ is already a very good choice), and let $\tau = \tau(p) > 0$ be given by Proposition 1. If we can find $r > 0$ such that $\omega_p(x, r) + \int_0^r h(t)^{1/2} \frac{dt}{t} \le \tau(p)$, then x is a point of low energy of K, by Proposition 1 and Lemma 9. In particular,

$$x \text{ is a point of low energy of } K \text{ as soon as } \liminf_{r \to 0} \omega_p(x, r) < \tau(p). \qquad (16)$$

Equivalently,

$$\liminf_{r \to 0} \omega_p(x, r) \ge \tau(p) \text{ when } x \text{ is a point of high energy of } K. \qquad (17)$$

And again Proposition 1 and Lemma 9 say that

$$x \text{ is a point of low energy of } K \text{ if and only if we can find } r > 0 \atop \text{such that } K \cap B(x, r) \text{ is a nice } C^1 \text{ curve or spider.} \qquad (18)$$

We shall denote by $K^\sharp$ the set of points of high energy of K, and we shall sometimes refer to it as the singular singular set. Note that $K^\sharp$ is closed, for instance because the set of points of K near which K is a nice C^1 curve or spider is open in K. And Theorem 51.20 says that (if $\int_0^{r_0} h(r)^{1/2} \frac{dr}{r} < +\infty$ for some $r_0 > 0$) the Hausdorff dimension of $K^\sharp$ is strictly less than 1. It is tempting to try to study (u, K) in terms of $K^\sharp$; we shall return to this point of view in Section 67.

61   Bonnet's theorem on connected global minimizers

The goal of this section is to prove the following theorem of A. Bonnet.

Theorem 1 [Bo]. *Let $(u, K) \in RGM(\mathbb{R}^2)$ be a reduced global minimizer in the plane, and assume that K is connected. Then (u, K) is one of the pairs defined in (54.14), (54.15), (54.16), and (54.17).*

Thus either K is empty, a line, or a propeller and u is locally constant, or else there is an orthonormal basis of $\mathbb{R}^2$ in which K is the half-line $\{(x, 0) \,;\, x \leq 0\}$, and there exist a constant sign $\pm$ and a constant C such that

$$u(r \cos \theta, r \sin \theta) = \pm \sqrt{\frac{2}{\pi}} \, r^{1/2} \sin \frac{\theta}{2} + C \ \text{ for } r > 0 \text{ and } -\pi < \theta < \pi. \quad (2)$$

Let us rapidly describe the general plan for the proof. First we shall show that for each $x \in \mathbb{R}^2$,

$$\omega_2(x, r) = \frac{1}{r} \int_{B(x,r) \setminus K} |\nabla u|^2 \ \text{ is a nondecreasing function of } r. \quad (3)$$

This will be an easy consequence of Proposition 47.7. A slightly closer look at the proof will also show that if u is not locally constant, the only case when $\omega_2(x, r)$ can be constant is when u is given by a formula like (2). The precise value of the constant $\sqrt{\frac{2}{\pi}}$ will have to be computed separately. It corresponds to the right balance between the two terms of the functional (length and energy). Thus the only possible constant values of $\omega_2(x, r)$ are 0 and 1 (where 1 corresponds to (2)).

Since $\omega_2(x, r)$ is nondecreasing, it has limits $l_0(x)$ at $r = 0$ and $l_\infty(x)$ at $r = +\infty$. Let us assume that (u, K) is not given by (54.14)–(54.17). Then $0 \leq l_0(x) < l_\infty(x) < 2\pi$, where the last inequality comes from the trivial energy estimate.

Next we shall consider any blow-up limit (u_0, K_0) of (u, K) at some $x \in K$. We shall prove that K_0 is connected, and that the analogue of $\omega_2(0, r)$ for (u_0, K_0) is constant, and equal to $l_0(x)$. Similarly, the analogue of $\omega_2(0, r)$ for any blow-in limit of (u, K) is constant, but this time equal to $l_\infty(x)$. From our discussion of the case when $\omega_2(0, r)$ is constant, we shall deduce that $l_0(x) = 0$ and $l_\infty(x) = 1$.

In particular, all points of K are low energy points, as in the last section. Thus K is locally composed of nice curves and spiders. We shall conclude with a small argument that shows that with such a local description, blow-in limits of K cannot be half-lines. This will use the fact that K contains no loops, by Proposition 55.1.

a. Monotonicity and the case when $\omega_2(x, r)$ is constant

Proof of (3). The following simple manipulation is meant to allow us to apply the statements of Lemma 47.4 and Proposition 47.7 directly, but we could also have observed that the proof goes through.

Let $\tau > 0$ be small, to be chosen soon. By invariance of RGM under transla-tions and dilations, it is enough to check that for all $(u, K) \in RGM$, the restriction of $\omega_2(0, \cdot)$ to $(0, \tau)$ is nondecreasing.

Set $\Omega = B(0, 1) \setminus K$, and consider the restriction of u to Ω. The conditions (47.1)–(47.3) (with K replaced with $K \cap B(0, 1)$) are satisfied by definition of RGM. Set

$$E(r) = \int_{B(0,r) \setminus K} |\nabla u|^2 = r \, \omega_2(0, r) \tag{4}$$

for $0 < r < 1$. Lemma 47.4 tells us that the derivative of E exists and is equal to $E'(r) = \int_{\partial B(0,r) \setminus K} |\nabla u|^2$ almost-everywhere, and that $E(r) = \int_0^r E'(t)dt$. In addition, Proposition 47.7 says that

$$E(r) \leq r\, \alpha(r)\, E'(r) \quad \text{for almost every } r \in \mathcal{B}, \tag{5}$$

where $\mathcal{B}$ is the set of radii $r \in (0, 1)$ that satisfy some appropriate condition (47.8), and $2\pi r\alpha(r)$ is the length of the longest connected component of $\partial B(0, r) \setminus K$.

Let us check that if τ is chosen small enough,

$$r \in \mathcal{B} \text{ for all } r < \tau \text{ such that } K \cap \partial B(0, r) \text{ is finite.} \tag{6}$$

Let r be as in (6), and y, $z \in K \cap \partial B(0, r)$ be given. Since K is connected, Proposition 57.1 says that there is a curve $\gamma_{y,z} \subset K$ that goes from y to z, and that the length of $\gamma_{y,z}$ is at most $C|y - z| \leq 2C\tau$. Let H denote the (finite) union of all the curves $\gamma_{y,z}$ that we get this way, when y and z vary. Then H is a compact connected subset of K that contains $K \cap \partial B(0, r)$, and it is contained in $B(0, (2C + 1)\tau) \subset B(0, 1)$ if τ is small enough. Hence $r \in \mathcal{B}$, and (6) holds.

Since $H^1(K \cap B(0, 1)) < +\infty$, Lemma 26.1 tells us that $K \cap \partial B(0, r)$ is finite for almost every $r < 1$. Then (6) says that almost all points of $(0, \tau)$ lie in $\mathcal{B}$, and (5) can be rewritten as

$$E(r) \leq r\, \alpha(r)\, E'(r) \leq r\, E'(r) \quad \text{for almost every } r \in (0, \tau), \tag{7}$$

because $\alpha(r) \leq 1$ by definition. Let us see what this means in terms of

$$f(r) = \omega_2(0, r) = E(r)/r. \tag{8}$$

First, $f'(r)$ exists wherever $E'(r)$ exists, and $f'(r) = E'(r)/r - E(r)/r^2$. Hence $f'(r) \geq 0$ almost-everywhere on $(0, \tau)$, by (7). Now (3) (i.e., the fact that f is nondecreasing) will follow as soon as we show that f is the integral of its derivative, i.e., that

$$f(s) - f(r) = \int_r^s f'(t)dt \text{ for } 0 < r < s < 1. \tag{9}$$

This follows rather easily from the analogous property for E. Let us first deduce it from general facts about $W^{1,1}$, and then give an essentially equivalent direct argument. Since E is the integral of its derivative, the proof of Proposition 9.1

shows that $E \in W^{1,1}_{\text{loc}}((0,1))$ (see (9.3) in particular). Then Exercise 9.15 shows that $f \in W^{1,1}_{\text{loc}}((0,1))$, with the derivative computed above, and Proposition 9.1 says that f is the integral of its derivative. We can also show this directly, by a soft integration by parts. Set $g(r) = 1/r$, so that $f = gE$. Then

$$f(s) - f(r) = g(s)E(s) - g(r)E(r) = g(s)[E(s) - E(r)] + [g(s) - g(r)]E(r), \quad (10)$$

$$g(s)[E(s) - E(r)] = \int_r^s g(s)E'(t)dt = \int_r^s g(t)E'(t)dt + \int_{t=r}^s \int_{u=t}^s g'(u)E'(t)dudt, \quad (11)$$

$$[g(s) - g(r)]E(r) = \int_r^s g'(u)E(r)du = \int_r^s g'(u)E(u)du - \int_{u=r}^s \int_{t=r}^u g'(u)E'(t)dtdu \quad (12)$$

and, since the two double integrals are equal (by Fubini, and because g' and E' are both integrable), we are left with $f(s) - f(r) = \int_r^s [g(t)E'(t) + g'(t)E(t)]dt = \int_r^s f'(t)dt$, as required. This proves (9), and we have already seen that (3) follows. $\square$

Next we want to see what happens when $\omega_2(x, \cdot)$ is constant. We already control the case when u is locally constant everywhere (by Proposition 58.10), so we may restrict to the case when $\omega_2(x, \cdot) > 0$. Also let us take $x = 0$ to simplify.

Lemma 13. *Let $(u, K) \in RGM(\mathbb{R}^2)$, assume that K is connected, and suppose that $f(r) = \omega_2(0, r)$ is constant and nonzero. Then there are constants A, C, and $\theta_0 \in \mathbb{R}$ such that $K = \{(-r\cos\theta_0, -r\sin\theta_0)\,;\, r \geq 0\}$ and*

$$u(r\cos\theta, r\sin\theta) = A\,r^{1/2}\sin\frac{\theta - \theta_0}{2} + C \ \text{ for } r > 0 \text{ and } \theta_0 - \pi < \theta < \theta_0 + \pi. \quad (14)$$

We shall see in Subsection b that $A^2 = 2/\pi$, but let us not worry for the moment. Also, one could prove the same thing with the only information that $f(r)$ is constant on an interval [see [BoDa], Lemma 35.29 on page 208], but this will not be needed here.

Let (u, K) be as in the lemma, and call $f > 0$ the constant value of $f(r)$. Thus $E(r) = fr$ for all r, $E'(r) = f$, and (7) says that $\alpha(r) = 1$ for almost all $r > 0$. [Recall that our restriction to $r \leq \tau$ could be removed by invariance of RGM under dilations.] By definition, $2\pi r\alpha(r)$ is the length of the longest connected component of $\partial B(0, r) \setminus K$, so we get that for almost every $r > 0$,

$$K \cap \partial B(0, r) \text{ has at most one point.} \quad (15)$$

We also want to know what u looks like on $\partial B(0, r) \setminus K$, and for this we need to look at our proof of Proposition 47.7. We start with the information that

(7) and (5) (or equivalently (47.9)) are equalities for almost all $r < \tau$. Let us only consider radii r for which the identities (47.10) and (47.13) hold. We know from Section 47 that this is the case for almost all r, and let us rapidly recall how (47.9) can be deduced from them. Set $I = \partial B(0,r) \setminus K$; because of (15) we can assume that I is an interval or the whole circle. Call m the mean value of u on J, and write that

$$E(r) = \int_I u \frac{\partial u}{\partial r} dH^1 = \int_I [u - m] \frac{\partial u}{\partial r} dH^1, \tag{16}$$

by (47.10) and (47.13). Here $\frac{\partial u}{\partial r}$ denotes the radial derivative of u, and we shall denote by $\frac{\partial u}{\partial \tau}$ the tangential derivative. By Wirtinger's inequality (Lemma 47.18),

$$\int_I (u - m)^2 dH^1 \leq \left(\frac{|I|}{\pi}\right)^2 \int_I \left(\frac{\partial u}{\partial \tau}\right)^2 dH^1 = 4r^2 \int_I \left(\frac{\partial u}{\partial \tau}\right)^2 dH^1. \tag{17}$$

Write $\int_I \left(\frac{\partial u}{\partial \tau}\right)^2 dH^1 = a^2$ and $\int_I \left(\frac{\partial u}{\partial r}\right)^2 dH^1 = b^2$. Then

$$E(r) \leq 2rab \leq r(a^2 + b^2) = r \int_J |\nabla u|^2 = rE'(r) \tag{18}$$

by (16), Cauchy–Schwarz, (17), and (47.5).

On the other hand, we know that $E(r) = fr$ and $E'(r) = f$, so all the inequalities in (18) must be identities. This forces $a^2 = b^2 = f/2$, and since $E(r) = 2rab$ and $E(r) \neq 0$, we must also have an equality in (17). That is, $u - m$ is one of the extremal functions in Wirtinger's inequality.

Let $\theta(r)$ be such that $K \cap \partial B(x,r) = \{(-r\cos\theta(r), -r\sin\theta(r))\}$ if this set has one point, and pick any $\theta(r)$ otherwise (but we shall see soon that this does not happen). By Remark 47.61 (and a small change of variables), we see that there are constants $A(r)$ and $C(r)$ such that

$$u(r\cos\theta, r\sin\theta) = A(r) \sin\frac{\theta - \theta(r)}{2} + C(r) \quad \text{for } \theta(r) - \pi < \theta < \theta(r) + \pi. \tag{19}$$

At this point it is tempting to try to use the fact that u is harmonic, compute everything, and show that $\theta(r)$ and $C(r)$ are constant. If the author is not mistaken, this would lead to a slightly unpleasant computation, and the result would be that there are actually harmonic functions of the form (19) for which $\theta(r)$ is not constant. So let us insist a little and derive additional information.

By (19),

$$\frac{\partial u}{\partial \tau} = r^{-1} \frac{\partial u}{\partial \theta} = (2r)^{-1} A(r) \cos\frac{\theta - \theta(r)}{2}, \quad \text{hence} \quad \int_I \left(\frac{\partial u}{\partial \tau}\right)^2 dH^1 = \frac{\pi}{4r} A(r)^2$$

and so

$$A(r)^2 = \frac{4r}{\pi} \int_I \left(\frac{\partial u}{\partial \tau}\right)^2 dH^1 = \frac{4r}{\pi} a^2 = \frac{2fr}{\pi}. \tag{20}$$

In particular, $A(r) \neq 0$, u has a jump at $(-r\cos\theta(r), -r\sin\theta(r))$, and $K \cap \partial B(0,r)$ has one point. Let us still refrain from computing (although we would have a better chance now) and use the minimality of (u, K) to conclude more rapidly.

Let r be as above, and let us define a new pair $(\widetilde{u}, \widetilde{K})$ as follows. First call $L(r) = \{(-t\cos\theta(r), -t\sin\theta(r)); 0 \leq t \leq r\}$ the line segment between the origin and the point of $K \cap \partial B(0,r)$, set $\widetilde{K} = L(r) \cup [K \setminus B(0,r)]$, keep $\widetilde{u}(y) = u(y)$ on $\mathbb{R}^2 \setminus [K \cup B(0,r)]$, and set

$$\widetilde{u}(t\cos\theta, t\sin\theta) = A(r)\left(\frac{t}{r}\right)^{1/2}\sin\frac{\theta - \theta(r)}{2} + C(r) \tag{21}$$

for $0 < t < r$ and $\theta(r) - \pi < \theta < \theta(r) + \pi$. It is easy to see that $(\widetilde{u}, \widetilde{K})$ is a topological competitor for (u, K) in $B(0,r)$. In particular, there is no difficulty with the definition of limits or restriction of u to $\partial B(0,r) \setminus K$ (because u is harmonic on $\mathbb{R}^2 \setminus K$), $\widetilde{u}$ is continuous across $\partial B(0,r) \setminus K$, and Corollary 14.28 says that $u \in W_{\text{loc}}^{1,2}(\mathbb{R}^2 \setminus \widetilde{K})$. For the topological condition, note that $\partial B(0,r) \setminus \widetilde{K}$ is equal to $\partial B(0,r) \setminus K$ and is connected, so Lemma 20.10 applies. Let us verify that

$$H^1(\widetilde{K} \cap \overline{B}(0,r)) = r \text{ and } \int_{B(0,r)\setminus\widetilde{K}} |\nabla\widetilde{u}|^2 = \frac{\pi}{2}A(r)^2 = fr. \tag{22}$$

The first part is obvious. For the second part, the same sort of computations as for (20), but on the circle $\partial B(0,t)$, give

$$\frac{\partial\widetilde{u}}{\partial\tau} = t^{-1}\frac{\partial\widetilde{u}}{\partial\theta} = \frac{1}{2}(tr)^{-1/2}A(r)\cos\frac{\theta-\theta(r)}{2}, \tag{23}$$

$$\frac{\partial\widetilde{u}}{\partial t} = \frac{1}{2}(tr)^{-1/2}A(r)\sin\frac{\theta-\theta(r)}{2}, \tag{24}$$

$$|\nabla\widetilde{u}|^2 = \left|\frac{\partial\widetilde{u}}{\partial\tau}\right|^2 + \left|\frac{\partial\widetilde{u}}{\partial t}\right|^2 = \frac{1}{4tr}A(r)^2, \tag{25}$$

$$\int_{B(0,r)\setminus\widetilde{K}} |\nabla\widetilde{u}|^2 = \int_0^r \int_{\partial B(0,t)\setminus L(r)} |\nabla\widetilde{u}|^2 dH^1 dt$$
$$= \int_0^r \frac{2\pi t}{4tr}A(r)^2 dt = \frac{\pi}{2}A(r)^2 = fr, \tag{26}$$

where the last part comes from (20). This proves (22).

Since (u, K) is a global minimizer, (54.5) says that

$$H^1(K \cap \overline{B}(0,r)) + \int_{B(0,r)\setminus K} |\nabla u|^2 \leq H^1(\widetilde{K} \cap \overline{B}(0,r)) + \int_{B(0,r)\setminus\widetilde{K}} |\nabla\widetilde{u}|^2 = r + fr. \tag{27}$$

Since $\int_{B(0,r)\setminus K}|\nabla u|^2 = E(r) = fr$ by definition, we get that $H^1(K \cap \overline{B}(0,r)) \leq r$.

On the other hand K is connected, contains the origin (because $K \cap \partial B(0,t)$ has one point for almost all $t > 0$) and a point on $\partial B(0,r)$, so $H^1(K \cap \overline{B}(0,r)) \geq r$. And the only possibility for equality is that $K \cap B(0,r) = L(r)$. [See Exercise 55.] Since we can do this for almost every $r > 0$, we get that

$$K \text{ is the half-line } \{(-r\cos\theta_0, -r\sin\theta_0)\,;\, r \geq 0\} \tag{28}$$

for some $\theta_0 \in \mathbb{R}$. Thus we can take $\theta(r) = \theta_0$ in the representation (19), and (20) forces us to take $A(r) = \pm\sqrt{\frac{2fr}{\pi}}$. We still need to check that $C(r)$ and the sign $\pm$ are constant. Note that $C(r) = u(r\cos\theta_0, r\sin\theta_0)$ by (19), so $C(r)$ is smooth. This forces the sign $\pm$ to be constant (by continuity and because $f > 0$).

Let us compute the radial derivative on $\partial B(0,r)$ one last time.

$$\frac{\partial u}{\partial r} = A'(r)\sin\frac{\theta - \theta_0}{2} + C'(r) = \pm\sqrt{\frac{f}{2\pi r}}\sin\frac{\theta - \theta_0}{2} + C'(r). \tag{29}$$

Since the functions $\sin\frac{\theta-\theta_0}{2}$ and 1 are orthogonal on $(\theta_0 - \pi, \theta_0 + \pi)$, (29) yields

$$\int_{\partial B(0,r)\backslash K} \left(\frac{\partial u}{\partial r}\right)^2 dH^1 = \frac{f}{2} + 2\pi r C'(r)^2. \tag{30}$$

But $\int_{\partial B(0,r)\backslash K} \left(\frac{\partial u}{\partial r}\right)^2 dH^1 = b^2 = \frac{f}{2}$ (see just below (17) for the definition, and three lines below (18) for the equality). Then $C(r)$ is constant, and Lemma 13 follows. $\qquad\square$

b. Why is the constant equal to $\sqrt{\frac{2}{\pi}}$?

Let us see why a representation of (u, K) as in Lemma 13 can only give a global minimizer if $A = \pm\sqrt{\frac{2}{\pi}}$. This fact is fairly simple, and was already known by Mumford and Shah [MuSh]. The converse, i.e., the fact that (14) actually gives a minimizer if $A = \pm\sqrt{\frac{2}{\pi}}$, is true, but much more delicate. See Section 62.

So far, we only know that $A^2 = 2f/\pi$, where f is the constant value of $\omega_2(0,r) = E(r)/r$. [Compare (20), (14), and (19).] The fact that $f = 1$ and $A^2 = 2/\pi$ will come from the balance between the length and energy terms in the functional. We shall see that if we picked $A^2 < 2/\pi$, (u, K) would not be a global minimizer because we could make K a little shorter and save more length than we lose in energy. While if $A^2 > 2/\pi$, we could extend K a little, and save more energy than what we lose in length.

Before we start the computations, let us remark that we shall see another way to compute the constant A^2. Indeed, Theorem 63.2 will give a formula for the square of the complex derivative of u, in terms of K, which will immediately give the form of u if K is a half-line. The two methods are not so different, because Theorem 63.2 will also be derived from variational principles.

We are now ready to compute A^2 when (u, K) is as in Lemma 13. Clearly it is enough to do this when $\theta_0 = C = 0$ in (14). We can also assume that $A > 0$ (because changing the sign of u does not modify the minimizing property of (u, K)). Thus $K = (-\infty, 0]$ is the negative real line (when we identify $\mathbb{R}^2$ with the complex plane) and

$$u(r \cos \theta, r \sin \theta) = A\, r^{1/2} \sin(\theta/2) \quad \text{for } r > 0 \text{ and } -\pi < \theta < \pi. \qquad (31)$$

We want to compare (u, K) with competitors $(u_\pm, K_\pm)$ obtained by adding or removing a little interval near the origin. It will be more convenient to do this with intervals of length 1 and look at our pairs from very far, rather than using tiny intervals and staying in the unit ball. Of course the two are equivalent, by dilation invariance. Set $K_+ = (-\infty, 1]$ and $K_- = (-\infty, -1]$, and let $R > 0$ be large. [It will be easier to do asymptotic estimates with R tending to $+\infty$.] Denote by $u_\pm$ the harmonic function on $\Omega_\pm = B(0, R) \setminus K_\pm$ whose boundary values on $\partial B(0, R) \setminus K$ coincide with those of u, and which minimizes $E_\pm = \int_{B(0,R) \setminus K_\pm} |\nabla u_\pm|^2$. One way to obtain $u_\pm$ is to send $\Omega_\pm$ conformally onto a half-disk (by a composition of $z \to (z \mp 1)^{1/2}$ and a smooth conformal mapping), and use the conformal invariance of our minimization problem (as in Lemma 17.11) to reduce to the case of Example 17.5 and Lemma 17.6. If we proceed this way it is clear that $u_\pm$ is smooth all the way up to the boundary on $\Omega_\pm$, except for a singularity like $(z \mp 1)^{1/2}$ at the point ± 1. This is good to know, because we intend to do integrations by parts.

Out of $B(0, R)$, we keep $u_\pm = u$. It is easy to see that $(u_\pm, K_\pm)$ is a topological competitor for (u, K) in $B(0, R)$. Since $(u, K) \in RGM$, we get that

$$\Delta_+ = \int_{B(0,R) \setminus K} |\nabla u|^2 - E_+ \le 1 \quad \text{and} \quad \Delta_- = E_- - \int_{B(0,R) \setminus K} |\nabla u|^2 \ge 1. \qquad (32)$$

We start with the computation of Δ_+, which incidentally is the same as in the beginning of Section 21. We apply Green's formula to the vector-valued function $h = (u - u_+)\nabla(u + u_+)$. Since

$$\operatorname{div} h = \nabla(u - u_+) \cdot \nabla(u + u_+) + (u - u_+)\Delta(u + u_+) = |\nabla u|^2 - |\nabla u_+|^2 \qquad (33)$$

because u and u_+ are harmonic, we get that

$$\Delta_+ = \int_{\Omega_+} |\nabla u|^2 - |\nabla u_+|^2 = \int_{\partial \Omega_+} (u - u_+) \frac{\partial(u + u_+)}{\partial n}. \qquad (34)$$

Recall that $\partial \Omega_+ = \partial B(0, R) \cup (-R, 1]$, with $(-R, 1]$ counted twice with different orientations. There is no contribution from $\partial B(0, R)$, because $(u - u_+)$ vanishes there. Similarly, $\frac{\partial(u + u_+)}{\partial n} = 0$ on $(-R, 0)$ by Proposition 17.15 (the Neumann condition), and we are left with the (double) contribution of $(0, 1)$. Since in addition $\frac{\partial u_+}{\partial n} = 0$ on $(0, 1)$, we are left with the integral of $(u - u_+)\frac{\partial u}{\partial n}$. Every point of $(0, 1)$

is counted twice, with opposite values of $\frac{\partial u}{\partial n}$. Thus the two contributions of $u\frac{\partial u}{\partial n}$ cancel, and we are left with

$$\Delta_+ = \int_{(0,1)} \mathrm{Jump}(u_+)\frac{\partial u}{\partial n}. \tag{35}$$

Here $\frac{\partial u}{\partial n}$ is just the derivative of u in the vertical direction, and $\mathrm{Jump}(u_+)$ is the difference of the two boundary values of u_+ (the one from above minus the one from below). Now (31) says that $\frac{\partial u}{\partial n}(r,0) = \frac{1}{2}Ar^{-1/2}$, so

$$\Delta_+ = \frac{A}{2}\int_0^1 \mathrm{Jump}(u_+)(r)\, r^{-1/2}dr. \tag{36}$$

The precise computation of $\mathrm{Jump}(u_+)$ is unpleasant (if even possible), but fortunately we shall see that $u_+(z)$ is quite close to $v(z) = u(z-1)$, whose jump at r is $2A(1-r)^{1/2}$ by (31). Let us compute

$$\Delta'_+ = \frac{A}{2}\int_0^1 \mathrm{Jump}(v)(r)\, r^{-1/2}dr = A^2\int_0^1 (1-r)^{1/2}\, r^{-1/2}dr. \tag{37}$$

Set $r = \sin^2 t$, $t \in [0,\pi/2]$, so that $dr = 2\sin t \cos t\, dt$ and $(1-r)^{1/2}r^{-1/2} = \cos t(\sin t)^{-1}$. Then

$$\Delta'_+ = 2A^2\int_0^{\pi/2} \cos^2 t\, dt = \frac{\pi}{2}A^2. \tag{38}$$

We still need to estimate $\Delta_+ - \Delta'_+$. Note that $v(z) = u(z-1)$ is a harmonic function on Ω_+ whose normal derivative $\frac{\partial v}{\partial n}$ vanishes on $(-R,1)$, just like u_+. The boundary values on $\partial B(0,R)$ are different, but not so much: $|v(z) - u_+(z)| = |u(z-1) - u(z)| \le CAR^{-1/2}$ on $\partial B(0,R)$, by direct calculation.

We have a maximum principle here. Indeed, set $\varphi(z) = (z-1)^{1/2}$ on Ω_+, where we choose the natural square root, so that $\Omega_1 = \varphi(\Omega_+)$ lies just on the right of the vertical axis (see Figure 1). Set $w = (v-u_+)\circ\varphi^{-1}$. This is a harmonic

Figure 1

function on Ω_1, and it has a smooth extension up to the boundary (except perhaps at the origin and the two corners $\pm i\sqrt{R+1}$), because v, u_+ and φ are smooth up to the boundary, except at 0 and (perhaps) the corners. In addition, the Neumann condition $\frac{\partial(v-u_+)}{\partial n} = 0$ on $(-R,1)$ is preserved by φ, so the normal derivative of w vanishes on the vertical piece of $\partial\Omega_1$, i.e., on $\varphi((-R,1))$.

Call Ω_2 the union of Ω_1, its symmetric image with respect to the vertical axis, and $\varphi((-R,1))$. Because of the Neumann condition, the symmetric extension of w to Ω_2 is harmonic. [See Exercise 56 or 57.] Note that $\partial\Omega_2$ is composed of $\varphi(\partial B(0,R))$ and its symmetric image, so $|w(z)| \leq CAR^{-1/2}$ on $\partial\Omega_2$. The maximum principle says that $|w(z)| \leq CAR^{-1/2}$ on Ω_2 , and so $|v(z) - u_+(z)| \leq CAR^{-1/2}$ on Ω_+. Then

$$|\Delta_+ - \Delta'_+| \leq \frac{A}{2} \int_0^1 |\operatorname{Jump}(u_+ - v)(r)|\, r^{-1/2} dr \leq C'A^2 R^{-1/2}, \qquad (39)$$

by (36) and (37). Altogether, $|\Delta_+ - \frac{\pi}{2}A^2| \leq C'A^2R^{-1/2}$ by (39) and (38), and since we can take R as large as we want, the first half of (32) says that $\frac{\pi}{2}A^2 \leq 1$.

The estimates for Δ_- will be similar. The same computations as for (35) yield

$$\Delta_- = \int_{(-1,0)} \operatorname{Jump}(u) \frac{\partial u_-}{\partial n} . \qquad (40)$$

Since $\operatorname{Jump}(u)(-r) = 2Ar^{1/2}$ by (31), we get that

$$\Delta_- = 2A \int_0^1 \frac{\partial u_-}{\partial y}(-r)\, r^{1/2} dr. \qquad (41)$$

This time we use the auxiliary function $v(z) = u(z+1)$. If we replace u_- with v in (41), we get

$$\Delta'_- = 2A \int_0^1 \frac{\partial v}{\partial y}(-r)\, r^{1/2} dr = A^2 \int_0^1 (1-r)^{-1/2}\, r^{1/2} dr = \frac{\pi}{2}A^2, \qquad (42)$$

by the same sort of computation as for (38).

To estimate $\Delta'_- - \Delta_-$, set $\Omega_- = B(0,R) \setminus (-R,-1]$, define $\varphi(z) = (z+1)^{1/2}$ on Ω_- and set $w = (v - u_-) \circ \varphi^{-1}$ on $\Omega_3 = \varphi(\Omega_-)$. As before, w has a harmonic extension to a larger domain Ω_4, obtained from Ω_3 by symmetrization with respect to the line segment $\varphi((-R,-1))$. Call $\widetilde{w}$ this extension. We still have that $|\widetilde{w}(\xi)| \leq CAR^{-1/2}$ on Ω_3, by the maximum principle. Then $|\nabla w(\xi)| \leq CAR^{-1}$ for $\xi \in B(0,1)$, because Ω_4 contains $B(0,\sqrt{R}/2)$. We are interested in $\xi = \varphi(-r)$, where $0 < r < 1$. Since $v - u_- = w \circ \varphi$, the chain rule gives $|\nabla(v - u_-)(-r)| = |\varphi'(-r)||\nabla w(\xi)| \leq CAR^{-1}(1-r)^{-1/2}$. Finally

$$|\Delta'_- - \Delta_-| = 2A \left| \int_0^1 \frac{\partial(v - u_-)}{\partial y}(-r)\, r^{1/2} dr \right|$$

$$\leq CA^2 R^{-1} \int_0^1 (1-r)^{-1/2}\, r^{1/2} dr \leq C'A^2 R^{-1} \qquad (43)$$

by (41) and (42). Since $\Delta'_- = \frac{\pi}{2}A^2$ (by (42)), $\Delta_- \geq 1$ (by (32)), and we can take R as large as we want, we get that $\frac{\pi}{2}A^2 \geq 1$. Altogether, $\frac{\pi}{2}A^2 = 1$, and this completes our verification. $\qquad \square$

Remark 44. If the pair (u, K) is given by (14) with $A = \pm\sqrt{\frac{2}{\pi}}$, then

$$\omega_2(0, r) = r^{-1} \int_{B(0,r) \setminus K} |\nabla u|^2 = 1. \tag{45}$$

This follows from our proof of (22) (take $A(r) = \pm\sqrt{\frac{2r}{\pi}}$ in (21)).

c. A local description of K and the conclusion

We want to prove Theorem 1 now. Let (u, K) be a reduced global minimizer in the plane, and suppose that K is connected. We already know from Proposition 58.10 that if u is locally constant, then K is empty, a line, or a propeller. Thus it will be enough to assume that u is not locally constant, and prove that (u, K) is as in (54.17).

Let x be any point of K. We know from (3) that $\omega_2(x, r)$ is a nondecreasing function of r, so we may set

$$l_0(x) = \lim_{r \to 0} \omega_2(x, r) \quad \text{and} \quad l_\infty(x) = \lim_{r \to +\infty} \omega_2(x, r). \tag{46}$$

We already know that (u, K) is as in (54.17) if $\omega_2(x, \cdot)$ is constant, so we may assume that $l_0(x) < l_\infty(x)$ for every $x \in K$. Our plan for the near future is to show that $l_0(x) = 0$ and $l_\infty(x) = 1$. We start with $l_\infty(x)$, and to this effect we consider blow-in limits of (u, K).

Let $\{r_k\}$ be a sequence of positive numbers that tends to $+\infty$, and set

$$K_k = r_k^{-1}[K - x] \quad \text{and} \quad u_k(y) = r_k^{-1/2} u(x + r_k y) \text{ for } y \in \mathbb{R}^2 \setminus K_k. \tag{47}$$

Let us assume that the sequence $\{(u_k, K_k)\}$ converges to some limit (v, G). We can certainly find sequences like this; in fact, Proposition 37.8 says that given any sequence $\{r_k\}$ that tends to $+\infty$, we can extract a subsequence for which $\{(u_k, K_k)\}$ converges. The limit (v, G) is a reduced global minimizer, by the dilation invariance of RGM and (54.13). Let us first check that

$$r^{-1} \int_{B(0,r) \setminus G} |\nabla v|^2 = l_\infty(x) \text{ for all } r > 0. \tag{48}$$

Fix $r > 0$ and set $a_k = r^{-1} \int_{B(0,r) \setminus K_k} |\nabla u_k|^2$. Then

$$a_k = r^{-1} r_k^{-1} \int_{B(x,rr_k) \setminus K} |\nabla u|^2 = \omega_2(x, rr_k) \text{ by (47)},$$

and so $\lim_{k \to +\infty} a_k = l_\infty(x)$. On the other hand, a_k tends to the left-hand side of (48) because $\{(u_k, K_k)\}$ converges to (v, G), by Proposition 37.18 (for the easy inequality), and Corollary 38.48 (with $h = 0$) for the hard part; (48) follows.

Next we claim that G is connected. We cannot simply say that G is a limit of connected sets, but the chord-arc property of K will do the job. Let y, z be points of G, and let $\{y_k\}$ and $\{z_k\}$ be sequences that converge to y and z respectively, with $y_k, z_k \in K_k$. Since K_k is connected, Proposition 57.1 says that we can find a curve $\gamma_k \subset K_k$, that goes from y_k to z_k, and whose length is at most $C|y_k - z_k| \le C'$. We can find parameterizations $f_k : [0, 1] \to K_k$ of the γ_k, that are uniformly Lipschitz. Then we can extract a subsequence that converges to some Lipschitz function f. Since $f(0) = y$, $f(1) = z$, and $f(t) \in G$ for $t \in [0, 1]$ (because $\{K_k\}$ converges to G), we found a curve in G that goes from y to z. Hence G is connected.

Now (48) says that the analogue of $\omega_2(0, r)$ for (v, G) is constant and equal to $l_\infty(x)$. Recall that $l_\infty(x) > 0$, since it is larger than $l_0(x)$. By Lemma 13 and Subsection b, (v, G) is as in (54.17), and Remark 44 says that $l_\infty(x) = 1$.

Another piece of information that we shall use is that $H^1(G \cap B(0, 1)) = 1$. By Corollary 38.48, it implies that $\limsup_{k \to +\infty} H^1(K_k \cap B(0, 1)) \le 1$, or equivalently

$$\limsup_{k \to +\infty} r_k^{-1} H^1(K \cap B(x, r_k)) \le 1. \tag{49}$$

Next consider blow-up limits. That is, define (u_k, K_k) as in (47), but this time with a sequence $\{r_k\}$ that tends to 0. As before, we can find $\{r_k\}$ so that (u_k, K_k) converges to some limit (w, H), and $(w, H) \in RGM(\mathbb{R}^2)$. Moreover,

$$r^{-1} \int_{B(0, r) \setminus H} |\nabla w|^2 = l_0(x) \text{ for all } r > 0, \tag{50}$$

by the same argument as for (48). Once again, the analogue of $\omega_2(0, r)$ for (w, H) is constant equal to $l_0(x)$. Since H is also connected (with the same proof), Lemma 13 and Subsection b tell us that (w, H) is as in (54.17), or else w is locally constant. The first option is impossible, by Remark 44 and because $l_0(x) < l_\infty(x) = 1$. So w is locally constant, and H is a line or a propeller. [The empty set is excluded because we started with $x \in K$.] And $l_0(x) = 0$.

Thus every point of K is a point of low energy, and Proposition 60.1 tells us that for each $x \in K$, there is a positive radius $r(x)$ such that

$$K \cap B(x, r(x)) \text{ is a nice } C^1 \text{ curve or a nice } C^1 \text{ spider.} \tag{51}$$

We shall use this to find a long chord-arc curve in K and contradict (49). The construction may be useful in different contexts, so let us give a general statement.

Lemma 52. *Let (u, K) be a reduced global minimizer, and let K_0 be a (nonempty) connected component of K. Suppose that every point $x \in K_0$ is a point of low energy (or equivalently, that we can find $r(x) > 0$ such that (51) holds). Then for each $x \in K_0$, we can find an injective Lipschitz mapping $z : \mathbb{R} \to K_0$ such that $z(0) = x$, $|z'(t)| = 1$ for almost every $t \in \mathbb{R}$, and*

$$\lim_{t \to \pm\infty} |z(t)| = +\infty. \tag{53}$$

Thus every $x \in K_0$ lies in a simple rectifiable curve in K_0, with two infinite branches. We also know from Proposition 57.1 that this curve is chord-arc, and it would be fairly easy to check that it is piecewise C^1.

Let $x \in K_0$ be given. Call $\mathcal{Z}$ the set of pairs (I, z), where I is an interval that contains a neighborhood of the origin, and $z : I \to K_0$ is an injective Lipschitz mapping such that $|z'(t)| = 1$ for almost every $t \in I$.

Because of (51), $\mathcal{Z}$ is not empty. We have a natural order on $\mathcal{Z}$: we say that $(I, z) \prec (J, w)$ when $I \subset J$ and w is an extension of z. By abstract nonsense, (and without the help of Zorn!) we can find a maximal pair (I, z). Let us check that $I = \mathbb{R}$ and z satisfies the properties required for the lemma.

First assume that I has a finite upper bound b. Note that since z is Lipschitz, we can define $z(b)$ even if $b \notin I$. Actually, if $b \notin I$, the extension of z to $I \cup \{b\}$ does not define a pair in $\mathcal{Z}$ (by maximality), and the only reason why this can be the case is that z is no longer injective, i.e., $z(b) \in z(I)$. Let $a \in I$ be such that $z(a) = z(b)$. Then the restriction of z to $[a, b]$ defines a simple loop in K (because the restriction of z to $[a, b)$ is injective). This is impossible, by (57.3). So $b \in I$.

Let us apply (51) to the point $y = z(b)$. Note that $K \cap B(y, r(y)) \setminus \{y\}$ is composed of two or three disjoint C^1 arcs, all contained in K_0 (because $y \in K_0$). Suppose for a second that $z(I \setminus \{b\})$ meets two of them, and let $s, t \in I \setminus \{b\}$ be such that $z(s)$ and $z(t)$ lie on two different branches of $K \cap B(y, r(y)) \setminus \{y\}$. Then we have two different simple curves in K that go from $z(s)$ to $z(t)$, the arc $z([s, t])$ (that does not contain y because z is injective and $b \notin [s, t]$), and another one that we get from (51), which is contained in $B(y, r(y))$ and goes through y. This contradicts Proposition 57.1.

So $z(I) \setminus \{b\}$ only meets one of the branches of $K \cap B(y, r(y)) \setminus \{y\}$. We can use this to extend z on the interval $I' = I \cup [b, b + r(y))$, in such a way that in particular z is injective on $[b, b + r(y))$ and $z((b, b + r(y)))$ is contained in another branch of $K \cap B(y, r(y)) \setminus \{y\}$ than the one that contains $z(I)$. Then the extension is injective, and the pair (I', z) lies in $\mathcal{Z}$. This contradiction with the maximality of (I, z) proves that I has no upper bound $b \in \mathbb{R}$. We can prove the same way that I is not bounded from below, hence $I = \mathbb{R}$.

Finally, suppose that (53) does not hold. Then there is a sequence $\{t_n\}$, that tends to $+\infty$ or $-\infty$, and for which $z(t_n)$ stays bounded. We can extract a subsequence so that $|t_m - t_n| \geq 1$ for $m \neq n$ and $\{z(t_n)\}$ converges to some $y \in K_0$. By (51), $K \cap B(y, r(y)) \setminus \{y\}$ is composed of two or three C^1 arcs γ_i, and for at least one i, there are infinitely many values of n such that $z(t_n) \in \gamma_i$. [Recall that at most one $z(t_n)$ is equal to y, by injectivity.] By definition of z (and if we choose $r(y)$ small enough), γ_i is contained in $z([t_n - 1/3, t_n + 1/3])$ for each such n, a contradiction with the injectivity of z. Lemma 52 follows. $\qquad \square$

We may now complete the proof of Theorem 1. Let (u, K) be our global minimizer in the plane, with K connected, and which we assumed not to be given by (54.14)–(54.17). Let z be as in Lemma 52, with $K_0 = K$ and any choice of

$x \in K$. Set $\gamma_+ = z((0, +\infty))$ and $\gamma_- = z((-\infty, 0))$. These curves are disjoint and contained in K, so

$$H^1(K \cap B(x, r)) \geq H^1(\gamma_+ \cap B(x, r)) + H^1(\gamma_- \cap B(x, r)) \geq 2r \qquad (54)$$

because $\gamma_\pm$ starts from x and goes to infinity (by (53)), and by Remark 2.11. Now this contradicts (49), and Theorem 1 follows. See Exercise 59 for another way to conclude from Lemma 52, but without (49). $\qquad \square$

Exercise 55. Let K be a connected set in the plane such that K contains the origin and a point of $\partial B(0, 1)$, and suppose that $H^1(K \cap B(0, 1)) = 1$. Show that $K \cap B(0, 1)$ is a line segment. Hint for a cheap way: you may verify first that for each $r < 1$, there is a simple curve in K that goes from 0 to $\partial B(0, r)$. Then check that all the points on the curve are aligned. Hint for a way without Proposition 30.14 or 30.17: first check that $K \cap \partial B(0, r)$ has exactly one point for almost all $r < 1$, and then that these points are aligned.

Exercise 56. [A reflection principle for harmonic functions.] Let L denote the vertical axis in $\mathbb{R}^2$ and let D_1, D_2 denote the two components of $B(0, 1) \setminus L$, with D_1 on the right. Suppose u is harmonic on D_1 and of class C^1 on $\overline{D}_1$, and that it satisfies the Neumann boundary condition $\frac{\partial u}{\partial n} = 0$ on $L \cap B(0, 1)$.

1. Call $a(r)$ the mean value of u on $\partial B(0, r) \cap D_1$. Show that

$$a'(r) = \frac{1}{\pi r} \int_{\partial B(0,r) \cap D_1} \frac{\partial u}{\partial r}.$$

2. Check that $a(r)$ is constant.

3. Extend u to $B(0, 1)$ by symmetry with respect to L. Show that (the extension of) u is harmonic. Check that this applies to w a little before (39).

Exercise 57. [A more complicated way to get the reflection principle.] Let D_1, D_2, and u be as in Exercise 56.

1. Show that u is the (harmonic) function on D_1 that minimizes $\int_{D_1} |\nabla u|^2$ with the given boundary values on $\partial B(0, 1) \cap \overline{D}_1$. Hint: (34) or Exercise 17.20.

2. Show that the symmetric extension of u to $B(0, 1)$ is harmonic. Hint: 17.6.

Exercise 58. Find a sequence of connected sets (and even Ahlfors-regular curves) in the plane that converges to $(-\infty, -1] \cup [0, +\infty)$. [Which is why we used Proposition 57.1 to prove that G is connected, a few lines after (48).]

Exercise 59. Show that the existence of a curve $z(\mathbb{R})$ as in Lemma 52 contradicts the fact that some blow-in limit of K is a half-line. Hint: you can do this with Proposition 57.1. You may also use Corollary 52.25 and Remark 52.27 to find large radii r such that $K \cap B(x, r)$ has only one point.

62   Cracktip is a global minimizer

To simplify some statements, we shall mean by cracktip a pair (u, K) such that, in some orthonormal basis of the plane,

$$K = \{(x, 0) \, ; x \leq 0\} \quad \text{and} \quad u(r \cos \theta, r \sin \theta) = \pm \sqrt{\frac{2}{\pi}} \, r^{1/2} \sin \frac{\theta}{2} + C \qquad (1)$$

for $0 < r < +\infty$ and $-\pi < \theta < \pi$ (as in (54.17)). In this section we would like to say why

$$\text{cracktips are reduced global minimizers in the plane.} \qquad (2)$$

We shall also discuss the two following connected facts:

> if (u, K) is a reduced global minimizer in the plane and if
> there is a connected component K_0 of K such that $K \setminus K_0$
> is bounded, then (u, K) is one of the minimizers of the
> standard list (54.14)–(54.17);
>
> $\qquad\qquad\qquad\qquad\qquad\qquad\qquad\qquad\qquad\qquad\qquad (3)$

> if (u, K) is a reduced global minimizer and we can find a
> sequence $\{r_k\}$ such that $\lim_{k \to +\infty} r_k = +\infty$ and $\{r_k^{-1} K\}$
> converges a half-line, then (u, K) is a cracktip.
>
> $\qquad\qquad\qquad\qquad\qquad\qquad\qquad\qquad\qquad\qquad\qquad (4)$

These results are proved in [BoDa], and since the proof is quite long and a little bit in the same style as this book, we shall only give a rapid sketch here, maybe even with a few tiny lies, and refer to [BoDa] for most proofs. See Section 2 of [BoDa] for a similar, a little shorter attempt.

It would be very nice if one could prove (1) with a simple calibration argument, but the author doubts that this will be possible.

So let us try to see why cracktips should be global minimizers. It is amusing (at least to the author) that the argument is quite elaborate and uses almost every known theorem about Mumford-Shah minimizers. Of course it is enough to consider the pair (u, K) where K is the negative real axis (i.e., without having to pick a new orthonormal basis), and $u(r \cos \theta, r \sin \theta) = \sqrt{\frac{2}{\pi}} \, r^{1/2} \sin \frac{\theta}{2}$ (i.e., with a positive sign and $C = 0$ in (1)). As often, it is simpler to argue by contradiction, so we assume that we can find a strictly better competitor $(\widetilde{u}, \widetilde{K})$ in some ball B. That is,

$$\Delta = H^1(K \cap \overline{B}) + \int_{B \setminus K} |\nabla u|^2 - H^1(\widetilde{K} \cap \overline{B}) - \int_{B \setminus \widetilde{K}} |\nabla \widetilde{u}|^2 > 0. \qquad (5)$$

The general idea is to try to find so much information about $(\widetilde{u}, \widetilde{K})$ that eventually we get a contradiction. And for this it would be very useful if $(\widetilde{u}, \widetilde{K})$ were a minimizer, or something like this. We have a small problem here, about lack of compactness. We cannot find $(\widetilde{u}, \widetilde{K})$ so that Δ is maximal, because the dilation of

$(\widetilde{u}, \widetilde{K})$ by a factor λ will give a competitor $(\widetilde{u}_\lambda, \widetilde{K}_\lambda)$ on λB for which the analogue of Δ in (5) is $\lambda \Delta$. We can also try to minimize Δ among competitors in a fixed ball B, but then we shall not get a minimizer in the plane, and $(\widetilde{u}, \widetilde{K})$ will be hard to study because of the boundary constraints on ∂B.

The functionals J_R. The solution that we shall use is to minimize a slightly different functional. Set $L = (-\infty, -1]$ (a strict subset of K), and fix a large radius R (for the moment). Consider only the class $\mathcal{U}_R$ of pairs (v, G) that are competitors for (u, K) in $B(0, R)$ and for which G contains L. Also let h be a smooth positive function on $(0, +\infty)$, to be chosen soon. This function has no relation with the gauge functions from the previous parts, we are just trying to keep some of the notation of [BoDa]. We want to minimize the functional

$$J_R(v, G) = h(H^1(G \setminus L)) + \int_{B(0,R) \setminus G} |\nabla v|^2 \tag{6}$$

on the class $\mathcal{U}_R$.

By dilation invariance, we can always find a competitor $(\widetilde{u}, \widetilde{K})$ as above, and such that $\overline{B} \subset B(0,1)$, $\widetilde{K} \supset L$, and (5) holds. Then $(\widetilde{u}, \widetilde{K}) \in \mathcal{U}_R$ for $R > 1$. Set

$$\eta(R) = \inf \{ J_R(v, G) ; (v, G) \in \mathcal{U}_R \}. \tag{7}$$

We shall choose h so that $h(x) = x$ for $x \leq H^1(\widetilde{K} \setminus L)$. Then (5) says that

$$J_R(u, K) \geq J_R(\widetilde{u}, \widetilde{K}) + \Delta \geq \eta(R) + \Delta \quad \text{for every } R > 1. \tag{8}$$

We shall need to find minimizers for J_R on $\mathcal{U}_R$. For this it is easier to choose h so that it varies very slowly. Let us take no chance and require that h be nondecreasing, convex, of class C^1, that $h(x) = x$ for $x \leq 2$, and that $h'(Bx) \leq 2h'(x)$ for some very large constant B. Probably much less would be needed. We also require that

$$h(x) = Ax \quad \text{for } x \text{ large enough}, \tag{9}$$

but this will only show up later, when we show that $G \setminus L$ is bounded (uniformly in R).

We need to study J_R a little bit. We start with regularity properties of G when $(v, G) \in \mathcal{U}_R$ is a reduced minimizer for J_R. In particular, we are interested in local Ahlfors-regularity and the concentration property. As long as we stay at positive distance from $L \cup \partial B(0, R)$, we can get them with almost the same proof as in Part C. We have to be a little careful about the effect of the nonlinearity of h, but there is no serious trouble because h is so mild. Near L, but far from -1, we can use a reflection argument with respect to L to reduce to the same estimates as far from L. [See Section 78 for similar arguments.] Here is the local Ahlfors-regularity result that we get: there is a constant $C_1 \geq 1$, that does not depend on A, B, or R, and such that if we set $G^- = G \setminus L$, then

$$C_1^{-1} r \leq H^1(G^- \cap B(x, r)) \leq 2\pi r \tag{10}$$

for all $x \in G^-$ and $0 < r < r_G = B(1 + H^1(G^-))$ such that $B(x,r) \subset B(0,R) \setminus \{-1\}$. These estimates deteriorate when we get close to -1 or $\partial B(0,R)$, but we do not need precise estimates there.

From (10) and the same sort of arguments as in Part C, we also get uniform concentration estimates far from L and $\partial B(0,R)$, which is enough for the proof of existence alluded to below.

Call $\mathcal{U}_{R,k}$ the set of pairs $(v,G) \in \mathcal{U}_R$ for which G has at most k connected components, and denote by $J_{R,k}$ the restriction of J_R to $\mathcal{U}_{R,k}$. All the estimates mentioned above still hold for minimizers of $J_{R,k}$ on $\mathcal{U}_{R,k}$, with almost the same proof. We can use this to prove the existence of minimizers for J_R on $\mathcal{U}_R$, with the same sort of argument as described in Section 36. That is, we find a pair $(v_{R,k}, G_{R,k})$ that minimizes $J_{R,k}$ on $\mathcal{U}_{R,k}$, then we extract a subsequence so that $(v_{R,k}, G_{R,k})$ converges to a limit (v_R, G_R), and show that (v_R, G_R) minimizes J_R. The uniform concentration property for the $(v_{R,k}, G_{R,k})$ away from $\partial B(0,R) \cup L$ is used to show that $H^1(G_R \setminus L) \leq \liminf_{k \to +\infty} H^1(G_{R,k} \setminus L)$ [it turns out that the $G_{R,k}$ do not get close to $\partial B(0,R)$ for R large, by the same argument as for (11) below]; the rest is straightforward.

So we get a pair (v_R, G_R) that minimizes J_R on $\mathcal{U}_R$. An important point now is that if A is chosen large enough, then there is a constant C_0 such that

$$G_R \setminus L \subset B(0, C_0) \quad \text{for all } R > 1. \tag{11}$$

The basic reason for this is that picking a large set $G_R \setminus L$ would cost more in length than the amount of energy that it would allow us to save. The argument goes roughly as follows. Set $l = H^1(G_R \setminus L)$. We first show that

$$G_R \setminus L \subset B(0, Cl), \tag{12}$$

where C depends on C_1 in (10), but not on A. To check this, we take a point $x \in G_R^- = G_R \setminus L$ and try to apply (10) with $r = 2C_1 l$. If we could, (10) would say that $H^1(G_R^-) \geq C_1^{-1} r = 2l$ (a contradiction with the definition of l), so we can't. We may as well have chosen our constant B larger than $2C_1$, because C_1 does not depend on B. Then $r \leq r_{G_R}$, and the only reason why (10) fails has to be that $B(x,r)$ meets -1 or ∂B. Thus either $x \in B(0, 2C_1 l)$ (and we are happy), or else x lies at distance less than $2C_1 l$ from $\partial B(0,R)$.

In other words $G_R^- = Y \cup Z$, with $Y \subset B(0, 2C_1 l)$ and $Z \subset \overline{B}(0,R) \setminus B(0, R - 2C_1 l)$. We may assume that $R >> C_1 l$, because otherwise (12) holds trivially. Then Z and Y are far away from each other. Note that the reason why we still need to show that Z is empty is that we did not want to prove an Ahlfors-regularity estimate near $\partial B(0,R)$. But the point will be the same: if we remove Z from G_R and replace v_R by the best harmonic function w on $B(0,R) \setminus (Y \cup L)$ that coincides with u and v_R on $\partial B(0,R)$, we save more length than we lose in energy. The computations are similar to what we did for Ahlfors-regularity; after an integration by parts, we are reduced to estimating quantities like $|\operatorname{Jump}(v_R)\nabla w| \leq C R^{-1/2}|\operatorname{Jump}(v_R)|$, which tend to be small when R is much larger than l. Eventually, one proves that Z is empty and (12) holds.

Next we show that putting a piece of G_R in $B(0, Cl)$ can only save so much energy. More precisely, denote by u_- the best harmonic function on $B(0, R) \setminus L$ (that coincides with u and v_R on $\partial B(0, R)$) and by u^+ the best harmonic function on $B(0, R) \setminus [L \cup \overline{B}((0, Cl)]$ with the same boundary constraint. Then

$$\int_{B(0,R)\setminus L} |\nabla u^-|^2 \leq \int_{B(0,R)\setminus[L\cup\overline{B}((0,Cl)]} |\nabla u^+|^2 + C'l. \tag{13}$$

This comes from a direct computation with an integration by parts, similar to what we did in Section 61.b. Since (v_R, G_R) is a minimizer, $J_R(v_R, G_R) \leq J_R(u^-, L) = \int_{B(0,R)\setminus L} |\nabla u^-|^2$. Then

$$\begin{aligned}
h(l) = h(H^1(G_R^-)) &= J_R(v_R, G_R) - \int_{B(0,R)\setminus G_R} |\nabla v_R|^2 \\
&\leq \int_{B(0,R)\setminus L} |\nabla u^-|^2 - \int_{B(0,R)\setminus G_R} |\nabla v_R|^2 \\
&\leq \int_{B(0,R)\setminus[L\cup\overline{B}((0,Cl)]} |\nabla u^+|^2 - \int_{B(0,R)\setminus G_R} |\nabla v_R|^2 + C'l \leq C'l,
\end{aligned} \tag{14}$$

by (13) and because the restriction of v_R to $B(0, R) \setminus [L \cup \overline{B}((0, Cl)]$ is a competitor in the definition of u^+. And C' does not depend on A (because C_1 does not).

We see that if we chose A larger than C', (14) is only possible when $h(l) < Al$, which forces l to be less than a constant (by (9)). Then (11) follows from (12).

The modified functional. The next stage of the proof is to show that some sequence of pairs (v_R, G_R), with R tending to $+\infty$, converges to a pair (v, G) and that (v, G) is a minimizer for some appropriate functional. Let us first describe which functional.

We start with the definition of a set of acceptable pairs. This is the set $\mathcal{U}$ of pairs (v, G) such that G is a closed subset of the plane, with

$$G \supset L, \quad G^- = G \setminus L \text{ is bounded, and } H^1(G^-) < +\infty, \tag{15}$$

and $v \in W^{1,2}_{\text{loc}}(\mathbb{R}^2 \setminus G)$ is such that $\int_{B\setminus G} |\nabla v|^2 < +\infty$ for every disk B.

Then we define competitors for a given pair $(v, G) \in \mathcal{U}$. Those are the pairs $(\widetilde{v}, \widetilde{G}) \in \mathcal{U}$ such that for R large enough, G^- and $\widetilde{G}^-$ are contained in $B(0, R)$, and more importantly

$$\widetilde{v}(x) = v(x) \quad \text{for all } x \in \mathbb{R}^2 \setminus [G \cup B(0, R)]. \tag{16}$$

There is no need for a topological condition here, because $\mathbb{R}^2 \setminus G$ has only one unbounded component.

Finally, a "modified global minimizer" is a pair $(v, G) \in \mathcal{U}$ such that

$$h(H^1(G^-)) + \int_{B(0,R)\backslash G} |\nabla v|^2 \leq h(H^1(\widetilde{G}^-)) + \int_{B(0,R)\backslash \widetilde{G}} |\nabla \widetilde{v}|^2 \qquad (17)$$

when $(\widetilde{v}, \widetilde{G})$ is a competitor for (v, G) and R is as above.

Return to the pairs (v_R, G_R) constructed above. The existence of a convergent subsequence goes as in Section 37, where we work on the domain $\mathbb{R}^2 \backslash L$. Call (v, G) the limit that we get. The fact that

$$(v, G) \text{ is a modified global minimizer} \qquad (18)$$

can be proved as in Section 38. The fact that our sets G_R stay in a fixed disk $B(0, C_0)$ (as in (11)) is important, not only because it allows us to get (15), but also because we do not have to worry about weird behaviors of the pairs (v_R, G_R) near $\partial B(0, R)$, that would come from the Dirichlet constraints there. Apart from this, we use the uniform concentration of the G_R far from L, and standard lower semicontinuity properties for the energy integrals.

Let us also mention that (v, G) is coral (as in Definition 8.24), because G^- is locally Ahlfors-regular away from -1, and that we also get an analogue of Corollary 38.48.

Many of the results of this book still work for modified global minimizers, with essentially the same proof. The nonlinearity of h plays almost no role, and we just have to be careful to apply our results to balls that do not meet L.

Because of (11) in particular, we can show that the (v_R, G_R), and then (v, G), are very close to the cracktip near infinity. This will be used later in the discussion; in particular we shall need to know that for every $x \in \mathbb{R}^2$,

$$\limsup_{r \to +\infty} \frac{1}{r} \int_{B(x,r)\backslash G} |\nabla v|^2 \leq 1 \qquad (19)$$

[Compare with Remark 61.44.]

The conjugate function w. Now we start a long study of the properties of our modified global minimizer (v, G). The long term goal is to show that (v, G) is a cracktip, and then get a contradiction. And an important tool will be a careful study of the level sets of the conjugate harmonic function of v.

First observe that $\Omega = \mathbb{R}^2 \backslash G$ is connected. Indeed, it has only one unbounded component (because $G \backslash L$ is bounded), and no bounded component (by the proof of Proposition 55.1).

Let us identify $\mathbb{R}^2$ with the complex plane. We claim that there is a harmonic function w on $\mathbb{R}^2 \backslash G$ for which $f = v + iw$ is holomorphic. This is a (probably very classic) consequence of the Neumann boundary conditions on v; let us try to explain why. Another way to demand that $f = v + iw$ be holomorphic is to require that the derivative of w in the direction $i\xi$ (where ξ is a vector in the plane and i is the square root of -1) be equal to the derivative of u in the direction ξ. In the case of the cracktip (u, K) in (1), w is easy to find: we just notice that $u(z) = \pm\sqrt{\frac{2}{\pi}}\,\mathrm{Im}\,(z^{1/2}) + C$ (with the obvious determination of square root on $\mathbb{R}^2 \setminus K$), and we can take $w(z) = \mp\sqrt{\frac{2}{\pi}}\,\mathrm{Re}\,(z^{1/2})$.

Return to our modified global minimizer (v, G). If there is such a function w, it is easy to see that it must be given by the following formula. Pick an origin $x_0 \in \Omega$, and set $a_0 = w(x_0)$. For each smooth path $\gamma : [0, 1] \to \Omega$, with $\gamma(0) = x_0$, we should have that

$$w(\gamma(1)) = a_0 - \int_0^1 \nabla v(\gamma(t)) \cdot i\gamma'(t)\, dt. \tag{20}$$

We can take this as a definition of w (modulo an additive constant), but we have to check that two different paths that go from x_0 to the same point $\gamma(1)$ give the same right-hand side in (20). This amounts to checking that

$$\int_0^1 \nabla v(\gamma(t)) \cdot i\gamma'(t)\, dt = 0 \ \text{ when } \gamma \text{ defines a simple loop in } \Omega. \tag{21}$$

Call V the domain bounded by (the image of) γ, and set $\Omega_1 = V \cap \Omega$. Then

$$\int_0^1 \nabla v(\gamma(t)) \cdot i\gamma'(t)\, dt = \int_{\partial V} \frac{\partial v}{\partial n}\, dH^1 = \int_{\partial \Omega_1} \frac{\partial v}{\partial n}\, dH^1 = \pm \int_{\Omega_1} \Delta v = 0, \tag{22}$$

with a unit normal pointing inwards if V lies on the left of γ (and outwards otherwise), because $\partial\Omega_1 \setminus \partial V \subset \partial\Omega = G$, by the Neumann condition on $\frac{\partial v}{\partial n} = 0$ on G, by Green, and because v is harmonic on Ω_1. See [BoDa] for details and a justification of the integrations by parts.

So we can define w, and (20) shows that it is unique modulo an additive constant. We call it the conjugated function of v. By the same argument, we can also define w when (v, G) is a reduced global minimizer, at least if $\mathbb{R}^2 \setminus G$ is connected. [Otherwise, we could also define w component by component, but we let us not worry because $\mathbb{R}^2 \setminus G$ is connected for all exotic reduced global minimizers, by Theorem 66.1 below.]

Notice that $|\nabla w| = |\nabla v|$ on Ω, by definition. So we would expect w to have the same singularities as v. Actually, w is a little nicer. Indeed,

$$w \text{ has a (unique) continuous extension to the plane} \tag{23}$$

(which we shall still denote by w), and

$$w \text{ is constant on every connected component of } G. \tag{24}$$

These two facts are related. Let us first see how we can get (24) for a component G_0 of G that is composed of finitely many $C^{1+\varepsilon}$ arcs Γ_i. On the interior of each

Γ_i, v has C^1 extensions up to Γ_i (one for each access), by Proposition 17.15. Since $\nabla w \cdot i\xi = \nabla v \cdot \xi$, w also has C^1 extensions. Moreover, the tangential derivative $\frac{\partial w}{\partial \tau}$ of w on Γ_i is equal to the normal derivative $\frac{\partial v}{\partial n}$, which vanishes. Thus each of the two boundary extensions of w is locally constant on Γ_i. In the present case, it is also easy to see that the boundary values of w (from a given access) do not jump at points where two Γ_i connect, and then to deduce (24) from our local information. Here we use the fact that G_0 has no loop, so we go through each point $x \in G_0$ and each access to x when we start from any given point of G_0 with a given access, and turn around G_0. Once we get (24) for our nice G_0, it is easy to show that w extends continuously across G_0.

The actual proof of (23) and (24) takes some time because even though it is true that G coincides with a $C^{1+\varepsilon}$ curve in a neighborhood of almost every $x \in G$, we need to control the variations of w near the other points. To get this control (and even uniform Hölder estimates on w), we use the fact that Ω is a John domain with center at infinity (see Section 56), and the trivial estimate on the energy.

The level sets of w. We start with a description of almost every level set of w that also holds for any reduced global minimizer (v, G) such that $\Omega = \mathbb{R}^2 \setminus G$ is connected. [This includes all exotic global minimizers, by Theorem 66.1 below.] We claim that

$$\text{for almost every } m \in \mathbb{R}, \text{ every connected component of} \tag{25}$$
$$\Gamma_m = \left\{ x \in \mathbb{R}^2 \, ; \, w(x) = m \right\} \text{ is a rectifiable tree with no finite branch.}$$

This description does not exclude the possibility that $\Gamma_m = \emptyset$, which already occurs in the case of the cracktip (u, K) for all $m > 0$ (if we choose $w(z) = -\sqrt{\frac{2}{\pi}}\operatorname{Re}\left(z^{1/2}\right)$). By rectifiable tree, we mean a connected set Γ such that $H^1(\Gamma \cap B) < +\infty$ for every ball B, with the property that for each choice of points x, $y \in \Gamma$, there is a unique simple curve in Γ that goes from x to y. In particular, Γ contains no loop. When we say that Γ has only infinite branches, we mean that every simple arc in Γ is contained in a rectifiable Jordan curve through ∞ contained in Γ (i.e., a simple rectifiable curve in Γ, without endpoint). Thus most level sets look like what is suggested in Figure 1.

The figure also suggests that more is true. For instance, it is likely that the components of Γ_m are almost always curves (with no branches). [In Ω, at least, branching only occurs at critical points of v.]

Let us say a few words about the proof of (25). First observe that w (or equivalently v) is not constant anywhere in Ω, because $\Omega = \mathbb{R}^2 \setminus G$ is connected and v is not constant near ∞. The description of Γ_m away from G is very simple: since w is harmonic, $\Gamma_m \cap \Omega$ is composed of pieces of analytic curves. These curves do not have endpoints in Ω. They may meet transversally at critical points of w, but this is all right. Since the critical points are zeroes of f' (where $f = v + iw$), they are isolated in Ω (even though they may perhaps accumulate on $\partial \Omega$). Thus the only complicated behavior comes from G, so let us see to what extent we can avoid it.

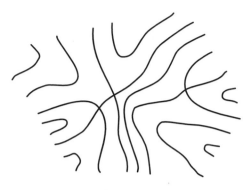

Figure 1

Because of (24), w is constant on every nontrivial component of G. There are only countably may such components, because each of them has a positive H^1 measure. So we may remove a countable set of numbers m, and we are left with level sets that do not meet any nontrivial component of G. Note that almost every point $x \in G$ is a C^1 point (i.e., $G \cap B(x,r)$ is a nice C^1 curve for r small). For points of $G^- = G \setminus L$, this follows from Theorem 51.20. For points of L, a small argument is needed, but this is true and relatively easy. Anyway, almost all points of G lie in a nontrivial connected component of G.

Denote by $\mathcal{S}$ the set of points $x \in G$ that do not lie on a nontrivial component of G. Thus $H^1(\mathcal{S}) = 0$, but $\mathcal{S}$ could still cause trouble. For instance, we do not know that $H^1(w(\mathcal{S})) = 0$, because w is not Lipschitz. So we cannot avoid G entirely. This causes difficulties in the proof, but let us not be too specific here. There is an averaging argument, based on the co-area formula and our estimates on $|\nabla v|^2$, that gives the following information for almost every m: Γ_m is rectifiable, v has a continuous extension to Γ_m, and we can compute the variations of v on (any simple curve in) Γ_m by integrating the gradient of v on $\Omega \cap \Gamma_m$, just as if $\mathcal{S}$ was not around. Let us only consider such m.

Let us try to see now why Γ_m does not contain any loop. Otherwise, $\mathbb{R}^2 \setminus \Gamma_m$ has at least one bounded component U. Notice that $w(x) \neq m$ on U, so $w - m$ has a constant sign on U. Denote by ∂ the set of points $x \in \partial U$ such that $x \in \Omega$ and $\nabla v(x) \neq 0$. Almost every point of ∂U lies in ∂, because $\Gamma_m \cap G \subset \mathcal{S}$ (by choice of m), $H^1(\mathcal{S}) = 0$, and the set of critical points of v is at most countable. Moreover, the normal derivative $\frac{\partial w}{\partial n}$ does not vanish on ∂ (because the tangential derivative vanishes), and it keeps a constant sign (because $w - m$ has a constant sign on U). By conjugation, the tangential derivative $\frac{\partial v}{\partial \tau}$ is equal to $\frac{\partial w}{\partial n}$ (maybe with the opposite sign, depending on a choice of orientation), so it is nonzero and has a constant sign on ∂ too. But we can get the variations of v on $\partial U \subset \Gamma_m$ by integrating $\frac{\partial v}{\partial \tau}$, so v is strictly monotonous along ∂U, a contradiction.

So Γ_m contains no loop. The fact that every component of a Γ_m is a rectifiable tree can be deduced from this, essentially as in Section 57. Then we need to know that Γ_m has no finite branch. If it did, we could find a point $x \in \Gamma_m \cap \Omega$ such that $\nabla v(x) \neq 0$ (so that Γ_m is an analytic curve near x) and a curve $\gamma \subset \mathbb{R}^2 \setminus \Gamma_m$ that connect points very close to x, and lying on different sides of Γ_m. This implication is not immediate, but it is not hard to deduce from the rectifiable tree structure; see Figure 2. Now $w - m$ keeps the same sign on γ, because γ does not meet

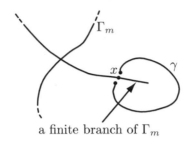

a finite branch of Γ_m

Figure 2

Γ_m. This is impossible, because the extremities of γ lie on different sides of Γ_m, where $w - m$ changes sign (because $|\nabla w(x)| = |\nabla v(x)| \neq 0$). This completes our description of the proof of (25).

All this was valid for any global minimizer such that Ω is connected. In the case of our modified global minimizer we can say a bit more, because (v, G) looks so much like (u, K) near infinity.

Set

$$w_0(z) = -\sqrt{\frac{2}{\pi}} \operatorname{Re}\left(z^{1/2}\right) \text{ on } \mathbb{R}^2 \setminus K.$$

Or, in polar coordinates,

$$w_0(r \cos \theta, r \sin \theta) = -\sqrt{\frac{2}{\pi}} r^{1/2} \cos(\theta/2).$$

Recall that w_0 is conjugate to the function u in the cracktip. We see that w_0 does not take positive values, and that for all $m < 0$, the level set $w_0^{-1}(m)$ is a parabola with the same asymptotic direction as L. The fact that it is a parabola is unimportant, but the point is that for $m < 0$ and $R > 0$ large enough, $w_0^{-1}(m) \cap \partial B(0, R)$ has exactly two points. One verifies that the same thing holds for our modified global minimizer. That is, let us normalize w so that $w(x) = 0$ for x in the unbounded connected component of G; this makes sense because of (24). Then Γ_m is empty for $m > 0$ and $\Gamma_m \cap \partial B(0, R)$ has exactly two points for $m < 0$ and R large enough (depending on m).

Let Γ_m be as in (25). Since every component of Γ_m has at least two branches that go to infinity, we see that there can be only one such component, and even that this component is a curve (i.e., with no branching). The details are fairly easy.

Thus Γ_m is empty for $m > 0$ and

Γ_m is a Jordan rectifiable curve through infinity for almost every $m < 0$.   (26)

By the same sort of discussion as when we checked that Γ_m contains no loop, we can also check that for almost all $m < 0$,

the continuous extension of v to Γ_m is strictly monotone.   (27)

Variations of v when we turn around a nontrivial component of G. Let G_0 be a connected component of G, other than the component that contains L and not reduced to one point. Recall that almost every point of G_0 is a C^1 point of G. At such a point x, there are two accesses to x from Ω, one from each side of G. And v has boundary values on G (one for each of the two accesses) that are C^1. [See Proposition 17.15 and the discussion after (55.5).]

Now there is a Lipschitz loop $z : \mathbb{S}^1 \to G_0$ that runs along G_0 so that each C^1 point of G_0 is visited exactly twice (once in each direction, corresponding to the two accesses from Ω). We get z as a limit of simple loops z_n that turn around G_0 at smaller and smaller distances, and are parameterized with constant speeds. Let us call z the tour of G_0. By construction (not detailed here), $z'(s) \neq 0$ and $z(s)$ is a C^1 point of G for almost every $s \in \mathbb{S}^1$, and we can define $f(s)$ as the boundary value of v at $x = z(s)$ corresponding to the access to x from the right of G_0. [The notion makes sense because $z'(s) \neq 0$.]

It turns out that f has a continuous extension to $\mathbb{S}^1$ (by the same sort of arguments as for the continuity of w, using the John property of Ω), and that it has the simplest possible variations. That is, $\mathbb{S}^1$ is the disjoint union of two intervals I^+ and I^-, and f is strictly increasing on I^+ and strictly decreasing on I^-. The proof relies on (26) and (27); we do not do it here, but hopefully Figure 3 will convince the reader that the two things are related. Recall that w is constant on G_0 (by (24)), so G_0 is contained in a level set Γ_{m_0}. And we can find level sets Γ_m, with m arbitrarily close to m_0 (and with both signs of $m - m_0$, because it can be proved that m_0 is *strictly* negative), that satisfy (26) and (27). Note that these Jordan curves do not meet, and that Γ_{m_0} lies between them. Since v is monotonous on the Γ_m, our description of its variations along the tour of G_0 is very plausible. The proof consists in checking that nothing goes wrong with the plane topology.

Monotonicity. Recall that $G \setminus L$ is bounded and w is continuous on the plane, so $m_0 = \inf_{x \in G} w(x)$ is finite. Set

$$\Omega_0 = \left\{ x \in \mathbb{R}^2 \,;\, w(x) < m_0 \right\}.   \tag{28}$$

Note that $\Omega_0 \subset \Omega$, by definition of m_0. Set

$$\Phi_x(r) = \frac{1}{r} \int_{\Omega_0 \cap B(x,r)} |\nabla v|^2 \quad \text{for } x \in \overline{\Omega}_0 \text{ and } r > 0.   \tag{29}$$

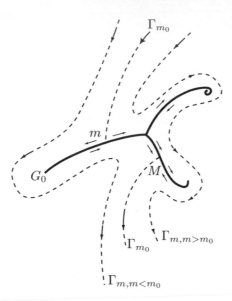

Figure 3.   The arrows indicate the variations of v on the level sets of w. On G_0, the minimum is m and the maximum is M.

This is not exactly the same thing as the normalized energy function $\omega_2(x,r)$ that we used in Sections 47 and 61, because we integrate on a smaller domain. We claim that

$$\Phi_x \text{ is nondecreasing.} \tag{30}$$

Let us first consider a simpler variant of Φ_x. Let $m < m_0$ be given, and suppose that (26) and (27) hold for m. We know that we can find m like this, arbitrarily close to m_0. Set $\Omega_m = \{x \in \mathbb{R}^2 \,;\, w(x) < m\}$ and $\Phi_{x,m}(r) = \frac{1}{r}\int_{\Omega_m \cap B(x,r)} |\nabla v|^2$. We shall only check that $\Phi_{x,m}$ is nondecreasing; (30) would follow by taking limits. The case of $\Phi_{x,m}$ is simpler, because it is easy to check that $\partial\Omega_m = \Gamma_m$, which is a rectifiable Jordan curve through infinity, and in addition is (locally) analytic (because $\Gamma_m \subset \Omega$). This will make integration by parts easier.

   The proof of monotonicity for $\Phi_{x,m}$ is essentially the same as for Proposition 47.7. What will make the proof work is that the normal derivative of v on Γ_m is zero, because it is the tangential derivative of the conjugate function w, and Γ_m is a level set of w.

   First,

$$r\Phi_{x,m}(r) = \int_{\Omega_m \cap B(x,r)} |\nabla v|^2$$

is the integral of its derivative $\int_{\Omega_m \cap \partial B(x,r)} |\nabla v|^2$, as in Lemma 47.4. Because of this, it is enough to check that $\Phi_{x,m}(r) \le \int_{\Omega_m \cap \partial B(x,r)} |\nabla v|^2$ for almost every r.

The proof relies on two integrations by parts. First,

$$r\Phi_{x,m}(r) = \int_{\Omega_m \cap B(x,r)} \nabla v \cdot \nabla v$$

$$= -\int_{\Omega_m \cap B(x,r)} v\Delta v + \int_{\partial(\Omega_m \cap B(x,r))} v\frac{\partial v}{\partial n} = \int_{\Omega_m \cap \partial B(x,r)} v\frac{\partial v}{\partial n} \quad (31)$$

by Green, because v is harmonic on $\Omega_m \subset \Omega$, and because $\frac{\partial v}{\partial n}$ vanishes on $\partial(\Omega_m \cap B(x,r)) \setminus \Omega_m \cap \partial B(x,r) \subset \Gamma_m$.

For the next formula, we only consider r such that $\Gamma_m \cap \partial B(x,r)$ is finite, with transverse intersections. This is the case for almost all r, by analyticity of Γ_m. For such r, we decompose $\Omega_m \cap \partial B(x,r)$ into its components I, and we need to know that $\int_I \frac{\partial v}{\partial n} = 0$ for every I. We shall prove this as before. We may assume that I has two extremities a and b; the cases when $I = \partial B(x,r)$ or $a = b$ are simpler and left to the reader. Since Γ_m is a simple curve and contains a and b, there is an arc of Γ_m from a to b. This arc does not meet I (which is contained in Ω_m), so we get a closed Jordan curve γ by adding I to our arc of Γ_m. Call U the bounded component of $\mathbb{R}^2 \setminus \gamma$. Then

$$0 = \int_U \Delta v = \int_{\partial U} \frac{\partial v}{\partial n} = \int_I \frac{\partial v}{\partial n} \quad (32)$$

because $\frac{\partial v}{\partial n}$ vanishes on the rest of $\partial U = \gamma$.

The rest of the proof is the same as in Section 47, between (47.16) and (47.19).

Thus the $\Phi_{x,m}$, and then Φ_x, are nondecreasing. The argument in [BoDa] is a little more complicated, because one also needs to know about the case when Φ_x is constant and nonzero (as in Section 61.a), and even about the case when it varies very slowly. This forces us to work directly on Ω_0, with a few complications, but the idea is the same.

Points of low energy. Let $x \in \partial\Omega_0$ be given. Since Φ_x is nondecreasing, it has limits

$$l(x) = \lim_{r \to 0} \Phi_x(r) \text{ and } l_\infty(x) = \lim_{r \to +\infty} \Phi_x(r). \quad (33)$$

Moreover $l_\infty(x) \leq 1$, by (19) and because $\Omega_0 \subset \mathbb{R}^2 \setminus G$. The next stage is to show that

$$l(x) = 0 \text{ for every point } x \in \partial\Omega_0 \setminus L. \quad (34)$$

The argument is based on a careful look at the proof (30). Let us assume that $l(x) > 0$. Then for each large $A > 0$, the variation of Φ_x on $[\rho, A\rho]$ tends to zero when ρ tends to zero. This forces $r\Phi'_x(r)$ to be very close to $\Phi_x(r)$ for most $r \in [\rho, A\rho]$. For these r, the length of the longest component of $\Omega_0 \cap \partial B(x,r)$ is very close to $2\pi r$, and the restriction of v to that component is very close to a sine function, by the same sort of arguments as in Section 61.a. After some amount of dirty work, one shows that some blow-up limit of (v, G) is $l(x)$ times a cracktip.

Set $\lambda = h'(H^1(G \setminus L))$, where h is as in the definition of the modified functional. We claim that

    if (v_0, G_0) is a blow-up limit of (v, G) at $x \notin L$,

        then $(\lambda^{-1/2} v_0, G_0)$ is a reduced global minimizer in the plane.   (35)

The proof is the same as for Proposition 40.9. We need to assume that $x \notin L$ so as not to get additional constraints from the fact that we forced $L \subset G$, and the multiplication by $\lambda^{-1/2}$ comes from the definition of our length term $h(H^1(G^-))$. More precisely, we are considering small perturbations of (v, G) near x, and small variations of $H^1(G^-)$ are essentially multiplied by λ. Thus (v_0, G_0) is a global minimizer in the plane, but for the variant of the usual functional where we would multiply the length terms by λ. Our multiplication of v_0 by $\lambda^{-1/2}$ compensates.

Let us return to the blow-up limit (v_0, G_0) that was $l(x)$ times a cracktip. By (35) and Section 61.b, $\lambda^{-1/2} v_0$ is a cracktip function (with the constant precisely equal to $\pm \sqrt{\frac{2}{\pi}}$). Hence $l(x) \leq \lambda^{1/2}$. We also get that $l(x) = \lambda^{1/2}$, i.e., that the contribution of $B(x, r) \setminus [\Omega_0 \cup G]$ is negligible near 0. Actually, for the proof of (2), we do not need to know this, because we are assuming that the cracktip (u, K) is not a global minimizer; thus we get a contradiction directly from (35) and Section 61.b, and (34) holds.

When we prove (3) and (4) (the results about global minimizers), we also get to a situation where we need to deduce (34) from (35), but we cannot conclude as fast (because the cracktip *is* a global minimizer). Then $\lambda = 1$ (there is no function h in this context), and we say that $l(x) = 1$, as before. Since we already know that $l(x) \leq l_\infty(x) \leq 1$ (see below (33)), this forces $l(x) = l_\infty(x) = 1$. Then Φ_x is constant, and a second look at the proof of (30) shows that (v, G) is a cracktip. This is exactly what we want in this case. So we have (34) in all cases.

Return to the proof of (2). Recall that $\Phi_x(r)$ may be strictly smaller than $\omega_2(x, r)$, because we integrate over a smaller domain. However, we can deduce from (34) that

    every $x \in G \cap \partial\Omega_0 \setminus L$ is a point of low energy.   (36)

There are two main cases for this. The simplest one is when we can find a sequence of radii r_k that tends to 0 and such that each $\Omega_0 \cap B(x, r_k)$ contains a ball of radius r_k/C. In this case, we pick a blow-up subsequence that converges to some limit (v_0, G_0). Then v_0 is constant on some ball of radius $1/C$ (by Proposition 37.18 on the lower semicontinuity of energy integrals) and since $(\lambda^{-1/2} v_0, G_0)$ is a global minimizer in the plane (by (35)), Proposition 58.19 says that G_0 is a line or a propeller, and v_0 is locally constant. In this case x is a point of low energy, by Corollary 38.48.

The other case is more painful. Let us just try to convince the reader that in this case also we should be able to find a blow-up limit (v_0, G_0) of (v, G) at x, for which v_0 is locally constant somewhere. [Which would allow us to conclude as before.] First observe that $w(x) \leq m_0$ because $x \in \partial\Omega_0$. Then there are level sets Γ_m, with $m < m_0$, that come as close to x as we want. By (26), we can even choose m so that Γ_m is a Jordan curve through infinity. Since we are not in the first case, Ω_0 looks like a very thin tube around Γ_m, that gets thinner and thinner as we approach x.

First suppose that we can find a sequence of radii r_k that tends to 0, and for which there is a ball B_k contained in $B(x, r_k)$, centered on Ω_0, of radius r_k/C, and that does not meet G. We get the strange situation of Figure 4 where $\Omega_0 \cap B_k$ is contained in a very thin tube, while the harmonic function $w - m_0$ is nonnegative on the rest of B_k. We also have a good control on the derivative of w on $\frac{1}{2}B_k$, and this forces w to be very close to m_0 on big parts of $\frac{1}{3}B_k \setminus \Omega_0$. Eventually, we can find a blow-up sequence of (v, G) at x that converges to a limit (v_0, G_0) for which v_0 is locally constant somewhere, and conclude as before.

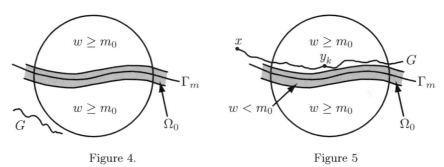

Figure 4.   Figure 5

The other possibility is that the points of Ω_0 get closer and closer to G. Then there are radii r_k that tend to zero and points $y_k \in G \cap B(x, r_k)$, for which $G \cap B(y_k, r_k/C)$ is a nice C^1 curve, and Ω_0 is still contained in a thin tube around Γ_m that crosses $B(y_k, r_k/C)$. [The existence of y_k uses Lemma 56.2.] This situation is also strange, because Ω_0 is still contained in a very thin tube around Γ_m, while $w \geq m_0$ on the boundary G and on the rest of the component of $B(y_k, r_k/C) \setminus G$ that contains the piece of Γ_m that we consider. See Figure 5. Here again, we eventually show that the only possibility is the existence of a blow-up limit (v_0, G_0) for which v_0 is locally constant somewhere. This completes our rough description of why (36) may be true. Obviously more precise limiting arguments and estimates on the Poisson kernel are needed for a true proof.

Return to the variations of v along G_0 and the final contradiction. Along the way, one also shows that $m_0 < 0$. This is a fairly straightforward consequence of our description of the level sets of w and the maximum principle, but let us not elaborate.

Recall that $m_0 = \inf_{x \in G} w(x)$. Since w is continuous and $G^- = G \setminus L$ is bounded, we can find $z_0 \in G$ such that $w(z_0) = m_0$. Call G_0 the connected component of z_0 in G. Note that G_0 does not meet L, because $w(z) = 0$ on L (by normalization of w) and $m_0 < 0$. We check that G_0 is not reduced to one point (another small argument that uses the maximum principle and our description of level sets), and so we can use our description of the variations of v along our tour of G_0.

Recall that we used a Lipschitz mapping $z : \mathbb{S}^1 \to G_0$ that was essentially two-to-one (our tour of G_0), and a continuous function f on $\mathbb{S}^1$, with the following property. For almost every $s \in \mathbb{S}^1$,

$$z(s) \text{ is a } C^1 \text{ point of } G, z \text{ is differentiable at } s, z'(s) \neq 0, \tag{37}$$

and then $f(s)$ is the limit of v at $z(s)$, with access from the right of the curve $z(\mathbb{S}^1)$. We also have a decomposition of $\mathbb{S}^1$ into disjoint intervals I^+, where f is strictly increasing, and I^-, where f is strictly decreasing. Call m and M the extremal values of f on $\mathbb{S}^1$. Then $m < M$ and none of the intervals $I^\pm$ has an empty interior. Also, we can take the convention that I^+ is closed and I^- open.

Notice that $z(s) \in \partial\Omega_0$ for almost all $s \in I^+$. Indeed let $s \in I^+$ satisfy (37), and suppose in addition that $\frac{\partial v}{\partial \tau}(z(s))$ (the tangential derivative of v on G) does not vanish. It is fairly easy to see that this is the case for almost all s, essentially because otherwise the full gradient of v would vanish on a set of positive measure in G. [Recall that $\frac{\partial v}{\partial n}$ vanishes almost-everywhere on G.] Then $\frac{\partial v}{\partial \tau}(z(s)) > 0$ (where we take an orientation compatible with $z'(s)$), because $s \in I^+$. Since w is conjugate to v, this yields $\frac{\partial w}{\partial n}(z(s)) = -\frac{\partial v}{\partial \tau}(z(s)) < 0$ (with a unit normal directed to the right). And the points of Ω directly to the right of $z(s)$ lie in Ω_0. Thus $z(s) \in \overline{\Omega}_0$. Since $w(z(s)) = m_0$ because $z(s) \in G_0$ and w is constant on G_0 (by (24)), we get that $z(s) \in \partial\Omega_0$, as needed.

Since $\partial\Omega_0$ is closed, $z(s) \in \partial\Omega_0$ for every $s \in I^+$. Then (36) tells us that $z(s)$ is a point of low energy, and Proposition 60.1 says that

$$z(s) \text{ is a nice } C^1 \text{ or spider point of } G \text{ for every } s \in I^+. \tag{38}$$

Thus we get a fairly nice description of the curve $z(I^+)$. The fact that we do not have a similarly nice description of $z(I^-)$ will not matter. Notice that spider points lie at positive distance from each other, so

$$I_s = \{s \in I^+; \ z(s) \text{ is a spider point}\}$$

is finite. Also, one can show that the restriction of z to I^+ is injective. This comes from (38) and the fact that G does not contain any loop (by the proof of (57.3)). In particular, since (38) forbids tips in $z(I^+)$, $z(s)$ cannot move along a segment and return along the same segment as long as s stays in I^+. Thus if G_0 is an arc, for instance, $z(I^+)$ stays strictly inside this arc, always with access from the same side. See Figure 6 for a symbolic description.

For almost every $s \in I^+$, $z(s)$ is a C^1 point of G and there is a unique $s^* \in \mathbb{S}^1$ such that $s^* \neq s$ and $z(s^*) = z(s)$. This is the point of $\mathbb{S}^1$ that corresponds to the other access to $z(s)$. Since the restriction of z to I^+ is injective, $s^* \in I^-$. By plane topology (and because z is the limit of a sequence of parameterizations z_n of simple curves around G_0), s^* is a decreasing function of s. Since f is decreasing on I^-, $f(s^*)$ is an increasing function of s (defined almost everywhere on I^+, to

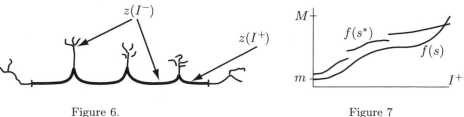

Figure 6.   Figure 7

start with). This function has a continuous extension to $I^+ \setminus I_s$, which has positive jumps at the points of I_s (because s^* jumps at these points).

We have that $f(s) = m \leq f(s^*)$ at the initial point of I^+ (with appropriate modifications if the initial point of I^+ lies in I_s), and $f(s) = M \geq f(s^*)$ at the endpoint. So we must have $f(s) = f(s^*)$ somewhere on I^+. [See Figure 7.] Let us see why this leads to a contradiction if $s \in I^+ \setminus I_s$; the case when $s \in I_s$ would be similar.

Recall that $z(s)$ is a nice C^1 point of G, by (38). So $f(s)$ and $f(s^*)$ are the two limits of v when we approach $z(s)$ from both sides of G. We just said that the jump vanishes, and it is easy to show directly that this is impossible. The point is that we would be able to get a better competitor than (v, G) by removing a tiny interval of G around $z(s)$, and modifying slightly the values of v near $z(s)$ so that they match. The computations are almost the same as for Proposition 55.1.

So we get our final contradiction, and this completes our description of the proof of (2). The reader probably has the same strange feeling about the construction as the author: the reason why we get a contradiction at the end is not so luminous, because we used so many little constructions. The central (and most delicate) part seems to be the monotonicity argument (i.e., the proof of (30)), which allows us to say that there are only two types of behaviors in the limit (depending on the constant value of Φ_x), one near infinity and a nice one near zero.   □

About the proof of (3) and (4). The argument above also shows that if (v, G) is a reduced global minimizer in the plane, with the same behavior as the cracktip (u, K) near infinity, then (v, G) is a cracktip. Let us not be too specific about what we mean by the same behavior, and refer to Theorem 39.10 in [BoDa].

If (v, G) is a reduced global minimizer in the plane such that *every* blow-in limit of (v, G) is a cracktip (possibly with different orientations), then Theorem 39.10 in [BoDa] applies and (v, G) itself is a cracktip. The verification is a little long, but fairly straightforward. One of the main points is to show that there is a curve $\Gamma \subset G$ such that $G \setminus \Gamma$ is bounded. This uses the flatness of $G \cap B(0, R)$ and the relative smallness of $|\nabla v|$ near $G \cap B(0, R) \setminus B(0, R/10)$ for R large, and Corollary 51.17, for instance.

Then we can prove (3). Let (u, K) be a reduced global minimizer such that $K \setminus K_0$ is bounded for some connected component K_0 of K, and let (v, G) be any blow-in limit of (u, K). Then G is connected. This follows from Proposition 57.1 (the fact that K_0 is a chord-arc tree), by the same argument as in Section 61.c (a little below (61.48)). Bonnet's theorem (Theorem 61.1) says that (v, G) is a cracktip, or else (v, G) is one of the trivial minimizers where v is locally constant. If any of the the blow-in limits of (u, K) is one of the trivial minimizers, Proposition 59.10 says that the same thing is true of (u, K). Otherwise, all blow-in limits of (u, K) are cracktips, and (u, K) is a cracktip too.

Finally let us say a few words about the proof of (4). So let (v, G) be a reduced global minimizer such that *some* blow-in limit of (v, G) is a cracktip. Thus there is a sequence $\{r_k\}$, that tends to $+\infty$, and for which

$$r_k^{-1}G \quad \text{is } \varepsilon_k\text{-close to a half-line in } B(0,1), \tag{39}$$

with ε_k tending to 0. This means that there is a half-line L_k such that $\text{dist}(x, L_k) \leq \varepsilon_k$ for $x \in r_k^{-1}G \cap B(0,1)$ and $\text{dist}(x, r_k^{-1}G) \leq \varepsilon_k$ for $x \in L_k \cap B(0,1)$. Let us also assume that some blow-in limits of G are different from half-lines. Then for some $\delta > 0$ we can find arbitrarily large radii s such that

$$s^{-1}G \quad \text{is not } \delta\text{-close to any half-line in } B(0,1). \tag{40}$$

We may assume that δ is very small. For k large enough, we can find $s_k \in (0, r_k)$ such that (40) holds with $s = s_k$. We choose s_k almost as large as possible. Then

$$s^{-1}G \quad \text{is } \delta\text{-close to a half-line in } B(0,1) \quad \text{for } 2s_k \leq s \leq r_k. \tag{41}$$

Since ε_k tends to 0, s_k cannot stay close to r_k, because otherwise (40) would contradict (39). More precisely, s_k/r_k tends to 0. Thus (41) concerns very long intervals.

Set $G_k = s_k^{-1}G$ and $v_k(x) = s_k^{-1/2}v(s_k x)$. We can replace $\{r_k\}$ with a subsequence for which $\{(v_k, G_k)\}$ converges to a limit (v', G'). Then

$$\rho G' \quad \text{is } \rho\delta\text{-close to a half-line in } B(0, \rho) \tag{42}$$

for every $\rho \geq 1$, by (41) and the fact that s_k/r_k tends to 0. Here again, we can use (42) to show that $G' \backslash B(0, 2)$ is connected. Given the way we started the argument, the simplest is to observe that if δ is small enough, Proposition 70.2 applies to any ball $B(x, r)$ centered on G' and such that $r = |x|/2$, so that $G' \cap B(x, r/C)$ is a nice C^1 curve for all such balls. There is no vicious circle here; Proposition 70.2 does not use this section.

But we could easily avoid Proposition 70.2 and use Corollary 51.17 instead. For this, we would incorporate closeness of v to cracktip functions in the closeness conditions (39)–(42). This would also give some control on v' in (42), and the desired control on $\int_{B(x,r)\backslash G'} |\nabla v'|^p$ when $r << |x|$ would ensue. This is what we did in [BoDa].

Once we get that $G' \backslash B(0, 2)$ is connected, (3) says that (v, G) is a cracktip, and we get a contradiction with (40).

This completes our discussion of (3) and (4), and the little summary of [BoDa]. □

63 The magic formula with $\left(\frac{\partial u}{\partial z}\right)^2$ in terms of K

Let (u, K) be a (topologically) reduced global minimizer in the plane (see Definition 54.4). The formula that follows gives a way to compute the gradient of u in terms of K, modulo the usual trivial modifications like adding a constant to, or changing the sign of u on a connected component of $\mathbb{R}^2 \setminus K$. In fact, the formula allows us to compute the square of the complex derivative $\frac{\partial u}{\partial z}$ as a constant times the Beurling transform of the restriction of H^1 to K.

We need additional notation. We identify the plane with $\mathbb{C}$, call $z = x + iy$ the generic point of $\mathbb{C}$, and set

$$F(z) = \left(\frac{\partial u}{\partial x} - i\frac{\partial u}{\partial y}\right)(z) = 2\frac{\partial u}{\partial z}(z) \quad \text{for } z \in \mathbb{C} \setminus K. \tag{1}$$

It is easy to see that F is holomorphic on $\mathbb{C} \setminus K$, because u is (real valued and) harmonic on $\mathbb{C} \setminus K$. Incidentally, recall that u has a conjugate function v, i.e., another harmonic function such that $u+iv$ is holomorphic. Its existence is very easy to check locally, and we can get its global existence from the Neumann condition $\frac{\partial u}{\partial n} = 0$ on K, as suggested in Section 62. Then F is also the complex derivative of $u + iv$. We shall not use this remark.

Theorem 2 [Lé2,3]. *If $(u, K) \in RGM(\mathbb{C})$, then*

$$F(z)^2 = -\frac{1}{2\pi} \int_K \frac{dH^1(w)}{(z - w)^2} \quad \text{for } z \in \mathbb{C} \setminus K. \tag{3}$$

See Notation 54.10 for RGM. The integral converges for $z \in \mathbb{C} \setminus K$, because K is Ahlfors-regular. That is, if we set $d = \text{dist}(z, K)$, then the contribution of the annulus $\{w \in K \, ; \, 2^{(k-1)}d \leq |w - z| < 2^k d\}$ to the integral on the right-hand side of (3) is at most $2^{-2(k-1)}d^{-2} \cdot H^1(K \cap B(z, 2^k d)) \leq C2^{-k}d^{-1}$. Thus the integral converges, and the right-hand side of (3) is less than $\leq Cd^{-1}$. Note that the corresponding estimate on $|F(z)|$ can be obtained directly from the trivial bound on $\int_{B(z,d)} |\nabla u|^2$, because u is harmonic on $B(z, d)$.

The reader may find it amusing to check that (3) yields $F = 0$ when K is a line or a propeller, and that $K = (-\infty, 0]$ gives $F(z)^2 = -\frac{1}{2\pi}\int_{-\infty}^0 (z-w)^{-2}dw = \frac{1}{2\pi z}$.

This last is coherent with the fact that $(u+iv)(z) = \pm i\sqrt{\frac{2}{\pi}}\, z^{1/2}+C$, as was checked in Section 61.

As was alluded to before, (3) allows us to compute u from K, modulo a sign (that comes from taking a square root of F^2) and an additive constant (that comes from integrating ∇u to get u) on each component of $\mathbb{C} \setminus K$. This is the best that we could hope, by Remark 54.7. In higher dimensions, we have no analogue of (3) so far, and no uniqueness result for u given K in general.

Notice that (3) only tells us what F is *if* (u, K) is a global minimizer. If you start from any K at random, (3) will give you a good function u to try, but most of the time (u, K) will not be a global minimizer.

As we shall see soon, (3) is an Euler-Lagrange formula, with respect to deformations of the pair (u, K) by vector fields. Probably many computations of the same type were tried before Léger, but the surprising thing here is the simplicity of the formula. It seems a little strange to the author that (3) did not yield enormous advances in the Mumford-Shah problem (so far). Possibly this is because the Beurling transform in (3) is not local, hence hard to use.

The proof below is almost the same as in [Lé2,3], except for a small improvement in the elimination of entire functions at the end, and a very leisurely computation of the variations of length and energy.

We give ourselves a smooth, compactly supported vector field $\varphi : \mathbb{C} \to \mathbb{C}$, and we set

$$\psi_t(z) = z + t\varphi(z) \text{ for } t \in \mathbb{R} \text{ (small) and } z \in \mathbb{C}. \tag{4}$$

For t small enough, ψ_t is a smooth diffeomorphism of the plane. We shall restrict our attention to such small t. We intend to compare (u, K) with the competitor (u_t, K_t) defined by

$$K_t = \psi_t(K) \text{ and } u_t(z) = u(\psi_t^{-1}(z)) \text{ for } z \in \mathbb{C} \setminus K_t. \tag{5}$$

Since φ is bounded and compactly supported, there is a ball B such that $\psi_t(z) = z$ for $z \in \mathbb{C} \setminus B$ and $|t| \leq 1$, say. Then (u_t, K_t) is a competitor for (u, K) in B. It is even a topological competitor (as in (54.3)), because ψ_t is a homeomorphism. Then the minimality condition (54.5) says that

$$L(t) + E(t) \geq L(0) + E(0) \text{ for } t \text{ small}, \tag{6}$$

where we set

$$L(t) = H^1(K_t \cap B) = H^1(\psi_t(K \cap B)) \text{ and } E(t) = \int_{B \setminus K_t} |\nabla u_t|^2. \tag{7}$$

Lemma 8. *The function L is differentiable near 0, and its derivative at the origin is*

$$L'(0) = \int_{K \cap B} \langle D\varphi(z) \cdot \tau(z), \tau(z) \rangle \, dH^1(z). \tag{9}$$

Here $D\varphi \cdot \tau$ denotes the differential of φ, applied to the vector τ, and $\tau(z)$ denotes a unit tangent vector to K at z. Note that we do not need $\tau(z)$ to be oriented. That is, we can replace $\tau(z)$ with $-\tau(z)$ without changing the scalar product in (9). We shall see soon how to define τ almost-everywhere on $K \cap B$. We shall also see later how to write (9) in terms of complex derivatives.

Let us prove the lemma quietly. We announced the formula formally to allow the experienced reader to go right to the next end-of-proof symbol.

We shall use the fact that $K \cap B$ is contained in $z(I)$, where I is a compact interval and z is a Lipschitz function, but the proof would also work if we needed more than one interval I, and in general if K was just rectifiable. The existence of z and I comes from Theorem 27.1, but again much less would be needed.

So let $z : I \to \mathbb{C}$ be a Lipschitz mapping such that $K \cap B \subset z(I)$. Our first goal is to write $H^1(K \cap B)$ in terms of z. Set

$$F = \{t \in I \,;\, z(t) \in K \cap B \text{ and } z(t) \neq z(s) \text{ for } s < t\}. \tag{10}$$

It is easy to see that F is a Borel set, and that the restriction of z to F is a bijection from F to $K \cap B$. Let μ denote the measure on I defined by pushing the restriction of H^1 to $K \cap B$ onto F, by the inverse of z. That is, set

$$\mu(A) = H^1(z(A \cap F)) \text{ for any Borel set } A \subset I. \tag{11}$$

It is clear that $\mu(A) \leq C|A|$ because z is Lipschitz (and by Remark 2.11), so μ is absolutely continuous with respect to dt (the Lebesgue measure on I). Call $\lambda(t)$ its density. We know that

$$\lambda(t) = \lim_{\varepsilon \to 0} \frac{1}{2\varepsilon} H^1(z(F \cap [t - \varepsilon, t + \varepsilon])) \tag{12}$$

for almost-every $t \in I$, by the Radon–Nikodym theorem. [See for instance [Ru].] In particular, λ is bounded (because z is Lipschitz), and $\lambda(z) = 0$ almost-everywhere on $I \setminus F$ (because $\frac{1}{2\varepsilon} H^1(F \cap [t - \varepsilon, t + \varepsilon])$ tends to zero at almost all such points, by the Lebesgue density theorem, and z is Lipschitz). Also, we claim that

$$\lambda(t) = \lim_{\varepsilon \to 0} \frac{1}{2\varepsilon} H^1(z([t - \varepsilon, t + \varepsilon])) = |z'(t)| \text{ for almost every } t \in F. \tag{13}$$

The first equality is easy, because $\frac{1}{2\varepsilon} H^1([t - \varepsilon, t + \varepsilon] \setminus F)$ tends to 0 almost-everywhere on F, again by the Lebesgue density theorem. The second one looks obvious, but let us check it anyway. Notice that $\frac{1}{2\varepsilon} H^1(z([t - \varepsilon, t + \varepsilon])) \geq \frac{1}{2\varepsilon}|z(t + \varepsilon) - z(t - \varepsilon)|$, which tends to $|z'(t)|$ when z is differentiable at t. So $\lambda(t) \geq |z'(t)|$ almost-everywhere on F.

We still need to check that $\lambda(t) \leq |z'(t)|$ almost-everywhere on F. Let us first verify that for every $\alpha > 0$,

$$\lambda(t) \leq \alpha \text{ almost-everywhere on } F_\alpha = \{t \in F \,;\, z'(t) \text{ exists and } |z'(t)| \leq \alpha\}. \tag{14}$$

Denote by μ_α the restriction of μ to F_α. If we show that $\mu_\alpha(A) \leq \alpha|A|$ for every Borel set $A \subset I$, we will get that the density of μ_α is $\leq \alpha$ almost-everywhere, and since this density is $\mathbf{1}_{F_\alpha} \lambda$, (14) will follow. So let $A \subset I$ be given, and let us estimate $\mu_\alpha(A) = H^1(z(A \cap F_\alpha))$. Set $A_{k,\varepsilon} = \{t \in A \cap F_\alpha \,;\, |z(s) - z(t)| \leq (\alpha + \varepsilon)|s - t| \text{ for all } s \in I \text{ such that } |s - t| \leq 2^{-k}\}$.

For each choice of k and ε, $H^1(z(A_{k,\varepsilon})) \leq (\alpha+\varepsilon)|A_{k,\varepsilon}| \leq (\alpha+\varepsilon)|A\cap F_\alpha|$. This is proved like Remark 2.11: for each covering of $A_{k,\varepsilon}$ by small sets (of diameters less than 2^{-k-1}, say), we get a covering of $z(A_{k,\varepsilon})$ by the images of the intersections with $A_{k,\varepsilon}$, with diameters at most $\alpha + \varepsilon$ times larger. For each fixed ε, $A \cap F_\alpha$ is the nondecreasing union of the $A_{k,\varepsilon}$, and so $H^1(z(A\cap F_\alpha)) \leq (\alpha + \varepsilon)|A \cap F_\alpha|$. This holds for every $\varepsilon > 0$, so $\mu_\alpha(A) = H^1(z(A\cap F_\alpha)) \leq \alpha|A \cap F_\alpha| \leq \alpha|A|$, as needed for (14).

Now it is easy to see that $\lambda(t) \leq |z'(t)|$ almost-everywhere on F. Otherwise, there would be a positive number η and a set of positive measure $Z \subset F$ such that $z'(t)$ exists and $\lambda(t) \geq |z'(t)| + \eta$ for every $t \in Z$. The set $Z_\alpha = \{t \in Z; \alpha - \eta/2 \leq |z'(t)| \leq \alpha\}$ would then have positive measure for some $\alpha > 0$, and since $Z_\alpha \subset F_\alpha$, (14) would say that $\lambda(t) \leq \alpha$ almost-everywhere on Z_α, which is impossible because $\lambda(t) \geq |z'(t)| + \eta \geq \alpha + \eta/2$ by definition. This contradiction completes our proof of (13).

Since $H^1(K \cap B) = H^1(z(F)) = \mu(F) = \int_F \lambda(s)ds$ by definition, (13) tells us that

$$H^1(K \cap B) = \int_F |z'(s)|ds. \tag{15}$$

Incidentally, we just checked something that was mentioned earlier: the equivalence between Hausdorff measure and length on a simple rectifiable curve (with a Lipschitz parameterization z). We even get the case when z itself is not necessarily injective but we restrict to a set F where z is injective. Of course all this is classical.

We may also do the computations above with the Lipschitz mapping $\tilde{z} = \psi_t \circ z$, where ψ_t is the same diffeomorphism as in (5), and get that

$$L(t) = H^1(K_t \cap B) = H^1(\psi_t(K \cap B)) = \int_F |\tilde{z}'(s)|ds = \int_F |D\psi_t(z(s)) \cdot z'(s)|ds, \tag{16}$$

by (7), because $\psi_t(B) = B$, and because F is the same for $\tilde{z}$ as for z (since ψ_t is bijective).

Next we want to differentiate under the integral sign. Note first that we may as well integrate on $F^* = \{s \in F; z'(s) \text{ exists and } z'(s) \neq 0\}$, because there is no contribution from $F \setminus F^*$ in (16). Fix $s \in F^*$ and set $a(s,t) = D\psi_t(z(s)) \cdot z'(s)$ for t small. Then $a(s,t) = z'(s) + tD\varphi(z(s)) \cdot z'(s)$, by (4). This is obviously a differentiable function of t, and $\frac{\partial a}{\partial t}(s,t) = D\varphi(z(s)) \cdot z'(s)$.

Notice that $a(s,t) = D\psi_t(z(s)) \cdot z'(s) \neq 0$ because $z'(s) \neq 0$ and $D\psi_t$ is bijective, so we can also differentiate the module. We get that

$$\frac{\partial}{\partial t}|a(s,t)| = \frac{\partial}{\partial t}\sqrt{|a(s,t)|^2} = |a(s,t)|^{-1}\langle\frac{\partial a}{\partial t}(s,t), a(s,t)\rangle$$
$$= |a(s,t)|^{-1}\langle D\varphi(z(s)) \cdot z'(s), a(s,t)\rangle. \tag{17}$$

Our function $|a(s,t)|$ and its derivative $\frac{\partial}{\partial t}|a(s,t)|$ are bounded (uniformly in s and t small), so the dominated convergence theorem applies and $L'(t) =$

$\int_{F^*} \frac{\partial}{\partial t} |a(s,t)| ds$. For $t = 0$, our expression in (17) simplifies, since $\psi_0(z) = z$ and hence $a(s,0) = z'(s)$. We are left with

$$L'(0) = \int_{F^*} |z'(s)|^{-1} \langle D\varphi(z(s)) \cdot z'(s), z'(s) \rangle\, ds. \tag{18}$$

This would be as easy to use as (9), but let us return to K and prove (9) anyway. Set $\tilde{\tau}(s) = |z'(s)|^{-1} z'(s)$ for $s \in F^*$ and $\tau(x) = \tilde{\tau}(z^{-1}(x))$ for $x \in z(F^*)$. [Recall that z is injective on F.] Then

$$L'(0) = \int_{F^*} |z'(s)|^{-1} \langle D\varphi(z(s)) \cdot z'(s), z'(s) \rangle\, ds \tag{19}$$

$$= \int_{F^*} \langle D\varphi(z(s)) \cdot \tilde{\tau}(s), \tilde{\tau}(s) \rangle |z'(s)| ds = \int_{z(F^*)} \langle D\varphi(x) \cdot \tau(x), \tau(x) \rangle\, dH^1(x).$$

This is almost the same thing as (9). There is no significant difference between $z(F^*)$ and $K \cap B$, because $K \cap B = z(F)$, $|F \setminus F^*| = 0$, and z is Lipschitz. Also, $\tau(x)$ is a tangent unit vector to $z(I)$ at x for all points $x \in z(F^*)$ where $z(I)$ has a tangent, and one could check that this is the case for almost every point of $z(I)$. The verification would be somewhat more painful than merely checking that $z'(s)$ exists for almost every s, and we shall find it simpler to use $\tau(x) = \tilde{\tau}(z^{-1}(x))$ as a definition of $\tau(x)$ (almost-everywhere on $K \cap B$). No vicious circle will ensue. Of course $\tau(x)$ is also a tangent unit vector to K at x when it is a tangent vector to $z(I)$. This issue is not serious anyway, because we know from Part E that almost all of $K \cap B$ is composed of C^1 arcs, for which there is a tangent everywhere. This completes our proof of Lemma 8.   □

Let us rewrite (9) in terms of complex derivatives. Set $\frac{\partial \varphi}{\partial z} = \frac{1}{2}(\frac{\partial \varphi}{\partial x} - i\frac{\partial \varphi}{\partial y})$ and $\frac{\partial \varphi}{\partial \bar{z}} = \frac{1}{2}(\frac{\partial \varphi}{\partial x} + i\frac{\partial \varphi}{\partial y})$. Then

$$D\varphi.(a, b) = a\frac{\partial \varphi}{\partial x} + b\frac{\partial \varphi}{\partial y} = (a + ib)\frac{\partial \varphi}{\partial z} + (a - ib)\frac{\partial \varphi}{\partial \bar{z}}, \tag{20}$$

by direct computation. Next

$$\langle D\varphi.\tau, \tau \rangle = \mathrm{Re}\left\{ \left[\tau\frac{\partial \varphi}{\partial z} + \bar{\tau}\frac{\partial \varphi}{\partial \bar{z}} \right] \bar{\tau} \right\} = \mathrm{Re}\left\{ \frac{\partial \varphi}{\partial z} + \bar{\tau}^2\frac{\partial \varphi}{\partial \bar{z}} \right\} \tag{21}$$

if $|\tau| = 1$, and (9) becomes

$$L'(0) = \int_{K \cap B} \mathrm{Re}\left\{ \frac{\partial \varphi}{\partial z} + \bar{\tau}(z)^2\frac{\partial \varphi}{\partial \bar{z}} \right\} dH^1(z). \tag{22}$$

Next we compute the energy $E(t)$.

Recall from (7) that $E(t) = \int_{B \setminus K_t} |\nabla u_t(x)|^2 dx$. Set $x = \psi_t(y)$. Then $dx = J(y)dy$, where $J(y)$ is the Jacobian determinant of $D\psi_t(y)$ (which stays positive for t small), and

$$E(t) = \int_{B \setminus K} |\nabla u_t(\psi_t(y))|^2 J(y)dy, \tag{23}$$

because $B \setminus K_t = \psi_t(B \setminus K)$. Since $u_t = u \circ \psi_t^{-1}$ by (5), $Du_t(\psi_t(y)) = Du(y) \circ D\psi_t(y)^{-1}$. Call $A = A_t(y)$ the matrix of $D\psi_t(y)^{-1}$, and write $A = \begin{pmatrix} a & b \\ c & d \end{pmatrix}$. Then $\frac{\partial u_t}{\partial x}(\psi_t(y)) = a \frac{\partial u}{\partial x} + c \frac{\partial u}{\partial y}$ and $\frac{\partial u_t}{\partial y}(\psi_t(y)) = b \frac{\partial u}{\partial x} + d \frac{\partial u}{\partial y}$, so

$$|\nabla u_t(\psi_t(y))|^2 = (a^2 + b^2)\left(\frac{\partial u}{\partial x}\right)^2 + (c^2 + d^2)\left(\frac{\partial u}{\partial y}\right)^2 + 2(ac + bd)\frac{\partial u}{\partial x}\frac{\partial u}{\partial y}. \tag{24}$$

Since $J(y) = (ad - bc)^{-1}$, (23) becomes

$$E(t) = \int_{B \setminus K} \left\{(a^2 + b^2)\left(\frac{\partial u}{\partial x}\right)^2 + (c^2 + d^2)\left(\frac{\partial u}{\partial y}\right)^2 + 2(ac + bd)\frac{\partial u}{\partial x}\frac{\partial u}{\partial y}\right\} \frac{dy}{(ad - bc)}. \tag{25}$$

Call a' the partial derivative of $a = a_t(y)$ with respect to t, and similarly for the functions b, c, d. The functions $a, b, c, d, a', b', c', d'$ are all bounded, with uniform bounds (if t stays small), and also $(ad - bc)^{-1}$ stays bounded. Since ∇u lies in $L^2(B \setminus K)$ by the trivial energy estimate, we can differentiate under the integral sign in (25), and in particular $E'(t)$ exists for t small.

We are only interested in the value of $E'(0)$, so let us compute our various functions at $t = 0$. Since $\psi_0(y) = y$ by (4), A is the identity matrix and $a = d = 1$, $b = c = 0$ for $t = 0$. Thus

$$E'(0) = \int_{B \setminus K} \left\{2a'\left(\frac{\partial u}{\partial x}\right)^2 + 2d'\left(\frac{\partial u}{\partial y}\right)^2 + 2(b' + c')\frac{\partial u}{\partial x}\frac{\partial u}{\partial y}\right.$$
$$\left. - (a' + d')\left[\left(\frac{\partial u}{\partial x}\right)^2 + \left(\frac{\partial u}{\partial y}\right)^2\right]\right\} dy$$
$$= \int_{B \setminus K} \left\{(a' - d')\left(\left(\frac{\partial u}{\partial x}\right)^2 - \left(\frac{\partial u}{\partial y}\right)^2\right) + 2(b' + c')\frac{\partial u}{\partial x}\frac{\partial u}{\partial y}\right\} dy, \tag{26}$$

where the functions a', b', c', d' are now computed at $t = 0$.

Recall that $\psi_t(y) = y + t\varphi(y)$, so $D\psi_t(y) = I + tD\varphi(y)$. Call M the matrix of $D\varphi(y)$; then $A = (I + tM)^{-1}$ and $A' = -M$ at $t = 0$. Thus a', b', c', d' are the coefficients of $-M$.

Next we want to write all this in terms of complex derivatives. Set $\frac{\partial \varphi}{\partial z} = \frac{1}{2}(\frac{\partial \varphi}{\partial x} + i\frac{\partial \varphi}{\partial y})$, and define $F = 2\frac{\partial u}{\partial z}$ as in (1). We claim that

$$E'(0) = -2\,\text{Re} \int_{B \setminus K} F^2 \frac{\partial \varphi}{\partial \bar{z}}. \tag{27}$$

Let us just cheat and compute the right-hand side of (27). With the notation above, $\frac{\partial \varphi}{\partial x} = -(a' + ic')$ and $\frac{\partial \varphi}{\partial y} = -(b' + id')$, so $\frac{\partial \varphi}{\partial z} = -\frac{1}{2}(a' - d' + ib' + ic')$. On the other hand, $F^2 = (\frac{\partial u}{\partial x} - i\frac{\partial u}{\partial y})^2 = \left(\frac{\partial u}{\partial x}\right)^2 - \left(\frac{\partial u}{\partial y}\right)^2 - 2i\frac{\partial u}{\partial x}\frac{\partial u}{\partial y}$, so

$$-2\,\mathrm{Re}\left(F^2 \frac{\partial \varphi}{\partial z}\right) = (a' - d')\left(\left(\frac{\partial u}{\partial x}\right)^2 - \left(\frac{\partial u}{\partial y}\right)^2\right) + 2(b' + c')\frac{\partial u}{\partial x}\frac{\partial u}{\partial y}. \qquad (28)$$

Thus (27) is the same thing as (26).

Recall from (6) that $L(t) + E(t)$ is minimal for $t = 0$, so $L'(0) + E'(0) = 0$. That is,

$$\mathrm{Re}\int_{K \cap B}\left\{\frac{\partial \varphi}{\partial z} + \overline{\tau}(z)^2 \frac{\partial \varphi}{\partial z}\right\} dH^1(z)\ -2\,\mathrm{Re}\int_{B \setminus K} F^2 \frac{\partial \varphi}{\partial z} = 0, \qquad (29)$$

by (22) and (27). Notice that we could also integrate on K and $\mathbb{C} \setminus K$ (rather than $K \cap B$ and $B \setminus K$), because φ is compactly supported in B anyway.

This holds for any smooth function $\varphi : \mathbb{C} \to \mathbb{C}$ with compact support. If we apply (29) to $i\varphi$, we get the same equation, but with the imaginary parts. Altogether,

$$\int_K \left\{\frac{\partial \varphi}{\partial z} + \overline{\tau}(z)^2 \frac{\partial \varphi}{\partial z}\right\} dH^1(z) = 2\int_{\mathbb{C} \setminus K} F^2 \frac{\partial \varphi}{\partial z} \qquad (30)$$

for every smooth compactly supported φ.

We can also state this in terms of distributions. Notice that F^2 is a distribution because $|\nabla u|^2 \in L^1_{\mathrm{loc}}(\mathbb{C})$, and so is $H^1_{|K}$ (the restriction of H^1 to K), because it is a Radon measure. Then (30) says that

$$2\frac{\partial}{\partial z}(F^2) = \frac{\partial}{\partial z}(H^1_{|K}) + \frac{\partial}{\partial z}\left(\overline{\tau}(z)^2 H^1_{|K}\right). \qquad (31)$$

Now we have our Euler–Lagrange equation, and to get (3) we shall just need to solve it in F^2. The main point will be to find a distribution G such that $\frac{\partial G}{\partial z} = \frac{\partial}{\partial z}(H^1_{|K})$, and for this we intend to convolve $H^1_{|K}$ with the appropriate fundamental solution.

Set $f = -p.v.\left(\frac{1}{\pi z^2}\right)$. This is the distribution on $\mathbb{C}$ defined by

$$\langle f, \varphi\rangle = -\lim_{\varepsilon \to 0}\int_{|z| > \varepsilon} \frac{\varphi(z)\,dz}{\pi z^2} \quad \text{for } \varphi \in C_c^\infty(\mathbb{C}). \qquad (32)$$

Let us check that the limit exists. For $0 < \varepsilon < \delta$,

$$\int_{\varepsilon < |z| < \delta} \frac{\varphi(z)}{\pi z^2}\,dz = \int_{\varepsilon < |z| < \delta} \frac{[\varphi(z) - \varphi(0)]}{\pi z^2}\,dz, \qquad (33)$$

because $\int_{\varepsilon < |z| < \delta} z^{-2} dz - 0$. The right-hand side of (33) tends to 0 when ε and δ tend to 0, because $\int_{|z| < 1} \frac{|\varphi(z) - \varphi(0)|}{\pi z^2} dz$ converges. So the limit in (32) exists, and the proof shows that f is a distribution.

Lemma 34. *Denote by δ_0 the Dirac mass at 0. Then $\frac{\partial f}{\partial \bar{z}} = \frac{\partial \delta_0}{\partial z}$.*

This is not surprising. We know that $\frac{\partial f}{\partial \bar{z}}$ is a distribution supported at the origin (because z^{-2} is holomorphic on $\mathbb{C} \setminus \{0\}$), and our formula is at least compatible with homogeneity. Let $\varphi \in C_c^\infty(\mathbb{C})$ be given. Then

$$\langle \frac{\partial f}{\partial \bar{z}}, \varphi \rangle = -\langle f, \frac{\partial \varphi}{\partial \bar{z}} \rangle = \lim_{\varepsilon \to 0} I(\varepsilon), \tag{35}$$

with $I(\varepsilon) = \int_{|z| > \varepsilon} \frac{\partial \varphi}{\partial \bar{z}} \frac{dz}{\pi z^2}$. We integrate by parts on the domain $\{|z| > \varepsilon\}$. Since $\frac{\partial}{\partial \bar{z}}(1/z^2) = 0$ on that domain, we are left with

$$I(\varepsilon) = \frac{1}{2} \int_{|z| = \varepsilon} \frac{\varphi(z)}{\pi z^2} (n_x + i n_y) \, dH^1(z), \tag{36}$$

where n_x and n_y denote the two coordinates of the unit tangent to the circle that points outward (i.e., in the direction of the origin). Thus $n_x + i n_y = -\varepsilon^{-1} z$ and

$$I(\varepsilon) = -\frac{1}{2\varepsilon} \int_{|z| = \varepsilon} \frac{\varphi(z)}{\pi z} \, dH^1(z). \tag{37}$$

Write $\varphi(z) = \varphi(0) + az + b\bar{z} + o(|z|)$ near 0, with $a = \frac{\partial \varphi}{\partial z}(0)$ and $b = \frac{\partial \varphi}{\partial \bar{z}}(0)$. The contributions of $\varphi(0)$ and $b\bar{z}$ to the integral in (37) vanish, and the contribution of az is exactly $-a$. Thus $I(\varepsilon) = -a + o(1)$, and $\langle \frac{\partial f}{\partial \bar{z}}, \varphi \rangle = -a = -\frac{\partial \varphi}{\partial z}(0)$, by (35). Since $-\frac{\partial \varphi}{\partial z}(0) = -\langle \frac{\partial \varphi}{\partial z}, \delta_0 \rangle = \langle \varphi, \frac{\partial \delta_0}{\partial z} \rangle$ by definition, Lemma 34 follows. □

Set $\mu = H^1_{|K}$. We want to define G by $G = f * \mu$ (so that $\frac{\partial G}{\partial \bar{z}} = \frac{\partial f}{\partial \bar{z}} * \mu = \frac{\partial \delta_0}{\partial z} * \mu = \frac{\partial \mu}{\partial z}$), but we need to be a little careful about the supports because neither f nor μ is compactly supported. So we cut f into two pieces before we do the convolution.

Let $h \in C_c^\infty(\mathbb{C})$ be such that $h(x) = 1$ for $|x| \leq 1$, $0 \leq h(x) \leq 1$ for $1 \leq |x| \leq 2$, and $h(x) = 0$ for $|x| \geq 2$. Set $f_1 = hf$ and $f_2 = (1 - h)f$. Now f_2 is a smooth function, and f_1 is a compactly supported distribution. Observe that

$$\frac{\partial f_2}{\partial \bar{z}} = \frac{1}{\pi z^2} \frac{\partial h}{\partial \bar{z}} \tag{38}$$

by direct computation (because $\frac{\partial f}{\partial \bar{z}} = 0$ near the support of $1 - h$), and then

$$\frac{\partial f_1}{\partial \bar{z}} = \frac{\partial \delta_0}{\partial z} - \frac{1}{\pi z^2} \frac{\partial h}{\partial \bar{z}} \tag{39}$$

by Lemma 34, and because $f = f_1 + f_2$.

Set $G_1 = f_1 * \mu$. Recall that this means that $\langle G_1, \varphi \rangle = \langle f_1, \varphi * \tilde{\mu} \rangle$, where $\tilde{\mu}$ is obtained from μ by symmetry with respect to the origin. Notice that $\varphi * \tilde{\mu}$

is smooth and f_1 is compactly supported, so $\langle f_1, \varphi * \widetilde{\mu} \rangle$ is defined and G_1 is a distribution. Moreover,

$$\frac{\partial G_1}{\partial \overline{z}} = \frac{\partial f_1}{\partial \overline{z}} * \mu = \frac{\partial \delta_0}{\partial \overline{z}} * \mu - \left(\frac{1}{\pi z^2} \frac{\partial h}{\partial \overline{z}}\right) * \mu = \frac{\partial \mu}{\partial \overline{z}} - \left(\frac{1}{\pi z^2} \frac{\partial h}{\partial \overline{z}}\right) * \mu, \qquad (40)$$

by standard results on derivatives of convolutions, (39), and because $\frac{\partial \delta_0}{\partial \overline{z}} * \mu = \delta_0 * \frac{\partial \mu}{\partial \overline{z}} = \frac{\partial \mu}{\partial \overline{z}}$.

We define $G_2 = f_2 * \mu$ by hand. That is, we set

$$G_2(z) = \int_{\mathbb{C}} f_2(z - w) d\mu(w) \quad \text{for } z \in \mathbb{C}. \qquad (41)$$

The integral converges, because $f_2(z - w) = 0$ for $|z - w| \le 1$ and hence

$$\int_{\mathbb{C}} |f_2(z - w)| d\mu(w) \le \frac{1}{\pi} \int_{|z-w| \ge 1} |z - w|^{-2} d\mu(w)$$

$$\le \frac{1}{\pi} \sum_{k=0}^{\infty} \int_{2^k \le |z-w| \le 2^{k+1}} 2^{-2k} \mu(B(z, 2^{k+1})) \le C \sum_{k=0}^{\infty} 2^{-k} < +\infty, \qquad (42)$$

where we also used the trivial part of the Ahlfors-regularity of K (see the first lines of Section 54.b, or Theorem 18.16). We even get that G_2 is bounded.

Next we want to differentiate G_2. Recall from (38) that $\frac{\partial f_2}{\partial \overline{z}} = \frac{1}{\pi z^2} \frac{\partial h}{\partial \overline{z}}$, which lies in $C_c^{\infty}(\mathbb{C})$ because $\frac{\partial h}{\partial \overline{z}}$ vanishes near 0. Thus we can differentiate under the integral sign, and we get that $\frac{\partial G_2}{\partial \overline{z}}$ is a continuous function, with

$$\frac{\partial G_2}{\partial \overline{z}}(z) = \int_{\mathbb{C}} \frac{1}{\pi(z - w)^2} \frac{\partial h}{\partial \overline{z}}(z - w) d\mu(w) = \left(\frac{1}{\pi z^2} \frac{\partial h}{\partial \overline{z}} * \mu\right)(z). \qquad (43)$$

Finally set $G = G_1 + G_2$, and note that $\frac{\partial G}{\partial \overline{z}} = \frac{\partial \mu}{\partial \overline{z}}$, by (40) and (43). Recall that $\mu = H^1_{|K}$; thus (31) says that $\frac{\partial}{\partial \overline{z}}\left[2F^2 - G - \overline{\tau}(z)^2 \mu\right] = 0$. In other words, there is an entire function H on $\mathbb{C}$ such that

$$2F^2 = G + \overline{\tau}(z)^2 \mu + 2H \qquad (44)$$

(in the sense of distributions). We want to see what this means on $\mathbb{C} \backslash K$. Recall that F^2 is a function: it was defined as the square of $2\frac{\partial u}{\partial z}$ on $\mathbb{C} \backslash K$, with no contribution from K. Thus we should not worry about $\overline{\tau}(z)^2 \mu$, which just compensates the contribution of K to G. Away from K, $G_1 = f_1 * \mu$ is a function, given by the usual formula

$$G_1(z) = \int_K f_1(z - w) d\mu(w) = -\int_K h(z - w) \frac{1}{\pi(z - w)^2} d\mu(w), \qquad (45)$$

where the integral converges because h is compactly supported and $z - w$ never gets close to 0. Then G also is a function on $\mathbb{C} \backslash K$, with

$$G(z) = G_1(z) + G_2(z) = \int_K f(z - w) d\mu(w) = -\frac{1}{\pi} \int_K \frac{d\mu(w)}{(z - w)^2}. \qquad (46)$$

And (44) says that

$$F(z)^2 = -\frac{1}{2\pi} \int_K \frac{d\mu(w)}{(z-w)^2} + H(z) \text{ for } z \in \mathbb{C} \setminus K. \tag{47}$$

To complete our proof of Theorem 2, we still need to check that $H = 0$. The idea is that otherwise ∇u would be too large on average. Set $A(z) = \int_K \frac{d\mu(w)}{(z-w)^2}$ for $z \in \mathbb{C} \setminus K$. Also let $1 < p < 2$ be given. First we want to check that for R large,

$$\int_{B(0,R) \setminus K} |A(z)|^{\frac{p}{2}} dz \le C_p R^{2-\frac{2}{p}}. \tag{48}$$

Fix $z \in \mathbb{C} \setminus K$ and set $d(z) = \text{dist}(z, K)$. The same estimates as for (42) yield

$$\begin{aligned}
|A(z)| &\le \int_{|z-w| \ge d(z)} |z-w|^{-2} d\mu(w) \\
&\le \sum_{k=0}^{\infty} \int_{2^k d(z) \le |z-w| \le 2^{k+1} d(z)} 2^{-2k} d(z)^{-2} \mu(B(z, 2^{k+1} d(z))) \tag{49} \\
&\le C \sum_{k=0}^{\infty} 2^{-k} d(z)^{-1} \le C d(z)^{-1},
\end{aligned}$$

by (the trivial part of the) Ahlfors-regularity of K. Then let $R > 0$ be given, and cut $B(0, R) \setminus K$ into the sets

$$E_k = \{ z \in B(0, R) \setminus K \,;\, 2^{-k} R \le d(z) \le 2^{-k+1} R \}. \tag{50}$$

Since K is Ahlfors-regular, Lemma 23.25 tells us that $|E_k| \le C 2^{-k} R^2$ for $k \ge 0$. [Recall that we get this by covering $K \cap B(0, R)$ by balls of radius $2^{-k} R$, and then counting how many balls we need. This time the full Ahlfors-regularity is used.] For $k < 0$ we have the more trivial estimate $|E_k| \le C R^2$. Then

$$\int_{E_k} |A(z)|^{\frac{p}{2}} dz \le C |E_k| (2^{-k} R)^{-\frac{p}{2}} \le C 2^{-k(1-\frac{p}{2})} R^{2-\frac{p}{2}} \tag{51}$$

for $k \ge 0$, and similarly $\int_{E_k} |A(z)|^{\frac{p}{2}} dz \le C 2^{k \frac{p}{2}} R^{2-\frac{p}{2}}$ for $k < 0$. Now (48) follows by summing over k.

On the other hand, $|F|^2 = \left| \frac{\partial u}{\partial x} - i \frac{\partial u}{\partial y} \right|^2 = |\nabla u|^2$, by (1), so

$$\begin{aligned}
\int_{B(0,R) \setminus K} |F(z)|^p dz &= \int_{B(0,R) \setminus K} |\nabla u|^p \\
&\le |B(0,R)|^{1-\frac{p}{2}} \left\{ \int_{B(0,R) \setminus K} |\nabla u|^2 \right\}^{\frac{p}{2}} \le C R^{2-\frac{p}{2}},
\end{aligned} \tag{52}$$

by the trivial energy estimate. Then

$$\int_{B(0,R)} |H(z)|^{\frac{p}{2}} dz \le \int_{B(0,R) \setminus K} \{ |F(z)|^2 + (2\pi)^{-1} |A(z)| \}^{\frac{p}{2}} dz \le C_p R^{2-\frac{p}{2}} \tag{53}$$

by (47), (48), and (52). Since H is entire, $|H(z)|^{\frac{p}{2}}$ is subharmonic (see Theorem 17.3 in [Ru]). But (53) says that the average of $|H(z)|^{\frac{p}{2}}$ on $B(0,R)$ tends to 0 when R tends to $+\infty$. By subharmonicity, $|H(z)|^{\frac{p}{2}} \equiv 0$. Then $H = 0$ too, and Theorem 2 follows from (47). $\qquad\qquad\square$

Remark 54. The longest proof of uniform rectifiability. At this point we cannot resist mentioning a different proof of uniform rectifiability for K when (u, K) is a reduced global minimizer in the plane. Let us assume that we already know that K is Ahlfors-regular (as in Section 18) and rectifiable. This last is a little unpleasant, but there are proofs of rectifiability that do not give uniform rectifiability, and we could also have this information directly if (u, K) was obtained from the SBV analogue of a global minimizer. Then the proof of Theorem 2 still works, and we get that (3) holds.

We also know that

$$\int_{B(x,r)\setminus K} |F(z)|^2 = \int_{B(x,r)\setminus K} |\nabla u(z)|^2 \leq 2\pi r \qquad (55)$$

for all $x \in \mathbb{C}$ and $r > 0$, by the remark just above (52) and the trivial estimate on energy. Let $z \in \mathbb{C} \setminus K$ be given, and set $d(z) = \mathrm{dist}(z, K)$ as before. Since F is holomorphic on $\mathbb{C} \setminus K$,

$$\begin{aligned}
|F(z)|^2 &\leq \left\{ |B(z, d(z))|^{-1} \int_{B(z,d(z))} |F(z)| \right\}^2 \\
&\leq |B(z, d(z))|^{-1} \int_{B(z,d(z))} |F(z)|^2 \leq 2d(z)^{-1},
\end{aligned} \qquad (56)$$

by the mean value property, Cauchy–Schwarz, and (55). Then

$$\int_{B(x,r)\setminus K} |F(z)|^4 d(z) \leq 2 \int_{B(x,r)\setminus K} |F(z)|^2 \leq 4\pi r \qquad (57)$$

for $x \in \mathbb{C}$ and $r > 0$, by (56) and (55). In the terminology of [DaSe3], $|F(z)|^4 d(z)dz$ is a Carleson measure on $\mathbb{C} \setminus K$ (see six lines above Lemma III.2.11 on page 220 of [DaSe3]). By (3) and because we also know that K is Ahlfors-regular, Lemma III.2.11 of [DaSe3] says that K satisfies the "usual square function estimate for the Cauchy integral" (USFE). Then Theorem I.2.41 on page 44 of [DaSe3] says that K is uniformly rectifiable (that is, is contained in a regular curve), as needed. Of course the proof looks short from here, but the proof of Theorem I.2.41 in [DaSe3] takes some time, even in dimension 1.

It is amusing that in [Lé2,3], Theorem I.2.41 in [DaSe3] was initially used in the other direction. That is, the author did not notice that it was enough to consider powers $p < 2$ in (48), and then used Theorem I.2.41 in [DaSe3] to estimate integrals of $A(z) = \int_K \frac{d\mu(w)}{(z-w)^2}$.

64   The case when K is contained in a countable union of lines

Theorem 1 [Lé3]. *Let $(u, K) \in RGM(\mathbb{R}^2)$ be a reduced global minimizer, and assume that we can find affine lines L_i, $i \in \mathbb{N}$, such that $H^1(K \setminus \bigcup_{i=0}^{\infty} L_i) = 0$. Then (u, K) is one of the standard minimizers defined by (54.14), (54.15), (54.16) or (54.17).*

In particular, we know all the reduced global minimizers for which K is contained in a line. In this special case, the result was also obtained (independently and almost simultaneously) by Chris Larsen [La].

The statement in [Lé3] is slightly different; the author does not assume that the collection of lines L_i is at most countable, but requires $H^1(K \cap L_i) > 0$ for all i. It is easy to see that this is equivalent. It would be interesting to know whether there is an analogue of this in higher dimensions. For instance, can we describe all the global minimizers in $\mathbb{R}^3$ for which K is contained in a finite union of planes (or even just one)?

We shall only prove Theorem 1 when K is contained in a line, and then give a very rapid sketch of the general case. See [Lé3] for details about the general case. The reader may also find there a generalization of Theorem 1 where we consider sets K that contain a fairly large piece of very flat Lipschitz graph in every ball. See Proposition 57 on page 456 of [Lé3]. This result is deduced from Theorem 1 by a compactness argument.

Let us prove Theorem 1 when K is contained in a line. Since we shall use the formula (62.3), let us still identify the plane with $\mathbb{C}$. Without loss of generality, we can assume that $K \subset \mathbb{R}$.

Set $F(z) = \left(\frac{\partial u}{\partial x} - i \frac{\partial u}{\partial y}\right)(z)$ for $z \in \mathbb{C} \setminus K$, as in (63.1). Then Theorem 63.2 says that

$$F(z)^2 = -\frac{1}{2\pi} \int_K \frac{dH^1(w)}{(z-w)^2} \quad \text{for } z \in \mathbb{C} \setminus K. \tag{2}$$

Let us first use a symmetry argument to show that

$$u \text{ is constant on } \mathbb{R} \setminus K. \tag{3}$$

There is nothing to prove if $K = \mathbb{R}$. Otherwise, notice that $F(z)^2 < 0$ on $\mathbb{R} \setminus K$, because of (2). Thus $F(z)$ is pure imaginary, i.e.,

$$F(z) = -\overline{F(\overline{z})} \tag{4}$$

(because $\overline{z} = z$). This holds for $z \in \mathbb{R} \setminus K$, but since both sides of (4) are analytic on $\mathbb{C} \setminus K$ (and also $\mathbb{C} \setminus K$ is connected, since K is strictly contained in $\mathbb{R}$), we get that (4) holds for all $z \in \mathbb{C} \setminus K$, by analytic continuation. Next

$$2\frac{\partial}{\partial z}(u(z) + u(\overline{z})) = \left(\frac{\partial}{\partial x} - i\frac{\partial}{\partial y}\right)(u(z) + u(\overline{z})) = F(z) + \overline{F(\overline{z})} = 0 \tag{5}$$

by definition of F, because u is real, and by (4). Thus $u(z)+u(\bar{z})$ is locally constant on $\mathbb{C} \setminus K$ and, since this set is connected, $u(z) + u(\bar{z})$ is constant on $\mathbb{C} \setminus K$. This proves (3).

For the second part of the argument, we want to study the variations of u on an interval of K. First observe that $\int_{\mathbb{R}} \frac{dH^1(w)}{(z-w)^2} = 0$ for $z \in \mathbb{C} \setminus \mathbb{R}$ (by direct computation), so (2) yields

$$F(z)^2 = \frac{1}{2\pi} \int_{\mathbb{R} \setminus K} \frac{dH^1(w)}{(z-w)^2} \quad \text{for } z \in \mathbb{C} \setminus \mathbb{R}. \tag{6}$$

If $K = \emptyset$ or $K = \mathbb{R}$, then $F = 0$ by (2) or (6), u is locally constant, and we are happy. Similarly, if K is a half-line, (2) gives the right formula for $F(z)$ and (54.17) follows by taking a square root and integrating. So we may assume that K is neither empty, nor equal to $\mathbb{R}$, nor to half a line. Let us also check that K cannot be the complement of a bounded interval I in $\mathbb{R}$.

Indeed, if this were the case, (6) would say that

$$F(z)^2 = \frac{1}{2\pi} \int_I \frac{dH^1(w)}{(z-w)^2} \sim \frac{|I|}{2\pi z^2} \tag{7}$$

near infinity, hence

$$\int_{B(0,R) \setminus K} |\nabla u|^2 = \int_{B(0,R) \setminus K} |F|^2 \geq c \operatorname{Log}(R) \tag{8}$$

for R large (by definition of F and (7)). On the other hand, (7) would also say that

$$\int_{\partial B(0,R) \setminus K} |\nabla u|^2 = \int_{\partial B(0,R) \setminus K} |F|^2 \leq C R^{-1}. \tag{9}$$

Let us construct a competitor to show that this is impossible. We take $\widetilde{K} = \mathbb{R}$, keep the same function u on $\mathbb{C} \setminus [B(0, R) \cup \mathbb{R}]$ (with R large), and define $\widetilde{u}$ on $B(0, R) \setminus \mathbb{R}$ so that it coincides with u on $\partial B(0, R)$. More precisely, denote by Ω_+ and Ω_- the two components of $B(0, R) \setminus \mathbb{R}$, and ∂_+ and ∂_- the corresponding pieces of $\partial B(0, R)$. We define $\widetilde{u}$ independently on each $\Omega_\pm$ by extending $u_{|\partial_\pm}$ to $\partial B(0, R)$ by symmetry, and then taking the Poisson extension to $B(0, R)$ and restricting to $\Omega_\pm$. By Lemma 22.16 (with $p = 2$), we get that

$$\int_{B(0,R) \setminus \mathbb{R}} |\nabla \widetilde{u}|^2 \leq C R \int_{\partial B(0,R) \setminus K} |\nabla u|^2 \leq C. \tag{10}$$

Actually, since we have a good control on $\int_{\partial B(0,R) \setminus K} |\nabla u|^2$ here, we do not even need to use the Poisson kernel. A simpler, essentially radial, extension formula would work; see Exercise 20.49.

Our pair $(\tilde{u}, \tilde{K})$ is a competitor for (u, K) in $B(0, R)$, which is clearly topological because $\mathbb{C} \setminus K$ is connected. But the comparison is favorable to $(\tilde{u}, \tilde{K})$, because we lose $|I|$ in the length term, while we win at least $c \log(R) - C$ in energy (by (8) and (10), and with constants that depend on I). Thus $K = \mathbb{R} \setminus I$ with I bounded is impossible.

Let us check that in the remaining case,

$$\text{we can find an interval } I = [a, b], \text{ with } b > a, \\ \text{such that } I \subset K \text{ and } a, b \in \mathbb{R} \setminus K. \tag{11}$$

First notice that $\mathbb{R} \setminus K$ is not empty or an interval, so it has at least two distinct connected components I_1 and I_2. Call J the compact interval between I_1 and I_2. Then J meets K. Since K is Ahlfors-regular (or directly because it is reduced), $H^1(K \cap J) > 0$. By Theorem 51.20, almost every point of $K \cap J$ is a C^1 point of K. Hence we can find at least one such point in $K \cap J$. By definition of C^1 points, and given the fact that $K \subset \mathbb{R}$, this just means that K contains a small open interval I_0 that contains this point. We get the interval I announced in (11) by replacing I_0 with a maximal interval I such that $I_0 \subset I \subset K$.

It is a little sad that we have to use Theorem 51.20 here, but the author did not find any sharp way to find an interval in K (that is, without tiny holes) without any work. Of course the proof of Theorem 51.20 becomes much simpler when $K \subset \mathbb{R}$, because most of our energy there was spent controlling the numbers $\beta(x, r)$. See Exercise 18 for a slightly more direct approach, though.

Because of (6), $F(z)^2$ has a continuous extension across (a, b), which we shall still call F^2. This is an abuse of notation, because F itself may not extend. Notice that $F^2(x) > 0$ for $x \in (a, b)$, by (6).

Consider the two domains $U_\pm = \{x + iy : x \in (a, b) \text{ and } \pm y > 0\}$. Because of what was just said and by simple connectedness, $F_{|U_\pm}$ has a unique continuous extension to $U_\pm \cup (a, b)$, which we shall call $F_\pm$. We even know that $F_\pm(x)^2 = F^2(x) > 0$ for $x \in (a, b)$, so $F_\pm(x)$ is real.

We may also integrate ∇u (whose coordinates are given by F) to compute u on $U_\pm$. Thus u also has a continuous extension to $U_\pm \cup (a, b)$, which we denote by $u_\pm$. [We already knew this from Section 17, but this is simpler to derive here.] We even get that the restriction of $u_\pm$ to (a, b) is of class C^1, with

$$\frac{\partial u_\pm}{\partial x}(x) = F_\pm(x) \quad \text{for } x \in (a, b). \tag{12}$$

The derivative may have singularities at a and b, but they are fairly mild. In fact, (6) says that

$$|F^2(z)| \leq \frac{1}{2\pi} \int_{\mathbb{R} \setminus K} \text{dist}(w, z)^{-2} dw$$
$$\leq C \, \text{dist}(z, \mathbb{R} \setminus K)^{-1} \quad \text{for } z \in (a, b) \cup [\mathbb{C} \setminus K], \tag{13}$$

so that in particular $u_\pm$ has a continuous extension to $I = [a, b]$.

Let us also check that $u_\pm(a) = u_\pm(b) = \lambda$, where λ is the constant value of u on $\mathbb{R} \setminus K$ given by (3). By definition of I (see (11)), we can find points $t \in \mathbb{R} \setminus K$, with $r = |a - t|$ as small as we want. Then

$$\int_{\partial B(a,r)} |\nabla u| dH^1 = \int_{\partial B(a,r)} |F^2(z)|^{1/2} dH^1$$

$$\leq C \int_{\partial B(a,r)} \text{dist}(z, \mathbb{R})^{-1/2} dH^1 \leq Cr^{1/2} \qquad (14)$$

by (13). The two intersections of $\partial B(a,r)$ with $\mathbb{R}$ are $a + r$, which lies on I, and $a - r$, which is therefore equal to t. Hence $|u_\pm(a+r) - \lambda| = |u_\pm(a+r) - u(t)| \leq Cr^{1/2}$ (by (14)), and since this holds with arbitrarily small values of r, $u_\pm(a) = \lambda$. We can get that $u_\pm(b) = \lambda$ with the same proof.

By Rolle, we can find $x_+ \in (a, b)$ such that $\frac{\partial u_+}{\partial x}(x) = 0$. But (12) says that $\frac{\partial u_+}{\partial x}(x) = F_+(x)$, and we observed earlier that $F_+(x)^2 = F^2(x) > 0$. This contradiction shows that our last case when $\mathbb{C} \setminus K$ is not an interval was impossible, and completes our proof of Theorem 1 when K is contained in a line. $\qquad \square$

Comments about the general case. The argument when K is almost contained in a countable union of lines is more complicated, but uses some of the same ideas.

First, one uses Theorem 51.20 to find a (countable) collection of open line intervals I, $I \in \mathcal{I}$, that are contained in K and such that $H^1(K \setminus \bigcup_{I \in \mathcal{I}} I) = 0$. For each $I \in \mathcal{I}$, the function F^2 has an analytic continuation across I, because if L_I denotes the line that contains I,

$$F(z)^2 = -\frac{1}{2\pi} \int_K \frac{dH^1(w)}{(z-w)^2} = -\frac{1}{2\pi} \int_{K \setminus I} \frac{dH^1(w)}{(z-w)^2} + \frac{1}{2\pi} \int_{L_I \setminus I} \frac{dH^1(w)}{(z-w)^2} \qquad (15)$$

for $z \in \mathbb{C} \setminus [K \cup L_I]$ (by (2), because $K = [K \setminus I] \cup I$, and because $\int_{L_I} \frac{dH^1(w)}{(z-w)^2} = 0$), and then the right-hand side of (15) extends across I. Thus F^2 has an analytic extension to $\Omega = \mathbb{C} \setminus (K \setminus \bigcup_{I \in \mathcal{I}} I)$.

Let $I \in \mathcal{I}$ be given, and suppose for simplicity that $I \subset \mathbb{R}$. The Neumann condition $\frac{\partial u}{\partial n} = 0$ on I implies that $F^2(z)$ is real on I, i.e., that

$$F^2(z) = \overline{F^2(\bar{z})} \qquad (16)$$

for $z \in I$. Observe that both sides of (16) are analytic on $\Omega \cap \Omega^*$, where Ω^* denotes the reflection of Ω with respect to $\mathbb{R}$. Recall that $H^1(\mathbb{C} \setminus \Omega) = H^1(K \setminus \bigcup_{I \in \mathcal{I}} I) = 0$, so $\Omega \cap \Omega^*$ is connected, and (16) holds for all $z \in \Omega \cap \Omega^*$ by analytic continuation.

If I is contained in some line L_I other than $\mathbb{R}$, we still get that F^2 has the appropriate symmetry with respect to L_I.

Because of all these symmetries, it turns out that one can get the following description of K (as a sort of star-like set). If K is not empty, there exist an integer

p and two (nonempty) closed sets K_0, K_1 contained in the positive real half-line $\mathbb{R}^+$, such that K is the image by a translation and a rotation of

$$\bigcup_{k=0}^{p-1} e^{2k\frac{i\pi}{p}} K_0 \ \cup \ \bigcup_{k=0}^{p-1} e^{(2k+1)\frac{i\pi}{p}} K_1. \tag{17}$$

[Note the symmetry of such a set with respect to the lines of direction $e^{2k\frac{i\pi}{p}}$ and $e^{(2k+1)\frac{i\pi}{p}}$.]

When $K_1 = K_0$, the situation is simpler, and a modification of the argument above (with the positivity of F^2 in (6)) gives the desired result (i.e., K_0 is a half-line, and then (u, K) is one of the standard global minimizers). See Proposition 28 in [Lé3].

We are left with the case when K is as in (17), with $K_1 \neq K_0$, where an additional argument is needed. Léger uses the formula (63.3) to show that $\omega_2(0, r) = r^{-1} \int_{B(0,r)\setminus K} |\nabla u|^2$ is nondecreasing [and even strictly increasing in some places, with some control], and then concludes with a blow-in argument. The argument is a little bit similar to the proof of Theorem 61.1 in [Bo]. See [Lé3] for details.

Exercise 18. Let (u, K) be a reduced global minimizer in the plane, with $K \subset \mathbb{R}$. We want to show that if $x_0 \in K$ is a density point for K, then K contains an open interval that contains x_0. Without loss of generality, we can assume that $x_0 = 0$.

1. Check that $|\nabla u(x + iy)|^2 = |F(x + iy)|^2 \leq C|y|^{-1}$ for $x + iy \in \mathbb{C} \setminus \mathbb{R}$.

2. Let $r > 0$ be given, suppose that $x + iy \in B(0, r)$, and that $H^1(K \cap B(0, Ar)) \leq \eta r$ for some large A and some small $\eta > 0$. Show that $|\nabla u(x + iy)|^2 \leq CA^{-1}r^{-1} + C\eta r|y|^{-2}$.

3. Let $1 < p < 2$ be given, and set $\omega_p^*(r) = r^{1-\frac{2}{p}} \left\{ \int_{\partial B(0,r)\setminus K} |\nabla u|^p dH^1 \right\}^{2/p}$ for $r > 0$. Show that $\lim_{r \to 0} \omega_p^*(r) = 0$. [Recall that $\rho^{-1} H^1(K \cap (-\rho, \rho))$ tends to 0 with ρ, because 0 is a point of density for K in $\mathbb{R}$.]

4. Show that for r small enough, $K \cap \partial B(0, r) = \{-r, r\}$. [Hint: use the argument near (25.16).] Conclude.

Note that this can be used to simplify the proof of (11), because the interval J there contains a density point for K.

5. Generalize the results of questions 1–4 to the case when $K \cap B(0, 1)$ is contained in a C^1 curve and 0 is a point of density of K (in the curve).

65 Another formula, with radial and tangential derivatives

The formula presented in this section was discovered independently by J.-C. Léger (who proved it for global minimizers in the plane, and deduced it from (63.3) and a computation with residues) and by F. Maddalena and S. Solimini (with a variational argument like the one presented in this section). It will be useful in the next sections because it will allow us to replace integrals of radial derivatives on a circle with integrals of tangential derivatives.

We are mostly interested in pairs $(u, K) \in RGM(\mathbb{R}^2)$, but let us give a more general statement, with local minimizers in a disk. Let $B(0, R)$ be a disk in the plane, and suppose that $(u, K) \in TRLAM(B(0, R), 0)$. Recall from Standard Assumption 41.1 that this means that (u, K) is a topologically reduced local topological almost-minimizer in $B(0, R)$, with gauge function $h = 0$, or a reduced local almost-minimizer in $B(0, R)$, again with gauge function $h = 0$. But an important special case is when $(u, K) \in RGM(\mathbb{R}^2)$.

Our formula will only make sense for almost every radius $r < R$. More precisely, let us restrict to radii r with the following properties:

$$K \cap \partial B(0, r) \text{ is finite} \tag{1}$$

and, for each $\xi \in K \cap \partial B(0, r)$, we can find $\rho > 0$ such that

$$K \cap B(\xi, \rho) \text{ is a curve of class } C^{1+\alpha} \text{ which is transverse to } \partial B(0, r) \text{ at } \xi. \tag{2}$$

The value of $\alpha > 0$ will not matter. Let us also require that

$$\int_{\partial B(0,r)} |\nabla u|^2 < +\infty \tag{3}$$

to be really safe, but one could check that it follows from the other conditions anyway. Before we state our result, we should observe that

$$\text{almost every } r < R \text{ satisfies (1), (2), and (3).} \tag{4}$$

Indeed $H^1(K \cap B(0, r)) + \int_{B(0,r)} |\nabla u|^2 \leq 2\pi r$ for $0 < r < R$, by the usual truncation argument (see Lemma 18.19). Then (1) holds for almost all $r < R$, by Lemma 26.1, and (3) holds for almost all r, by Fubini. Notice that we do not need to worry about the precise definition of ∇u on $\partial B(0, r)$, because u is harmonic on $B_0 \setminus K$.

Next, Theorem 51.20 says that H^1-almost every point $\xi \in K$ is a C^1 point of K, i.e., there is a $\rho > 0$ such that $K \cap B(\xi, \rho)$ is a nice C^1 curve. The proof even gives that it is a curve of class $C^{1+\alpha}$. Call K' the set of points of K that are not like this. Then $H^1(K') = 0$, and the set of $r < R$ such that $\partial B(0, r)$ contains a point of K' has vanishing measure too. [It is the image of K' by $z \to |z|$, so we can apply Remark 2.11.] This takes care of (2), except for the transversality condition. But the set of radii r for which K is tangent to $\partial B(0, r)$ at some C^1 point of K has measure 0, by Sard's theorem (applied to the same function $z \to |z|$). The verification is the same as for (58.5), so we skip the details. This proves (4).

Proposition 5. *Let* $(u, K) \in TRLAM(B(0, R), 0)$ *be a local minimizer in the disk* $B(0, R)$ *(as above), and* $r < R$ *be such that* (1), (2), *and* (3) *hold. For each* $\xi \in K \cap \partial B(0, r)$, *call* $\theta_\xi \in [0, \frac{\pi}{2}]$ *the (nonoriented) angle between the tangent line to* K *at* ξ *and the radius* $[0, \xi]$. *Also call* $\frac{\partial u}{\partial r}$ *and* $\frac{\partial u}{\partial \tau}$ *the radial and tangential derivatives of* u *at a point of* $\partial B(0, r) \setminus K$. *Then*

$$\int_{\partial B(0,r) \setminus K} \left(\frac{\partial u}{\partial r}\right)^2 dH^1 = \int_{\partial B(0,r) \setminus K} \left(\frac{\partial u}{\partial \tau}\right)^2 dH^1$$

$$+ \sum_{\xi \in K \cap \partial B(0,r)} \cos \theta_\xi - \frac{1}{r} H^1(K \cap B(0, r)). \tag{6}$$

See Remark 40 for a generalization to minimizers in a cone.

We shall prove the proposition by a variational argument. That is, we shall deform (u, K) with a one-parameter family of homeomorphisms φ_t, the minimality of (u, K) will say that some derivative vanishes at $t = 0$, and this will yield (6). The φ_t will move points in the whole ball $B(0, r)$, so our proof really needs (u, K) to be a local minimizer, and not merely an almost-minimizer (even with a very small gauge function).

Our homeomorphisms φ_t are radial functions defined by

$$\varphi_t(\rho \cos \theta, \rho \sin \theta) = (g_t(\rho) \cos \theta, g_t(\rho) \sin \theta) \tag{7}$$

for $0 < \rho < R$ and $\theta \in \mathbb{R}$, where

$$g_t(\rho) = \rho + t f(\rho) \tag{8}$$

and f is the following continuous function on $[0, R)$. We pick $r_0 < r$ (close to r) and set

$$\begin{aligned} f(\rho) &= \rho & \text{for } 0 \leq \rho \leq r_0, \\ f(\rho) &= \frac{r_0(r - \rho)}{r - r_0} & \text{for } r_0 \leq \rho \leq r, \\ f(\rho) &= 0 & \text{for } r \leq \rho \leq R. \end{aligned} \tag{9}$$

We shall restrict our attention to $|t| < (2\|f'\|_\infty)^{-1}$ so that

$$\frac{1}{2} \leq g_t'(\rho) \leq 2 \quad \text{and} \quad \frac{\rho}{2} \leq g_t(\rho) \leq 2\rho \quad \text{for } 0 \leq \rho < R. \tag{10}$$

Then g is a piecewise affine bijection of $[0, R)$, and φ_t is a piecewise smooth diffeomorphism of $B(0, R)$. Set

$$K_t = \varphi_t(K) \quad \text{and} \quad u_t(x) = u(\varphi_t^{-1}(x)) \quad \text{for } x \in B(0, R) \setminus \varphi_t(K). \tag{11}$$

It is easy to check that $u_t \in W_{\text{loc}}^{1,2}(B(0, R) \setminus K_t)$. Since $\varphi_t(x) = x$ for $x \notin B(0, r)$, (u_t, K_t) is a competitor for (u, K) in $B(0, r)$. It is even a topological competitor, because φ_t is a homeomorphism. Thus, if we set

$$a(t) = H^1(K_t \cap \overline{B}(0, r)) \quad \text{and} \quad e(t) = \int_{B(0,r) \setminus K_t} |\nabla u_t|^2, \tag{12}$$

the minimality of (u, K) yields

$$a(0) + e(0) \leq a(t) + e(t) \quad \text{for } |t| \text{ small.} \tag{13}$$

We want to check that a and e have derivatives at $t = 0$, and compute the derivatives. Set

$$B = B(0, r), B_0 = B(0, r_0), \text{ and } A = B \setminus B_0. \tag{14}$$

First notice that

$$\varphi_t(x) = (1 + t)x \quad \text{for } x \in B_0, \tag{15}$$

so that

$$H^1(\varphi_t(K \cap B_0)) = (1 + t)H^1(K \cap B_0). \tag{16}$$

Next consider the contribution of $K \cap A$. Recall from (1) and (2) that $K \cap \partial B$ is finite, and each point of $K \cap \partial B$ is a C^1 point of K, with transverse intersection with ∂B. If we take r_0 close enough to r, $K \cap A$ is a finite union of C^1 arcs. For $x \in K \cap A$, let $\tau(x)$ denote a unit tangent vector to K at x. It is all right if we only define $\tau(x)$ modulo multiplication by ± 1. By direct computation,

$$H^1(\varphi_t(K \cap A)) = \int_{K \cap A} |D\varphi_t(x) \cdot \tau(x)| dH^1(x) = \int_{K \cap A} h_t(x) dH^1(x), \tag{17}$$

where we set $h_t(x) = |D\varphi_t(x) \cdot \tau(x)|$.

Fix $x \in K \cap A$, and let us first see what $D\varphi_t(x)$ looks like. Set $\rho = |x|$, $p = \rho^{-1}x$ (the unit vector with direction x), and call q a unit vector perpendicular to p, so that (p, q) is an orthonormal basis. Then

$$D\varphi_t(x) \cdot p = g'_t(\rho) p \quad \text{and} \quad D\varphi_t(x) \cdot q = \rho^{-1} g_t(\rho) q, \tag{18}$$

by (7). If we write $\tau(x) = \tau_p p + \tau_q q$, (18) yields

$$h_t(x)^2 = |D\varphi_t(x) \cdot \tau(x)|^2 = g'_t(\rho)^2 \tau_p^2 + \rho^{-2} g_t(\rho)^2 \tau_q^2. \tag{19}$$

In particular, (10) says that

$$\frac{1}{2} \leq h_t(x) \leq 2. \tag{20}$$

Recall from (8) that $g_t(\rho) = \rho + tf(\rho)$. Then $h_t(x)^2$ is differentiable in t, with

$$\frac{\partial}{\partial t}[h_t(x)^2] = 2g'_t(\rho)\frac{\partial}{\partial t}[g'_t(\rho)] \tau_p^2 + 2\rho^{-2}g_t(\rho)\frac{\partial}{\partial t}[g_t(\rho)] \tau_q^2$$
$$= 2g'_t(\rho)f'(\rho) \tau_p^2 + 2\rho^{-2}g_t(\rho)f(\rho) \tau_q^2. \tag{21}$$

Then $h_t(x)$ also is differentiable,

$$\frac{\partial h_t}{\partial t} = \frac{\partial}{\partial t}[h_t^2(x)^{1/2}] = \frac{1}{2h_t(x)}\frac{\partial}{\partial t}[h_t^2(x)]$$
$$= \frac{1}{h_t(x)}\left\{ g'_t(\rho)f'(\rho) \tau_p^2 + \rho^{-2}g_t(\rho)f(\rho) \tau_q^2 \right\}, \tag{22}$$

and (20) says that $\frac{\partial h_t}{\partial t}$ is bounded, uniformly in x and t. Thus we can differentiate (17) under the integral sign, and

$$\frac{\partial}{\partial t}[H^1(\varphi_t(K \cap A))] = \int_{K \cap A} \frac{\partial}{\partial t}[h_t(x)] \, dH^1(x). \tag{23}$$

We are interested in the value of this at $t = 0$. Then $g_t(\rho) = \rho$, $g_t'(\rho) = 1$, $h_t(x)^2 = \tau_p^2 + \tau_q^2 = 1$ by (19), and

$$\frac{\partial}{\partial t}[H^1(\varphi_t(K \cap A))]_{|t=0} = \int_{K \cap A} \{f'(\rho) \tau_p^2 + \rho^{-1} f(\rho) \tau_q^2\} dH^1, \tag{24}$$

by (22). Now

$$\begin{aligned} a(t) &= H^1(K_t \cap \overline{B}(0,r)) = H^1(K_t \cap B(0,r)) \\ &= H^1(\varphi_t(K \cap B_0)) + H^1(\varphi_t(K \cap A)) \\ &= (1+t)H^1(K \cap B_0) + H^1(\varphi_t(K \cap A)) \end{aligned} \tag{25}$$

by (12), (1), (14), and (16), so (24) yields

$$a'(0) = H^1(K \cap B_0) + \int_{K \cap A} \{f'(\rho) \tau_p^2 + \rho^{-1} f(\rho) \tau_q^2\} dH^1. \tag{26}$$

Now we want to compute $e'(0)$. Recall from (12) and (11) that

$$e(t) = \int_{B(0,r) \setminus K_t} |\nabla u_t|^2(y) \, dy = \int_{\varphi_t(B \setminus K)} |\nabla(u \circ \varphi_t^{-1})|^2(y) \, dy \tag{27}$$

We split $\nabla u(x)$ into its radial part $\frac{\partial u}{\partial r}(x)$ and its tangential part $\frac{\partial u}{\partial \tau}(x)$, and similarly for u_t. Then (18) says that

$$\frac{\partial u_t}{\partial r}(\varphi_t(x)) = g_t'(\rho)^{-1} \frac{\partial u}{\partial r}(x) \quad \text{and} \quad \frac{\partial u_t}{\partial \tau}(\varphi_t(x)) = \rho \, g_t(\rho)^{-1} \frac{\partial u}{\partial \tau}(x), \tag{28}$$

where we still set $\rho = |x|$. Next set $y = \varphi_t(x)$ in (27). The Jacobian determinant of φ_t is $J_t(x) = \rho^{-1} g_t'(\rho) g_t(\rho)$, again by (18). Then

$$\begin{aligned} e(t) &= \int_{B \setminus K} |\nabla(u \circ \varphi_t^{-1})|^2(\varphi_t(x)) \, J_t(x) \, dx \\ &= \int_{B \setminus K} \left\{ g_t'(\rho)^{-2} \left(\frac{\partial u}{\partial r}(x)\right)^2 + \rho^2 \, g_t(\rho)^{-2} \left(\frac{\partial u}{\partial \tau}(x)\right)^2 \right\} J_t(x) \, dx \\ &= \int_{B \setminus K} \left\{ \frac{g_t(\rho)}{\rho g_t'(\rho)} \left(\frac{\partial u}{\partial r}(x)\right)^2 + \frac{\rho g_t'(\rho)}{g_t(\rho)} \left(\frac{\partial u}{\partial \tau}(x)\right)^2 \right\} dx. \end{aligned} \tag{29}$$

Set $w_t(\rho) = \frac{g_t(\rho)}{\rho g_t'(\rho)}$ for $0 < \rho < r$. Then (8) says that

$$w_t(\rho) = \frac{g_t(\rho)}{\rho g_t'(\rho)} = \frac{\rho + tf(\rho)}{\rho + t\rho f'(\rho)} = \frac{1 + t\rho^{-1} f(\rho)}{1 + tf'(\rho)}. \tag{30}$$

Since we restricted to $|t|$ small, w_t and its inverse are bounded, and their partial derivative with respect to t as well. Then we can differentiate in t inside the integral in (29) (because $\nabla u \in L^2(B \setminus K)$), and

$$e'(t) = \int_{B\setminus K} \left\{ \frac{\partial}{\partial t}(w_t(\rho)) \left(\frac{\partial u}{\partial r}(x) \right)^2 + \frac{\partial}{\partial t}\left(\frac{1}{(w_t(\rho))} \right) \left(\frac{\partial u}{\partial \tau}(x) \right)^2 \right\} dx. \qquad (31)$$

As before, we just want to compute this when $t = 0$. Then $g_t(\rho) = \rho$ by (8), and (30) yields that $w_t(\rho) = 1$, $\frac{\partial}{\partial t}(w_t(\rho)) = \rho^{-1}f(\rho) - f'(\rho)$, and $\frac{\partial}{\partial t}\left(\frac{1}{(w_t(\rho))} \right) = -\frac{\partial}{\partial t}(w_t(\rho))$. Thus

$$e'(0) = \int_{B\setminus K} \left\{ \rho^{-1}f(\rho) - f'(\rho) \right\} \left\{ \left(\frac{\partial u}{\partial r}(x) \right)^2 - \left(\frac{\partial u}{\partial \tau}(x) \right)^2 \right\} dx. \qquad (32)$$

It is time to use our definition (9) of f. Since $f(\rho) = \rho$ for $\rho \leq r_0$, we get $\rho^{-1}f(\rho) - f'(\rho) = 0$ there, and B_0 gives no contribution to the integral in (32). This is not surprising, because φ_t is conformal on B_0, and hence $\int_{\varphi(B_0\setminus K)} |\nabla u_t|^2 = \int_{B_0\setminus K} |\nabla u|^2$ by conformal invariance of energy integrals. We are left with the contribution of A, where $f(\rho) = \frac{r_0(r-\rho)}{r-r_0}$, $f'(\rho) = \frac{-r_0}{r-r_0}$, and $\rho^{-1}f(\rho) - f'(\rho) = \frac{r_0(r-\rho)}{\rho(r-r_0)} + \frac{\rho r_0}{\rho(r-r_0)} = \frac{r_0 r}{\rho(r-r_0)}$. Thus

$$e'(0) = \int_{A\setminus K} \frac{r_0 r}{\rho(r - r_0)} \left\{ \left(\frac{\partial u}{\partial r}(x) \right)^2 - \left(\frac{\partial u}{\partial \tau}(x) \right)^2 \right\} dx. \qquad (33)$$

So we proved that $a(t)$ and $e(t)$ are differentiable, and (13) says that $a'(0) + e'(0) = 0$. That is,

$$H^1(K \cap B_0) + \int_{K\cap A} \left\{ f'(\rho)\,\tau_p^2 + \rho^{-1}f(\rho)\,\tau_q^2 \right\} dH^1$$
$$+ \int_{A\setminus K} \frac{r_0 r}{\rho(r - r_0)} \left\{ \left(\frac{\partial u}{\partial r}(x) \right)^2 - \left(\frac{\partial u}{\partial \tau}(x) \right)^2 \right\} dx = 0, \qquad (34)$$

by (26) and (33). Now we want to let r_0 tend to r from below and see what we get. Obviously,

$$\lim_{r_0 \to r^-} H^1(K \cap B_0) = H^1(K \cap B). \qquad (35)$$

Next we check that

$$\lim_{r_0 \to r^-} \int_{K\cap A} \rho^{-1}f(\rho)\,\tau_q^2 dH^1 = 0. \qquad (36)$$

Recall from (9) that $f(\rho) = \frac{r_0(r - \rho)}{r - r_0}$ for $x \in A$, and so $0 \leq \rho^{-1}f(\rho) \leq \rho^{-1}r_0 \leq 1$. Moreover, (1) and (2) say that for r_0 close to r, $K \cap A$ is composed of a finite number of C^1 arcs that are transverse to ∂B. In particular, $H^1(K \cap A)$ tends to 0 when r_0 tends to r, and (36) follows.

For the contribution of $f'(\rho)\,\tau_p^2$, we need to look a little more closely. For each $\xi \in K \cap \partial B$, call K_ξ the little arc of K near ξ promised by (2). Recall that for r_0 close enough to r, $K \cap A$ is the (disjoint) union of the $K_\xi \cap A$, by (1) and (2).

When r_0 tends to r, $H^1(K_\xi \cap A)$ is equivalent to $(r - r_0)(\cos\theta_\xi)^{-1}$, where $\theta_\xi \in [0, \frac{\pi}{2})$ still denotes the angle of the tangent to K at ξ with the radius $[0, \xi]$. This angle is different from $\frac{\pi}{2}$ by the transversality in (2).

When $x \in K_\xi$, $\tau_p(x)$ is the radial part of $\tau(x)$, a unit tangent to K_ξ at x. Then $\tau_p(x)^2$ tends to $\cos^2\theta_\xi$, uniformly in $x \in K_\xi$, when r_0 tends to r.

Finally, (9) says that $f'(\rho) = -\frac{r_0}{r-r_0}$ for $x \in A$. Altogether, the contribution of $K_\xi \cap A$ to $\int_{K \cap A} f'(\rho)\,\tau_p^2 dH^1$ is equivalent to $-\frac{r}{r-r_0} \cdot \cos^2\theta_\xi \cdot (r-r_0)(\cos\theta_\xi)^{-1} = -r\cos\theta_\xi$, and

$$\lim_{r_0 \to r^-} \int_{K \cap A} f'(\rho)\,\tau_p^2\, dH^1 = -r \sum_{\xi \in K \cap \partial B} \cos\theta_\xi. \qquad (37)$$

Finally, we claim that

$$\lim_{r_0 \to r^-} \int_{A \setminus K} \frac{r_0 r}{\rho(r - r_0)} \left\{ \left(\frac{\partial u}{\partial r}(x)\right)^2 - \left(\frac{\partial u}{\partial \tau}(x)\right)^2 \right\} dx$$
$$= r \int_{\partial B \setminus K} \left\{ \left(\frac{\partial u}{\partial r}\right)^2 - \left(\frac{\partial u}{\partial \tau}\right)^2 \right\} dH^1. \qquad (38)$$

The main point here is that by minimality of (u, K), the function u is harmonic on $B(0, R) \setminus K$, hence ∇u is continuous there. We also required in (2) that K be of class $C^{1+\alpha}$ near ∂B; then ∇u has continuous extensions up to K (by Proposition 17.15), and in particular it stays bounded near the points of $K \cap \partial B$. Since $\frac{r_0 r}{\rho(r - r_0)}$ is equivalent to $\frac{r}{(r - r_0)}$, we get (38) easily by integrating in polar coordinates.

We deduce from (34)–(38) that

$$H^1(K \cap B) - r \sum_{\xi \in K \cap \partial B} \cos\theta_\xi + r \int_{\partial B \setminus K} \left\{ \left(\frac{\partial u}{\partial r}\right)^2 - \left(\frac{\partial u}{\partial \tau}\right)^2 \right\} dH^1 = 0, \qquad (39)$$

which is the same thing as (6). This proves Proposition 5. $\qquad\qquad\qquad\square$

Remark 40. Proposition 5 also works when we replace $B(0, R)$ with its intersection with an open cone. That is, let $\mathcal{C}$ be an open cone centered at the origin, and set $\Omega = \mathcal{C} \cap B(0, R)$. Also let $(u, K) \in TRLAM(\Omega, 0)$ be given. Then

$$\int_{\mathcal{C} \cap \partial B(0,r) \setminus K} \left(\frac{\partial u}{\partial r}\right)^2 dH^1 = \int_{\mathcal{C} \cap \partial B(0,r) \setminus K} \left(\frac{\partial u}{\partial \tau}\right)^2 dH^1$$
$$+ \sum_{\xi \in K \cap \partial B(0,r)} \cos\theta_\xi - \frac{1}{r} H^1(K \cap B(0,r)) \qquad (41)$$

for almost every $r < R$ (with the same notation as above). The proof is the same; the main point (and also the reason why the proof would not extend to other domains Ω) is that our homeomorphisms φ_t preserve $\mathcal{C} \cap B$.

66 $\mathbb{R}^2 \setminus K$ is connected when (u, K) is an exotic global minimizer

In this section we want to discuss, but not prove entirely, the following result.

Theorem 1 [DaLé]. *Let $(u, K) \in RGM(\mathbb{R}^2)$ be a reduced global minimizer in the plane, and suppose that K is neither a line nor a propeller. Then $\mathbb{R}^2 \setminus K$ is connected.*

See the second part of this section for a few slightly more general or related results, and the (short) last part for a description of the global minimizers in some cones.

The proof of Theorem 1 is based on a monotonicity argument, a little like for Bonnet's Theorem 61.1, but with a different function. Let $(u, K) \in RGM(\mathbb{R}^2)$ and $x \in \mathbb{R}^2$ be given, and set

$$l(r) = H^1(K \cap B(x,r)) \quad \text{and} \quad E(r) = \int_{B(x,r) \setminus K} |\nabla u|^2 \tag{2}$$

for $r > 0$. We consider the function

$$F(r) = \frac{l(r) + 2E(r)}{r} \tag{3}$$

and show that it is nondecreasing on $[\rho, +\infty)$ as soon as $\partial B(x,r)$ meets at least two connected components of $\mathbb{R}^2 \setminus K$. Since there are different estimates involved here, and some of them extend to other situations, it is probably worthwhile to be more specific.

a. Monotonicity estimates

Let us take $x = 0$ to simplify, and merely assume that $(u,K) \in TRLAM(B(0,R),0)$ for some $R > 0$, as in Section 65. This means that (u, K) is a reduced local minimizer in $B(0, R)$ (topological or not, this will not matter), and with the gauge function $h = 0$. We are only interested in $F(r)$ for $r < R$. The first verification is that

F is differentiable almost everywhere on $(0, R)$, and

$$F(b) - F(a) \geq \int_a^b F'(r)\, dr \quad \text{for } 0 \leq a < b < R. \tag{4}$$

Let us check this and compute $F'(r)$ at the same time. Call $\mathcal{R}$ the set of radii $r \in (0, R)$ such that $K \cap \partial B(0,r)$ is finite, and every point $\xi \in K \cap \partial B(0,r)$ has a small neighborhood where K is a C^1 curve that is transverse to $\partial B(0,r)$ at ξ. Note that $\mathcal{R}$ is open, and (65.4) says that almost every point of $(0, R)$ lies in $\mathcal{R}$. It is easy to see that l is continuously differentiable on $\mathcal{R}$, with

$$l'(r) = \sum_{\xi \in K \cap \partial B(0,r)} \frac{1}{\cos \theta_\xi}, \tag{5}$$

where we still denote by $\theta_\xi \in [0, \frac{\pi}{2})$ the (nonoriented) angle between the tangent line to K at ξ and the radius $[0, \xi]$. Then $l(r)/r$ also is continuously differentiable on $\mathcal{R}$, with the derivative $\frac{rl'(r)-l(r)}{r^2}$. And the analogue of (4) for $l(r)/r$ can be deduced from this, because l is nondecreasing. See Exercise 42 for details. Notice that $l(r)$ may have positive jumps on $(0, R) \setminus \mathcal{R}$, which is the reason why we only wrote an inequality in (4).

The computations for $E(r)$ are easier. When we compute $E(r)$ in polar coordinates, we get that E is differentiable almost everywhere, with $E'(r) = \int_{\partial B(0,r)\setminus K} |\nabla u|^2$, and that E is the integral of its derivative. The analogue of (4) for $E(r)/r$ follows easily; see the proof of (61.9), where we checked this a little more carefully.

This completes our verification of (4). We also get that for almost every $r < R$,

$$
F'(r) = \frac{l'(r)}{r} - \frac{l(r)}{r^2} + \frac{2}{r}\int_{\partial B(0,r)\setminus K} |\nabla u|^2 - \frac{2E(r)}{r^2}
$$

$$
= \frac{2}{r}\int_{\partial B(0,r)\setminus K} |\nabla u|^2 + \frac{1}{r}\sum_{\xi \in K \cap \partial B(0,r)} \frac{1}{\cos\theta_\xi} - \frac{2E(r)+l(r)}{r^2}. \tag{6}
$$

Let us say more about how we get lower bounds for $F'(r)$. We write

$$
\int_{\partial B(0,r)\setminus K} |\nabla u|^2 \, dH^1 = I_r + I_\tau , \tag{7}
$$

where

$$
I_r = \int_{\partial B(0,r)\setminus K} \left(\frac{\partial u}{\partial r}\right)^2 dH^1 \quad \text{and} \quad I_\tau = \int_{\partial B(0,r)\setminus K} \left(\frac{\partial u}{\partial \tau}\right)^2 dH^1. \tag{8}
$$

Proposition 65.5 says that for $r \in \mathcal{R}$,

$$
I_r = I_\tau + \sum_{\xi \in K \cap \partial B(0,r)} \cos\theta_\xi - \frac{1}{r}l(r). \tag{9}
$$

Denote by $N(r)$ the number of points in $K \cap \partial B(0,r)$ (which is finite for $r \in \mathcal{R}$). Most of the time, we use the following estimates. We first notice that since $a + a^{-1} \geq 2$ for $a > 0$,

$$
\sum_{\xi \in K \cap \partial B(0,r)} \cos\theta_\xi + \sum_{\xi \in K \cap \partial B(0,r)} \frac{1}{\cos\theta_\xi} \geq 2N(r), \tag{10}
$$

and then use (6) and (9) to get that

$$
rF'(r) = 2I_r + 2I_\tau + \sum_{\xi \in K \cap \partial B(0,r)} \frac{1}{\cos\theta_\xi} - \frac{2E(r)+l(r)}{r}
$$

$$
\geq I_r + 3I_\tau + 2N(r) - \frac{2E(r)+2l(r)}{r}. \tag{11}
$$

This is often easy to use. Notice in particular that since $I_r \geq 0$, $F'(r) \geq 0$ as soon as

$$E(r) + l(r) \leq \frac{3}{2} r I_\tau + r N(r). \tag{12}$$

This is what we prove in a number of cases. We do not want to give the details here, but the point of (12) is that we can hope to prove it with our standard methods. That is, we can use the values of K and u on $\partial B(0, r)$ to construct new competitors (v, G) for (u, K) in $B(0, r)$. The constraints on (v, G) are the following. We want $G \cap \overline{B}(0, r)$ to separate the $N(r)$ connected components of $\partial B(0, r) \setminus K$ from each other in $\overline{B}(0, r)$, because this way we can apply Lemma 45.8 to show that (v, G) is a topological competitor for (v, G), and also because otherwise it would be impossible to estimate $\int_{B(0,r) \setminus G} |\nabla v|^2$ in terms of $N(r)$, I_r, and I_τ anyway. We also need v to coincide with u on $\partial B(0, r) \setminus G$, as usual, and the best is to extend harmonically on each component of $B \setminus G$ independently.

If we can construct a pair (v, G) like this, and for which

$$H^1(G \cap \overline{B}(0, r)) + \int_{B(0,r) \setminus G} |\nabla v|^2 \leq \frac{3}{2} r I_\tau + r N(r), \tag{13}$$

then we get (12) because $E(r) + l(r) \leq H^1(G \cap \overline{B}(0, r)) + \int_{B(0,r) \setminus G} |\nabla v|^2$. This happens in quite a few cases: we try a few natural competitors (v, G), get estimates on the left-hand side of (13), and often (13) holds for one of them. Figure 1 suggests three reasonable choices for G when $N(r) = 2$; the best one will mostly depend on the angle made by the two points of $K \cap \partial B(0, \rho)$. But if the integral I_τ is large, it may be preferable to choose G as in Figure 2, because the norm of our extension operator from $\partial B \setminus G$ to $B \setminus G$ will be smaller if $B \setminus G$ is composed of smaller sectors. Things get easier when $N(r) \geq 3$, because it is often quite profitable to use forks as in Figure 3, since this makes $H^1(G \cap \overline{B}(0, r))$ much smaller than $r N(r)$. Let us refer to [DaLé] for the constructions, and merely record the estimates that we get.

In what follows, we merely assume that $(u, K) \in TRLAM(B(0, R), 0)$ (as above), but restrict our attention to $r \in \mathcal{R}$ (which is the case for almost every $r \in (0, R)$). Recall that $N(r)$ is the number of points in $K \cap \partial B(0, r)$.

14. If $N(r) = 0$, then $r F'(r) \geq \int_{\partial B(0,r)} |\nabla u|^2$.

15. If $N(r) \geq 3$, then $r F'(r) \geq 10^{-10} \int_{\partial B(0,r) \setminus K} |\nabla u|^2$.

16. If $N(r) \geq 4$, then we also have that $r F'(r) \geq 10^{-3}$.

17. If $N(r) \geq 2$ and all the connected components of $\partial B(0, r) \setminus K$ have lengths smaller than or equal to $\omega \pi r$ for some $\omega \leq \frac{3}{2}$, then

$$r F'(r) \geq \text{Min}\{1, 3 - 2\omega\} \int_{\partial B(0,r) \setminus K} |\nabla u|^2.$$

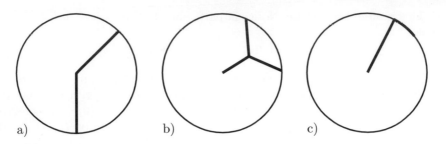

Figure 1. Three reasonable choices of G when $N(r) = 2$

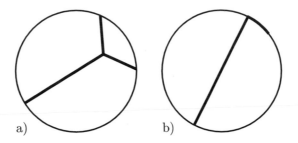

Figure 2. Two other choices when $N(r) = 2$

These were the relatively easy, or at least straightforward estimates. Note that we do not get good estimates when $N(r) = 1$, and there is a good reason for this. It turns out that already when (u, K) is a cracktip for which K contains 0 but the tip is not the origin, F may be strictly decreasing on some intervals. This is possible because $l(r)/r$ decreases from 2 to 1. When $N(r) = 2$ and $\omega \geq 3/2$, we can still get a good estimate, but only with an additional condition.

18. If $N(r) = 2$, and the two points of $K \cap \partial B(0, r)$ lie on the same connected component of K, then $rF'(r) \geq 10^{-10} \int_{\partial B(0,r) \setminus K} |\nabla u|^2$.

For this last one, we need something else than (11) and the comparison with a reasonable competitor. We also need to know that

$$E(r) \leq r\,\omega(r)I_r^{1/2}I_\tau^{1/2}, \tag{19}$$

where $\pi r\omega(r)$ denotes the length of the longest component of $\partial B(0, r) \setminus K$. Let us check (19). The proof is almost the same as for Bonnet's lower bound on $E'(r)$ (Proposition 47.7). Denote by J_1 and J_2 the two components of $\partial B(0, r) \setminus K$, and recall from (47.12) that

$$E(r) = \int_{\partial B(0,r) \setminus K} u\frac{\partial u}{\partial r}dH^1 = \int_{J_1} u\frac{\partial u}{\partial r}dH^1 + \int_{J_2} u\frac{\partial u}{\partial r}dH^1. \tag{20}$$

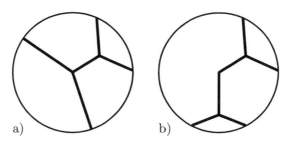

a)   b)

Figure 3. Two cases with $N(r) = 4$

Then observe that $\int_{J_j} \frac{\partial u}{\partial r} dH^1 = 0$ for $j = 1, 2$, by (47.13). This is where we use the fact that the two points of $K \cap \partial B(0, r)$ lie on the same component of K. Then we may subtract the mean value m_j of u on J_j and get that

$$\left| \int_{J_j} u \frac{\partial u}{\partial r} dH^1 \right| = \left| \int_{J_j} (u - m_j) \frac{\partial u}{\partial r} dH^1 \right| \leq \left\{ \int_{J_j} (u - m_j)^2 \right\}^{1/2} \left\{ \int_{J_j} \left(\frac{\partial u}{\partial r} \right)^2 \right\}^{1/2}$$

$$\leq \frac{H^1(J_j)}{\pi} \left\{ \int_{J_j} \left(\frac{\partial u}{\partial \tau} \right)^2 \right\}^{1/2} \left\{ \int_{J_j} \left(\frac{\partial u}{\partial r} \right)^2 \right\}^{1/2},$$

$$(21)$$

by the Wirtinger inequality in Lemma 47.18.

Set $a_j = \left\{ \int_{J_j} \left(\frac{\partial u}{\partial \tau} \right)^2 \right\}^{1/2}$, $b_j = \left\{ \int_{J_j} \left(\frac{\partial u}{\partial r} \right)^2 \right\}^{1/2}$, and $\lambda = I_\tau^{-1/2} I_r^{1/2}$ (if $I_\tau = 0$, we get $E(r) = 0$ directly). Also recall that $H^1(J_j) \leq \pi r \omega(r)$ by definition of $\omega(r)$. Then

$$\left| \int_{J_j} u \frac{\partial u}{\partial r} dH^1 \right| \leq r \omega(r) a_j b_j \leq r \omega(r) \left(\frac{\lambda}{2} a_j^2 + \frac{1}{2\lambda} b_j^2 \right)$$

$$(22)$$

by (21), and now (20) yields

$$E(r) \leq \frac{\lambda}{2} r \omega(r) \left(a_1^2 + a_2^2 \right) + \frac{1}{2\lambda} r \omega(r) \left(b_1^2 + b_2^2 \right)$$

$$= \frac{\lambda}{2} r \omega(r) I_\tau + \frac{1}{2\lambda} r \omega(r) I_r = r \omega(r) I_\tau^{1/2} I_r^{1/2},$$

$$(23)$$

as needed for (19).

For the proof of **18**, it is convenient to use the following other combination of (6) and (9) (in addition to (11)):

$$r F'(r) = 2I_r + 2I_\tau + \sum_{\xi \in K \cap \partial B(0, r)} \frac{1}{\cos \theta_\xi} - \frac{2E(r) + l(r)}{r} \geq 3I_r + I_\tau - \frac{2E(r)}{r}. \quad (24)$$

Once we get all these estimates, it is just a (slightly unpleasant) matter of brute force to see that they lead to **18**. We refer to [DaLé] for details on the construction of the appropriate competitors and the other verifications.

b. Proof of Theorem 1 and related results

We may now say how to deduce Theorem 1 from our estimates **14–18**. Let $(u, K) \in RGM(\mathbb{R}^2)$ be given, and assume that $\mathbb{R}^2 \setminus K$ is not connected. We want to show that K is a line or a propeller. Without loss of generality, we may assume that K contains the origin. Let r_0 be such that $B(0, r_0)$ meets at least two different components of $\mathbb{R}^2 \setminus K$. Then for every $r > r_0$, $\partial B(0, r)$ meets both components (because they are unbounded, by Proposition 55.1). This forces $N(r) \geq 2$, but in addition we claim that

$$\text{if } N(r) = 2, \text{ then the two points of } \partial B(0, r) \cap K$$

$$\text{lie on the same component of } K. \tag{25}$$

Indeed let x_1, x_2 be points of $\partial B(0, r) \setminus K$ that lie in different components of $\mathbb{R}^2 \setminus K$. Then K separates x_1 from x_2, and Theorem 14.3 on page 123 of [Ne] says that there is a connected component of K that still separates them. This component contains the two points of $\partial B(0, r) \cap K$, because otherwise we could go from x_1 to x_2 along an arc of $\partial B(0, r)$.

Because of (25), we can apply **15** and **18** and we get that

$$rF'(r) \geq 10^{-10} \int_{\partial B(0,r) \setminus K} |\nabla u|^2 \text{ for almost every } r > r_0. \tag{26}$$

In particular, F is nondecreasing on $(r_0, +\infty)$, and since $F(r) = \frac{l(r) + 2E(r)}{r} \leq 4\pi$ by the trivial estimate, it has a limit l at infinity. Let $\varepsilon > 0$ be small, and choose $r_1 > r_0$ be so large that $F(r_1) \geq l - \varepsilon$. Also let $a > 0$ be small, to be chosen soon. For $R > a^{-1}r_1$,

$$\int_{B(0,R) \setminus [K \cup B(0,aR)]} |\nabla u|^2 = \int_{aR}^{R} \left\{ \int_{\partial B(0,r) \setminus K} |\nabla u|^2 \right\} dr \leq 10^{10} \int_{aR}^{R} rF'(r) dr \tag{27}$$

$$\leq 10^{10} R \int_{aR}^{R} F'(r) dr \leq 10^{10} R[F(R) - F(ar)] \leq 10^{10} \varepsilon R$$

by Fubini, (26), and (4). Since $\int_{B(0,aR) \setminus K} |\nabla u|^2 \leq 2\pi aR$ by the trivial estimate, we get that

$$\int_{B(0,R) \setminus K} |\nabla u|^2 \leq (10^{10}\varepsilon + 2\pi a)R. \tag{28}$$

If a and ε are chosen small enough, $10^{10}\varepsilon + 2\pi a$ is smaller than the constant τ in Proposition 60.1, and that proposition says that $K \cap B(0, R/2)$ is a nice C^1 curve or spider.

At this point, the simplest way to conclude is to say that $K \cap B(0, R/2)$ is connected for R large, so K is connected and Theorem 61.1 says that K is a line or a propeller.

We do not really need to use Bonnet's theorem here. We may also apply Proposition 59.10, because our proof of (28) shows that $\int_{B(0,R)\setminus K} |\nabla u|^2$ tends to 0 when R tends to $+\infty$, and hence any blow-in limit of (u, K) must be locally constant, by Proposition 37.18. Or we may conclude directly from the fact that $K \cap B(0, R/2)$ is a nice C^1 curve or spider, with uniform estimates in R which imply that on a given ball B, $K \cap B$ looks more and more like a line or a propeller. See the end of the proof of Proposition 59.10, starting at (59.12). At any rate, we get that K is a line or a propeller, as needed for Theorem 1. $\qquad\square$

We continue with consequences of Theorem 1 and its proof.

Corollary 29. *For each small $\tau > 0$ there is a constant $a_1 > 0$ with the following property. Suppose $(u, K) \in TRLAM(B(0, R), 0)$ is a coral minimizer in $B(0, R) \subset \mathbb{R}^2$ (that is, a coral almost-minimizer with gauge function $h = 0$; see Standard Assumption 41.1), and suppose that*

$$\text{we can find two connected components of } B(0, R) \setminus K \tag{30}$$
$$\text{that meet } \partial B(x, a_1 R) \text{ and } \partial B(x, R/2),$$

or that

$$\text{we can find two disjoint connected sets } \gamma_1 \text{ and } \gamma_2 \tag{31}$$
$$\text{that are contained in } K \text{ and meet } \partial B(0, a_1 R) \text{ and } \partial B(0, R/2).$$

Then $K \cap B(0, a_1 R)$ is a nice C^1 curve or spider, and $\int_{B(0,r)\setminus K} |\nabla u|^2 \leq \tau r$ for $0 \leq r \leq a_1 R$.

See Definitions 51.2 and 53.3 for the definition of C^1 curves and spiders. Let us first assume that (30) holds, or that (31) holds with sets γ_1 and γ_2 that lie in a single component of K. We claim that for every $r \in (a_1 R, R/2)$, $N(r) \geq 2$ and (25) holds.

Indeed, if (30) holds, the two components of $B(0, R) \setminus K$ meet $\partial B(x, r)$, so $N(r) \geq 2$ because otherwise the components would communicate through $\partial B(x, r)$. If in addition $N(r) = 2$, first observe that $K_r = K \cap \overline{B}(x, r)$ separates the two components of $\partial B(x, r)$ in $\overline{B}(x, r)$. Then the union of K_r with two half-lines that emanate from the points of $K \cap \partial B(x, r)$ still separates the two components of $\partial B(x, r)$ in $\mathbb{R}^2$. The same separation theorem as before (on page 123 in [Ne]) gives a connected subset that still separates, and (25) follows. The details are the same as for Lemma 29.51.

If (31) holds and both γ_i are contained in the same component of K, then $N(r) \geq 2$ because each γ_i meets $\partial B(x, r)$, and (25) holds trivially. This proves our claim.

We can deduce (26) from (25), **15**, and **18**, as before. In particular, F is nondecreasing. Then let $a > 0$ and $\varepsilon > 0$ be small, to be chosen soon. We also have that $F(R) \leq 4\pi$, by the trivial estimate, so if the new constant $a_0 > 0$ is small enough, depending on a and ε, we can find ρ such that $a_0 R \leq \rho \leq R/2$ and $F(\rho) - F(a\rho) \leq \varepsilon$. If $a_1 \leq a_0 a$, then $a\rho \geq aa_0 R \geq a_1 R$, (25) holds for $a\rho \leq r \leq \rho$, and the proof of (28) yields

$$\int_{B(x,\rho)\setminus K} |\nabla u|^2 \leq (10^{10}\varepsilon + 2\pi a)\rho. \tag{32}$$

If a and ε are small enough, we can apply Proposition 60.1 (with $p = 2$ and $h = 0$) and we get that $K \cap B(x, \eta\rho)$ is a nice C^1 curve or spider. Our definitions of nice curves and spiders are such that $K \cap B(x, r)$ is also a nice C^1 curve or spider for every $r \leq \eta\rho$.

We still need an estimate for the energy in smaller disks. The simplest is probably to notice that since $K \cap B(x, \eta\rho)$ is a nice C^1 curve or spider, Proposition 47.7 says that $\omega_2(x, r) = \frac{1}{r}\int_{B(x,r)\setminus K} |\nabla u|^2$ is a nondecreasing function of $r \leq \eta\rho$. Then

$$\omega_2(x, r) \leq \omega_2(x, \eta\rho) \leq \eta^{-1}\omega_2(x, \rho) \leq \eta^{-1}(10^{10}\varepsilon + 2\pi a) \leq \tau \quad \text{for } r \leq \eta\rho, \tag{33}$$

by (32), and if a and ε are small enough. This gives the conclusions of the corollary, as soon as we take $a_1 \leq \eta a_0$. This completes our proof in the special case when (31) holds, or (32) holds and the γ_i lie in the same component of K. Let us call $a(\tau)$ the value of a_1 that we get in this case.

We are now left with the general case when (32) holds. We shall deduce this case from the previous one, by contradiction and compactness. Suppose that the corollary fails for some $\tau > 0$. Then for each $k \geq 0$ there is a minimizer (u_k, K_k) that contradicts its statement with $a_1 = 2^{-k-1}$. By invariance under translations and dilations, we can normalize things so that $x = 0$ and $R = 2$ for each (u_k, K_k). Then $(u_k, K_k) \in TRLAM(B(0, 2), 0)$, and we can find disjoint curves $\gamma_{1,k}$ and $\gamma_{2,k}$ that are contained in K_k and meet $\partial B(0, 2^{-k})$ and $\partial B(0, 1)$. But $K_k \cap B(0, 2^{-k})$ is neither a nice C^1 arc nor a spider, or else we can find $r \leq 2^{-k}$ such that $\int_{B(0,r)\setminus K} |\nabla u|^2 > \tau r$.

By Proposition 37.8, Theorem 38.3, and Remark 38.64, we can replace $\{(u_k, K_k)\}$ with a subsequence that converges to some limit (u, K), and $(u, K) \in TRLAM(B(0, 2), 0)$.

By Proposition 30.17, each $\gamma_{i,k}$ contains a simple arc $\gamma'_{i,k}$ that goes from $\partial B(0, 2^{-k})$ to $\partial B(0, 1)$, and that is contained in $\overline{B}(0, 1) \setminus B(0, 2^{-k})$. By the trivial estimate, the length of $\gamma'_{i,k}$ is at most $H^1(K_k \cap \overline{B}(0, 1)) \leq 2\pi$, so we can find uniformly Lipschitz parameterizations $z_{i,k} : [0, 1] \to K_k$ of the $\gamma'_{i,k}$. Thus we can take a subsequence for which each $\{z_{i,k}\}$, $i = 1, 2$, converges to a limit z_i. Then $z_i([0, 1]) \cap B(0, 1) \subset K$, $z_1(0) = z_2(0) = 0$, and both $z_i(1)$ lie on $\partial B(0, 1)$. Set $\gamma_i = z_i([0, 1]) \cap B(0, 1)$; we claim that

$$\gamma_1 \cap \gamma_2 = \{0\}. \tag{34}$$

Suppose not. Let $x \in \gamma_1 \cap \gamma_2 \setminus \{0\}$, and set $r = \frac{1}{2} \operatorname{Min}(|x|, 1 - |x|)$. Also denote by $x_{i,k}$ the point of $\gamma'_{i,k}$ that lies closest to x. Set $B_k = B(x_{1,k}, r)$. For k large enough, $B_k \subset B(0, 1)$, and it contains two disjoint subarcs of $\gamma'_{1,k}$ that go from its center to ∂B_k. We can apply Corollary 29 in this case (with $\tau = 10^{-1}$, say), because the subarcs lie in the same component of K_k. We get that $K_k \cap a(10^{-1}) B_k$ is a nice curve or spider. But this is incompatible with the existence of two disjoint simple curves $\gamma'_{1,k}$ and $\gamma'_{2,k}$ in K_k that come from 0, go through the $x_{i,k}$ (that both lie well inside in $a(10^{-1}) B_k$ for k is large enough), and go out B_k again. So (34) holds.

Because of (34), we can apply apply the special case of Corollary 29 that we already proved to the pair (u, K) (in the disk $B(0, 2)$, with the disjoint arcs $\gamma_i \setminus \{0\}$, and with a constant τ_0 that we shall choose soon). We get that $\int_{B(0,r) \setminus K} |\nabla u|^2 \leq \tau_0 r$ for $r = a(\tau_0)$. Then $\int_{B(0,r) \setminus K_k} |\nabla u_k|^2 \leq 2 \tau_0 r$ for k large, by Corollary 38.48.

If τ_0 is small enough, Proposition 60.1 says that for k large enough, $K_k \cap B(0, \eta r)$ is a nice C^1 curve or spider. In addition, the proof of (33) shows that for $\rho \leq \eta r$,

$$\omega_2(0, \rho) \leq \omega_2(0, \eta r) \leq \eta^{-1} \omega_2(0, r) \leq 2 \eta^{-1} \tau_0 \leq \tau, \tag{35}$$

where the $\omega_2(0, \cdot)$ are computed in terms of u_k, and if τ_0 is small enough. Altogether, (u_k, K_k) satisfies the conclusions of the corollary for k large, in contradiction with its definition. This completes our proof of Corollary 29. $\qquad \square$

The following is an immediate consequence of Corollary 29 and Proposition 58.10.

Corollary 36. *If $(u, K) \in RGM(\mathbb{R}^2)$ is a reduced global minimizer, and if K contains two disjoint unbounded connected sets, then K is a line or a propeller.*

Fix $r > 0$ fairly large, and apply Corollary 29 with very small values of τ and radii $R = a_1(\tau)^{-1} r$. $\qquad \square$

Another consequence of Corollary 29 is one of the main estimates in the description of a nontrivial connected component of K in Proposition 57.26. That is, if K_0 is a component of K for some $(u, K) \in RGM(\mathbb{R}^2)$, if $x \in K_0$, and if we can find two simple curves $\gamma_i \subset K$ of length $r > 0$ such that $\gamma_1 \cap \gamma_2 = \{x\}$, then $K \cap B(x, \frac{a_1 r}{2\pi})$ is a nice C^1 curve or spider. Indeed, γ_i meets $\partial B(x, \rho)$ for $\rho < \frac{r}{2\pi}$, by connectedness and because $H^1(K \cap B(x, \frac{r}{2\pi})) \leq r$ by the trivial estimate. Then we can apply Corollary 29 with $R = \frac{r}{2\pi}$.

There are other circumstances where the estimates **14–18** may be useful. Here is a simple example; one could certainly come up with other ones of the same type.

Corollary 37. *If $(u, K) \in RGM(\mathbb{R}^2)$ and if K is symmetric with respect to the origin, then K is a line or the empty set.*

This is fairly easy. The point is that $N(r)$, the number of points in $K \cap \partial B(0,r)$, is always even, and in particular is never equal to 1. In addition, if $N(r) \geq 2$, the length of the longest component of $\partial B(0,r) \setminus K$ is at most πr. Thus we can apply **14** or **16**, with $\omega = 1$. This yields $rF'(r) \geq \frac{1}{2} \int_{\partial B(0,r) \setminus K} |\nabla u|^2$ for almost all $r > 0$, and we can conclude as in the proof of Theorem 1. $\qquad\square$

c. Global minimizers in a cone

Surprisingly, the analogue of $RGM(\mathbb{R}^2)$ in a cone is easier to study, at least if the aperture of the cone is not too large. Here we only consider the cones

$$\mathcal{C}_\alpha = \{ \rho e^{i\theta} \, ; \, \rho > 0 \text{ and } 0 < \theta < \alpha \}, \text{ with aperture } \alpha < \frac{3\pi}{2}. \tag{38}$$

We denote by $RGM(\mathcal{C}_\alpha)$ the set of topologically reduced (or equivalently, coral, see Remark 54.8) topological minimizers in $\mathcal{C}_\alpha$ (that is, almost-minimizers with gauge function $h = 0$).

Theorem 39. *If $\alpha < \frac{3\pi}{2}$ and $(u, K) \in RGM(\mathcal{C}_\alpha)$, then K is either empty or a half-line with its origin in $\partial \mathcal{C}_\alpha$, and u is locally constant on $\mathcal{C}_\alpha \setminus K$. See Definitions 7.17, 8.7, and 8.24.*

There are additional constraints on the half-line. If its origin lies in $\partial \mathcal{C}_\alpha \setminus \{0\}$, then K must be perpendicular to $\partial \mathcal{C}_\alpha$. If K starts from the origin, its angle with each of the branches of $\partial \mathcal{C}_\alpha$ must be $\geq \frac{\pi}{2}$ (which forces $\alpha \geq \pi$). Finally, $\alpha \geq \pi$ is needed in all cases.

It is easy to see that these constraints are necessary and sufficient for $(0, K)$ to be a global minimizer in $\mathcal{C}_\alpha$, if K is a half-line.

The proof of Theorem 39 is based on the analogue of Proposition 65.5 for $\mathcal{C}_\alpha$ (see Remark 65.40), which leads to an analogue of **17** with

$$F(r) = \frac{1}{r} \left\{ H^1(K \cap B(0,r)) + 2 \int_{\mathcal{C}_\alpha \cap B(0,r) \setminus K} |\nabla u|^2 \right\}. \tag{40}$$

Here $\omega \leq \frac{\alpha}{\pi} < \frac{3}{2}$, so we get that

$$rF'(r) \geq \text{Min}\{1, 3 - \frac{2\alpha}{\pi}\} \int_{\mathcal{C}_\alpha \cap \partial B(0,r) \setminus K} |\nabla u|^2 \geq 0. \tag{41}$$

Thus the reason why there are more restrictions on global minimizers in $\mathcal{C}_\alpha$ than in the plane is that there is less room around, or equivalently that when we extend functions from an arc $\partial B(0,r) \cap \mathcal{C}_\alpha$ to the conic sector $B(0,r) \cap \mathcal{C}_\alpha$, we can get better estimates on the energy of the extension.

It is not as easy to conclude from here as it was for Theorem 1, because (41) only holds with balls centered at the origin. However, after a blow-in argument and an additional monotonicity argument, one can deduce Theorem 39 from (41). We refer to Section 12 of [DaLé] for the proof. $\qquad\square$

Exercise 42. Suppose $I = [0, R]$ is an interval, l is a nondecreasing function defined on I, with $l(0) = 0$, and $\mathcal{R}$ is an open set of I, with $H^1(I \setminus \mathcal{R}) = 0$. Also suppose that l is continuously differentiable on $\mathcal{R}$.

1. Show that $l(b) - l(a) \geq \int_{\mathcal{R} \cap [a,b]} l'(r)dr$ for $0 \leq a \leq b \leq R$. [Start with the intersection of [a,b] with a finite union of connected components of $\mathcal{R}$.] Set $f(r) = r^{-1}l(r)$ for $r \in (0, R]$. We want to show that

$$f(b) - f(a) \geq \int_{\mathcal{R} \cap [a,b]} f'(r)dr \quad \text{for } 0 < a < b \leq R. \tag{43}$$

2. Let a and b be given, with $0 < a < b \leq R$, and set $E = [a, b] \setminus \mathcal{R}$. Let $\varepsilon > 0$ be given. Check that we can find intervals (a_j, b_j), $1 \leq j \leq k$, such that $a \leq a_1 < b_1 < a_2 < \cdots < a_k < b_k \leq R$, $E \subset \{a_1\} \cup \{b_k\} \cup \bigcup_{1 \leq j \leq k} (a_j, b_j)$, and $\sum_{1 \leq j \leq k} (b_j - a_j) \leq \varepsilon$.

3. Show that $\sum_{1 \leq j \leq k} (f(b_j) - f(a_j)) \geq -\varepsilon a^{-2} l(R)$.

4. Set $U_\varepsilon = [a, b] \setminus \bigcup_{1 \leq j \leq k} (a_j, b_j)$. Show that $f(b) - f(a) \geq \int_{U_\varepsilon} f'(r)dr - \varepsilon a^{-2} l(R)$.

5. Prove (43). Note that this applies to $f(r) = l(r) + 2E(r)$ in the proof of (4).

67 Points of high energy, the singular singular set $K^\sharp$

Let $(u, K) \in TRLAM(\Omega, h)$ be a local almost-minimizer in a planar domain Ω [see Standard Assumption 41.1.], and with a gauge function $h(r)$ such that $\int_0^{r_0} h(t)^{1/2} \frac{dt}{t} < +\infty$ for some $r_0 > 0$. We may think in particular about the case when $(u, K) \in RGM(\mathbb{R}^2)$ is a reduced global minimizer. We shall denote by $K^\sharp$ the set of points of high energy of K. That is,

$$K^\sharp = \{x \in K \, ; \, \limsup_{r \to 0} \frac{1}{r} \int_{B(x,r) \setminus K} |\nabla u|^2 > 0\}. \tag{1}$$

The set $K^\sharp$ was introduced by Léger [Lé3], with the idea that studying it could help classify the global minimizers. We may also refer to it as the singular singular set.

Recall that there are two types of points in K: the points of $K^\sharp$ and the points of low energy. The dichotomy is rather clear, because for $1 < p \leq 2$, there is a positive constant $\tau(p)$ such that if $x \in K$ is such that $\omega_p(x, r) + \int_0^r h(t)^{1/2} \frac{dt}{t} \leq \tau(p)$ for some $r > 0$, then x is a point of low energy. See Remark 60.15 (and (23.5) for the definition of the normalized L^p-energy $\omega_p(x, r)$). Note that it is still possible to define points of high and low energy in higher dimensions, but so far we do not have a very good description of the behavior of K near points of low energy, even in dimension 3.

Notice that $K^\sharp$ is the set of points of K that are neither C^1 points nor spider points, by Proposition 60.1 and Lemma 60.9. In particular, $K^\sharp$ is closed.

It is very tempting to try to use the set $K^\sharp$ to get interesting information about (u, K). So far, there are only a few results about $K^\sharp$, that we shall describe here, and the author's impression is that it will be easier to study $K^\sharp$ *after* we know about (u, K), and not so much the other way around. We start with the stability of $K^\sharp$ under limits.

Proposition 2. *Let $\{(u_k, K_k)\}$ be a sequence in $RGM(\mathbb{R}^2)$, and suppose that it converges to some limit (u, K). Then the singular singular sets $K_k^\sharp$ converge to $K^\sharp$.*

See Definition 37.6 for the convergence, and recall that $(u, K) \in RGM$ by (54.13) (so we can talk about $K^\sharp$). The first half of the proposition (that is, (8) below) was proved in [Lé3], and extends to almost-minimizers with the same proof. The second part comes from [DaLé].

Before we prove the proposition itself, it will be convenient to prove two of the inequalities in Proposition 57.26.

Lemma 3. *Let $(u, K) \in RGM(\mathbb{R}^2)$ and $x \in K$ be such that $r_3(x) = \text{dist}(x, K^\sharp) > 0$. Then we can find an injective Lipschitz mapping $f : (-r_3(x), r_3(x)) \to K$ such that $f(0) = x$ and $|f'(t)| = 1$ almost everywhere.*

With the notation of Section 57, Lemma 3 says that $r_1(x) \geq r_3(x)$; see (57.21) and (57.23).

The proof is almost the same as for Lemma 61.52. Let us review how it goes, and refer to the proof of Lemma 61.52 for details. Let (u, K) and x be as in the lemma. First observe that all the points of $K \cap B(x, r_3(x))$ are C^1 or spider points of K, by Proposition 60.1. Recall that in Lemma 61.52 we assumed this for all points of K, but we proved the conclusion with $r_3(x) = +\infty$.

Denote by $\mathcal{Z}$ the set of pairs (I, f), where $I = (-\rho, \rho)$ is an open symmetric interval around the origin and $f : I \to K$ is injective, Lipschitz, and such that $f(0) = x$ and $|f'(t)| = 1$ almost everywhere. We know that such a pair exists, because x is a C^1 or spider point of K. We are happy if we can take $\rho = r_3(x)$, so let us assume that we cannot. By general principles, we can find a pair $(I, f) \in \mathcal{Z}$, which is maximal in the sense that f cannot be extended to a strictly larger open symmetric interval, with the properties above.

Note that $f(-\rho)$ and $f(\rho)$ can be defined, because f is Lipschitz, and they both lie in $B(x, r_3(x))$ because $\rho < r_3(x)$. Then $f(-\rho)$ and $f(\rho)$ are C^1 or spider points of K. Because of this, we can extend f to a slightly larger interval $(-\rho', \rho')$, in such a way that we still have $|f'(t)| = 1$ almost everywhere, and that f is injective near $-\rho$ and $-\rho'$. By maximality of $f_{|(-\rho,\rho)}$, the extension cannot be injective on $(-\rho', \rho')$, and this forces K to contain a loop, in contradiction with (57.3). See the proof of Lemma 61.52 for more detail on the construction of the extension and the loop.

So we can take $\rho = r_3(x)$, and the lemma follows. $\square$

Remark 4. There is a constant $a > 0$ such that if $f : (-\rho, \rho) \to K$ is injective, Lipschitz, and such that $f(0) = x$ and $|f'(t)| = 1$ almost everywhere, then $K \cap B(x, a\rho)$ is a nice C^1 curve or spider.

To see this, set $\gamma_1 = f((-\rho, 0))$ and $\gamma_2 = f((0, \rho))$. First observe that the length of γ_i is at least ρ. For each $r < \rho/(2\pi)$, $H^1(K \cap B(x, r)) \leq 2\pi r$ by the trivial estimate, hence γ_i cannot be contained in $B(x, r)$, and so it meets $\partial B(x, r)$. Of course the γ_i are essentially disjoint, so we can apply Corollary 66.29, with any $R < \rho/(2\pi)$, and we get the result. [In this case, we only need the simpler special case of Corollary 66.29 where (66.31) holds and both γ_i lie in the same component of K.]

Corollary 66.29 also says that for each small $\tau > 0$, we can find $a_\tau > 0$ such that if $f : (-\rho, \rho) \to K$ is as above, then $\int_{B(x,t) \backslash K} |\nabla u|^2 \leq \tau t$ for all $t \leq a_\tau \rho$.

We may now start the proof of Proposition 2. Let $\{(u_k, K_k)\}$ and its limit (u, K) be as in the proposition, and let us first show that

$$\lim_{k \to +\infty} \mathrm{dist}(x, K_k^\sharp) = 0 \quad \text{for } x \in K^\sharp. \tag{5}$$

Suppose not. Then we can find $x \in K^\sharp$, $\rho > 0$, and an increasing sequence $\{k_j\}$ such that $\mathrm{dist}(x, K_{k_j}^\sharp) \geq 2\rho$ for all j. Since K is the limit of the K_k, we can find points $x_k \subset K_k$ such that $\{x_k\}$ tends to x. Then

$$\mathrm{dist}(x_{k_j}, K_{k_j}^\sharp) > \rho \quad \text{for } j \text{ large enough.} \tag{6}$$

By Lemma 3, we can find an injective Lipschitz function $f_j : (-\rho, \rho) \to K_{k_j}$ such that $f_j(0) = x_{k_j}$ and $|f_j'(t)| = 1$ almost everywhere. Let $\tau > 0$ be small, to be chosen very soon, and let a_τ be as in Remark 4. Then $\int_{B(x_{k_j}, a_\tau \rho) \backslash K_{k_j}} |\nabla u_{k_j}|^2 \leq \tau a_\tau \rho$, and

$$\int_{B(x, a_\tau \rho/2) \backslash K} |\nabla u|^2 \leq \liminf_{j \to +\infty} \int_{B(x, a_\tau \rho/2) \backslash K_{k_j}} |\nabla u_{k_j}|^2 \leq \tau a_\tau \rho \tag{7}$$

by Proposition 37.18, and because $\{x_k\}$ tends to x. If τ is small enough, we can apply Proposition 60.1, and we get that $K \cap B(x, \eta a_\tau \rho/2)$ is a nice C^1 curve or spider. In particular, $x \notin K^\sharp$. This contradiction proves (5).

Now we check that

$$\text{if } x \in K \backslash K^\sharp, \text{ there is a radius}$$
$$\rho > 0 \text{ such that } K_k^\sharp \cap B(x, \rho) = \emptyset \text{ for } k \text{ large.} \tag{8}$$

This will be easier. Let $x \in K \backslash K^\sharp$ be given. By definition (1),

$$\lim_{r \to 0} \tfrac{1}{r} \int_{B(x,r) \backslash K} |\nabla u|^2 = 0.$$

Let $\varepsilon > 0$ be small, to be chosen soon, and let $r > 0$ be such that $\int_{B(x,r)\setminus K} |\nabla u|^2 \le \varepsilon r$. By Corollary 38.48, $\int_{B(x,r)\setminus K_k} |\nabla u_k|^2 \le 2\varepsilon r$ for k large enough.

Call x_k the point of K_k that lies closest to x. Since $\{K_k\}$ converges to K, x_k tends to x, and $\int_{B(x_k,r/2)\setminus K_k} |\nabla u_k|^2 \le \int_{B(x,r)\setminus K_k} |\nabla u_k|^2 \le 2\varepsilon r$ for k large enough.

If ε is small enough, Proposition 60.1 says that $K_k \cap B(x_k, \eta r/2)$ is a nice C^1 curve or spider, and in particular $B(x_k, \eta r/2)$ does not contain any point of $K_k^\sharp$. Since $B(x, \eta r/4) \subset B(x_k, \eta r/2)$ for k large, this gives (8), with $\rho = \eta r/4$. Notice that (8) is still true for almost-minimizers (with a gauge function h such that $\int_0^{r_0} h(t)^{1/2} \frac{dt}{t} < +\infty$ for some $r_0 > 0$, so that we can apply Proposition 60.1). The proof is the same.

The convergence of the $K_k^\sharp$ to $K^\sharp$ is now easy. Recall from Definition 34.4 that we just need to show that

$$\lim_{k \to 0} d_S(K_k^\sharp, K^\sharp) = 0 \tag{9}$$

for each (sufficiently large) compact set $S \subset \mathbb{R}^2$, where

$$d_S(K_k^\sharp, K^\sharp) = \sup_{x \in S \cap K_k^\sharp} \operatorname{dist}(x, K^\sharp) + \sup_{x \in S \cap K^\sharp} \operatorname{dist}(x, K_k^\sharp). \tag{10}$$

If $\sup_{x \in S \cap K_k^\sharp} \operatorname{dist}(x, K^\sharp)$ does not tend to 0, we can find $\varepsilon > 0$, a sequence $\{k_j\}$ that tends to $+\infty$, and for each j a point $x_{k_j} \in S \cap K_{k_j}^\sharp$ such that $\operatorname{dist}(x_{k_j}, K^\sharp) \ge \varepsilon$. Since we can replace $\{k_j\}$ with a subsequence, we may assume that $\{x_{k_j}\}$ converges to some $x \in S$. Then $x \in K \setminus K^\sharp$, because $\{K_k\}$ converges to K and $\operatorname{dist}(x_{k_j}, K^\sharp) \ge \varepsilon$. By (8), we can find $\rho > 0$ such that $B(x, \rho)$ does not meet $K_k^\sharp$ for k large. This is impossible because $x_{k_j} \in K_{k_j}^\sharp$ and tends to x.

If $\sup_{x \in S \cap K^\sharp} \operatorname{dist}(x, K_k^\sharp)$ does not tend to 0, we can find $\varepsilon > 0$, a sequence $\{k_j\}$ that tends to $+\infty$, and a sequence $\{x_j\}$ in $S \cap K^\sharp$ such that $\operatorname{dist}(x_j, K_{k_j}^\sharp) \ge \varepsilon$ for all j. Since we can replace $\{k_j\}$ with a subsequence, we may assume that $\{x_j\}$ converges to some $x \in K$. But then $x \in K^\sharp$ (because $K^\sharp$ is closed), and (5) says that $\operatorname{dist}(x, K_k^\sharp)$ tends to 0. This is impossible, because $\operatorname{dist}(x_j, K_{k_j}^\sharp) \ge \varepsilon$ for all j.

Altogether, (9) holds, $\{K_k^\sharp\}$ converges to $K^\sharp$, and this proves Proposition 2. $\square$

We end this section with a few simple results on $K^\sharp$. These are taken from [Lé3], except for the last one which comes from [DaLé]. As usual, we assume that $(u, K) \in RGM(\mathbb{R}^2)$. First recall that by Theorem 51.20, $H^1(K^\sharp) = 0$, and even the Hausdorff dimension of $K^\sharp$ is strictly smaller than 1.

Next let us check that if K is not a line or a propeller,

$$\text{every connected component of } K \text{ contains at least one point of } K^\sharp. \tag{11}$$

For components that are reduced to one point, this follows from Proposition 60.1, which says that points of low energy are C^1 or spider points. Then let K_0 be a component of K that is not reduced to a point, and assume that K_0 does not meet $K^\sharp$. Then Lemma 61.52 says that for each point $x \in K_0$ we can find an injective Lipschitz mapping $f : \mathbb{R} \to K_0$ such that $f(0) = x$, $|f'(t)| = 1$ almost everywhere, and $\lim_{t \to \pm\infty} |f(t)| = +\infty$. The two sets $f((-\infty, 0))$ and $f((0, +\infty))$ are disjoint unbounded connected subsets of K, and Corollary 66.36 says that K is a line or a propeller. This proves (11).

Let us return to the general case of $(u, K) \in RGM(\mathbb{R}^2)$. If $K^\sharp$ is empty, then K is empty, a line, or a propeller (by (11)). If $K^\sharp$ is reduced to one point, (11) says that K is connected, and Bonnet's Theorem 61.1 says that (u, K) is a cracktip minimizer like the one in (54.17).

Let us assume that $K^\sharp$ has at least two points, or equivalently that (u, K) is an exotic global minimizer. Then

$$K^\sharp \text{ is unbounded.} \tag{12}$$

Indeed let $(u, K) \in RGM$ be such that $K^\sharp$ is bounded, and let (v, G) be a blow-in limit of (u, K). Recall from (54.13) that $(v, G) \in RGM$, and call $G^\sharp$ its singular singular set. By Proposition 2, $G^\sharp = \{0\}$, so (11) says that (v, G) is a cracktip minimizer, and (62.4) says that (u, K) is a cracktip minimizer as well. The author does not know any simpler proof (i.e., one that would not use the results of Section 62).

An amusing fact is that $K^\sharp$ cannot be invariant under a nontrivial translation, because otherwise the singular singular set of any blow-in limit of (u, K) would contain a line, which is impossible because it has vanishing H^1 measure. See Proposition 42 in [Lé3] for a similar result with any rotation of angle $\alpha \notin A$ (mod 2π), where A a finite set of rational multiples of π.

Let us also prove that there is a constant $C > 1$ such that if $K^\sharp$ has at least two points, then

for each $x \in K^\sharp$ and every $\lambda > \text{dist}(x, K^\sharp \setminus \{x\})$,

$$K^\sharp \text{ meets the annulus } A_C(x, \lambda) = \{z \in \mathbb{R}^2 \,;\, \lambda < |z - x| < C\lambda\}. \tag{13}$$

Thus $K^\sharp$ does not have big gaps near ∞.

Let us check this by contradiction and compactness. That is, assume that for each integer $k \geq 2$, we can find an exotic minimizer (u_k, K_k), a point $x_k \in K_k^\sharp$, and a positive number λ_k such that

$$\lambda_k \geq \text{dist}(x_k, K_k^\sharp \setminus \{x_k\}) \quad \text{and} \quad A_k(x_k, \lambda_k) \text{ does not meet } K_k^\sharp, \tag{14}$$

where $A_k(x_k, \lambda_k) = \{z \in \mathbb{R}^2 \,;\, \lambda_k < |z - x_k| < k\lambda_k\}$ as above. By translation invariance, we may assume that $x_k = 0$. We may also replace λ_k with the infimum of all the λ_k that satisfy (14), because this infimum still satisfies (14). Then $K_k^\sharp$

meets $\partial B(x_k, \lambda_k)$ (either by minimality, or because $\lambda_k = \text{dist}(x_k, K_k^\sharp \setminus \{x_k\})$). Let y_k be a point of $\partial B(x_k, \lambda_k) \cap K_k^\sharp$. By dilation invariance, we may assume that $|y_k| = \lambda_k = 1$.

Now let us extract a sequence $\{k_j\}$ so that the (u_{k_j}, K_{k_j}) converge to a limit (u, K), and $\{y_{k_j}\}$ converges to a limit y. We know that $(u, K) \in RGM$, by (54.13), and that $\{K_{k_j}^\sharp\}$ converges to $K^\sharp$, by Proposition 2. In particular, 0 and y lie on $K^\sharp$, so (u, K) is an exotic minimizer. But also, $K^\sharp$ does not meet $\{z \in \mathbb{R}^2 \, ; \, |z| > 1\}$, by (14). Then $K^\sharp$ is bounded, in contradiction with (12). This completes our proof of (13).

Exercise 15 [Lé3]. Show that there is a constant $C > 1$ such that if $\Gamma \subset \mathbb{R}^2$ is a chord-arc curve, with constant C, and if $(u, K) \in RGM(\mathbb{R}^2)$ is such that $K \subset \Gamma$, then K is a line or a half-line. Here chord-arc means that Γ is an (unbounded) simple curve, and that for each choice of x, $y \in \Gamma$, the length of the arc of Γ from x to y is at most $C|x - y|$. Hint: otherwise, construct a sequence of exotic minimizers with $\{0, 1\} \subset K^\sharp$. Use Section 64 too.

G. Applications to Almost-Minimizers $(n = 2)$

In this small part we want to use some of the results of Part F to get additional information on local almost-minimizers in a domain.

We shall assume in this part that $(u, K) \in TRLAM(\Omega, h)$, where Ω is a domain in $\mathbb{R}^2$ and h is a gauge function. Recall from Standard Assumption 41.1 that this means that (u, K) is a reduced local almost-minimizer in Ω, with gauge function h (see Definitions 7.14 and 8.1), or a topologically reduced local topological almost-minimizer in Ω, with gauge function h (as in Definitions 7.17 and 8.7). As before, we can equivalently assume that (u, K) is coral instead of reduced (see Definition 8.24).

68 No small loop in K, the chord-arc and the John conditions

We start with rather straightforward adaptations of some results of Sections 55–57. None of these is really surprising, but it is probably worth stating the results and checking that nothing goes wrong with the boundary of Ω. The first one gives a lower bound on the diameter of connected components of $\Omega \setminus K$ that do not get close to the boundary.

Proposition 1. *There is a constant $\tau > 0$ such that if $(u, K) \in TRLAM(\Omega, h)$, $B(x, r)$ is a disk such that $B(x, 2r) \subset \Omega$, and Ω_0 is a connected component of $\Omega \setminus K$ that is contained in $B(x, r)$, then $h(2r) \geq \tau$.*

The proof is almost the same as for Proposition 55.1, but we have to be a little careful about uniformity. The constant 2 could be replaced with any number (strictly) larger than 1, with only minor modifications to the proof. Let (u, K), $B(x, r)$, and Ω_0 be as in the statement, and suppose in addition that $h(2r) \leq \tau$ for some very small τ. First observe that

$$H^1(\partial \Omega_0) \geq \operatorname{diam}(\Omega_0). \tag{2}$$

This is easy to check: if y, z are two points of Ω_0 and $\gamma \subset \Omega_0$ is a curve from y to z, then every line perpendicular to the segment $[y, z]$ meets γ, so it also meets $\partial \Omega_0$ twice, because both of its infinite branches lie outside of Ω_0. Then (2) follows from Remark 2.11, applied to the orthogonal projection on $[y, z]$.

Let $\varepsilon > 0$ be very small, to be chosen later. Also choose $p \in (1, 2)$, for instance $p = 3/2$. We claim that if τ is small enough, depending on ε, we can find $y \in \partial\Omega_0$ and $t \in [C_\varepsilon^{-1} \operatorname{diam}(\Omega_0), \operatorname{diam}(\Omega_0)/10]$ such that

$$\beta_K(y, t) + \omega_p(y, t) \le \varepsilon. \tag{3}$$

Here $\beta_K(y, t)$ and $\omega_p(y, t)$ are still as in (41.2) and (23.5). The proof uses the Carleson measure estimates (41.15) and (41.16) (coming from Section 23 and the local uniform rectifiability of K), and is the same as for (56.5) in Lemma 56.2, applied to a ball of radius $\operatorname{diam}(\Omega_0)/2$ that contains a big piece of $\partial\Omega_0$. Such a ball exists, by (2).

If the two main components of $\partial B(y, t) \setminus K$ lie in different components of $\overline{B}(y, t) \setminus K$, Remark 50.62 allows us to apply Theorem 50.1 with $J(y, t)^{-1} = 0$, and we get that

$$K \cap B(y, \rho) \text{ is a } (1 + \varepsilon')\text{-chord-arc curve with endpoints in } \partial B(y, \rho) \tag{4}$$

for some $\rho \in (t/2, t)$, and where ε' is as small as we want, provided that we take ε and τ accordingly small.

Before we continue with this case, let us start with the other case when the two main components of $\partial B(y, t) \setminus K$ lie in the same component of $\overline{B}(y, t) \setminus K$. Let τ_1 be small, and let η_1 be as in Lemma 51.5 (with $\tau = \tau_1$). If our constants ε and τ are small enough, we get that the normalized jump $J(x, \eta_1 t)$ is larger than τ_1^{-1}. Then we can apply Theorem 50.1, and find $\rho \in (\eta_1 t/2, \eta_1 t)$ such that (4) holds. Thus we get a radius ρ in both cases.

By (3) and Fubini, we can find $\rho_1 \in (\rho/2, \rho)$ such that

$$\omega_p^*(y, \rho_1) \le C\omega_p(y, \rho) \le C(\eta_1)\,\varepsilon. \tag{5}$$

[Compare the definitions (42.5) and (23.5). Also recall that $\omega_p^*(y, \rho_1) < +\infty$ also has implications on the radial limits of u almost-everywhere on $\partial B(y, \rho_1)$, as in Convention 42.6.]

We are ready to construct a competitor for (u, K). Recall from (3) that $\beta_K(y, \rho_1) \le 4\eta^{-1}\beta_K(y, t) \le 4\eta^{-1}\varepsilon \le 10^{-2}$, so there is a line L through y such that all points of $K \cap \overline{B}(y, \rho_1)$ lie within $10^{-2}\rho_1$ of L. Set

$$B_1 = B(y, \rho_1) \text{ and } Z = \{z \in \partial B_1 \,;\, \operatorname{dist}(z, L) \le 10^{-1}\rho_1\}. \tag{6}$$

Also call ∂_0 and ∂_1 the two components of $\partial B_1 \setminus Z$. By (4) and since $y \in \partial\Omega_0$, at least one of the ∂_i lies in Ω_0. So we may assume that $\partial_0 \subset \Omega_0$.

The most interesting case is when ∂_1 lies in some other component Ω_1. By (5) and a minor variant of Lemma 22.16, we can find a continuous function v on $B_1 \setminus Z$ that coincides with u on ∂_1, is constant on ∂_0, and for which

$$\int_{B_1} |\nabla v|^2 \le C\omega_p^*(y, \rho_1)\rho_1 \le C(\eta_1)\varepsilon\rho_1 \le 10^{-1}\rho_1 \tag{7}$$

if ε is small enough. See Exercise 20 for more detail.

Set $G = [K \setminus B_1] \cup Z$. We also want to define v outside of $G \cup B_1$. Let m denote the constant value of v on ∂_0. We set $v(z) = m$ on $\Omega_0 \setminus B_1$, and keep $v(z) = u(z)$ on $\Omega \setminus [G \cup B_1 \cup \Omega_0]$.

It is easy to see that G is closed in Ω and $v \in W^{1,2}_{\mathrm{loc}}(\Omega \setminus G)$. Recall that $\Omega_0 \subset B(x, r)$ (by definition of Ω_0), and also $B_1 \subset B(x, \frac{3r}{2})$ (because $y \in \partial\Omega_0$). So (v, G) is a competitor for (u, K) in $B(x, \frac{3r}{2})$.

If $\mathrm{diam}(\Omega_0) \geq r/10$, say, we are happy with this information and set $B = B(x, \frac{3r}{2})$. Otherwise, we set $B = B(y, 2\,\mathrm{diam}(\Omega_0))$, and note that B is compactly contained in Ω and (v, G) is a competitor for (u, K) in B. In both cases, (v, G) is also a topological competitor for (u, K) in B; the verification is the same as in Section 55, just before (55.8). Call $r(B)$ the radius of B. Then

$$H^1(K \cap \overline{B}) + \int_{B \setminus K} |\nabla u|^2 \leq H^1(G \cap \overline{B}) + \int_{B \setminus G} |\nabla v|^2 + h(r(B))\, r(B), \qquad (8)$$

by (7.7). But

$$\begin{aligned}
H^1(G \cap \overline{B}) &= H^1(K \cap \overline{B}) - H^1(K \cap B_1) + H^1(Z) \\
&\leq H^1(K \cap \overline{B}) - 2\rho_1 + H^1(Z) \leq H^1(K \cap \overline{B}) - \rho_1
\end{aligned} \qquad (9)$$

by (4) and the definition of Z and ρ_1. Also,

$$\int_{B \setminus G} |\nabla v|^2 \leq \int_{B \setminus K} |\nabla u|^2 + \int_{B_1} |\nabla v|^2 \leq \int_{B \setminus K} |\nabla u|^2 + 10^{-1}\rho_1, \qquad (10)$$

by (7). Finally, $\rho_1 \geq \rho/2 \geq \eta_1 t/4 \geq \eta_1 C_\varepsilon^{-1}\,\mathrm{diam}(\Omega_0)/4 \geq \eta_1 C_\varepsilon^{-1} r(B)/60$ by construction, and (9) and (10) contradict (8) if τ (our upper bound for $h(2r)$) is small enough.

We are left with the easier case when ∂_1 and ∂_0 are both contained in Ω_0. In this case, we take $G = [K \setminus B_1] \cup Z$ as before, replace u by any constant on $\Omega_0 \cup B_1$, and keep its values on the rest of $\Omega \setminus G$. This still gives a topological competitor for (u, K) in B, and since $\int_{B \setminus G} |\nabla v|^2 \leq \int_{B \setminus K} |\nabla u|^2$, (9) contradicts (10) as before. This completes our proof of Proposition 1 by contradiction. $\qquad \square$

Remark 11. Unlike Proposition 55.1, it is not clear to the author that Proposition 1 generalizes to higher dimensions. Our proof does not (because (2) is hard to generalize), and it seems hard to exclude the case of a long and thin component Ω_0 for which we would not be able to find a pair (y, t) where the analogue of (3) holds. There is a similar problem if we try to deduce the proposition from Proposition 55.1 by a compactness argument: we need to know that a sequence of components Ω_0 will not become thinner and thinner and tend to a set with empty interior. In dimension 2, we can proceed as in the proof of Lemma 57.4, but again this uses lower bounds on the length of $\partial\Omega_0$.

Remark 12. If (u, K) is an almost-minimizer (no longer local), then we do not need to restrict to components Ω_0 that stay away from $\partial\Omega$. That is, there is a constant

τ such that if (u, K) is an almost-minimizer in Ω with gauge function h and if Ω_0 is a component of $\Omega \setminus K$ that is contained in a ball of radius r and contains a point x such that $\mathrm{dist}(x, \partial\Omega) \geq C^{-1}r$, then $h(2r) \geq \tau$. The proof is the same; the point x is used to prove the analogue of (2) and find (y, t) such that (3) holds (using the local almost-minimality). We probably don't need this when $\partial\Omega$ is not too wild, but let us not elaborate.

Let us now turn to local John and chord-arc conditions. We start with a local version of Proposition 56.1.

Lemma 13. *There are constants $C_1 \geq 1$ and $\tau > 0$ with the following property. Suppose that $(u, K) \in TRLAM(\Omega, h)$, $x \in \Omega \setminus K$, $B(x, r) \subset \Omega$, and $h(r) \leq \tau$. Call Ω_0 the connected component of $\Omega \setminus K$ that contains x. Then $\Omega_0 \cap B(x, r/2)$ contains a disk of radius r/C_1.*

The proof is almost the same as for Proposition 56.1. We know from Proposition 1 that if τ is small enough, $\mathrm{diam}(\Omega_0) \geq r/2$. If $B(x, r/10) \subset \Omega_0$, we are happy. Otherwise $B(x, r/10)$ meets $\partial\Omega_0$.

Let us check that $H^1(\partial\Omega_0 \cap B(x, r/3)) \geq r/C$. If $B(x, r/5)$ does not meet any other component of $\Omega \setminus K$, then $K \cap B(x, r/5) \subset \partial\Omega_0$ and hence $H^1(\partial\Omega_0 \cap B(x, r/5)) \geq C^{-1}r$ by local Ahlfors-regularity (see Proposition 18.16). Otherwise, $B(x, r/5)$ meets some other component Ω_1, and Proposition 1 says that $\mathrm{diam}(\Omega_1) \geq r/3$. Then $B(x, r/2)$ contains two curves of diameters $r/10$, one starting from x and contained in Ω_0 and the other one contained in Ω_1. It is straightforward (if perhaps painful, because one has to go through a few cases) to see that $H^1(\partial\Omega_0 \cap B(x, r/3)) \geq r/100$.

Next pick $p \in (1, 2)$ and let ε be small. We can find $y \in \partial\Omega_0 \cap B(x, r/3))$ and $t \in [C_\varepsilon^{-1}r, r/10]$ such that $\beta_K(y, t) + \omega_p(y, t) \leq \varepsilon$; the proof is the same as for (3). And the same argument as before gives a radius $\rho \in (\eta_1 t/2, t)$ such that (4) holds. Since $y \in \partial\Omega_0$, at least one of the two components of $B(y, \rho) \setminus K$ is contained in Ω_0, and since this component contains a disk of radius $\rho/3$ (because $\beta_K(y, t) \leq \varepsilon$), we see that $\Omega_0 \cap B(x, r/2)$ contains a disk of radius $C_\varepsilon^{-1}\eta_1 r/6$. The lemma follows. $\square$

Our next step is a generalization of Lemma 56.10.

Lemma 14. *There are constants $C_2 \geq 1$ and $\tau > 0$, with the following property. Let $(u, K) \in TRLAM(\Omega, h)$ and $x \in \Omega \setminus K$ be given. Set $d = \mathrm{dist}(x, K)$, and suppose that $B(x, 2C_2 d) \subset \Omega$ and $h(2C_2 d) \leq \tau$. Then there is a curve γ such that*

$$\gamma \subset [\Omega \setminus K] \cap B(x, C_2 d), \quad \mathrm{dist}(\gamma, K) \geq C_2^{-1} d, \tag{15}$$

and γ goes from x to a point x_1 such that $\mathrm{dist}(x_1, K) \geq 2d$.

We prove the lemma by contradiction and compactness.

If the lemma fails with $C_2 = 2^k$ and $\tau = 2^{-k}$, we can find a domain Ω_k, a pair $(u_k, K_k) \in TRLAM(\Omega_k, h_k)$, and a point $x_k \in \Omega_k \setminus K_k$ such that $d_k =$

$\mathrm{dist}(x, K_k) \leq 2^{-k-1} \mathrm{dist}(x_k, \partial\Omega_k)$ and $h_k(2^{k+1}d_k) \leq 2^{-k}$. But there is no path γ as above.

We may normalize so that $x_k = 0$ and $d_k = 1$. Then $B(0, 2^{k+1}) \subset \Omega_k$, and $\{\Omega_k\}$ converges to $\mathbb{R}^2$. By Proposition 37.8, we may replace $\{(u_k, K_k)\}$ with a subsequence that converges to some limit (u, K). Since $d_k = \mathrm{dist}(0, K_k) = 1$ for all k, we also get that $\mathrm{dist}(0, K) = 1$.

Moreover, set $h_k^*(r) = \sup_{l \geq k} h_l(r)$ for $r > 0$. Each $h_k^*(r)$ tends to 0 when r tends to 0, because the h_k are gauge functions and $h_k(2^{k+1}) \leq 2^{-k}$. So we may apply Theorem 38.3 to the pairs (u_l, K_l), on the domain $B(0, 2^k)$ and with the gauge function h_k^*. We get that (u, K) is a coral topological almost-minimizer on $B(0, 2^k)$, with gauge function $h_k^{*,+}(r) = \lim_{\rho \to r+} h_k^*(\rho)$. Since we can do this with any k and $h_k^{*,+}(r) \leq 2^{-k}$ for $r < 2^k$, we also get that (u, K) is a coral topological almost-minimizer on any ball, with gauge function $h = 0$. In other words, $(u, K) \in RGM(\mathbb{R}^2)$ (see Definition 54.4 and Remark 54.8).

Now Proposition 56.1 says that we can find a ball of radius 3 in the component of 0 in $\mathbb{R}^2 \setminus K$. Call x_1 the center of that ball, and let γ be a path from 0 to x_1 in $\mathbb{R}^2 \setminus K$. Then the following properties hold for k large: $\mathrm{dist}(x_1, K_k) > 2$, $\mathrm{dist}(\gamma, K_k) > 2^{-k}$, and $\gamma \subset B(0, 2^k)$. This contradicts the definition of u_k, K_k, and x_k, and Lemma 14 follows. $\qquad\square$

The following is a generalization of the existence of escape paths (i.e., the John condition). It can be deduced from Lemma 14 exactly as Proposition 56.7 follows from Lemma 56.10.

Proposition 16. *There are constants $C_3 \geq 1$ and $\tau > 0$ such that whenever $(u, K) \in TRLAM(\Omega, h)$, $x \in \Omega \setminus K$, and $A \in (0, \mathrm{dist}(x, \partial\Omega))$ is also such that $h(A) \leq \tau$, then there is a 1-Lipschitz mapping $z : [0, A] \to \Omega \setminus K$ such that*

$$\mathrm{dist}(z(t), K) \geq C_3^{-1}[t + \mathrm{dist}(x, K)] \ for \ t \in [0, A]. \tag{17}$$

We end this section with a local version of the chord-arc condition (Lemma 57.4).

Proposition 18. *There are constants $C_4 \geq 1$ and $\tau > 0$ with the following property. Let $(u, K) \in TRLAM(\Omega, h)$, and let $\gamma \subset K$ be a simple curve in K. Suppose that $\gamma \subset B(x, r)$ for some disk $B(x, r)$ such that $B(x, 2r) \subset \Omega$, and that $h(2r) \leq \tau$. Call y and z the endpoints of γ. Then $H^1(\gamma) \leq C_4|y - z|$.*

We could prove Proposition 18 with Proposition 16 (see Exercise 21), but instead let us sketch an argument that follows our main proof of Lemma 57.4, and refer to that proof for details. As before, we proceed by contradiction and compactness. So we suppose that the proposition fails with $C_4 = 2^k$ and $\tau = 2^{-k}$ and find a counterexample $(u_k, K_k) \in TRLAM(\Omega_k, h_k)$, with a disk $B(x_k, r_k)$, and a curve γ_k with endpoints y_k and z_k. We normalize things so that $y_k = 0$ and $\mathrm{diam}(\gamma_k) = 1$. Then z_k tends to 0. Also note that $H^1(\gamma_k)$ stays bounded, by the trivial estimate, so we can find a Lipschitz parameterization z_k of γ_k by the unit interval I, with uniform Lipschitz bounds.

Then we extract a subsequence for which $\{z_k\}$ converges to a Lipschitz function z, and also $B(x_k, r_k)$ converges to a limit B. It may happen that B is the whole plane, or even a half-plane, but this does not matter. Since $(u_k, K_k) \in TRLAM(B(x_k, 2r_k), h_k)$ (because $B(x_k, 2r_k) \subset \Omega_k)$ and by restriction), we can extract a subsequence for which $\{(u_k, K_k)\}$ converges to a limit (u, K), and (u, K) is a reduced local minimizer on $2B$ (i.e., with gauge function $h = 0$). [If B is the plane or a half-plane, take $2B = \mathbb{R}^2$.] The verification is the same as in Lemma 14.

Now we need a (slightly painful) argument to show that $z(I)$ contains a loop. The argument is the same as for Lemma 57.4, so we skip it. Since $z(I) \subset K$ by definition, this contradicts Proposition 1 (or directly the proof of Proposition 55.1, since (u, K) is a local minimizer). Proposition 18 follows. $\square$

Remark 19. In our statement of Proposition 18, we require γ to stay far from the boundary, but the author is not completely convinced that this is needed. If we deal with local almost-minimizers, there will be difficulties in the proof. Figure 1 suggests a situation where K contains a small loop γ, but it is hard to remove

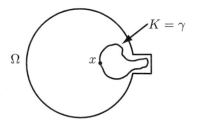

Figure 1

it because there is no ball B that contains γ and is contained in Ω (so that it is harder than usual to construct competitors). But this does not seem to give a true counterexample easily; the curve of Figure 1 can still be deformed locally, so the pair (u, K) is probably not a local almost-minimizer with small h.

Our constraint on γ seems even less justified when we deal with plain (not local) almost-minimizers, but the proofs suggested above don't seem to work.

Notice that loops are the real enemy; at least if $|y - z|$ is substantially smaller than $\operatorname{dist}(y, \partial\Omega)$, there is a short arc from y to z. This is a fairly easy consequence of Corollary 66.29 (and a compactness argument).

Exercise 20. We want to check the existence of v in (7). Let (u, K), $B_1 = B(y, \rho_1)$, and Z be as in Proposition 1, and call ∂_0 and ∂_1 the two components of $\partial B_1 \setminus Z$. Denote by a and b the extremities of ∂_1.

1. Show that $|u(b) - u(a)| \leq C\omega_p^*(y, \rho_1)^{1/2}\rho_1^{1/2}$.

2. Find an extension w of $u_{|\partial_1}$ to ∂B_1 such that $w(z) = u(a)$ on ∂_0 and

$$\int_{\partial B_1} \left|\frac{\partial w}{\partial \tau}\right|^p \leq C \int_{\partial_1} |\nabla u|^p.$$

3. Check that w has a continuous extension v to $\overline{B}_1$ for which (7) holds.

Exercise 21. Show that Exercise 57.29 can be adapted to give a proof of Proposition 18 that relies on Proposition 16.

69 Spiral points

In this section we consider the places where a local almost-minimizer looks a lot like the cracktip minimizers studied in Sections 61 and 62. We define spiral points as the points where all blow-up limits of (u, K) are cracktips, show that $x \in K$ is a spiral point as soon as at least one blow-up limit of K at x is a half-line, and then give a description of K near x as a slowly turning spiral (Theorem 29). We shall also see sufficient conditions for the existence of spiral points; see Theorems 2 and 29 in this section, and Propositions 70.2 and 70.5 in the next one.

Definition 1. *Let $(u, K) \in TRLAM(\Omega, h)$ be a local almost-minimizer in a planar domain Ω. We say that the point $x \in K$ is a* spiral point *if for each blow-up limit (v, G) of (u, K) at the point x, G is a half-line starting at the origin and there is an orthonormal basis of the plane in which (v, G) is the cracktip minimizer described in (54.17).*

Recall that blow-up limits at x are limits of sequences $\{(u_k, K_k)\}$ as in (40.2)–(40.3), with $y_k = x$ and a sequence t_k that tends to 0. See Standard Assumption 41.1 for TRLAM and Theorem 29 below for more information on (u, K) when x is a spiral point. We could have used a less restrictive definition, because of the following result.

Theorem 2. *Let $(u, K) \in TRLAM(\Omega, h)$ and $x \in K$ be such that $\{r_k^{-1}[K - x]\}$ converges to a half-line for some sequence $\{r_k\}$ of positive numbers that tends to 0. Then x is a spiral point.*

This is essentially proved in [Lé3]; we shall try to give a slightly more constructive (if less sharp) argument. The basic idea behind the theorem is that all the global minimizers in the plane that are close to a cracktip are themselves cracktips (obtained from the original one by small translations and rotations). The fact that all the blow-up limits of (u, K) at x are cracktips thus comes from the connectedness of the set of blow-up limits of (u, K) at x. But we do not want to formalize this for the moment.

It will be more convenient to prove Theorem 2 a little later in this section. We start with yet another regularity result. See Theorem 70.2 below for a generalization where lines are replaced with connected sets.

Lemma 3. *For each $\varepsilon_0 > 0$, there are positive constants τ_0, η_0, and a_0 with the following property. Let $\Omega \subset \mathbb{R}^2$, $(u, K) \in TRLAM(\Omega, h)$, $x \in K$, and $r > 0$ be such that $B(x, r) \subset \Omega$. Suppose that $h(r) \leq \tau_0$ and that we can find a line P through x such that*

$$\text{dist}(y, P) \leq \eta_0 r \text{ for } y \in K \cap B(x, r) \text{ and } \text{dist}(y, K) \leq \eta_0 r \text{ for } y \in P \cap B(x, r). \tag{4}$$

Then $\omega_2(x, a_0 r) = \dfrac{1}{a_0 r} \displaystyle\int_{B(x, a_0 r) \setminus K} |\nabla u|^2 \leq \varepsilon_0$.

We called this a regularity result, because once we get flatness (as in the hypothesis) and small energy (as in the conclusion), we can apply Theorem 50.1 or one of its corollaries to get regularity. See the end of the proof of Sublemma 14.

Let us prove the lemma by contradiction and compactness. If it fails, we can find $\varepsilon_0 > 0$ and, for each $k \geq 0$, a local almost-minimizer (u_k, K_k) and a disk $B(x_k, r_k)$ that satisfy the hypotheses of Lemma 3 with $a_0 = \tau_0 = 2^{-k}$ and $\eta_0 = 2^{-2k}$, but not the conclusion. By invariance under translations and dilations, we may assume that $x_k = 0$ and $r_k = 2^k$. Then

$$B(0, 2^k) = B(x_k, r_k) \subset \Omega_k \text{ and } h_k(2^k) \leq 2^{-k}. \tag{5}$$

By (5) and Proposition 37.8, we can replace $\{(u_k, K_k)\}$ with a subsequence that converges to a limit (u, K) in the plane. We want to show that (u, K) is a global minimizer.

Set $h_k^* = \sup_{l \geq k} h_k$. Each $h_k^*(r)$ tends to 0 when r tends to 0, by (5) and because this is the case for the h_k. By Theorem 38.3, (u, K) is a coral local topological almost-minimizer in $B(0, 2^k)$, with gauge function $h_k^{*+}(r) = \lim_{\rho \to r^+} h_k^*(\rho)$. Since this is true for all k, and the h_k^* converge to $h = 0$ uniformly on every $(0, r]$, we get that (u, K) is a reduced global minimizer in the plane (see Remark 54.8 concerning reduction).

Now (4) says that K_k is 2^{-k}-close to a line in $B(0, 2^k)$, so the limit K is a line. Hence u is locally constant, by Lemma 59.6, and $\lim_{k \to +\infty} \int_{B(0, 1) \setminus K_k} |\nabla u_k|^2 = 0$, by Corollary 38.48. This contradicts our assumption that $\omega_2(x_k, 2^{-k} r_k) = \omega_2(0, 1) \geq \varepsilon_0$ for each k, and Lemma 3 follows. $\qquad\square$

Our next goal is an analogue of Lemma 3 with approximation by half-lines, but we shall first try to describe $K \cap B(0, 1)$ when (u, K) is an almost-minimizer in $B(0, 1)$ such that $K \cap B(x, r)$ looks a lot like a line segment for all disks $B(x, r) \subset B(0, 1)$ such that $r \geq 2^{-l_0}$ for some (large) l_0. Let us be more specific. We consider a pair $(u, K) \in TRLAM(\Omega, h)$, where Ω is a planar domain that contains $B(0, 1)$ and h is a gauge function that will rapidly be assumed to be small. We assume that there is a (small) constant $\eta > 0$ and an integer $l_0 > 5$ with the following properties. There is a half-line L_0 starting from 0 such that

$$\text{dist}(y, L_0) \leq \eta \text{ for } y \in K \cap B(0,1) \quad \text{and} \quad \text{dist}(y, K) \leq \eta \text{ for } y \in L_0 \cap B(0,1), \quad (6)$$

and, for every $x \in K \cap B(0, 1/2)$ and $1 \leq l \leq l_0$, there is a half-line $L_{x,l}$ such that

$$\text{dist}(y, L_{x,l}) \leq \eta \, 2^{-l} \text{ for } y \in K \cap B(x, 2^{-l})$$
$$\text{and } \text{dist}(y, K) \leq \eta \, 2^{-l} \text{ for } y \in L_{x,l} \cap B(x, 2^{-l}). \quad (7)$$

Notice that we do not require $L_{x,l}$ to start from x; in most cases it will have its origin outside of $B(x, 2^{-l})$, and (7) will be a condition of approximation by a line.

Lemma 8. *If the constant $\eta > 0$ is small enough, we have the following. Let $(u, K) \in TRLAM(B(0,1), h)$ be an almost-minimizer in the unit disk, assume that $h(1) \leq \eta$ and that we can find half-lines L_0 and $L_{x,l}$, $x \in K \cap B(0, 1/2)$ and $1 \leq l \leq l_0$, such that (6) and (7) hold. Then there is a point $x_0 \in K \cap B(0, 2^{-5})$ and a connected compact set H such that*

$$K \cap [B(0, 1/3) \setminus B(x_0, 2^{-l_0})] \subset H \subset K \cap [B(0, 1/2) \setminus B(x_0, 2^{-l_0-1})]. \quad (9)$$

It may help the reader to think about K as a spiral, and then x_0 is a good approximation of the center of the spiral (only at scales larger than 2^{-l_0}, which is natural with our hypotheses). We expect to find x_0 in the sets K_l^*, $1 \leq l \leq l_0$, defined by

$$K_l^* = \{ x \in K \cap B(0, 1/2) \, ; \text{ the origin of } L_{x,l} \text{ lies in } B(x, 2^{-l-2}) \}. \quad (10)$$

Our definition of K_l^* makes it depend on our choices of good half-lines $L_{x,l}$, but it would be easy to check that other choices give almost the same set. Let us first verify that

$$K_l^* \subset B(0, 2^{-l-1}) \text{ for } 1 \leq l \leq 5. \quad (11)$$

Indeed, if $x \in K_l^*$ and η is small enough, K *cannot* be very close to a line in $B(x, 2^{-l-1})$, because it is already close to a half-line with its extremity in $B(x, 2^{-l-2})$. But if $l \leq 5$ and $x \in K \setminus B(0, 2^{-l-1})$, (6) says that K is η-close to L_0 in $B(x, 2^{-l-1})$, and L_0 coincides with a line on $B(x, 2^{-l-1})$. Then $x \notin K_l^*$, as announced.

It will also be useful to know that

$$K_{l+1}^* \subset K_l^* \text{ for } l \geq 1. \quad (12)$$

Call $\xi_{x,l}$ the endpoint of $L_{x,l}$. If $x \in K_{l+1}^*$, $\xi_{x,l+1}$ lies in $B(x, 2^{-l-3})$. Inside $B(x, 2^{-l-1})$, K is at the same time $\eta \, 2^{-l-1}$-close to $L_{x,l+1}$ and $\eta \, 2^{-l}$-close to $L_{x,l}$, and this forces $\xi_{x,l}$ to lie in $B(x, 2^{-l-2})$, as needed. Next we verify that

$$\text{diam}(K_l^*) \leq 2^{-l} \text{ for } 1 \leq l \leq l_0. \quad (13)$$

Let $x, z \in K_l^*$ be given. If $l \leq 5$, (11) says that $|x - z| \leq 2^{-l}$, and we are happy. Otherwise, we still have that $|x - z| \leq 2^{-5}$. Suppose that $|x - z| > 2^{-l}$, and let m be such that $2^{-m-1} \leq |x - z| < 2^{-m}$. Then $4 \leq m < l$. By (12), x and y lie inside K_{m+1}^*. Note that $B(x, 2^{-m+1})$ contains $B(z, 2^{-m})$. Thus, on $B(z, 2^{-m})$, K is very close both to $L_{z,m+1}$ and $L_{x,m-1}$. Because of this, and also since $\xi_{z,m+1}$ lies well inside $B(z, 2^{-m-1})$, we get that $|\xi_{x,m-1} - \xi_{z,m+1}| \leq 10\eta 2^{-m}$, say. The same argument yields $|\xi_{x,m-1} - \xi_{x,m+1}| \leq 10\eta 2^{-m}$, and hence $|\xi_{x,m+1} - \xi_{z,m+1}| \leq 20\eta 2^{-m}$. On the other hand, $|\xi_{x,m+1} - \xi_{z,m+1}| \geq |x-z| - |x - \xi_{x,m+1}| - |z - \xi_{z,m+1}| \geq 2^{-m-1} - 2 \cdot 2^{-m-3} \geq 2^{-m-2}$, by definition of m and since $x, z \in K_{m+1}^*$. This contradiction proves (13).

Let us check that the sets K_l^*, $1 \leq l \leq l_0$ are not empty. For $l \leq 5$, we first notice that by (6), there is a point $z_0 \in K \cap \overline{B}(0, \eta)$. This point lies in K_l^* for $l \leq 5$, because if K is very close to any half-line L in $B(z_0, 2^{-l})$, (6) says that the extremity of L is close to z_0.

Now let $l \in [5, l_0 - 1]$ be given, suppose that K_l^* is not empty, and pick $x \in K_l^*$. Thus $|\xi_{x,l} - x| \leq 2^{-l-2}$. By (7), there is a point $z \in K$ such that $|z - \xi_{x,l}| \leq \eta 2^{-l} < 2^{-l-3}$. In particular, $z \in K \cap B(0, 1/2)$, by (11), and since $l < l_0$ there is a half-line $L_{z,l+1}$ such that the analogue of (7) holds. As before, the extremity of this half-line cannot be far from $\xi_{x,l}$ (because of (7) with $L_{x,l}$), and $z \in K_{l+1}^*$.

So we can pick x_0 in $K_{l_0}^*$, and (12) says that $x_0 \in K_l^*$ for $l \leq l_0$. We still need to find a connected set H as in Lemma 8.

Sublemma 14. *For each constant $C_1 > 1$, we can find $a_1 \in (0, 10^{-2})$ such that if η is small enough, then for every $x \in K \cap [B(0, 3/4) \setminus B(x_0, 2^{-l_0})]$, we can find a radius t such that $a_1|x - x_0| \leq t \leq 10^{-2}|x - x_0|$, and*

$$K \cap B(x, t) \text{ is a chord-arc (open) curve with constant } \leq C_1, \qquad (15)$$
$$\text{and whose endpoints lie on } \partial B(x, t).$$

Fix $x \in K \cap [B(0, 3/4) \setminus B(x_0, 2^{-l_0})]$, set $r = 10^{-2}|x - x_0|$, and call $D(x_0, x)$ the line through x_0 and x. Let us check that

$$\text{dist}(y, D(x_0, x)) \leq C\eta r \text{ for } y \in K \cap B(x, r) \qquad (16)$$
$$\text{and } \text{dist}(y, K) \leq C\eta r \text{ for } y \in D(x_0, x) \cap B(x, r).$$

If $|x - x_0| \geq \frac{3}{20}$, notice that x_0 and x are both within η of L_0. Since $x_0 \in B(0, 2^{-5})$ and L_0 starts at the origin, L_0 is $C\eta$-close to $D(x_0, x)$ in $B(x, 2r)$. Moreover, $B(x, r) \subset B(0, 1)$, so (16) follows from (6).

If $|x - x_0| < \frac{3}{20}$, let l denote the integer such that $\frac{3}{10} 2^{-l} \leq |x - x_0| < \frac{3}{5} 2^{-l}$. Then $l > 1$ (because $\frac{3}{10} 2^{-l} \leq |x - x_0| < \frac{3}{20}$) and $l < l_0$ (because $2^{-l} > \frac{3}{5} 2^{-l} > |x - x_0| \geq 2^{-l_0}$). By definition of r and l, $B(x, r)$ lies well inside $B(x_0, 2^{-l}) \setminus B(x_0, 2^{-l-2})$. Since $x_0 \in K_l^*$, the extremity of $L_{x_0,l}$ lies inside $B(x_0, 2^{-l-2})$, and so $L_{x_0,l}$ coincides with a line near $B(x,r)$. Also, x_0 and x both lie within $\eta 2^{-l}$ of $L_{x_0,l}$ (by (7)), and now (16) follows from (7).

Let ε_0 be small, to be chosen soon, and let τ_0, η_0 and a_0 be the constants provided by Lemma 3. If η is small enough (compared to τ_0 and η_0; recall that η also controls $h(1)$), we can apply Lemma 3, and we get that $\omega_2(x, a_0 r) \leq \varepsilon_0$. We also have that $\beta_K(x, a_0 r) \leq C a_0^{-1} \eta \leq \varepsilon_0$, by (16) and if η is small enough.

Set $B_0 = B(x, a_0 r)$. If the two main components of $\partial B_0 \setminus K$ are contained in different components of $\overline{B}_0 \setminus K$, Remark 50.62 allows us to apply Theorem 50.1 with $J(x, a_0 r)^{-1} = 0$. If ε_0 and $h(1)$ are small enough (i.e., if η is small enough), we get a radius $t \in (a_0 r/2, a_0 r)$ such that (15) holds.

If the two main components of $\partial B_0 \setminus K$ lie in the same component of $\overline{B}_0 \setminus K$, we need to control a jump before we apply Theorem 50.1. Let us apply Lemma 51.5. For any given $\varepsilon_1 > 0$, this lemma gives us a new constant $\eta_1 > 0$ such that, in the present situation, and if ε_0 and $h(1) \leq \eta$ are small enough, then $J(x, \eta_1 a_0 r)^{-1} \leq \varepsilon_1$. The previous estimates still give that $\omega_2(x, \eta_1 a_0 r) \leq \eta_1^{-1} \omega_2(x, a_0 r) \leq \eta_1^{-1} \varepsilon_0$ (which is still as small as we want; we can let ε_0 depend on η_1), and $\beta_K(x, \eta_1 a_0 r) \leq C \eta_1^{-1} a_0^{-1} \eta \leq \eta_1^{-1} \varepsilon_0$. Now we can apply Theorem 50.1, and we find $t \in (\eta_1 a_0 r/2, \eta_1 a_0 r)$ such that (15) holds. Sublemma 14 follows. $\square$

Remark 17. If we assume that $\int_0^1 h(t)^2 \frac{dt}{t}$ is sufficiently small, then we can apply Corollary 51.17 instead of Theorem 50.1, and we get that $K \cap B(x,t)$ is a nice C^1 curve (instead of (15)). This does not make the rest of the proof of Lemma 8 significantly simpler, but it will make our later description of K nicer.

To complete our proof of Lemma 8, we still need to put together the various chord-arc curves provided by Sublemma 14.

Set $\lambda = 1 + 10^{-2} a_1$, say, and consider the radii $r_j = \lambda^j 2^{-l_0}$, where we restrict to the integers such that $j \geq 1$ and $r_j \leq \frac{4}{10}$. For each such j, call l_j the integer such that $2^{-l_j-1} \leq r_j < 2^{-l_j}$. Note that $1 \leq l_j \leq l_0$, and set $L_j = L_{x_0, l_j}$ to simplify the notation. Recall that the origin of L_j lies in $B(x_0, 2^{l_j-2})$, because $x_0 \in K_{l_j}^*$. Then L_j meets $\partial B(x_0, r_j)$. Call ξ_j the intersection. By (7), we can find $x_j \in K$ such that $|x_j - \xi_j| \leq \eta 2^{-l_j} \leq 2\eta r_j$. We apply Sublemma 14 to x_j, and get a radius that we call t_j. We claim that

$$K \cap [B(x_0, 1/3) \setminus B(x_0, 2^{-l_0})] \subset \bigcup_j B(x_j, t_j). \tag{18}$$

Set $A_j = \{z \in B(0,1) \,;\, \lambda^{-2} r_j \leq |z - x_0| \leq \lambda^2 r_j\}$. Clearly, $[B(x_0, 1/3) \setminus B(x_0, 2^{-l_0})]$ is contained in the union of the A_j, and it is enough to check that

$$K \cap A_j \subset B(x_j, t_j) \quad \text{for each } j. \tag{19}$$

So let $y \in K \cap A_j$ be given. By (7), there is a point $\xi \in L_j$ such that $|\xi - y| \leq \eta 2^{-l_j} \leq 2\eta r_j$. In particular, $\lambda^{-3} r_j \leq |\xi - x_0| \leq \lambda^3 r_j$. Since the origin of L_j lies inside $B(x_0, 2^{l_j-2}) \subset B(x_0, r_j/2)$ and both ξ and ξ_j lie on L_j, $|\xi - \xi_j| \leq 3\|\xi -$

$x_0| - |\xi_j - x_0|| = 3||\xi - x_0| - r_j| \leq 10^{-1} a_1 r_j$, say. Also recall that $|x_j - \xi_j| \leq 2\eta r_j$. Then $|x_j - y| \leq 10^{-1} a_1 r_j + 4\eta r_j$, and $y \in B(x_j, t_j)$. This proves (18) and (19).

Recall from (15) that $\gamma_j = K \cap B(x_j, t_j)$ is a chord-arc curve. By (19), γ_j contains x_{j-1} if $j > 1$. Thus $H = \bigcup_j \bar{\gamma}_j$ is connected. By construction, H can barely get out of $B(x_0, \frac{4}{10}) \setminus B(x_0, 2^{-l_0})$, so $H \subset K \cap [B(0, 1/2) \setminus B(x_0, 2^{-l_0-1})]$, (because $x_0 \in B(0, 2^{-5})$). Moreover, (18) says that it contains $K \cap [B(0, 1/3) \setminus B(x_0, 2^{-l_0})]$. In other words, (9) holds. This completes our proof of Lemma 8. $\square$

Remark 20. If the hypotheses of Lemma 8 hold for every $l_0 < +\infty$, then we can choose $x_0 \in \cap_l K_l^*$. Indeed each K_l^* is nonempty, $\overline{K}_{l+1}^* \subset K_l^*$ by the proof of (12), and $\mathrm{diam}(K_l^*)$ tends to 0 by (13), so we can use completeness. Then our proof of Lemma 8 also gives a compact connected set H such that

$$K \cap B(0, 1/3) \subset H \subset K \cap B(0, 1/2). \tag{21}$$

The next lemma says that if the almost-minimizer (u, K) looks a lot like a cracktip at the scale of the unit ball, this has to stay true at smaller scales too. We shall see later that Theorem 2 follows easily. To simplify our notation slightly, we shall say that $\alpha_K(x, r) \leq \eta$ if $x \in K$ and there is a half-line $L(x, r)$ such that

$$\begin{aligned} &\mathrm{dist}(y, L(x, r)) \leq \eta r \text{ for } y \in K \cap B(x, r) \\ &\text{and} \quad \mathrm{dist}(y, K) \leq \eta r \text{ for } y \in L(x, r) \cap B(x, r). \end{aligned} \tag{22}$$

Lemma 23. *For each small constant $\eta > 0$, we can find $\varepsilon > 0$ such that if $(u, K) \in TRLAM(B(0, 1), h)$, if $h(1) \leq \varepsilon$, and if we can find a half-line L starting from the origin such that*

$$\mathrm{dist}(y, L) \leq \varepsilon \text{ for } y \in K \cap B(0, 1) \quad \text{and} \quad \mathrm{dist}(y, K) \leq \varepsilon \text{ for } y \in L \cap B(0, 1), \tag{24}$$

then $\alpha_K(x, r) \leq \eta$ for $x \in K \cap B(0, 1/4)$ and $0 < r < 1/2$.

We shall prove this by contradiction and compactness, so let us assume that we can find $\eta > 0$ for which the lemma fails. Then the lemma also fails for $\eta' < \eta$, so we may take η as small as we want. Since $\varepsilon = 2^{-k}$ does not work, we can find a gauge function h_k such that $h_k(1) \leq 2^{-k}$, a pair $(u_k, K_k) \in TRLAM(B(0, 1), h_k)$, and a half-line L_k starting from 0, so that (24) holds but $\alpha_{K_k}(x_k, t_k) > \eta$ for some pair (x_k, r_k).

Let us choose (x_k, r_k) so that r_k is almost maximal. Then

$$\alpha_{K_k}(x, r) \leq \eta \text{ for } x \in K_k \cap B(0, 1/4) \text{ and } 2r_k \leq r < 1/2. \tag{25}$$

Notice that (22) holds for K_k with $L(x, r) = L_k$ as soon as $\varepsilon = 2^{-k} \leq \eta r$. This is not the case for the pair (x_k, r_k), so $r_k < \eta^{-1} 2^{-k}$. In particular, r_k tends to 0.

Define a new pair (v_k, G_k) by

$$G_k = r_k^{-1}[K_k - x_k] \quad \text{and} \quad v_k(y) = r_k^{-1/2}u_k(r_k y + x_k).$$

Then $(v_k, G_k) \in TRLAM(\Omega_k, g_k)$, with

$$\Omega_k = r_k^{-1}[B(0,1) - x_k] = B(-r_k^{-1}x_k, r_k^{-1}) \quad \text{and} \quad g_k(t) = h_k(r_k t).$$

Notice that $\{\Omega_k\}$ converges to the whole plane (because $|x_k| < 1/4$ and r_k tends to 0), and $\{g_k\}$ converges to 0, uniformly on every $(0, R]$ (because $g_k(r_k^{-1}) = h_k(1) \leq 2^{-k}$). By Proposition 37.8, we can replace $\{(v_k, G_k)\}$ with a subsequence that converges to a limit (v, G) in the plane, and then Theorem 38.3 says that (v, G) is a reduced global minimizer. The details of the verification are the same as in the proof of Lemma 3, just after (5).

We want to use (62.3) to show that (v, G) is one of the usual minimizers, so let us show that

$$\text{there is a connected component } H \text{ of } G \text{ such that } G \setminus H \text{ is bounded.} \qquad (26)$$

Notice that for k large, (u_k, K_k) satisfies the assumption of Lemma 8, with an integer $l_0 = l_0(k)$ such that $2^{-l_0} \leq 4r_k \leq 4\eta^{-1}2^{-k}$. So we can find a point $y_k \in K_k \cap B(0, 2^{-5})$ and a compact connected set H_k such that

$$K_k \cap [B(0, 1/3) \setminus B(y_k, 2^{-l_0})] \subset H_k \subset K_k \cap [B(0, 1/2) \setminus B(y_k, 2^{-l_0-1})]. \qquad (27)$$

Set $z_k = r_k^{-1}[y_k - x_k]$. A first possibility is that the z_k stay in a fixed disk $B(0, R)$.

Then we claim that there is a component of G that contains $G \setminus B(0, R+5)$. Indeed let ξ, ζ lie on $G \setminus B(0, R+5)$. Since G is the limit of the G_k, we can find sequences $\{\xi_k\}$ and $\{\zeta_k\}$ of points of G_k, that converge to ξ and ζ respectively. Set $\widetilde{\xi}_k = r_k \xi_k + x_k$ and $\widetilde{\zeta}_k = r_k \zeta_k + x_k$. Notice that $\widetilde{\xi}_k \in B(0, 1/3)$ for k large, because r_k tends to 0 and $x_k \in B(0, 1/4)$. Also, $|\widetilde{\xi}_k - y_k| = r_k|\xi_k - z_k| \geq 5r_k > 2^{-l_0(k)}$, so $\widetilde{\xi}_k \in H_k$ by (27). Similarly, $\widetilde{\zeta}_k \in H_k$ for k large. Since H_k is connected and $H^1(H_k) < +\infty$ by (27), Proposition 30.14 gives us a simple curve $\gamma_k \in H_k$ that goes from $\widetilde{\xi}_k$ to $\widetilde{\zeta}_k$.

Note that $\gamma_k \subset B(0, 1/2)$, so Proposition 68.16 says that for k large (so that $h(1)$ is small enough), $H^1(\gamma_k) \leq C|\widetilde{\xi}_k - \widetilde{\zeta}_k|$. Equivalently, $\Gamma_k = r_k^{-1}[\gamma_k - x_k]$ is a simple arc in G_k that goes from ξ_k to ζ_k, and $H^1(\Gamma_k) \leq C|\xi_k - \zeta_k|$. We can find uniformly Lipschitz parameterizations of the Γ_k defined on the unit interval, then take a limit, and we get a parameterization of some arc in G from ξ to ζ. Thus ξ and ζ lie in the same component of G, and (26) holds in this case.

The other case is when the z_k do not stay in a fixed disk. Then we can replace (v_k, G_k) with a subsequence for which $|z_k|$ goes to infinity. We claim that in this case G is connected. Indeed, if ξ and ζ lie on G and we define ξ_k, ζ_k, $\widetilde{\xi}_k$, and $\widetilde{\zeta}_k$ as above, then $\widetilde{\xi}_k$ and $\widetilde{\zeta}_k$ lie on H_k for k large (because $2^{l_0}|\widetilde{\xi}_k - y_k| \geq$

$(4r_k)^{-1}|\widetilde{\xi}_k - y_k| = \frac{1}{4}|\xi_k - z_k|$ tends to $+\infty$, and similarly for $\widetilde{\zeta}_k$). So we get arcs γ_k and Γ_k as above, and ξ and ζ lie in the same component of G, as needed for (26).

Since (v, G) is a global minimizer, (26) and (62.3) say that G is empty, a half-line, a line, or a propeller. But G is also the limit of the sets $G_k = r_k^{-1}[K_k - x_k]$. Thus the first case is impossible, because $x_k \in K_k$. The second and third cases also, because $\alpha_{G_k}(0, 1) = \alpha_{K_k}(x_k, r_k) > \eta$ by definition of r_k. And the last case is impossible (if η is small enough), this time because $\alpha_{G_k}(0, 2) = \alpha_{K_k}(x_k, 2r_k) \leq \eta$ (by (25)). This gives the desired contradiction and proves Lemma 23. $\qquad\square$

Remark 28. Let (u, K) be as in Lemma 23. Then we can apply Lemma 8 (to the slightly rescaled restriction of (u, K) to $B(0, 1/2)$) with any l_0, and Remark 20 says that we can find $x_0 \in \cap_l K_l^*$ and a compact connected set H such that $K \cap B(0, 1/6) \subset H \subset K \cap B(0, 1/4)$. With a little more work, we could check that H is a chord-arc curve with one extremity at x_0, but we shall find it more pleasant to prove a little more, under the additional assumption that $\int_0^1 h(t)^{1/2}\, \frac{dt}{t} < +\infty$. See Theorem 29.

Let us show that x_0 is a spiral point. Let (v, G) be any blow-up limit of (u, K) at x_0. Thus G is the limit of some sequence $\{G_j\}$, where $G_j = \rho_j^{-1}[K - x_0]$, and $\{\rho_j\}$ tends to 0. We want to show that G is a half-line that starts from the origin, and (v, G) is a cracktip minimizer. Let us first check that G is connected.

Set $H_j = \rho_j^{-1}[H - x_0]$, with H as above, and $B_j = \rho_j^{-1}[B(0, 1) - x_0]$. Note that $\{\frac{1}{6}B_j\}$ converges to the whole plane (because x_0 is very close to 0), and $G_j \cap \frac{1}{6}B_j \subset H_j \subset G_j \cap \frac{1}{4}B_j$. For each choice of $y, z \in G$, we can find points $y_j, z_j \in G_j$, so that the sequences $\{y_j\}$ and $\{z_j\}$ converge to y and z respectively. For j large, y_j and z_j lie in H_j, so there is a simple rectifiable curve γ_j in $H_j \subset G_j$ that goes from y_j to z_j. [See Proposition 30.14.] Also $\rho_j \gamma_j + x_0 \subset H \subset B(0, 1/2)$, hence we can apply Proposition 68.18 (if $h(1) \leq \varepsilon$ is small enough), and we get that $H^1(\gamma_j) \leq C|z_j - y_j| \leq C(1 + |z - y|)$ for j large. As before, we can find uniformly Lipschitz parameterizations of the γ_j defined on the unit interval, take a limit, and get an arc in G from y to z. Thus G is connected.

But (v, G) is a blow-up limit of (u, K), so it is a reduced global minimizer. By Theorem 61.1, G is a half-line, a line, or a propeller (the empty set is excluded because $0 \in G$). But $x_0 \in K_l^*$ for every $l \geq 0$, which exclude lines and propellers (compare with (7) and (10)). So G is a half-line. Moreover, if the endpoint of G was not the origin, it would be easy to construct another blow-up sequence for which G is a line, by multiplying ρ_j by some sequence λ_j that tends slowly to 0. We just saw that this is impossible, so G starts from the origin. Finally, v is as in (54.17), by the discussion in Section 61, and altogether x_0 is a spiral point.

Proof of Theorem 2. Let $(u, K) \in TRLAM(\Omega, h)$ and $x \in K$ be given, and suppose that we can find a sequence $\{r_k\}$ that tends to 0 and for which $\{r_k^{-1}[K - x]\}$ converges to a half-line. Note that we may assume that this half-line starts from the origin, because otherwise we may replace $\{r_k\}$ with a new sequence $\{\lambda_k r_k\}$; if

λ_k tends sufficiently slowly to $+\infty$, $\{(\lambda_k r_k)^{-1}[K - x]\}$ will converge to a half-line that starts from 0.

Set $\Omega_k = r_k^{-1}[\Omega - x]$, $K_k = r_k^{-1}[K - x]$, and $u_k(y) = r_k^{-1/2} u(r_k y + x)$. Then $(u_k, K_k) \in TRLAM(\Omega_k, h_k)$, with $h_k(t) = h(r_k t)$.

Fix a small η. For k large enough, (u_k, K_k) satisfies the hypotheses of Lemma 23. Fix such a k. Then Remark 28 gives us a point $x_0 \in \cap_l K_l^*$ (where the sets K_l^* are now defined as in (10), but in terms of K_k). Note that since the $r_j^{-1}[K - x] = (r_j^{-1} r_k) K_k$ converge to a half-line, we can find arbitrarily large values of l such that 0 lies in K_l^*. Then $x_0 = 0$, by (13), and $0 \in K_l^*$ for every $l \geq 0$.

If (v, G) is a blow-up limit of (u, K) at x, it is also a blow-up limit of (u_k, K_k) at 0, by definition. But Remark 28 says that $x_0 = 0$ is a spiral point for (u_k, K_k), so G is a half-line starting from 0 and v is as in (54.17). Theorem 2 follows. $\square$

Let us give now a little more information about the behavior of almost-minimizers near a spiral point. We shall keep the same hypotheses as in Lemma 23, except that we find it more convenient to assume that the gauge function $h(r)$ decays reasonably fast and get local C^1 regularity.

Theorem 29. *There exists $\varepsilon > 0$ such that if $(u, K) \in TRLAM(B(0, 1), h)$, with a gauge function h such that $\int_0^1 h(t)^{1/2} \frac{dt}{t} \leq \varepsilon$, and if there is a half-line L starting from the origin such that*

$$\mathrm{dist}(y, L) \leq \varepsilon \ \text{for } y \in K \cap B(0, 1) \quad and \quad \mathrm{dist}(y, K) \leq \varepsilon \ \text{for } y \in L \cap B(0, 1), \quad (30)$$

then there is a point $x_0 \in K \cap B(0, 2^{-5})$ and a C^1 function $f : (0, 1/2) \to \mathbb{R}$ such that

$$K \cap B(x_0, 1/2) = \{x_0\} \cup \{x_0 - (r \cos f(r), r \sin f(r)); \ 0 < r < 1/2\}, \quad (31)$$

$$|f'(r)| \leq \frac{1}{8r}, \quad and \quad \lim_{r \to 0}[r f'(r)] = 0. \quad (32)$$

In addition, there exist a sign $\pm$ and a constant C such that

$$u(x_0 + (r \cos \theta, r \sin \theta)) = \pm\sqrt{\frac{2}{\pi}} r^{1/2} \sin(\frac{\theta - f(r)}{2}) + C + o(r^{1/2}) \quad (33)$$

for $0 < r < 1/2$ and $f(r) - \pi < \theta < f(r) + \pi$, and where $o(r^{1/2})$ denotes the product of $r^{1/2}$ with a bounded function of r and θ that tends to 0 (uniformly in θ) when r tends to 0. Finally

$$\lim_{r \to 0} r^{-1} \int_{B(x_0, r) \setminus K} |\nabla u|^2 = \lim_{r \to 0} r^{-1} H^1(K \cap B(x_0, r)) = 1. \quad (34)$$

The reader should not be alarmed by the constant 8 in (32); it is just easier to get in the proof, but one could get any other large constant.

As we shall see, all these properties follow from the information that we already have, including the fact that all the blow-up limits of (u, K) at x_0 are cracktips. See the end of the section for comments.

Pick a small $\eta > 0$, and choose ε so small that Lemma 23 applies. By Remark 28, we can find $x_0 \in \cap_l K_l^* \subset K \cap B(0, 2^{-5})$, and x_0 is a spiral point. Let us first establish the description in (31)–(32). We proceed as in Sublemma 14, except that we use Corollary 51.17 instead of Theorem 50.1.

Pick any radius $\rho \in (0, 1/2)$. If $\rho < 1/3$, choose $l \geq 1$ such that $2^{-l-1} \leq \frac{11\rho}{10} < 2^{-l}$, and then set $L = L_{x_0,l}$, where $L_{x_0,l}$ is the half-line for which (7) holds. Recall that the extremity of L lies in $B(x_0, 2^{-l-2})$, because $x_0 \in K_l^*$. If $\rho \geq 1/3$, set $l = 0$, but take for L the half-line from (6). In both cases, call ξ_ρ the point of intersection of L with $\partial B(x_0, \rho)$.

By (6) or (7), we can find $x_\rho \in K$ such that $|x_\rho - \xi_\rho| \leq \eta 2^{-l}$. Then (16) says that if we set $r = 10^{-2}|x_\rho - x_0|$, K is $C\eta r$-close inside $B(x_\rho, r)$ to the line $D(x_0, x_\rho)$ through x_0 and x_ρ. If η is small enough, we can apply Lemma 3 and get that $\omega_2(x_\rho, a_0 r) \leq \varepsilon_0$, where ε_0 is as small as we want. Then (if ε is small enough), Corollary 51.17 says that $K_\rho = K \cap B(x_\rho, a_0 r/2)$ is a nice C^1 curve. In particular, it is a 10^{-1}-Lipschitz graph in some set of coordinates (see Definition 51.2). By (16), the first axis in these coordinates makes an angle less than $1/9$ with the line $D(x_0, x_\rho)$. Hence K_ρ has a representation as in (31), with a C^1 function f such that $f'(r) \leq (8r)^{-1}$.

In addition, set $\lambda = 1 + \frac{a_0}{10}$, say, and $A_\rho = B(x_0, \lambda\rho) \setminus B(x_0, \lambda^{-1}\rho)$. Then $K \cap A_\rho \subset K \cap B(x_\rho, a_0 r/2) = K_\rho$, by (6) or (7) and our definition of x_ρ. Since the A_ρ cover $K \cap B(x_0, 1/2) \setminus \{x_0\}$ nicely, we get (31) and the first half of (32).

For the second part of (32), we shall need to know that the K_ρ above are C^1, with uniform estimates. That is, we obtained their C^1 regularity through Corollary 51.17, which itself relies directly on Corollary 50.51. The proof gives a continuous choice of unit tangent $e(y)$ to K_ρ at $y \in K_\rho$ such that

$$|e(y) - e(z)| \leq C a_\rho^*(4|y - z|) \quad \text{for } y, z \in K_\rho, \tag{35}$$

at least if $|y - z| \geq C^{-1}\rho$, and with a function $a_\rho^*(t)$ that tends to 0 when t tends to 0. See (50.60) and the two lines above it, and (50.56), (50.38), and (49.15) for the definition of a_ρ^*. Now a_ρ^* depends also on ρ, through the term $\tau_0^{1/2} \left(\frac{t}{r}\right)^{b/2}$ in (50.38), so we need to be a little more careful. In fact, the computation in (50.38) leads to $a_\rho^*(t) \leq a_1(t) + C \left(\frac{t}{\rho}\right)^{b/2}$ for $t \leq \rho$, where $a_1(t)$ is a nondecreasing function of t that tends to 0 at $t = 0$ and no longer depends on ρ.

Thus (35) is good when $|y - z| \ll \rho$, and we need to complete this with an estimate on flatness. Define $\alpha_K(x_0, r)$ as in (22). Then $\lim_{r \to 0} \alpha_K(x_0, r) = 0$, because all blow-up limits of K at x_0 are half-lines starting from 0. And there is a line L_r through x_0 such that $\text{dist}(y, L_r) \leq 2\alpha_K(x_0, r)r$ for $y \in K \cap B(x_0, r)$. We apply this with $r = 2\rho$ and get that

$$\text{dist}(y, L_r) \leq 4\alpha_K(x_0, 2\rho)\rho \text{ for } y \in K_\rho. \tag{36}$$

Call π the orthogonal projection on the vector line perpendicular to L_r; then $|\pi(y) - \pi(z)| \leq 4\alpha_K(x_0, 2\rho)\rho$ for $y, z \in K_\rho$. However, if γ denotes the arc of K_ρ between y and z,

$$|\pi(y) - \pi(z)| = \left| \int_\gamma \pi(e(\zeta))dH^1(\zeta) \right| \geq \left| \int_\gamma \pi(e(y))dH^1 \right| - Ca_\rho^*(4|y - z|)H^1(\gamma)$$

$$\geq |y - z||\pi(e(y))| - C\left[a_1(4|y - z|) + \left(\frac{|y - z|}{\rho}\right)^{b/2}\right]|y - z|, \tag{37}$$

by (35) and because K_ρ is a Lipschitz curve. Hence

$$|\pi(e(y))| \leq C\left[a_1(4|y - z|) + \left(\frac{|y - z|}{\rho}\right)^{b/2}\right] + |y - z|^{-1}|\pi(y) - \pi(z)|$$

$$\leq Ca_1(\rho) + C\left(\frac{|y - z|}{\rho}\right)^{b/2} + 4\alpha_K(x_0, 2\rho)\rho|y - z|^{-1}. \tag{38}$$

We choose z such that $|y - z| = \alpha_K(x_0, 2\rho)^{1/2}\rho$, say, and get that

$$|\pi(e(y))| \leq Ca_1(\rho) + C\alpha_K(x_0, 2\rho)^{b/4} + C\alpha_K(x_0, 2\rho)^{1/2} \text{ for every } y \in K_\rho. \tag{39}$$

The main point of this is that the right-hand side tends to 0 when ρ tends to 0. Thus the angle of the tangent to K at y with the line L_r tends to 0. The same thing holds if we replace L_r with the line through x_0 and y, because (36) says that it makes an angle $\leq C\alpha_K(x_0, r)$ with L_r. Translated in terms of f, this means that $rf'(r)$ tends to 0, as required for (32).

For (33) and (34), it will be convenient to identify the plane with $\mathbb{C}$. Set

$$K_r = (re^{if(r)})^{-1}[K - x_0] \text{ and } u_r(y) = r^{-1/2}u(x_0 + re^{if(r)}y) \tag{40}$$

for r small and $x_0 + re^{if(r)}y \in B(0, 1) \setminus K$. Note that $\{K_r\}$ converges to the negative real axis, by (31) and (32). We want to show that

$$(u_r, K_r) \text{ converges to one of the cracktip minimizers in (54.17).} \tag{41}$$

Since we did not really define a topology on the set of pairs (u_r, K_r), (41) is a little ambiguous a priori. Let us define it to mean that there is a minimizer (v, G) as in (54.17) such that for each sequence $\{r_k\}$ that tends to 0, the sequence $\{(u_{r_k}, K_{r_k})\}$ converges to (v, G) (as in Definition 37.6).

Let $\{r_k\}$ be a sequence that tends to 0. By Proposition 37.8, we can find a subsequence of $\{(u_{r_k}, K_{r_k})\}$ that converges to some global minimizer (v, G). Since we know from (31)–(32) that G is the negative real axis, the results of Section 61 (or Theorem 63.2) say that v is given by (54.17), with a choice of sign and a constant that may a priori depend on our choice of subsequence.

In particular we can find a sequence $\{t_k\}$ that tends to 0, for which $\{u_{t_k}\}$ converges to a function v as in (54.17). For definiteness, let us assume that v comes with the $+$ sign in (54.17). Recall from Definition 37.6 that the convergence means that we can find constants c_k such that $\{u_{t_k} - c_k\}$ converges to v in $L^1(A)$ for each compact set $A \subset \mathbb{R}^2 \setminus G$. Normally, we should restrict to compact sets A contained in any given component of $\mathbb{R}^2 \setminus G$ and let c_k depend on that component, but here the issue does not arise because $\mathbb{R}^2 \setminus G$ is connected.

Now let us assume that (41) does not hold. That is, assume that there is a sequence $\{r_k\}$ that tends to 0, and for which $\{u_{r_k}\}$ does not converge to v. In other words, there is a compact set $A \subset \mathbb{R}^2 \setminus G$ such that $a_k = \inf_{c \in \mathbb{R}} \{\|u_{r_k} - c - v\|_{L^1(A)}\}$ does not tend to 0. We may assume that A contains a small disk, because taking A larger only makes things worse. Notice that (when A and v are fixed) a_k is a continuous function of r_k. Since we know that the analogue of a_k for our initial sequence $\{t_k\}$ tends to 0, we can replace $\{r_k\}$ with a new sequence that tends to 0 and for which $a_k = \varepsilon$ for all k, where $\varepsilon > 0$ is any small enough constant (smaller than the lim sup of a_k).

As above, we can extract a subsequence for which $\{(u_{r_k}, K_{r_k})\}$ converges to some pair (w, H), and we know that $H = G$ (because K_r tends to G) and w is as in (54.17) (because (w, G) is a global minimizer, and by Section 61). But also $\inf_{c \in \mathbb{R}} \{\|w - c - v\|_{L^1(A)}\} = \varepsilon$ (by definition of convergence), and this is incompatible with the formulae for v and w if ε is small enough (and because A contains a disk). This proves (41).

Since the proof of (33) will be a little longer, let us first check (34).

Call (v, G) the limit of (u_r, K_r) (given by (41)). Recall from Remark 61.44 that $\int_{B(0,1) \setminus G} |\nabla v|^2 = 1$. Then $\lim_{r \to 0} \int_{B(0,1) \setminus K_r} |\nabla u_r|^2 = 1$, by Proposition 37.18 and Corollary 38.48 (which we can apply with any gauge function $t \to h(r_0 t)$, $r_0 > 0$). The energy part of (34) follows, by (40). Similarly, $\lim_{r \to 0} H^1(K_r \cap B(0,1)) = H^1(G \cap B(0,1)) = 1$ by Lemma 38.6 and Corollary 38.48, which proves the rest of (34).

We are left with (33) to prove. We still want to use the convergence of (u_r, K_r) to a cracktip minimizer (v, G), as in (41). Recall from (31)–(32) that G is the negative real axis, and then v must be as in (54.17). The choice of constant C does not matter (by definition of convergence), so we may assume that $C = 0$. Let us also assume that v is given with a positive sign in (54.17) (otherwise, replace u with $-u$). We shall need to work a little because the convergence given by (41) is a little too weak. It only tells us that for each compact set $A \subset \mathbb{R}^2 \setminus G$ and each sequence $\{r_k\}$ that tends to 0, we can find constants c_k such that $\{u_{r_k} - c_k\}$ converges to v in $L^1(A)$. We have a little more, because Lemma 37.25 says that

$$\{u_{r_k} - c_k\} \text{ converges to } v \text{ uniformly on } A. \tag{42}$$

But even this is not enough yet, because we want some form of uniform convergence near G as well. This will not come from general principles only, we need to use the relative smoothness of K_{r_k}. Recall from (31) and (40) that

$$K_{r_k} \cap B(0, (2r_k)^{-1}) = \{0\} \cup \{ -re^{if_k(r)} \, ; \, 0 < r < (2r_k)^{-1} \}, \qquad (43)$$

with $f_k(r) = f(r_k r) - f(r_k)$. Note that $|f'_k(r)| \leq r^{-1}$ and $\{rf'_k(r)\}$ converges to 0 uniformly on compact sets, by (32). Also $f_k(1) = 0$ by definition.

Set $H = \overline{B}(0,1) \setminus [B(0, 10^{-1}) \cup K_{r_k}]$, and let $x \in H$ be given. Write $x = re^{i\theta}$, with $f_k(r) - \pi < \theta < f_k(r) + \pi$. Let us show that

$$\varepsilon_k = \sup \left\{ |u_{r_k}(x) - c_k - v(re^{i(\theta - f_k(r))})| \, ; \, x \in H \right\} \qquad (44)$$

tends to 0 when k tends to $+\infty$. Set $A_\tau = \{ x \in \overline{B}(0,1) \setminus B(0, 10^{-1}) \, ; \, \mathrm{dist}(x, G) \geq \tau \}$ for each small $\tau > 0$. Since A_τ is a compact subset of $\mathbb{R}^2 \setminus G$, (42) says that $\varepsilon'_k = \sup \left\{ |u_{r_k}(x) - c_k - v(x)| \, ; \, x \in A_\tau \right\}$ tends to 0. In addition, $v(x) - v(re^{i(\theta - f_k(r))})$ tends to zero uniformly on A_τ, because v is uniformly continuous in a neighborhood of A_τ and $\{f_k\}$ converges to 0 uniformly on $[10^{-1}, 1]$. Then $\varepsilon''_k = \sup \left\{ |u_{r_k}(x) - c_k - v(re^{i(\theta - f_k(r))})| \, ; \, x \in A_\tau \right\}$ also tends to 0.

Next let $x \in H \setminus A_\tau$ be given. Let γ_x be the arc of the circle $\partial B(0, r)$ that starts from x, does not meet K_{r_k}, and whose length is 2τ. Call z the other extremity of γ_x, and notice that $z \in A_\tau$ (by (32)). Let us check that

$$|u_{r_k}(z) - u_{r_k}(x)| \leq C\tau^{1/2}. \qquad (45)$$

Let ξ be any point of γ_x, and set $d(\xi) = |\xi - x|/10$. If τ is small enough, $B_\xi = B(\xi, d(\xi))$ does not meet K_{r_k}, by (43) and because $f'_k(r) \leq r^{-1}$. Also, $(u_{r_k}, K_{r_k}) \in TRLQ(B_\xi)$ for k large, with $r_0 = 1$ and uniform constants M and a. This comes from Remark 7.22 (also see Standard Assumption 18.14 for the notation), and the fact that $h(r_k)$ tends to 0. Thus Lemma 18.22 says that

$$|u_{r_k}(\xi) - u_{r_k}(\zeta)| \leq Cd(\xi)^{1/2} \ \text{ for } \zeta \in B_\xi. \qquad (46)$$

And (45) follows from (46), applied to a collection of points ξ, chosen along γ_x in a geometric progression.

Now write $z = re^{i\theta'}$, with $f_k(r) - \pi < \theta' < f_k(r) + \pi$. Recall that x is quite close to K_{r_k}, because it lies outside of A_τ. Also, z lies on the same side of K_{r_k} as x, by definition of γ_x. Thus θ and θ' are both very close to $-\pi$, or both close to π, and $|\theta' - \theta| = 2\tau/r \leq 20\tau$. Hence $|v(re^{i(\theta - f_k(r))}) - v(re^{i(\theta' - f_k(r))})| \leq C\tau^{1/2}$ (by direct computation; recall that we have a formula for v). Finally,

$$
\begin{aligned}
|u_{r_k}(x) - c_k - v(re^{i(\theta - f_k(r))})| &\leq |u_{r_k}(x) - u_{r_k}(z)| + |u_{r_k}(z) - c_k - v(re^{i(\theta - f_k(r))})| \\
&\leq C\tau^{1/2} + |u_{r_k}(z) - c_k - v(re^{i(\theta - f_k(r))})| \\
&\leq C\tau^{1/2} + |u_{r_k}(z) - c_k - v(re^{i(\theta' - f_k(r))})| \\
&\leq C\tau^{1/2} + \varepsilon''_k, \qquad (47)
\end{aligned}
$$

in particular by (45) and because $z \in A_\tau$. Since this holds for all $x \in H \setminus A_\tau$, and ε_k'' already controls what happens on A_τ, we get that $\varepsilon_k \leq C\tau^{1/2} + \varepsilon_k'' \leq (C+1)\tau^{1/2}$ for k large. Since we can prove this for every small $\tau > 0$, we get that ε_k tends to 0, as needed.

We are now ready to check (33). We shall use our estimate of ε_n with the sequence $\{r_k\}$ given by $r_k = 2^{-k}$ for k large. Recall from (44), (40), and the line below (43) that

$$
\begin{aligned}
\varepsilon_k &= \sup \left\{ |u_{r_k}(x) - c_k - v(xe^{-if_k(|x|)})| \, ; \, x \in H \right\} \\
&= \sup \left\{ |r_k^{-1/2} u(x_0 + r_k e^{if(r_k)} x) - c_k - v(xe^{-if_k(|x|)})| \, ; \, x \in H \right\} \\
&= \sup \left\{ |r_k^{-1/2} u(x_0 + r_k e^{if(r_k)} x) - c_k - v(xe^{-if(|r_k x|)} e^{if(r_k)})| \, ; \, x \in H \right\} \\
&= \sup \left\{ |r_k^{-1/2} u(x_0 + r_k e^{if(r_k)} x) - c_k - r_k^{-1/2} v(r_k x e^{-if(|r_k x|)} e^{if(r_k)})| \, ; \, x \in H \right\}.
\end{aligned}
\tag{48}
$$

Set $y = r_k e^{if(r_k)} x$ and $H_k = [r_k e^{if(r_k)}] H = \overline{B}(0, r_k) \setminus [B(0, 10^{-1} r_k) \cup (K - x_0)]$. Then

$$
\begin{aligned}
\varepsilon_k &= \sup \left\{ |r_k^{-1/2} u(x_0 + y) - c_k - r_k^{-1/2} v(ye^{-if(|y|)})| \, ; \, y \in H_k \right\} \\
&= r_k^{-1/2} \sup \left\{ |u(x_0 + y) - r_k^{1/2} c_k - v(ye^{-if(|y|)})| \, ; \, y \in H_k \right\}.
\end{aligned}
\tag{49}
$$

Set $d_k = r_k^{1/2} c_k$, and apply (49) to $y = r_k e^{if(r_k)}$. Since $v(ye^{-if(|y|)}) = v(r_k) = 0$, we get that $|u(x_0 + r_k e^{if(r_k)}) - d_k| \leq r_k^{1/2} \varepsilon_k$. We can also apply (49) to the same y, but seen as a point of H_{k-1}, and we get that $|u(x_0 + r_k e^{if(r_k)}) - d_{k-1}| \leq r_{k-1}^{1/2} \varepsilon_{k-1}$. In particular, $|d_k - d_{k-1}| \leq r_k^{1/2} \varepsilon_k + r_{k-1}^{1/2} \varepsilon_{k-1}$, and since $r_k = 2^{-k}$ for k large, the series $\sum_k |d_k - d_{k-1}|$ converges, we can define $d_\infty = \lim_{k \to +\infty} d_k$, and even $r_k^{-1/2}[d_\infty - d_k]$ tends to 0. Now (49) says that

$$
r_k^{-1/2} \sup \left\{ |u(x_0 + y) - d_\infty - v(ye^{-if(|y|)})| \, ; \, y \in H_k \right\} \leq \varepsilon_k + r_k^{-1/2} |d_\infty - d_k|, \tag{50}
$$

with a right-hand side that tends to 0. And (33) (with $C = d_\infty$) follows from (50), because every point $(r \cos \theta, r \sin \theta)$, $0 < r < 1/2$ and $f(r) - \pi < \theta < f(r) + \pi$, can be written as $y \in H_k$, for some k such that $r_k/2 \leq r \leq r_k$. This completes our proof of Theorem 29. □

Let us rapidly summarize the situation so far. Theorem 2 and Theorem 29 give reasonably simple sufficient conditions to have a spiral point, and Theorem 29 gives a fairly nice description of K and u near a spiral point. It should probably be stressed here that the description of K and u in Theorem 29 is not the most precise one; it is just a first step. At this time, we do not know the precise regularity of K and u near spiral points; for instance, we do not know whether K is necessarily of class C^1 up to the tip (i.e., with a modulus of continuity of the tangent direction that does not explode near the tip).

In the next section, we shall see other conditions that imply the presence of a spiral point; the additional information (compared with Theorem 29) will be that we know that even locally, K can only look like a limited number of objects.

70 Two last regularity results

We want to give two last sufficient conditions for an almost-minimizer (u, K) to have a nice C^1 or spider point, or a spiral point, in a ball. As we shall see, this is a straightforward consequence of the sufficient conditions that we already know about, put together with our partial knowledge of global minimizers with a standard compactness argument.

To simplify the discussion, we restrict to pairs $(u, K) \in TRLAM(B, h)$, where B is the unit disk in the plane and h is a gauge function such that

$$\int_0^1 h(t)^{1/2} \frac{dt}{t} \leq \varepsilon \tag{1}$$

for some $\varepsilon > 0$, and $TRLAM(\Omega, h)$ is still as in Standard Assumption 41.1.

Proposition 2. *There are constants $\varepsilon > 0$ and $\tau > 0$ such that if $(u, K) \in TRLAM(B, h)$, B is the unit disk, h satisfies (1), $0 \in K$, and there is a compact connected set H such that*

$$\mathrm{dist}(x, H) \leq \varepsilon \text{ for } x \in K \cap B(0, 1) \text{ and } \mathrm{dist}(x, K) \leq \varepsilon \text{ for } x \in H \cap B(0, 1), \tag{3}$$

then $K \cap B(0, \tau)$ is a nice C^1 curve, or $K \cap B(0, \tau)$ is a nice spider, or there is a point $x \in B(0, 2\tau)$ such that $K \cap B(x, 10\tau)$ is a nice local spiral.

We already defined nice C^1 curves and spiders (see Definitions 51.2 and 53.3), but not yet nice local spirals. Let us say that $K \cap B(x, \tau)$ is a nice local spiral if there is a C^1 function $f : (0, \tau) \to \mathbb{R}$ such that $|rf'(r)| \leq 1/8$ on $(0, \tau)$, $\lim_{r \to 0} rf'(r) = 0$, and

$$K \cap B(x, \tau) = \{x\} \cup \{x - (r \cos f(r), r \sin f(r)); \, 0 < r < \tau\}. \tag{4}$$

Note the similarity with the conclusion of Theorem 69.29, which says in particular that if x is a spiral point, there is a $\tau > 0$ such that $K \cap B(x, \tau)$ is a nice local spiral.

Let us give a second statement and a few comments before we prove Proposition 2.

Proposition 5. *For each integer $N \geq 1$, there are constants $\varepsilon > 0$ and $\tau > 0$ such that if $(u, K) \in TRLAM(B, h)$, B is the unit disk, h satisfies (1), $0 \in K$, and we can find N lines $L_1, \ldots, L_N$ such that*

$$\mathrm{dist}\Big(x, \bigcup_{j=1}^N L_j\Big) \leq \varepsilon \quad \text{for } x \in K \cap B(0, 1), \tag{6}$$

then $K \cap B(0, \tau)$ is a nice C^1 curve, or $K \cap B(0, \tau)$ is a nice spider, or there is a point $x \in B(0, 2\tau)$ such that $K \cap B(x, 10\tau)$ is a nice local spiral.

We decided to state Propositions 2 and 5 for almost-minimizers in the unit disk to simplify the statement, but of course the statement applies to other domains, by dilation invariance and because $TRLAM(\Omega, h) \subset TRLAM(B, h)$ when $B \subset \Omega$ (see Exercise 7.58). Also, there should be similar statements, with a slightly weaker notion of nice curves, spiders, and spirals, for gauge functions h that tend to 0 more slowly.

The proof of the propositions will be based on a small compactness argument, plus the fact that the only global minimizers (v, G) for which G is connected, or contained in a finite union of lines, are the usual ones. As the reader may easily guess, other similar facts on global minimizers would give other analogues of Propositions 2 and 5. And we already used similar arguments before.

Let us prove Proposition 2 first. Suppose we cannot find ε and τ as in the statement. Then taking $\varepsilon_k = 4^{-k}$ and $\tau_k = 2^{-k}$ does not work. Then there is a pair $(u_k, K_k) \in TRLAM(B, h_k)$ that satisfies the hypotheses of the proposition, but not the conclusion.

Let us change scales, set $G_k = 2^k K_k$, $v_k(x) = 2^{k/2} u_k(2^{-k}x)$ for $x \in B(0, 2^k) \setminus G_k$, and $\widetilde{h}_k(r) = h_k(2^{-k}r)$. Then $(v_k, G_k) \in TRLAM(B(0, 2^k), \widetilde{h}_k)$ and

$$\int_0^{2^k} \widetilde{h}_k(t)^{1/2} \, \frac{dt}{t} \leq \varepsilon_k = 4^{-k}. \tag{7}$$

In particular, $\widetilde{h}_k(2^{k-1}) \leq Log(2)^{-1} 4^{-k}$, which tends to 0.

Since (u_k, K_k) does not satisfy the conclusion of the proposition, $G_k \cap B(0, 1)$ is not a nice C^1 curve or spider, and there is no point $x \in B(0, 2)$ such that $G_k \cap B(x, 10)$ is a nice spiral.

Since $\{B(0, 2^k)\}$ converges to the plane and $\widetilde{h}_k$ converges to 0 uniformly on each $(0, R)$, we can find a subsequence (v_{k_j}, G_{k_j}) that converges to a pair (v, G), and (v, G) is a global minimizer. The argument relies on Proposition 37.8 and Theorem 38.3, but we do not give it in detail here because it is the same as for Lemma 69.3, just below (69.5). Now the main point here is to show that G is connected.

Our hypothesis gives us a compact connected set that is 4^{-k}-close to K_k in $B(0, 1)$. We dilate it by a factor 2^k and get a compact connected set H_k such that

$$\text{dist}(x, H_k) \leq 2^{-k} \text{ for } x \in G_k \cap B(0, 2^k)$$
$$\text{and } \text{dist}(x, G_k) \leq 2^{-k} \text{ for } x \in H_k \cap B(0, 2^k). \tag{8}$$

Since the G_{k_j} converge to G, (8) shows that the H_{k_j} converge to G too. That is, for every $R > 0$ there is a sequence $\{\delta_j\}$ that tends to 0 and such that

$$\text{dist}(x, H_{k_j}) \leq \delta_j \text{ for } x \in G \cap B(0, R)$$
$$\text{and } \text{dist}(x, G) \leq \delta_j \text{ for } x \in H_{k_j} \cap B(0, R). \tag{9}$$

At this point we have the usual difficulty: even though G is the limit of the H_{k_j} we cannot conclude immediately that it is connected (see Exercise 61.58). There are probably ways to deal with this difficulty directly (by showing that the sets H_k have some sort of chord-arc property at large scales), but we shall find it easier to use a trick.

Let us first check that G has no bounded connected component. First observe that G is not empty (because $0 \in G_k$ for all k), hence it is unbounded (because it is Ahlfors-regular). Let $R > 1$ be given, and pick any $x \in B(0, R-1)$. Because of (9) and since H_{kj} is connected, we can find a chain of points of G, starting with x, such two consecutive points of the chain lie at distance at most $2\delta_j$ from each other, and so that the last point of the chain lies outside of $B(0, R - \delta_j)$. Since we can do this for all j, we get that the connected component of x in G is not contained in $B(0, R-1)$. [Apply Exercise 12 to $[G \cap B(0, R-1)] \cup \partial B(0, R-1)$ or, if you want to avoid thinking about topology, see the proof of Proposition 30.1.] Since we could do this with any $R > 0$, we get that G has no bounded connected component.

Now G is connected, because otherwise it would contain at least two disjoint unbounded connected sets (take different components of G). But Corollary 66.36 says that in this case, G is a line or a propeller (and hence connected, a contradiction). So G is connected, and by Theorem 61.1, G is a half-line, a line, or a propeller.

If G is a line or a propeller, v is constant and Corollary 38.48 says that

$$\lim_{j \to +\infty} \int_{B(0,R) \setminus G_{k_j}} |\nabla v_{k_j}|^2 = 0 \quad \text{for every } R > 0. \tag{10}$$

First assume that G is a propeller whose center x_0 lies in $B(0, 2)$. Call x_j the point of G_{k_j} that lies closest to x_0. Then x_j tends to x_0. Since for j large, G_{k_j} is as close as we want to a propeller centered at x_j, (7) and (10) allow us to apply Theorem 53.4 to G_{k_j} in the ball $B(x_j, 8)$. We get that $G_{k_j} \cap B(x_j, 4)$ is a nice spider. Then $G_{k_j} \cap B(0, 1)$ is a C^1 curve or spider (depending on the position of x_j relative to $\partial B(0, 1)$), a contradiction.

If G is a line or a propeller centered outside of $B(0, 2)$, then $G \cap B(0, 2)$ coincides with a line. Recall that $0 \in K$, so 0 lies in all G_k, and G goes through the origin. For j large, we can apply Corollary 51.17 to G_{k_j} inside the ball $B(0, 2)$ (again by (7) and (10)). We get that $G_{k_j} \cap B(0, 1)$ is a nice C^1 curve, a contradiction.

We are left with the most interesting case when G is a half-line. Call x_0 its extremity. If $x_0 \in B(0, 3/2)$, call x_j the point of G_{k_j} that lies closest to x_0. Again x_j tends to x_0, and for j large enough we can apply Theorem 69.29 to G_j (after a linear change of variables that maps $B(x_j, 20)$ to the unit disk). We get that there is a point z_j near x_j such that $G_{k_j} \cap B(z_j, 10)$ is a nice local spiral (see the definition near (4)). Thus we get a contradiction in this case too.

Our last case is when the extremity x_0 lies outside of $B(0, 3/2)$. Then $G \cap B(0, 3/2)$ coincides with a line through the origin, but we cannot apply Corollary 51.17 as fast as before, because we do not have (10). So let us estimate energies. Let $r > 0$ be small, to be chosen soon, and let x be any point of $G \cap B(0, 1)$. Notice that since ∇v is bounded on $B(0, 5/4) \setminus G$, Corollary 38.48 says that

$$\limsup_{j \to +\infty} \int_{B(x, 3r) \setminus G_{k_j}} |\nabla v_{k_j}|^2 \leq \int_{B(x, 3r) \setminus G} |\nabla v|^2 \leq Cr^2. \tag{11}$$

Call x_j the point of G_{k_j} that lies closest to x. If r is small enough, (11) says that for j large, $\omega_2(x_j, 2r) \leq Cr$ is very small, and since $G_{k_j} \cap B(x_j, 2r)$ also stays very close to G, we can apply Corollary 51.17 to (G_{k_j}, v_{k_j}) in $B(x_j, 2r)$. We get that $G_{k_j} \cap B(x_j, r)$ is a nice C^1 curve. The proof even gives that it is a 10^{-2}-Lipschitz graph (instead of just 10^{-1}-Lipschitz) in some coordinates. This is true for every $x \in G \cap B(0, 1)$, and G_{k_j} is as close to G as we want, so a small gluing argument allows us to deduce that $G_{k_j} \cap B(0, 1)$ is a nice C^1 curve. See the end of the proof of Corollary 51.17 for a similar argument. This last contradiction with the definition of (u_{k_j}, K_{k_j}) completes our proof of Proposition 2. $\square$

To prove Proposition 5 we fix N and do as for Proposition 2. That is, we proceed by contradiction, find contradicting pairs (u_k, K_k) and dilate them by a factor of 2^k to get new pairs (v_k, G_k), and then extract a subsequence that converges to a global minimizer (v, G). This time, we do not have a connected set H_k as before, but we know that each of the K_k is contained in a union of at most N lines. The same thing holds for each G_k, and then also for G. By Theorem 64.1, G is a half-line, a line, or a propeller, and we can conclude as before. $\square$

The reader may be slightly unhappy about Proposition 5, because we did not prove it entirely in this book. Note that for most applications, we expect to have better information on $K \cap B(0, 1)$, such as the fact that it is very close to a line or a propeller, and then Proposition 5 becomes accordingly easy. We did not list all such results because we do not want to bore the reader to death.

Exercise 12. Let E be a compact subset of $\mathbb{R}^n$ such that for each $\varepsilon > 0$ and each choice of x, y in E, there is a chain of points of E that goes from x to y, and such that two consecutive points in the chain lie at distance at most ε. Show that E is connected.

Exercise 13. Prove that there are constants $\varepsilon > 0$ and $\tau > 0$ such that, if $(u, K) \in TRLAM(B(0, 1), h)$ (in the plane), h satisfies (1), and the Hausdorff distance between K and its reflection with respect to the origin is less than ε, then the conclusion of Propositions 2 and 5 holds.

71   From the Mumford-Shah conjecture in the plane to its local version

In this short section we show how Conjecture 54.18 (the analogue of the Mumford-Shah conjecture for global minimizers) implies the original Mumford-Shah conjecture (6.3)–(6.4). Recall that this conjecture says that if Ω is a smooth domain and (u, K) is a reduced minimizer of the Mumford-Shah functional (given by (2.13), with $a = b = c = 1$, and with a bounded function g), then

$$K \text{ is a finite union of } C^1 \text{ curves, that can only meet} \tag{1}$$
$$\text{at their ends, by sets of three and with } 120° \text{ angles.}$$

Note that the conjecture allows the existence of free ends for the curves, and (as far as I know) does not necessarily require that the curves be C^1 all the way to the tips. Incidentally, this last question is open too.

In this section we shall not require any regularity for Ω, but we shall only prove (1) locally (on compact subsets of Ω). In fact, we shall prove the following slightly more precise (local) result. [See the last lines of this section and Section 79 for the boundary regularity.]

Proposition 2. *Let Ω be a domain in the plane, and let $(u, K) \in TRLAM(\Omega, h)$ be a coral local almost-minimizer (topological or not, see Standard Assumption 41.1). Assume that the gauge function is such that $\int_0^{r_0} h(t)^{1/2} \frac{dt}{t} < +\infty$ for some $r_0 > 0$, and that for every point $x \in K$, at least one of the blow-up limits of K at x is a line, a propeller, or a half-line. Then, for each relatively compact subdomain $\Omega' \subset\subset \Omega$, K coincides on Ω' with a set that satisfies (1).*

Recall that a blow-up limit of K at x is a set that can be obtained as the limit (as in Definition 34.4) of $\{r_k^{-1}[K - x]\}$, where $\{r_k\}$ is a sequence that tends to 0.

Since Proposition 2 is easy to prove, let us do this first; we shall see later why it applies to Mumford-Shah minimizers if Conjecture 54.18 holds. So let Ω and $(u, K) \in TRLAM(\Omega, h)$ be as in the statement, and let $x \in K$ be given. Let G be a blow-up limit of K at x, and let $\{r_k\}$ denote a sequence that tends to 0 such that G is the limit of $\{r_k^{-1}[K - x]\}$. Let us first assume that G is a line or a propeller, and let us rapidly say why x is a C^1 or spider point in that case. [The argument is the same as usual.] Set $K_k = r_k^{-1}[K - x]$ and $u_k(y) = r_k^{-1/2}u(x + r_k y)$. By Proposition 37.8, we can replace $\{r_k\}$ by a subsequence for which the pairs (u_k, K_k) converge to a limit. Call v the limit of $\{u_k\}$ (we already know that K_k tends to G). By Proposition 40.9, (v, G) is a global minimizer. By Lemma 59.6, v is locally constant. By Corollary 38.48, $\int_{B(0,1) \setminus K_k} |\nabla u_k|^2$ tends to 0, so for k large we can apply Corollary 51.17 or Theorem 53.4 to (u_k, K_k) in $B(0, 1)$, and get that $B(x, r) \cap K$ is a nice C^1 or spider disk for some $r > 0$.

If G is a half-line, then for k large we can apply Theorem 69.29 and get a small ball $B(x, r)$ where K is a nice local spider. In all cases we can find a small ball near x where K behaves as in (1). The proposition follows by compactness.
□

Let us see why Proposition 2 may apply to Mumford-Shah minimizers. First recall from Proposition 7.8 that if (u, K) is a reduced Mumford-Shah minimizer in Ω (associated to some bounded function g), then $(u, K) \in TRLAM(\Omega, h)$, with $h(r) = Cr$. In addition, if x is any point of K and G is any blow-up limit of K at x (and Proposition 34.6 says that we can always find such limits), then there is a function v on $\mathbb{R}^2 \setminus G$ such that (v, G) is a reduced global minimizer. This comes from Proposition 37.8 (which says that we can extract a subsequence, so that the blow-up sequence (u_k, r_k) associated to $\{r_k\}$ as above converges) and Proposition 40.9 (which says that the limit is a reduced global minimizer). If Conjecture 54.18 is true, then G is a line, a propeller, or a half-line (the empty set is excluded, since $0 \in G$). Then we can apply Proposition 2 to get the local variant of (1).

The reader should not worry too much about what happens near the boundary of Ω. We shall see in Section 79 that if Ω is bounded and C^1, $\int_0^{r_0} h(t)^{1/2} \frac{dt}{t} < +\infty$ for some $r_0 > 0$, and $(u, K) \in TRLAM(\Omega, h)$, then there is a neighborhood of $\partial\Omega$ where K is a finite union of C^1 curves (including up to $\partial\Omega$), and that only touch $\partial\Omega$ perpendicularly.

H. Quasi- and Almost-Minimizers in Higher Dimensions

In this part we want to discuss some specific aspects of Mumford-Shah theory in dimensions larger than 2. Many of the proofs will not be done entirely, but we shall try to give an idea of what happens.

We shall start with the local Ahlfors-regularity and uniform rectifiability of K when (u, K) is a reduced quasiminimizer, as in Part C. This will also be an excuse for giving some information on uniform rectifiability. Then we shall say a few words about the C^1 regularity almost-everywhere of K when (u, K) is a reduced almost-minimizer (i.e., about the analogue of the results of Part E). We shall conclude this part with a discussion of what could possibly be analogues of the Mumford-Shah conjecture and Conjecture 54.18 in dimension 3.

72 Local Ahlfors-regularity for quasiminimizers

Recall from Theorem 18.16 that if Ω is a domain in $\mathbb{R}^n$ and $(u, K) \in TRLQ(\Omega)$, with constants (r_0, M, a), and if a is small enough (depending on n and M), then K is locally Ahlfors-regular in Ω. More precisely, there is a constant $C \geq 1$, that depends only on n and M, such that

$$C^{-1}r^{n-1} \leq H^{n-1}(K \cap B(x,r)) \leq Cr^{n-1} \tag{1}$$

for $x \in K$ and $0 < r < r_0$ such that $B(x, r) \subset \Omega$.

Here as before, $TRLQ(\Omega)$ denotes the set of reduced local quasiminimizers in Ω, topological or not (and reduced accordingly); see Standard Assumption 18.14.

As we said earlier, in dimensions $n > 2$ this result comes from [CaLe2] but was written in the context of functions of bounded variation and for almost-minimizers. The generalization to (topological) quasiminimizers can be found in [Si]; the proof is similar, but a certain number of adaptations are needed. Among these, the need for some Poincaré estimates for u in a good part of the complement of K, as in Lemmas 5 and 9. These seem different, but yet look a little like the Hölder estimates in [So2], that were probably devised for a similar purpose.

Recall that there is another proof of local Ahlfors-regularity for almost-minimizers in [So2]. Probably this proof also extends to quasiminimizers as here, but the truth is that I did not check.

We want to give a detailed sketch of proof for (1) in this section. Also, we shall say later that the proof works as well for quasiminimizers near the boundary, so the experienced reader may already start checking that nothing special will happen when balls hit the boundary of Ω (and Ω is a nice domain).

First observe that the second inequality of (1) holds (by Lemma 18.19), so we may concentrate on the lower bound. Let $(u, K) \in TRLQ(\Omega)$ be given, with quasiminimality constants (r_0, M, a), and let $B \subset \Omega$ be given; without loss of generality, we may assume that $B = B(0, 1)$. Assume that $r_0 \geq 1$, that a is extremely small (depending on n and M if needed), and that

$$H^{n-1}(K \cap B) \leq \eta, \tag{2}$$

where η is another very small constant that may depend on n and M. We want to get a contradiction, but not directly. Instead it is easier to try to produce an at least equally surprising situation in the smaller ball $B(0, 10^{-1})$. For instance, imagine that we could prove by induction that $H^{n-1}(K \cap B(0, 10^{-j})) < 10^{-j(n-1)}\eta$ for all $j \geq 0$. The same argument, starting with any $x \in K \cap B(0, 1/2)$ (and maybe with $\eta' < \eta$ in (2)), would also yield that $H^{n-1}(K \cap B(x, 10^{-j})) \leq \eta 10^{-(n-1)j}$ for $j \geq 0$. Then the upper density $d(x) = \limsup_{r \to 0} \{r^{1-n} H^{n-1}(K \cap B(x, r))\}$ would be less than $10^{(n-1)}\eta$ for every $x \in K \cap B(0, 1/2)$. By Lemma 19.1 (the same density result that we used to conclude the argument in dimension 2), $H^{n-1}(K \cap B(0, 1/2)) = 0$. Then Proposition 10.1 would say that u has an extension in $W^{1,2}_{\mathrm{loc}}(B(0, 1/2))$, and that (u, K) cannot be reduced (even topologically, since $K \cap B(0, 1/2)$) does not help separating anything). So we would get the desired contradiction.

As usual, the only way to get information is to construct new competitors. Here we want to remove $K \cap B_1$, where $B_1 = B(0, 1/2)$, and see whether we can replace u with a smooth function on B_1 without losing too much energy.

A first difficulty (compared with the two-dimensional case) is that we cannot isolate $K \cap B_1$ from the rest of K as neatly as before. That is, even when (2) holds, it could be that $\partial B(0, \rho)$ meets K for every $\rho \in (1/2, 1)$, so u is not smooth there and the restriction of u to $\partial B(0, \rho) \setminus K$ could be hard to extend.

Here we shall produce a function $w \in W^{1,1}(3B/4)$ (by a limiting argument that will be alluded to later) and do a smooth transition on a thin annulus $A \subset B \setminus B_1$. That is, we shall pick a smooth radial function φ such that $\varphi = 1$ on the inside component of $B \setminus A$ and $\varphi = 0$ on the outside component. Then we shall set

$$v(x) = \varphi(x)\, w(x) + (1 - \varphi(x))\, \overline{u}(x), \tag{3}$$

where $\overline{u}$ is a minor truncation of u (that will be discussed soon). The singular set G that goes with v does not need to meet B_1 (or the inside component of $B \setminus A$), because $v = w$ there, and $w \in W^{1,1}(4B/5)$. On the other hand, $\overline{u}$ will have a few additional jumps along a set $G_0 \subset \overline{A}$, due to the truncation, and we can take $G = [K \setminus B_1] \cup G_0$. Write

$$\nabla v = \nabla \overline{u} + \varphi \nabla(w - \overline{u}) + (w - \overline{u})\nabla \varphi. \tag{4}$$

Somewhere in the argument, we shall need to estimate the L^2 norm of $w - \overline{u}$, and in particular it is good to make sure that $\overline{u} \in L^2$.

About the truncation. If we chose $\overline{u} = u$, it would be difficult to control its L^2 norm, even if we had good estimates for the gradients. Indeed, we don't know how complicated the domain $B \setminus K$ will be, and then Poincaré estimates will be essentially impossible to prove. The point of taking a truncation $\overline{u}$ is precisely to restrict to a subregion of $A \setminus K$ where u is controlled. Thus the first ingredient of the proof in [Si] consists in proving analogues of the Poincaré estimates for $u \in W^{1,2}(B \setminus K)$, but on slightly smaller domains.

Let us be a little more specific about these Poincaré estimates. Forget for a moment that (u, K) is a quasiminimizer, and just recall that $H^{n-1}(K)$ is small and $u \in W^{1,2}(B \setminus K)$. We want a control on u on most of $B \setminus K$.

Lemma 5 [Si]. *Set $D = B \setminus \overline{B}(0, 1/4)$. There exist constants C_1 and C_2 (that depend only on n) such that if $K \subset D$ is closed, with $H^{n-1}(K) \leq C_1^{-1} \varepsilon^{n-1}$, and if $u \in W^{1,1}(D \setminus K)$, then we can find a constant λ and a measurable set $E \subset D \setminus K$ such that*

$$|D \setminus E| \leq \varepsilon \quad \text{and} \quad |u(x) - \lambda| \leq C_2 \varepsilon^{-n} \int_{D \setminus K} |\nabla u| \text{ for } x \in E. \tag{6}$$

The reader should not pay too much attention to the fact that we work with an annulus (the lemma would work on B too); this is just what we use later. Also, we shall never need the precise dependence on ε of the estimates in (6).

Let us describe the proof of Lemma 5. Let us assume that u is smooth to simplify the discussion; the general case could be obtained by density, as for the Poincaré inequality (see near (12.15)), or we could just follow the argument below, be a little more careful, and in particular use the absolute continuity of u on almost all lines. The idea is still to apply the fundamental theorem of calculus on many curves γ, to get that $|u(x) - u(y)| \leq \int_\gamma |\nabla u|$ for $x, y \in \gamma$, but we have to restrict to curves that do not meet K.

For instance, let us first assume that $n = 2$. For each $\theta \in \partial B$, call L_θ the intersection of D with the line through 0 and θ. Call X_1 the set of $\theta \in \partial B$ for which L_θ meets K. Then $H^1(X_1) \leq 4 H^1(K) \leq \varepsilon/4$, because the radial projection onto ∂B is 4-Lipschitz on D, and if $C_1 \geq 16$. We also want to throw out the set X_2 of $\theta \in \partial B \setminus X_1$ for which $\int_{L_\theta} |\nabla u| \geq 10 \varepsilon^{-1} \int_{D \setminus K} |\nabla u|$. Note that $H^1(X_2) \leq \varepsilon/4$, by Fubini and Chebyshev, and that the fundamental theorem of calculus says that for each $\theta \in \partial B \setminus [X_1 \cup X_2]$,

$$|u(x) - u(y)| \leq 10 \varepsilon^{-1} \int_{D \setminus K} |\nabla u| \text{ for } x, y \in L_\theta. \tag{7}$$

Next consider the circles $N_r = \partial B(0, r)$, call Y_1 the set of $r \in (1/4, 1)$ such that N_r meets K, and Y_2 the set of $r \in (1/4, 1) \setminus Y_1$ such that $\int_{N_r} |\nabla u| \geq$

$100\varepsilon^{-1}\int_{D\setminus K}|\nabla u|$. The same argument as before shows that $H^1(Y_1\cup Y_2)\leq(4\pi)^{-1}\varepsilon$, and that

$$|u(y)-u(z)|\leq100\varepsilon^{-1}\int_{D\setminus K}|\nabla u|\quad\text{for }y,z\in N_r \tag{8}$$

when $r\in(1/4,1)\setminus[Y_1\cup Y_2]$. Set $Z=[\bigcup_{\theta\in X_1\cup X_2}L_\theta]\cup[\bigcup_{r\in Y_1\cup Y_2}N_r]$ and $E=D\setminus Z$. Then $|Z|\leq\varepsilon$, and it is easy to see that $|u(x)-u(z)|\leq110\varepsilon^{-1}\int_{D\setminus K}|\nabla u|$ for $x,y\in E$, by (7) and (8). The lemma for $n=2$ follows.

In higher dimensions, we have to replace (8) with a similar estimate valid on most hypersurfaces $\partial B(0,r)$. Let us just sketch the argument. First we check that for most $r\in(1/4,1)$, $H^{n-2}(K\cap\partial B(0,r))$ is very small, and $\int_{\partial B(0,r)\setminus K}|\nabla u|$ is not too large. Call F the set of these good radii. For $r\in F$, we can use an induction hypothesis to find a very large subset E_r of $\partial B(0,r)\setminus K$ such that $|u(y)-u(z)|\leq C(\varepsilon)\int_{D\setminus K}|\nabla u|$ for $x,y\in E_r$.

Fix some $r_0\in F$, and call H the union of all the good radii $L_\theta,\theta\in\partial B\setminus[X_1\cup X_2]$ which meet E_{r_0}. Notice that most of D still lies in H. Moreover, for every $r\in F$, most of $\partial B(0,r)$ lies in $E_r\cap H$.

Now if $x\in E_r\cap H$ and $y\in E_{r'}\cap H$, the line L_θ that goes through x meets $\partial B(0,r_0)$ at a point $x'\in E_{r_0}$, and similarly the radius that goes through y meets $\partial B(0,r_0)$ at some $y'\in E_{r_0}$. We have good estimates for $|u(x)-u(x')|$ and $|u(y)-u(y')|$, by (7), and $|u(x')-u(y')|$ is not too large either, by definition of E_{r_0}. The lemma follows rather easily, but we leave the details. In particular, the reader should not worry about the measurability of the set E that we get, since we can always replace it with a bigger measurable set defined by (6). □

Notice that Lemma 5 extends, with almost the same proof, to domains $D'=D\setminus H$, where H is any closed half-plane that does not contain the origin. One can even extend it to any bilipschitz image of such a D', by bilipschitz invariance of the statement.

Lemma 9 [Si]. *For $K\subset B$ (closed) and $u\in W^{1,2}(B\setminus K)$, we can find a closed set $S\subset B$ and a constant λ such that*

$$H^{n-1}(\partial S)\leq CH^{n-1}(K),\qquad|S|\leq CH^{n-1}(K)^{\frac{n}{n-1}}, \tag{10}$$

and

$$\left\{\int_{B(0,3/4)\setminus S}|u-\lambda|^p\right\}^{1/p}\leq C_p\left\{\int_{B\setminus K}|\nabla u|^2\right\}^{1/2}\quad\text{for }2\leq p<\frac{2n}{n-2}. \tag{11}$$

Of course the constants C and C_p do not depend on K or u. Hopefully the reader will find this new lemma believable. First notice that the lemma is trivial unless $H^{n-1}(K)$ is very small (because we can take $S=B$), and that the second part of (10) essentially follows from the first part by the isoperimetric inequality (the proof gives both at the same time). When K is empty, we can take $S=\emptyset$ and

Lemma 9 reduces to the standard Sobolev embedding. Finally, Lemma 9 would follow at once from Lemma 5 (on the unit ball) if we replaced (10) with the weaker demand that $|S|$ be as small as we want (if $H^{n-1}(K)$ is small enough).

So the point of Lemma 9 is that we get a better control on the shape of the set where u is very different from λ. It is obtained by a combination of Lemma 5 and a maximal function (or a covering) argument. For each $x \in B(0, 3/4)$ and $j \geq 2$, set $D_j(x) = B(x, 2^{-j}) \setminus \overline{B}(x, 2^{-j-2})$. Fix a small $\varepsilon < 0$, let C_1 be as in Lemma 5, call Z_j the set of (bad) points $x \in B(0, 3/4)$ such that $H^{n-1}(K \cap D_j(x)) \geq C_1^{-1} \varepsilon^{n-1} 2^{-(n-1)j}$, and set $Z = \bigcup_{j \geq 2} Z_j$.

If $x \in B(0, 3/4) \setminus Z$ and $j \geq 2$, we can apply Lemma 5 to u on the annulus $D_j(x)$. We get a constant $\lambda_j(x)$ and a set $E_j(x)$ such that $|D_j(x) \setminus E_j(x)| \leq \varepsilon 2^{-nj}$ and

$$|u(y) - \lambda_j(x)| \leq C(\varepsilon) 2^{(n-1)j} \int_{D_j(x) \setminus K} |\nabla u| \quad \text{for } y \in E_j(x) \qquad (12)$$

(where the extra $2^{(n-1)j}$ comes from the scaling). We have a similar result for $D_{j+1}(x)$, and since $E_{j+1}(x)$ meets $E_j(x)$ if ε is small enough, the comparison yields

$$|\lambda_{j+1}(x) - \lambda_j(x)| \leq C(\varepsilon) 2^{(n-1)j} \int_{[D_j(x) \cup D_{j+1}(x)] \setminus K} |\nabla u|. \qquad (13)$$

If in addition $x \in B(0, 3/4) \setminus K$ and x is a Lebesgue point for u, the $\lambda_j(x)$ tend to $u(x)$ by (12), and iterations of (13) yield

$$\begin{aligned}
|u(x) - \lambda_2(x)| &\leq C(\varepsilon) \sum_{j \geq 2} 2^{(n-1)j} \int_{[D_j(x) \cup D_{j+1}(x)] \setminus K} |\nabla u| \\
&\leq C \int_{B(x, 2^{-2}) \setminus K} |\nabla u(y)| |x - y|^{1-n}.
\end{aligned} \qquad (14)$$

Thus $\mathbf{1}_{B(0,3/4) \setminus [K \cup Z]} [u(x) - \lambda_2(x)]$ is dominated by the convolution of ∇u with $|x|^{1-n}$, and it lies in $L^p(B(0, 3/4))$ for $p < \frac{2n}{n-2}$; the computations are the same as for the Sobolev inequality in Section 12. Notice that $|\lambda_2(x) - \lambda_2(x')| \leq C \int_{B \setminus K} |\nabla u|$ (if $H^{n-1}(K)$ is small enough), by the proof of Lemma 5. So (11) will hold for any set S that contains $K \cup Z$.

To complete the proof of Lemma 9, we just need to show that we can find a closed set S that contains $K \cup Z$ and for which (10) holds. The question reduces to showing that we can cover $K \cup Z$ with balls $B(x_i, r_i)$, with $\sum_i r_i^{n-1} \leq CH^{n-1}(K)$. Indeed, then we may take for S the closure of $\cup_i B(x_i, r_i)$; it is easy to see that any point of S that does not already lie on the closure of some $B(x_i, r_i)$ has to lie on K (because the r_i tend to 0). Also note that $\sum_i r_i^n \leq \sup_i r_i \cdot \sum_i r_i^{n-1} \leq CH^{n-1}(K)$, so we do not even need to invoke the isoperimetric inequality.

The existence of a covering for K is no more than the definition of $H^{n-1}(K)$, so we just need to cover Z. For each $x \in Z$, we can find $j = j(x)$ such that $H^{n-1}(K \cap D_j(x)) \geq C_1^{-1} \varepsilon^{n-1} 2^{-(n-1)j}$. Then we set $B_x = B(x, 2^{-j})$. This gives a covering of Z, and the first Vitali covering (Lemma 33.1) gives a collection $\{B_x\}_{x \in I}$ such that the B_x, $x \in I$, are disjoint and the $5B_x$ cover Z. Since

$$\sum_{x \in I} (5 \cdot 2^{-j})^{n-1} \leq 5^{n-1} C_1 \varepsilon^{1-n} \sum_{x \in I} H^{n-1}(K \cap D_j(x)) \tag{15}$$
$$\leq 5^{n-1} C_1 \varepsilon^{1-n} H^{n-1}(K),$$

the $5B_x$ provide the desired covering of Z, and Lemma 9 follows. $\square$

First estimates on the competitor (v, G). All this work about Sobolev estimates was merely done to replace u with a suitable truncation $\bar{u}$, to be used in (3). As far as the author knows, this part of the argument is somewhat simpler in the context of SBV functions, but even there something needs to be done.

We may now return to the competitor (v, G) promised near (3). Let A be a thin (closed) annulus contained in $B(0, 3/4) \setminus B(0, 1/2)$; this is the same transition annulus as in (3), and it will be chosen soon. Call A_- the bounded component of $B \setminus A$, and set $A_+ = A \cup A_-$. [Thus A_+ is the convex hull of A.] Set

$$\bar{u}(x) = u(x) \text{ for } x \in B \setminus [A_+ \cap S] \quad \text{and} \quad \bar{u}(x) = \lambda \text{ for } x \in A_+ \cap S, \tag{16}$$

where S and λ are as in Lemma 9. Finally define v as in (3). Then we may take

$$G = [K \setminus A_-] \cup [S \cap \partial A] \cup [\partial S \cap A] \tag{17}$$

as the singular set associated with v. That is, G is closed and $v \in W^{1,2}(B \setminus G)$.

There is a little complication with the topological condition, though. The set G in (17) does not necessarily give a topological competitor for (u, K) in B (or a smaller ball). For instance, $B \setminus K$ may have a very thin connected component that goes from near 0 to ∂B [think about a very thin cone with its tip at the origin]. Then removing $K \cap A_-$ will put this component in contact with the rest of A_-, and adding $[S \cap \partial A] \cup [\partial S \cap A]$ will not necessarily improve the situation. However, this is rather easy to fix. Call $\mathcal{O}$ the union of all the components of $B \setminus K$, except the largest one; then $|\mathcal{O}| \leq C |\partial \mathcal{O}|^{\frac{n}{n-1}} \leq C H^{n-1}(K \cap B)^{\frac{n}{n-1}}$, by the isoperimetric inequality. We may choose $\rho \in (1/2, 3/4)$ such that $H^{n-1}(\partial B(0, \rho) \cap \mathcal{O}) \leq 4|\mathcal{O}| \leq C H^{n-1}(K \cap B)^{\frac{n}{n-1}}$, and one can check that if we add $\partial B(0, \rho) \cap \mathcal{O}$ to the set G in (17), (v, G) becomes a topological competitor for (u, K) in $B(0, 3/4)$, by Lemma 20.10. Adding $\partial B(0, \rho) \cap \mathcal{O}$ does not disturb the estimates below (see (19) and (22) in particular).

Lemma 18. Let $\tau > 0$ be given (very small). Set $A = A_r = \overline{B}(0, r + \tau) \setminus B(0, r)$ for $r \in (1/2, 2/3)$, and call $\mathcal{R}$ the set of radii $r \in (1/2, 2/3)$ such that

$$H^{n-1}([S \cap \partial A_r] \cup [\partial S \cap A_r]) \leq C [H^{n-1}(K \cap B)^{\frac{1}{n-1}} + \tau] H^{n-1}(K \cap B). \tag{19}$$

Then $|\mathcal{R}| \geq 1/10$, if the absolute constant C is large enough.

Indeed we easily deduce from (10) that

$$\int_{1/2}^{2/3} H^{n-1}(S \cap \partial A_r) \, dr \le 2|S| \le C H^{n-1}(K \cap B)^{\frac{n}{n-1}}, \tag{20}$$

while (a trivial discrete variant of) Fubini easily yields

$$\int_{1/2}^{2/3} H^{n-1}(\partial S \cap A_r) \, dr \le 2\tau H^{n-1}\partial S \le C\tau H^{n-1}(K \cap B); \tag{21}$$

Lemma 18 follows from Chebyshev. □

If we choose any $r \in \mathcal{R}$, then (19) holds and so

$$H^{n-1}(G \setminus K) \le C \left[H^{n-1}(K \cap B)^{\frac{1}{n-1}} + \tau \right] H^{n-1}(K \cap B). \tag{22}$$

It will also be useful to estimate energies. Recall from (3) that $v(x) = \varphi(x) w(x) + (1 - \varphi(x)) \overline{u}(x)$, where φ is a smooth function such that $\varphi(x) = 1$ near A_- and $\varphi(x) = 0$ out of A_+, and $w \in W^{1,1}(3B/4)$ will be chosen later. Near A, we shall use the brutal estimate

$$|\nabla v(x)| \le \varphi(x)|\nabla w(x)| + (1 - \varphi(x))|\nabla \overline{u}(x)| + |\nabla \varphi(x)||w(x) - \overline{u}(x)|, \tag{23}$$

which leads to

$$|\nabla v(x)|^2 \le 2|\nabla w(x)|^2 + 2|\nabla \overline{u}(x)|^2 + C\tau^{-2}|w(x) - \overline{u}(x)|^2. \tag{24}$$

For technical reasons, it will be easier to cut some integrals along a slightly different annulus $A' = \overline{B}(0, r + \tau) \setminus B(0, r')$, where r is still as in the definition of $A = A_r$ and $r - \tau \le r' \le r$. Set $\delta_1 = \int_{A'} |\nabla v|^2$. Then

$$\begin{aligned} \delta_1 &\le \int_{A'} \left\{ 2|\nabla w|^2 + 2|\nabla \overline{u}|^2 + C\tau^{-2}|w - \overline{u}|^2 \right\} \\ &\le \int_{A'} \left\{ 2|\nabla w|^2 + 2|\nabla u|^2 + C\tau^{-2}|w - \overline{u}|^2 \right\} \end{aligned} \tag{25}$$

by (24) and (16). We shall also need the function $\widetilde{u}$ such that $\widetilde{u}(x) = u(x)$ for $x \in B \setminus S$ and $\widetilde{u}(x) = \lambda$ for $x \in S$. Note that $\widetilde{u}(x) = \overline{u}$ on $A' \subset A_+$, so we can replace $\overline{u}$ with $\widetilde{u}$ in (25). The same discrete Fubini argument as above (i.e., cutting $B(0, 1/2)$ into annuli of width τ and counting how many may have large contributions) shows that, except if we choose r in some exceptional set of measure $\le 10^{-2}$, we have that

$$\delta_1 \le C\tau \int_{B(0,3/4) \setminus K} |\nabla u|^2 + C\tau \int_{B(0,3/4)} |\nabla w|^2 + C\tau^{-1} \int_{B(0,3/4)} |w - \widetilde{u}|^2 \tag{26}$$

no matter how we choose $r' \in [r - \tau, r]$.

Recall that $v = \bar{u} = u$ out of A_+ (by (16)), and $v = w$ on A_-. Hence

$$\int_{B\setminus G} |\nabla v|^2 - \int_{B\setminus K} |\nabla u|^2 \leq \int_{B(0,r')} |\nabla w|^2 - \int_{B(0,r')\setminus K} |\nabla u|^2 + \delta_1. \qquad (27)$$

Construction of $w \in W^{1,2}(B(0,3/4))$. Before we get to use (22) and (27) to prove estimates on (u, K), we need to find an acceptable function w. Of course it will be important to use the fact that K is very small, and this will be easier to do with a compactness argument. Thus, instead of considering only one pair (u, K), we now assume that we have a whole sequence of pairs $(u_k, K_k) \in TRLQ(B)$. We suppose that

$$\lim_{k \to +\infty} H^{n-1}(K_k \cap B) = 0 \qquad (28)$$

and that $(u_k, K_k) \in TRLQ(B)$ with the same constants $r_0 \geq 1$ and $M \geq 1$, and a fixed small a. The basic idea is to make some subsequence of $\{u_k\}$ converge to some limit, and take w as the limit. But since we expect the energies $\int_{B\setminus K_k} |\nabla u_k|^2$ to tend to 0, we shall get a little more information if we first normalize the u_k. So we set

$$E_k = \int_{B\setminus K_k} |\nabla u_k|^2 \quad \text{and} \quad u_k^* = E_k^{-1/2} u_k. \qquad (29)$$

The reader may be worried about the case when $E_k = 0$ for infinitely many k. If this happens, we do not get a contradiction directly, but a simple modification of the argument below works. More precisely, we can show that $H^{n-1}(K \cap B(0,1/2)) \ll H^{n-1}(K \cap B)$ as in the case when $E_k \ll H^{n-1}(K \cap B)$ below, and then iterate. So let us assume that $E_k \neq 0$ for k large.

Apply Lemma 9 to each u_k^*; this gives small sets S_k and constants λ_k. Without loss of generality, we can subtract a constant from each u_k, so as to get $\lambda_k = 0$ for all k and simplify the notation.

Lemma 30 [Si]. *We can find a subsequence (still denoted by $\{(u_k, K_k)\}$ for simplicity) and a function u_∞^* such that*

$$\text{the functions } \mathbf{1}_{B\setminus S_k} u_k^* \text{ converge to } u_\infty^* \text{ in } L^2(B(0,3/4)), \qquad (31)$$

$$u_\infty^* \in W^{1,2}(B(0,3/4)), \quad \text{and} \qquad (32)$$

$$\nabla u_\infty^* \text{ is the weak limit in } L^2(B(0,3/4)) \text{ of the } \nabla(\mathbf{1}_{B\setminus S_k} u_k^*) = \mathbf{1}_{B\setminus S_k} \nabla u_k^*. \qquad (33)$$

Let us start with the convergence in L^2. Since we may extract subsequences whenever we wish, we may assume that $|S_k| \leq 2^{-k}$, say. We want to use the Cauchy criterion, so we wish to estimate

$$I_{k,l} = \int_{B(0,3/4)} |\mathbf{1}_{B\setminus S_k} u_k^* - \mathbf{1}_{B\setminus S_l} u_l^*|^2 \qquad (34)$$

for some subsequence that we want to extract now. For each $k_0 \geq 0$, set $\widehat{S}(k_0) = \bigcup_{k \geq k_0} S_k$ and $E(k_0) = B(0, 3/4) \backslash \widehat{S}(k_0)$. Then $|\widehat{S}(k_0)| \leq 2^{-k_0+1}$. Select a countable dense set $\mathcal{D}(k_0)$ in each $E(k_0)$. For each $x \in \mathcal{D}(k_0)$ and each $k \geq k_0$, we know that $x \notin S_k$ (because $S_k \subset \widehat{S}(k_0)$), so the construction of Lemma 9 gives annuli $D_{j,k}(x)$, $j \geq 2$, and constants $\lambda_{j,k}(x)$, so that (12) holds (with u replaced by u_k^*).

For each x the $\lambda_{j,k}(x)$ stay uniformly bounded, by the proof of Lemma 9, so we can use the diagonal process to find a subsequence $\{(u_{k_m}, K_{k_m})\}$ such that for every $k_0 \geq 0$, all the $\{\lambda_{j,k_m}(x)\}$, $x \in \mathcal{D}(k_0)$, $k_m \geq k$, and $j \geq 2$ converge. Let us prove that $\lim_{m,l \to +\infty} I_{k_m,k_l} = 0$ for this subsequence.

Fix an integer $k_0 \geq 0$, and let $m, l \geq k_0$ be given. Cut the domain of integration for I_{k_m,k_l} into $B(0, 3/4) \cap \widehat{S}(k_0)$ and $E(k_0) = B(0, 3/4) \backslash \widehat{S}(k_0)$. This yields $I_{k_m,k_l} = I(m,l) + J(m,l)$, with

$$I(m,l) = \int_{B(0,3/4) \cap \widehat{S}(k_0)} |\mathbf{1}_{B \backslash S_{k_m}} u_{k_m}^* - \mathbf{1}_{B \backslash S_{k_l}} u_{k_l}^*|^2, \tag{35}$$

$$J(m,l) = \int_{E(k_0)} |\mathbf{1}_{B \backslash S_{k_m}} u_{k_m}^* - \mathbf{1}_{B \backslash S_{k_l}} u_{k_l}^*|^2. \tag{36}$$

Fix $p \in (2, \frac{2n}{n-2})$, and apply Lemma 9 to $u_{k_m}^*$. Recall that we can take $\lambda = 0$ (because we removed a constant from u_{k_m}), and that $\int_{B \backslash K_k} |\nabla u_{k_m}^*|^2 \leq 1$, by (29). This yields

$$\int_{B(0,3/4)} |\mathbf{1}_{B \backslash S_{k_m}} u_{k_m}^*|^p \leq C_p. \tag{37}$$

There is a similar estimate for $u_{k_l}^*$, and then

$$\begin{aligned} I(m,l) &\leq |\widehat{S}(k_0)|^{\frac{p-2}{p}} \left\{ \int_{B(0,3/4)} |\mathbf{1}_{B \backslash S_{k_m}} u_{k_m}^* - \mathbf{1}_{B \backslash S_{k_l}} u_{k_l}^*|^p \right\}^{2/p} \\ &\leq C |\widehat{S}(k_0)|^{\frac{p-2}{p}} \leq C 2^{-\frac{p-2}{p} k_0}, \end{aligned} \tag{38}$$

because $|\widehat{S}(k_0)| \leq 2^{-k_0+1}$.

We also need to control $J(m,l)$. Cover $E(k_0)$ by balls $B_i = B(x_i, 2^{-k_0-2})$, $i \in I(k_0)$, so that $x_i \in \mathcal{D}(k_0)$ and $|x_i - x_j| \geq 2^{-k_0-3}$ for $i, j \in I(k_0)$, $i \neq j$. Thus

$$J(m,l) \leq \sum_{i \in I(k_0)} \int_{E(k_0) \cap B_i} |\mathbf{1}_{B \backslash S_{k_m}} u_{k_m}^* - \mathbf{1}_{B \backslash S_{k_l}} u_{k_l}^*|^2 =: \sum_{i \in I(k_0)} a_i. \tag{39}$$

Recall that for each $x_i \in \mathcal{D}(k_0)$ and each $k \geq k_0$, we have annuli $D_{j,k}(x_i)$, $j \geq 2$, and numbers $\lambda_{j,k}(x_i)$, and that the sequence $\{\lambda_{j,k_m}(x_i)\}$ converges for each $j \geq 2$.

[But we shall only use this for $j = k_0 + 2$.] Write $a_i = b_i + c_i + d_i$, with

$$b_i = \int_{E(k_0) \cap B_i} |\mathbf{1}_{B \setminus S_{k_m}} u_{k_m}^* - \lambda_{j,k_m}(x_i)|^2, \tag{40}$$

$$c_i = \int_{E(k_0) \cap B_i} |\mathbf{1}_{B \setminus S_{k_l}} u_{k_l}^* - \lambda_{j,k_l}(x_i)|^2, \tag{41}$$

$$d_i = |E(k_0) \cap B_i| |\lambda_{j,k_m}(x_i) - \lambda_{j,k_l}(x_i)|$$
$$\leq C2^{-nk_0} |\lambda_{j,k_m}(x_i) - \lambda_{j,k_l}(x_i)|. \tag{42}$$

Notice that

$$\lim_{m,l \to +\infty} \sum_{i \in I(k_0)} d_i = 0 \tag{43}$$

because the $\{\lambda_{j,k_m}(x_i)\}$ are uniformly bounded and convergent, and $I(k_0)$ has at most $C2^{-nk_0}$ elements. Now let $x \in B_i \cap E(k_0)$ be given. Since x does not lie in any S_k, $k \geq k_0$, the construction of Lemma 9 gives sets $E_{j,k_m}(x)$ and numbers $\lambda_{j,k_m}(x)$, $j \geq 2$, so that the analogue of (12) for $u_{k_m}^*$ holds. That is,

$$|u_{k_m}^*(y) - \lambda_{j,k_m}(x)| \leq C2^{(n-1)j} \int_{D_{j,k_m}(x)} |\nabla u_{k_m}^*| \quad \text{for } y \in E_{j,k_m}(x). \tag{44}$$

We may use this for $j \geq k_0 + 2$, and we get the analogue of (14). That is, if in addition x is a density point for $u_{k_m}^*$,

$$|u_{k_m}^*(x) - \lambda_{k_0+2,k_m}(x)| \leq C \int_{B(x,2^{-k_0-2})} |\nabla u_{k_m}^*(y)| |x-y|^{1-n} \tag{45}$$

and hence

$$\int_{E(k_0) \cap B_i} |u_{k_m}^*(x) - \lambda_{k_0+2,k_m}(x)|^2 \leq C2^{-2k_0} \int_{B(x_i,2^{-k_0-2}) \setminus K_{k_m}} |\nabla u_{k_m}^*|^2, \tag{46}$$

by an easy estimate on convolutions. Since $x \in B_i = B(x_i, 2^{-k_0-2})$, B_i has a large intersection with $B(x, 2^{-k_0-2})$ and Lemma 5 says that

$$|\lambda_{k_0+2,k}(x) - \lambda_{k_0+2,k}(x_i)|^2 \leq C2^{-2k_0} 2^{nk_0} \int_{B_i \setminus K_{k_m}} |\nabla u_{k_m}^*|^2. \tag{47}$$

Altogether

$$b_i = \int_{E(k_0) \cap B_i} |u_{k_m}^* - \lambda_{k_0+2,k_m}(x_i)|^2$$
$$\leq 2 \int_{E(k_0) \cap B_i} |u_{k_m}^*(x) - \lambda_{k_0+2,k_m}(x)|^2 + C2^{-nk_0} |\lambda_{k_0+2,k_m}(x) - \lambda_{k_0+2,k_m}(x_i)|^2$$
$$\leq C2^{-2k_0} \int_{B(x_i,2^{-k_0-1}) \setminus K_{k_m}} |\nabla u_{k_m}^*|^2 \tag{48}$$

by (40), because $\mathbf{1}_{B\backslash S_{km}} = 1$ on $E(k_0)$ (since $S_{km} \subset \widehat{S}(k_0)$ for $m \geq k_0$), and by (46) and (47). Thus

$$\sum_{i \in I(k_0)} b_i \leq C 2^{-2k_0} \int_{B\backslash K_{km}} |\nabla u_{km}^*|^2 = C 2^{-2k_0} \tag{49}$$

because the $B(x_i, 2^{-k_0-1}) = 2B_i$ have bounded overlap. Similarly, $\sum_{i \in I(k_0)} c_i \leq C 2^{-2k_0}$. Altogether,

$$J(m,l) \leq \sum_{i \in I(k_0)} a_i = \sum_{i \in I(k_0)} (b_i + c_i + d_i) \leq C 2^{-2k_0} + \sum_{i \in I(k_0)} d_i \leq C 2^{-2k_0} \tag{50}$$

for m and l large enough, by (39) and (43). And then $I_{k_m,k_l} = I(m,l) + J(m,l) \leq C(2^{-\frac{p-2}{p}k_0} + 2^{-2k_0})$, by (38). Since we could do this with any $k_0 > 0$, we get that I_{k_m,k_l} tends to 0, $\{\mathbf{1}_{B\backslash S_{km}} u_{km}^*\}$ is a Cauchy sequence in $L^2(B(0,3/4))$, and (31) holds.

Let us now turn to (32) and (33). Since $\{\nabla u_k^*\}$ (and hence also $\mathbf{1}_{B\backslash S_k}\{\nabla u_k^*\}$) is a bounded sequence in $L^2(B(0,3/4))$, we can extract a new subsequence so that the $\mathbf{1}_{B\backslash S_k} \nabla u_k^*$ converge weakly in $L^2(B(0,3/4))$ to some limit V. We still need to show that $u_\infty^* \in W^{1,2}(B(0,3/4))$ and its derivative is V. That is, we need to check that

$$\int_{B(0,3/4)} u_\infty^* \frac{\partial \varphi}{\partial x_i} = -\int_{B(0,3/4)} V_i\, \varphi = -\lim_{k \to +\infty} \int_{B(0,3/4)\backslash S_k} \frac{\partial u_k^*}{\partial x_i}\, \varphi \tag{51}$$

for every smooth function φ with support in $B(0,3/4)$ and $1 \leq i \leq n$. Let us do this for $i = 1$; the verification for the other variables would be the same. Also, since we may cut φ into (finitely many) pieces, we may assume that its support is contained in a small cube $Q \subset B(0,3/4)$, with faces parallel to the axes. Observe that

$$\int_{B(0,3/4)} u_\infty^* \frac{\partial \varphi}{\partial x_1} = \lim_{k \to +\infty} \int_{B(0,3/4)\backslash S_k} u_k^* \frac{\partial \varphi}{\partial x_1}, \tag{52}$$

by (31). Set

$$J_k = \int_{B(0,3/4)\backslash S_k} \left[u_k^* \frac{\partial \varphi}{\partial x_1} + \frac{\partial u_k^*}{\partial x_1} \varphi \right] = \int_{Q\backslash S_k} \left[u_k^* \frac{\partial \varphi}{\partial x_1} + \frac{\partial u_k^*}{\partial x_1} \varphi \right];$$

we want to show that J_k tends to 0. Write $Q = I \times Q'$, call $\pi : Q \to Q'$ the orthogonal projection parallel to the x_1-axis, and set $F_k = \pi(Q \cap [S_k \cup K_k])$. Notice that S_k does not go as far as ∂B, by construction. Hence $F_k \subset \pi(\partial S_k) \cup \pi(K_k)$, and so

$$H^{n-1}(F_k) \leq H^{n-1}(\partial S_k) + H^{n-1}(K_k \cap B) \leq C H^{n-1}(K_k \cap B), \tag{53}$$

by (10). Proposition 9.5 says that for almost every $y \in Q' \setminus F_k$, $u_k^*(\cdot, y) \in W^{1,1}(I)$, and its derivative is $\frac{\partial u_k^*}{\partial x_1}(\cdot, y)$. Which means that

$$\int_I [u_k^*(x, y) \frac{\partial \varphi}{\partial x_1}(x, y) + \frac{\partial u_k^*}{\partial x_1}(x, y) \, \varphi(x, y)] \, dx = 0.$$

Also notice that $I \times (Q' \setminus F_k)$ does not meet S_k. Thus the contribution of $I \times (Q' \setminus F_k)$ to J_k vanishes, and

$$J_k = \int \int_{(I \times F_k) \setminus S_k} [u_k^* \frac{\partial \varphi}{\partial x_1} + \frac{\partial u_k^*}{\partial x_1} \varphi] \le C \int \int_{(I \times F_k) \setminus S_k} |u_k^*| + |\frac{\partial u_k^*}{\partial x_1}|$$

$$\le C |I|^{1/2} |F_k|^{1/2} \{ \int_{Q \setminus S_k} |u_k^*|^2 + |\frac{\partial u_k^*}{\partial x_1}|^2 \}^{1/2} \le C |F_k|^{1/2}, \tag{54}$$

with constants C that may depend on φ and Q, but not on k, and where we used (29) and (31) to control $\int_{Q \setminus S_k} |u_k^*|^2 + |\frac{\partial u_k^*}{\partial x_1}|^2$. Now $|F_k|$ tends to 0, by (53), and so does J_k. This completes our proof of (51), (32) and (33) follow, and so does Lemma 30. $\qquad\square$

We shall need to know that for every open set $U \subset B(0, 3/4)$,

$$\int_U |\nabla u_\infty^*|^2 \le \liminf_{k \to +\infty} \int_{U \setminus S_k} |\nabla u_k^*|^2 = \liminf_{k \to +\infty} \{ E_k^{-1} \int_{U \setminus S_k} |\nabla u_k|^2 \}. \tag{55}$$

And indeed, (33) says that for every $g \in L^2_{\mathbb{R}^n}(U)$,

$$\int_U \nabla u_\infty^* \cdot g = \lim_{k \to +\infty} \int_{U \setminus S_k} \nabla u_k^* \cdot g \le ||g||_2 \liminf_{k \to +\infty} \{ \int_{U \setminus S_k} |\nabla u_k^*|^2 \}^{1/2}, \tag{56}$$

and then

$$\{ \int_U |\nabla u_\infty^*|^2 \}^{1/2} = \sup \{ \int_U \nabla u_\infty^* \cdot g \; ; \; g \in L^2_{\mathbb{R}^n}(U) \text{ and } ||g||_2 \le 1 \}$$

$$\le \liminf_{k \to +\infty} \{ \int_{U \setminus S_k} |\nabla u_k^*|^2 \}^{1/2}, \tag{57}$$

which proves the first half of (55). The second half follows from (29).

An intermediate statement. As we said before, the main point of the proof is to show that (under some additional conditions) if (2) holds, then an even better estimate holds in smaller balls. Here is a precise statement.

Lemma 58. *For each constant $M \ge 1$, we can find $\eta > 0$, $\alpha \in (0, 1/3)$, and $C_M \ge 1$ such that if $(u, K) \in TRLQ(B)$, with quasiminimality constants $r_0 \ge 1$, $M \ge 1$, $a > 0$, and if $H^{n-1}(K \cap B) \le \eta$, then*

$$H^{n-1}(K \cap B(0, \alpha)) + \int_{B(0, \alpha) \setminus K} |\nabla u|^2 \le \frac{\alpha^{n-1}}{2} \{ H^{n-1}(K \cap B) + \int_{B \setminus K} |\nabla u|^2 \} + C_M \, a. \tag{59}$$

We took $a > 0$ here just for convenience of notation, to avoid adding an arbitrary small constant to the right-hand side of (59).

We shall see later how to conclude once we have Lemma 58, but let us first discuss the proof. We prove the lemma with r_0, M, and α fixed, but we shall see later that α depends only on n, and η and C_M depend only on n and M, not on a or r_0.

Here we shall find it easier to distinguish between two cases. In [CaLe2] and [Si], the authors proceed slightly differently; they introduce functionals of the form $\int |\nabla u|^2 + c H^{n-1}(K)$, with c variable, and prove a single (synthetic) estimate.

The case when energy is much smaller. We start with the slightly easier case when

$$\int_{B\backslash K} |\nabla u|^2 \leq \frac{a^{n-1}}{4} H^{n-1}(K \cap B). \tag{60}$$

We prove (59) by compactness. We consider a sequence (u_k, K_k) in $TRLQ(B)$, (with the same constants r_0, M, and a) for which $H^{n-1}(K_k)$ tends to 0 and (60) holds, but not (59), and we try to get contradicting information for some large values of k. First we define the u_k^* as in (29), extract a subsequence, and find a limit u_∞^* as in Lemma 30. This allows us to define competitors (v_k, G_k) for (u_k, K_k), as in (3), (16), and (17). That is, we choose an annulus A_k and a corresponding function φ_k as in Lemma 18, set

$$\overline{u}_k = \mathbf{1}_{B\backslash[A_{k,+}\cap S_k]} u_k \tag{61}$$

(where S_k is still obtained from Lemma 9, $A_{k,+}$ is the union of A_k and its interior complementary component, and we subtracted a constant from u_k to be able to take $\lambda = 0$ in (16), as we did for Lemma 30), then set $w_k = E_k^{1/2} u_\infty^*$,

$$v_k(x) = \varphi_k(x) w_k(x) + (1 - \varphi_k(x)) \overline{u}_k(x), \tag{62}$$

and

$$G_k = [K_k \backslash A_{k,-}] \cup [S_k \cap \partial A_k] \cup [\partial S_k \cap A_k] \tag{63}$$

(maybe plus a tiny additional set in the case of topological quasiminimizers, as discussed just after (17)). This way, all the estimates proved above will apply.

Let us extract one last subsequence, so that the A_k converge to some annulus $A = \overline{B}(0, r+\tau)\backslash B(0, \tau)$. This will allow us to apply (25)–(27) for all k large enough, and with the fixed radius $r' = \text{Max}(1/2, r - \tau/2)$. Recall from (55) that

$$\int_{B(0,r')} |\nabla u_\infty^*|^2 \leq \liminf_{k \to +\infty} \left\{ E_k^{-1} \int_{B(0,r')\backslash S_k} |\nabla u_k|^2 \right\}. \tag{64}$$

Hence, by definition of w_k, there is a sequence $\{\varepsilon_k\}$ that tends to 0 and for which

$$\int_{B(0,r')} |\nabla w_k|^2 \leq \int_{B(0,r')\backslash S_k} |\nabla u_k|^2 + \varepsilon_k E_k. \tag{65}$$

Next set $\widetilde{E}_k = \int_{B\backslash G_k} |\nabla v_k|^2$; then (27) and (65) say that $\widetilde{E}_k - E_k \leq \varepsilon_k E_k + \delta_{1,k}$, with a quantity $\delta_{1,k}$ that can be estimated as in (26). Set $\delta_k E = \text{Max}\{\widetilde{E}_k - E_k, M(\widetilde{E}_k - E_k)\}$, as in (7.18) and (7.19). Then $\delta_k E \leq M\varepsilon_k E_k + M\delta_{1,k}$. The quasiminimality condition is written

$$H^{n-1}(K_k \setminus G_k) \leq MH^{n-1}(G_k \setminus K_k) + \delta_k E + a$$
$$\leq MH^{n-1}(G_k \setminus K_k) + M\varepsilon_k E_k + M\delta_{1,k} + a. \tag{66}$$

Recall from (26) that

$$\delta_{1,k} \leq C\tau \int_{B(0,3/4)\backslash K_k} |\nabla u_k|^2 + C\tau \int_{B(0,3/4)} |\nabla w_k|^2 + C\tau^{-1} \int_{B(0,3/4)} |w_k - \widetilde{u}_k|^2, \tag{67}$$

where $\widetilde{u}_k = u_k \mathbf{1}_{B\backslash S_k}$ because we can take $\lambda = 0$ (see above (26)).

Next, $\int_{B(0,3/4)\backslash K_k} |\nabla u_k|^2 \leq E_k$ by definition of E_k, and $\int_{B(0,3/4)} |\nabla w_k|^2 \leq E_k + \varepsilon_k E_k$ (or with some other sequence that tends to 0) by the proof of (65). Finally, $\int_{B(0,3/4)} |w_k - \widetilde{u}_k|^2 = E_k \int_{B(0,3/4)} |u_\infty^* - u_k^* \mathbf{1}_{B\backslash S_k}|^2$, by definition of w_k and (29), and this tends to 0 by (31) and because $\{E_k\}$ is bounded. So $\delta_{1,k} \leq 2C\tau E_k + \varepsilon_k'$, where $\{\varepsilon_k'\}$ is another sequence that tends to 0. Now

$$H^{n-1}(K_k \cap B(0,1/2)) \leq H^{n-1}(K_k \setminus G_k) \tag{68}$$
$$\leq CM[H^{n-1}(K_k \cap B)^{\frac{1}{n+1}} + \tau] H^{n-1}(K_k \cap B) + M\varepsilon_k E_k + M\delta_{1,k} + a,$$
$$\leq CM[\eta^{\frac{1}{n+1}} + \tau] H^{n-1}(K_k \cap B) + M\varepsilon_k E_k + 2MC\tau E_k + M\varepsilon_k' + a$$

by (63), (66), (22), and (2). We choose η and τ so small that $CM[\eta^{\frac{1}{n+1}} + \tau] \leq \frac{\alpha^{n-1}}{4}$ in the first term of the right-hand side and $2MC\tau \leq \frac{\alpha^{n-1}}{4}$ in the third term. Then

$$H^{n-1}(K_k \cap B(0,1/2)) \leq \frac{\alpha^{n-1}}{4} H^{n-1}(K_k \cap B) + \frac{\alpha^{n-1}}{4} E_k + 2a \tag{69}$$

for k large. [We required $a > 0$ in Lemma 58 only to be able to throw away the sequences that tend to 0 and pay with an extra a.]

So far we did not use the extra assumption (60).

If (60) holds, then $\int_{B(0,\alpha)\backslash K} |\nabla u_k|^2 \leq \frac{\alpha^{n-1}}{4} H^{n-1}(K_k \cap B)$, and (59) follows from (69) [in contradiction with the definition of $\{(u_k, K_k)\}$].

The case when energy is not much smaller. So we are left with the case when (60) does not hold. Since we still have (69), it is enough to show that

$$\int_{B(0,\alpha)\backslash K} |\nabla u|^2 \leq \frac{\alpha^{n-1}}{4} \int_{B\backslash K} |\nabla u|^2 + (C_M - 2)a. \tag{70}$$

As before, we proceed by compactness, suppose that we have a sequence $\{(u_k, K_k)\}$ in $TRLQ(B)$ for which $H^{n-1}(K_k)$ tends to 0 and (60) and (70) never hold, and we try to reach a contradiction. We still pick annuli A_k and define $\bar{u}_k$ as before, but we shall try to use a better function than $w_k = E_k^{1/2} u_\infty^*$ to define our new competitor (v_k, G_k).

Choose a radius $\rho \in (1/3, 1/2)$ such that the restriction of u_∞^* to $\partial B(0, \rho)$ lies in $W^{1,2}$, and is equal almost-everywhere on $\partial B(0, \rho)$ to the radial limits of u_∞^*; we know from Corollary 14.28 that this is the case for almost every $\rho \in (1/3, 1/2)$. Then let w^* denote the function in $W^{1,2}(B(0, \rho))$ that coincides with u_∞^* (in the sense of radial limits again) on $\partial B(0, \rho)$ and minimizes $\int_{B(0,\rho)} |\nabla w^*|^2$ (see Proposition 15.6). We know from Lemma 15.17 that w^* is harmonic on $B(0, \rho)$. Since $\int_{B(0,\rho)} |\nabla w^*|^2 \leq \int_{B(0,3/4)} |\nabla u_\infty^*|^2 \leq 1$ by (55) and the normalization (29), we get that

$$\int_{B(0,\alpha)} |\nabla w^*|^2 \leq C\alpha^n \leq \frac{\alpha^{n-1}}{10}, \tag{71}$$

where C does not depend on α, and where we choose $\alpha < 1/3$ (depending only on the dimension) so that the second inequality holds. Here we used a trivial estimate on harmonic functions in a ball, but we shall have to be be careful about its analogue when we consider quasiminimizers near the boundary, and B is replaced with another domain.

We set $w_k(x) = E_k^{1/2} w^*(x)$ for $x \in B(0, \rho)$ and $w_k(x) = E_k^{1/2} u_\infty^*$ otherwise, and then define v_k and G_k as in (62) and (63). We continue the estimates as in the previous case. Set

$$\Delta_E = \int_{B(0,\rho)} |\nabla u_\infty^*|^2 - |\nabla w^*|^2. \tag{72}$$

Notice that we can improve our energy estimate (65) by $E_k \Delta_E$. Then the same computations as for (68) yield

$$0 \leq H^{n-1}(K_k \cap B(0, 1/2)) \tag{73}$$

$$\leq CM[\eta^{\frac{1}{n+1}} + \tau] H^{n-1}(K_k \cap B) + M\varepsilon_k E_k + 2MC\tau E_k + M\varepsilon_k' + a - E_k \Delta_E.$$

Now $H^{n-1}(K_k \cap B) \leq 4\alpha^{1-n} E_k$ because (60) does not hold for (u_k, K_k), and so (73) says that $E_k\{\Delta_E - 4CM[\eta^{\frac{1}{n+1}} + \tau]\alpha^{1-n} - M\varepsilon_k - 2MC\tau\} \leq M\varepsilon_k' + a$, which implies that

$$E_k\{\Delta_E - C'M[\eta^{\frac{1}{n+1}} + \tau]\alpha^{1-n}\} \leq 2a \quad \text{for } k \text{ large.} \tag{74}$$

Since we assume that (70) does not hold, $E_k \geq \int_{B(0,\alpha)\backslash K_k} |\nabla u_k|^2 \geq (C_M - 2)a$, and (74) says that

$$\Delta_E \leq C'M[\eta^{\frac{1}{n+1}} + \tau]\alpha^{1-n} + \frac{2}{C_M - 2}. \tag{75}$$

Of course the precise expression in the right-hand side does not matter; the point is that we can make it as small as we want. Also note that this was the place where distinguishing cases helps. If we had not excluded the case when the energy E_k is much smaller than $H^{n-1}(K \cap B)$, we would find it hard to use (73) to get information on the u_k.

Next we want to use the smallness of Δ_E to estimate $\int_{B(0,\alpha)} |\nabla u_\infty^*|^2$. Set $f = u_\infty^* - w^*$ on $B(0,\rho)$. Recall that w^* minimizes $\int_{B(0,\rho)} |\nabla w^*|^2$ under the constraint that $w^* = u_\infty^*$ on $\partial B(0,\rho)$. Then $w^* + tf$ is a competitor in the definition of w^* for every small $t \in \mathbb{R}$, and the comparison yields that ∇f is orthogonal to ∇w^* in $L^2(B(0,\rho))$. Hence

$$\int_{B(0,\rho)} |\nabla u_\infty^*|^2 = \int_{B(0,\rho)} |\nabla w^*|^2 + \int_{B(0,\rho)} |\nabla f|^2, \tag{76}$$

by Pythagoras, and $\int_{B(0,\rho)} |\nabla f|^2 = \Delta_E$ (by (72)). This yields

$$\int_{B(0,\alpha)} |\nabla u_\infty^*|^2 \leq 2 \int_{B(0,\alpha)} |\nabla w^*|^2 + 2 \int_{B(0,\alpha)} |\nabla f|^2 \leq \frac{\alpha^{n-1}}{5} + 2\Delta_E, \tag{77}$$

by (71). Next set

$$\Delta_E' = \liminf_{k \to +\infty} \int_{B(0,\alpha)\setminus K_k} |\nabla u_k^*|^2 - \int_{B(0,\alpha)} |\nabla u_\infty^*|^2. \tag{78}$$

Notice that $\Delta_E' \geq 0$, by (55), and that if r' is as near (64),

$$\int_{B(0,r')} |\nabla u_\infty^*|^2 = \int_{B(0,\alpha)} |\nabla u_\infty^*|^2 + \int_{B(0,r')\setminus \overline{B}(0,\alpha)} |\nabla u_\infty^*|^2$$

$$\leq \liminf_{k \to +\infty} \int_{B(0,\alpha)\setminus K_k} |\nabla u_k^*|^2 - \Delta_E' + \liminf_{k \to +\infty} \int_{B(0,r')\setminus[\overline{B}(0,\alpha)\cup K_k]} |\nabla u_k^*|^2$$

$$\leq \liminf_{k \to +\infty} \int_{B(0,r')\setminus K_k} |\nabla u_k^*|^2 - \Delta_E'. \tag{79}$$

Thus we can win an extra Δ_E' in (64), and the same computations as before lead to an analogue of (73), with Δ_E replaced with Δ_E'. Then the proof of (75) yields

$$\Delta_E' \leq C'M[\eta^{\frac{1}{n+1}} + \tau]\alpha^{1-n} + \frac{2}{C_M - 2}. \tag{80}$$

We can find arbitrarily large integers k for which

$$\int_{B(0,\alpha)\setminus K_k} |\nabla u_k^*|^2 \leq \liminf_{k \to +\infty} \int_{B(0,\alpha)\setminus K_k} |\nabla u_k^*|^2 + a = \int_{B(0,\alpha)} |\nabla u_\infty^*|^2 + \Delta_E' + a$$

$$\leq \frac{\alpha^{n-1}}{5} + 2\Delta_E + \Delta_E' + a$$

$$\leq \frac{\alpha^{n-1}}{5} + 3C'M[\eta^{\frac{1}{n+1}} + \tau]\alpha^{1-n} + \frac{6}{C_M - 2} + a \tag{81}$$

by (78), (77), (75), and (80). If we choose η, τ, and C_M^{-1} small enough (depending on α, which was already chosen and depends only on the dimension), we get that

$$\int_{B(0,\alpha)\setminus K_k} |\nabla u_k|^2 = E_k \int_{B(0,\alpha)\setminus K_k} |\nabla u_k^*|^2 \leq \frac{\alpha^{n-1}}{4} E_k + a E_k \leq \frac{\alpha^{n-1}}{4} E_k + Ca,$$

$$(82)$$

which contradicts the fact that (u_k, K_k) does not satisfy (70) (by construction). This completes our proof of (70) and (59) in this second case, and Lemma 58 follows. □

How to conclude from there. Now let (u, K) be as in the beginning of this section. That is, assume that $(u, K) \in TRLQ(\Omega)$ for some domain Ω that contains B, and that both $H^{n-1}(K \cap B)$ and the quasiminimality constant a are very small (depending on M). Set

$$\theta_m = \alpha^{m(1-n)} \left[H^{n-1}(K \cap B(0, \alpha^m)) + \int_{B(0,\alpha^m)\setminus K} |\nabla u|^2 \right]$$

$$(83)$$

for $m \geq 0$. Lemma 58 (applied to a dilation of (u, K) by a factor α^{-m}) applies as soon as $\alpha^{m(1-n)} H^{n-1}(K \cap B(0, \alpha^m))$ is small enough, and yields

$$\theta_{m+1} \leq \frac{1}{2}\theta_m + C_M a.$$

$$(84)$$

If we start with $H^{n-1}(K \cap B)$ and a small enough, we can apply Lemma 58 a few times, and find m_0 such that $\theta_{m_0} \leq \eta$, where η is the constant in Lemma 58. If $\theta_{m_0} \geq 3C_M a$, Lemma 58 and (84) apply and shows that $\theta_{m_0+1} \leq \frac{5}{6}\theta_{m_0}$. In fact, as long as $m \geq m_0$ is such that $\theta_m \geq 3C_M a$, we can repeat the application of Lemma 58 and (84) and get that $\theta_{m+1} \leq \frac{5}{6}\theta_m$. Eventually, we get a first $m_1 \geq m_0$ such that $\theta_{m_1} < 3C_M a$. Of course this is still true if $\theta_{m_0} < 3C_M a$.

We still have (84), but now it says that $\theta_{m_1+1} < 3C_M a$, and this stays true for all subsequent values of m. Thus $\theta_m \leq 3C_M a$ for m large. If we define the upper density $d(x)$ by

$$d(x) = \limsup_{r\to 0} \left\{ r^{1-n} H^{n-1}(K \cap B(x, r)) \right\},$$

$$(85)$$

then $d(0) \leq Ca$. The constant C depends on M and α, but this does not matter. The point is that if a is small enough, Lemma 19.1 says that $d(x) > Ca$ for H^{n-1}-almost every point of K, so 0 was a very exceptional point.

Now suppose that the first inequality in (1) fails for some reduced quasi-minimizer (u, K) and some ball $B(x, r)$. If C is large enough, every point $x \in K \cap B(x, r/2)$ will satisfy our initial condition (after translation by $-x$ and dilation by a factor $2/r$), hence $d(x) \leq Ca$. By Lemma 19.1, $H^{n-1}(K \cap B(x, r/2)) = 0$. By Proposition 10.1, u has an extension to $B(x, r/2)$ that lies in $W^{1,2}$, and hence (u, K) is not be reduced (or coral), a contradiction. [The argument is the same as in Section 20; also see the first lines of this section.] This completes our description of Siaudeau's proof of Ahlfors-regularity. □

Remark 86. The argument above, and hence local Ahlfors-regularity, extend to quasiminimizers with linear distortion (see Definition 7.52). That is, given n, $M \geq 1$, and $L \geq 1$, we can find constants $a > 0$ and $C \geq 1$ such that, if $(u, K) \in TRLQ(\Omega)$ with constants (r_0, M, a) and linear distortion $\leq L$, then (1) holds for all pairs (x, r) such that $x \in K$, $0 < r < r_0$, and $B(x, r) \subset \Omega$.

Unless the author missed something, the argument needs surprisingly little modification; the energy computations are just done after a linear change of variable (which we can make constant, modulo errors that tend to 0, by extracting a subsequence for which the matrices A in Definition 7.52 converge), and the main point is again that we have enough decay for the energy of harmonic functions in a ball.

73   Uniformly rectifiable sets of all dimensions

In this section we want to give a fairly rapid overview of uniform rectifiability, essentially with no proofs. The excuse for doing this is the result of uniform rectifiability described in the next section, but many of the results quoted here are probably completely irrelevant to the theory of Mumford-Shah minimizers or quasiminimizers. So this will be mostly be a section of advertisement for quantitative geometric measure theory. See [Da9] for a slightly more detailed survey, and [DaSe3] and its references for lots of additional information on uniform rectifiability.

In what follows, E is an Ahlfors-regular set of dimension d in $\mathbb{R}^n$. This means that E is closed, not reduced to one point, and that there is a constant $C \geq 1$ such that

$$C^{-1} r^d \leq H^d(E \cap B(x, r)) \leq C r^d \text{ for every } x \in E \text{ and } 0 < r < \text{diam}(E). \quad (1)$$

Here d will be an integer. The two definitions that follow do not make sense otherwise, but some equivalent definitions (like with (16) or (17) below, or the boundedness of singular integral operators) can be generalized to noninteger dimensions. However, no Ahlfors-regular set satisfies them when d is not an integer. This is not surprising; recall that rectifiability behaves just the same way.

The definitions of uniform rectifiability (there are many equivalent ones) are all a little more complicated than in dimension 1. We start with the definition that is closest to our definition of Section 27 (requiring that E be contained in an Ahlfors-regular curve). So we first define ω-regular surfaces. We still use parameterizations z, but they will no longer be Lipschitz. Instead they will merely be controlled by A_1-weights, as follows.

Recall that an $\underline{A_1\text{-weight}}$ is a positive, locally integrable function ω on $\mathbb{R}^d$ such that

$$\frac{1}{|B|} \int_B \omega \leq C \text{ ess.inf}_B \omega \text{ for every ball } B, \quad (2)$$

for some constant $C \geq 1$. It is well known that every A_1-weight is also an A_∞-weight, which means that for each $\varepsilon > 0$ we we can find $\delta > 0$ such that $\int_A \omega(x)dx \leq \varepsilon \int_B \omega(x)dx$ for each ball B and each measurable subset $A \subset B$ such that $|A| \leq \delta|B|$. Also, A_∞-weights are characterized by the existence of constants $C \geq 1$ and $\eta > 0$ such that

$$\frac{1}{C}\left(\frac{|A|}{|B|}\right)^\eta \leq \left\{\int_A \omega(x)dx\right\}\left\{\int_B \omega(x)dx\right\}^{-1} \leq C\left(\frac{|A|}{|B|}\right)^{1/\eta} \qquad (3)$$

whenever B is a ball and A is a measurable subset of B. See [Jou] or [GaRu] for more information.

Definition 4. *An $\underline{\omega\text{-regular parameterization}}$ is a function $z \in W^{1,1}_{\mathrm{loc}}(\mathbb{R}^d)$, with values in $\mathbb{R}^n$, such that there is an A_1-weight ω on $\mathbb{R}^d$ for which*

$$|\nabla z(y)| \leq \omega(y)^{1/d} \ \text{almost-everywhere}, \qquad (5)$$

and

$$\int_{\{y \in \mathbb{R}^d \,;\, z(y) \in B(x,r)\}} \omega(y)\,dy \leq Cr^d \ \text{ for } x \in \mathbb{R}^n \text{ and } r > 0. \qquad (6)$$

An $\underline{\omega\text{-regular surface}}$ of dimension d is a subset of $\mathbb{R}^n$ of the form $E = z(\mathbb{R}^d)$, where z is an ω-regular parameterization.

Notice that the conditions (5) and (6) tend to go in different directions and that when $d = 1$, regular parameterizations are ω-regular (as in Definition 4), with $\omega = 1$. The two notions of regular curve coincide, though, because we may compose z in Definition 4 with the inverse of a change of variable with derivative $\omega(y)$ to reduce to $\omega = 1$. When $d > 1$, it is much more difficult to find quasiconformal changes of variables in $\mathbb{R}^d$ that have a given Jacobian, even up to a constant, and this is why we need weights in Definition 4.

It is reasonably easy to show that ω-regular surfaces are automatically Ahlfors-regular (as in (1)), the main point being that we can compare $H^d(z(B))$ with $\int_B \omega(y)dy$. The reader is asked to believe that ω-regular surfaces keep many of the good properties of Lipschitz graphs, with more flexibility.

Definition 7. *Let E be a d-dimensional Ahlfors-regular set in $\mathbb{R}^n$. We say that E is $\underline{\text{uniformly rectifiable}}$ when we can find an ω-regular parameterization $z : \mathbb{R}^d \to \mathbb{R}^n$ such that $E \subset z(\mathbb{R}^d)$.*

Recall that a rectifiable set of dimension d is a set $F \subset \mathbb{R}^n$ that can be covered, up to a set of H^d-measure zero, by a countable number of d-dimensional surfaces Γ_j of class C^1. We also get an equivalent definition if we merely ask the Γ_j to be Lipschitz images of $\mathbb{R}^d$, or even images of $\mathbb{R}^d$ by applications z_j that satisfy (5). [This fact is based on Lusin's theorem.] Thus uniform rectifiability implies rectifiability, but we get a better control on the set, because we need only one mapping z and in addition we have some uniform control on the parameterization z.

The existence of an ω-regular surface that contains E is often hard to prove directly, but fortunately a few other characterizations of uniform rectifiability are available. Let us start with an analogue of Definition 31.2 and Theorem 31.5.

Theorem 8. *[DaSe1] Let E be a d-dimensional Ahlfors-regular set in $\mathbb{R}^n$. Then E is uniformly rectifiable if and only if it <u>contains big pieces of Lipschitz images of</u> $\mathbb{R}^d$, i.e., if there are constants $M \geq 1$ and $\theta > 0$ such that, for each ball $B(x,r)$ centered on E and with $0 < r < \text{diam}(E)$, we can find an M-Lipschitz mapping $f_{x,r} : B(0,r) \subset \mathbb{R}^d \to \mathbb{R}^n$ such that*

$$H^d(E \cap f_{x,r}(B(0,r)) \cap B(x,r)) \geq \theta r^d. \tag{9}$$

The direct implication is rather easy; the converse is much more complicated, but the construction is a little bit in the same spirit as in Section 31. However we cannot use a minimizing surface, so we have to construct our ω-regular surface little by little, starting from large scales and using a stopping time construction.

There are a few intermediate equivalent conditions. For instance, E also contains big pieces of bilipschitz images of subsets of $\mathbb{R}^d$, which means that for each ball $B(x,r)$ centered on E, we can find a closed set $F_{x,r} \subset \mathbb{R}^d$ and an M-bilipschitz mapping $f_{x,r} : F_{x,r} \to \mathbb{R}^n$ such that $H^d(E \cap f_{x,r}(F_{x,r}) \cap B(x,r)) \geq \theta r^d$. We may even find fewer than C such sets $F_{x,r,i}$ and M-bilipschitz mappings $f_{x,r,i}$ so that the $f_{x,r,i}(F_{x,r,i})$ cover 99% of $E \cap B(x,r)$. See [DaSe3] for this and other similar characterizations.

The first excuse for studying uniform rectifiability came from singular integral operators. It turns out that the uniformly rectifiable sets are the Ahlfors-regular sets in $\mathbb{R}^n$ such that all the kernels K in a reasonable class $\mathcal{K}_{d,n}$ define bounded operators on $L^2(E)$ by a formula similar to $Tf(x) = \int_E K(x-y)f(y)dH^d(y)$. The class $\mathcal{K}_{d,n}$ contains all the odd kernels K that are defined and smooth on $\mathbb{R}^d \setminus \{0\}$ and are homogeneous of degree $-d$, and a few other ones with similar estimates. When $d = 1$ and $n = 2$, Mattila, Melnikov and Verdera [MatMV] proved that the continuity of the Cauchy operator alone is enough. This was generalized to general sets and measures (not necessarily Ahlfors-regular) by Tolsa [To2] and Nazarov, Treil, Volberg [NTV]. When $n > 2$ and $d = n - 1$, it is not known whether the continuity of the Riesz transforms alone implies uniform rectifiability (because there is no available analogue of Menger curvature).

Uniformly rectifiable sets have a few interesting regularity properties. Let us just mention a Carleson measure estimate on the Peter Jones numbers $\beta_{E,q}(x,r)$. For $x \in E$, $0 < r < \text{diam}(E)$, and $1 \leq q < +\infty$, set

$$\beta_{E,q}(x,r)^q = \inf \left\{ r^{-d} \int_{E \cap B(x,r)} [r^{-1} \text{dist}(y,P)]^q dH^d(y) \, ; \, P \text{ is an affine } d\text{-plane} \right\}. \tag{10}$$

This is the analogue of $\beta_E(x,r)$ above, but with an L^q norm.

Theorem 11. *[DaSe1] If E is uniformly rectifiable, then for each $q \in [1, \frac{2d}{d-2})$, there is a constant C_q such that*

$$\int_{y \in E \cap B(x,r)} \int_{0<t<r} \beta_{E,q}(x,r)^2 \frac{dH^d(y)dt}{t} \leq C_q r^d \tag{12}$$

for $x \in E$ and $0 < r < \text{diam}(E)$.

Actually, this is also a characterization of uniform rectifiability: even for $q = 1$ alone, the Carleson estimate (12) is enough to imply uniform rectifiability.

Also, it is easy to see that (12) implies the following weaker estimate (called the Weak Geometric Lemma): for each $\varepsilon > 0$, there is a constant $C_\varepsilon > 0$ such that if we set $\mathcal{B}(\varepsilon) = \{(x,r) \in E \times (0, \text{diam}(E)) \, ; \, \beta_E(x,r) > \varepsilon\}$, then $\mathcal{B}(\varepsilon)$ is a Carleson set. This means that

$$\int_{y \in E \cap B(x,r)} \int_{0<t<r} \mathbf{1}_{\mathcal{B}(\varepsilon)}(y,t) \frac{dH^d(y)dt}{t} \leq C(\varepsilon) \, r^d \tag{13}$$

for $x \in E$ and $0 < r < \text{diam}(E)$. The proof of (13) goes as in Corollary 23.38, except that we also need to use the fact that

$$\beta_E(x,r) \leq C \beta_{E,q}(x,r)^{\frac{q}{d+q}} \quad \text{for } x \in E \text{ and } 0 < r < \text{diam}(E), \tag{14}$$

with a constant C that depends on q and the regularity constant for E. The proof of (14) is the same as for (41.12).

Recall that if (13) holds, then for each $x \in E$ and $r > 0$ we can find a point $y \in E \cap B(x, r/2)$ and a radius $t \in (C_\varepsilon^{-1} r, r/2)$ such that $\beta_E(y,t) \leq \varepsilon$. The proof is as in Corollary 23.38 and Proposition 41.17.

It is very amusing that apparently minor variants of the weak geometric lemma are enough to imply uniform rectifiability. Let us just give three examples. Set

$$
\begin{aligned}
b\mathcal{G}(\varepsilon) = \big\{ & (x,r) \in E \times (0, \text{diam}(E)); \text{there is an affine } d\text{-plane } P \\
& \text{such that } \text{dist}(y, P) \leq \varepsilon r \text{ for } y \in E \cap B(x,r) \text{ and} \\
& \text{dist}(y, E) \leq \varepsilon r \text{ for } y \in P \cap B(x,r) \big\};
\end{aligned} \tag{15}
$$

$$
\begin{aligned}
c\mathcal{G}(\varepsilon) = \big\{ & (x,r) \in E \times (0, \text{diam}(E)); \\
& \text{dist}(\frac{y+z}{2}, E) \leq \varepsilon r \text{ for } y, z \in E \cap B(x,r) \big\};
\end{aligned} \tag{16}
$$

$$
\begin{aligned}
s\mathcal{G}(\varepsilon) = \big\{ & (x,r) \in E \times (0, \text{diam}(E)); \\
& \text{dist}(2y - z, E) \leq \varepsilon r \text{ for } y, z \in E \cap B(x,r) \big\}.
\end{aligned} \tag{17}
$$

These correspond to balls where E is very close to a d-plane, or close to being convex, or nearly symmetric with respect to its points. Then there is an $\varepsilon > 0$ (that depends on n, d, and the Ahlfors-regularity constant for E) such that if the complement in $E \times (0, \text{diam}(E))$ of $b\mathcal{G}(\varepsilon)$, $c\mathcal{G}(\varepsilon)$, or $s\mathcal{G}(\varepsilon)$ is a Carleson set, then E is uniformly rectifiable. See [DaSe3] for this (look for the BWLG, WLCV, and WLS) and many other conditions of the same type.

We end this section with two sufficient conditions for uniform rectifiability in codimension 1. Both rely on the idea that E almost separates $\mathbb{R}^n$ into more than one piece in most balls.

The first one (Condition B) was introduced by S. Semmes [Se1] as a sufficient condition for the L^2-boundedness on E of some singular integral operators. We say that the Ahlfors-regular set E of dimension $n-1$ in $\mathbb{R}^n$ satisfies Condition B if there is a constant $C > 0$ such that, for each $x \in E$ and $0 < r < \text{diam}(E)$, we can find two balls B_1 and B_2 with the same radius $C^{-1}r$, that are both contained in $B(0,r) \setminus E$, and that lie in different connected components of $\mathbb{R}^n \setminus E$. It was shown in [DaSe2] or [DaJe] that Ahlfors-regular sets that satisfy Condition B contain big pieces of Lipschitz graphs. This means that we can find constants $M > 0$ and $\theta > 0$ such that for each $x \in E$ and $0 < r < \text{diam}(E)$, we can find a set $\Gamma_{x,r}$, which is the graph of an M-Lipschitz function in some orthogonal set of coordinates, and for which $H^{n-1}(E \cap \Gamma_{x,r} \cap B(x,r)) \geq \theta r^{n-1}$. Then E is uniformly rectifiable too, by Theorem 8.

In the case of Mumford-Shah minimizers, for instance, Condition B is a little too restrictive to be used directly. Indeed, if E is the singular set of a Mumford-Shah minimizer (u, E), it may be that the complement of E is connected (and hence E does not satisfy Condition B), even though there are many balls in which E appears to separate space into at least two pieces. [Think about a half-line in the plane, for instance.] In such situations, the following result is more helpful.

Let $E \subset \mathbb{R}^n$ be Ahlfors-regular of dimension $n-1$, and let the (large positive) constants N, k, C_0 be given. We want to define a notion of (bad) balls $B(x,r)$, where we have reasonably good Poincaré estimates on the complement of E. Call $\mathcal{B}(C_0, k, N)$ the set of pairs (x, r) such that $x \in E$, $0 < r < \text{diam}(E)$, and we have

$$\left| \frac{1}{|B_1|} \int_{B_1} f(y) dy - \frac{1}{|B_2|} \int_{B_2} f(y) dy \right| \leq N r^{1-n} \int_{B(x,kr) \setminus E} |\nabla f(y)| dy \qquad (18)$$

whenever B_1 and B_2 are two balls contained in $B(x,r) \setminus E$ and with radii larger than $C_0^{-1}r$, and $f \in W^{1,1}(B(x,kr) \setminus E)$.

The point of this definition is that if $(x, r) \in \mathcal{B}(C_0, k, N)$, then E does not separate balls too well near $B(x,r)$. If E satisfies Condition B, then for each $B(x,r)$ we can find B_1 and B_2 as in the definition of Condition B, then take $f = 1$ on the component of $\mathbb{R}^n \setminus E$ that contains B_1 and $f = 0$ elsewhere, and it is easy to see that (18) is violated. Thus $\mathcal{B}(C_0, k, N)$ is empty (if C_0 is large enough) when E satisfies Condition B.

Theorem 19 [DaSe6]. *If $E \subset \mathbb{R}^n$ is Ahlfors-regular of dimension $n-1$, and if there is a constant $C_0 \geq 1$ such that all the $\mathcal{B}(C_0, k, N)$, $k \geq 1$ and $N \geq 1$ are Carleson sets, then E is uniformly rectifiable.*

See (13) for the definition of Carleson sets. The reader may find the statement a little strange. What is proved is that there are (large) constants k and N, that can be computed in terms of C_0, the regularity constant for E, and n, such that if $\mathcal{B}(C_0, k, N)$, is a Carleson sets, then E is uniformly rectifiable. But in practice it is just as easy to check that all $\mathcal{B}(C_0, k, N)$ are Carleson sets. See the last lines of this section too.

Remark 20. The proof of Theorem 19 also yields that E contains big pieces of Lipschitz graphs. See the definition above (18) and Theorem 3.42 on page 407 of [DaSe6].

In addition to the original proof of [DaSe6], there is a slightly more natural (but not much simpler) proof that we want to mention here. It is based on the following result.

Theorem 21 [DaSe8], [JKV]. *For each integer $n \geq 1$ and $C_1 \geq 1$ we can find $C_2 \geq 1$ such that the following holds. Let A be a closed subset of the annulus $B(0, 2) \setminus B(0, 1)$. Assume that A separates $B(0, 1)$ from $\mathbb{R}^n \setminus B(0, 2)$ (which means that these two sets lie in different components of $\mathbb{R}^n \setminus A$), and that $H^{n-1}(A) \leq C_1$. Then there is a Lipschitz graph Γ, with Lipschitz constant $\leq C_2$, such that $H^{n-1}(A \cap \Gamma) \geq C_2^{-1}$.*

In [JKV], this is proved with a very nice combinatorial argument (with checkerboards). In [DaSe8], the idea of the proof consists in trying to find compact sets $F \subset \overline{B}(0, 2) \setminus B(0, 1)$ that still separate $B(0, 1)$ from infinity, and that minimize

$$J(F) = H^{n-1}(F \cap A) + C_3 H^{n-1}(F \setminus A), \qquad (22)$$

where C_3 is a large constant. If C_3 is large enough, any minimizer for J will have to have a big intersection with A. Indeed $H^{n-1}(F) \geq H^{n-1}(\partial B(0, 1))$ by Remark 2.11, and because the radial projection from F to $\partial B(0, 1)$ is onto, by the separation condition; also $H^{n-1}(F \setminus A) \leq C_3^{-1} J(F) \leq C_3^{-1} H^{n-1}(A)$, by (22) and because A itself is a competitor, so that $H^{n-1}(F \cap A) \geq \frac{1}{2} H^{n-1}(\partial B(0, 1))$ if C_3 is large enough. The main point of the argument is that any minimizer of J (or even any quasiminimizer for H^{n-1}) under such topological constraint must be uniformly rectifiable, which allows us to find a big piece of Lipschitz graph in $F \cap A$. The delicate problem of existence of minimizers for J is not dealt with directly; instead we go to the BV setting and consider an analogous functional on sets of finite perimeter, for which existence results are easier.

Once we have Theorem 21, it is not hard to prove Theorem 19. First we observe that if $\mathcal{B}(C_0, k, N)$ is a Carleson set, then for each $x \in E$ and $r < \operatorname{diam} E$, we can find $y \in E \cap B(x, r/2)$ and t such that $\frac{r}{C} \leq t \leq \frac{r}{2k}$ and $(y, t) \notin \mathcal{B}(C_0, k, N)$. The argument is the same as in Corollary 23.38. This means that we can find B_1 and B_2 contained in $B(y, t)$, not too small, and $f \in W^{1,1}(B(y, kt) \setminus E)$ for which the analogue of (18) fails. After a small modification of f, we can reduce to the case where f is Lipschitz, and also constant on B_1 and B_2. Then call $m_i = |B_i|^{-1} \int_{B_i} f(z) dz$ the mean (and now constant) value of f on B_i. Thus

$|m_1 - m_2| \geq Nt^{1-n} \int_{B(y,kt)\backslash E} |\nabla f|$. Without loss of generality, we may assume that $m_1 < m_2$.

We use the co-area formula (see Section 28) to find a level set $\gamma_\xi = f^{-1}(\xi)$ such that $\xi \in (m_1, m_2)$ and

$$H^{n-1}(\gamma_\xi) \leq C(m_2 - m_1)^{-1} \int_{B(y,kt)\backslash E} |\nabla f| \leq CN^{-1}t^{n-1}. \qquad (23)$$

Notice that γ_ξ separates $\frac{1}{2}B_1$ from $\frac{1}{2}B_2$ in $B(y,kt)\backslash E$, because f is now continuous and equal to m_i on B_i. This means that $\widehat{E} = E \cup \gamma_\xi$ separates $B_1/2$ from $B_2/2$ in $B(y,kt)$.

We have an upper bound on $t^{1-n}H^{n-1}(\widehat{E} \cap B(y,kt))$ that does not depend on $N \geq 1$, so a minor variant of Theorem 21 (where the annulus $B(0,2)\backslash B(0,1)$ is replaced with a slightly different set) gives a Lipschitz graph Γ such that $H^{n-1}(\Gamma \cap \widehat{E} \cap B(y,kt)) \geq \theta t^{n-1}$. The constant $\theta > 0$ may depend on k and the regularity constant for E, but it does not depend on N. So if we choose N large enough, (23) says that $H^{n-1}(\gamma_\xi) < \theta/2$, and we still have that $H^{n-1}(\Gamma \cap E \cap B(y,kt)) \geq \theta t^{n-1}/2$. Note that $B(y,kt) \subset B(x,r)$ and $t \geq C^{-1}r$ by definition, so $E \cap B(x,r)$ contains a comparatively large piece of Lipschitz graph.

Thus E contains big pieces of Lipschitz graphs, hence is uniformly rectifiable (by our second definition with Lipschitz images). We even see that in the statement of Theorem 19 it is enough to require that $\mathcal{B}(C_0, k, N)$ be a Carleson set for some $C_0 \geq 1$ and some $k > 1$, provided that we take N large enough (depending on C_0 and k). This is a little improvement over the statement in [DaSe6].

74   Uniform rectifiability for quasiminimizers

Local quasiminimizers are also locally uniformly rectifiable in dimensions larger than 2. The precise statement is as follows.

Theorem 1 [DaSe6]. *For each dimension $n \geq 2$ and each constant $M \geq 1$, there is a constant $a > 0$ such that if $\Omega \subset \mathbb{R}^n$ is open, $(u, K) \in TRLQ(\Omega)$ with quasiminimality constants (r_0, M, a), and if $x \in K$ and $r \leq r_0$ are such that $B(x, 2r) \subset \Omega$, then there is an Ahlfors-regular, uniformly rectifiable set E of dimension $n - 1$ such that $K \cap B(x, r) \subset E$.*

See Standard Assumption 18.14 for the definition of $TRLQ(\Omega)$. The constants for the Ahlfors-regularity and uniform rectifiability of E depend only on n, M, and a. The proof also gives that E contains big pieces of Lipschitz graphs (see Remark 73.20 and the definition above (73.18)), and a tiny bit more precisely that K contains big pieces of Lipschitz graphs locally. This is not a very big deal anyway; it is known that uniformly rectifiable sets with big projections contain big pieces of Lipschitz graphs (see for instance [DaSe2]), and here the big projections would be easy to get (see the proof of (75.5) below).

As usual, the statement in [DaSe6] does not mention general quasiminimizers, but only minor modifications of the original argument are needed.

Let us give an idea of the proof. The point is to show that K satisfies the local version of the Weak Sobolev Components condition in Theorem 73.19. That is, define the bad sets $\mathcal{B}(C_0, k, N)$ as we did just above (73.18), but with E replaced with K. Then there is a $C_0 \geq 1$ such that every $\mathcal{B}(C_0, k, N)$, $k \geq 1$ and $N \geq 1$, is a local Carleson set. This last means that

$$\int_{y \in K \cap B(x,r)} \int_{0 < t < r} \mathbf{1}_{\mathcal{B}(C_0, k, N)}(y, t) \frac{dH^{n-1}(y) dt}{t} \leq C r^{n-1} \qquad (2)$$

when $x \in K$ and $0 < r < r_0$ are such that $B(x, 2r) \subset \Omega$. Here C may depend on n, M, a, but also on k and N.

Once we get (2), Theorem 1 follows from the proof of Theorem 73.19, or by applying Theorem 73.19 to $[K \cap B(x,r)] \cup \partial B(x,r)$. The argument is very similar to what we did for Theorem 27.1; we skip it and refer to [DaSe6].

So let us concentrate on the proof of (2). Let k and N be given, and let C_0 be large, to be chosen soon (independent of k and N). Pick any $p \in (\frac{2(n-1)}{n}, 2)$, and let ε be very small, to be chosen later (this time depending on k and N). We want to use the fact that

$$\{(x, r) \in K \times (0, r_0) \, ; \, B(x, 2r) \subset \Omega \ \text{and} \ \omega_p(x, r) > \varepsilon\} \ \text{is a local Carleson set,} \qquad (3)$$

which follows at once from Theorem 23.8, Remark 23.10, and Chebyshev's inequality. We shall check that if C_0 and ε are chosen correctly, then

$$\begin{aligned} &\omega_p(x, kr) > \varepsilon \ \text{for all} \ x \in K \ \text{and} \ 0 < r \leq r_0 \\ &\text{such that} \ B(x, 2kr) \subset \Omega \ \text{and} \ (x, r) \in \mathcal{B}(C_0, k, N). \end{aligned} \qquad (4)$$

It is easy to see that (2) follows as soon as we have (3) and (4). In particular, the reader should not worry about the set $\{(x, r) \in K \times (0, r_0) \, ; \, B(x, r) \subset \Omega \ \text{but} \ B(x, 2kr) \not\subset \Omega\}$, which is automatically a Carleson set. We leave the verifications to the reader and start directly with the proof of (4).

So let (x, r) be as in (4). In particular, $(x, r) \in \mathcal{B}(C_0, k, N)$, and we have the Poincaré type estimate (73.18) whenever B_1 and B_2 are two balls of radius $C_0^{-1} r$ contained in $B(x, r) \setminus K$. Since we can apply (73.18) with $f = u$, we get the existence of a constant $\alpha \in \mathbb{R}$ such that

$$\left| \alpha - \frac{1}{|B|} \int_B u \right| \leq N r^{1-n} \int_{B(x, kr) \setminus K} |\nabla u| \qquad (5)$$

for every ball B of radius $C_0^{-1} r$ that is contained in $B(x, r) \setminus K$. Incidentally,

$$\int_{B(x, kr) \setminus K} |\nabla u| \leq C (kr)^{n(1 - \frac{1}{p})} \left\{ \int_{B(x, kr) \setminus K} |\nabla u|^p \right\}^{1/p} \qquad (6)$$

by Hölder, so (5) says that

$$\left| \alpha - \frac{1}{|B|} \int_B u \right|^p \leq C(N,k)\, r^{p-n} \int_{B(x,kr)\setminus K} |\nabla u|^p. \tag{7}$$

Our plan is to construct a competitor for (u, K) in $B(x, r)$, and (7) will help us control its energy. In fact, if (4) fails for (x, r), the right-hand side of (7) is very small, and we shall see that this is enough to contradict the quasiminimality of (u, K).

Set $r_1 = C_0^{-1} r$ and let y_j, $j \in J$, be a maximal collection of points of $K \cap B(x, r)$ that lie at distances at least r_1 from each other. Observe that J has at most CC_0^{n-1} elements because the balls $B(y_j, r_1/2)$ are disjoint, have masses $H^{n-1}(K \cap B(y_j, r_1/2)) \geq C^{-1} C_0^{1-n} r^{n-1}$ by (72.1), while their total mass is at most Cr^{n-1}, by the easy part of (72.1). For $r/2 < \rho < r$, set $J(\rho) = \{j \in J;\ \overline{B}(y_j, 2r_1)$ meets $\partial B(x, \rho)\}$. Then

$$\int_{r/2}^r \sharp J(\rho)\, d\rho = \sum_{j \in J} \int_{r/2}^r \mathbf{1}_{\{|\rho - |y_j - x|| \leq 2r_1\}}(\rho)\, d\rho \leq \sum_{j \in J} 4C_0^{-1} r \leq CC_0^{n-2} r, \tag{8}$$

so we can choose $\rho \in (r/2, 2r/3)$ such that $\sharp J(\rho) \leq CC_0^{n-2}$. We choose ρ such that in addition

$$\int_{\partial B(x,\rho)\setminus K} |\nabla u|^p \leq Cr^{-1} \int_{B(x,r)\setminus K} |\nabla u|^p, \tag{9}$$

and $u \in W^{1,2}(\partial B(x, \rho) \setminus K)$, with a derivative that comes from the restriction of ∇u, and also such that the restriction of u to $\partial B(x, \rho) \setminus K$ coincides almost-everywhere with the radial limits of u from inside and outside $B(x, \rho)$. We know that this is possible, because our last constraints are satisfied for almost every ρ; see (the proof of) Corollary 14.28.

Set $Z = \partial B(x, \rho) \cap \left[\bigcup_{j \in J(\rho)} \overline{B}(y_j, 2r_1) \right]$. Then

$$H^{n-1}(Z) \leq C\, [\sharp J(\rho)]\, r_1^{n-1} \leq CC_0^{-1} r^{n-1}. \tag{10}$$

Call v the restriction of u to $\partial B(x, \rho) \setminus Z$. We want to show that v has a nice extension to $\overline{B}(x, \rho)$, and we start with an extension to $\partial B(x, \rho)$. First observe that

$$\operatorname{dist}(\partial B(x, \rho) \setminus Z, K) \geq r_1, \tag{11}$$

because otherwise we could find $z \in \partial B(x, \rho) \setminus Z$ and $y \in K$ such that $|y - z| \leq r_1$, and then $|y - y_j| \geq r_1$ for all j because $z \notin Z$. This is impossible, because $y \in K \cap B(x, r)$ and by maximality of our collection $\{y_j\}$.

Now let φ be a smooth function on $\partial B(x, \rho)$ such that $0 \leq \varphi \leq 1$ everywhere, $\varphi(z) = 1$ in a neighborhood of $\partial B(x, \rho) \setminus Z$, $\varphi(z) = 0$ when $\operatorname{dist}(z, \partial B(x, \rho) \setminus Z) > r_1/4$, and $|\nabla \varphi(z)| \leq Cr_1^{-1}$ everywhere. Our extension of v to $\partial B(x, \rho)$ is

$$w(z) = v(z)\varphi(z) + \alpha(1 - \varphi(z)), \tag{12}$$

where α is the same constant as in (5) and (7). Set

$$U = \{z \in \partial B(x,\rho)\,;\; \mathrm{dist}(z, \partial B(x,\rho) \setminus Z) < r_1/2\}. \tag{13}$$

Note that $\varphi(z) = 0$ on a neighborhood of $\partial B(x,\rho) \setminus U$, so $w(z) = \alpha$ and $\nabla w = 0$ there. Thus it is enough to estimate ∇w on U. By (11), v is defined on U, and

$$|\nabla w(z)| \le |\nabla v(z)| + |\nabla \varphi(z)||v(z) - \alpha| \le |\nabla v(z)| + Cr_1^{-1}|v(z) - \alpha| \tag{14}$$

for $z \in U$. We want to compute $\int_U |\nabla w|^p$, and for this we cover U by less than CC_0^{n-1} balls B_i of radius r_1 centered on $\partial B(x,\rho) \setminus Z$. [This is possible, by the definition (13).] For each B_i, call $m_i = |B_i|^{-1} \int_{B_i} u$ the mean value of u on B_i. Since $B_i \subset B(x,r) \setminus K$ by (11) and because $\rho \le 2r/3$, we can apply (7) and get that

$$|m_i - \alpha|^p \le C(N,k)\, r^{p-n} \int_{B(x,kr)\setminus K} |\nabla u|^p. \tag{15}$$

Also set $n_i = H^{n-1}(B_i \cap \partial B(x,\rho))^{-1} \int_{B_i \cap \partial B(x,\rho))} u(z)\, dH^{n-1}(z)$. Then

$$|n_i - m_i|^p \le C\left\{\frac{r_1}{|B_i|} \int_{B_i} |\nabla u|\right\}^p \le Cr_1^{p-n} \int_{B_i} |\nabla u|^p \tag{16}$$

by Lemma 13.20 and Hölder. Finally,

$$\int_{B_i \cap \partial B(x,\rho)} |v(z) - n_i|^p \le Cr_1^p \int_{B_i \cap \partial B(x,\rho)} |\nabla v(z)|^p \le CC_0^{-p} r^{p-1} \int_{B(x,r)\setminus K} |\nabla u|^p \tag{17}$$

by Exercise 12.29 (or the proof of Proposition 12.23), a smooth change of variable (to go from $B_i \cap \partial B(x,\rho)$ to a flat disk), and (9). Altogether,

$$\int_{B_i \cap \partial B(x,\rho)} |v(z) - \alpha|^p \le C(N,k)\, r^{p-1} \int_{B(x,kr)\setminus K} |\nabla u|^p \tag{18}$$

by (15), (16), and (17). We sum this over i and get that

$$\int_{\partial B(x,\rho)} |\nabla w|^p \le C \int_U |\nabla v|^p + Cr_1^{-p} \int_U |v(z) - \alpha|^p$$

$$\le Cr^{-1} \int_{B(x,r)\setminus K} |\nabla u|^p + C(N,k,C_0)\, r^{-1} \int_{B(x,kr)\setminus K} |\nabla u|^p, \tag{19}$$

by (14), (9), and (18). Now we may extend w to the ball $\overline{B}(x,\rho)$. Lemma 22.32 allows us to do this so that

$$\int_{B(x,\rho)} |\nabla w|^2 \le Cr^{n - \frac{2n}{p} + \frac{2}{p}} \left\{\int_{\partial B(x,\rho)} |\nabla w|^p\right\}^{2/p}$$

$$\le C(N,k,C_0)\, r^{n - \frac{2n}{p}} \left\{\int_{B(x,kr)\setminus K} |\nabla u|^p\right\}^{2/p} \tag{20}$$

$$= C(N,k,C_0)\, r^{n-1}\, \omega_p(x,kr),$$

by (19) and (23.5).

We are now ready to define a competitor for (u, K). Set $G = [K \setminus B(x, \rho)] \cup Z$, and extend our new function w to $\Omega \setminus G$ by taking $w(z) = u(z)$ out of $\overline{B}(x, \rho) \cup K$. As usual, G is closed in Ω and $w \in W^{1,2}_{loc}(\Omega \setminus G)$, by Corollary 14.28. Thus (w, G) is a competitor for (u, K) in $B(x, \rho)$. If (u, K) is a topological quasiminimizer, we still need to check that (w, G) is a topological competitor for (u, K) in $B(x, \rho)$ before we apply the quasiminimality condition (7.20). Let us postpone this for the moment and apply (7.20).

The analogue of $\widetilde{E} - E$ here is

$$\int_{B(x,\rho)} |\nabla w|^2 - \int_{B(x,\rho) \setminus K} |\nabla u|^2 \leq C(N, k, C_0) \, r^{n-1} \, \omega_p(x, kr), \tag{21}$$

by (20), so (7.20) says that

$$\begin{aligned} H^{n-1}(K \setminus G) &\leq M H^{n-1}(G \setminus K) + MC(N, k, C_0) \, r^{n-1} \, \omega_p(x, kr) + a\rho^{n-1} \\ &\leq MCC_0^{-1} r^{n-1} + MC(N, k, C_0) \, r^{n-1} \, \omega_p(x, kr) + ar^{n-1}, \end{aligned} \tag{22}$$

by (10). On the other hand, we saw in Section 72 that if the constant a is small enough (depending on n and M), then K is locally Ahlfors regular. Thus

$$H^{n-1}(K \setminus G) = H^{n-1}(K \cap B(x, \rho)) \geq C_1^{-1} r^{n-1}, \tag{23}$$

with a constant C_1 that depends only on n and M. We may now choose C_0 so large that $C_1^{-1} \geq 3MCC_0^{-1}$ (with C as in (22)), and also require that $a < \frac{1}{3}C_1^{-1}$. Then (22) and (23) imply that

$$\omega_p(x, kr) > \frac{1}{3}[M \, C(N, k, C_0) \, C_1]^{-1}. \tag{24}$$

That is, if we set $\varepsilon = \frac{1}{3}[M \, C(N, k, C_0) \, C_1]^{-1}$, then we get that $\omega_p(x, kr) > \varepsilon$, as required for (4).

We still need to check that (w, G) is a topological competitor for (u, K) in $B(x, \rho)$. If not, Lemma 20.10 says that $\partial B(x, \rho) \setminus G = \partial B(x, \rho) \setminus Z$ is *not* contained in a single component of $\Omega \setminus K$. Thus we can find points $z_1, z_2 \in \partial B(x, \rho) \setminus Z$ that lie in different components of $\Omega \setminus K$. Note that K still separates z_1 from z_2 in $B(x, kr)$ (which is contained in Ω by definition of (x, r) in (4)), and that the $B_i = B(z_i, r_1)$ do not meet K, by (11). Then B_1 and B_2 lie in different components of $B(x, kr) \setminus K$, and we can find a locally constant function f on $B(x, kr) \setminus K$ that is equal to 0 on B_1 and to 1 on B_2. But this contradicts (73.18) and our initial assumption that $(x, r) \in \mathcal{B}(C_0, k, N)$. So (w, G) was a topological competitor for (u, K), and this completes our proof of (4); (2) and Theorem 1 follow, as we said earlier. $\qquad\square$

Remark 25. When $\partial\Omega$ is itself Ahlfors-regular and uniformly rectifiable and (u, K) is as in Theorem 1, we also get that $K \cup \partial\Omega$ is uniformly rectifiable. The proof is the same as in Remark 32.5, but once again this is not much more than what Theorem 1 tells us.

Remark 26. Theorem 1 extends to quasiminimizers with linear distortion, and the proof is almost the same. [The main point was to get local Ahlfors-regularity; see Remark 72.86.]

75 C^1 regularity almost-everywhere for almost-minimizers

The main results of Part E generalize to higher dimensions, although with a slightly more delicate proof. In what follows, (u, K) is a reduced local almost-minimizer in a domain $\Omega \subset \mathbb{R}^n$ with a gauge function h such that

$$h(r) = C_0 r^\alpha \text{ for } r > 0 \tag{1}$$

for some constants $\alpha > 0$ and $C_0 \geq 0$. Probably the results stay true with topologically reduced or coral topological local almost-minimizers (i.e., when $(u, K) \in TRLAM(\Omega, h)$ as in Standard Assumption 41.1), but this is not stated in [AFP] and [Am], and I did not check. Similarly, (1) can probably be replaced by some Dini conditions, and even almost-minimizers with linear distortion should be allowed. We start with an analogue of Corollary 51.17.

Theorem 2 [AmPa],[AFP1],[Ri1]. *For each $p \in (\frac{2(n-1)}{n}, 2]$, we can find constants $\varepsilon > 0$ and $\eta > 0$ such that if (u, K) is a reduced local almost-minimizer in Ω, with gauge function h as in (1), $x \in K$, $B(x, r) \subset \Omega$, and*

$$\beta_K(x, r) + \omega_p(x, r) + h(r) \leq \varepsilon, \tag{3}$$

then we can find coordinates in $\mathbb{R}^n$ in which $K \cap B(x, \eta r) = \Gamma \cap B(x, \eta r)$, where Γ is the graph of some C^1 and 10^{-1}-Lipschitz function.

See (41.22) and (23.5) for the definitions of $\beta_K(x, r)$ and $\omega_p(x, r)$. The reader should not worry too much about the constant η; a closer look at the proof, or putting the statement back into the machine, would show that we can take $\eta = 1/2$.

The most important part of Theorem 2 is the case when $p = 2$, which was done in [AmPa] and [AFP1]. The statement may be slightly different, because the authors use the L^2-version $\beta_{K,2}(x, r)$ of $\beta_K(x, r)$ (see (73.10) for a definition), but the two are equivalent, since K is locally Ahlfors-regular and hence $C^{-1}\beta_{K,2}(x, r) \leq \beta_K(x, r) \leq C\beta_{K,2}(x, 2r)^{\frac{2}{n+1}}$ by (73.14) or the proof of (41.12). It is in this main part of the proof that I did not check that things extend to topological almost-minimizers (or almost-minimizers with linear distortion).

The proof of this part of Theorem 2 is quite elaborate; we shall only say a few words about it and refer to Chapter 8 of [AFP3] for a detailed and improved proof. The contribution of S. Rigot was to show that the hypotheses of Theorem 2 imply those of [AFP1] for a slightly smaller radius, and we shall discuss this in a moment. See [MaSo2] for a different way to conclude from [AFP1].

The proof of Ambrosio, Fusco, and Pallara uses the same sort of general method as in Part E. That is, they introduce quantities that measure how nice $K \cap B(x,r)$ is, then prove various inequalities between these quantities, and eventually show that they all decay to 0 when r tends to 0. An important new quantity (compared to Part E) is the tilt. If P is a vector hyperplane, set

$$T(x,r,P) = r^{1-n} \int_{K \cap B(x,r)} |\pi(\nu(y))|^2 dH^{n-1}(y), \tag{4}$$

where $\pi(\nu(y))$ denotes the orthogonal projection on P of a unit tangent vector to K at y. [Recall that K is rectifiable, so we can define $\nu(y)$ almost-everywhere, and the orientation does not matter. In [AFP1], almost-minimizers are given by a function $u \in SBV$, and $\nu(y)$ comes naturally from the definition of the jump part of ∇u.]

In Part E, one could say that the main point of the argument was that if K looks a lot like a line L in $B(x,r)$, and u looks like a constant on each component of $B(x,r) \setminus L$, then ∇u has the same sort of decay near x as the gradient of a harmonic function on $B(x,r) \setminus L$. Some care is needed, for instance to show that the holes in K stay sufficiently small to allow the comparison, but the main points are that the normalized energy $\omega_2(x,r)$ decays almost as it would do for a harmonic function in a disk, and that it controls all the other quantities that need to stay small.

In the argument of [AFP1,3], one also uses the fact that ∇u decays reasonably fast near x when K looks like a plane and u looks like two constants in $B(x,r) \setminus K$, but this is no longer enough. In Part E we used the (trivial) fact that the shortest path between two points a, b is the line segment $[a,b]$, and also that a connected set that contains a and b and whose length is barely larger than $|b-a|$ must stay quite close to $[a,b]$. The analogue in higher dimensions is much less trivial, and some control is needed on sets like K that almost minimize surface measure under some boundary constraints (like the fact that the intersection of K with $\partial B(x,r)$ is given, and the requirement that K almost separates $B(x,r)$ into at least two large pieces, because the normalized jump is not too small).

Ambrosio, Fusco, and Pallara use the tilt to control the behavior of K, and show that $\omega_2(x,r)$ and $T(x,r) = \inf_P T(x,r,P)$ both decay like powers. As far as the author can tell, part of the point here is that for minimal surfaces, for instance, an integration by parts shows that the normalized tilt would decay sufficiently fast.

It would be interesting to find out whether a modification of the argument in [AFP1,3] still works in $\mathbb{R}^3$ when $\omega_2(x,r)$ is very small and $K \cap B(x,r)$ lies very close to the product of a propeller by a line (instead of a plane), and yields that $K \cap B(x,\eta r)$ is a C^1 version of the product of a propeller by a line. We don't know either about the case when $K \cap B(x,r)$ is very close to the minimal set composed of six faces bounded by four half-lines starting from x and making maximal angles with each other.

The reduction of Theorem 2 to the case when $p = 2$ is easier. If (u, K) and (x, r) are as in the theorem, we can find $r_1 \in (r/2, r)$ such that $\omega_p^*(x, r) \leq C\varepsilon$ (see (42.5) and Convention 42.6 for the definition, and just apply Chebyshev to get r_1). Call $J(x, r_1)$ the normalized jump of u and $J^*(x, r_1)$ its version on the sphere $\partial B(x, r_1)$, as in (42.4) and (42.7), except that we shall use the convenient convention that $J^*(x, r_1) = J(x, r_1) = +\infty$ when (u, K) is a topological almost-minimizer and the two main components of $\partial B(x, r_1) \setminus K$ lie in different components of $\Omega \setminus K$. Proposition 42.10 says that $J^*(x, r_1) \geq \tau_0$, and then $J(x, r_1) \geq \tau_0/2$, by (42.8) and if ε is small enough. Next we claim that

$$H^{n-1}(K \cap B(x, r_1)) \geq \alpha r_1^{n-1} - C\varepsilon^{1/2} r_1^{n-1}, \tag{5}$$

where α denotes the Lebesgue measure of the unit ball in $\mathbb{R}^{n-1}$. The idea of the proof is the same as in Section 44, except that since we allow our estimate here to depend on $\beta_K(x, r)$ we can be a little more rapid, and in particular we do not need to use the co-area theorem.

Let us prove (5) anyway. By (3), we can find an affine hyperplane P through x such that $\text{dist}(y, P) \leq \varepsilon r$ for $y \in K \cap B(x, r)$. Let us assume, without loss of generality, that P is horizontal. Set $P_1 = P \cap B(x, r_1 - \varepsilon r)$. For $\xi \in P_1$, denote by L_ξ the intersection of $\overline{B}(x, r_1)$ with the (vertical) line through ξ and perpendicular to P, and call $a_-(\xi)$ and $a_+(\xi)$ the two extremities of L_ξ, with $a_+(\xi)$ above P. Also call $\Sigma_\pm$ the set of points $a_\pm(\xi)$, $\xi \in P_1$. Thus $\Sigma_\pm$ is a nice cap, a little smaller than a hemisphere, and $\Sigma_\pm \subset \partial B(x, r_1) \setminus K$ by definition of P and P_1. Also, the other cap $\Sigma_\pm(x, r_1)$ that was used to define $J^*(x, r_1)$ is contained in $\Sigma_\pm$. Since $u \in W^{1,p}(\Sigma_\pm)$ by choice of r_1, we can apply Poincaré's inequality (Proposition 12.11 and a bilipschitz change of variable) and we get that

$$\int_{\Sigma_\pm} |u(y) - m_{\Sigma_\pm(x,r_1)} u| \, dH^{n-1}(y) \leq Cr \int_{\Sigma_\pm} |\nabla u| \, dH^{n-1}$$
$$\leq Cr^{n-\frac{1}{2}} \omega_p^*(x, r_1)^{1/2} \leq C\varepsilon^{1/2} r^{n-\frac{1}{2}}, \tag{6}$$

by (42.5) and our definition of r_1. Since the image of the measure dH^{n-1} on $\Sigma_\pm$ by the orthogonal projection on P is larger than dH^{n-1}, this yields

$$\int_{P_1} |u(a_\pm(\xi)) - m_{\Sigma_\pm(x,r_1)} u| \, dH^{n-1}(\xi) \leq C\varepsilon^{1/2} r^{n-\frac{1}{2}}. \tag{7}$$

Recall that $|m_{\Sigma_+(x,r_1)} u - m_{\Sigma_-(x,r_1)} u| = r_1^{1/2} J^*(x, r_1) \geq \tau_0 r_1^{1/2}$, by (42.7). If we set

$$E = \left\{ \xi \in P_1 \, ; \, |u(a_+(\xi)) - u(a_-(\xi))| \geq \frac{1}{2} \tau_0 r_1^{1/2} \right\}, \tag{8}$$

then $|u(a_+(\xi)) - m_{\Sigma_+(x,r_1)} u| \geq \tau_0 r_1^{1/2}/4$ or $|u(a_-(\xi)) - m_{\Sigma_-(x,r_1)} u| \geq \tau_0 r_1^{1/2}/4$ for $\xi \in P_1 \setminus E$, so

$$H^{n-1}(P_1 \setminus E) \leq C\tau_0^{-1} r^{n-1} \varepsilon^{1/2}, \tag{9}$$

by (7) and Chebyshev. Next let E_1 be the set of points $\xi \in E$ such that L_ξ does not meet K. For almost every $\xi \in E_1$, u is absolutely continuous on L_ξ and

$$\frac{1}{2}\tau_0 r_1^{1/2} \leq |u(a_+(\xi)) - u(a_-(\xi))| \leq \int_{L_\xi} |\nabla u|, \tag{10}$$

where the first inequality holds by definition of E. So

$$H^{n-1}(E_1) \leq \int_{E_1} 2\tau_0^{-1} r_1^{-1/2} \int_{L_\xi} |\nabla u| \leq 2\tau_0^{-1} r_1^{-1/2} \int_{B(x,r_1)\setminus K} |\nabla u| \tag{11}$$
$$\leq C\tau_0^{-1} r_1^{n-1} \omega_p(x, r_1)^{1/2} \leq C\tau_0^{-1} \varepsilon^{1/2} r_1^{n-1}$$

by Fubini, the definition (23.5) of $\omega_p(x, r_1)$, and our assumption (3). Finally,

$$\begin{aligned} H^{n-1}(K \cap B(x, r_1)) &\geq H^{n-1}(\pi(K \cap B(x, r_1))) \geq H^{n-1}(E \setminus E_1) \\ &\geq H^{n-1}(P_1) - H^{n-1}(P_1 \setminus E) - H^{n-1}(E_1) \\ &\geq \alpha r_1^{n-1} - C\tau_0^{-1} \varepsilon^{1/2} r_1^{n-1}, \end{aligned} \tag{12}$$

where π denotes the orthogonal projection on P, and by (9) and (11). This proves (5).

We may now use Proposition 45.1, which says that

$$H^{n-1}(K \cap \overline{B}(x, r_1)) + \int_{B(x,r_1)\setminus K} |\nabla u|^2$$
$$\leq \alpha r_1^{n-1} + C_p r_1^{n-1} \omega_p^*(x, r_1) + C r_1^{n-1} \beta_K(x, r_1) + r_1^{n-1} h(r) \leq \alpha r_1^{n-1} + C r_1^{n-1} \varepsilon, \tag{13}$$

by (3). When we compare this with (5), we get that

$$\omega_2(x, r_1) = r_1^{1-n} \int_{B(x,r_1)\setminus K} |\nabla u|^2 \leq C\varepsilon^{1/2}. \tag{14}$$

Thus the pair (x, r_1) satisfies the hypotheses of Ambrosio, Fusco, and Pallara's theorem (with ε replaced with $C\varepsilon^{1/2}$), and we have completed our reduction. $\square$

Many of the little improvements of the C^1 regularity theorem that we explained in dimension 2 still work in higher dimensions, with almost the same proof. For instance, we may replace our assumption that $\omega_p(x, r)$ is very small in Theorem 2 with the requirement that the normalized jump $J(x, r)$ be very large, as in Corollary 52.25. Or with the assumption that not only $\beta_K(x, r)$ is very small, so that all the points of $K \cap B(x, r)$ lie very close to some hyperplane P, but also that all the points of $P \cap B(x, r)$ lie very close to K. The proof is then almost the same as in the special case of Proposition 70.2 where H is a line.

There is also an analogue of Theorem 2, where (3) is replaced with the assumption that $h(r) + \omega_p(x, r) + r^{1-n}|H^{n-1}(K \cap B(x, r)) - H^{n-1}(P \cap B(x, r))| \leq \varepsilon$, where P is any hyperplane through x. See [AFH].

Let us only state analogues of Theorems 51.4 and 51.20 here.

We shall say that $B(x, r)$ is a C^1 ball for K when $K \cap B(x, r) = \Gamma \cap B(x, r)$ for some image under a rotation of a C^1 and 10^{-1}-Lipschitz graph (as in the statement of Theorem 2). And a C^1 point of K will be a point $x \in K$ such that $B(x, r)$ is a C^1 ball for K for some $r > 0$.

Corollary 15 [Ri1]. *For each dimension n and each gauge function h such that (1) holds, we can find constants $r_0 > 0$ and $C_0 > 0$ such that if (u, K) is a reduced local almost-minimizer in Ω, with gauge function h, $0 < r \leq r_0$, and $B(x, r) \subset \Omega$, we can find $y \in K \cap B(x, r/2)$ and $t \in (C_0^{-1} r, r/2)$ such that $B(y, t)$ is a C^1 ball for K.*

Indeed, it is enough to find $y \in K \cap B(x, r/2)$ and $t \in (C_0^{-1} \eta^{-1} r, r/2)$ that satisfy the hypotheses of Theorem 2, and then the pair $(y, \eta t)$ will do. If r_0 is small enough, we shall get that $h(t) \leq \varepsilon/2$ automatically, so it is enough to have that $\beta_K(x, r) + \omega_p(x, r) \leq \varepsilon/2$.

We know from Theorem 23.8, the trivial bound on energy, and the local Ahlfors-regularity result in Section 72, that $\omega_p(y, t) \, dH^{n-1}(y) \frac{dt}{t}$ is a local Carleson measure on $K \times (0, r_0)$. In addition, Theorem 74.1 says that K is locally uniformly rectifiable. Thus we can apply Theorem 73.11 to the uniformly rectifiable sets E given by Theorem 74.1; we get that $\beta_{K,2}(y, t)^2 \, dH^{n-1}(y) \frac{dt}{t}$, for instance, is also a local Carleson measure. Here $\beta_{K,2}(y, t)$ is still the L^2-version of $\beta_K(y, t)$, as in (73.10), and we just used the fact that $\beta_{K,2}(y, t) \leq \beta_{E,2}(y, t)$ when $K \subset E$. We still have that $\beta_K(x, r) \leq C\beta_{K,2}(x, 2r)^{\frac{2}{n+1}}$ by local Ahlfors-regularity of K and (73.14) or the proof of (41.12), so $\beta_K(y, t)^{n+1} \, dH^{n-1}(y) \frac{dt}{t}$ is also a local Carleson measure.

Thus $[\omega_p(y, t) + \beta_K(y, t)^{n+1}] \, dH^{n-1}(y) \frac{dt}{t}$ is a local Carleson measure, and the proof of Corollary 23.38 gives the desired pair (y, t).   □

Corollary 16 [Ri1]. *If (u, K) is a reduced local almost-minimizer in Ω, with a gauge function h such that (1) holds, then the set of points of K that are not C^1 points has vanishing H^{n-1}-measure, and is even of Hausdorff dimension $\leq d$, for some $d < n - 1$ that does not depend on h.*

The proof is the same as for Theorem 51.20.   □

Remark 17. The argument above also gives the uniform concentration property for K when $(u, K) \in TRLAM(\Omega, h)$ and h satisfies (1), as a consequence of the local uniform rectifiability of K. [See Remark 25.5 for the definition.]

To check this, fix a small $\varepsilon > 0$.

We just saw that $[\omega_p(y, t) + \beta_K(y, t)^{n+1}] \, dH^{n-1}(y) \frac{dt}{t}$ is a local Carleson measure. Thus if $x \in K$, r is small enough, and in addition $B(x, r) \subset \Omega$, we can find $y \in K \cap B(x, r/2)$ and $t \in [C(\varepsilon)^{-1} r, r/2]$, such that $\omega_p(y, t) + \beta_K(y, t)^{n+1} \leq \varepsilon$. By the same argument as in the reduction of Theorem 2 to the case when $p = 2$, we can find $r_1 \in (t/2, t)$ such that (5) holds. Thus, the pair (y, r_1) satisfies (25.6) with ε replaced with $C\varepsilon^{1/2}$, which is enough.

Remark 18. Although it is quite probable that the arguments in [AmPa] and [AFP1] still work with almost-minimizers, and even with linear distortion, I did not check this. At least, the reduction argument given in this section (i.e., the proof of (5) and (14)) extends, with the same sort of modifications as in Remarks 45.14 and 52.28. In particular, we may even replace (5) with

$$H^{n-1}(A(K \cap B(x, r_1))) \geq H^{n-1}(A(P(x, r) \cap B(x, r_1))) - C\varepsilon^{1/2}r_1^{n-1}, \qquad (19)$$

where $P(x, r)$ is a hyperplane through x such that $\operatorname{dist}(y, P(x, r)) \leq r\varepsilon$ for $y \in K \cap B(x, r)$, and A is any bijective affine mapping. Naturally the constant C depends on A.

The argument of Remark 17 also extends to quasiminimizers with linear distortion, by Remark 74.26 and because all we use is the local uniform rectifiability of K and the proof of (5) or (19). This gives the uniform concentration property for K when (u, K) is a quasiminimizer with linear distortion (and a is small enough, depending on M and L). Because we even get (19) it also gives the uniform concentration property for every $A(K)$, as announced in Remark 25.28.

76   What is a Mumford-Shah conjecture in dimension 3?

Recall from Definition 54.4 that a global minimizer in $\mathbb{R}^n$ is a topologically reduced (or coral) local topological almost-minimizer in $\Omega = \mathbb{R}^n$, with gauge function $h = 0$. But the definition simplifies. It is very unlikely that we shall ever have a complete description of all global minimizers in $\mathbb{R}^n$, because it would at least contain a description of all minimal sets of codimension 1 in $\mathbb{R}^n$ (i.e., global minimizers for which u is locally constant). In this section we restrict to $n = 3$, and try to show that the following question is interesting (and almost untouched): what should be the list of global minimizers in $\mathbb{R}^n$?

a. Minimal sets

We start with the description of the global minimizers (u, K) for which u is constant on each component of $\mathbb{R}^n \setminus K$. In this case, Definition 54.4 simplifies a little and says that K is a minimal set in the following sense:

$$H^{n-1}(K \cap B) \leq H^{n-1}(G \cap B) \qquad (1)$$

when G is another closed set in $\mathbb{R}^n$, B is a ball, $G \setminus B = K \setminus B$, and in addition G separates x from y in $\mathbb{R}^n$ whenever $x, y \in \mathbb{R}^n \setminus [B \cup K]$ lie in different components of $\mathbb{R}^n \setminus K$.

The author believes that, thanks to Jean Taylor [Ta], we have a complete description of the minimal sets in $\mathbb{R}^3$. The two simplest examples are the empty set, and a plane. A third example is the product $K = P \times L$, where P is a propeller in a plane and L is a line perpendicular to that plane. That is, K is the union of three half-planes with a line parallel to L as their common boundary, and that make $120°$ angles with each other. The verification of minimality in these cases is easy, and goes as in dimension 2.

There is a fourth minimal set in $\mathbb{R}^3$, which can be obtained as follows. First pick four half-lines L_j that start from the origin and make equal angles with each other. Thus the points of $L_j \cap \partial B(0,1)$ are the four vertices of a regular tetrahedron centered at 0. Then K is the union of the six angular sectors bounded by the pairs of lines L_j. The fact that K is a minimal set is more complicated to establish than for the previous examples; the best proof uses simple integrations by parts [Morg3].

The four minimal sets above are easy to spot as tangent objects in soap films; the last two will be seen, for instance, if you dip a cube (that is, a wire sculpture with 12 edges) into the correct liquid. Apparently there is no other minimal set of codimension 1 than the four above. To the author's knowledge, this is not written explicitly, but this should follow from Jean Taylor's results in [Ta]. More specifically, Jean Taylor proved that all minimal cones of codimension 1 are as above, and also that minimal sets are locally C^1-equivalent to one of the four examples above. [Part of the point is that there is a monotonicity formula, so the tangent objects are cones.] Then a small argument is needed to exclude some behaviors near infinity and conclude, which the author plans to write down somewhere soon. There is also an argument in [Morg2], which the author does not trust entirely.

b. Cracktip times a line

Next we focus on the global minimizers for which u is not locally constant. The only one we know so far is the product of the cracktip with a line. That is, let K be the half-plane $K = \{(x,0,z)\,;\, x \leq 0 \text{ and } z \in \mathbb{R}\}$, and then set $u(x,y,z) = u_0(x,y)$, where u_0 is the function associated with the cracktip minimizer (54.17) in the plane. That is, $u_0(r\cos\theta, r\sin\theta) = \sqrt{\frac{2}{\pi}}\, r^{1/2} \sin\left(\frac{\theta}{2}\right)$ for $r > 0$ and $|\theta| < \pi$. Of course we could have chosen a different sign and added a constant.

We claim that (u, K) is a global minimizer in $\mathbb{R}^3$. Indeed, let (v, G) be a competitor, and assume that (v, G) coincides with (u, K) out of $B(0, R)$. For each choice of $z \in \mathbb{R}$, set $G_z = \{(x,y) \in \mathbb{R}^2 \,;\, (x,y,z) \in G\}$ and define v_z on $\mathbb{R}^2 \setminus G_z$ by $v_z(x,y) = v(x,y,z)$. The reader may be worried because we only know that $v \in W^{1,2}_{\text{loc}}(\mathbb{R}^3 \setminus G)$, but this is not a serious issue. It is enough to consider almost every $z \in \mathbb{R}$ in the argument below, and also we could have replaced v with a harmonic function on $B(0, R) \setminus G$ with the same values outside $B(0, 2R)$ and that minimizes the energy.

Now (v_z, G_z) is a competitor for the cracktip (u_0, K_0) in the plane, where we set $K_0 = \{(x,0)\,;\, x \leq 0\}$. Note that we do not need to worry about the topological condition, because $\mathbb{R}^2 \setminus K_0$ is connected. Since (u_0, K_0) is a global minimizer in

$\mathbb{R}^2$ (see Section 62),

$$H^1(K_0 \cap B(0,R)) + \int_{B(0,R) \setminus K_0} |\nabla u_0|^2 \leq H^1(G_z \cap B(0,R)) + \int_{B(0,R) \setminus G_z} |\nabla v_z|^2. \quad (2)$$

But $|\nabla v_z(x,y)| \leq |\nabla v(x,y,z)|$ almost everywhere, so

$$\int_{[-R,R]} \int_{B(0,R) \setminus G_z} |\nabla v_z(x,y)|^2 dx dy \, dz \leq \int_{[B(0,R) \times [-R,R]] \setminus G} |\nabla v|^2 dx dy dz. \quad (3)$$

Similarly,

$$\int_{[-R,R]} H^1(G_z \cap B(0,R)) \, dz \leq H^2\big(G \cap [B(0,R) \times [-R,R]]\big). \quad (4)$$

Actually, there is a small problem with this, but assume that (4) holds for the moment. Then

$$H^2(K \cap [B(0,R) \times [-R,R]]) + \int_{[B(0,R) \times [-R,R]] \setminus K} |\nabla u|^2 dx dy dz$$

$$= R H^1(K_0 \cap B(0,R)) + R \int_{B(0,R) \setminus K_0} |\nabla u_0|^2$$

$$\leq H^2(G \cap [B(0,R) \times [-R,R]]) + \int_{[B(0,R) \times [-R,R]] \setminus G} |\nabla v|^2 dx dy dz \quad (5)$$

by (2), (3), and (4), and hence

$$H^2(K \cap \overline{B}(0,R)) + \int_{B(0,R) \setminus K} |\nabla u|^2 dx dy dz$$

$$\leq H^2(G \cap \overline{B}(0,R)) + \int_{B(0,R) \setminus G} |\nabla v|^2 dx dy dz \quad (6)$$

because (v,G) coincides with (u,K) out of $B(0,R)$. Thus (u,K) is a global minimizer on $\mathbb{R}^3$, as announced.

Unfortunately, (4) is not as straightforward as it looks. It is true if G is rectifiable; see for instance Theorem 2.93 in [AFP3] or Theorem 3.2.22 in [Fe]. In fact, this is then the easy part of the co-area formula, and it is not too hard to check directly: we can prove it first for measurable subsets of C^1 surfaces (essentially by Fubini), and then take countable disjoint unions to go to the general case (see Exercise 14); we shall also see another proof (based on almost the same principle) in a moment.

The author does not know whether (4) remains true for nonrectifiable sets G. The converse inequality is known to fail in some cases: there are compact sets G in the plane, with $H^1(G) < +\infty$, such that $H^2(G \times [0,1]) > H^1(G)$. See [BM]. The difficulty comes from the fact that Hausdorff measure is defined in terms of covering by sets, and summing powers of their diameters. If we replace H^2 by its variant $\mathcal{L}^2$ where we only consider coverings by balls, the analogue of (4) is easy to prove; let us see how. Given $\varepsilon > 0$, find a covering of $G' = G \cap [B(0,R) \times [-R,R]]$ by balls $B(x_j, r_j)$, with $r_j \le \varepsilon$ and $\pi \sum r_j^2 \le \mathcal{L}^2(G') + \varepsilon$ [compare with Definition 2.6]. For each $r \in [-R,R]$, the traces of the $B(x_j, r_j)$ give a covering of $G'_r = G' \cap \{z = r\}$ by sets of diameter $\le 2\varepsilon$. Then

$$\int_{[-R,R]} H^1_{2\varepsilon}(G_z \cap B(0,R))\, dz \le \int_{[-R,R]} \sum_j \operatorname{diam}\left[B(x_j, r_j) \cap \{z = r\}\right] dz$$

$$= \sum_j \int_{[-R,R]} \operatorname{diam}\left[B(x_j, r_j) \cap \{z = r\}\right] dz. \tag{7}$$

Since for each j,

$$\int_{-\infty}^{+\infty} \operatorname{diam}\left[B(x_j, r_j) \cap \{z = r\}\right] dz = \int_{-r_j}^{r_j} 2(r_j^2 - t^2)^{1/2} dt = 4r_j^2 \int_0^1 (1 - t^2)^{1/2} dt$$

$$= 4r_j^2 \int_0^{\pi/2} \cos^2 u \, du = \pi r_j^2 \tag{8}$$

(where we set $t = \sin u$, $dt = \cos u \, du$), we get that

$$\int_{[-R,R]} H^1_{2\varepsilon}(G_z \cap B(0,R))\, dz \le \pi \sum_j r_j^2 \le \mathcal{L}^2(G') + \varepsilon. \tag{9}$$

We now let ε tend to 0 in (9), and Fatou's Lemma yields the analogue of (4). Next observe that

$$\mathcal{L}^d(E) = H^d(E) \text{ when } E \text{ is rectifiable (of dimension } d). \tag{10}$$

This is a consequence of the isodiametric inequality; see Exercise 35.19 or directly [Fe], 2.10.30 and 3.2.26. Then (10) shows that (4) holds for rectifiable sets, as promised.

We could have defined the Mumford-Shah functional and all our other notions in terms of $\mathcal{L}^2$ instead of H^2. This would not have changed much of the theory, by (10) and because the reduced quasiminimizers are rectifiable; in the analogous theory with SBV functions, things would have been the same too, because the singular sets are rectifiable by definition. Thus the proof of minimality of our pair (u, K) above in the SBV setting is easier and only uses the argument above.

Of course all this looks like a bad excuse. If we really want to show that (u, K) is a global minimizer, we can. The point is that we do not need to consider all the competitors (v, G) as above, but only those for which G is rectifiable (and v is harmonic on $B(0, R) \setminus G$, for that matter). Indeed, we can always minimize $H^2(G \cap \overline{B}(0,R)) + \int_{B(0,R) \setminus G} |\nabla v|^2$ under the boundary conditions that $(v, G) =$

(u, K) out of $\overline{B}(0, R)$; the best argument for the existence of a minimizer uses the SBV setting, but we could also proceed as suggested in Section 36. Then it is enough to check (5) for the minimizer (v, G), and the argument above (with (7)) works because G is rectifiable.

It would be good to check that (u, K) is the only global minimizer for which K is a half-plane (modulo the usual symmetries). At least, $\pm(2/\pi)^{1/2}$ is the only possible constant in the definition of u_0 above, essentially for the same reason as in the plane, but a priori there could be minimizers for which u is not invariant by translations.

Remark 11. What we did for Cracktip generalizes to other global minimizers: if (u, K) is a global minimizer in $\mathbb{R}^n$, the product of (u, K) by a line is a global minimizer in $\mathbb{R}^{n+1}$, at least if we use $\mathcal{L}^{n-1}$ instead of H^{n-1} or if $\mathbb{R}^n \backslash K$ is connected. See Exercise 16. Conversely, all translation invariant global minimizers are of that type; see Exercise 17.

c. Why there should be at least another minimizer

First let us say that we have no idea about how many other global minimizers there are in $\mathbb{R}^3$. We do not even know that $\mathbb{R}^3 \backslash K$ is connected, or even that it has no thin connected component, when (u, K) is a global minimizer for which u is not locally constant (compare with Sections 56 and 66).

Next we wish to convince the reader that there is probably at least one global minimizer that we did not mention yet. Consider the usual Mumford-Shah functional J in (1.4), in the domain $\Omega = B(0, 1) \times (-N, N)$, with $a = b = c = 1$. We take N relatively large ($N = 10$ should be enough) to make sure that the image g will have some influence. Take a function g of the form $g(x, y, z) = \varphi(z) g_0(x, y)$, where φ is a smooth function such that $0 \le \varphi \le 1$, $\varphi(z) = 1$ for $z \le -1$, and $\varphi(z) = 0$ for $z \ge 1$. To define g_0, pick a propeller P in $\mathbb{R}^2$ centered at the origin, denote by D_1, D_2, and D_3 the three connected components of $B(0, 1) \backslash P$, and set $g_0(x, y) = 10j - 20$ on D_j. The constant 10 was chosen a little bit at random, but it should be larger than $\sqrt{2}$, because then (g_0, P) is the only minimizer of the 2-dimensional analogue of J where $g = g_0$, by [ABD].

We know that the functional J has at least one minimizer (u, K), and let us try to guess what it looks like. If we restricted to the smaller domain $B(0, 1) \times (-N, -1)$, (u, K) would coincide with $(g, P \times (-N, -1))$. This comes from the minimizing property of (g_0, P) in $B(0, 1)$, and the argument of reduction to this is the same as for the cracktip minimizer just above. It is reasonable to think that $K \cap [B(0, 1) \times (-N, -3/2)]$, say, looks a lot like $P \times (-N, -3/2)$, with three almost vertical walls that meet along a curve γ with 120° angles. If there are no other global minimizers than the five mentioned above, all the blow-up limits of (u, K) are among the five. If in addition we have a perturbation theorem like Theorem 53.4, but associated to the product of a propeller by a line, we should be able to follow the curve γ, where the blow-up limits of K are products of propellers by lines, up until some other type of blow-up limit shows up. One possibility is that this does not happen, and γ goes all the way to $\partial\Omega$. For instance, the three walls could go all the way up to the top of Ω. This does not seem very likely, because since u should look like g, and g is constant near the top, it does not really look useful to build a wall that high.

It would also seem a little strange if γ turned and went to the side part of $\partial\Omega$, mostly because it seems that u would have to stay far from g for too long, and it is not clear that it would make us save so much surface. Even though we expect γ to turn a little because g is not entirely symmetric.

The other option would be that γ has an end inside Ω, and that at this end (call it x_0) the blow-up limit of K is our fourth minimal set (the one with four edges and six faces). Assuming we also have a perturbation theorem for this minimal set, we would be able to follow the three new curves like γ that start from x_0, and this until they reach $\partial\Omega$ or some other junction like x_0. In this case also, it looks like K separates Ω into a few different regions, and this does not seem to be the optimal way to proceed. This is a place where numerical experiments could help, since the main decision seems to be at a large scale, between enclosing four regions into six walls, and keeping only three walls and letting u change slowly with altitude.

What seems the most likely to the author is that the three almost vertical walls will stop somewhere between the bottom of Ω and its top. At the top boundary of the walls, there should be three edges where the blow-up limits of K are half-planes (as in Subsection b), and one central point where the curves meet, and where another type of global minimizer shows up. Of course the reader can probably imagine lots of other ways to continue the walls, but hopefully he or she will find the description above to be the simplest and most likely.

It would be amusing to be able to prove that there is an additional global minimizer with some abstract argument like the one above (as opposed to coming up with some concrete pair (u, K) and showing that it is a minimizer). The author tried to do something like this in dimension 2, to show that if the cracktip pair in (54.17) is not a global minimizer, then there is an even uglier exotic global minimizer. With no success.

d. What should an extra global minimizer look like?

We shall continue this discussion under the assumption that the list of five global minimizers given above is not complete, and the next goal will be to try to guess what is the simplest form that an additional global minimizer (u, K) would take.

Let us first try to work with K, with the hope that u will be reasonably easy to find once we have K. If we believe in the argument in Subsection c, we should look for a set K that looks like $P \times \mathbb{R}$ (where P is a propeller centered at the origin in $\mathbb{R}^2$) in the region where $z \ll 0$, and like the empty set in the region where $z \gg 0$.

The set of global minimizers is stable under the usual dilations [i.e., replacing K with λK and u with $\lambda^{1/2}u(x/\lambda)$], and although this does not mean that there should exist minimizers that are invariant under dilations, let us decide that we are looking for an invariant minimizer anyway. Thus, K is a cone and $u(x) = |x|^{1/2}u_1(x/|x|)$ for some function u_1 on $\partial B(0,1) \setminus K$. Of course, nothing really forces this, and the reader may find it more pleasant to claim that there is a minimizer that is only invariant under dilations by $\lambda_k = \left(\sqrt{7\pi}\right)^k$, $k \in \mathbb{Z}$.

Let us go one more step into speculation, and decide that K should be contained in $P \times \mathbb{R}$. Thus $K_1 = K \cap \partial B(0,1)$ is contained in $\partial B(0,1) \cap [P \times \mathbb{R}]$, which is the union of three arcs of great circles. Let us even decide that K_1 is composed of three arcs γ_j of these great circles, that start at the bottom. That is, the γ_j, $1 \le j \le 3$, are vertical arcs of circles that start at the south pole of $\partial B(0,1)$ and make $120°$ angles there. Call l_j, $1 \le j \le 3$, the length of γ_j.

Before we continue our discussion of K, we should say a few words about u. We know that u is harmonic on $\mathbb{R}^3 \setminus K$ and satisfies the Neumann condition $\frac{\partial u}{\partial n} = 0$ on K, which is just a way of saying that it minimizes the energy locally. We also assumed that u is homogeneous of degree $1/2$, so we can rewrite $\Delta u = 0$ in terms of $u_1 = u_{|\partial B(0,1)}$. After some computations (see Exercise 18), we get that

$$\Delta_1 u_1 + \frac{3}{4}u_1 = 0, \tag{12}$$

where Δ_1 is the Laplace operator on $\partial B(0,1)$. [To compute $\Delta_1 u_1$ at a point $\xi \in \partial B(0,1)$, we may choose two orthogonal great circles through ξ, and sum the second derivatives of u_1 along these two circles.] In other words, u_1 is an eigenvector of $-\Delta_1$ on $\partial B(0,1) \setminus K_1$, with the Neumann boundary condition, corresponding to the eigenvalue $3/4$. This does not necessarily determine u, but we only have a vector space of finite dimension left (at most).

The first thought that comes to mind is that we should try to take $l_1 = l_2 = l_3 = \pi/2$, or equivalently $K = P \times (-\infty, 0]$, but it is not clear that this conjecture is more serious than lazy. Here are a few potential problems with this.

The main one is that apparently $3/4$ is not an eigenvalue for $-\Delta_1$ on $\partial B(0,1) \setminus K_1$ for this choice of K_1. More generally, the fact that we want $3/4$ to be an eigenvalue should put some constraint on the choice of K_1. But since the author has no clue about the precise value of the eigenvalues of $-\Delta_1$, depending on the three lengths l_j, this information will perhaps not be so easy to use.

We could even be more ambitious and decide that $3/4$ should be the smallest eigenvalue of $-\Delta_1$. There are excuses for doing this, but none of them is very serious. The first one is that the lowest eigenfunctions usually have a simpler variation pattern. Also, the lowest eigenvalue and eigenfunctions may be easier to compute, because they correspond to optimal functions for the 2-Poincaré inequality. That is, there is a smallest constant γ such that

$$\int_{\partial B(0,1)\setminus K_1} |f|^2 \le \gamma^2 \int_{\partial B(0,1)\setminus K_1} |\nabla f|^2 \tag{13}$$

for all functions $f \in W^{1,2}(\partial B(0,1) \setminus K_1)$ with integral 0. And the functions f for which (13) is an equality are solutions of $\Delta_1 f + \lambda f = 0$, where λ is the smallest eigenvalue of $-\Delta_1$. This boils down to the fact that $\|\nabla f\|_2^2 = -\langle \Delta_1 f, f \rangle$, by Green. Thus $\lambda = \gamma^{-2}$. Observe that since $W^{1,2}(\partial B(0,1) \setminus K_1)$ is an increasing function of K_1, γ is a nondecreasing function of K_1, and λ is nonincreasing. When K_1 is empty, λ is equal to 2 (and the eigenfunctions are the restriction to $\partial B(0,1) \setminus K_1$ of affine functions). When two of the lengths l_j tend to π, γ tends to $+\infty$ (bad Poincaré inequalities) and λ tends to 0.

The second excuse is that this is what happens in dimension 2. Indeed, the cracktip function in the plane is homogeneous of degree $1/2$, and its restriction to $\partial B(0,1) \setminus \{-1\}$ is $u_1(e^{i\theta}) = (2/\pi)^{1/2} \sin(\theta/2)$. And $\sin(\theta/2)$ is a solution of $v'' + v/4 = 0$, corresponding to the lowest eigenvalue of $-\Delta$ on $(-\pi, \pi)$. [See Wirtinger's Lemma 47.18.] The difference between $3/4$ and $1/4$ comes from the computation of Δ in polar coordinates. When we think about this and the proof of monotonicity in the connected 2-dimensional case, assuming that $3/4$ is the lowest eigenvalue also looks like a pleasant economy of means.

Let us return to $K = P \times (-\infty, 0]$. The main reason why it is tempting to take this choice of K, or at least $l_1 = l_2 = l_3$ if $3/4$ is not an eigenvalue when $K = P \times (-\infty, 0]$, is because it has more symmetry. But this is a bad reason. Indeed, call ψ a rotation of $120°$ around the vertical axis. Our function u cannot be invariant under ψ, or even such that $u(\psi(x)) = \pm u(x) + C$; this is easy to see, because the three jumps of u at the south pole are nonzero (see for instance the proof of Proposition 42.10), and their sum is 0. See Exercise 20 for a hint. Of course, this does not give a contradiction: the set of eigenvectors is invariant by ψ, each u is not, so we only know that $3/4$ is a multiple eigenvalue. The only thing that we would clearly lose here is the essential uniqueness of u given K (which we had in dimension 2 because of the magic formula in Theorem 63.2). But the idea of requiring K to be invariant when we know that u will not be, does not seem so satisfactory. After all, if u and u_1 are not invariant by rotation, there will probably be a branch γ_j along which the jump of u_1 (or its gradient near the top) is larger, and maybe it would make sense to make this γ_j a little longer or the other ones a little shorter.

So it seems more tempting to consider sets K that are not invariant under ψ. We can still decide that $l_1 = l_2 \neq l_3$, and the objections above do not apply. It may seem that this leaves a lot of choice on the lengths l_j, but there are also a few constraints.

As we said earlier, $3/4$ should be an eigenvalue for $-\Delta_1$ and u_1 should be an eigenvector, with the Neumann condition $\frac{\partial u}{\partial n} = 0$. Also, the asymptotic behavior of u near each of the three extremities of K_1 should be equivalent to the behavior of cracktip functions (times a line, as in Subsection b), with precise constants $\pm(2/\pi)^{1/2}$ in particular. We get this by looking at blow-up limits of (u, K) at such points. Finally, the fact that the mean curvature of K (away from a few lines) is zero shows that the boundary values of $|\nabla u|^2$ on either side of K coincide. This is proved like (58.28) above and, translated in terms of u_1, this just means that the boundary values of $\left|\frac{\partial u_1}{\partial \tau}\right|^2 + \frac{u_1^2}{4}$ on either side of each arc γ_j coincide. Note that symmetry may actually diminish the number of conditions that we get this way, and help with the jump condition.

Our discussion about symmetry leads us to another interesting question: could it be that the set of global minimizers with complex values is fundamentally different from the set of global minimizers with real values (which we have been considering so far)? It is easy to see that if (u, K) is a real global minimizer, multiplying u by a complex number of modulus 1 will give a complex global minimizer. So far, all the complex global minimizers that we know are obtained this way, but we do not know for sure that there is no other one. Even in dimension 2, one could imagine the existence of an intrinsically complex exotic global minimizer! If we allow u to be complex-valued in the discussion above, u may be such that $u(\psi(x)) = ju(x)$, for instance, where j is a third root of 1. It would be very amusing to have a truly different set of global minimizers when u is allowed to be complex.

The following paragraph was added with the galley proofs, and replaces a too optimistic one where the author guessed that there should be a minimizer as in the discussion above, with $l_1 = l_2 < l_3$. In the mean time, the situation started to move rapidly, as (much needed) numerical experiments were undertaken. First, Benoît Merlet [Me] computed the first eigenvalues for the Laplacian on a sphere minus three arcs of great circles, and estimated the leading term in the asymptotics near the tips. Apparently no combination of l_1, l_2 and l_3 gives an eigenvalue equal to 3/4 and at the same time exact cracktip asymptotics at the tips. So the author may have to forget some of his dreams. Perhaps the smallest modification would consist in replacing the two short arcs of great circles with curves that deviate slightly from great circles; note that this is not incompatible with homogeneity, but it is somewhat unpleasant to have to choose a curve here. One could also imagine two pieces of surface leaving tangentially from a cracktip-like minimizer, as suggested by Merlet. Hopefully very recent software by Blaise Bourdin will allow us to come up with better guesses.

The reader may be worried that the existence of a dilation-invariant global minimizer (u, K) as in the discussion above is incompatible with the fact the minimizer of Subsection b (cracktip times a line) is probably the only global minimizer for which K is a half-plane. Let us check that there is no apparent problem with this. Let (u, K) be our dilation-invariant minimizer, and let $x_0 \neq 0$ be a point of the boundary of K. Any blow-up limit of u at x_0 is a global minimizer whose singular set is a half-plane, so it should be like the minimizer of Subsection b. Thus we expect an expansion of u near x_0 of the type $u(x) = u(x_0) + u_0(\pi(x - x_0)) + o(|x - x_0|^{1/2})$, where π is the orthogonal projection on a plane perpendicular to the vector line through x_0, and u_0 is essentially the same function on the plane as in Subsection b. Then the homogeneity of u and u_0 yields $u(\lambda x) = \lambda^{1/2}[u(x_0) + u_0(\pi(x - x_0)) + o(|x - x_0|^{1/2})] =$

$u(\lambda x_0) + u_0(\pi(\lambda x - \lambda x_0)) + o(|x - x_0|^{1/2})$, which is compatible with the fact that blow-up limits of u at λx_0 should also be global minimizers.

Exercise 14. We want to show that if G is a rectifiable set of dimension 2 in $\mathbb{R}^3$ and if we set $G_z = G \cap \{x \in \mathbb{R}^3 ; x_3 = z\}$ for $z \in \mathbb{R}$, then

$$\int_{\mathbb{R}} H^1(G_z) \, dz \leq H^2(G). \tag{15}$$

1. First let $f : \mathbb{R}^2 \to \mathbb{R}$ be of class C^1, and set $\Gamma_f = \{x \in \mathbb{R}^3 ; x_1 = f(x_2, x_3)\}$. Prove (15) when Q is a square and $G = \{x \in \Gamma_f ; (x_2, x_3) \in Q\}$.

2. Prove (15) when G is a measurable subset of Γ_f.

[This will take care of the pieces of C^1 surfaces that are not too horizontal; the next question deals with almost horizontal surfaces.]

3. Now let $h : \mathbb{R}^2 \to \mathbb{R}$ be Lipschitz with constant less than 10^{-3}, and set $H_h = \{x \in \mathbb{R}^3 ; x_3 = h((x_1, x_2))\}$ Show that $\int_{\mathbb{R}} H^1(G_z) \, dz \leq 10^{-1} H^2(G)$ for every measurable set $G \subset H_h$. [You may use coverings.]

4. Show that $\int_{\mathbb{R}} H^1(G_z) \, dz \leq 100 H^2(G)$ for every measurable set $G \subset \mathbb{R}^3$. [Use coverings again.]

5. Prove (15) for all rectifiable sets.

Exercise 16. (Product of a global minimizer by a line.) Let (u_0, K_0) be a reduced global minimizer in $\mathbb{R}^n$, and set $K = K_0 \times \mathbb{R} \subset \mathbb{R}^{n+1}$ and $u(x, t) = u_0(x)$ for $x \in \mathbb{R}^n \setminus K_0$ and $t \in \mathbb{R}$. We want to prove that (u, K) is a reduced global minimizer in $\mathbb{R}^{n+1}$, but we shall assume that $\mathbb{R}^n \setminus K_0$ is connected, or that we used the spherical variant $\mathcal{L}^{n-1}$ of H^{n-1} in the definition of minimizers.

1. Let (v, G) be a topological competitor for (u, K) in some large ball $B(0, R)$. For $z \in \mathbb{R}$, call (v_z, G_z) the section of (v, G) at $t = z$ (see before (2)). Show that (v_z, G_z) is a topological competitor for (u_0, K_0). Hint: K separates points if K_0 separates their projections.

2. Check that the argument of Subsection b gives the conclusion. Why did we add the unnatural hypothesis?

Exercise 17. (Translation invariant global minimizers.) We want to show that every reduced global minimizer (u, K) in $\mathbb{R}^{n+1}$ that is invariant under all translations by vectors $(0, t)$, $t \in \mathbb{R}$, is the product of a global minimizer by a line, as in the previous exercise. Call (u_0, K_0) the section of (u, K) by the hyperplane $\{t = 0\}$, and let (v_0, G_0) be a topological competitor for (u_0, K_0) in some $B(0, R)$.

1. Set $T = B(0, R) \times [-M, M]$, and construct a topological competitor (v, G) for (u, K) in T, such that G is the union of $G_0 \times [-M, M]$, $K_0 \times [\mathbb{R} \setminus [-M, M]]$, and $\overline{B}(0, R) \times \{-M, M\}$.

2. Show that if (v_0, G_0) does strictly better than (u_0, K_0), then for M large (v, G) does strictly better than (u, K). Conclude.

Exercise 18. (Computations in spherical coordinates in $\mathbb{R}^3$.) Write

$$x_1 = \rho \sin \varphi \sin \theta, \qquad x_2 = \rho \sin \varphi \cos \theta, \qquad x_3 = \rho \cos \varphi,$$

and show that

$$\Delta u = \frac{\partial^2 u}{\partial \rho^2} + \frac{2}{\rho} \frac{\partial u}{\partial \rho} + \frac{1}{\rho^2 \sin \varphi} \left\{ \frac{\partial^2 u}{\partial \theta^2} + \sin \varphi \frac{\partial^2 u}{\partial \varphi^2} + \cos \varphi \frac{\partial u}{\partial \varphi} \right\}, \qquad (19)$$

(or go to the library). Then check (12).

Exercise 20. Check the assertion (about one page below (12)) that we cannot find a minimizer (u, K) such that $u(\psi(x)) = \pm u(x) + C$. [Hint: in case of $-$, use ψ^2.]

Exercise 21. (Poincaré estimates and spherical harmonics.) Denote by S the unit sphere in $\mathbb{R}^n$ and, for each integer $k \geq 0$, denote by E_k the set of restrictions to S of polynomials on $\mathbb{R}^n$ that are homogeneous of degree k and harmonic. Also denote by $\mathcal{H}$ the linear span of the E_k, i.e., the set of functions on S that can be written

$$f = \sum_{k=0}^{l} h_k, \quad \text{with } h_k \in E_k \text{ for all } k. \qquad (22)$$

We shall need to know (and admit) that $\mathcal{H}$ is dense in $L^2(S)$. Finally denote by Δ_1 the Laplacian on S.

1. "Check" that for $k \geq 1$, $\Delta_1 f = -k(n + k - 2)f$ for $f \in E_k$ (or return to the library).

2. Show that $\|\nabla f\|_2^2 = - \int_S f \Delta_1 f$ when f is smooth on S.

3. Show that $\|f\|_2^2 = \sum_{k=0}^{l} \|h_k\|_2^2$ when (22) holds (use **1** and the proof of **2**).

4. Show that

$$\int_S (f - m_S f)^2 \leq \frac{1}{n-1} \int_S |\nabla f|^2 \qquad (23)$$

   for all $f \in \mathcal{H}$, where $m_S f = \frac{1}{|S|} \int_S f$.

5. Extend (23) to $f \in W^{1,2}(S)$. Notice that (23) is an identity for $f \in E_1$.

6. Denote by $S^+ = \{x \in S \,;\, x_n > 0\}$ the upper half-sphere. For each $f \in W^{1,2}(S^+)$, extend f to S (minus the great circle) by symmetry. Show that the extension $\tilde{f}$ lies in $W^{1,2}(S)$.

7. Show that the analogue of (23) for $f \in W^{1,2}(S^+)$ holds, and that the constant $\frac{1}{n-1}$ is optimal there too.

8. Show that (23) also holds on the quarter of sphere $\{x \in S \,;\, x_{n-1} > 0 \text{ and } x_n > 0\}$. It is no longer optimal, though, even when $n = 2$.

I. Boundary Regularity

What can we say about minimizers, almost-minimizers, and quasiminimizers in a domain Ω when we approach the boundary? The reader probably expects some complications, but it turns out that if $\partial\Omega$ is C^1 (to simplify), most of the general regularity results that we proved inside Ω stay true near the boundary, with similar proofs. To the author's knowledge, S. Solimini is the first person who observed that it should be as easy to deal with minimizers near the boundary as inside Ω.

As we shall see in the next section, quasiminimizers in Ω are still Ahlfors-regular near $\partial\Omega$. This also gives uniform rectifiability results, but those are no real improvement over the results of Parts C and H. A more interesting consequence is that $H^{n-1}(\overline{K} \cap \partial\Omega) = 0$ (where we take the closure of K in $\mathbb{R}^n$), which allows nice reflection tricks, in particular along flat parts of $\partial\Omega$; see Section 78.

Maybe we should recall here that our theorem about limits of almost-minimizers goes all the way to the boundary; see Section 39.

When $n = 2$ we are lucky, and we can get a complete description of K near $\partial\Omega$ when Ω is a bounded smooth domain and (u, K) is an almost-minimizer in Ω with a sufficiently small gauge function h: there is a neighborhood of $\partial\Omega$ where K is a finite union of C^1 curves that end on $\partial\Omega$ perpendicularly. See Section 79.

77 Ahlfors-regularity near the boundary

We start with the local Ahlfors-regularity of K near Ω when (u, K) is a quasiminimizer with small enough constant a. We shall assume in this section that

$$(u, K) \text{ is a reduced (or coral) } (r_0, M, a)\text{-quasiminimizer in } \Omega \qquad (1)$$

(this time, no longer merely local!) or that

$$
\begin{aligned}
&(u, K) \text{ is a topologically reduced or coral}\\
&\text{topological } (r_0, M, a)\text{-quasiminimizer in } \Omega.
\end{aligned} \qquad (2)
$$

See Definitions 7.21, 8.1, and 8.24 for (1); since topological quasiminimizers were not mentioned in Definition 7.21, let us say that (2) means that the quasiminimality condition (7.20) holds when $(\widetilde{u}, \widetilde{K})$ is a topological competitor for (u, K) in some ball $B(x, r)$ such that $r \leq r_0$ (see Definition 7.15), and that in addition (u, K) is topologically reduced (as in Definition 8.7) or coral (Definition 8.24).

We shall state our results with locally Lipschitz domains, even though we do not have immediate applications for this generality. We shall say that $B(x,r)$ is a λ-Lipschitz ball for $\partial\Omega$ when $x \in \partial\Omega$ and we can find orthonormal coordinates in $\mathbb{R}^n$ and a Lipschitz function $A : \mathbb{R}^{n-1} \to \mathbb{R}$ such that $||\nabla A||_\infty \leq \lambda$, so that in the given coordinates,

$$\Omega \cap B(x,r) = \{(\xi,\zeta) \in B(x,r) ; \zeta > A(\xi)\}. \tag{3}$$

Let us give two different statements, starting with the case of dimension 2.

Theorem 4 ($n = 2$). *For each choice of $\lambda > 0$ and $M \geq 1$, we can find constants $a > 0$ and $C \geq 1$ such that if $\Omega \subset \mathbb{R}^2$, (u,K) satisfies (1) or (2), $x \in \partial\Omega$, $0 < r \leq r_0$, $B(x,r)$ is a λ-Lipschitz ball for $\partial\Omega$, and $K \cap B(x,r/2)$ is not empty, then $H^1(K \cap B(x,r)) \geq C^{-1}r$.*

In higher dimensions we shall restrict to small Lipschitz constants; I don't know whether the restriction is needed.

Theorem 5. *For each choice of n, we can find $\lambda > 0$ such that the following holds. For each $M \geq 1$, there are constants $a > 0$ and $C \geq 1$ such that, if (u,K) satisfies (1) or (2), $x \in \partial\Omega$, $0 < r \leq r_0$, $B(x,r)$ is a λ-Lipschitz ball for $\partial\Omega$, and $K \cap B(x,r/2)$ is not empty, then $H^{n-1}(K \cap B(x,r)) \geq C^{-1}r^{n-1}$.*

In the case of Mumford-Shah minimizers on C^1 domains, Theorems 4 and 5 are due to Solimini [So2]; it is quite probable that the proof extends to the present situation, but I did not check.

Because of Lemma 18.19, we also have the upper bound $H^{n-1}(K \cap B(x,r)) \leq Cr^{n-1}$. Altogether, we get that if every ball of radius r_0 centered on $\partial\Omega$ is a λ-Lipschitz ball and (u,K) is as in the theorems, then

$$C^{-1}r^{n-1} \leq H^{n-1}(K \cap B(x,r)) \leq Cr^{n-1} \tag{6}$$

whenever $x \in \overline{K}$ and $r \leq r_0$, because we can apply Theorem 4 or 5 to a slightly different ball when $B(x,r/4)$ meets $\partial\Omega$, and apply (72.1) otherwise.

We shall try to prove the two theorems more or less at the same time. The proof will follow the same scheme as for (72.1), and we shall not give all the details.

So let n, $M \geq 1$, and λ be given. When $n > 2$, we shall be allowed to put an additional constraint on λ (depending on n only), but let us not worry for the moment.

We want to prove an analogue of Lemma 72.58. That is, we want to find constants $\eta > 0$, $\alpha \in (0, 1/3)$, and $C_M \geq 1$ such that if (u,K) satisfies (1) or (2), $x \in \overline{\Omega}$, $r \leq r_0$, $B(x,r)$ is either contained in Ω or contained in a λ-Lipschitz ball for $\partial\Omega$, and $H^{n-1}(K \cap B(x,r)) \leq \eta r^{n-1}$, then

$$H^{n-1}(K \cap B(x,\alpha r)) + \int_{B(x,\alpha r)\setminus K} |\nabla u|^2$$
$$\leq \frac{\alpha^{n-1}}{2}\left\{ H^{n-1}(K \cap B(x,r)) + \int_{B(x,r)\setminus K} |\nabla u|^2 \right\} + C_M \, ar^{n-1}. \tag{7}$$

Once we have (7), we can conclude by the same density argument as in Section 72; see below (72.82).

We want to prove (7) by contradiction and compactness, so we give ourselves a small $\alpha > 0$ (to be chosen later) and assume that there are sequences $\{(u_k, K_k)\}$ $\{x_k\}$, $\{r_k\}$, for which the hypotheses hold with $\eta_k = 2^{-k}$, say, and yet (7) fails, and we try to find a contradiction. By dilation invariance, we can assume that $x_k = 0$ and $r_k = 1$. If $B(x_k, r_k) = B(0,1)$ is contained in Ω_k for infinitely many k, we can replace our sequence with a subsequence for which this happens for all k, and then we are exactly in the situation of Section 72. In this case the argument of Section 72 gives a contradiction, provided that α is small enough, depending only on n.

So we may assume that $B(0,1)$ is never contained in Ω_k, and hence is contained in a λ-Lipschitz ball for $\partial\Omega_k$. We can even extract a subsequence so that the axes relative to the Lipschitz description of $\partial\Omega_k$ converge to a limit. Without loss of generality, we can assume that the limit corresponds to the usual coordinates in $\mathbb{R}^n$. Then we have a description of every $\Omega_k \cap B$, k large enough as in (3), maybe with λ replaced with 2λ. We can even extract a second subsequence, to make sure that the $\Omega_k \cap B(0,1)$ converge to the intersection of $B(0,1)$ with some λ-Lipschitz domain Ω (again in the usual coordinates).

We define the normalized functions $u_k^* = E_k^{-1/2} u_k$ as in (72.29), and we would like to use Lemma 72.9 and Lemma 72.30 to produce truncations and take limits, as we did in Lemma 72.30. But observe that if λ is not too small (which can only happen when $n = 2$), the domains $\Omega_k \cap B$ can be a little complicated; for instance, they do not need to be connected (see Figure 1 on page 545). For this reason, we consider a cylinder R centered at the origin, with horizontal circular base and vertical walls, such that $B(0, 1/2) \subset R \subset B$. The point is that the $\Omega_k \cap R$, k large, and $\Omega \cap R$ are all bilipschitz equivalent to intersections $B \cap H$, where H is a half-space that contains the origin, and with bilipschitz mappings which preserve the origin and have uniform bilipschitz bounds (that depend on λ, but this is all right).

As was alluded to in Section 72, Lemmas 72.5, and 72.9 still work with B replaced with the $\Omega_k \cap R$. We apply them after subtracting constants from the u_k, so as to be able to take $\lambda_k = 0$ in Lemma 72.9 (as we did before) and simplify notation. We get small sets $S_k \subset \Omega_k \cap R$ where we shall be able to truncate.

Then we apply Lemma 72.30, with B replaced by $\Omega_k \cap R$ and the constant $3/4$ (as in $B(0, 3/4)$) replaced by 10^{-1}. The statement should be slightly different also because we now have domains $\Omega_k \cap R$ that converge to $\Omega \cap R$ (instead of a fixed domain B), but our proof is mainly local. More precisely, only (72.31) (i.e., the convergence in L^2 of the $1_{\Omega_k \cap R \setminus S_k} u_k^*$ to a limit u_∞^*) looks dangerous, but the proof gives the convergence in $L^2(H)$ for every compact set $H \subset \Omega \cap B(0, 10^{-1}) \subset \Omega \cap R$, say, and we should not worry about the part of $\Omega_k \cap B(0, 10^{-1})$ that lives close to $\partial\Omega$, because (72.11) gives a uniform control on the $1_{\Omega_k \cap B(0, 10^{-1}) \setminus S_k} u_k^*$ in L^p for some $p > 2$.

Recall that Lemma 72.30 was used to produce a limiting function u_∞^*, and then functions $w_k = E_k^{1/2} u_\infty^*$ that were used in the formula (72.62) to produce competitors v_k for the u_k. So we have an additional bit of trouble, because we want w_k to be defined on $\Omega_k \cap R$, and not only on $\Omega \cap R$. The simpler fix seems to use the fact that $\Omega_k \cap R$ is Lipschitz, and extend u_∞^* and w_k across $\partial\Omega$, using for instance a bilipschitz reflection across the Lipschitz graph given by (3). This does not give harmonic functions u_∞^* and w_k on $\Omega_k \setminus \Omega$, but we shall not care because this set is very small and so

$$\lim_{k \to +\infty} \int_{\Omega_k \cap B(0,10^{-1}) \setminus \Omega} |\nabla u_\infty^*|^2 = 0. \tag{8}$$

Indeed, the definition with a bilipschitz reflection yields

$$\int_{\Omega_k \cap B(0,10^{-1}) \setminus \Omega} |\nabla u_\infty^*|^2 \leq C \int_{D_k} |\nabla u_\infty^*|^2, \tag{9}$$

where D_k is a part of Ω that lives very close to the boundary, and the right-hand side tends to 0 by the dominated convergence theorem.

We can follow the computations of Section 72 quietly, and the only effect of the extra energy term coming from $\Omega_k \setminus \Omega$ is to make the error terms ε_k and ε_k' in (72.65) and (72.68), for instance, a little larger. We leave the details, let the reader check that nothing goes wrong, and we go directly to the place where we need to be careful.

Recall that in our second case where the energy is not so small, we used a trivial estimate on the decay of energy for harmonic functions in a ball, to get (72.71). More precisely, we wanted to get a normalized decay estimate for our function u_∞^*, and we introduced an energy minimizing function w^* for which it would be easier to prove. The fact that the energy for u_∞^* was very close to the energy for w^* (and hence we could deduce decay for u_∞^* from decay for w^*) was obtained because otherwise replacing u_∞^* by w^* in the comparison argument would be too profitable. But for the decay of w^*, we can forget about our sequence.

We shall do something similar here, but first we want to select a radius $\rho < 10^{-1}$ such that $\partial B(0,\rho) \cap \Omega$ is reasonably nice, so that we can cut and paste along $\partial B(0,\rho)$.

If $\mathrm{dist}(0, \partial\Omega) > 10^{-2}$, we choose $\rho \in (1/200, 1/100)$ such that the restriction of u_∞^* to $\partial B(0,\rho)$ lies in $W^{1,2}$, and is equal almost-everywhere to the radial limits of u_∞^* from outside and inside. This case is not too interesting, because we shall be able to proceed exactly as in Section 72.

Next suppose that $\mathrm{dist}(0, \partial\Omega) \leq 10^{-2}$ and $n > 2$. If λ is small enough (depending on n only), then $\partial\Omega \cap B(0,1)$ is a Lipschitz graph (i.e., $\partial\Omega$ does not get out and in the ball) and for every choice of $\rho \in (1/20, 1/10)$, $\Omega \cap \partial B(0,\rho)$ is the piece of $\partial B(0,\rho)$ that lives on one side of the Lipschitz graph $\partial B(0,\rho) \cap \partial\Omega$. We choose $\rho \in (1/20, 1/10)$ so that the restriction of u_∞^* to $\Omega \cap \partial B(0,\rho)$ lies in $W^{1,2}$, and is equal almost-everywhere to the radial limits of u_∞^*. We know (by Proposition 9.5, a localization argument, and a change of variables) that this is the case for almost every ρ.

If $\operatorname{dist}(0, \partial\Omega) \leq 10^{-2}$ and $n = 2$, we choose $\rho \in (1/20, 1/10)$ so that $\partial B(0, \rho)$ only meets $\partial\Omega$ finitely many times, and each time transversally at a point where $\partial\Omega$ has a tangent. Then $\partial(\Omega \cap B(0, \rho))$ is composed of a finite number of intervals of $\partial\Omega \cap B(0, \rho)$, plus a finite number of arcs of $\Omega \cap \partial B(0, \rho)$. We modify ρ slightly, if needed, so that the restriction of u_∞^* to these arcs lies in $W^{1,2}$, and is equal to the radial limits of u_∞^* from both sides. See Figure 1.

With our careful choice of ρ, we can apply Proposition 15.6 to find a function $w^* \in W^{1,2}(\Omega \cap B(0, \rho))$, whose boundary values on $\Omega \cap \partial B(0, \rho)$ coincide with u_∞^* almost-everywhere, and for which $\int_{\Omega \cap B(0, \rho)} |\nabla w^*|^2$ is minimal.

Lemma 10. *We can find $\tau \in (0, 1)$ and $C \geq 0$ such that in the present situation*

$$\int_{\Omega \cap B(0, \alpha)} |\nabla w^*|^2 \leq C\alpha^{n-1+\tau} \int_{\Omega \cap B(0, \rho)} |\nabla w^*|^2 \qquad (11)$$

for $\alpha \leq \rho$.

When $n = 2$, τ and C will depend only on λ. When $n > 2$, we shall be able to take any $\tau \in (0, 1)$, but we shall need to take λ small, depending on n and τ, and C also will depend on n and τ.

Once we have this lemma, (11) gives an analogue of (72.71), which we can use as in Section 72 to conclude. [That is, we can use the decay of ∇w^* and the fact that u_∞^* is almost minimizing to prove a similar decay for ∇u_∞^*, get (7), and deduce the theorems.] So it will be enough to prove Lemma 10.

Set $d = \operatorname{dist}(0, \partial\Omega)$, and let $z \in \partial\Omega$ be such that $|z| = d$. Let us check that if $\alpha \leq d$,

$$\int_{B(0, \alpha)} |\nabla w^*|^2 \leq C\frac{\alpha^n}{d^n} \int_{B(0, d)} |\nabla w^*|^2. \qquad (12)$$

This is clear when $d/2 \leq \alpha \leq d$. Otherwise, we just need to say that w^* is harmonic in $B(0, d)$, hence $|\nabla w^*|$ is bounded by $C\{d^{-n} \int_{B(0,d)} |\nabla w^*|^2\}^{1/2}$ on $B(0, d/2)$; (12) follows.

If $d > \rho/4$ and $\alpha > d$, (11) is immediate. If $d > \rho/4$ and $\alpha \leq d$, (11) follows from (12). So we may assume that $d \leq \rho/4$.

Suppose that we can prove that

$$\int_{\Omega \cap B(z, r)} |\nabla w^*|^2 \leq Cr^{n-1+\tau} \int_{\Omega \cap B(z, \rho/2)} |\nabla w^*|^2 \quad \text{for } 2d \leq r \leq \rho/2. \qquad (13)$$

Then (11) will follow. Indeed, if $\alpha < d$,

$$\int_{\Omega \cap B(0,\alpha)} |\nabla w^*|^2 \leq C \frac{\alpha^n}{d^n} \int_{\Omega \cap B(0,d)} |\nabla w^*|^2 \leq C \frac{\alpha^n}{d^n} \int_{\Omega \cap B(z,2d)} |\nabla w^*|^2$$

$$\leq C \frac{\alpha^n}{d^n} d^{n-1+\tau} \int_{\Omega \cap B(z,\rho/2)} |\nabla w^*|^2$$

$$\leq C \alpha^{n-1+\tau} \int_{\Omega \cap B(0,\rho)} |\nabla w^*|^2 \tag{14}$$

by (12) and (13), while if $d \leq \alpha \leq \rho/4$ we simply say that

$$\int_{\Omega \cap B(0,\alpha)} |\nabla w^*|^2 \leq \int_{\Omega \cap B(z,2\alpha)} |\nabla w^*|^2 \leq C (2\alpha)^{n-1+\tau} \int_{\Omega \cap B(z,\rho/2)} |\nabla w^*|^2 \tag{15}$$

by (13), which yields (11) too. The case when $\alpha \geq \rho/4$ is trivial.

So we are left with (13) to prove. Let us do a first argument, based on Bonnet's computations from Section 47, but which will only give the desired result when $n \leq 4$. Then we shall give another proof, which works for all n (but only with small λ). Set

$$E(r) = \int_{\Omega \cap B(z,r)} |\nabla w^*|^2 \tag{16}$$

for $d < r \leq \rho/2$. By the proof of Lemma 47.4 (i.e., polar coordinates and Fubini), $E(r)$ is the indefinite integral of its derivative $E'(r) = \int_{\Omega \cap \partial B(z,r)} |\nabla w^*|^2$. We want to compare these two quantities.

The boundary of $\Omega \cap B(z,r)$ is composed of $\Omega \cap \partial B(z,r)$, plus the inside boundary $\partial\Omega \cap B(z,r)$ where $\frac{\partial w^*}{\partial n} = 0$. Hence

$$E(r) = \int_{\Omega \cap B(z,r)} \nabla w^* \cdot \nabla w^*$$

$$= -\int_{\Omega \cap B(z,r)} w^* \Delta w^* + \int_{\partial(\Omega \cap B(z,r))} w^* \frac{\partial w^*}{\partial n} = \int_{\Omega \cap \partial B(z,r)} w^* \frac{\partial w^*}{\partial n}, \tag{17}$$

as in (47.11). We only checked the integration by parts in Section 47 when $n = 2$, but the case when $n > 2$ works essentially the same way, and we omit the details.

Next we decompose $\Omega \cap \partial B(z,r)$ into its connected components I_j, and we want to prove that

$$\int_{I_j} \frac{\partial w^*}{\partial n} = 0 \quad \text{for all } j \tag{18}$$

as in (47.15). When $n > 2$ this is easier, because we know that $\partial\Omega \cap B(0,1)$ is a Lipschitz graph with small constant λ, and there is only one component $I_j = \Omega \cap \partial B(z,r)$. Then

$$\int_{I_j} \frac{\partial w^*}{\partial n} = \int_{\Omega \cap \partial B(z,r)} \frac{\partial w^*}{\partial n} = \int_{\partial(\Omega \cap B(z,r))} \frac{\partial w^*}{\partial n} = -\int_{\Omega \cap B(z,r)} \Delta w^* = 0 \tag{19}$$

because $\frac{\partial w^*}{\partial n} = 0$ on the rest of $\partial(\Omega \cap B(z,r))$, and by an integration by parts that we leave to the reader. [See Section 47 for a hint.]

When $n = 2$, there may be more than one I_j (if λ is large), but if $r \leq \beta$, where β is a constant that depends only on λ, all the points of $\partial\Omega \cap \partial B(z,r)$ lie in a same compact arc of $\partial\Omega \cap B(0,\rho)$. In this case, we can apply the proof of (47.13) and get (18) for almost every $r \leq \beta$.

Call m_j the mean value of w^* on I_j. When (18) holds, (17) yields

$$E(r) = \sum_j \int_{I_j} w^* \frac{\partial w^*}{\partial n} = \sum_j \int_{I_j} (w^* - m_j) \frac{\partial w^*}{\partial n}$$
$$\leq \sum_j \left\{ \int_{I_j} (w^* - m_j)^2 \right\}^{1/2} \left\{ \int_{I_j} \left(\frac{\partial w^*}{\partial n} \right)^2 \right\}^{1/2}. \tag{20}$$

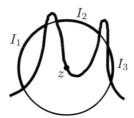

Figure 1

Let us continue the argument when $n = 2$. In this case, each I_j is an arc of circle of length at most $2\pi r(1-\varepsilon)$, where $\varepsilon > 0$ depends on the Lipschitz constant λ. (See Figure 1.) Wirtinger's Lemma 47.19 says that

$$\left\{ \int_{I_j} (w^* - m_j)^2 \right\}^{1/2} \leq 2r(1-\varepsilon) \left\{ \int_{I_j} \left(\frac{\partial w^*}{\partial \tau} \right)^2 \right\}^{1/2}, \tag{21}$$

where $\frac{\partial w^*}{\partial \tau}$ denotes the tangential derivative of w^*. Hence

$$E(r) \leq 2r(1-\varepsilon) \sum_j \left\{ \int_{I_j} \left(\frac{\partial w^*}{\partial \tau} \right)^2 \right\}^{1/2} \left\{ \int_{I_j} \left(\frac{\partial w^*}{\partial n} \right)^2 \right\}^{1/2}$$
$$\leq r(1-\varepsilon) \sum_j \int_{I_j} \left(\frac{\partial w^*}{\partial \tau} \right)^2 + \left(\frac{\partial w^*}{\partial n} \right)^2 \tag{22}$$
$$= r(1-\varepsilon) \int_{\Omega \cap \partial B(z,r)} |\nabla w^*|^2 = r(1-\varepsilon)E'(r)$$

for almost every $r \leq \beta$.

We are now ready to prove (13) when $n = 2$. If $r \geq \beta$, (13) is trivial (because β is a constant). If $E(r) = 0$, we are happy too. Otherwise, set $W(r) = \text{Log}(E(r))$;

the same argument as for (48.16) shows that $W(r)$ is the integral of its derivative $W'(r) = E'(r)/E(r)$. Hence

$$
W(\beta) - W(r) = \int_r^\beta W'(r)dr = \int_r^\beta \frac{E'(r)}{E(r)}dr
$$

$$
\geq \int_r^\beta \frac{dr}{(1-\varepsilon)\,r} = (1-\varepsilon)^{-1}[\mathrm{Log}(\beta) - \mathrm{Log}(r)],
$$

(23)

and then

$$
\int_{\Omega \cap B(z,r)} |\nabla w^*|^2 = E(r) = E(\beta)\,e^{W(r)-W(\beta)} \leq E(\beta)\,e^{(1-\varepsilon)^{-1}[\mathrm{Log}(r)-\mathrm{Log}(\beta)]}
$$

$$
\leq CE(\beta)r^{(1-\varepsilon)^{-1}} \leq Cr^{(1-\varepsilon)^{-1}} \int_{\Omega \cap B(z,\rho/2)} |\nabla w^*|^2,
$$

(24)

by (16) and (23). This is the same as (13), with $\tau = (1-\varepsilon)^{-1} - 1 > 0$. This completes our proof of (13) when $n = 2$; Theorem 4 follows. $\qquad\square$

We are left with the case when $n > 2$. Let us first try to adapt the proof above. We still have (20), and we would like to have an analogue of (21). Call $\gamma(r)$ the best constant in the 2-Poincaré inequality on $\Omega \cap \partial B(x,r)$. That is, $\gamma(r)$ is the smallest constant such that

$$
\left\{ \int_{\Omega \cap \partial B(x,r)} (w^* - m)^2 \right\}^{1/2} \leq \gamma(r) \left\{ \int_{\Omega \cap \partial B(x,r)} |\nabla_t w^*|^2 \right\}^{1/2}
$$

(25)

for every $w^* \in W^{1,2}(\Omega \cap \partial B(x,r))$, where we denote by m the mean value of w^* on $\Omega \cap \partial B(x,r)$ and by $\nabla_t w^*$ its gradient on $\Omega \cap \partial B(x,r)$ (to avoid any confusion with the full gradient ∇w^* above). Then (20) yields

$$
E(r) \leq \left\{ \int_{\Omega \cap \partial B(x,r)} (w^* - m)^2 \right\}^{1/2} \left\{ \int_{\Omega \cap \partial B(x,r)} \left(\frac{\partial w^*}{\partial n}\right)^2 \right\}^{1/2}
$$

$$
\leq \gamma(r) \left\{ \int_{\Omega \cap \partial B(x,r)} |\nabla_t w^*|^2 \right\}^{1/2} \left\{ \int_{\Omega \cap \partial B(x,r)} \left(\frac{\partial w^*}{\partial n}\right)^2 \right\}^{1/2}
$$

$$
\leq \frac{\gamma(r)}{2} \int_{\Omega \cap \partial B(x,r)} \left[|\nabla_t w^*|^2 + \left(\frac{\partial w^*}{\partial n}\right)^2 \right] = \frac{\gamma(r)}{2} E'(r)
$$

(26)

because there is only one set I_j, equal to $\Omega \cap \partial B(x,r)$. If we want to get (13), it will be enough to prove that

$$
\gamma(r) \leq 2(n - 1 + \tau)^{-1}r,
$$

(27)

and then our proof of (24) will apply.

Call S the unit half-sphere; the best constant for the 2-Poincaré inequality on S is $(n-1)^{-1/2}$; see for instance Exercise 76.21. Recall that $\partial\Omega$ is a λ-Lipschitz graph that contains z. Then $r^{-1}(\Omega \cap \partial B(x,r))$ is bilipschitz-equivalent to S, with a constant C that is as close to 1 as we want, provided that we take λ small enough. We can use this to deduce the 2-Poincaré inequality on $r^{-1}(\Omega \cap \partial B(x,r))$ from the same thing on S and we get that the best constant for the 2-Poincaré inequality on $r^{-1}(\Omega \cap \partial B(x,r))$ is as close as we want to $(n-1)^{-1/2}$ (if λ is small enough). By homogeneity, the best constant on $\Omega \cap \partial B(x,r)$ is r times the previous one, hence we can get (27) if $(n-1)^{-1/2} < 2(n-1+\tau)^{-1}$, or equivalently $(n-1+\tau)^2 < 4(n-1)$. We can find $\tau > 0$ like this when $n \leq 4$, and this gives a proof of Theorem 5 in these cases.

Maybe it is worth analyzing why the proof does not work for n large. Notice that it does not work even if Ω is a half-space. On the other hand, (13) holds for a half-space, with $\tau = 1$, because in this case w^* has a symmetric extension which is harmonic in $B(z, \rho/2)$, and by the proof of (12).

When we follow the argument above, the worst case seems to be when $\Omega = \{x_n > 0\}$ and w^* is the first coordinate, say. Notice that the Neumann condition holds in this case (because w^* does not depend on x_n), and also that w^* essentially has the worst behavior with respect to (13). The inequality in (25) is an identity for this function, with $\gamma(r) = (n-1)^{-1/2}$, so we do not lose anything there. We do not lose anything in (20) either, when we apply Cauchy–Schwarz, because the function w^* is proportional to its radial derivative (a coincidence?). But we lose something important in (26), when we say that $ab \leq (a^2 + b^2)/2$, where a and b are the L^2 norms of the radial and tangential derivatives and are quite different from each other. [The tangential derivative is larger close to the equator $\{x_1 = 0\}$, and gives a much larger contribution when n is large, because most of the mass of the sphere is there.] There is not so much to complain about here, the surprising thing is that the argument of Bonnet works so well in dimension 2, where the same function is extremal three times.

We still need to prove Theorem 5 in large dimensions. As before we just need to prove (13), and we shall even be able to take any given $\tau \in (0,1)$.

First notice that it is enough to find a (small) constant $\eta > 0$ such that, if Ω is a λ-Lipschitz domain as before, $z \in \partial\Omega$, $\rho > 0$, $w_0 \in W^{1,2}(\partial B(z,\rho) \cap \Omega)$, and if $w \in W^{1,2}(\Omega \cap B(z,\rho))$ minimizes the energy $\int_{\Omega \cap B(z,\rho)} |\nabla w|^2$ among all functions of $W^{1,2}(\Omega \cap B(z,\rho))$ with boundary values equal to w_0, then

$$\int_{\Omega \cap B(z,\eta\rho)} |\nabla w|^2 \leq \eta^{n-1+\tau} \int_{\Omega \cap B(z,\rho)} |\nabla w|^2. \tag{28}$$

Indeed, if r is as in (13) and if $j \geq 0$ is such that $\eta^{j+1}\rho/2 \leq r \leq \eta^j \rho/2$, multiple applications of (28) (with the radii $\eta^i \rho/2$, $i < j$), yield

$$\int_{\Omega \cap B(z,r)} |\nabla w^*|^2 \leq \int_{\Omega \cap B(z,\eta^j \rho/2)} |\nabla w^*|^2 \leq \eta^{j(n-1+\tau)} \int_{\Omega \cap B(z,\rho/2)} |\nabla w^*|^2$$
$$\leq Cr^{n-1+\tau} \int_{\Omega \cap B(z,\rho/2)} |\nabla w^*|^2. \tag{29}$$

Denote by H the upper half-space, and let us first check (28) when $\Omega = H$. In this case, w is obtained from w_0 by first extending it to the sphere $\partial B(z, \rho)$ by symmetry, then taking the harmonic extension to $B(z, \rho)$, and finally restricting to H. The proof is the same as for Example 17.5. Easy estimates on harmonic functions (as in the proof of (12)) then yield

$$\int_{B(z, t\rho) \cap H} |\nabla w|^2 \leq C_0 t^n \int_{B(z, \rho) \cap H} |\nabla w|^2 \qquad (30)$$

for $t < 1$, and with a constant C_0 that depends only on n.

This is slightly better than (28) (when $\Omega = H$ and if $\eta = t$ is chosen small enough), and we want to deduce the general case from this by an approximation argument.

So let Ω, ρ, and w be given, as above. Since $\partial \Omega$ is a λ-Lipschitz graph, we can find a bilipschitz bijection φ from $\overline{\Omega} \cap B(z, \rho)$ to $\overline{H} \cap B(0, \rho)$, such that $\varphi(z) = 0$, and φ maps $\partial \Omega \cap B(z, \rho)$ onto $\partial H \cap B(0, \rho)$ and $\Omega \cap \partial B(z, \rho)$ onto $H \cap \partial B(0, \rho)$. We can even choose φ to be bilipschitz with constant less than $1 + C\lambda$.

Set $v_0 = w_0 \circ \varphi^{-1}$ on $H \cap \partial B(0, \rho)$ and $v = w \circ \varphi^{-1}$ on $H \cap B(0, \rho)$. Then $v_0 \in W^{1,2}(H \cap \partial B(0, \rho))$ and $v \in W^{1,2}(H \cap B(0, \rho))$, by bilipschitz invariance. The boundary values of v are equal to v_0 almost-everywhere on $H \cap \partial B(0, \rho)$; the argument suggested in Remark 13.16 still works here (because the bilipschitz constant is close to 1), and anyway we could even construct φ so that it is C^1 away from $\partial \Omega$.

Call $\widetilde{v}$ the function of $W^{1,2}(H \cap B(0, \rho))$ whose boundary values on $H \cap \partial B(0, \rho)$ are equal to v_0 almost-everywhere, and for which $\widetilde{E} = \int_{H \cap B(0, \rho)} |\nabla \widetilde{v}|^2$ is minimal. Notice that v is a competitor for $\widetilde{v}$, so

$$\widetilde{E} \leq \int_{H \cap B(0, \rho)} |\nabla v|^2 \leq (1 + C\lambda) \int_{\Omega \cap B(z, \rho)} |\nabla w|^2, \qquad (31)$$

where the last inequality comes from a bilipschitz change of variable. Set $E = \int_{\Omega \cap B(z, \rho)} |\nabla w|^2$; thus (31) says that $\widetilde{E} \leq (1 + C\lambda) E$.

Now set $\widetilde{w} = \widetilde{v} \circ \varphi$. Then $\widetilde{w}$ is a competitor in the definition of w as above, and so

$$E = \int_{\Omega \cap B(z, \rho)} |\nabla w|^2 \leq \int_{\Omega \cap B(z, \rho)} |\nabla \widetilde{w}|^2 \leq (1 + C\lambda)\widetilde{E}, \qquad (32)$$

again by a change of variables. In other words, all these numbers are comparable.

Next observe that $\nabla(\widetilde{w} - w)$ is orthogonal to ∇w in $L^2(\Omega \cap B(z, \rho))$, because w has minimal energy (consider competitors of the form $w + t(\widetilde{w} - w)$). Pythagoras says that

$$\int_{\Omega \cap B(z, \rho)} |\nabla \widetilde{w} - \nabla w|^2 = \int_{\Omega \cap B(z, \rho)} \left[|\nabla \widetilde{w}|^2 - |\nabla w|^2 \right]$$
$$\leq (1 + C\lambda)\widetilde{E} - E \leq C'\lambda E \qquad (33)$$

by (32) and (31). Since

$$\int_{\Omega \cap B(z,\eta\rho)} |\nabla \widetilde{w}|^2 \leq (1+C\lambda) \int_{H \cap B(0,2\eta\rho)} |\nabla \widetilde{v}|^2 \tag{34}$$

$$\leq (1+C\lambda)C_0(2\eta)^n \int_{H \cap B(0,\rho)} |\nabla \widetilde{v}|^2 = C_0(1+C\lambda)(2\eta)^n \widetilde{E} \leq C\eta^n E$$

by a change of variables, then (30) and (31), we get that

$$\int_{\Omega \cap B(z,\eta\rho)} |\nabla w|^2 \leq 2 \int_{\Omega \cap B(z,\eta\rho)} |\nabla \widetilde{w} - \nabla w|^2 + 2 \int_{\Omega \cap B(z,\eta\rho)} |\nabla \widetilde{w}|^2$$

$$\leq 2C'\lambda E + 2C\eta^n E = [2C'\lambda + 2C\eta^n] \int_{\Omega \cap B(z,\rho)} |\nabla w|^2 \tag{35}$$

by (33) and (34); (28) follows by choosing η small enough, and then λ very small, so that the expression in brackets is less than $\eta^{n-1+\tau}$. This completes our proof of (28), (13), and finally Theorem 5. $\qquad\square$

Remark 36. The author does not know whether Theorem 5 holds for general Lipschitz domains Ω. The decay estimate (28) fails, though; see Exercise 40.

Notice however that domains with angles or corners are a different issue, since we may always hope for a better behavior of (28) on a specific domain. For instance, if Ω is a product of intervals and z is one of the corners (or more generally lies on some edge), we can extend w by successive reflections, and eventually get a harmonic function on a small ball. Thus (28) holds in this case, and this leads to a generalization of Theorem 5 to C^1 domains with square corners, for instance.

It seems to the author that (28) holds when Ω is a cone over some open set $V \subset \partial B(0,1)$, provided that the isoperimetric constant γ for V (defined as in (25)) is smaller than $\sqrt{\frac{4}{2n-3}}$, which leaves some room because the half-sphere corresponds to $\sqrt{\frac{1}{n-1}}$. See Exercise 41 for suggestions, but be aware that I did not check this seriously. Then Theorem 5 will still hold for such domains Ω, and even small Lipschitz perturbations of them (as in our last proof).

Remark 37. The author claims that Theorems 4 and 5 extend to quasiminimizers with linear distortion $\leq L$ (see Section 7.c), and that the proof is still a straightforward modification of the argument above. The point is that we still need the same decay as in Lemma 10 or (13), but for functions that minimize $\int |A\nabla w^*|^2$ instead of harmonic functions. Thus we have to make a linear change of variables, but fortunately our assumptions on Ω in both theorems are invariant under linear transformations (except for the fact that constants get worse). The other modifications should be the same ones as in Section 72.

Remark 38. Once we have the local Ahlfors-regularity of K near the boundary, our interior regularity results give apparently more global results. For instance, if

Ω is a bounded smooth domain, we get that if (u, K) is a coral quasiminimizer in Ω, with small enough constant a, then $\overline{K}$ is a uniformly rectifiable Ahlfors-regular set. The proof is rather simple once we have the other results. For instance we could deduce this from Theorem 74.1 and the fact that any Ahlfors-regular set that contains big pieces of uniformly rectifiable sets is uniformly rectifiable. To get the big pieces, one would use the local Ahlfors-regularity near the boundary (to get enough mass), Theorem 78.1 below (which says that $H^{n-1}(\overline{K} \setminus K) = 0$, so this mass does not lie on the boundary), and a small argument to say that most of the mass is not too close to $\partial\Omega$ (if not, we can deform (u, K) and push a big part of K out of Ω; see Lemma 79.3 below). One could also save some time in the verification of the big pieces theorem by observing that K contains big pieces of Lipschitz graphs locally (by Theorem 75.2, to be lazy). We do not want to elaborate, because no new element really comes in; see Remark 32.5 for a similar argument and Exercise 42 for a few hints.

Exercise 39. Prove (28) for a product of intervals, and then for the image of such a set by a bilipschitz image with constant close enough to 1.

Exercise 40. We want to show (or at least convince the reader) that (28) fails for a general Lipschitz domain. Call S the unit sphere in $\mathbb{R}^n$, with $n \geq 3$, and let V be a domain in S such that $\Omega = \{r\theta\,;\, r > 0$ and $\theta \in V\}$ is a Lipschitz domain. Call γ the best Poincaré constant for V, as in (25).

1. Show that we can choose V so that γ is as large as we want. Suggestion: make V look like the upper half-sphere, minus two arcs of a vertical great circle that go from the equator to points very close to the north pole. You may even choose a smooth V (but the Lipschitz constant has to be large).

2. Show that there is a function $w^* \in W^{1,2}(V)$ for which the Poincaré inequality (25) is an identity. Hint: restrict to w^* with integral 0, notice that $||w^*||_2$ is a Lipschitz function of w^*, and look near (36.8).

3. Let w^* be as above. Show that $\Delta_1 f = -\gamma^{-2} f$ on V, where Δ_1 is the Laplacian on the sphere. Then check that w^* satisfies the Neumann boundary condition $\partial w^*/\partial n = 0$ at ∂V. [Sorry, all this is quite unpleasant and actually I did not check the details, but I did not see any simpler way either. For the last part, you may use the smoothness of ∂V, but hopefully a weak form as in (15.22) will be enough for Question 5.]

4. Let $\lambda > 0$ be such that $\lambda(\lambda + n - 2) = \gamma^{-2}$. Set $w(r\theta) = r^\lambda w^*(\theta)$ for $r > 0$ and $\theta \in V$. Show that w is harmonic on Ω and $\partial w/\partial n = 0$ on $\partial\Omega$. [Use a generalization of (76.19).]

5. Show that w is locally energy-minimizing, as in the statement of (28). [See Exercise 17.20.] But $|\nabla w(r\theta)|^2$ behaves like $r^{2\lambda-2}$ near the origin, and the left-hand side of (28) looks like $\eta^{n+2\lambda-2}$. Since we can take λ as close to 0 as we want, (28) fails.

Exercise 41. Warning: this one was checked even less carefully than the previous one. Still consider an open domain V in the unit sphere, and $\Omega = \{r\theta \, ; \, r > 0$ and $\theta \in V\}$ as before. Suppose that the best constant for the Poincaré inequality in V is $\gamma < \sqrt{\frac{4}{2n-3}}$, and let w be as in (28) (in particular, w is harmonic on Ω and $\partial u/\partial n = 0$ on $\partial\Omega$).

Set $E(r) = \int_{\Omega \cap B(z,r)} |\nabla w|^2$ for r small, and also $J_t(r) = \int_{\Omega \cap \partial B(z,r)} |\nabla_t w|^2$ and $J_r(r) = \int_{\Omega \cap \partial B(z,r)} |\frac{\partial w}{\partial r}|^2$, corresponding to the tangential and radial parts of the gradient.

1. Check that $E(r) \leq \gamma \, r \, J_t(r)^{1/2} J_r(r)^{1/2}$ for almost-every small r. [Hint: (26).]

2. Check that $E'(r) = J_t(r) + J_r(r)$.

3. Show that $r J_t(r) = r J_r(r) + (n-2)E(r)$. This may take a while, but the proof of Proposition 65.5 should give something like this. Perhaps also the appropriate integration by parts. This is the place where we need Ω to be a cone.

4. Check that (28) will follow if we prove that $rE'(r) \geq (n-1+\tau)E(r)$.

5. Set $a = r J_r(r)/E(r)$. Then $rE'(r) = (2a+n-2)E(r)$ and $\gamma^2 a(a+n-2) \geq 1$. Let a_0 be such that $a_0(a_0 + n - 2) = \gamma^{-2}$. Check that (28) holds as soon as $2a_0 \geq 1 + \tau$.

6. Check that (28) holds for some $\tau \in (0,1)$ as soon as $\gamma^2 < \frac{4}{2n-3}$.

Exercise 42. Let (u, K) be as in Remark 38. Thus Ω is a smooth bounded domain, and (u, K) is a reduced quasiminimizer on Ω, with a small. Also let $B(x, r)$ be a small ball centered on $\partial\Omega \cap \overline{K}$.

1. Set $E(\tau) = \{y \in \Omega \cap B(x,r) \, ; \, \text{dist}(y, \partial\Omega) \leq \tau r\}$. Show that if τ is small enough, $K \cap B(x,r)$ cannot be contained in $E(\tau)$. Hint: use a function φ such that $\varphi(y) = y$ outside of $B(x,r)$ and $\varphi(E(\tau) \cap B(x, r/2))$ lies outside of Ω. This is not so easy, but Lemma 79.3 will give more than a hint.

2. Take τ small, and use a point $y \in K \cap B(x,r) \setminus E(\tau)$ to find a uniformly rectifiable set (or even a Lipschitz graph) G such that $H^{n-1}(K \cap B(x,2) \cap G) \geq C^{-1}r^{n-1}$.

78 A reflection trick

The main point of this section is that if (u, K) is a Mumford-Shah minimizer in a domain Ω, and $\partial\Omega$ has a flat piece, we can extend (u, K) by symmetry across that flat piece, and the extension is also a Mumford-Shah minimizer on the larger domain. The same thing works with almost-minimizers or quasiminimizers and, to a limited extent, with reflections across smooth pieces of boundary.

As far as the author knows, the first place where this reflection trick appears is in [DMS2].

a. Points of accumulation of K on $\partial\Omega$

Before we do the extension by reflection, we need to know that the trace of K on $\partial\Omega$ has vanishing H^{n-1} measure.

Theorem 1. *Let Ω be a domain in $\mathbb{R}^n$, and suppose that (u, K) is a coral (r_0, M, a)-quasiminimizer in Ω (i.e., satisfies (77.1) or (77.2)). Also let $x \in \partial\Omega$ and $r > 0$ be such that $B(x, r)$ is a λ-Lipschitz ball for $\partial\Omega$ (see (77.3)). Then*

$$H^{n-1}(\overline{K} \cap \partial\Omega \cap B(x, r)) = 0 \qquad (2)$$

if $n = 2$ and a is sufficiently small, depending on M and λ, or if $n \geq 3$ and λ and a are sufficiently small, depending on n and M.

Thus, if Ω is locally Lipschitz (with small constant if $n \geq 3$) and a is small enough (depending on all other constants), then $H^{n-1}(\overline{K} \cap \partial\Omega) = 0$.

Notice that our quasiminimality hypotheses hold automatically when (u, K) is a coral almost-minimizer on Ω (topological or not).

It is also possible to control the mass of K very near $\partial\Omega$; see Lemma 79.3 below.

Our proof of Theorem 1, and perhaps Theorem 1 itself, comes from some work with S. Semmes, but unfortunately the author does not recall where.

Let u, K, x, and r be as in the statement. Obviously it is enough to prove that $H^{n-1}(\overline{K} \cap \partial\Omega \cap B(x, \rho)) = 0$ for every $\rho < r$, so let $\rho < r$ be given. For each small $\tau > 0$, let Z_τ be a maximal subset of $H = \overline{K} \cap \partial\Omega \cap \overline{B}(x, \rho)$ such that the balls $B(y, \tau)$, $y \in Z_\tau$, are disjoint. If $\tau < r - \rho$, each $B(y, \tau)$ is a λ-Lipschitz ball for $\partial\Omega$. If in addition $\tau < r_0$ and λ and a are small enough (as in the statement), we can apply Theorem 77.4 or 77.5, and we get that $H^{n-1}(K \cap B(y, \tau)) \geq C^{-1}\tau^{n-1}$. Then

$$\sharp Z_\tau \leq C\tau^{1-n} \sum_{y \in Z_\tau} H^{n-1}(K \cap B(y, \tau)) \leq C\tau^{1-n} H^{n-1}(H_\tau), \qquad (3)$$

where $H_\tau = \{z \in K ; \operatorname{dist}(z, H) \leq \tau\}$. By maximality of Z_τ, the balls $B(y, 2\tau)$, $y \in Z_\tau$, cover H. Then, with the notation of (2.8), $H_{4\tau}^{n-1}(H) \leq C \sum_{y \in Z_\tau} (2\tau)^{n-1} \leq CH^{n-1}(H_\tau)$ for τ small enough, by (3). Since $H^{n-1}(H_\tau)$ tends to 0 when τ tends to 0 (because H_τ decreases to $\emptyset$), we get that $H^{n-1}(H) = 0$, as needed for Theorem 1. $\qquad \square$

Note that Theorem 1 generalizes to quasiminimizers with linear distortion, because local Ahlfors-regularity does. See Remarks 72.86 and 75.18.

b. Reflection across a plane

Let us state our reflection result in the simple case of a reflection across a hyperplane. See Subsection **c** for slight generalizations.

Let Ω be a domain in $\mathbb{R}^n$, and assume that Ω lies above $P = \{x \in \mathbb{R}^n ; x_n = 0\}$. The interesting case is when P contains a fairly large piece of $\partial\Omega$, but we do not need to assume this explicitly. Call σ the reflection across P (defined by $\sigma((x_1, \ldots, x_n)) = (x_1, \ldots, -x_n)$), and set $\widetilde{\Omega} = \Omega \cup \sigma(\Omega) \cup D$, where D is the set of points $x \in P$ such that Ω contains $\{y \in B(x,r) \, ; \, y_n > 0\}$ for r small. It is easy to see that $\widetilde{\Omega}$ is open.

Let K be closed in Ω, and let $u \in W^{1,2}_{loc}(\Omega \setminus K)$ be such that $\int_{\Omega \cap B(x,r) \setminus K} |\nabla u|^2 < +\infty$ for $x \in D$ and r small; this will be the case for quasiminimizers, by the proof of (18.20). Set

$$\widetilde{K} = K \cup \sigma(K) \cup (\overline{K} \cap D) \subset \widetilde{\Omega}; \tag{4}$$

it is easy to see that $\widetilde{K}$ is closed in $\widetilde{\Omega}$. Define $\widetilde{u}$ almost-everywhere on $\widetilde{\Omega} \setminus \widetilde{K}$ by

$$\widetilde{u}(y) = u(y) \text{ for } y \in \Omega \setminus K, \quad \text{and} \quad \widetilde{u}(y) = u(\sigma(y)) \text{ for } y \in \sigma(\Omega \setminus K). \tag{5}$$

Let us check that

$$\widetilde{u} \in W^{1,2}_{loc}(\widetilde{\Omega} \setminus \widetilde{K}). \tag{6}$$

We did not define $\widetilde{u}$ on $D \setminus \overline{K}$, but we could have. Indeed, if $x \in D$ and $r < \text{dist}(x, K)$ is as in the definition of D (two lines above (4)), Section 13 tells us that u has vertical limits almost everywhere on $D \cap B(x,r)$. In addition, $\widetilde{u}$ has the same limits from the other side too (by symmetry), and now Welding Lemma 14.4 says that $\widetilde{u} \in W^{1,1}(B(x,r))$. Since $u \in W^{1,2}(\Omega \cap B(x,r))$ by assumption and $|\nabla \widetilde{u}(\sigma(y))| = |\nabla u(y)|$ on $\Omega \cap B(x,r)$, we get that $\widetilde{u} \in W^{1,2}(B(x,r))$. So $\widetilde{u} \in W^{1,2}$ near points of $D \setminus \overline{K}$; since there is no problem away from P, we get (6).

Let us give a first statement with Mumford-Shah minimizers. Let $g \in L^\infty(\Omega)$ be given, and set

$$J(u, K) = H^{n-1}(K) + \int_{\Omega \setminus K} |\nabla u|^2 + \int_{\Omega \setminus K} |u - g|^2 \tag{7}$$

for K closed in Ω and $u \in W^{1,2}(\Omega \setminus K)$. Extend g to $\widetilde{\Omega}$ by symmetry (that is, set $\widetilde{g}(y) = g(\sigma(y))$ for $y \in \sigma(\Omega)$), and let $\widetilde{J}$ be the analogue of J on $\widetilde{\Omega}$, defined with the function $\widetilde{g}$.

Theorem 8 [DMS2]. *Keep the same notation; if (u, K) is a reduced minimizer for J, then $(\widetilde{u}, \widetilde{K})$ is a reduced minimizer for $\widetilde{J}$.*

We first prove that $(\widetilde{u}, \widetilde{K})$ is a minimizer. Let (v, G) be a competitor for $\widetilde{J}$, and let us check that $\widetilde{J}(\widetilde{u}, \widetilde{K}) \leq \widetilde{J}(v, G)$. We may assume that $\widetilde{J}(v, G) < +\infty$. Set

$$A = H^{n-1}(G \cap \Omega) + \int_{\Omega \setminus G} \left[|\nabla v|^2 + |v - \widetilde{g}|^2 \right] \quad \text{and}$$

$$B = H^{n-1}(G \cap \sigma(\Omega)) + \int_{\sigma(\Omega) \setminus G} \left[|\nabla v|^2 + |v - \widetilde{g}|^2 \right]. \tag{9}$$

First observe that $J(u, K) \leq A$, because (u, K) minimizes J and the restriction of (v, G) to Ω is a competitor for J. Similarly, $J(u, K) \leq B$, this time because the restriction to Ω of the pair (v^*, G^*) defined by $G^* = \sigma(G)$ and $v^*(y) = v(\sigma(y))$ is a competitor for J. Altogether, $2J(u, K) \leq A + B \leq \widetilde{J}(v, G)$.

On the other hand,

$$\widetilde{J}(\widetilde{u}, \widetilde{K}) = 2J(u, K) + H^{n-1}(D \cap \widetilde{K}) = 2J(u, K) + H^{n-1}(D \cap \overline{K}),$$

by the analogue of (7), (5), and (4). But Theorem 1 says that $H^{n-1}(D \cap \overline{K}) = 0$, so $\widetilde{J}(\widetilde{u}, \widetilde{K}) = 2J(u, K) \leq \widetilde{J}(v, G)$, as needed.

Now we check that $(\widetilde{u}, \widetilde{K})$ is reduced, and the simplest is to use the definition. Suppose that $\widetilde{u}$ extends to a function of $W^{1,2}(\widetilde{\Omega} \backslash Z)$ for some closed set Z contained in $\widetilde{K}$. Then $Z \cap \Omega = K$, because (u, K) is reduced. Similarly, $Z \cap \sigma(\Omega) = \sigma(K)$. Finally, Z contains $\overline{K} \cap D$, because it is closed in $\widetilde{\Omega}$ and contains K. Thus $Z = \widetilde{K}$, and $(\widetilde{u}, \widetilde{K})$ is reduced. Theorem 8 follows. $\square$

It is tempting to state the analogue of Theorem 8 for almost- and quasimin- imizers. We shall do this next, but first let us warn the reader not to focus too much on specifics. The main message of this section is that Theorem 8 is easy to extend to other situations and functionals, and the central point should always be that Theorem 1 allows us to extend u across most of D. We keep the same notation and assumptions on Ω as above, but forget about g.

Theorem 10. *If (u, K) is a reduced almost-minimizer on Ω, with gauge function h, then $(\widetilde{u}, \widetilde{K})$ is a reduced almost-minimizer on $\widetilde{\Omega}$, with the gauge function $2h$. If (u, K) is a reduced (r_0, M, a)-quasiminimizer in Ω and a is small enough, de- pending on n and M, then $(\widetilde{u}, \widetilde{K})$ is a reduced $(r_0, M^2, (M+1)a)$-quasiminimizer in $\widetilde{\Omega}$. The analogous statements for coral topological almost-minimizers and quasi- minimizers hold as well.*

Let us do the proof for topological quasiminimizers; the other cases are sim- pler. Let (v, G) be a topological competitor for $(\widetilde{u}, \widetilde{K})$ in a ball B, and let us try to prove the quasiminimality condition (7.20). We shall need to know that the restriction of (v, G) to Ω is a topological competitor for (u, K).

If $\overline{B}$ does not meet Ω, then (v, G) coincides with (u, K) on Ω, and there is nothing to prove. Otherwise, it is obvious that the restriction of (v, G) is a competitor for (u, K) in B (see Definition 7.2), so we just need to worry about the topological condition (7.16). We want to check that if y and $z \in \Omega \backslash [K \cup \overline{B}]$ lie in the same component of $\Omega \backslash G$, then they lie in the same component of $\Omega \backslash K$ too.

Notice that y and z lie in the same component of $\widetilde{\Omega} \backslash G$ (because $\Omega \subset \widetilde{\Omega}$), hence they lie in the same component of $\widetilde{\Omega} \backslash \widetilde{K}$, because (v, G) is a topological competitor for $(\widetilde{u}, \widetilde{K})$ in B. Let γ be a path from y to z in $\widetilde{\Omega} \backslash \widetilde{K}$. See Figure 1, where we decided to let P be vertical (to save some space). Since $\widetilde{\Omega} \backslash \widetilde{K}$ is open, we can replace γ with a polygonal line with the same properties. We can even make sure, by moving slightly the extremities of the segments that compose γ,

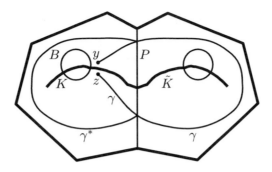

Figure 1

that none of them lies in D. Then γ only meets D on a finite set, and each time transversally.

Define σ^* on $\widetilde{\Omega}$ by $\sigma^*(x) = x$ for $x \in \Omega \cup D$ and $\sigma^*(x) = \sigma(x)$ for $x \in \sigma(\Omega)$, and set $\gamma^* = \sigma^*(\gamma)$. Then $\gamma^* \subset \gamma \cup \sigma(\gamma)$ does not meet $\widetilde{K}$, because $\widetilde{K}$ is symmetric and does not meet γ. Of course γ^* is not necessarily contained in Ω, because it can have a finite number of points on D. But each of these points lies on $D \setminus \widetilde{K} = D \setminus \overline{K}$, and we can modify γ^* near each such point so that it stays in $\Omega \setminus K$. Then we get a path in $\Omega \setminus K$ that goes from y to z, as needed.

Thus the restriction of (v, G) to Ω is a topological competitor for (u, K) in B. Call r the radius of B; if $r \leq r_0$, we can apply (7.20) to (u, K). We get that

$$H^{n-1}(K \setminus G) \leq M H^{n-1}([G \cap \Omega] \setminus K) + \partial E + ar^{n-1}, \tag{11}$$

where $\partial E = \mathrm{Max}\{F - E, M(F - E)\}$, $E = \int_{\Omega \cap B \setminus K} |\nabla u|^2$, and $F = \int_{\Omega \cap B \setminus G} |\nabla v|^2$.

Now set $G_1 = \sigma(G)$ and $v_1(z) = v(\sigma(z))$ on $\widetilde{\Omega} \setminus G_1$. By symmetry of $(\widetilde{u}, \widetilde{K})$, (v_1, G_1) is a topological competitor for $(\widetilde{u}, \widetilde{K})$ in $\sigma(B)$, and the argument above yields the analogue of (11) for the restriction of (v_1, G_1) to Ω. That is,

$$H^{n-1}(K \setminus G_1) \leq M H^{n-1}([G_1 \cap \Omega] \setminus K) + \partial_1 E + ar^{n-1}, \tag{12}$$

where we set $\partial_1 E = \mathrm{Max}\{F_1 - E_1, M(F_1 - E_1)\}$, $E_1 = \int_{\Omega \cap \sigma(B) \setminus K} |\nabla u|^2$, and $F_1 = \int_{\Omega \cap \sigma(B) \setminus \sigma(G)} |\nabla v_1|^2 = \int_{\sigma(\Omega) \cap B \setminus G} |\nabla v|^2$. We want to add (11) and (12). Observe that $K \setminus G_1 = K \setminus \sigma(G) = \widetilde{K} \cap \Omega \setminus \sigma(G) = \sigma(\widetilde{K} \cap \sigma(\Omega) \setminus G)$, so

$$H^{n-1}(K \setminus G) + H^{n-1}(K \setminus G_1)$$
$$= H^{n-1}(\widetilde{K} \cap \Omega \setminus G) + H^{n-1}(\sigma(\widetilde{K} \cap \sigma(\Omega) \setminus G)) \tag{13}$$
$$= H^{n-1}(\widetilde{K} \cap \Omega \setminus G) + H^{n-1}(\widetilde{K} \cap \sigma(\Omega) \setminus G) = H^{n-1}(\widetilde{K} \setminus G),$$

where the last equality comes from the fact that $H^{n-1}(\widetilde{K} \cap D) = H^{n-1}(\overline{K} \cap D) = 0$ by Theorem 1, and requires a to be small enough. Next

$$H^{n-1}([G \cap \Omega] \setminus K) + H^{n-1}([G_1 \cap \Omega] \setminus K) \tag{14}$$
$$= H^{n-1}([G \cap \Omega] \setminus \widetilde{K}) + H^{n-1}([\sigma(G) \cap \Omega] \setminus \widetilde{K})$$
$$= H^{n-1}([G \cap \Omega] \setminus \widetilde{K}) + H^{n-1}([G \cap \sigma(\Omega)] \setminus \widetilde{K}) \le H^{n-1}(G \setminus \widetilde{K}).$$

For ∂E and $\partial_1 E$, things will be less pleasant and it will be easier to distinguish cases. Set

$$E' = E + E_1 = \int_{\Omega \cap B \setminus K} |\nabla u|^2 + \int_{\Omega \cap \sigma(B) \setminus K} |\nabla u|^2$$
$$= \int_{\Omega \cap B \setminus K} |\nabla u|^2 + \int_{\sigma(\Omega) \cap B \setminus \widetilde{K}} |\nabla \widetilde{u}|^2 = \int_{\widetilde{\Omega} \cap B \setminus \widetilde{K}} |\nabla \widetilde{u}|^2 \tag{15}$$

because $\widetilde{u}$ is symmetric. Also set

$$F' = F + F_1 = \int_{\Omega \cap B \setminus G} |\nabla v|^2 + \int_{\sigma(\Omega) \cap B \setminus G} |\nabla v|^2 = \int_{\widetilde{\Omega} \cap B \setminus G} |\nabla v|^2; \tag{16}$$

thus

$$F' - E' = \int_{\widetilde{\Omega} \cap B \setminus G} |\nabla v|^2 - \int_{\widetilde{\Omega} \cap B \setminus \widetilde{K}} |\nabla \widetilde{u}|^2. \tag{17}$$

Also set $\partial' E = \text{Max}\{F' - E', M(F' - E')\}$.

If $F - E$ and $F_1 - E_1$ are both nonnegative, $\partial E + \partial_1 E = M(F - E) + M(F_1 - E_1) = M(F' - E') = \partial' E$. If $F - E$ and $F_1 - E_1$ are both nonpositive, $\partial E + \partial_1 E = (F - E) + (F_1 - E_1) = F' - E' = \partial' E$. In both cases, we add (11) and (12) and get that

$$H^{n-1}(\widetilde{K} \setminus G) \le M H^{n-1}(G \setminus \widetilde{K}) + \partial' E + 2a r^{n-1}, \tag{18}$$

by (13) and (14). This is the quasiminimality condition (7.20) that we want, by (17).

Next assume that $F - E$ is positive and $F_1 - E_1$ is negative. Then $\partial E = M(F - E)$ and $\partial_1 E = F_1 - E_1$. In this case we multiply (12) by M and add it to (11). This yields

$$H^{n-1}(\widetilde{K} \setminus G) \le H^{n-1}(K \setminus G) + M H^{n-1}(K \setminus G_1)$$
$$\le M H^{n-1}([G \cap \Omega] \setminus K) + \partial E + a r^{n-1}$$
$$\quad + M^2 H^{n-1}([G_1 \cap \Omega] \setminus K) + M \partial_1 E + M a r^{n-1}$$
$$\le M^2 H^{n-1}(G \setminus \widetilde{K}) + M(F - E + F_1 - E_1) + (1 + M) a r^{n-1}$$
$$\le M^2 H^{n-1}(G \setminus \widetilde{K}) + M(F' - E') + (1 + M) a r^{n-1} \tag{19}$$

by (13), (11), (12), and (14). Now $M(F' - E') \le \text{Max}\{F' - E', M^2(F' - E')\}$, so (19) is the desired quasiminimality condition, with constants M^2 and $(M + 1)a$.

The last case when $F - E$ is negative and $F_1 - E_1$ positive goes the same way, except that we multiply (11) by M and add to (12).

Thus $(\widetilde{u}, \widetilde{K})$ is a topological $(r_0, M^2, (M+1)a)$-quasiminimizer in $\widetilde{\Omega}$. We still need to check that it is coral. But if a is small enough $\widetilde{K}$ is locally Ahlfors-regular, by Sections 72 and 77, and hence it is coral. [See Definition 8.24.] Actually, we did not need the local Ahlfors-regularity away from D, because we already knew that K is coral in Ω, but for D we seem to need something.

This completes the case of coral topological quasiminimizers. The case of reduced quasiminimizers is simpler (we do not need to worry about topology, and reduction is dealt with as in Theorem 8). Finally, almost-minimizers are treated like quasiminimizers, except that the algebra is simpler (and almost the same as for Theorem 8). This completes our proof of Theorem 10. □

Remark 20. We did not mention topologically reduced topological almost- and quasiminimizers in Theorem 10, because it is not clear that this is the right notion here. In Definition 8.7, one only compares K with sets $K' \subset K$ that coincide with K out of a ball $B \subset\subset \Omega$; this does not allow a trivial proof of topological reduction like the one in Theorem 8, because we could also remove pieces of $\widetilde{K}$ near D. But anyway Theorem 10 also holds with topologically reduced almost- and quasiminimizers. The point is that Sections 72 and 77 still say that $\widetilde{K}$ is locally Ahlfors-regular, and then we can get the topological reduction as in Remark 8.31: if we had a topological competitor for an extension of $\widetilde{u}$ to $\widetilde{\Omega} \setminus H$, with $H \subset \widetilde{K}$, we could use (7.20) to show that $H^{n-1}(\widetilde{K} \setminus H) = 0$, and then conclude that $H = K$.

c. Other reflections, linear distortion

It is tempting to extend the results of the previous subsection to less rigid reflections. The general setup would be the following. Let Ω be a locally Lipschitz domain, and suppose we have a bilipschitz reflection σ defined on $\overline{\Omega}$, such that $\overline{\Omega} \cap \sigma(\overline{\Omega}) \subset \partial\Omega$ and $\sigma(x) = x$ on $\overline{\Omega} \cap \sigma(\overline{\Omega})$. Call D the open subset of $\partial\Omega$ of points $x \in \partial\Omega$ such that for r small, $B(x,r) \setminus \partial\Omega$ is the disjoint union of $\Omega \cap B(x,r)$ and $\sigma(\Omega) \cap B(x,r)$.

Given a pair (u, K) as above, we introduce the extended pair $(\widetilde{u}, \widetilde{K})$ on $\widetilde{\Omega} = \Omega \cup \sigma(\Omega) \cup D$, which we still define by (4) and (5), and we consider minimality and quasiminimality properties of $(\widetilde{u}, \widetilde{K})$.

Theorem 8 generalizes easily to this context, except that now the functional $\widetilde{J}$ on $\widetilde{\Omega}$ is no longer of Mumford-Shah type. Its simplest expression is

$$\widetilde{J}(v, G) = H^{n-1}(G \cap [\Omega \cup D]) + H^{n-1}(\sigma(G \cap \sigma(\Omega))) \tag{21}$$

$$+ \int_{\Omega \setminus G} \left[|\nabla v|^2 + |v - g|^2 \right] + \int_{\Omega \setminus \sigma(G)} \left[|\nabla v(\sigma(x))|^2 + |v(\sigma(x)) - g(\sigma(x))|^2 \right] dx,$$

but of course one could write it as a functional on $\widetilde{\Omega}$, using the differential of σ and a change of variables. When we do this, we get a more complicated type of surface term (where Hausdorff measure is multiplied by a factor that also depends on the direction of the tangent plane), and similarly for the energy integral.

Notice already that $(\widetilde{u}, \widetilde{K})$ is no longer an almost-minimizer in $\widetilde{\Omega}$, or even a quasiminimizer, because $(\nabla v) \circ \sigma$ is obtained from $\nabla (v \circ \sigma)$ by composition with a linear mapping which does not need to be conformal. If our reflection σ is sufficiently smooth, though, $(\widetilde{u}, \widetilde{K})$ is an almost-minimizer with distortion, as in Section 7.c (see Exercise 7.60).

Let us now try to generalize Theorem 10, and see how far we can get easily. The simplest case is when

$$n = 2, \text{ and } \sigma \text{ is bilipschitz on } \overline{\Omega} \text{ and conformal on } \Omega. \tag{22}$$

In this case Theorem 10 for quasiminimizers generalizes rather directly; we just lose factors k in the quasiminimality constants M and a, where k is the bilipschitz constant for σ, and also r_0 has to be replaced with $k^{-1}r_0$. The main point is that the energy integrals are invariant by σ because σ is conformal, while Hausdorff measures are at most multiplied or divided by k because σ is bilipschitz.

There is no direct analogue for almost-minimizers, even with smooth conformal mappings, because when we compose by σ, the length term is locally multiplied by the derivative, while the energy stays the same; this creates a problem because almost-minimizers stop being almost-minimizers when we multiply the length term by a factor. Or, put in another way, if σ looks like a dilation by λ somewhere, the right way to get invariance would be to use $\lambda^{-1/2}u(\sigma(y))$ instead of $u(\sigma(y))$.

In higher dimensions, we still get an analogue of Theorem 10 for quasiminimizers when σ is both C^1 and conformal; this time the energy is not exactly invariant, but it is multiplied by $|D\sigma(y)|^{n-2}$, which is bounded and continuous. See Proposition 7.27 for hints on the proof.

Now it is not so pleasant to restrict to conformal reflections, and the discussion above seems to lead to two conclusions.

The first one is that maybe we should not try to prove that the extended pair $(\widetilde{u}, \widetilde{K})$ is an almost-minimizer, or at least ask what we want to do with such a result. If we just want to prove some regularity, maybe knowing that we started from an almost-minimizer (u, K), and using regularity results up to the boundary (including Theorem 1) will be enough. In this case it is perhaps enough to know that (u, K) is a quasiminimizer (even, with linear distortion). The same comment applies if we just want to show that $(\widetilde{u}, \widetilde{K})$ minimizes, or almost minimizes some functional. In this case Theorem 1 should be enough.

The second conclusion is that, if we really want analogues of Theorem 10, we should consider almost-minimizers (or quasiminimizers) with linear distortion, as in Section 7.b. The notions are nicely invariant under C^1 changes of variables with uniform bounds on the derivative and its inverse (see Exercise 7.60). In addition, Theorem 1 is still valid for quasiminimizers with linear distortion $\leq M$, because we still have local Ahlfors-regularity at the boundary (see Remark 77.37). Then the proof of Theorem 10 should go through.

Once we take the decision to allow linear distortion, we also have the option of trying to reduce to reflection across a hyperplane, by first mapping Ω to a simpler domain with a flat piece. The two methods should be roughly equivalent.

79 Boundary regularity for almost-minimizers in dimension 2

The main result of this section is the following.

Theorem 1. *Let Ω be a bounded C^1 domain in $\mathbb{R}^2$, and let (u, K) be a reduced almost-minimizer on Ω, or a topologically reduced (or coral) topological almost-minimizer on Ω, with a gauge function h such that $\int_0^1 h(t)^{1/2} \frac{dt}{t} < +\infty$. Then we can find $\tau > 0$ such that if we set*

$$\Omega(\tau) = \left\{ x \in \Omega \,;\, \mathrm{dist}(x, \partial\Omega) \leq \tau \right\}, \tag{2}$$

then $\overline{K} \cap \Omega(\tau)$ is a finite, disjoint union of C^1 curves that cross $\Omega(\tau)$ and end on $\partial\Omega$ perpendicularly.

This result was first proved in [MaSo4], and merely announced in [DaLé] at about the same time. The proofs are somewhat different, but both rely on blow-ups and monotonicity. Here we give the proof suggested in [DaLé], both because it is simpler once we trust Section 66c and because we promised in [DaLé] that we would do the missing details here.

Theorem 1 can easily be improved; we shall not try to give optimal statements to this effect, but state our lemmas so that it should be reasonably easy to formulate improvements. See Remark 41 for small weakenings of the C^1 assumption on Ω, and Remark 42 for a rapid discussion about the better regularity of K near $\partial\Omega$ and the easy adaptation of the results of Part E. We start the proof with a few lemmas.

In the following, we shall assume that Ω is a domain in $\mathbb{R}^2$, and (u, K) is a reduced almost-minimizer on Ω, or a topologically reduced (or coral) topological almost-minimizer on Ω. Suitable assumptions on the regularity of $\partial\Omega$ will come along the way.

Lemma 3. *For each $\lambda > 0$ we can find constants $\varepsilon > 0$ and $\eta > 0$ such that if $x \in \overline{K} \cap \partial\Omega$ and $r > 0$ are such that*

$$B(x, r) \text{ is a } \lambda\text{-Lipschitz ball for } \partial\Omega \tag{4}$$

(see (77.3)) and $h(r) \leq \varepsilon$, then we can find $y \in K \cap B(x,r)$ such that $\mathrm{dist}(y, \partial\Omega) > \eta r$.

Suppose instead that

$$K \cap B(x,r) \subset \Omega(\eta r), \tag{5}$$

where $\Omega(\eta r)$ is as in (2). We want to construct a competitor (v, G) and contradict the almost-minimality of (u, K); the idea will be to push K out of Ω and save some length. Let us first choose coordinates, as in (77.3), so that $x = 0$ and

$$\Omega \cap B(x,r) = \big\{(\xi, \zeta) \in B(x,r) \, ; \, \zeta > F(\xi)\big\} \tag{6}$$

for some λ-Lipschitz function F such that $F(0) = 0$.

Let us first choose a box B and an annulus A. Set $r_1 = \frac{1}{3}(1 + \lambda)^{-1}r$, $B = (-r_1, r_1) \times (-(1+\lambda)r_1, (1+\lambda)r_1)$ and then $B_+ = (1+\tau)B$ and $A = \overline{B}_+ \setminus B$, where $\tau \in (0, 10^{-2})$ will be chosen soon. Notice that $B_+ \subset B(x, 2r/3)$.

We shall also need a security zone $D \subset A$, near the graph, so we set

$$D_0 = \big\{(\xi, \zeta) \in B(x,r) \, ; \, F(\xi) < \zeta \leq F(\xi) + (1+\lambda)\eta r\big\} \tag{7}$$

and $D = D_0 \cap A$ (see Figure 1). Let us check that

$$K \cap B^+ \subset \Omega(\eta r) \cap B^+ \subset D_0. \tag{8}$$

The first inclusion comes from (5); for the second one, let $y = (\xi, \zeta)$ lie in $\Omega(\eta r)$. Thus there is $y_1 = (\xi_1, \zeta_1) \in \partial\Omega$ such that $|y_1 - y| \leq \eta r$. Hence $F(\xi_1) = \zeta_1$ and $|\xi_1 - \xi| \leq \eta r$. Now $\zeta \leq \zeta_1 + \eta r = F(\xi_1) + \eta r \leq F(\xi) + \lambda|\xi_1 - \xi| + \eta r \leq F(\xi) + (1+\lambda)\,\eta r$; (8) follows.

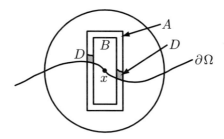

Figure 1

We shall use a smooth bijection ψ to push $K \cap B$ out of Ω. The simplest is to take ψ so that it only moves the points vertically. That is, if we write $y = (\xi, \zeta)$, we take $\psi(y) = (\xi, \zeta - \varphi(y))$ for some function φ. We choose φ such that $\varphi(y) = 0$ out of B_+, $\varphi(y) = (1 + \lambda)\,\eta r$ on B, $0 \leq \varphi(y) \leq (1 + \lambda)\,\eta r$ on A, and $||D\psi - I|| \leq C(\lambda, \tau)\,\eta$ everywhere (which, incidentally, is a way to make sure that ψ is bijective when η is small enough). Now set

$$G = [K \setminus B_+] \cup [\partial D \cap \Omega], \tag{9}$$

and observe that G is closed in Ω because $K \cap A \subset D$, by (8). We want to define v by $v(y) = 0$ for $y \in D$ and

$$v(y) = u(\psi^{-1}(y)) \text{ for } y \in \Omega \setminus [G \cup D], \tag{10}$$

so let us check that this makes sense.

For $y \in \Omega \setminus B^+$, $\psi^{-1}(y) = y$ and (10) says that $v(y) = u(y)$. In the remaining cases when $y \in B^+$, $\psi^{-1}(y) \in B^+$ too, because ψ is bijective and leaves $\mathbb{R}^2 \setminus B^+$ alone.

If $y \in A \setminus D$, $\psi^{-1}(y)$ lies above y, hence $\psi^{-1}(y) \in A \setminus D \subset \Omega \setminus K$, by (8).

If $y \in B \cap \Omega$ and $\psi^{-1}(y) \in B$, then $\psi^{-1}(y)$ lies at least $(1 + \lambda)\eta r$ above y (by definition of φ), so $\psi^{-1}(y) \in \Omega \setminus K$, again by (8).

Finally, if $y \in B \cap \Omega$ but $\psi^{-1}(y) \notin B$, $\psi^{-1}(y)$ lies in (the top part of) A, hence in $\Omega \setminus K$ again. Thus (10) makes sense in all cases, and it even defines a function $v \in W^{1,2}_{\text{loc}}(\Omega \setminus K)$.

Thus (v, G) is a competitor for (u, v) in $B_+ \subset B(x, r)$. It is even a topological competitor, by the proof of Lemma 20.10. Notice that

$$H^1(G \cap B(x, r)) \leq H^1(K \cap B(x, r) \setminus B^+) + H^1(\partial D \cap \Omega)$$
$$\leq H^1(K \cap B(x, r)) - C_0^{-1} r_1 + (1 + \lambda)(4\eta r + 2\tau r_1) \tag{11}$$

by (9), (7), and Theorem 77.4. Here C_0 may depend on λ, but not on τ or η. Next

$$\int_{\Omega \cap B_+ \setminus G} |\nabla v|^2 \leq \left(1 + C(\lambda, \tau)\eta\right) \int_{\Omega \cap B_+ \setminus D} |\nabla u(\psi^{-1}(y))|^2$$
$$\leq \left(1 + C(\lambda, \tau)\eta\right) \int_{\Omega \cap B_+ \setminus K} |\nabla u|^2 \leq \int_{\Omega \cap B_+ \setminus K} |\nabla u|^2 + C(\lambda, \tau)\eta r \tag{12}$$

because $\|DF - I\| \leq C(\lambda, \tau)\eta$ everywhere, $\psi^{-1}(\Omega \cap B_+ \setminus D) \subset \Omega \cap B_+ \setminus K$, and by a change of variables. Altogether,

$$H^1(G \cap B(x, r)) + \int_{\Omega \cap B(x,r) \setminus G} |\nabla v|^2 \leq H^1(K \cap B(x, r)) + \int_{\Omega \cap B(x,r) \setminus K} |\nabla u|^2$$
$$- C_0^{-1} r_1 + (1 + \lambda)(4\eta r + 2\tau r_1) + C(\lambda, \tau)\eta r. \tag{13}$$

On the other hand, the almost-minimality of (u, K) says that

$$H^1(K \cap B(x, r)) + \int_{\Omega \cap B(x,r) \setminus K} |\nabla u|^2 \leq H^1(G \cap B(x, r)) + \int_{\Omega \cap B(x,r) \setminus G} |\nabla v|^2 + h(r)r, \tag{14}$$

so $(1 + \lambda)(4\eta r + 2\tau r_1) + h(r)r \leq C_0^{-1} r_1$. Recall that $r_1 = \frac{1}{3}(1 + \lambda)^{-1}r$; now we get the desired contradiction if $h(r)$ is small enough and if we choose τ, and then η small enough (all depending on λ). Thus (5) was impossible and Lemma 3 holds. $\qquad \square$

Remark 15. Lemma 3 still holds for quasiminimizers (with a sufficiently small, depending on M and λ). The proof is almost the same. Also, Lemma 3 extends to $\mathbb{R}^n$, but we have to take λ small enough (depending on n and M), because we need to apply Theorem 77.5.

Lemma 16. *For every choice of $\alpha < 3\pi/2$, $\lambda > 0$, and $\delta > 0$, we can find $\varepsilon > 0$ and $M \geq 1$, with the following property. Let $x \in \partial\Omega$ and $r > 0$ be such that $K \cap B(x, r) \neq \emptyset$ and $h(Mr) \leq \varepsilon$. Also assume that*

$$B(x, 10r) \text{ is a } \lambda\text{-Lipschitz ball for } \partial\Omega \tag{17}$$

and that there is an open cone $\mathcal{C} = \mathcal{C}(x, Mr)$, with vertex x and aperture at most α, such that

$$\begin{aligned} \sup\{\operatorname{dist}(y, \mathcal{C})\,;\, y \in \Omega \cap B(x, Mr)\} \\ + \sup\{\operatorname{dist}(y, \Omega)\,;\, y \in \mathcal{C} \cap B(x, Mr)\} \leq \varepsilon Mr. \end{aligned} \tag{18}$$

Then there is a half-line $L \subset \mathcal{C}$, with its extremity on $B(x, \frac{3r}{2}) \cap \partial\mathcal{C}$, such that

$$\begin{aligned} \operatorname{dist}(y, L) \leq \delta r \text{ for } y \in K \cap B(x, 10r) \setminus \Omega(\delta r) \\ \text{and } \operatorname{dist}(y, K) \leq \delta r \text{ for } y \in L \cap B(x, 10r). \end{aligned} \tag{19}$$

See (77.3) for the definition of λ-Lipschitz balls and (2) for $\Omega(\delta r)$. There are additional constraints on L and the aperture of $\mathcal{C}(x, Mr)$, but we did not write them down yet to simplify the statement; see (23) and (24). We shall also see later a few improvements on the conclusion, but let us start with this. We shall prove the lemma by contradiction and compactness, so let us fix α, λ, and δ, assume that there are pairs (u_k, K_k) and balls $B(x_k, r_k)$ that satisfy the hypotheses with $M_k = 2^n$ and $\varepsilon_k = 4^{-n}$, but not the conclusions, and try to derive a contradiction.

By dilation invariance, we may assume that $x_k = 0$ and $r_k = 1$. We may extract a first subsequence, so that (after extraction) $\mathcal{C}(x_k, M_k r_k)$ converges to some cone $\mathcal{C}$. Then (18) says that the domains Ω_k converge to $\mathcal{C}$, as in (39.5) and (39.32).

By Proposition 37.8, we can extract a new subsequence so that $\{(u_k, K_k)\}$ converges to some pair (v, L). Then we want to apply Theorem 39.34. Remark 39.42 allows us to do this with the cone $\mathcal{C}$ as our limiting domain. We also need to choose a gauge function h that works for every pair (u_k, K_k), k large enough, so we set $h(t) = \sup_{k \geq k_0} h_k(t)$, where h_k corresponds to (u_k, K_k). Since each h_k is nondecreasing and tends to 0 at 0, and $h_k(t) \leq 4^{-k}$ for $t \leq 2^k$ by construction, it is easy to see that $\lim_{t \to 0^+} h(t) = 0$. Obviously, h is nondecreasing too, so it is a gauge function. Thus Theorem 39.34 says that (v, L) is a coral topological almost-minimizer on $\mathcal{C}$, with gauge function $h(4t)$. Since $h(t) \leq 4^{-k_0}$ for $t \leq 2^{k_0}$ and we can take k_0 as large as we want, we even get that (v, G) is a minimizer. With the notation of Section 66c, $(v, G) \in RGM(\mathcal{C})$.

Now the aperture of $\mathcal{C}$ is at most $\alpha < 3\pi/2$, so Theorem 66.39 says that either L is empty and v is constant, or else L is a half-line with its origin on $\partial\mathcal{C}$, and v is locally constant.

We want to get additional information on L, and it will be useful to know that for k large,

$$K_k \cap B(0, \frac{12}{10})) \text{ is not contained in } \Omega_k(\eta), \tag{20}$$

where η is a small positive constant that depends on λ. Indeed K_k meets $B(0,1)$, and (17) allows us to apply (72.1) or Theorem 77.4. So we get that

$$H^1(K_k \cap B(0, \frac{11}{10})) \geq C(\lambda)^{-1}. \tag{21}$$

Then we can use the proof of Lemma 3 to get (20). The main point is that in the lemma, we only used our assumption that $x \in \overline{K}$ to get the analogue of (21) and use it in (11). In the present situation, in order to deduce (20) as in Lemma 3, we would also need to replace $B = B(0, \frac{11}{10})$ by a ball B' of radius $(1+\lambda)^{-1}/100$ contained in B and such that $H^1(K_k \cap B')$ is still reasonably large, so that there is enough room to construct the boxes, but this is only a cosmetic change. So (20) holds.

Return to our minimizer (v, L). Choose a point y_k in $K_k \cap B(0, \frac{12}{10}) \setminus \Omega_k(\eta)$ for k large; then there is a subsequence for which $\{y_k\}$ converges to a point $y \in \overline{B}(0, \frac{12}{10})$. Since $\mathcal{C}$ is the limit of the cones in (18), we get that $y \in \mathcal{C}$ and $\text{dist}(y, \partial\mathcal{C}) \geq \eta$. Then $y \in L$, by definition of convergence for the pairs (u_k, K_k). Thus L is a half-line, and it passes through $\overline{B}(0, \frac{12}{10})$.

Recall from Section 69c that we have the following additional constraints on L and $\mathcal{C}$. First, the aperture of $\mathcal{C}$ is at least π. Call x_0 the extremity of L. If $x_0 = 0$, L makes angles at least $\pi/2$ with the two branches of $\partial\mathcal{C}$. If not, L is perpendicular to the branch of $\partial\mathcal{C}$ that contains x_0.

Call $\mathcal{C}_k$ the cone associated to Ω_k in (18), and let L_k be a half-line in $\mathcal{C}_k$ that starts from $\partial\mathcal{C}_k$ and is $\delta/2$-close to L in $B(0, 10)$. Such a half-line exists for k large, because $\mathcal{C}_k$ tends to $\mathcal{C}$. Now (19) holds for k large, with this choice of L_k and by definition of the convergence of K_k to L. In particular, this is why, in the first half of (19), we restricted to $y \in B(x, 10r) \setminus \Omega(\delta r)$, which is contained in a fixed compact subset of $\mathcal{C}$ for k large. We do not need to be so careful for the second part, as all points of $L \cap B(x, 10r)$ lie close to some point of $L \cap B(x, 11r) \setminus \Omega(\delta r/2)$. Now recall that (19) was not supposed to hold, by definition (u_k, K_k); this contradiction completes our proof of Lemma 16. $\qquad\square$

Remark 22. Since we have additional information on L, we can use it in Lemma 16. For instance, the fact that the aperture of $\mathcal{C}$ is at least π implies that

$$\text{the aperture of } \mathcal{C}(x, Mr) \text{ is at least } \pi - \delta \tag{23}$$

in the proof of Lemma 16. Also, if x_0 lies in $B(0, \delta/10)$, then we can choose L_k so that it starts from the origin and makes angles at least $\frac{\pi}{2} - \delta$ with the two branches of $\mathcal{C}_k$. If $x_0 \notin B(0, \delta/10)$, we can choose L_k so that it starts on one branch of $\partial \mathcal{C}_k$ and is perpendicular to that branch. Translated in terms of Lemma 16 itself, this means that we can pick L such that

$$L \text{ starts from } x \text{ and makes angles} \geq \tfrac{\pi}{2} - \delta \text{ with the two}$$
$$\text{branches of } \mathcal{C}(x, Mr), \text{ or else is perpendicular} \qquad (24)$$
$$\text{to the branch of } \partial\mathcal{C}(x, Mr) \text{ that contains its extremity.}$$

Let us continue with the hypotheses of Lemma 16, and derive a little more information. First observe that we could replace (19) with

$$\operatorname{dist}(y, L) \leq \delta r \text{ for } y \in K \cap B(x, 9r)$$

and

$$\operatorname{dist}(y, K) \leq \delta r \text{ for } y \in L \cap B(x, 10r). \qquad (25)$$

That is, we can also control $K \cap B(x, 9r) \cap \Omega(\delta r)$. Indeed, pick δ' much smaller than δ, and apply Lemma 16. We get a half-line L for which the analogue of (19) holds, with δ replaced with δ'.

Now suppose that (25) does not hold, and let $y \in K \cap B(x, 9r)$ be such that $\operatorname{dist}(y, L) > \delta r$. Set $B = B(y, \frac{\delta}{2(1+\lambda)})$. Since B lies inside $B(0, 10)$ and far from L, (19) says that $K \cap B \subset \Omega(\delta' r)$ (i.e., very close to $\partial \Omega$). On the other hand, $H^1(K \cap B) \geq C^{-1}\delta$, by (17) and Theorem 77.4 or (72.1). If δ'/δ is sufficiently small, the proof of Lemma 3 gives a contradiction, as before; (25) follows.

Lemma 26. *Keep the notation, quantifiers, and assumptions of Lemma* 16. *Then*

$$\int_{\Omega \cap B(x, 9r) \setminus K} |\nabla u|^2 \leq \delta r. \qquad (27)$$

Let us prove this like Lemma 16, by contradiction and compactness. Choose pairs (u_k, K_k) as above, except that our contradiction assumption is now that

$$\int_{\Omega_k \cap B(0,9) \setminus K_k} |\nabla u_k|^2 \geq \delta. \qquad (28)$$

As before, we extract a subsequence so that $\{(u_k, K_k)\}$ converges to (L, g), and L is a half-line and v is locally constant on $\mathcal{C} \setminus L$. The general idea is that we should get a contradiction because of Corollary 38.48, but since the corollary was stated for limits of local minimizers in a fixed domain, let us give a rapid proof. The main difference with the local case is that we also need to worry about energy pockets that would go to the boundary.

Pick a small constant $\eta > 0$, and set $H = \{y \in \overline{B}(0, 10) \, ; \, \operatorname{dist}(y, L \cup \partial\mathcal{C}) \geq \eta\}$. This is a compact subset of $\mathcal{C} \setminus L$, so $H \subset \Omega_k \setminus K_k$ for k large. In addition,

Proposition 37.25 says that if H_1 and H_2 denote the two connected components of H, we can find constants $a_{k,1}$ and $a_{k,2}$ such that

$$\{u_k - a_{k,i}\} \text{ converges uniformly to } v \text{ on } H_i \tag{29}$$

for $i = 1, 2$. Since we may always replace our sequence with a subsequence, we can assume that

$$|u_k - a_{k,i} - v| \leq \varepsilon_k \text{ on } H_i, \tag{30}$$

where the rapidly decreasing sequence $\{\varepsilon_k\}$ will be chosen soon.

We want to construct a competitor for (u_k, K_k) in $B(0,10)$, so we choose a radius $\rho \in (9,10)$ such that each u_k coincides, almost everywhere on $\partial B(0,\rho) \cap H$, with its radial limits on both sides (recall that this property holds for almost every ρ because $H \subset \Omega_k \setminus K_k$), and in addition

$$\int_{\partial B(0,\rho) \setminus K_k} |\nabla u_k|^2 \leq C2^k \tag{31}$$

for k large. This is possible, because $\int_{B(0,10) \setminus K_k} |\nabla u_k|^2 \leq 100$ by (the proof of) the trivial estimate (18.8). [We lose 2^k because we want a single ρ that works for all k.] Now set $B = B(0, \rho)$ and

$$G_k = [K_k \setminus B] \cup [L' \cap \Omega_k \cap B] \cup Z, \tag{32}$$

where L' is the line that contains L and

$$Z = \partial B \cap \Omega_k \setminus \text{int}(H) = \{z \in \partial B \cap \Omega_k \,;\, \text{dist}(z, L \cup \partial C) \leq \eta\} \tag{33}$$

(see Figure 2).

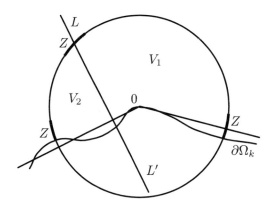

Figure 2

We keep $v_k = u_k$ out of B, and define v_k independently on each of the two components of $\Omega \cap B \setminus L'$. Call V_1 and V_2 these components, and set $I_j =$

$\partial V_j \cap [\Omega \setminus G_k]$. Notice that I_j is one of the two components of $\partial B \cap \Omega \setminus Z$, and that it is contained in one of the two H_i; let us choose our names so that $I_j \subset H_j$.

To define v_k on V_j, we take the restriction of u_k to I_j, extend it to $\overline{B}$, and restrict to V_j. As usual, this defines a pair (v_k, G_k) which is a topological competitor for (u_k, K_k) in B (by Lemma 45.8). For the energy estimate, we use Lemma 38.26. Recall that this lemma says that for each $\tau > 0$ we can find $C(\tau)$ such that in the present situation

$$\int_{V_j} |\nabla v_k|^2 \leq \tau \int_{I_j} |\nabla u_k|^2 + C(\tau) \left\{ \int_{I_j} |u_k - m_{k,j}| \right\}^2, \qquad (34)$$

where $m_{k,j}$ denotes the mean value of u_k on I_j. Here we take $\tau = \tau_k = 4^{-k}$. Since $|u_k - m_{k,j}| \leq \varepsilon_k$, by (30), we get that

$$\int_{\Omega_k \cap B \setminus L'} |\nabla v_k|^2 = \int_{V_1 \cup V_2} |\nabla v_k|^2 \leq C 2^{-k} + C\, C(\tau_k)\, \varepsilon_k^2 \leq C 2^{-k} \qquad (35)$$

by (31) and if ε_k is sufficiently small.

We also need to estimate lengths. First,

$$H^1(G_k \cap \overline{B}) = H^1(L' \cap \Omega_k \cap B) + H^1(Z) \leq H^1(L' \cap \Omega_k \cap B) + \delta/10 \qquad (36)$$

by (32), (33), and if we choose η small enough. The main point of the argument is that, thanks to the uniform concentration property for the K_k, the convergence of K_k to L yields

$$H^1(L \cap B) \leq \liminf_{k \to +\infty} H^1(K_k \cap B). \qquad (37)$$

See the proof of Lemma 38.6, and notice that here the local result is enough. Thus

$$H^1(L' \cap \Omega_k \cap B) \leq H^1(L \cap B) + \delta/10 \leq H^1(K_k \cap B) + 2\delta/10 \qquad (38)$$

for k large, by definition of L', and by (37). We now add (35), (36), and (38) to get that

$$\int_{\Omega_k \cap B \setminus L'} |\nabla v_k|^2 + H^1(G_k \cap \overline{B}) \leq C 2^{-k} + H^1(K_k \cap B) + 3\delta/10$$

$$\leq \int_{\Omega_k \cap B \setminus K_k} |\nabla u_k|^2 + H^1(K_k \cap B) + C 2^{-k} - 7\delta/10, \qquad (39)$$

by (28). This obviously contradicts the almost-minimality of (u_k, K_k) for k large; Lemma 26 follows. $\square$

Next we claim that if the assumptions of Lemma 16 are satisfied, with δ small enough, and in addition $\int_0^r h(t)^{1/2} \frac{dt}{t}$ is small enough, then

$$K \cap B(z, r/2) \text{ is a nice } C^1 \text{ curve for every } z \in K \cap B(x, 8r) \setminus \Omega(2r). \qquad (40)$$

Indeed $B(z, r) \subset B(x, 9r) \setminus \Omega(r)$, so (19) says that $\beta_K(z, r) \leq \delta$ and Lemma 26 says that $\omega_2(z, r) \leq \delta$. Since we control $\int_0^r h(t)^{1/2} \frac{dt}{t}$ by assumption, Corollary 51.17 applies and gives (40).

We are now ready to prove Theorem 1. Let Ω be a bounded C^1 domain and (u, K) an almost-minimizer, as in the statement. The main point will be that for every small $\delta > 0$, there is a radius $r_0 > 0$ such that the assumptions of Lemma 16, and also the extra assumption for (40), are fulfilled for every choice of $x \in \overline{K} \cap \partial\Omega$ and $r \leq r_0$.

Let $x \in \overline{K} \cap \partial\Omega$ be given. We easily construct a locally C^1 curve $\Gamma \subset K \cap B(x, 8r_0)$ by putting together lots of good curves given by (40), applied to the pairs $(x, 2^{-k}r_0)$, $k \geq 0$. There is no problem for the gluing, because the half-lines L associated to successive radii $2^{-k}r_0$, $k \geq 0$, are close to each other, by (19). Then we also get that $K \cap B(x, r_0) \subset \Gamma$ (indeed, if we want to show that $y \in K \cap B(x, r_0)$ lies on Γ, we apply (24) to the pair $(x, 2^{-k}r_0)$, with $|y-x|/6 \leq 2^{-k}r_0 \leq |y-x|/3$, say).

Now we promised that $\overline{K} \cap B(x, r_0)$ would be a C^1 curve, and we merely showed that Γ is (locally) C^1 in Ω, so we need a little more information at x. First notice that when r tends to 0, we can apply Lemma 16 with values of δ that tend to 0. Then (19) and (24) show that Γ has a tangent at x. Now we still need to show that the direction of the tangent to Γ is continuous at x, i.e., that it does not oscillate much. We can get it from the fact that (40) can be obtained with applications of Corollary 17 with constants that tend to 0, so that in particular the sets $K \cap B(z, r/2)$ are Lipschitz graphs with constants that tend to 0. We do not want to insist too much, because it is possible to get much better estimates, at least if h decays sufficiently fast. See Remark 42 below.

So Γ is C^1 up to x, and (24) also shows that it is perpendicular to $\partial\Omega$ at r. Finally note that x is an isolated point of $\overline{K} \cap \partial\Omega$; in fact it is the only point of $\overline{K} \cap \partial\Omega \cap B(x, r_0)$. This completes our proof of Theorem 1. $\qquad\square$

Remark 41. We can still say things when Ω is not C^1. For instance, if $\partial\Omega$ is merely Lipschitz and x is a point of $\overline{K} \cap \partial\Omega$ where Ω has a tangent cone (i.e., $\partial\Omega$ has two half-tangent lines), we still get that $K \cap B(x, r_0)$ is a locally C^1 curve, provided that the aperture of the cone is $< 3\pi/2$. We can even get a C^1 control, or better, if h decays fast enough; see the discussion below. In particular, we get an analogue of Theorem 1 for piecewise C^1 domains, as soon as $\partial\Omega$ does not have corners with (inside) angles $\geq 3\pi/2$.

Note that K does not even get close to corners where Ω looks like a sector with angle $< \pi/2$, by (23).

Our Lipschitz assumption was used to get local Ahlfors-regularity, by Theorem 77.4. Probably it can be weakened, maybe even dropped if we know that Ω looks a lot like a sector at all points of $\partial\Omega$. Then $\partial\Omega$ could have slowly turning spirals too.

Remark 42. Once we know that $K \cap B(x, r_0)$ is a (locally) C^1 curve that goes to x, it is reasonably easy to prove better regularity estimates on K and u. The arguments are similar to what we did in Part E, but simpler.

Let (u, K) be a reduced almost-minimizer (topological or not) on $\Omega \subset \mathbb{R}^2$, let $x \in \overline{K} \cap \partial\Omega$ be given, and assume that (x, r) satisfies the hypotheses of Lemma

16 for $r \leq r_0$. This is the case, for instance, if $\partial\Omega$ is C^1 near x. Also suppose that

$$h(r) \leq Cr^\gamma \text{ for } r \leq r_0 \tag{43}$$

for some $\gamma > 0$. We want to show that K is a $C^{1+\varepsilon}$ curve all the way to x.

We already know from Lemma 16 or (25) that for $r \leq r_0$, $K \cap B(x, 9r)$ is close to a half-line L. In addition, $\int_{\Omega \cap B(x,9r) \setminus K} |\nabla u|^2$ is small, by Lemma 26, and we may even use the fact that for r small, $K \cap B(x, r)$ is a locally C^1 curve that ends at x. All this makes it easier to check that the main quantities that control the good behavior of u and K decrease fairly fast. In particular, we do not really need to control jumps and holes, because we already know that K is locally connected. Thus the main quantities to consider are $\beta_K(x, r)$ and $\omega_2(x, r)$ (see the definitions (41.2) and (23.5)). Let us first say why

$$\beta_K(x, r/2) \leq C\omega_2(x, r)^{1/2} + Ch(r)^{1/2}. \tag{44}$$

We can use the good description of $K \cap B(x, r)$ to choose $\rho \in (r/2, r)$ such that $K \cap \partial B(x, \rho)$ has only one point (call it z), the restriction of u to $\Omega \cap \partial B(x, \rho) \setminus K$ is equal to the radial limit of u almost-everywhere and lies in $W^{1,2}$ with a derivative given by the restriction of ∇u, and

$$\int_{\Omega \cap \partial B(x,\rho) \setminus K} |\nabla u|^2 \leq 10\,\omega_2(x, r). \tag{45}$$

Now construct a competitor (v, G) such that $G = [K \setminus B(x, \rho)] \cup [z, x)$. As usual, we keep $v = u$ out of $B(x, \rho)$, and we choose it to minimize the energy on each of the two components of $\Omega \cap B(x, \rho) \setminus G$ with the given values on $\Omega \cap \partial B(x, \rho) \setminus K$. Then

$$\int_{\Omega \cap B(x,\rho) \setminus G} |\nabla v|^2 \leq C \int_{\Omega \cap \partial B(x,\rho) \setminus K} |\nabla u|^2 \leq Cr\,\omega_2(x, r) \text{ by (45),}$$

and on the other hand

$$H^1(G \cap B(x, \rho)) = |z - x| = \rho \leq H^1(K \cap B(x, \rho)) - C^{-1}\beta_K(x, r/2)^2 r,$$

by Pythagoras and because K goes from z to x through any given point of $K \cap \overline{B}(x, r/2)$; (44) follows.

But the main point again is to show that for some constant $\tau > 0$,

$$\omega_2(x, \tau r) \leq \frac{1}{2}\omega_2(x, r) + Ch(r). \tag{46}$$

Choose ρ as above, but this time keep K and merely replace u on $\Omega \cap B(x, \rho) \setminus K$ by the function v with the same values on $\Omega \cap \partial B(x, \rho) \setminus K$ as u, and which minimizes the energy under this constraint. Then

$$\int_{\Omega \cap B(x,\rho) \setminus K} |\nabla u|^2 \leq \int_{\Omega \cap B(x,\rho) \setminus K} |\nabla v|^2 + rh(r), \tag{47}$$

by almost-minimality. Since $\nabla(u-v)$ is orthogonal to ∇v in $L^2(\Omega \cap \partial B(x,\rho) \setminus K)$ (by minimality of v and because $v + t(u-v)$ is a competitor for all t),

$$\int_{\Omega \cap B(x,\rho) \setminus K} |\nabla(u-v)|^2 = \int_{\Omega \cap B(x,\rho) \setminus K} |\nabla u|^2 - \int_{\Omega \cap B(x,\rho) \setminus K} |\nabla v|^2 \leq rh(r). \quad (48)$$

Then we want to apply Bonnet's argument from Section 47. Set $E(t) = \int_{\Omega \cap B(x,t) \setminus K} |\nabla v|^2$ for $0 < t < \rho$, and try to show that $tE'(t) \geq 3E(t)/2$ for almost every $t < \rho$. We observe that for $t \leq \rho/C$, where C depends on the Lipschitz constant for $\partial\Omega$ near x, all the points of $\partial B(x,t) \cap [K \cup \partial\Omega]$ lie on a same connected compact subset of $B(x,\rho) \cap [K \cup \partial\Omega]$; this allows us to estimate $E(t)$ as we did in Section 47, i.e., integrate by parts, cut $\partial B(x,t) \setminus [K \cup \partial\Omega]$ into intervals I, replace $\int_I v \frac{\partial v}{\partial \tau}$ with $\int_I (v - m_I v) \frac{\partial v}{\partial \tau}$, and use Cauchy–Schwarz and Wirtinger. The argument is just the same as for (77.13) in dimension 2, so we skip the details. Eventually, we get that $E(t) \leq \theta t E'(t)$, where $2\pi\theta t$ is the length of the longest component I of $\Omega \cap \partial B(x,t) \setminus K$ (the intervals in $\partial B(x,t) \setminus \Omega$ do not matter, we take $v = 0$ there). Here $2\pi\theta$ can barely be larger than $3\pi/2 - \pi/2 = \pi$ (when Ω looks like a sector of aperture $3\pi/2$, and by (24)) so we can make sure that $\theta \leq 2/3$. Thus $tE'(t) \geq 3E(t)/2$, and when we integrate we get that

$$\int_{\Omega \cap B(x,t) \setminus K} |\nabla v|^2 = E(t) \leq C(\rho^{-1}t)^{3/2} E(\rho) \leq C(\rho^{-1}t)^{3/2} \int_{\Omega \cap B(x,\rho) \setminus K} |\nabla u|^2,$$

$$(49)$$

where the last inequality comes from the minimality of v. Finally,

$$\int_{\Omega \cap B(x,t) \setminus K} |\nabla u|^2 \leq 2 \int_{\Omega \cap B(x,t) \setminus K} |\nabla v|^2 + 2 \int_{\Omega \cap B(x,t) \setminus K} |\nabla(u-v)|^2$$

$$\leq C(\rho^{-1}t)^{3/2} \int_{\Omega \cap B(x,\rho) \setminus K} |\nabla u|^2 + 2rh(r) \quad (50)$$

$$\leq C(r^{-1}t)^{3/2} \int_{\Omega \cap B(x,r) \setminus K} |\nabla u|^2 + 2rh(r)$$

by (48) and because $r/2 \leq \rho \leq r$. Then divide by t to get that

$$w_2(x,t) = t^{-1} \int_{\Omega \cap B(x,t) \setminus K} |\nabla u|^2 \leq C\left(\frac{t}{r}\right)^{1/2} w_2(x,\rho) + 2\frac{r}{t} h(r). \quad (51)$$

This proves (46); we just have to pick τ small enough and take $t = \tau r$ in (51).

Once we have (44) and (46), we know that $w_2(x,r)$ and $\beta_K(x,r)$ decay like a power, and it is very easy to get $C^{1+\varepsilon}$ estimates on K for some $\varepsilon > 0$. If γ in (43) is large enough, we should even get all the way to $C^{1,1}$.

Even if we do not have the assumptions for Lemma 16, or in dimension $n \geq 2$, the arguments of Part E for C^1-regularity (when we already control flatness and energy, for instance) seem to go through near the boundary. For instance, jumps and holes are apparently as easy to control near Ω as inside. Of course the reader is advised to check all this before using it.

80 A few questions

It seems natural to end this book with a few questions. Naturally most of them were already mentioned in the text, so this short section is mainly here for easier access. Also see the entry "Questions" in the index for a few sporadic ones.

We start with the situation in dimension 2. The main question is still the Mumford-Shah conjecture (if (u, K) is a reduced minimizer of the functional J (see (1.4)), in a smooth domain, can we say that K is a finite union of C^1 curves, with the usual constraint of $120°$ angles where they meet?). It even seems reasonable to ask for more: is there a constant $\tau > 0$, that depends only on Ω and $||g||_\infty$, such that K is a union of C^1 curves of diameters at least τ? See Section 6 for more details about this.

As was observed in Section 54 and proved in Section 71, the Mumford-Shah conjecture would follow if we proved that the list of four global minimizers in (54.14)–(54.17) is complete (Conjecture 54.18). See also Section 79 for the boundary behavior. It would be very amusing if Conjecture 54.18 was false, and the corresponding exotic global minimizer in the plane did not show up as a blow-up limit of a Mumford-Shah minimizer in a domain. But the most likely is that both conjectures are true, and Conjecture 54.18 will be proved first.

Since we know the answer to Conjecture 54.18 when K separates $\mathbb{R}^2$ into at least two components, or even has two or more unbounded components, the next case could be when K has one unbounded component. It seems to help slightly in monotonicity arguments, but (so far) not enough.

One can also ask whether knowing that K is contained in a chord-arc curve helps. There is a positive answer if the chord-arc constant is small enough (see Exercise 67.15), but not in general.

It could be amusing to consider global minimizers in other planar domains. We were lucky and got a characterization in the case of sectors with aperture $< 3\pi/2$ (see Section 66c), but other shapes could be interesting, independently of any potential use for the boundary behavior on nonsmooth domains. One can also imagine that there are exotic global minimizers in an appropriate surface of dimension 2 (like a sector with aperture $\alpha > 2\pi$), because there is more room.

It seems that variants of the Mumford-Shah functional (and conjecture) where we keep length as it is, but replace energy integrals $\int |\nabla u|^2$ with less homogeneous variants like $\int \left(\frac{\partial u}{\partial x}\right)^2 + a\left(\frac{\partial u}{\partial y}\right)^2$, $a \neq 1$, were not studied. Maybe the list of global minimizers for such variants is fundamentally different. This has to do with the local behavior of almost-minimizers with linear distortion; see Definition 7.45 and Remark 40.12.

Return to the Mumford-Shah functional on a domain. Is it possible to find Ω and $g \in L^\infty(\Omega)$ such that some blow-up limit of (u, K) at $x \in \Omega$ is a cracktip (i.e., such that x is a spiral point, see Section 69)? The answer is not obvious; the fact that the cracktip pair from (54.17) is a global minimizer in the plane only gives direct information on problems with a Dirichlet constraint at the boundary. So far, no calibration argument works for this.

We do not know the precise regularity of K at a spiral point (for the Mumford-Shah functional in a domain, if spiral points exist, or more generally for almost-minimizers). For instance, does K remain C^1 all the way to the tip, or can it turn infinitely many times around it?

Can we recognize a global minimizer from a small piece? Of course the answer is yes if Conjecture 54.18 holds, but even otherwise one could hope to have enough rigidity to prove something like this. This is not such a stupid question, because we know that minimizers have some rigidity. E.g., it was proved in [LeMo] that K is real-analytic far from its singular points. [This is even true in higher dimensions, see [KLM].] One could also take the question less strictly and ask to which extent two global minimizers that are very close in a ball need to stay close in larger balls.

As a special case of this, we do not know how to conclude when K contains a line segment, or equivalently (see Exercise 64.18) meets one on a set of positive measure (compare with Section 64). This assumption should force K to be symmetric with respect to the corresponding line, for instance, but so far we were not able to exploit analytic continuation and prove this.

We may also ask whether we can recognize a global minimizer, or at least get very useful information, from its singular singular set (defined in Section 67).

A similar question concerns the connectedness of the set RGM of global minimizers. We have a few perturbation results that tend to say that points of RGM are isolated, but we do not know about possible exotic minimizers. Here one should state results and work modulo translations and rotations.

Some of our proofs do not generalize if the energy $\int |\nabla u|^2$ is replaced with $\int |\nabla u|^p$, for instance. This is so not only because p-harmonic functions are harder to study, but because precise formulae were used. It would be nice to have generalizations of Bonnet's monotonicity argument, or Léger's magic formula, or at least the essential uniqueness of u given K, which otherwise seem a little miraculous. This could be a first step towards similar questions in higher dimensions, but it could also be harder.

The case of functions with values in a sphere (as opposed to $\mathbb{R}$ or $\mathbb{R}^m$) was considered by De Giorgi, Carriero, and Leaci [DCL]; some results stay true (like existence and local Ahlfors-regularity), and of course there are specifics (e.g., harmonic mappings with values in a sphere are very different). There are probably lots of other interesting variants of this type, but let us not mention those (out of ignorance).

In dimensions 3 and higher (one can always dream), many of the theorems of dimension 2 become open questions. As was discussed in Section 76, we still do not have a closed list of putative global minimizers in $\mathbb{R}^3$, but there are simpler questions that we cannot answer.

Can we at least show that all the global minimizers in $\mathbb{R}^3$ (i.e., pairs $(u, K) \in RGM(\mathbb{R}^3)$, see Section 54) for which K is contained in a plane are the product of a cracktip by an orthogonal line? We do not even know this when K is a plane sector and u is homogeneous of degree $1/2$.

Actually, could it be that every global minimizer in $\mathbb{R}^3$ is homogeneous of degree $1/2$?

What if K contains a hyperplane? We know the answer to the analogous question in dimension 2, by Section 64 or 66. It could also be that global minimizers in a half-space, or a slightly smaller domain, are easier to study.

Is it true that if $(u, K) \in RGM(\mathbb{R}^3)$ and u is constant somewhere, or $\mathbb{R}^3 \setminus K$ is not connected, then u is constant everywhere (and hence K is a minimal set)? Can we at least say that the components of $\mathbb{R}^3 \setminus K$ are not thin (i.e., that Proposition 56.1 generalizes), hence are John domains (as in Proposition 56.7)? At least we know that $\mathbb{R}^3 \setminus K$ has no bounded component.

We are also missing analogues of some of our perturbation results. For instance, we would like to know that if $\int_{B(x,r) \setminus K} |\nabla u|^2$ is very small and $K \cap B(0, 1)$ is very close to one of the minimal sets in $\mathbb{R}^3$ (namely, a plane, the product of a propeller with a line, or the minimal set with six faces and four edges), then $K \cap B(x, r/2)$ is a C^1 version of this minimal set. [Unless we only get a slightly less regular parameterization of K, as in Reifenberg's theorem.] We only know this in dimension 2 [Da5] and when the minimal set is a plane [AFP1]. In this direction, it would be good to have a different proof of Ambrosio, Fusco, and Pallara's result in [AFP1], that would not use tilt. Or to understand J. Taylor's result on minimizing sets in $\mathbb{R}^3$ better.

Suppose $(u, K) \in RGM(\mathbb{R}^3)$, and some blow-in limit of (u, K) is the product of a cracktip by a line. Can we say that (u, K) itself is the product of a cracktip by a line? [This is the case in dimension 2, see (62.4).] Or locally, is there is an analogue of Section 69 on spiral points when (u, K) looks a lot like the product of a cracktip by a line in a given ball?

It seems hard to adapt Bonnet's argument to dimension 3 and get monotonicity results, but it would be very nice if we could do that.

Can we determine u (modulo the obvious symmetries) when we know K and $(u, K) \in RGM(\mathbb{R}^3)$? In dimension 2, the answer is yes, because of Léger's formula (63.3), but apparently we have no such formula in $\mathbb{R}^3$. This could be linked to the question of the missing minimizer in Section 76, because if K has a symmetry of order 3, and u cannot have such a symmetry, u will not be unique. See Section 76.

This question of symmetry leads to a simple question that we already do not know how to answer in the plane: could it be that the set of global minimizers in $\mathbb{R}^n$ really changes if we consider functions u with values in $\mathbb{R}^2$, or $\mathbb{R}^m$? Taking functions in $\mathbb{R}^2$ could make it easier for u to have a symmetry of order 3, as suggested in Section 76, but this is probably not a good enough reason.

Finally, is it true that for generic choices of $g \in L^\infty(\Omega)$, there is a unique minimizer for the Mumford-Shah functional? And even, that the functional never has two distinct minimizers that are very close to each other? See Section 5.d for a rapid discussion.

References

[Ah] L. Ahlfors, Zur Theorie der Überlagerungsflächen, Acta Math. 65 (1935), 157–194.

[ABD1] G. Alberti, G. Bouchitté, and G. Dal Maso, The calibration method for the Mumford-Shah functional, C. R. Acad. Sci. Paris Sér. I Math. 329 (1999), no. 3, 249–254.

[ABD2] G. Alberti, G. Bouchitté, and G. Dal Maso, The calibration method for the Mumford-Shah functional and free-discontinuity problems, Calc. Var. Partial Differential Equations 16, No. 3 (2003), 299–333.

[Al] F.J. Almgren, Existence and regularity almost everywhere of solutions to elliptic variational problems with constraints, Memoirs of the Amer. Math. Soc. 165, volume 4 (1976), i–199.

[Am] L. Ambrosio, Existence theory for a new class of variational problems, Arch. Rational Mech. Anal. 111 (1990), 291–322.

[AFH] L. Ambrosio, N. Fusco, and J. Hutchinson, Higher integrability of the gradient and dimension of the singular set for minimizers of the Mumford-Shah functional, Calc. Var. Partial Differential Equations 16 (2003), no. 2, 187–215.

[AFP1] L. Ambrosio, N. Fusco, and D. Pallara, Partial regularity of free discontinuity sets II., Ann. Scuola Norm. Sup. Pisa Cl. Sci. (4), 24 (1997), 39–62.

[AFP2] L. Ambrosio, N. Fusco and D. Pallara, Higher regularity of solutions of free discontinuity problems. Differential Integral Equations 12 (1999), no. 4, 499–520.

[AFP3] L. Ambrosio, N. Fusco and D. Pallara, Functions of bounded variation and free discontinuity problems, Oxford Mathematical Monographs, Clarendon Press, Oxford 2000.

[AmPa] L. Ambrosio and D. Pallara. Partial regularity of free discontinuity sets I., Ann. Scuola Norm. Sup. Pisa Cl. Sci. (4), 24 (1997), 1–38.

[BM] A.S. Besicovitch and P.A.P. Moran, The measure of product and cylinder sets, J. London Math. Soc. 20, (1945). 110–120.

[Bo] A. Bonnet, On the regularity of edges in image segmentation, Ann. Inst. H. Poincaré, Analyse non linéaire, Vol. 13, 4 (1996), 485–528.

[BoDa] A. Bonnet and G. David, Cracktip is a global Mumford-Shah minimizer, Astérisque 274, Société Mathématique de France 2001.

[CaLe1] M. Carriero and A. Leaci, Existence theorem for a Dirichlet problem with free discontinuity set, Nonlinear Anal. 15 (1990), 661–677.

[CaLe2] M. Carriero and A. Leaci, S^k-valued maps minimizing the L^p-norm of the gradient with free discontinuities, Ann. Scuola Norm. Sup. Pisa Cl. Sci. (4), 18 (1991), 321–352.

[CMM]   R.R. Coifman, A. McIntosh and Y. Meyer, L'intégrale de Cauchy définit un opérateur borné sur L^2 pour les courbes lipschitziennes, Ann. of Math. 116 (1982), 361–387.

[DMS1]   G. Dal Maso, J.-M. Morel, and S. Solimini, Une approche variationnelle en traitement d'images: résultats d'existence et d'approximation, C. R. Acad. Sci. Paris Sér. I Math. 308 (1989), no. 19, 549–554.

[DMS2]   G. Dal Maso, J.-M. Morel, and S. Solimini, A variational method in image segmentation: Existence and approximation results, Acta Math. 168 (1992), no. 1-2, 89–151.

[Da1]   G. David, Opérateurs intégraux singuliers sur certaines courbes du plan complexe, Ann. Sci. Ec. Norm. Sup. 17 (1984), 157–189.

[Da2]   G. David, Opérateurs d'intégrale singulière sur les surfaces régulières, Ann. Sci. Ec. Norm. Sup., série 4, t.21 (1988), 225–258.

[Da3]   G. David, Morceaux de graphes lipschitziens et intégrales singulières sur une surface, Revista Matematica Iberoamericana, vol. 4, 1 (1988), 73–114.

[Da4]   G. David, Wavelets and singular integrals on curves and surfaces, Lecture Notes in Math. 1465, Springer-Verlag 1991.

[Da5]   G. David, C-1 arcs for minimizers of the Mumford-Shah functional, SIAM. Journal of Appl. Math. Vol. 56, No. 3 (1996), 783–888.

[Da6]   G. David, Unrectifiable 1-sets have vanishing analytic capacity, Revista Matematica Iberoamericana 14, 2 (1998), 369–479.

[Da7]   G. David, Analytic capacity, Calderón-Zygmund operators, and rectifiability, Publicacions Matemàtiques 43 (1999), 3–25.

[Da8]   G. David, Limits of Almgren-quasiminimal sets, Proceedings of the conference on Harmonic Analysis, Mount Holyoke, A.M.S. Contemporary Mathematics series, Vol. 320 (2003), 119–145.

[Da9]   G. David, uniformly rectifiable sets, notes of a summer school on Harmonic Analysis at Park City, to be published, IAS/Park City Mathematics series, AMS; available in the mean time at http://www.math.u-psud.fr/~gdavid.

[DaJe]   G. David and D. Jerison, Lipschitz approximation to hypersurfaces, harmonic measure, and singular integrals, Indiana Univ. Math. J. 39 (1990), no. 3, 831–845.

[DaLé]   G. David and J.-C. Léger, Monotonicity and separation for the Mumford-Shah problem, Annales de l'Inst. Henri Poincaré, Analyse non linéaire 19, 5, 2002, pages 631–682.

[DaSe1] G. David and S. Semmes, Singular integrals and rectifiable sets in $\mathbb{R}^n$: au-delà des graphes lipschitziens, Astérisque 193, Société Mathématique de France 1991.

[DaSe2] G. David and S. Semmes, Quantitative rectifiability and Lipschitz mappings, Trans. Amer. Math. Soc. 337 (1993), no. 2, 855–889.

[DaSe3] G. David and S. Semmes, Analysis of and on uniformly rectifiable sets, A.M.S. series of Mathematical surveys and monographs, Volume 38, 1993.

[DaSe4] G. David and S. Semmes, On the singular sets of minimizers of the Mumford-Shah functional, Journal de Math. Pures et Appl. 75 (1996), 299–342.

[DaSe5] G. David and S. Semmes, On a variational problem from image processing, proceedings of the conference in honor of J.-P. Kahane, special issue of the Journal of Fourier Analysis and Applications, 1995, 161–187.

[DaSe6] G. David and S. Semmes, Uniform rectifiability and Singular sets, Annales de l'Inst. Henri Poincaré, Analyse non linéaire, Vol. 13, No. 4 (1996), p. 383–443.

[DaSe7] G. David and S. Semmes, Fractured fractals and broken dreams: Self-similar geometry through metric and measure, Oxford Lecture series in Mathematics and its applications 7, Oxford University Press 1997.

[DaSe8] G. David and S. Semmes, Quasiminimal surfaces of codimension 1 and John domains, Pacific Journal of Mathematics Vol. 183, No. 2 (1998), 213–277.

[DaSe9] G. David and S. Semmes, Uniform rectifiability and quasiminimizing sets of arbitrary codimension, Memoirs of the A.M.S. Number 687, volume 144, 2000.

[DG]   E. De Giorgi, Problemi con discontinuità libera, Int. Symp. Renato Caccioppoli, Napoli, Sept. 20–22, 1989. Ricerche Mat. 40 (1991), suppl. 203–214.

[DCL]   E. De Giorgi, M. Carriero, and A. Leaci, Existence theorem for a minimum problem with free discontinuity set, Arch. Rational Mech. Anal. 108 (1989), 195–218.

[Di]   F. Dibos, Uniform rectifiability of image segmentations obtained by a variational method, Journal de Math. Pures et Appl. 73, 1994, 389–412.

[DiKo]   F. Dibos and G. Koepfler, Propriété de régularité des contours d'une image segmentée, Comptes Rendus Acad. Sc. Paris 313 (1991), 573–578.

[DiSé]   F. Dibos and E. Séré, An approximation result for Minimizers of the Mumford-Shah functional, Boll. Un. Mat. Ital. A(7), 11 (1997), 149–162.

[Du]   P. Duren, Theory of Hardy spaces, Academic Press 1970.

[Fa]   K. Falconer, The Geometry of fractal sets, Cambridge University Press 1984.

[Fe]   H. Federer, Geometric measure theory, Grundlehren der Mathematischen Wissenschaften 153, Springer Verlag 1969.

[GaRu]   J. García-Cuerva and J.-L. Rubio de Francia, Weighted norm inequalities and related topics, North-Holland Mathematics Studies, 116, North-Holland Publishing Co., Amsterdam, 1985.

[Ga]   J. Garnett, Bounded analytic functions, Academic Press 1981.

[Gi]   E. Giusti, Minimal surfaces and functions of bounded variation, Monographs in Mathematics, 80. Birkhäuser Verlag, Basel-Boston, Mass., 1984.

[HaKo]   P. Hajlasz, and P. Koskela, Sobolev met Poincaré, Mem. Amer. Math. Soc. 145 (2000), no. 688.

[HaLiPo] G. Hardy, J.E. Littlewood, and G. Pólya, Inequalities, Second Edition, Cambridge University Press 1952.

[Jo1]   P. Jones, Square functions, Cauchy integrals, analytic capacity, and harmonic measure, Proc. Conf. on Harmonic Analysis and Partial Differential Equations, El Escorial 1987 (ed. J. García-Cuerva), p. 24–68, Lecture Notes in Math. 1384, Springer-Verlag 1989.

[Jo2]   P. Jones, Rectifiable sets and the traveling saleseman problem, Inventiones
           Mathematicae 102, 1 (1990), 1–16.

[JKV]   P. Jones, N. Katz, and A. Vargas, Checkerboards, Lipschitz functions and uni-
           form rectifiability, Rev. Mat. Iberoamericana 13 (1997), no. 1, 189–210.

[Jou]   J.-L. Journé, Calderón-Zygmund operators, pseudodifferential operators and
           the Cauchy integral of Calderón, Lecture Notes in Mathematics 994, Springer-
           Verlag, Berlin, 1983.

[KLM]   H. Koch, G. Leoni, and M. Morini, On optimal regularity of free boundary
           problems and a conjecture of De Giorgi, preprint.

[La]   C. Larsen, personal communication.

[Lé1]   J.-C. Léger, Une remarque sur la régularité d'une image segmentée, Journal de
           Math. pures et appliquées 73, 1994, 567–577.

[Lé2]   J.-C. Léger, Courbure de Menger et rectifiabilité et sur la fonctionnelle de
           Mumford-Shah, thèse, Paris-Sud Orsay, January 1997.

[Lé3]   J.-C. Léger, Flatness and finiteness in the Mumford-Shah problem, J. Math.
           Pures Appl. (9) 78 (1999), no. 4, 431–459.

[LeMo]   G. Leoni and M. Morini, Some remarks on the analyticity of minimizers of free
           discontinuity problems, J. Math. Pures Appl. (9) 82 (2003), no. 5, 533–551.

[LMS]   F.A. Lops, F. Maddalena, and S. Solimini, Hölder continuity conditions for the
           solvability of Dirichlet problems involving functionals with free discontinuities,
           Ann. Inst. H. Poincaré, Anal. Non Linéaire 18 (2001), no. 6, 639–673.

[MaSo1]   F. Maddalena and S. Solimini, Concentration and flatness properties of the
           singular set of bisected balls, Ann. Scuola Norm. Sup. Pisa Cl. Sci. (4) 30 (2001),
           no. 3-4, 623–659 (2002).

[MaSo2]   F. Maddalena and S. Solimini, Regularity properties of free discontinuity sets,
           Ann. Inst. H. Poincaré, Anal. Non Linéaire 18 (2001), no. 6, 675–685.

[MaSo3]   F. Maddalena and S. Solimini, Lower semicontinuity properties of functionals
           with free discontinuities, Arch. Ration. Mech. Anal. 159 (2001), no. 4, 273–294.

[MaSo4]   F. Maddalena and S. Solimini, Blow-up techniques and regularity near the
           boundary for free discontinuity problems, Adv. Nonlinear Stud. 1 (2001), no. 2,
           1–41.

[Mar]   D. Marr, Vision, Freeman and Co. 1982.

[Mat1]   P. Mattila, Cauchy singular integrals and rectifiability of measures in the plane,
           Advances in Math. 115, 1 (1995), 1–34.

[Mat2]   P. Mattila, Geometry of sets and measures in Euclidean space, Cambridge
           Studies in Advanced Mathematics 44, Cambridge University Press 1995.

[MatMV]   P. Mattila, M. Melnikov, and J. Verdera, The Cauchy integral, analytic capac-
           ity, and uniform rectifiability, Ann. of Math. 144, 1 (1996), 127–136.

[MatP]   P. Mattila and D. Preiss, Rectifiable measures in $\mathbb{R}^n$ and existence of principal
           values for singular integrals, J. London Math. Soc. (2) 52 (1995), no. 3, 482–496.

[Me]   B. Merlet, Sur qeulques équations aux dérivées partielles et leur analyse
           numérique, thesis, Université de Paris-Sud 2004.

[MoSo1] J.-M. Morel and S. Solimini, Estimations de densité pour les frontières de segmentations optimales, C. R. Acad. Sci. Paris Sér. I Math. 312 (1991), no. 6,
429–432.

[MoSo2] J.-M. Morel and S. Solimini, Variational methods in image segmentation, Progress in nonlinear differential equations and their applications 14, Birkhäuser
1995.

[Morg1] F. Morgan, Geometric measure theory, A beginner's guide, Academic Press
1988.

[Morg2] F. Morgan, Size-minimizing rectifiable currents, Invent. Math. 96 (1989), no. 2,
333–348.

[Morg3] F. Morgan, Minimal surfaces, crystals, shortest networks, and undergraduate
research, Math. Intelligencer 14 (1992), no. 3, 37–44.

[MuSh1] D. Mumford and J. Shah, Boundary detection by minimizing functionals, IEEE
Conference on computer vision and pattern recognition, San Francisco 1985.

[MuSh2] D. Mumford and J. Shah, Optimal approximations by piecewise smooth functions and associated variational problems, Comm. Pure Appl. Math. 42 (1989),
577–685.

[NTV] F. Nazarov, S. Treil, and A. Volberg, Cauchy integral and Calderón-Zygmund
operators on non homogeneous spaces, International Math. Res. Notices 1997,
15, 703–726.

[Ne] M.H.A. Newman, Elements of the topology of plane sets of points, Second edition, reprinted, Cambridge University Press, New York 1961.

[Ok] K. Okikiolu, Characterization of subsets of rectifiable curves in $\mathbb{R}^n$, J. of the
London Math. Soc. 46 (1992), 336–348.

[Pa] H. Pajot, Analytic capacity, rectifiability, Menger curvature and the Cauchy
integral, Lecture Notes in Mathematics, 1799, Springer-Verlag, Berlin, 2002.

[Po] C. Pommerenke, Boundary behavior of conformal maps, Grundlehren der Mathematischen Wissenschaften 299, Springer-Verlag 1992.

[Pr] D. Preiss, Geometry of measures in R^n: distribution, rectifiability, and densities,
Ann. of Math. (2) 125 (1987), no. 3, 537–643.

[Re] E.R. Reifenberg, Solution of the Plateau problem for m-dimensional surfaces of
varying topological type, Acta Math. 104 (1960), 1–92.

[Ri1] S. Rigot, Big Pieces of $C^{1,\alpha}$-Graphs for Minimizers of the Mumford-Shah Functional, Ann. Scuola Norm. Sup. Pisa Cl. Sci. (4) 29 (2000), no. 2, 329–349.

[Ri2] S. Rigot, Uniform partial regularity of quasi minimizers for the perimeter, Cal.
Var. Partial Differential Equations 10 (2000), no. 4, 389–406.

[Ri3] S. Rigot, Ensembles quasi-minimaux avec contrainte de volume et rectifiabilité
uniforme, Mém. Soc. Math. Fr. (N.S.) 82 (2000), v+104pp.

[Ru] W. Rudin, Real and complex analysis, Third edition. McGraw Hill Book Co.,
New York, 1987.

[Se1] S. Semmes, A criterion for the boundedness of singular integrals on hypersurfaces, Trans. Amer. Math. Soc. 311 (1989), no. 2, 501–513.

[Se2]   S. Semmes, Differentiable function theory on hypersurfaces in $\mathbb{R}^n$ (without bounds on their smoothness), Indiana Univ. Math. Journal 39 (1990), 985–1004.

[Se3]   S. Semmes, Analysis vs. geometry on a class of rectifiable hypersurfaces in $\mathbb{R}^n$, Indiana Univ. Math. Journal 39 (1990), 1005–1036.

[Se4]   S. Semmes, Finding structure in sets with little smoothness, Proceedings of the International Congress of Mathematicians, Vol. 1, 2 (Zürich, 1994), 875–885, Birkhäuser, Basel, 1995.

[Se5]   S. Semmes, Finding curves on general spaces through quantitative topology, with applications to Sobolev and Poincaré inequalities. Selecta Math. (N.S.) 2 (1996), no. 2, 155–295.

[Si]   A. Siaudeau, Alhfors-régularité des quasi minima de Mumford-Shah, J. Math. Pures Appl. (9) 82 (2003), no. 12, 1697–1731.

[Sim1]   L. Simon, Lectures on geometric measure theory, Proceedings of the Centre for Mathematical Analysis, Australian National University, 3, Australian National University, Centre for Mathematical Analysis, Canberra, 1983.

[Sim2]   L. Simon, Recent developments in the theory of minimal surfaces, Proceedings of the International Congress of Mathematicians, Vol. 1, 2 (Warsaw, 1983), 579–584, PWN, Warsaw, 1984.

[So1]   S. Solimini, Functionals with surface terms on a free singular set, Nonlinear partial differential equations and their applications, Collège de France Seminar, Vol. XII (Paris, 1991–1993), 211–225, Pitman Res. Notes Math. Ser., 302, Longman Sci. Tech., Harlow, 1994.

[So2]   S. Solimini, Simplified excision techniques for free discontinuity problems in several variables, J. Funct. Anal. 151 (1997), no. 1, 1–34.

[St]   E.M. Stein, Singular integrals and differentiability properties of functions, Princeton University Press 1970 .

[Ta]   J. Taylor, The structure of singularities in soap-bubble-like and soap-film-like minimal surfaces, Ann. of Math. (2) 103 (1976), no. 3, 489–539.

[To1]   X. Tolsa, L^2-boundedness of the Cauchy integral operator for continuous measures. Duke Math. J. 98 (1999), no. 2, 269–304.

[To2]   X. Tolsa, Painlevé's problem and the semiadditivity of analytic capacity, Acta Math. 190 (2003), no. 1, 105–149.

[Zi]   W.P. Ziemer, Weakly differentiable functions, Graduate texts in Mathematics 120, Springer-Verlag 1989.

Index

We try to give closest numbers; ± may be an additional position indication.
C stands for Corollary, D for Definition, E for Exercise or Example, L. for Lemma,
P. for Proposition, R. for Remark, S. for Section, T. for Theorem. We try to put
definitions first.

Monografie Matematyczne

New Series

Your Specialized Publisher in Mathematics

Birkhäuser

For orders originating from all over the world except USA/Canada/Latin America:

Birkhäuser Verlag AG
c/o Springer GmbH & Co
Haberstrasse 7
D-69126 Heidelberg
Fax: +49 / 6221 / 345 4 229
e-mail: birkhauser@springer.de
http://www.birkhauser.ch

For orders originating in the USA/Canada/Latin America:

Birkhäuser
333 Meadowland Parkway
USA-Secaucus
NJ 07094-2491
Fax: +1 201 348 4505
e-mail: orders@birkhauser.com

Managing Editor:
Przemysław Wojtaszczyk, IMPAN and Warsaw University, Poland

Starting in the 1930s with volumes written by such distinguished mathematicians as Banach, Saks, Kuratowski, and Sierpinski, the original series grew to comprise 62 excellent monographs up to the 1980s. In cooperation with the Institute of Mathematics of the Polish Academy of Sciences (IMPAN), Birkhäuser now resumes this tradition to publish high quality research monographs in all areas of pure and applied mathematics.

◼ **Vol. 63: Schürmann, J.**, Westfälische Wilhelms-Universität Münster, Germany

Topology of Singular Spaces and Constructible Sheaves

2003. 464 pages. Hardcover.
ISBN 3-7643-2189-X

Assuming that the reader is familiar with sheaf theory, the book gives a self-contained introduction to the theory of constructible sheaves related to many kinds of singular spaces, such as cell complexes, triangulated spaces, semialgebraic and subanalytic sets, complex algebraic or analytic sets, stratified spaces, and quotient spaces. The relation to the underlying geometrical ideas are worked out in detail, together with many applications to the topology of such spaces. All chapters have their own detailed introduction, containing the main results and definitions, illustrated in simple terms by a number of examples. The technical details of the proof are postponed to later sections, since these are not needed for the applications.

◼ **Vol. 64: Walczak, P.**, University of Łódź, Poland

Dynamics of Foliations, Groups and Pseudogroups

2004. 240 pages. Hardcover.
ISBN 3-7643-7091-2

Foliations, groups and pseudogroups are objects which are closely related via the notion of holonomy. In the 1980s they became considered as general dynamical systems. This book deals with their dynamics. Since "dynamics" is a very extensive term, we focus on some of its aspects only. Roughly speaking, we concentrate on notions and results related to different ways of measuring complexity of the systems under consideration. More precisely, we deal with different types of growth, entropies and dimensions of limiting objects. Invented in the 1980s (by E. Ghys, R. Langevin and the author) geometric entropy of a foliation is the principal object of interest among all of them. Throughout the book, the reader will find a good number of inspiring problems related to the topics covered.

◼ **Vol. 65: Badescu, L.**, Università degli Studi di Genova, Italy

Projective Geometry and Formal Geometry

2004. 228 pages. Hardcover.
ISBN 3-7643-7123-4

The aim of this monograph is to introduce the reader to modern methods of projective geometry involving certain techniques of formal geometry. Some of these methods are illustrated in the first part through the proofs of a number of results of a rather classical flavor, involving in a crucial way the first infinitesimal neighbourhood of a given subvariety in an ambient variety. Motivated by the first part, in the second formal functions on the formal completion X/Y of X along a closed subvariety Y are studied, particularly the extension problem of formal functions to rational functions.

O b e r w o l f a c h
S e m i n a r s

The workshops organized by the
Mathematisches Forschungsinstitut
Oberwolfach are intended to introduce students and young mathematicians to current fields of research. By means of these well-orga-
nized seminars, also scientists from other fields will be introduced to
new mathematical ideas.
The publication of these workshops in the series Oberwolfach
Seminars (formerly DMV Seminar) makes the material available to
an even larger audience.

Your Specialized
Publisher in
Mathematics
Birkhäuser

For orders originating from all over the world
except USA/Canada/Latin America:
All countries excluding those listed below:
Birkhäuser Verlag AG
c/o Springer Auslieferungs-Gesellschaft (SAG)
Customer Service
Haberstrasse 7, D-69126 Heidelberg
Tel.: +49 / 6221 / 345 0
Fax: +49 / 6221 / 345 42 29
e-mail: orders@birkhauser.ch

For orders originating in the
USA/Canada/Latin America:

Birkhäuser
333 Meadowland Parkway
USA-Secaucus
NJ 07094-2491
Fax: +1 201 348 4505
e-mail: orders@birkhauser.com